DIE GRUNDLEHREN DER

MATHEMATISCHEN WISSENSCHAFTEN

IN EINZELDARSTELLUNGEN MIT BESONDERER BERÜCKSICHTIGUNG DER ANWENDUNGSGEBIETE

HERAUSGEGEBEN VON

J. L. DOOB · E. HEINZ · F. HIRZEBRUCH
E. HOPF · H. HOPF · W. MAAK · S. MAC LANE
W. MAGNUS · F. K. SCHMIDT · K. STEIN

GESCHÄFTSFÜHRENDE HERAUSGEBER

B. ECKMANN UND B. L. VAN DER WAERDEN
ZÜRICH

BAND 123

SPRINGER-VERLAG
BERLIN · HEIDELBERG · NEW YORK
1966

FUNCTIONAL ANALYSIS

BY

KÔSAKU YOSIDA
PROFESSOR OF MATHEMATICS
AT THE UNIVERSITY OF TOKYO

2nd PRINTING CORRECTED

SPRINGER-VERLAG
BERLIN · HEIDELBERG · NEW YORK
1966

Library of Congress Catalog Card Number 64-8025

Printed in Germany

Titel Nr. 5106

Preface

The present book is based on lectures given by the author at the University of Tokyo during the past ten years. It is intended as a textbook to be studied by students on their own or to be used in a course on Functional Analysis, i.e., the general theory of linear operators in function spaces together with salient features of its application to diverse fields of modern and classical analysis.

Necessary prerequisites for the reading of this book are summarized, with or without proof, in Chapter 0 under titles: Set Theory, Topological Spaces, Measure Spaces and Linear Spaces. Then, starting with the chapter on Semi-norms, a general theory of Banach and Hilbert spaces is presented in connection with the theory of generalized functions of S. L. SOBOLEV and L. SCHWARTZ. While the book is primarily addressed to graduate students, it is hoped it might prove useful to research mathematicians, both pure and applied. The reader may pass e.g. from Chapter IX (Analytical Theory of Semi-groups) directly to Chapter XIII (Ergodic Theory and Diffusion Theory) and to Chapter XIV (Integration of the Equation of Evolution). Such materials as "Weak Topologies and Duality in Locally Convex Spaces" and "Nuclear Spaces" are presented in the form of the appendices to Chapter V and Chapter X, respectively. These might be skipped for the first reading by those who are interested rather in the application of linear operators.

In the preparation of the present book, the author has received valuable advice and criticism from many friends. Especially, Mrs. K. HILLE has kindly read through the manuscript as well as the galley and page proofs. Without her painstaking help, this book could not have been printed in the present style in the language which was not spoken to the author in the cradle. The author owes very much to his old friends, Professor E. HILLE and Professor S. KAKUTANI of Yale University and Professor R. S. PHILLIPS of Stanford University for the chance to stay in their universities in 1962, which enabled him to polish the greater part of the manuscript of this book, availing himself of their valuable advice. Professor S. ITO and Dr. H. KOMATSU of the University of Tokyo kindly assisted the author in reading various parts of the galley proof, correcting errors and improving the presentation. To all of them, the author expresses his warmest gratitude.

Thanks are also due to Professor F. K. SCHMIDT of Heidelberg University and to Professor T. KATO of the University of California at Berkeley who constantly encouraged the author to write up the present book. Finally, the author wishes to express his appreciation to Springer-Verlag for their most efficient handling of the publication of this book.

Tokyo, September 1964

KÔSAKU YOSIDA

Contents

0. Preliminaries

It is the purpose of this chapter to explain certain notions and theorems used throughout the present book. These are related to *Set Theory*, *Topological Spaces*, *Measure Spaces* and *Linear Spaces*.

1. Set Theory

Sets. $x \in X$ means that x is a *member* or *element* of the set X; $x \bar{\in} X$ means that x is not a member of the set X. We denote the set consisting of all x possessing the property P by $\{x; P\}$. Thus $\{y; y = x\}$ is the set $\{x\}$ consisting of a single element x. The *void set* is the set with no members, and will be denoted by $\emptyset$. If every element of a set X is also an element of a set Y, then X is said to be a *subset* of Y and this fact will be denoted by $X \subseteqq Y$, or $Y \supseteqq X$. If $\mathfrak{X}$ is a set whose elements are sets X, then the set of all x such that $x \in X$ for some $X \in \mathfrak{X}$ is called the *union* of sets X in $\mathfrak{X}$; this union will be denoted by $\bigcup_{X \in \mathfrak{X}} X$. The *intersection* of the sets X in $\mathfrak{X}$ is the set of all x which are elements of every $X \in \mathfrak{X}$; this intersection will be denoted by $\bigcap_{X \in \mathfrak{X}} X$. Two sets are *disjoint* if their intersection is void. A family of sets is disjoint if every pair of distinct sets in the family is disjoint. If a sequence $\{X_n\}_{n=1,2,\ldots}$ of sets is a disjoint family, then the union $\bigcup_{n=1}^{\infty} X_n$ may be written in the form of a sum $\sum_{n=1}^{\infty} X_n$.

Mappings. The term *mapping*, *function* and *transformation* will be used synonymously. The symbol $f: X \rightarrow Y$ will mean that f is a single-valued function whose *domain* is X and whose *range* is contained in Y; for every $x \in X$, the function f assigns a uniquely determined element $f(x) = y \in Y$. For two mappings $f: X \rightarrow Y$ and $g: Y \rightarrow Z$, we can define their *composite mapping* $gf: X \rightarrow Z$ by $(gf)(x) = g(f(x))$. The symbol $f(M)$ denotes the set $\{f(x); x \in M\}$ and $f(M)$ is called the *image* of M under the mapping f. The symbol $f^{-1}(N)$ denotes the set $\{x; f(x) \in N\}$ and $f^{-1}(N)$ is called the *inverse image* of N under the mapping f. It is clear that

$$Y_1 = f(f^{-1}(Y_1)) \text{ for all } Y_1 \subseteqq f(X), \text{ and } X_1 \subseteqq f^{-1}(f(X_1)) \text{ for all } X_1 \subseteqq X.$$

If $f: X \to Y$, and for each $y \in f(X)$ there is only one $x \in X$ with $f(x) = y$, then f is said to have an *inverse* (mapping) or to be *one-to-one*. The inverse mapping then has the domain $f(X)$ and range X; it is defined by the equation $x = f^{-1}(y) = f^{-1}(\{y\})$.

The domain and the range of a mapping f will be denoted by $D(f)$ and $R(f)$, respectively. Thus, if f has an inverse then

$$f^{-1}(f(x)) = x \text{ for all } x \in D(f), \text{ and } f(f^{-1}(y)) = y \text{ for all } y \in R(f).$$

The function f is said to map X *onto* Y if $f(X) = Y$ and *into* Y if $f(X) \subseteqq Y$. The function f is said to be an *extension* of the function g and g a *restriction* of f if $D(f)$ contains $D(g)$, and $f(x) = g(x)$ for all x in $D(g)$.

Zorn's Lemma

Definition. Let P be a set of elements $a, b, \ldots$ Suppose there is a binary relation defined between certain pairs (a, b) of elements of P, expressed by $a \prec b$, with the properties:

$$\begin{cases} a \prec a, \\ \text{if } a \prec b \text{ and } b \prec a, \text{ then } a = b, \\ \text{if } a \prec b \text{ and } b \prec c, \text{ then } a \prec c \text{ (transitivity).} \end{cases}$$

Then P is said to be *partially ordered* (or *semi-ordered*) by the relation $\prec$.

Examples. If P is the set of all subsets of a given set X, then the set inclusion relation ($A \subseteqq B$) gives a partial ordering of P. The set of all complex numbers $z = x + iy$, $w = u + iv, \ldots$ is partially ordered by defining $z \prec w$ to mean $x \leqq u$ and $y \leqq v$.

Definition. Let P be a partially ordered set with elements $a, b, \ldots$ If $a \prec c$ and $b \prec c$, we call c an *upper bound* for a and b. If furthermore $c \prec d$ whenever d is an upper bound for a and b, we call c the *least upper bound* or the *supremum* of a and b, and write $c = \sup(a, b)$ or $a \vee b$. This element of P is unique if it exists. In a similar way we define the *greatest lower bound* or the *infimum* of a and b, and denote it by $\inf(a, b)$ or $a \wedge b$. If $a \vee b$ and $a \wedge b$ exist for every pair (a, b) in a partially ordered set P, P is called a *lattice*.

Example. The totality of subsets M of a fixed set B is a lattice by the partial ordering $M_1 \prec M_2$ defined by the set inclusion relation $M_1 \subseteqq M_2$.

Definition. A partially ordered set P is said to be *linearly ordered* (or *totally ordered*) if for every pair (a, b) in P, either $a \prec b$ or $b \prec a$ holds. A subset of a partially ordered set is itself partially ordered by the relation which partially orders P; the subset might turn out to be linearly ordered by this relation. If P is partially ordered and S is a subset of P, an $m \in P$ is called an upper bound of S if $s \prec m$ for every $s \in S$. An $m \in P$ is said to be *maximal* if $p \in P$ and $m \prec p$ together imply $m = p$.

Zorn's Lemma. Let P be a non-empty partially ordered set with the property that every linearly ordered subset of P has an upper bound in P. Then P contains at least one maximal element.

It is known that Zorn's lemma is equivalent to Zermelo's axiom of choice in set theory.

2. Topological Spaces

Open Sets and Closed Sets

Definition. A system τ of subsets of a set X defines a *topology* in X if τ contains the void set, the set X itself, the union of every one of its subsystems, and the intersection of every one of its finite subsystems. The sets in τ are called the *open sets* of the *topological space* (X, τ); we shall often omit τ and refer to X as a topological space. Unless otherwise stated, we shall assume that a topological space X satisfies *Hausdorff's axiom of separation*:

For every pair (x_1, x_2) of distinct points x_1, x_2 of X, there exist disjoint open sets G_1, G_2 such that $x_1 \in G_1$, $x_2 \in G_2$.

A *neighbourhood* of the point x of X is a set containing an open set which contains x. A neighbourhood of the subset M of X is a set which is a neighbourhood of every point of M. A point x of X is an *accumulation point* or *limit point* of a subset M of X if every neighbourhood of x contains at least one point $m \in M$ different from x.

Definition. Any subset M of a topological space X becomes a topological space by calling "open" the subsets of M which are of the form $M \cap G$ where G's are open sets of X. The induced topology of M is called the *relative topology* of M as a subset of the topological space X.

Definition. A set M of a topological space X is *closed* if it contains all its accumulation points. It is easy to see that M is closed iff[1] its *complement* $M^C = X - M$ is open. Here $A - B$ denotes the totality of points $x \in A$ not contained in B. If $M \subseteq X$, the intersection of all closed subsets of X which contain M is called the *closure* of M and will be denoted by M^a (the superscript "a" stands for the first letter of the German: abgeschlossene Hülle).

Clearly M^a is closed and $M \subseteq M^a$; it is easy to see that $M = M^a$ iff M is closed.

Metric Spaces

Definition. If X, Y are sets, we denote by $X \times Y$ the set of all ordered pairs (x, y) where $x \in X$ and $y \in Y$; $X \times Y$ will be called the *Cartesian product* of X and Y. X is called a *metric space* if there is defined a func-

[1] iff is the abbreviation for "if and only if".

tion d with domain $X \times X$ and range in the real number field R^1 such that

$$\begin{cases} d(x_1, x_2) \geqq 0 \text{ and } d(x_1, x_2) = 0 \text{ iff } x_1 = x_2, \\ d(x_1, x_2) = d(x_2, x_1), \\ d(x_1, x_3) \leqq d(x_1, x_2) + d(x_2, x_3) \text{ (the triangle inequality).} \end{cases}$$

d is called the *metric* or the *distance function* of X. With each point x_0 in a metric space X and each positive number r, we associate the set $S(x_0; r) = \{x \in X; d(x, x_0) < r\}$ and call it the *open sphere* with centre x_0 and radius r. Let us call "open" the set M of a metric space X iff, for every point $x_0 \in M$, M contains a sphere with centre x_0. Then the totality of such "open" sets satisfies the axiom of open sets in the definition of the topological space.

Hence a metric space X is a topological space. It is easy to see that a point x_0 of X is an accumulation point of M iff, to every $\varepsilon > 0$, there exists at least one point $m \neq x_0$ of M such that $d(m, x_0) < \varepsilon$. The n-dimensional Euclidean space R^n is a metric space by

$$d(x, y) = \left(\sum_{i=1}^{n} (x_i - y_i)^2\right)^{1/2}, \text{ where } x = (x_1, \ldots, x_n) \text{ and } y = (y_1, \ldots, y_n).$$

Continuous Mappings

Definition. Let $f: X \to Y$ be a mapping defined on a topological space X into a topological space Y. f is called *continuous at a point* $x_0 \in X$ if to every neighbourhood U of $f(x_0)$ there corresponds a neighbourhood V of x_0 such that $f(V) \subseteq U$. The mapping f is said to be *continuous* if it is continuous at every point of its domain $D(f) = X$.

Theorem. Let X, Y be topological spaces and f a mapping defined on X into Y. Then f is continuous iff the inverse image under f of every open set of Y is an open set of X.

Proof. If f is continuous and U an open set of Y, then $V = f^{-1}(U)$ is a neighbourhood of every point $x_0 \in X$ such that $f(x_0) \in U$, that is, V is a neighbourhood of every point x_0 of V. Thus V is an open set of X. Let, conversely, for every open set $U \ni f(x_0)$ of Y, the set $V = f^{-1}(U)$ be an open set of X. Then, by the definition, f is continuous at $x_0 \in X$.

Compactness

Definition. A system of sets G_α, $\alpha \in A$, is called a *covering* of the set X if X is contained as a subset of the union $\bigcup_{\alpha \in A} G_\alpha$. A subset M of a topological space X is called *compact* if every system of open sets of X which covers M contains a finite subsystem also covering M.

In view of the preceding theorem, *a continuous image of a compact set is also compact.*

Proposition 1. Compact subsets of a topological space are necessarily closed.

Proof. Let there be an accumulation point x_0 of a compact set M of a topological space X such that $x_0 \bar{\in} M$. By Hausdorff's axiom of separation, there exist, for any point $m \in M$, disjoint open sets G_{m,x_0} and $G_{x_0,m}$ of X such that $m \in G_{m,x_0}$, $x_0 \in G_{x_0,m}$. The system $\{G_{m,x_0};\ m \in M\}$ surely covers M. By the compactness of M, there exists a finite subsystem $\{G_{m_i,x_0};\ i = 1, 2, \ldots, n\}$ which covers M. Then $\bigcap_{i=1}^{n} G_{x_0,m_i}$ does not intersect M. But, since x_0 is an accumulation point of M, the open set $\bigcap_{i=1}^{n} G_{x_0,m_i} \ni x_0$ must contain a point $m \in M$ distinct from x_0. This is a contradiction, and M must be closed.

Proposition 2. A closed subset M_1 of a compact set M of a topological space X is compact.

Proof. Let $\{G_\alpha\}$ be any system of open sets of X which covers M_1. M_1 being closed, $M_1^C = X - M_1$ is an open set of X. Since $M_1 \subseteqq M$, the system of open sets $\{G_\alpha\}$ plus M_1^C covers M, and since M is compact, a properly chosen finite subsystem $\{G_{\alpha_i};\ i = 1, 2, \ldots, n\}$ plus M_1^C surely covers M. Thus $\{G_{\alpha_i};\ i = 1, 2, \ldots, n\}$ covers M_1.

Definition. A subset of a topological space is called *relatively compact* if its closure is compact. A topological space is said to be *locally compact* if each point of the space has a compact neighbourhood.

Theorem. Any locally compact space X can be embedded in another compact space Y, having just one more point than X, in such a way that the relative topology of X as a subset of Y is just the original topology of X. This Y is called a *one point compactification* of X.

Proof. Let y be any element distinct from the points of X. Let $\{U\}$ be the class of all open sets in X such that $U^C = X - U$ is compact. We remark that X itself $\in \{U\}$. Let Y be the set consisting of the points of X and the point y. A set in Y will be called open if either (i) it does not contain y and is open as a subset of X, or (ii) it does contain y and its intersection with X is a member of $\{U\}$. It is easy to see that Y thus obtained is a topological space, and that the relative topology of X coincides with its original topology.

Suppose $\{V\}$ be a family of open sets which covers Y. Then there must be some member of $\{V\}$ of the form $U_0 \cup \{y\}$, where $U_0 \in \{U\}$. By the definition of $\{U\}$, U_0^C is compact as a subset of X. It is covered by the system of sets $V \cap X$ with $V \in \{V\}$. Thus some finite subsystem: $V_1 \cap X, V_2 \cap X, \ldots, V_n \cap X$ covers U_0^C. Consequently, $V_1, V_2, \ldots, V_n$ and $U_0 \cup \{y\}$ cover Y, proving that Y is compact.

Tychonov's Theorem

Definition. Corresponding to each α of an index set A, let there be given a topological space X_α. The Cartesian product $\prod_{\alpha \in A} X_\alpha$ is, by definition, the set of all functions f with domain A such that $f(\alpha) \in X_\alpha$ for every $\alpha \in A$. We shall write $f = \prod_{\alpha \in A} f(\alpha)$ and call $f(\alpha)$ the α-th coordinate of f. When A consists of integers $(1, 2, \ldots, n)$, $\prod_{k=1}^{n} X_k$ is usually denoted by $X_1 \times X_2 \times \cdots \times X_n$. We introduce a (*weak*) *topology* in the *product space* $\prod_{\alpha \in A} X_\alpha$ by calling "open" the sets of the form $\prod_{\alpha \in A} G_\alpha$, where the open set G_α of X_α coincides with X_α for all but a finite set of α.

Tychonov's Theorem. The Cartesian product $X = \prod_{\alpha \in A} X_\alpha$ of a system of compact topological spaces X_α is also compact.

Remark. As is well known, a closed bounded set on the real line R^1 is compact with respect to the topology defined by the distance $d(x, y) = |x - y|$ (the Bolzano-Weierstrass theorem). By the way, a subset M of a metric space is said to be *bounded*, if M is contained in some sphere $S(x_0, r)$ of the space. Tychonoff's theorem implies, in particular, that a *parallelopiped*:

$$-\infty < a_i \leqq x_i \leqq b_i < \infty \qquad (i = 1, 2, \ldots, n)$$

of the n-dimensional Euclidean space R^n is compact. From this we see that R^n is locally compact.

Proof of Tychonov's Theorem. A system of sets has the *finite intersection property* if its every finite subsystem has a non-void intersection. It is easy to see, by taking the complement of the open sets of a covering, that a topological space X is compact iff, for every system $\{M_\alpha; \alpha \in A\}$ of its subsets with finite intersection property, the intersection $\bigcap_{\alpha \in A} M_\alpha^a$ is non-void.

Let now a system $\{S\}$ of subsets S of $X = \prod_{\alpha \in A} X_\alpha$ have the finite intersection property. Let $\{N\}$ be a system of subsets N of X with the following properties:

(i) $\{S\}$ is a subsystem of $\{N\}$,
(ii) $\{N\}$ has the finite intersection property,
(iii) $\{N\}$ is maximal in the sense that it is not a proper subsystem of other systems having the finite intersection property and containing $\{S\}$ as its subsystem.

The existence of such a maximal system $\{N\}$ can be proved by Zorn's lemma or transfinite induction.

For any set N of $\{N\}$ we define the set $N_\alpha = \{f(\alpha); f \in N\} \subseteq X_\alpha$. We denote then by $\{N_\alpha\}$ the system $\{N_\alpha; N \in \{N\}\}$. Like $\{N\}$, $\{N_\alpha\}$ enjoys the finite intersection property. Thus, by the compactness of X_α, there exists at least one point $p_\alpha \in X_\alpha$ such that $p_\alpha \in \bigcap_{N \in \{N\}} N_\alpha^a$. We have to show that the point $p = \prod_{\alpha \in A} p_\alpha$ belongs to the set $\bigcap_{N \in \{N\}} N^a$.

But since p_{α_0} belongs to the intersection $\bigcap_{N\in\{N\}} N^a_{\alpha_0}$, any open set G_{α_0} of X_{α_0} which contains p_{α_0} intersects every $N_{\alpha_0} \in \{N_{\alpha_0}\}$. Therefore the open set

$$G^{(\alpha_0)} = \{x;\, x = \prod_{\alpha\in A} x_\alpha \text{ with } x_{\alpha_0} \in G_{\alpha_0}\}$$

of X must intersect every N of $\{N\}$. By the maximality condition (iii) of $\{N\}$, $G^{(\alpha_0)}$ must belong to $\{N\}$. Thus the intersection of a finite number of sets of the form $G^{(\alpha_0)}$ with $\alpha_0 \in A$ must also belong to $\{N\}$ and so such a set intersect every set $N \in \{N\}$. Any open set of X containing p being defined as a set containing such an intersection, we see that $p = \prod_{\alpha\in A} p_\alpha$ must belong to the intersection $\bigcap_{N\in\{N\}} N^a$.

Urysohn's Theorem

Proposition. A compact space X is *normal* in the sense that, for any disjoint closed sets F_1 and F_2 of X, there exist disjoint open sets G_1 and G_2 such that $F_1 \subseteqq G_1$, $F_2 \subseteqq G_2$.

Proof. For any pair (x, y) of points such that $x \in F_1$, $y \in F_2$, there exist disjoint open sets $G(x, y)$ and $G(y, x)$ such that $x \in G(x, y)$, $y \in G(y, x)$. F_2 being compact as a closed subset of the compact space X, we can, for fixed x, cover F_2 by a finite number of open sets $G(y_1, x)$, $G(y_2, x), \ldots, G(y_{n(x)}, x)$. Set

$$G_x = \bigcup_{j=1}^{n(x)} G(y_j, x) \quad \text{and} \quad G(x) = \bigcap_{j+1}^{n(x)} G(x, y_j).$$

Then the disjoint open sets G_x and $G(x)$ are such that $F_2 \subseteqq G_x$, $x \in G(x)$. F_1 being compact as a closed subset of the compact space X, we can cover F_1 by a finite number of open sets $G(x_1), G(x_2), \ldots, G(x_k)$. Then

$$G_1 = \bigcup_{j=1}^{k} G(x_j) \quad \text{and} \quad G_2 = \bigcap_{j=1}^{k} G_{x_j}$$

satisfy the condition of the proposition.

Corollary. A compact space X is *regular* in the sense that, for any non-void open set G_1' of X, there exists a non-void open set G_2' such that $(G_2')^a \subseteqq G_1'$.

Proof. Take $F_1 = (G_1')^C$ and $F_2 = \{x\}$ where $x \in G_1'$. We can then take for G_2' the open set G_2 obtained in the preceding proposition.

Urysohn's Theorem. Let A, B be disjoint closed sets in a normal space X. Then there exists a real-valued continuous function $f(t)$ on X such that

$$0 \leqq f(t) \leqq 1 \text{ on } X, \text{ and } f(t) = 0 \text{ on } A,\ f(t) = 1 \text{ on } B.$$

Proof. We assign to each rational number $r = k/2^n$ $(k = 0, 1, \ldots, 2^n)$, an open set $G(r)$ such that (i) $A \subseteqq G(0)$, $B = G(1)^c$, and (ii) $G(r)^a \subseteqq G(r')$ whenever $r < r'$. The proof is obtained by induction with respect to n.

For $n = 0$, there exist, by the normality of the space X, disjoint open sets G_0 and G_1 with $A \subseteqq G_0$, $B \subseteqq G_1$. We have only to set $G_0 = G(0)$. Suppose that $G(r)$'s have been constructed for r of the form $k/2^{n-1}$ in such a way that condition (ii) is satisfied. Next let k be an odd integer > 0. Then, since $(k-1)/2^n$ and $(k+1)/2^n$ are of the form $k'/2^{n-1}$ with $0 \leqq k' \leqq 2^{n-1}$, we have $G((k-1)/2^n)^a \subseteqq G((k+1)/2^n)$. Hence, by the normality of the space X, there exists an open set G which satisfies $G((k-1)/2^n)^a \subseteqq G$, $G^a \subseteqq G((k+1)/2^n)$. If we set $G(k/2^n) = G$, the induction is completed.

Define $f(t)$ by

$$f(t) = 0 \text{ on } G(0), \text{ and } f(t) = \sup_{t \bar{\in} G(r)} r \text{ whenever } t \in G(0)^C.$$

Then, by (i), $f(t) = 0$ on A and $f(t) = 1$ on B. We have to prove the continuity of f. For any $t_0 \in X$ and positive integer n, we take r with $f(t_0) < r < f(t_0) + 2^{-n-1}$. Set $G = G(r) \cap G(r - 2^{-n})^{aC}$ (we set, for convention, $G(s) = \emptyset$ if $s < 0$ and $G(s) = X$ if $s > 1$). The open set G contains t_0. For, $f(t_0) < r$ implies $t_0 \in G(r)$, and $(r - 2^{-n-1}) < f(t_0)$ implies $t_0 \in G(r - 2^{-n-1})^C \subseteqq G(r - 2^{-n})^{aC}$. Now $t \in G$ implies $t \in G(r)$ and so $f(t) \leqq r$; similarly $t \in G$ implies $t \in G(r - 2^{-n})^{aC} \subseteqq G(r - 2^{-n})^C$ so that $r - 2^{-n} \leqq f(t)$. Therefore we have proved that $|f(t) - f(t_0)| \leqq 1/2^n$ whenever $t \in G$.

The Stone-Weierstrass Theorem

Weierstrass' Polynomial Approximation Theorem. Let $f(x)$ be a real-valued (or complex-valued) continuous function on the closed interval $[0, 1]$. Then there exists a sequence of polynomials $P_n(x)$ which converges, as $n \to \infty$, to $f(x)$ uniformly on $[0, 1]$. According to S. Bernstein, we may take

$$P_n(x) = \sum_{p=0}^{n} {}_nC_p\, f(p/n)\, x^p (1-x)^{n-p}. \tag{1}$$

Proof. Differentiating $(x+y)^n = \sum_{p=0}^{n} {}_nC_p\, x^p y^{n-p}$ with respect to x and multiplying by x, we obtain $nx(x+y)^{n-1} = \sum_{p=0}^{n} p\, {}_nC_p\, x^p y^{n-p}$. Similarly, by differentiating the first expression twice with respect to x and multiplying by x^2, we obtain $n(n-1)x^2(x+y)^{n-2} = \sum_{p=0}^{n} p(p-1)\, {}_nC_p\, x^p y^{n-p}$. Thus, if we set

$$r_p(x) = {}_nC_p\, x^p (1-x)^{n-p}, \tag{2}$$

we have

$$\sum_{p=0}^{n} r_p(x) = 1, \quad \sum_{p=0}^{n} p\, r_p(x) = nx, \quad \sum_{p=0}^{n} p(p-1)\, r_p(x) = n(n-1)x^2. \tag{3}$$

Hence

$$\sum_{p=0}^{n} (p - nx)^2 r_p(x) = n^2 x^2 \sum_{p=0}^{n} r_p(x) - 2nx \sum_{p=0}^{n} p r_p(x) + \sum_{p=0}^{n} p^2 r_p(x)$$
$$= n^2 x^2 - 2nx \cdot nx + (nx + n(n-1)x^2)$$
$$= nx(1-x). \tag{4}$$

We may assume that $|f(x)| \leqq M < \infty$ on $[0, 1]$. By the uniform continuity of $f(x)$, there exists, for any $\varepsilon > 0$, a $\delta > 0$ such that

$$|f(x) - f(x')| < \varepsilon \text{ whenever } |x - x'| < \delta. \tag{5}$$

We have, by (3),

$$\left| f(x) - \sum_{p=0}^{n} f(p/n) r_p(x) \right| = \left| \sum_{p=0}^{n} (f(x) - f(p/n)) r_p(x) \right|$$
$$\leqq \left| \sum_{|p-nx| \leqq \delta n} \right| + \left| \sum_{|p-nx| > \delta n} \right| .$$

For the first term on the right, we have, by $r_p(x) \geqq 0$ and (3),

$$\left| \sum_{|p-nx| \leqq \delta n} \right| \leqq \varepsilon \sum_{p=0}^{n} r_p(x) = \varepsilon .$$

For the second term on the right, we have, by (4) and $|f(x)| \leqq M$,

$$\left| \sum_{|p-nx| > \delta n} \right| \leqq 2M \sum_{|p-nx| > \delta n} r_p(x) \leqq \frac{2M}{n^2 \delta^2} \sum_{p=0}^{n} (p - nx)^2 r_p(x)$$
$$= \frac{2Mx(1-x)}{n\delta^2} \leqq \frac{M}{2\delta^2 n} \to 0 \quad (\text{as } n \to \infty).$$

The Stone-Weierstrass Theorem. Let X be a compact space and $C(X)$ the totality of real-valued continuous functions on X. Let a subset B of $C(X)$ satisfy the three conditions: (i) if $f, g \in B$, then the function product $f \cdot g$ and linear combinations $\alpha f + \beta g$, with real coefficients α, β, belong to B, (ii) the constant function 1 belongs to B, and (iii) the uniform limit f_∞ of any sequence $\{f_n\}$ of functions $\in B$ also belongs to B. Then $B = C(X)$ iff B *separates the points of* X, i.e. iff, for every pair (s_1, s_2) of distinct points of X, there exists a function x in B which satisfies $x(s_1) \neq x(s_2)$.

Proof. The necessity is clear, since a compact space is normal and so, by Urysohn's theorem, there exists a real-valued continuous function x such that $x(s_1) \neq x(s_2)$.

To prove the sufficiency, we introduce the *lattice notations*:

$$(f \vee g)(s) = \max(f(s), g(s)), \ (f \wedge g)(s) = \min(f(s), g(s)), \ |f|(s) = |f(s)|.$$

By the preceding theorem, there is a sequence $\{P_n\}$ of polynomials such that

$$\big| |t| - P_n(t) \big| < 1/n \quad \text{for} \quad -n \leqq t \leqq n.$$

Hence $||f(s)| - P_n(f(s))| < 1/n$ if $-n \leqq f(s) \leqq n$. This proves, by (iii), that $|f| \in B$ if $f \in B$, because any function $f(s) \in B \subseteqq C(X)$ is bounded on the compact space X. Thus, by

$$f \vee g = \frac{f+g}{2} + \frac{|f-g|}{2} \quad \text{and} \quad f \wedge g = \frac{f+g}{2} - \frac{|f-g|}{2},$$

we see that B is closed under the lattice operations $\vee$ and $\wedge$.

Let $h \in C(X)$ and $s_1, s_2 \in X$ be arbitrarily given such that $s_1 \neq s_2$. Then we can find an $f_{s_1 s_2} \in B$ with $f_{s_1 s_2}(s_1) = h(s_1)$ and $f_{s_1 s_2}(s_2) = h(s_2)$. To see this, let $g \in B$ be such that $g(s_1) \neq g(s_2)$, and take real numbers α and β so that $f_{s_1 s_2} = \alpha g + \beta$ satisfies the conditions: $f_{s_1 s_2}(s_1) = h(s_1)$ and $f_{s_1 s_2}(s_2) = h(s_2)$.

Given $\varepsilon > 0$ and a point $t \in X$. Then, for each $s \in X$, there is a neighbourhood $U(s)$ of s such that $f_{st}(u) > h(u) - \varepsilon$ whenever $u \in U(s)$. Let $U(s_1), U(s_2), \ldots, U(s_n)$ cover the compact space X and define

$$f_t = f_{s_1 t} \vee \cdots \vee f_{s_n t}.$$

Then $f_t \in B$ and $f_t(u) > h(u) - \varepsilon$ for all $u \in X$. We have, by $f_{s_j t}(t) = h(t)$, $f_t(t) = h(t)$. Hence there is a neighbourhood $V(t)$ of t such that $f_t(u) < h(u) + \varepsilon$ whenever $u \in V(t)$. Let $V(t_1), V(t_2), \ldots, V(t_k)$ cover the compact space X, and define

$$f = f_{t_1} \wedge \cdots \wedge f_{t_k}.$$

Then $f \in B$ and $f(u) > h(u) - \varepsilon$ for all $u \in X$, because $f_{t_j}(u) > h(u) - \varepsilon$ for $u \in X$. Moreover, we have, for an arbitrary point $u \in X$, say $u \in V(t_i)$, $f(u) \leqq f_{t_i}(u) < h(u) + \varepsilon$.

Therefore we have proved that $|f(u) - h(u)| < \varepsilon$ on X.

We have incidentally proved the following two corollaries.

Corollary 1 (Kakutani-Krein). Let X be a compact space and $C(X)$ the totality of real-valued continuous functions on X. Let a subset B of $C(X)$ satisfy the conditions: (i) if $f, g \in B$, then $f \vee g$, $f \wedge g$ and the linear combinations $\alpha f + \beta g$, with real coefficients α, β, belong to B, (ii) the constant function 1 belongs to B, and (iii) the uniform limit f_∞ of any sequence $\{f_n\}$ of functions $\in B$ also belongs to B. Then $B = C(X)$ iff B separates the points of X.

Corollary 2. Let X be a compact space and $C(X)$ be the totality of complex-valued continuous functions on X. Let a subset B of $C(X)$ satisfy the conditions: (i) if $f, g \in B$, then the function product $f \cdot g$ and the linear combinations $\alpha f + \beta g$, with complex coefficients α, β, belong to B, (ii) the constant function 1 belongs to B, and (iii) the uniform limit f_∞ of any sequence $\{f_n\}$ of functions $\in B$ also belongs to B. Then $B = C(X)$ iff B satisfies the conditions: (iv) B separates points of X and (v) if $f(s) \in B$, then its complex conjugate function $\bar{f}(s)$ also belongs to B.

Weierstrass' Trigonometric Approximation Theorem. Let X be the circumference of the unit circle of R^2. It is a compact space by the usual topology, and a complex-valued continuous function on X is represented by a continuous function $f(x)$, $-\infty < x < \infty$, of period 2π. If we take, in the above Corollary 2, for B the set of all functions representable by linear combinations, with complex coefficients of the trigonometric functions

$$e^{inx} \ (n = 0, \pm 1, \pm 2, \ldots)$$

and by those functions obtainable as the uniform limit of such linear combinations, we obtain *Weierstrass' trigonometric approximation theorem*: Any complex-valued continuous function $f(x)$ with period 2π can be approximated uniformly by a sequence of *trigonometric polynomials* of the form $\sum_n c_n e^{inx}$.

Completeness

A sequence $\{x_n\}$ of elements in a metric space X converges to a limit point $x \in X$ iff $\lim_{n\to\infty} d(x_n, x) = 0$. By the triangle inequality $d(x_n, x_m) \leqq d(x_n, x) + d(x, x_m)$, we see that a convergent sequence $\{x_n\}$ in X satisfies *Cauchy's convergence condition*

$$\lim_{n,m\to\infty} d(x_n, x_m) = 0. \tag{1}$$

Definition. Any sequence $\{x_n\}$ in a metric space X satisfying the above condition (1) is called a *Cauchy sequence*. A metric space X is said to be *complete* if every Cauchy sequence in it converges to a limit point $\in X$.

It is easy to see, by the triangle inequality, that the limit point of $\{x_n\}$, if it exists, is uniquely determined.

Definition. A subset M of a topological space X is said to be *non-dense* in X if the closure M^a does not contain a non-void open set of X. M is called *dense* in X if $M^a = X$. M is said to be of *the first category* if M is expressible as the union of a countable number of sets each of which is non-dense in X; otherwise M is said to be of *the second category*.

Baire's Category Argument

The Baire-Hausdorff Theorem. A non-void complete metric space is of the second category.

Proof. Let $\{M_n\}$ be a sequence of closed sets whose union is a complete metric space X. Assuming that no M_n contains a non-void open set, we shall derive a contradiction. Thus M_1^C is open and $M_1^{Ca} = X$, hence M_1^C contains a closed sphere $S_1 = \{x; d(x_1, x) \leqq r_1\}$ whose centre x_1 may be taken arbitrarily near to any point of X. We may assume that $0 < r_1 < 1/2$. By the same argument, the open set M_2^C contains a closed sphere

$S_2 = \{x; d(x_2, x) \leqq r_2\}$ contained in S_1 and such that $0 < r_2 < 1/2^2$. By repeating the same argument, we obtain a sequence $\{S_n\}$ of closed spheres $S_n = \{x; d(x_n, x) \leqq r_n\}$ with the properties:

$$0 < r_n < 1/2^n, \ S_{n+1} \subseteqq S_n, \ S_n \cap M_n = \emptyset \qquad (n = 1, 2, \ldots).$$

The sequence $\{x_n\}$ of the centres forms a Cauchy sequence, since, for any $n < m$, $x_m \in S_n$ so that $d(x_n, x_m) \leqq r_n < 1/2^n$. Let $x_\infty \in X$ be the limit point of this sequence $\{x_n\}$. The completeness of X guarantees the existence of such a limit point x_∞. By $d(x_n, x_\infty) \leqq d(x_n, x_m) + d(x_m, x_\infty) \leqq r_n + d(x_m, x_\infty) \to r_n$ (as $m \to \infty$), we see that $x_\infty \in S_n$ for every n. Hence x_∞ is in none of the sets M_n, and hence x_∞ is not in the union $\bigcup_{n=1}^{\infty} M_n = X$, contrary to $x_\infty \in X$.

Baire's Theorem 1. Let M be a set of the first category in a compact topological space X. Then the complement $M^C = X - M$ is dense in X.

Proof. We have to show that, for any non-void open set G, M^C intersects G. Let $M = \bigcup_{n=1}^{\infty} M_n$ where each M_n is a non-dense closed set. Since $M_1 = M_1^a$ is non-dense, the open set M_1^C intersects G. Since X is regular as a compact space, there exists a non-void open set G_1 such that $G_1^a \subseteqq G \cap M_1^C$. Similarly, we can choose a non-void open set G_2 such that $G_2^a \subseteqq G_1 \cap M_2^C$. Repeating the process, we obtain a sequence of non-void open sets $\{G_n\}$ such that

$$G_{n+1}^a \subseteqq G_n \cap M_{n+1}^C \qquad (n = 1, 2, \ldots).$$

The sequence of closed sets $\{G_n^a\}$ enjoys, by the monotony in n, the finite intersection property. Since X is compact, there is an $x \in X$ such that $x \in \bigcap_{n=1}^{\infty} G_n^a$. $x \in G_1^a$ implies $x \in G$, and from $x \in G_{n+1}^a \subseteqq G_n \cap M_{n+1}^C$ ($n = 0, 1, 2, \ldots;\ G_0 = G$), we obtain $x \in \bigcap_{n=1}^{\infty} M_n^C = M^C$. Therefore we have proved that $G \cap M^C$ is non-void.

Baire's Theorem 2. Let $\{x_n(t)\}$ be a sequence of real-valued continuous functions defined on a topological space X. Suppose that a finite limit:

$$\lim_{n\to\infty} x_n(t) = x(t)$$

exists at every point t of X. Then the set of points at which the function x is discontinuous constitutes a set of the first category.

Proof. We denote, for any set M of X, by M^i the union of all the open sets contained in M; M^i will be called the *interior* of M.

Put $P_m(\varepsilon) = \{t \in X;\ |x(t) - x_m(t)| \leqq \varepsilon, \varepsilon > 0\}$, $G(\varepsilon) = \bigcup_{m=1}^{\infty} P_m^i(\varepsilon)$.

Then we can prove that $C = \bigcap_{n=1}^{\infty} G(1/n)$ coincides with the set of all points at which $x(t)$ is continuous. Suppose $x(t)$ is continuous at $t = t_0$.

We shall show that $t_0 \in \bigcap_{n=1}^{\infty} G(1/n)$. Since $\lim_{n\to\infty} x_n(t) = x(t)$, there exists an m such that $|x(t_0) - x_m(t_0)| \leqq \varepsilon/3$. By the continuity of $x(t)$ and $x_m(t)$ at $t = t_0$, there exists an open set $U_{t_0} \ni t_0$ such that $|x(t) - x(t_0)| \leqq \varepsilon/3$ $|x_m(t) - x_m(t_0)| \leqq \varepsilon/3$ whenever $t \in U_{t_0}$. Thus $t \in U_{t_0}$ implies

$$|x(t) - x_m(t)| \leqq |x(t) - x(t_0)| + |x(t_0) - x_m(t_0)| + |x_m(t_0) - x_m(t)| < \varepsilon,$$

which proves that $t_0 \in P_m^i(\varepsilon)$ and so $t_0 \in G(\varepsilon)$. Since $\varepsilon > 0$ was arbitrary, we must have $t_0 \in \bigcap_{n=1}^{\infty} G(1/n)$.

Let, conversely, $t_0 \in \bigcap_{n=1}^{\infty} G(1/n)$. Then, for any $\varepsilon > 0$, $t_0 \in G(\varepsilon/3)$ and so there exists an m such that $t_0 \in P_m^i(\varepsilon/3)$. Thus there is an open set $U_{t_0} \ni t_0$ such that $t \in U_{t_0}$ implies $|x(t) - x_m(t)| \leqq \varepsilon/3$. Hence, by the continuity of $x_m(t)$ and the arbitrariness of $\varepsilon > 0$, $x(t)$ must be continuous at $t = t_0$.

After these preparations, we put

$$F_m(\varepsilon) = \{t \in X;\ |x_m(t) - x_{m+k}(t)| \leqq \varepsilon \quad (k = 1, 2, \ldots)\}.$$

This is a closed set by the continuity of the $x_n(t)$'s. We have $X = \bigcup_{m=1}^{\infty} F_m(\varepsilon)$ by $\lim_{n\to\infty} x_n(t) = x(t)$. Again by $\lim_{n\to\infty} x_n(t) = x(t)$, we have $F_m(\varepsilon) \subseteqq P_m(\varepsilon)$. Thus $F_m^i(\varepsilon) \subseteqq P_m^i(\varepsilon)$ and so $\bigcup_{m=1}^{\infty} F_m^i(\varepsilon) \subseteqq G(\varepsilon)$. On the other hand, for any closed set F, $(F - F^i)$ is a non-dense closed set. Thus $X - \bigcup_{m=1}^{\infty} F_m^i(\varepsilon)$ $= \bigcup_{m=1}^{\infty} (F_m(\varepsilon) - F_m^i(\varepsilon))$ is a set of the first category. Thus its subset $G(\varepsilon)^C = X - G(\varepsilon)$ is also a set of the first category. Therefore the set of all the points of discontinuity of the function $x(t)$, which is expressible as $X - \bigcap_{n=1}^{\infty} G(1/n) = \bigcup_{n=1}^{\infty} G(1/n)^C$, is a set of the first category.

Theorem. A subset M of a complete metric space X is relatively compact iff it is *totally bounded* in the sense that, for every $\varepsilon > 0$, there exists a finite system of points $m_1, m_2, \ldots, m_n$ of M such that every point m of M has a distance $< \varepsilon$ from at least one of $m_1, m_2, \ldots, m_n$. In other words, M is totally bounded if, for every $\varepsilon > 0$, M can be covered by a finite system of spheres of radii $< \varepsilon$ and centres $\in M$.

Proof. Suppose M is not totally bounded. Then there exist a positive number ε and an infinite sequence $\{m_n\}$ of points $\in M$ such that $d(m_i, m_j) \geqq \varepsilon$ for $i \neq j$. Then, if we cover the compact set M^a by a system of open spheres of radii $< \varepsilon$, no finite subsystem of this system can cover M^a. For, this subsystem cannot cover the infinite subset $\{m_i\} \subseteqq M \subseteqq M^a$. Thus a relatively compact subset of X must be totally bounded.

Suppose, conversely, that M is totally bounded as a subset of a complete metric space X. Then the closure M^a is complete and is totally bounded with M. We have to show that M^a is compact. To this purpose, we shall first show that any infinite sequence $\{p_n\}$ of M^a contains a subsequence $\{p_{(n)'}\}$ which converges to a point of M^a. Because of the total boundedness of M, there exist, for any $\varepsilon > 0$, a point $q_\varepsilon \in M^a$ and a subsequence $\{p_{n'}\}$ of $\{p_n\}$ such that $d(p_{n'}, q_\varepsilon) < \varepsilon/2$ for $n = 1, 2, \ldots$; consequently, $d(p_{n'}, p_{m'}) \leqq d(p_{n'}, q_\varepsilon) + d(q_\varepsilon, p_{m'}) < \varepsilon$ for $n, m = 1, 2, \ldots$ We set $\varepsilon = 1$ and obtain the sequence $\{p_{i'}\}$, and then apply the same reasoning as above with $\varepsilon = 2^{-1}$ to this sequence $\{p_{i'}\}$. We thus obtain a subsequence $\{p_{n''}\}$ of $\{p_{n'}\}$ such that

$$d(p_{n'}, p_{m'}) < 1, \; d(p_{n''}, p_{m''}) < 1/2 \qquad (n, m = 1, 2, \ldots).$$

By repeating the process, we obtain a subsequence $\{p_{n^{(k+1)}}\}$ of the sequence $\{p_{n^{(k)}}\}$ such that

$$d(p_{n^{(k+1)}}, p_{m^{(k+1)}}) < 1/2^k \qquad (n, m = 1, 2, \ldots).$$

Then the subsequence $\{p_{(n)'}\}$ of the original sequence $\{p_n\}$, defined by the diagonal method:

$$p_{(n)'} = p_{n^{(n)}},$$

surely satisfies $\lim\limits_{n,m\to\infty} d(p_{(n)'}, p_{(m)'}) = 0$. Hence, by the completeness of M^a, there must exist a point $p \in M^a$ such that $\lim\limits_{n\to\infty} d(p_{(n)'}, p) = 0$.

We next show that the set M^a is compact. We remark that there exists a countable family $\{F\}$ of open sets F of X such that, if U is any open set of X and $x \in U \cap M^a$, there is a set $F \in \{F\}$ for which $x \in F \subseteqq U$. This may be proved as follows. M^a being totally bounded, M^a can be covered, for any $\varepsilon > 0$, by a finite system of open spheres of radii ε and centres $\in M^a$. Letting $\varepsilon = 1, 1/2, 1/3, \ldots$ and collecting the countable family of the corresponding finite systems of open spheres, we obtain the desired family $\{F\}$ of open sets.

Let now $\{U\}$ be any open covering of M^a. Let $\{F^*\}$ be the subfamily of the family $\{F\}$ defined as follows: $F \subseteqq \{F^*\}$ iff $F \subseteqq \{F\}$ and there is some $U \in \{U\}$ with $F \subseteqq U$. By the property of $\{F\}$ and the fact that $\{U\}$ covers M^a, we see that this countable family $\{F^*\}$ of open sets covers M^a. Now let $\{U^*\}$ be a subfamily of $\{U\}$ obtained by selecting just one $U \in \{U\}$ such that $F \subseteqq U$, for each $F \in \{F^*\}$. Then $\{U^*\}$ is a countable family of open sets which covers M^a. We have to show that some finite subfamily of $\{U^*\}$ covers M^a. Let the sets in $\{U^*\}$ be indexed as U_1, $U_2, \ldots$ Suppose that, for each n, the finite union $\bigcup\limits_{j=1}^{n} U_j$ fails to cover M^a. Then there is some point $x_n \in \left(M - \bigcup\limits_{k=1}^{n} U_k\right)$. By what was proved above, the sequence $\{x_n\}$ contains a subsequence $\{x_{(n)'}\}$ which converges to a point, say x_∞, in M^a. Then $x_\infty \in U_N$ for some index N, and so $x_n \in U_N$ for infinitely many values of n, in particular for an $n > N$. This contradicts the fact that x_n was chosen so that $x_n \in \left(M - \bigcup_{k=1}^{n} U_k\right)$. Hence we have proved that M^a is compact.

3. Measure Spaces

Measures

Definition. Let S be a set. A pair $(S, \mathfrak{B})$ is called a *σ-ring* or a *σ-additive family* of sets $\subseteqq S$ if $\mathfrak{B}$ is a family of subsets of S such that

$$S \in \mathfrak{B}, \tag{1}$$

$$B \in \mathfrak{B} \text{ implies } B^C = (S - B) \in \mathfrak{B}, \tag{2}$$

$$B_j \in \mathfrak{B} \ (j = 1, 2, \ldots) \text{ implies that } \bigcup_{j=1}^{\infty} B_j \in \mathfrak{B} \ (\sigma\textit{-additivity}). \tag{3}$$

Let $(S, \mathfrak{B})$ be a σ-ring of sets $\in S$. Then a triple $(S, \mathfrak{B}, m)$ is called a *measure space* if m is a non-negative, σ-additive measure defined on $\mathfrak{B}$:

$$m(B) \geqq 0 \text{ for every } B \in \mathfrak{B}, \tag{4}$$

$$m\left(\sum_{j=1}^{\infty} B_j\right) = \sum_{j=1}^{\infty} m(B_j) \text{ for any disjoint sequence } \{B_j\} \text{ of sets } \in \mathfrak{B}$$

$$(\textit{countable- or } \sigma\textit{-additivity of } m), \tag{5}$$

S is expressible as a countable union of sets $B_j \in \mathfrak{B}$ such that $m(B_j) < \infty$ $(j = 1, 2, \ldots)$ (*σ-finiteness* of the measure space $(S, \mathfrak{B}, m)$). (6)
This value $m(B)$ is called the *m-measure* of the set B.

Measurable Functions

Definition. A real- (or complex-) valued function $x(s)$ defined on S is said to be $\mathfrak{B}$-*measurable* or, in short, *measurable* if the following condition is satisfied:

For any open set G of the real line R^1 (or complex plane C^1), the set $\{s; x(s) \in G\}$ belongs to $\mathfrak{B}$. (7)

It is permitted that $x(s)$ takes the value ∞.

Definition. A property P pertaining to points s of S is said to hold *m-almost everywhere* or, in short *m-a. e.*, if it holds except for those s which form a set $\in \mathfrak{B}$ of m-measure zero.

A real- (or complex-) valued function $x(s)$ defined m-a.e. on S and satisfying condition (7) shall be called a *$\mathfrak{B}$-measurable function defined m-a.e. on S* or, in short, a *$\mathfrak{B}$-measurable function*.

Egorov's Theorem. If B is a $\mathfrak{B}$-measurable set with $m(B) < \infty$ and if $\{f_n(s)\}$ is a sequence of $\mathfrak{B}$-measurable functions, finite m-a. e. on B, that converges m-a. e. on B to a finite $\mathfrak{B}$-measurable function $f(s)$, then there exists, for each $\varepsilon > 0$, a subset E of B such that $m(B-E) \leqq \varepsilon$ and on E the convergence of $f_n(s)$ to $f(s)$ is uniform.

Proof. By removing from B, if necessary, a set of m-measure zero, we may suppose that on B, the functions $f_n(s)$ are everywhere finite, and converge to $f(s)$ on B.

The set $B_n = \bigcap_{k=n+1}^{\infty} \{s \in B;\ |f(s) - f_k(s)| < \varepsilon\}$ is $\mathfrak{B}$-measurable and $B_n \subseteqq B_k$ if $n < k$. Since $\lim_{n\to\infty} f_n(s) = f(s)$ on B, we have $B = \bigcup_{n=1}^{\infty} B_n$. Thus, by the σ-additivity of the measure m, we have

$$\begin{aligned} m(B) &= m\{B_1 + (B_2 - B_1) + (B_3 - B_2) + \cdots\} \\ &= m(B_1) + m(B_2 - B_1) + m(B_3 - B_2) + \cdots \\ &= m(B_1) + (m(B_2) - m(B_1)) + (m(B_3) - m(B_2)) + \cdots \\ &= \lim_{n\to\infty} m(B_n)\,. \end{aligned}$$

Hence $\lim_{n\to\infty} m(B - B_n) = 0$, and therefore, from a sufficiently large k_0 on, $m(B - B_k) < \eta$ where η is any given positive number.

Thus there exist, for any positive integer k, a set $C_k \subseteqq B$ such that $m(C_k) \leqq \varepsilon/2^k$ and an index N_k such that

$$|f(s) - f_n(s)| < 1/2^k \ \text{ for } n > N_k \text{ and for } s \in B - C_k.$$

Let us set $E = B - \bigcup_{k=1}^{\infty} C_k$. Then we find

$$m(B-E) \leqq \sum_{k=1}^{\infty} m(C_k) \leqq \sum_{k=1}^{\infty} \varepsilon/2^k = \varepsilon,$$

and the sequence $f_n(s)$ converges uniformly on E.

Integrals

Definition. A real- (or complex-) valued function $x(s)$ defined on S is said to be *finitely-valued* if it is a finite non-zero constant on each of a finite number, say n, of disjoint $\mathfrak{B}$-measurable sets B_j and equal to zero on $S - \bigcup_{j=1}^{n} B_j$. Let the value of $x(s)$ on B_j be denoted by x_j. Then $x(s)$ is *m-integrable* or, in short, *integrable* over S if $\sum_{j=1}^{n} |x_j|\, m(B_j) < \infty$, and the value $\sum_{j=1}^{n} x_j\, m(B_j)$ is defined as the *integral* of $x(s)$ over S with respect to the measure m; the integral will be denoted by $\int_S x(s)\, m(ds)$

or, in short, by $\int_S x(s)$ or simply by $\int x(s)$ when no confusion can be expected. A real- (or complex-) valued function $x(s)$ defined m-a. e. on S is said to be *m-integrable* or, in short, *integrable* over S if there exists a sequence $\{x_n(s)\}$ of finitely-valued integrable functions converging to $x(s)$ m-a. e. and such that

$$\lim_{n,k\to\infty} \int_S |x_n(s) - x_k(s)|\, m(ds) = 0.$$

It is then proved that a finite $\lim_{n\to\infty} \int_S x_n(s)\, m(ds)$ exists and the value of this limit is independent of the choice of the approximating sequence $\{x_n(s)\}$. The value of the integral $\int_S x(s)\, m(ds)$ over S with respect to the measure m is, by definition, given by $\lim_{n\to\infty} \int_S x_n(s)\, m(ds)$. We shall sometimes abbreviate the notation $\int_S x(s)\, m(ds)$ to $\int x(s)\, m(ds)$ or to $\int x(s)$.

Properties of the Integral

i) If $x(s)$ and $y(s)$ are integrable, then $\alpha x(s) + \beta y(s)$ is integrable and $\int_S (\alpha x(s) + \beta y(s))\, m(ds) = \alpha \int_S x(s)\, m(ds) + \beta \int_S y(s)\, m(ds)$.

ii) $x(s)$ is integrable iff $|x(s)|$ is integrable.

iii) If $x(s)$ is integrable and $x(s) \geqq 0$ a. e., then $\int_S x(s)\, m(ds) \geqq 0$, and the equality sign holds iff $x(s) = 0$ a. e.

iv) If $x(s)$ is integrable, then the function $X(B) = \int_B x(s)\, m(ds)$ is σ-additive, that is, $X\left(\sum_{j=1}^{\infty} B_j\right) = \sum_{j=1}^{\infty} X(B_j)$ for any disjoint sequence $\{B_j\}$ of sets $\in \mathfrak{B}$. Here $\int_B x(s)\, m(ds) = \int_S C_B(s)\, x(s)\, m(ds)$, where $C_B(s)$ is the *defining function* of the set B, that is,

$$C_B(s) = 1 \text{ for } s \in B \text{ and } C_B(s) = 0 \text{ for } s \in S - B.$$

v) $X(B)$ in iv) is *absolutely continuous* with respect to m in the sense that $m(B) = 0$ implies $X(B) = 0$. This condition is equivalent to the condition that $\lim_{m(B)\to 0} X(B) = 0$ uniformly in $B \in \mathfrak{B}$.

The Lebesgue-Fatou Lemma. Let $\{x_n(s)\}$ be a sequence of real-valued integrable functions. If there exists a real-valued integrable function $x(s)$ such that $x(s) \geqq x_n(s)$ a. e. for $n = 1, 2, \ldots$ (or $x(s) \leqq x_n(s)$ a. e. for $n = 1, 2, \ldots$), then

$$\int_S \left(\overline{\lim_{n\to\infty}}\, x_n(s)\right) m(ds) \geqq \overline{\lim_{n\to\infty}} \int_S x_n(s)\, m(ds)$$

$$\left(\text{or } \int_S \left(\underline{\lim_{n\to\infty}}\, x_n(s)\right) m(ds) \leqq \underline{\lim_{n\to\infty}} \int_S x_n(s)\, m(ds)\right),$$

under the convention that if $\overline{\lim}_{n\to\infty} x_n(s)$ $\left(\text{or } \underline{\lim}_{n\to\infty} x_n(s)\right)$ is not integrable, we understand that $\overline{\lim}_{n\to\infty} \int_S x_n(s)\, m(ds) = -\infty$ $\left(\text{or } \underline{\lim}_{n\to\infty} \int_S x_n(s)\, m(ds) = \infty\right)$.

Definition. Let $(S, \mathfrak{B}, m)$ and $(S', \mathfrak{B}', m')$ be two measure spaces. We denote by $\mathfrak{B}\times\mathfrak{B}'$ the smallest σ-ring of subsets of $S\times S'$ which contains all the sets of the form $B\times B'$, where $B\in\mathfrak{B}$, $B'\in\mathfrak{B}'$. It is proved that there exists a uniquely determined σ-finite, σ-additive and non-negative measure $m\times m'$ defined on $\mathfrak{B}\times\mathfrak{B}'$ such that

$$(m\times m')\,(B\times B') = m(B)\, m'(B').$$

$m\times m'$ is called the *product measure* of m and m'. We may define the $\mathfrak{B}\times\mathfrak{B}'$-measurable functions $x(s, s')$ defined on $S\times S'$, and the $m\times m'$-integrable functions $x(s, s')$. The value of the integral over $S\times S'$ of an $m\times m'$-integrable function $x(s, s')$ will be denoted by

$$\iint_{S\times S'} x(s, s')\,(m\times m')\,(ds\, ds') \quad \text{or} \quad \iint_{S\times S'} x(s, s')\, m(ds)\, m'(ds').$$

The Fubini-Tonelli Theorem. A $\mathfrak{B}\times\mathfrak{B}'$-measurable function $x(s, s')$ is $m\times m'$-integrable over $S\times S'$ iff at least one of the iterated integrals

$$\int_{S'}\left\{\int_S |x(s, s')|\; m(ds)\right\} m'(ds') \quad \text{and} \quad \int_S\left\{\int_{S'} |x(s, s')|\; m'(ds')\right\} m(ds)$$

is finite; and in such a case we have

$$\begin{aligned}\iint_{S\times S'} x(s, s')\, m(ds)\, m'(ds') &= \int_{S'}\left\{\int_S x(s, s')\, m(ds)\right\} m'(ds')\\ &= \int_S\left\{\int_{S'} x(s, s')\, m'(ds')\right\} m(ds).\end{aligned}$$

Topological Measures

Definition. Let S be a locally compact space, e.g., an n-dimensional Euclidean space R^n or a closed subset of R^n. The *Baire subsets* of S are the members of the smallest σ-ring of subsets of S which contains every compact G_δ-set, i.e., every compact set of X which is the intersection of a countable number of open sets of S. The *Borel subsets* of S are the members of the smallest σ-ring of subsets of S which contains every compact set of S.

If S is a closed subset of a Euclidean space R^n, the Baire and the Borel subsets of S coincide, because in R^n every compact (closed bounded) set is a G_δ-set. If, in particular, S is a real line R^1 or a closed interval on R^1, the Baire (= Borel) subsets of S may also be defined as the members of the smallset σ-ring of subsets of S which contains half open intervals $(a, b]$.

Definition. Let S be a locally compact space. Then a non-negative *Baire* (*Borel*) *measure* on S is a σ-additive measure defined for every Baire

(Borel) subset of S such that the measure of every compact set is finite. The Borel measure m is called *regular* if for each Borel set B we have

$$m(B) = \inf_{U \supseteq B} m(U)$$

where the infimum is taken over all open sets U containing B. We may also define the regularity for Baire measures in a similar way, but it turns out that a Baire measure is always regular. It is also proved that each Baire measure has a uniquely determined extension to a regular Borel measure. Thus we shall discuss only Baire measures.

Definition. A complex-valued function $f(s)$ defined on a locally compact space S is a *Baire function* on S if $f^{-1}(B)$ is a Baire set of S for every Baire set B in the complex plane C^1. Every continuous function is a Baire function if S is a countable union of compact sets. A Baire function is measurable with respect to the σ-ring of all Baire sets of S.

The Lebesgue Measure

Definition. Suppose S is the real line R^1 or a closed interval of R^1. Let $F(x)$ be a monotone non-decreasing function on S which is continuous from the right: $F(x) = \inf_{x<y} F(y)$. Define a function m on half closed intervals $(a, b]$ by $m((a, b]) = F(b) - F(a)$. This m has a uniquely determined extension to a non-negative Baire measure on S. The extended measure m is finite, i.e., $m(S) < \infty$ iff F is bounded. If m is the Baire measure induced by the function $F(s) = s$, then m is called the *Lebesgue measure*. The Lebesgue measure in R^n is obtained from the n-tuple of the one-dimensional Lebesgue measures through the process of forming the product measure.

Concerning the Lebesgue measure and the corresponding *Lebesgue integral*, we have the following two important theorems:

Theorem 1. Let M be a Baire set in R^n whose Lebesgue measure $|M|$ is finite. Then, if we denote by $B \ominus C$ the *symmetric difference* of the set B and C: $B \ominus C = B \cup C - B \cap C$, we have

$$\lim_{|h| \to 0} |(M + h) \ominus M| = 0, \text{ where } M + h = \{x \in R^n;\ x = m + h,\ m \in M\}.$$

Here $m + h = (m_1 + h_1, \ldots, m_n + h_n)$ for $m = (m_1, \ldots, m_n)$, $h = (h_1, \ldots, h_n)$ and $|h| = \left(\sum_{j=1}^{n} h_j\right)^{1/2}$.

Theorem 2. Let G be an open set of R^n. For any Lebesgue integrable function $f(x)$ in G and $\varepsilon > 0$, there exists a continuous function $C_\varepsilon(x)$ in G such that $\{x \in G;\ C_\varepsilon(x) \neq 0\}^a$ is a compact subset of G and

$$\int_G |f(x) - C_\varepsilon(x)|\, dx < \varepsilon.$$

Remark. Let m be a Baire measure on a locally compact space S. A subset Z of S is called a set of *m-measure zero* if, for each $\varepsilon > 0$, there is a Baire set B containing Z with $m(B) < \varepsilon$. One can extend m to the class of *m-measurable sets,* such a set being one which differs from a Baire set by a set of m-measure zero. Any property pertaining to a set of m-measure zero is said to hold m-almost everywhere (m-a. e.). One can also extend integrability to a function which coincides m-a. e. with a Baire function.

4. Linear Spaces

Linear Spaces

Definition. A set X is called a *linear space* over a field K if the following conditions are satisfied:

X is an abelian group (written additively), (1)

A *scalar multiplication* is defined: to every element $x \in X$ and each $\alpha \in K$ there is associated an element of X, denoted by αx, such that we have

$$\left.\begin{aligned} &\alpha(x+y) = \alpha x + \alpha y \quad (\alpha \in K;\ x, y \in X),\\ &(\alpha + \beta)\, x = \alpha x + \beta x \quad (\alpha, \beta \in K;\ x \in X),\\ &(\alpha\beta)\, x = \alpha(\beta x) \quad (\alpha, \beta \in K;\ x \in X),\\ &1 \cdot x = x \ (1 \text{ is the unit element of the field } K). \end{aligned}\right\} \qquad (2)$$

In the sequel we consider linear spaces only over the real number field R^1 or the complex number field C^1. A linear space will be said to be *real* or *complex* according as the field K of coefficients is the real number field R^1 or the complex number field C^1. Thus, in what follows, we mean by a linear space a real or complex linear space. We shall denote by Greek letters the elements of the field of coefficients and by Roman letters the elements of X. The zero of X (= the unit element of the additively written abelian group X) and the number zero will be denoted by the same letter 0, since it does not cause inconvenience as $0 \cdot x = (\alpha - \alpha)\, x = \alpha x - \alpha x = 0$. The *inverse element* of the additively written abelian group X will be denoted by $-x$; it is easy to see that $-x = (-1)x$.

Definition. The elements of a linear space X are called vectors (of X). The vectors $x_1, x_2, \ldots, x_n$ of X are said to be *linearly independent* if the equation $\sum_{j=1}^{n} \alpha_j x_j = 0$ implies $\alpha_1 = \alpha_2 = \cdots = 0$. They are *linearly dependent* if such an equation holds where at least one coefficient is different from 0. If X contains n linearly independent vectors, but every system of $(n+1)$ vectors is linearly dependent, then X is said to be of *n-dimension*. If the number of linearly independent vectors is not finite, then X is said to be of infinite dimension. Any set of n linearly independent vectors in an n-dimensional linear space constitutes a *basis* for X and each vector x of X has a unique representation of the form $x = \sum_{j=1}^{n} \alpha_j y_j$ in terms of the basis $y_1, y_2, \ldots, y_n$. A subset M of a linear space X is called a *linear subspace* or, in short, a *subspace*, if whenever $x, y \in M$, the linear combinations $\alpha x + \beta y$ also belong to M. M is thus a linear space over the same coefficient field as X.

Linear Operators and Linear Functionals

Definition. Let X, Y be linear spaces over the same coefficient field K. A mapping $T : x \to y = T(x) = Tx$ defined on a linear subspace D of X and taking values in Y is said to be *linear*, if

$$T(\alpha x_1 + \beta x_2) = \alpha(Tx_1) + \beta(Tx_2).$$

The definition implies, in particular,

$$T \cdot 0 = 0, \ T(-x) = -(Tx).$$

We denote

$$D = D(T), \{y \in Y; y = Tx, x \in D(T)\} = R(T), \{x \in D(T); Tx = 0\} = N(T)$$

and call them the *domain*, the *range* and the *null space* of T, respectively. T is called a *linear operator* or *linear transformation* on $D(T) \subseteqq X$ into Y, or somewhat vaguely, a linear operator from X into Y. If the range $R(T)$ is contained in the scalar field K, then T is called a *linear functional* on $D(T)$. If a linear operator T gives a one-to-one map of $D(T)$ onto $R(T)$, then the inverse map T^{-1} gives a linear operator on $R(T)$ onto $D(T)$:

$$T^{-1}Tx = x \text{ for } x \in D(T) \quad \text{and} \quad T\,T^{-1}y = y \text{ for } y \in R(T).$$

T^{-1} is the *inverse operator* or, in short, the *inverse* of T. By virtue of $T(x_1 - x_2) = Tx_1 - Tx_2$, we have the following.

Proposition. A linear operator T admits the inverse T^{-1} iff $Tx = 0$ implies $x = 0$.

Definition. Let T_1 and T_2 be linear operators with domains $D(T_1)$ and $D(T_2)$ both contained in a linear space X, and ranges $R(T_1)$ and $R(T_2)$ both contained in a linear space Y. Then $T_1 = T_2$ iff $D(T_1) = D(T_2)$ and $T_1 x = T_2 x$ for all $x \in D(T_1) = D(T_2)$. If $D(T_1) \subseteqq D(T_2)$ and $T_1 x = T_2 x$ for all $x \in D(T_1)$, then T_2 is called an *extension* of T_1, and T_1 a *restriction* of T_2; we shall then write $T_1 \subseteqq T_2$.

Convention. The value $T(x)$ of a linear functional T at a point $x \in D(T)$ will sometimes be denoted by $\langle x, T \rangle$, i. e.

$$T(x) = \langle x, T \rangle.$$

Factor Spaces

Proposition. Let M be a linear subspace in a linear space X. We say that two vectors $x_1, x_2 \in X$ are *equivalent modulo* M if $(x_1 - x_2) \in M$ and write this fact symbolically by $x_1 \equiv x_2 \pmod{M}$. Then we have:

(i) $x \equiv x \pmod{M}$,

(ii) if $x_1 \equiv x_2 \pmod{M}$, then $x_2 \equiv x_1 \pmod{M}$,

(iii) if $x_1 \equiv x_2 \pmod{M}$ and $x_2 \equiv x_3 \pmod{M}$, then $x_1 \equiv x_3 \pmod{M}$.

Proof. (i) is clear since $x - x = 0 \in M$. (ii) If $(x_1 - x_2) \in M$, then $(x_2 - x_1) = -(x_1 - x_2) \in M$. (iii) If $(x_1 - x_2) \in M$ and $(x_2 - x_3) \in M$, then $(x_1 - x_3) = (x_1 - x_2) + (x_2 - x_3) \in M$.

We shall denote the set of all vectors $\in X$ equivalent modulo M to a fixed vector x by ξ_x. Then, in virtue of properties (ii) and (iii), all vectors in ξ_x are mutually equivalent modulo M. ξ_x is called a *class* of equivalent (modulo M) vectors, and each vector in ξ_x is called a *representative* of the class ξ_x. Thus a class is completely determined by any one of its representatives, that is, $y \in \xi_x$ implies that $\xi_y = \xi_x$. Hence, two classes ξ_x, ξ_y are either disjoint (when $y \bar{\in} \xi_x$) or coincide (when $y \in \xi_x$). Thus the entire space X decomposes into classes ξ_x of mutually equivalent (modulo M) vectors.

Theorem. We can consider the above introduced classes (modulo M) as vectors in a new linear space where the operation of addition of classes and the multiplication of a class by a scalar will be defined through

$$\xi_x + \xi_y = \xi_{x+y}, \quad \alpha \xi_x = \xi_{\alpha x}.$$

Proof. The above definitions do not depend upon the choice of representatives x, y of the classes ξ_x, ξ_y respectively. In fact, if $(x_1 - x) \in M$, $(y_1 - y) \in M$, then

$$(x_1 + y_1) - (x + y) = (x_1 - x) + (y_1 - y) \in M,$$
$$(\alpha x_1 - \alpha x) = \alpha (x_1 - x) \in M.$$

We have thus proved $\xi_{x_1+y_1} = \xi_{x+y}$ and $\xi_{\alpha x_1} = \xi_{\alpha x}$, and the above definitions of the class addition and the scalar multiplication of the classes are justified.

Definition. The linear space obtained in this way is called the *factor space* of X modulo M and is denoted by X/M.

References

Topological Space: P. Alexandroff-H. Hopf [1], N. Bourbaki [1], J. L. Kelley [1].

Measure Space: P. R. Halmos [1], S. Saks [1].

I. Semi-norms

The *semi-norm* of a vector in a linear space gives a kind of length for the vector. To introduce a topology in a linear space of infinite dimension suitable for applications to classical and modern analysis, it is sometimes necessary to make use of a system of an infinite number of semi-norms. It is one of the merits of the Bourbaki group that they stressed the importance, in functional analysis, of *locally convex spaces* which are defined through a system of semi-norms satisfying the *axiom of separation.* If the system reduces to a single semi-norm, the corresponding linear space is called a *normed linear space.* If, furthermore, the space is complete with respect to the topology defined by this semi-norm, it is called a *Banach space.* The notion of complete normed linear spaces was introduced around 1922 by S. BANACH and N. WIENER independently of each other. A modification of the norm, the *quasi-norm* in the present book, was introduced by M. FRÉCHET. A particular kind of limit, the *inductive limit,* of locally convex spaces is suitable for discussing the *generalized functions* or the *distributions* introduced by L. SCHWARTZ, as a systematic development of S. L. SOBOLEV's generalization of the notion of functions.

1. Semi-norms and Locally Convex Linear Topological Spaces

As was stated in the above introduction, the notion of semi-norm is of fundamental importance in discussing linear topological spaces. We shall begin with the definition of the semi-norm.

Definition 1. A real-valued function $p(x)$ defined on a linear space X is called a *semi-norm* on X, if the following conditions are satisfied:

$$p(x+y) \leqq p(x) + p(y) \quad (\textit{subadditivity}), \tag{1}$$

$$p(\alpha x) = |\alpha| \, p(x). \tag{2}$$

Example 1. The n-dimensional Euclidean space R^n of points $x = (x_1, \ldots, x_n)$ with coordinates $x_1, x_2, \ldots, x_n$ is an n-dimensional linear space by the operations:

$$x + y = (x_1 + y_1, x_2 + y_2, \ldots, x_n + y_n),$$
$$\alpha x = (\alpha x_1, \alpha x_2, \ldots, \alpha x_n).$$

In this case $p(x) = \max\limits_{1 \leqq j \leqq n} |x_j|$ is a semi-norm. As will be proved later, $p(x) = \left(\sum\limits_{j=1}^{n} |x_j|^q\right)^{1/q}$ with $q \geqq 1$ is also a semi-norm on R^n.

Proposition 1. A semi-norm $p(x)$ satisfies

$$p(0) = 0, \tag{3}$$

$$p(x_1 - x_2) \geqq |p(x_1) - p(x_2)|, \text{ in particular, } p(x) \geqq 0. \tag{4}$$

Proof. $p(0) = p(0 \cdot x) = 0 \cdot p(x) = 0$. We have, by the subadditivity, $p(x_1 - x_2) + p(x_2) \geqq p(x_1)$ and hence $p(x_1 - x_2) \geqq p(x_1) - p(x_2)$. Thus $p(x_1 - x_2) = |-1| \cdot p(x_2 - x_1) \geqq p(x_2) - p(x_1)$ and so we obtain (4).

Proposition 2. Let $p(x)$ be a semi-norm on X, and c any positive number. Then the set $M = \{x \in X; p(x) \leqq c\}$ enjoys the properties:

$$M \ni 0, \tag{5}$$

M is *convex*: $x, y \in M$ and $0 < \alpha < 1$ implies

$$\alpha x + (1 - \alpha) y \in M, \tag{6}$$

M is *balanced* (équilibré in Bourbaki's terminology):

$$x \in M \text{ and } |\alpha| \leqq 1 \text{ imply } \alpha x \in M, \tag{7}$$

M is *absorbing*: for any $x \in X$, there exists $\alpha > 0$

$$\text{such that } \alpha^{-1} x \in M, \tag{8}$$

$$p(x) = \inf_{\alpha > 0, \alpha^{-1} x \in M} \alpha c \quad (\inf = \text{infimum} = \text{the greatest lower bound}). \tag{9}$$

Proof. (5) is clear from (3). (7) and (8) are proved by (2). (6) is proved by the subadditivity (1) and (2). (9) is proved by observing the equivalence of the three propositions below:

$$[\alpha^{-1} x \in M] \rightleftarrows [p(\alpha^{-1} x) \leqq c] \rightleftarrows [p(x) \leqq \alpha c].$$

Definition 2. The functional

$$p_M(x) = \inf_{\alpha > 0, \alpha^{-1} x \in M} \alpha \tag{9'}$$

is called the *Minkowski functional* of the convex, balanced and absorbing set M of X.

Proposition 3. Let a family $\{p_\gamma(x); \gamma \in \Gamma\}$ of semi-norms of a linear space X satisfy the axiom of separation:

For any $x_0 \neq 0$, there exists $p_{\gamma_0}(x)$ in the family such that $p_{\gamma_0}(x_0) \neq 0$. (10)

Take any finite system of semi-norms of the family, say $p_{\gamma_1}(x), p_{\gamma_2}(x), \ldots, \ldots, p_{\gamma_n}(x)$ and any system of n positive numbers $\varepsilon_1, \varepsilon_2, \ldots, \varepsilon_n$, and set

$$U = \{x \in X; p_{\gamma_j}(x) \leqq \varepsilon_j \qquad (j = 1, 2, \ldots, n)\}. \tag{11}$$

U is a convex, balanced and absorbing set. Consider such a set U as a neighbourhood of the vector 0 of X, and defined a neighbourhood of any vector x_0 by the set of the form

$$x_0 + U = \{y \in X; y = x_0 + u, u \in U\}. \tag{12}$$

Consider a subset G of X which contains a neighbourhood of each of its point. Then the totality $\{G\}$ of such subsets G satisfies the axiom of open sets, given in Chapter 0, Preliminaries, 2.

Proof. We first show that the set G_0 of the form $G_0 = \{x \in X; p_\gamma(x) < c\}$ is open. For, let $x_0 \in G_0$ and $p_\gamma(x_0) = \beta < c$. Then the neighbourhood of x_0, $x_0 + U$ where $U = \{x \in X; p_\gamma(x) \leqq 2^{-1}(c - \beta)\}$, is contained in G_0, because $u \in U$ implies $p_\gamma(x_0 + u) \leqq p_\gamma(x_0) + p_\gamma(u) < \beta + (c - \beta) = c$.

Hence, for any point $x_0 \in X$, there is an open set $x_0 + G_0$ which contains x_0. It is clear, by the above definition of open sets, that the union of open sets and the intersection of a finite number of open sets are also open.

Therefore we have only to prove Hausdorff's axiom of separation: If $x_1 \neq x_2$, then there exist disjoint open sets G_1 and G_2 such that

$$x_1 \in G_1, x_2 \in G_2. \tag{13}$$

In view of definition (12) of the neighbourhood of a general point x_0, it will be sufficient to prove (13) for the case $x_1 = 0$, $x_2 \neq 0$. We choose, by (10), $p_{\gamma_2}(x)$ such that $p_{\gamma_2}(x_2) = \alpha > 0$. Then $G_1 = \{x \in X; p_{\gamma_2}(x) < \alpha/2\}$ is open, as proved above. Surely $G_1 \ni 0 = x_1$. We have to show that G_1 and $G_2 = x_2 + G_1$ have no point in common. Assume the contrary and let there exist a $y \in G_1 \cap G_2$. $y \in G_2$ implies $y = x_2 + g = x_2 - (-g)$ with some $g \in G_1$ and so, by (4), $p_{\gamma_2}(y) \geqq p_{\gamma_2}(x_2) - p(-g) \geqq \alpha - 2^{-1}\alpha = \alpha/2$, because $-g$ belongs to G_1 with g. This contradicts the inequality $p_{\gamma_2}(y) < \alpha/2$ implied by $y \in G_1$.

Proposition 4. By the above definition of open sets, X is a *linear topological space*, that is, X is a linear space and at the same time a topological space such that the two mappings $X \times X \to X : (x, y) \to x + y$ and $K \times X \to X : (\alpha, x) \to \alpha x$ are both continuous. Moreover, each semi-norm $p_\gamma(x)$ is a continuous function on X.

Proof. For any neighbourhood U of 0, there exists a neighbourhood V of 0 such that

$$V + V = \{w \in X; w = v_1 + v_2 \text{ where } v_1, v_2 \in V\} \subseteqq U,$$

since the semi-norm is subadditive. Hence, by writing

$$(x + y) - (x_0 + y_0) = (x - x_0) + (y - y_0),$$

we see that the mapping $(x, y) \to x + y$ is continuous at $x = x_0$, $y = y_0$. For any neighbourhood U of 0 and any scalar $\alpha \neq 0$, the set $\alpha U = \{x \in X; x = \alpha u, u \in U\}$ is also a neighbourhood of 0. Thus, by writing

$$\alpha x - \alpha_0 x_0 = \alpha(x - x_0) + (\alpha - \alpha_0) x_0,$$

we see by (2) that $(\alpha, x) \to \alpha x$ is continuous at $\alpha = \alpha_0$, $x = x_0$.

The continuity of the semi-norm $p_\gamma(x)$ at the point $x = x_0$ is proved by $|p_\gamma(x) - p_\gamma(x_0)| \leqq p_\gamma(x - x_0)$.

Definition 3. A linear topological space X is called a *locally convex, linear topological space*, or, in short, a *locally convex space*, if any of its open sets $\ni 0$ contains a convex, balanced and absorbing open set.

Proposition 5. The Minkowski functional $p_M(x)$ of the convex, balanced and absorbing subset M of a linear space X is a semi-norm on X.

Proof. By the convexity of M, the inclusions

$$x/(p_M(x)+\varepsilon)\in M,\ y/(p_M(y)+\varepsilon)\in M \text{ for any } \varepsilon>0$$

imply

$$\frac{p_M(x)+\varepsilon}{p_M(x)+p_M(y)+2\varepsilon}\cdot\frac{x}{p_M(x)+\varepsilon}+\frac{p_M(y)+\varepsilon}{p_M(x)+p_M(y)+2\varepsilon}\cdot\frac{y}{p_M(y)+\varepsilon}\in M$$

and so $p_M(x+y)\leqq p_M(x)+p_M(y)+2\varepsilon$. Since $\varepsilon>0$ was arbitrary, we obtain the subadditivity of $p_M(x)$. Similarly we obtain $p_M(\alpha x)=|\alpha|\,p_M(x)$ since M is balanced.

We have thus proved

Theorem. A linear space X, topologized as above by a family of semi-norms $p_\gamma(x)$ satisfying the axiom of separation (10), is a locally convex space in which each semi-norm $p_\gamma(x)$ is continuous. Conversely, any locally convex space is nothing but the linear topological space, topologized as above through the family of semi-norms obtained as the Minkowski functionals of convex balanced and absorbing open sets of X.

Definition 4. Let $f(x)$ be a complex-valued function defined in an open set Ω of R^n. By the *support* (or *carrier*) of f, denoted by supp(f), we mean the smallest closed set (of the topological space Ω) containing the set $\{x\in\Omega; f(x)\neq 0\}$. It may equivalently be defined as the smallest closed set of Ω outside which f vanishes identically.

Definition 5. By $C^k(\Omega)$, $0\leqq k\leqq\infty$, we denote the set of all complex-valued functions defined in Ω which have continuous partial derivatives of order up to and including k (of order $<\infty$ if $k=\infty$). By $C_0^k(\Omega)$, we denote the set of all functions $\in C^k(\Omega)$ *with compact support*, i.e., those functions $\in C^k(\Omega)$ whose supports are compact subsets of Ω. A classical example of a function $\in C_0^\infty(R^n)$ is given by

$$\begin{aligned} f(x) &= \exp((|x|^2-1)^{-1}) \text{ for } |x|=|(x_1,\ldots,x_n)|=\Big(\sum_{j=1}^n x_j^2\Big)^{1/2}<1, \\ &= 0 \ \text{ for } |x|\geqq 1. \end{aligned} \tag{14}$$

The Space $\mathfrak{E}^k(\Omega)$

$C^k(\Omega)$ is a linear space by

$$(f_1+f_2)(x)=f_1(x)+f_2(x),\ (\alpha f)(x)=\alpha f(x).$$

For any compact subset K of Ω and any non-negative integer $m\leqq k$ ($m<\infty$ when $k=\infty$), we define the semi-norm

$$p_{K,m}(f)=\sup_{|s|\leqq m, x\in K}|D^s f(x)|,\ f\in C^k(\Omega),$$

where sup = supremum = the least upper bound and

$$D^s f(x) = \frac{\partial^{s_1+s_2+\cdots+s_n}}{\partial x_1^{s_1}\,\partial x_2^{s_2}\cdots\partial x_n^{s_n}} f(x_1, x_2, \ldots, x_n),$$

$$|s| = |(s_1, s_2, \ldots, s_n)| = \sum_{j=1}^{n} s_j.$$

Then $C^k(\Omega)$ is a locally convex space by the family of these semi-norms. We denote this locally convex space by $\mathfrak{E}^k(\Omega)$. The convergence $\lim_{h\to\infty} f_h = f$ in this space $\mathfrak{E}_k(\Omega)$ is exactly the uniform convergence $\lim_{h\to\infty} D^s f_h(x) = D^s f(x)$ on every compact subset K of Ω, for each s with $|s| \leqq k$ ($|s| < \infty$ if $k = \infty$).

Proposition 6. $\mathfrak{E}^k(\Omega)$ is a metric space.

Proof. Let $K_1 \subseteqq K_2 \subseteqq \cdots \subseteqq K_n \subseteqq \cdots$ be a monotone increasing sequence of compact subsets of Ω such that $\Omega = \bigcup_{n=1}^{\infty} K_n$. Define, for each positive integer h, the distance

$$d_h(f, g) = \sum_{m=0}^{k} 2^{-m} p_{K_h,m}(f-g) \cdot (1 + p_{K_h,m}(f-g))^{-1}.$$

Then the convergence $\lim_{s\to\infty} f_s = f$ in $\mathfrak{E}_k(\Omega)$ is defined by the distance

$$d(f, g) = \sum_{h=1}^{\infty} 2^{-h} d_h(f, g) \cdot (1 + d_h(f, g))^{-1}.$$

We have to show that $d_h(f, g)$ and $d(f, g)$ satisfy the triangle inequality. The triangle inequality for $d_h(f, g)$ is proved as follows: by the subadditivity of the semi-norm $p_{K_h,m}(f)$, we easily see that $d_h(f, g) =$ satisfies the triangle inequality $d_h(f, g) \leqq d_h(f, k) + d_h(k, g)$, if we can prove the inequality

$$|\alpha-\beta| \cdot (1 + |\alpha-\beta|)^{-1} \leqq |\alpha-\gamma|\,(1 + |\alpha-\gamma|)^{-1} + |\gamma-\beta|\,(1 + |\gamma-\beta|)^{-1}$$

for complex numbers α, β and γ; the last inequality is clear from the inequality valid for any system of non-negative numbers α, β and γ:

$$(\alpha+\beta)\,(1+\alpha+\beta)^{-1} \leqq \alpha(1+\alpha)^{-1} + \beta(1+\beta)^{-1}.$$

The triangle inequality for $d(f, g)$ may be proved similarly.

Definition 6. Let X be a linear space. Let a family $\{X_\alpha\}$ of linear subspaces X_α of X be such that X is the union of X_α's. Suppose that each X_α is a locally convex linear topological space such that, if $X_{\alpha_1} \subseteqq X_{\alpha_2}$, then the topology of X_{α_1} is identical with the relative topology of X_{α_1} as a subset of X_{α_2}. We shall call "open" every convex balanced and absorbing set U of X iff the intersection $U \cap X_\alpha$ is an open set of X_α containing the zero vector 0 of X_α, for all X_α. If X is a locally convex linear topological space whose topology is defined in the stated way, then X is called the (strict) *inductive limit* of X_α's.

Remark. Take, from each X_α, a convex balanced neighbourhood U_α of 0 of X_α. Then the *convex closure* U of the union $V = \bigcup_\alpha U_\alpha$, i.e.,

$$U = \left\{ u \in X;\, u = \sum_{j=1}^{n} \beta_j v_j,\ v_j \in V,\ \beta_j \geqq 0\ (j = 1, 2, \ldots, n),\ \sum_{j=1}^{n} \beta_j = 1 \right.$$
$$\left. \text{with arbitrary finite } n \right\}$$

surely satisfies the condition that it is convex balanced and absorbing in such a way that $U \cap X_\alpha$ is a convex balanced neighbourhood of 0 of X_α, for all X_α. The set of all such U's corresponding to an arbitrary choice of U_α's is a *fundamental system of neighbourhood* of 0 of the (strict) inductive limit X of X'_αs, i.e., every neighbourhood of 0 of the (strict) inductive limit X of X'_αs contains one of the U's obtained above. This fact justifies the above definition of the (strict) inductive limit.

The Space $\mathfrak{D}(\Omega)$

$C_0^\infty(\Omega)$ is a linear space by

$$(f_1 + f_2)(x) = f_1(x) + f_2(x), \quad (\alpha f)(x) = \alpha f(x).$$

For any compact subset K of Ω, let $\mathfrak{D}_K(\Omega)$ be the set of all functions $f \in C_0^\infty(\Omega)$ such that $\operatorname{supp}(f) \subseteqq K$. Define a family of semi-norms on $\mathfrak{D}_K(\Omega)$ by

$$p_{K,m}(f) = \sup_{|s| \leqq m,\, x \in K} |D^s f(x)|, \text{ where } m < \infty.$$

$\mathfrak{D}_K(\Omega)$ is a locally convex linear topological space, and, if $K_1 \subseteqq K_2$, then it follows that the topology of $\mathfrak{D}_{K_1}(\Omega)$ is identical with the relative topology of $\mathfrak{D}_{K_1}(\Omega)$ as a subset of $\mathfrak{D}_{K_2}(\Omega)$. Then the (strict) inductive limit of $\mathfrak{D}_K(\Omega)$'s, where K ranges over all compact subsets of Ω, is a locally convex, linear topological space. Topologized in this way, $C_0^\infty(\Omega)$ will be denoted by $\mathfrak{D}(\Omega)$. It is to be remarked that,

$$p(f) = \sup_{x \in \Omega} |f(x)|$$

is one of the semi-norms which defines the topology of $\mathfrak{D}(\Omega)$. For, if we set $U = \{f \in C_0^\infty(\Omega);\, p(f) \leqq 1\}$, then the intersection $U \cap \mathfrak{D}_K(\Omega)$ is given by $U_K = \{f \in \mathfrak{D}_K(\Omega);\, p_K(f) = \sup_{x \in K} |f(x)| \leqq 1\}$.

Proposition 7. The convergence $\lim_{h\to\infty} f_h = 0$ in $\mathfrak{D}(\Omega)$ means that the following two conditions are satisfied: (i) there exists a compact subset K of Ω such that $\operatorname{supp}(f_h) \subseteqq K$ $(h = 1, 2, \ldots)$, and (ii) for any differential operator D^s, the sequence $\{D^s f_h(x)\}$ converges to 0 uniformly on K.

Proof. We have only to prove (i). Assume the contrary, and let there exist a sequence $\{x^{(k)}\}$ of points $\in \Omega$ having no accumulation points in Ω and a subsequence $\{f_{h_j}(x)\}$ of $\{f_h(x)\}$ such that $f_{h_j}(x^{(j)}) \neq 0$. Then the semi-norm

$$\begin{cases} p(f) = \sum_{k=1}^{\infty} 2 \sup_{x \in K_k - K_{k-1}} |f(x)/f_{h_k}(x^{(k)})|, \text{ where the mono-} \\ \text{tone increasing sequence of compact subsets } K_j \text{ of} \\ \Omega \text{ satisfies } \bigcup_{j=1}^{\infty} K_j = \Omega \text{ and } x^{(k)} \in K_k - K_{k-1} \\ (k = 1, 2, \ldots), K_0 = \emptyset \end{cases}$$

defines a neighbourhood $U = \{f \in C_0^\infty(\Omega); p(f) \leqq 1\}$ of 0 of $\mathfrak{D}(\Omega)$. However, none of the f_{h_k}'s is contained in U.

Corollary. The convergence $\lim_{h\to\infty} f_h = f$ in $\mathfrak{D}(\Omega)$ means that the following two conditions are satisfied: (i) there exists a compact subset K of Ω such that $\text{supp}(f_h) \subseteqq K$ $(h = 1, 2, \ldots)$, and (ii) for any differential operator D^s, the sequence $D^s f_h(x)$ converges to $D^s f(x)$ uniformly on K.

Proposition 8 (A theorem of approximation). Any continuous function $f \in C_0^0(R^n)$ can be approximated by functions of $C_0^\infty(R^n)$ uniformly on R^n.

Proof. Let $\theta_1(x)$ be the function introduced in (14) and put

$$\theta_a(x) = h_a^{-1} \theta_1(x/a), \text{ where } a > 0 \text{ and } h_a > 0 \text{ are such that} \\ \int_{R^n} \theta_a(x)\, dx = 1. \tag{15}$$

We then define the *regularization* f_a of f:

$$f_a(x) = \int_{R^n} f(x-y)\, \theta_a(y)\, dy = \int_{R^n} f(y)\, \theta_a(x-y)\, dy, \text{ where} \\ x - y = (x_1 - y_1, x_2 - y_2, \ldots, x_n - y_n). \tag{16}$$

The integral is convergent since f and θ_a have compact support. Moreover, since

$$f_a(x) = \int_{\text{supp}(f)} f(y)\, \theta_a(x-y)\, dy,$$

the support of f_a may be taken to be contained in any neighbourhood of the supp (f) if we take $a > 0$ sufficiently small. Next, by differentiating under the integral sign, we have

$$D^s f_a(x) = D_x^s f_a(x) = \int_{R^n} f(y)\, D_s^x \theta_a(x-y)\, dy, \tag{17}$$

and so f_a is in $C_0^\infty(R^n)$. Finally we have by $\int\limits_{R^n} \theta_a(x-y)\,dy = 1$,

$$|f_a(x) - f(x)| \leq \int\limits_{R^n} |f(y) - f(x)|\,\theta_a(x-y)\,dy$$

$$\leq \int\limits_{|f(y)-f(x)|\leq\varepsilon} |f(y) - f(x)|\,\theta_a(x-y)\,dy$$

$$+ \int\limits_{|f(y)-f(x)|>\varepsilon} |f(y) - f(x)|\,\theta_a(x-y)\,dy.$$

The first term on the right is $\leq \varepsilon$; and the second term on the right equals 0 for sufficiently small $a > 0$, because, by the uniform continuity of the function f with compact support, there exists an $a > 0$ such that $|f(y) - f(x)| > \varepsilon$ implies $|y - x| > a$. We have thus proved our Proposition.

2. Norms and Quasi-norms

Definition 1. A locally convex space is called a *normed linear space*, if its topology is defined by just one semi-norm.

Thus a linear space X is called a normed linear space, if for every $x \in X$, there is associated a real number $\|x\|$, the *norm* of the vector x, such that

$$\|x\| \geq 0 \text{ and } \|x\| = 0 \text{ iff } x = 0, \tag{1}$$

$$\|x + y\| \leq \|x\| + \|y\| \text{ (\textit{triangle inequality})}, \tag{2}$$

$$\|\alpha x\| = |\alpha| \cdot \|x\|. \tag{3}$$

The topology of a normed linear space X is thus defined by the distance

$$d(x, y) = \|x - y\|. \tag{4}$$

In fact, $d(x, y)$ satisfies the *axiom of distance*:

$$d(x, y) \geq 0 \text{ and } d(x, y) = 0 \text{ iff } x = y,$$

$$d(x, y) \leq d(x, z) + d(z, y) \text{ (triangle inequality)},$$

$$d(x, y) = d(y, x).$$

For, $d(x, y) = \|x - y\| = \|y - x\| = d(y, x)$ and $d(x, y) = \|x - y\| = \|x - z + z - y\| \leq \|x - z\| + \|z - y\| = d(x, z) + d(z, y)$ by (1), (2), (3) and (4).

The convergence $\lim\limits_{n\to\infty} d(x_n, x) = 0$ in a normed linear space X will be denoted by $s\text{-}\lim\limits_{n\to\infty} x_n = x$ or simply by $x_n \to x$, and we say that the sequence $\{x_n\}$ *converges strongly* to x. The adjective "strong" is introduced to distinguish it from the "weak" convergence to be introduced later.

Proposition 1. In a normed linear space X, we have

$$\lim_{n\to\infty} \|x_n\| = \|x\| \text{ if } \operatorname*{s-lim}_{n\to\infty} x_n = x, \tag{5}$$

$$\operatorname*{s-lim}_{n\to\infty} \alpha_n x_n = \alpha x \text{ if } \lim_{n\to\infty} \alpha_n = \alpha \text{ and } \operatorname*{s-lim}_{n\to\infty} x_n = x, \tag{6}$$

$$\operatorname*{s-lim}_{n\to\infty} (x_n + y_n) = x + y \text{ if } \operatorname*{s-lim}_{n\to\infty} x_n = x \text{ and } \operatorname*{s-lim}_{n\to\infty} y_n = y. \tag{7}$$

Proof. (5), (6) and (7) are already proved, since X is a locally convex space topologized by just one semi-norm $p(x) = \|x\|$. However, we shall give a direct proof as follows. As a semi-norm, we have

$$\|x - y\| \geqq |\|x\| - \|y\|| \tag{8}$$

and hence (5) is clear. (7) is proved by $\|(x + y) - (x_n + y_n)\| = \|(x - x_n) + (y - y_n)\| \leqq \|x - x_n\| + \|y - y_n\|$. From $\|\alpha x - \alpha_n x_n\| \leqq \|\alpha x - \alpha_n x\| + \|\alpha_n x - \alpha_n x_n\| \leqq |\alpha - \alpha_n| \cdot \|x\| + |\alpha_n| \cdot \|x - x_n\|$ and the boundedness of the sequence $\{\alpha_n\}$ we obtain (6).

Definition 2. A linear space X is called a *quasi-normed linear space*, if, for every $x \in X$, there is associated a real number $\|x\|$, the *quasi-norm* of the vector x, which satisfies (1), (2) and

$$\|-x\| = \|x\|, \lim_{\alpha_n\to 0} \|\alpha_n x\| = 0 \text{ and } \lim_{\|x_n\|\to 0} \|\alpha x_n\| = 0. \tag{3'}$$

Proposition 2. In a quasi-normed linear space X, we have (5), (6) and (7).

Proof. We need only prove (6). The proof in the preceding Proposition shows that we have to prove

$$\lim_{n\to\infty} \|x_n\| = 0 \text{ implies that } \lim_{n\to\infty} \|\alpha x_n\| = 0 \text{ uniformly in } \alpha \text{ on any bounded set of } \alpha. \tag{9}$$

The following proof of (9) is due to S. Kakutani (unpublished). Consider the functional $p_n(\alpha) = \|\alpha x_n\|$ defined on the linear space R^1 of real numbers normed by the absolute value. By the triangle inequality of $p_n(\alpha)$ and (3'), $p_n(\alpha)$ is continuous on R^1. Hence, from $\lim_{n\to\infty} p_n(\alpha) = 0$ implied by (3') and Egorov's theorem (Chapter 0, Preliminaries, 3. Measure Spaces), we see that there exists a Baire measurable set A on the real line R^1 with the property:

$$\text{the Lebesgue measure } |A| \text{ of } A \text{ is} > 0 \text{ and } \lim_{n\to\infty} p_n(\alpha) = 0 \text{ uniformly on } A. \tag{10}$$

Since the Lebesgue measure on the real line is continuous with respect to translations, we have, denoting by $B \ominus C$ the symmetric difference $B \cup C - B \cap C$,

$$|(A + \sigma) \ominus A| \to 0 \quad \text{as} \quad \sigma \to 0.$$

Thus there exists a positive number σ_0 such that

$|\sigma| \leqq \sigma_0$ implies $|(A+\sigma) \ominus A| < |A|/2$, in particular, $|(A+\sigma) \cap A| > 0$.

Hence, for any real number σ with $|\sigma| \leqq \sigma_0$, there is a representation

$$\sigma = \alpha - \alpha' \text{ with } \alpha \in A,\ \alpha' \in A.$$

Therefore, by $p_n(\sigma) = p_n(\alpha - \alpha') \leqq p_n(\alpha) + p_n(\alpha')$, we see that

$$\lim_{n\to\infty} p_n(\sigma) = 0 \text{ uniformly in } \sigma \text{ when } |\sigma| \leqq \sigma_0.$$

Let M be any positive number. Then, taking a positive integer $k \geqq M/\sigma_0$ and remembering $p_n(k\sigma) \leqq k p_n(\sigma)$, we see that (9) is true for $|\alpha| \leqq M$.

Remark. The above proof may naturally be modified so as to apply to complex quasi-normed linear spaces X as well.

As in the case of normed linear spaces, the convergence $\lim_{n\to\infty} \|x - x_n\| = 0$ in a quasi-normed linear space will be denoted by $s\text{-}\lim_{n\to\infty} x_n = x$, or simply by $x_n \to x$; we shall then say that the sequence $\{x_n\}$ *converges strongly* to x.

Example. Let the topology of a locally convex space X be defined by a countable number of semi-norms $p_n(x)$ $(n = 1, 2, \ldots)$. Then X is a quasi-normed linear space by the quasi-norm

$$\|x\| = \sum_{n=1}^{\infty} 2^{-n} p_n(x) (1 + p_n(x))^{-1}.$$

For, the convergence $\lim_{h\to\infty} p_n(x_h) = 0$ $(n = 1, 2, \ldots)$ is equivalent to $s\text{-}\lim_{h\to\infty} x_h = 0$ with respect to the quasi-norm $\|x\|$ above.

3. Examples of Normed Linear Spaces

Example 1. $C(S)$. Let S be a topological space. Consider the set $C(S)$ of all real-valued (or complex-valued), bounded continuous functions $x(s)$ defined on S. $C(S)$ is a normed linear space by

$$(x+y)(s) = x(s) + y(s),\ (\alpha x)(s) = \alpha x(s),\ \|x\| = \sup_{s \in S} x(s).$$

In $C(S)$, $s\text{-}\lim_{n\to\infty} x_n = x$ means the uniform convergence of the functions $x_n(s)$ to $x(s)$.

Example 2. $L^p(S, \mathfrak{B}, m)$, or, in short, $L^p(S)$ $(1 \leqq p < \infty)$. Let $L^p(S)$ be the set of all real-valued (or complex-valued) $\mathfrak{B}$-measurable functions $x(s)$ defined m-a. e. on S such that $|x(s)|^p$ is m-integrable over S. $L^p(S)$ is a linear space by

$$(x+y)(s) = x(s) + y(s),\ (\alpha x)(s) = \alpha x(s).$$

For, $(x(s) + y(s))$ belongs to $L^p(S)$ if $x(s)$ and $y(s)$ both belong to $L^p(S)$, as may be seen from the inequality $|x(s) + y(s)|^p \leqq 2^p(|x(s)|^p + |y(s)|^p)$. We define the norm in $L^p(S)$ by

$$\|x\| = \Big(\int_S |x(s)|^p\, m(ds)\Big)^{1/p}. \tag{1}$$

The subadditivity

$$\Big(\int_S |x(s) + y(s)|^p\, m(ds)\Big)^{1/p} \leqq \Big(\int_S |x(s)|^p\, m(ds)\Big)^{1/p} + \Big(\int_S |y(s)|^p\, m(ds)\Big)^{1/p}, \tag{2}$$

called *Minkowski's inequality*, is clear for the case $p = 1$. To prove the general case $1 < p < \infty$, we need

Lemma 1. Let $1 < p < \infty$ and let the *conjugate exponent* p' of p be defined by

$$\frac{1}{p} + \frac{1}{p'} = 1. \tag{3}$$

Then, for any pair of non-negative numbers a and b, we have

$$ab \leqq \frac{a^p}{p} + \frac{b^{p'}}{p'}, \tag{4}$$

where the equality is satisfied iff $a = b^{1/(p-1)}$.

Proof. The minimum of the function $f(c) = \dfrac{c^p}{p} + \dfrac{1}{p'} - c$ for $c \geqq 0$ is attained only at $c = 1$, and the minimum value is 0. By taking $c = ab^{-1/(p-1)}$ we see that the Lemma is true.

The proof of (2). We first prove *Hölder's inequality*

$$\int |x(s)\, y(s)| \leqq \Big(\int |x(s)|^p\Big)^{1/p} \cdot \Big(\int |y(s)|^{p'}\Big)^{1/p'} \tag{5}$$

$\Big($for convenience, we write $\int z(s)$ for $\int_S z(s)\, m(ds)\Big)$.

To this end, we assume that $A = \big(\int |x(s)|^p\big)^{1/p}$ and $B = \big(\int |y(s)|^{p'}\big)^{1/p'}$ are both $\neq 0$, since otherwise $x(s)\, y(s) = 0$ a.e. and so (5) would be true. Now, by taking $a = |x(s)|/A$ and $b = |y(s)|/B$ in (4) and integrating, we obtain $\dfrac{\int |x(s)\, y(s)|}{AB} \leqq \dfrac{1}{p}\dfrac{A^p}{A^p} + \dfrac{1}{p'}\dfrac{B^{p'}}{B^{p'}} = 1$ which implies (5).

Next, by (5), we have

$$\begin{aligned}
\int |x(s) + y(s)|^p &\leqq \int |x(s) + y(s)|^{p-1} \cdot |x(s)| \\
&\quad + \int |x(s) + y(s)|^{p-1} \cdot |y(s)| \\
&\leqq \Big(\int |x(s) + y(s)|^{p'(p-1)}\Big)^{1/p'} \Big(\int |x(s)|^p\Big)^{1/p} \\
&\quad + \Big(\int (|x(s) + y(s)|^{p'(p-1)}\Big)^{1/p'} \Big(\int (|y(s)|^p\Big)^{1/p}\Big),
\end{aligned}$$

which proves (2) by $p'(p-1) = p$.

Remark 1. The equality sign in (2) holds iff there exists a non-negative constant c such that $x(s) = cy(s)$ m-a.e. (or $y(s) = cx(s)$ m-a.e.). This is implied from the fact that, by Lemma 1, the equality sign in Hölder's inequality (5) holds iff $|x(s)| = c \cdot |y(s)|^{1/(p-1)}$ (or $|y(s)| = c \cdot |x(s)|^{1/p-1}$) are satisfied m-a.e.

Remark 2. The condition $\|x\| = \left(\int |x(s)|^p\right)^{1/p} = 0$ is equivalent to the condition that $x(s) = 0$ m-a.e. We shall thus consider two functions of $L^p(S)$ as equivalent if they are equal m-a.e. By this convention, $L^p(S)$ becomes a normed linear space. The limit relation $s\text{-}\lim_{n\to\infty} x_n = x$ in $L^p(S)$ is sometimes called the *mean convergence of p-th order* of the sequence of functions $x_n(s)$ to the function $x(s)$.

Example 3. $L^\infty(S)$. A $\mathfrak{B}$-measurable function $x(s)$ defined on S is said to be *essentially bounded* if there exists a constant α such that $|x(s)| \leqq \alpha$ m-a.e. The infimum of such constants α is denoted by

$$\operatorname*{vrai\,max}_{s\in S} |x(s)| \text{ or } \operatorname*{essential\,sup}_{s\in S} |x(s)|.$$

$L^\infty(S, \mathfrak{B}, m)$ or, in short, $L^\infty(S)$ is the set of all $\mathfrak{B}$-measurable, essentially bounded functions defined m-a.e. on S. It is a normed linear space by

$$(x+y)(s) = x(s) + y(s),\ (\alpha x)(s) = \alpha x(s),\ \|x\| = \operatorname*{vrai\,max}_{s\in S} |x(s)|,$$

under the convention that we consider two functions of $L^\infty(S)$ as equivalent if they are equal m-a.e.

Theorem 1. Let the total measure $m(S)$ of S be finite. Then we have

$$\lim_{p\to\infty} \left(\int_S |x(s)|^p\, m(ds)\right)^{1/p} = \operatorname*{vrai\,max}_{s\in S} |x(s)| \quad \text{for} \quad x(s) \in L^\infty(S). \quad (6)$$

Proof. It is clear that $\left(\int_S |x(s)|^p\, m(ds)\right)^{1/p} \leqq m(S)^{1/p} \operatorname*{vrai\,max}_{s\in S} |x(s)|$ so that $\overline{\lim}_{p\to\infty} \left(\int_S |x(s)|^p\right)^{1/p} \leqq \operatorname*{vrai\,max}_{s\in S} |x(s)|$. By the definition of the vrai max, there exists, for any $\varepsilon > 0$, a set B of m-measure > 0 at each point of which $|x(s)| \geqq \operatorname*{vrai\,max}_{s\in S} |x(s)| - \varepsilon$. Hence $\left(\int_S |x(s)|^p\, m(ds)\right)^{1/p}$ $\geqq m(B)^{1/p} \left(\operatorname*{vrai\,max}_{s\in S} |x(s)| - \varepsilon\right)$. Therefore $\underline{\lim}_{p\to\infty} \left(\int |x(s)|^p\right)^{1/p} \geqq \operatorname*{vrai\,max}_{s\in S}$ $|x(s)| - \varepsilon$, and so (6) is true.

Example 4. Let, in particular, S be a *discrete topological space* consisting of countable points denoted by $1, 2, \ldots$; the term *discrete* means that each point of $S = \{1, 2, \ldots\}$ is itself open in S. Then as linear subspaces of $C(\{1, 2, \ldots\})$, we define (c_0), (c) and (l^p), $1 \leqq p < \infty$.

(c_0): Consider a bounded sequence of real or complex numbers $\{\xi_n\}$. Such a sequence $\{\xi_n\}$ defines a function $x(n) = \xi_n$ defined and continuous on the discrete space $S = \{1, 2, \ldots\}$; we shall call $x = \{\xi_n\}$ a vector with components ξ_n. The set of all vectors $x = \{\xi_n\}$ such that $\lim_{n\to\infty} \xi_n = 0$ constitutes a normed linear space (c_0) by the norm

$$||x|| = \sup_n |x(n)| = \sup_n |\xi_n|.$$

(c): The set of all vectors $x = \{\xi_n\}$ such that finite $\lim_{n\to\infty} \xi_n$ exist, constitutes a normed linear space (c) by the norm $||x|| = \sup_n |x(n)|$ $= \sup_n |\xi_n|$.

(l^p), $1 \leqq p < \infty$: The set of all vectors $x = \{\xi_n\}$ such that $\sum_{n=1}^{\infty} |\xi_n|^p < \infty$ constitutes a normed linear space (l^p) by the norm $||x|| = \left(\sum_{n=1}^{\infty} |\xi_n|^p\right)^{1/p}$. As an abstract linear space, it is a linear subspace of $C(\{1, 2, \ldots\})$. It is also a special case of $L^p(S, \mathfrak{B}, m)$ in which $m(\{1\}) = m(\{2\}) = \cdots = 1$.

$(l^\infty) = (m)$: As in the case of $L^\infty(S)$, we shall denote by (l^∞) the linear space $C(\{1, 2, \ldots\})$, normed by $||x|| = \sup_n |x(n)| = \sup_n |\xi_n|$. (l^∞) is also denoted by (m).

The Space of Measures. Let $\mathfrak{B}$ be a σ-ring of subsets of S. Consider the set $A(S, \mathfrak{B})$ of all real- (or complex-) valued functions $\varphi(B)$ defined on $\mathfrak{B}$ such that

$$|\varphi(B)| \neq \infty \text{ for every } B \in \mathfrak{B}, \tag{7}$$

$$\varphi\left(\sum_{j=1}^{\infty} B_j\right) = \sum_{j=1}^{\infty} \varphi(B_j) \text{ for any disjoint sequence } \{B_j\} \text{ of sets } \in \mathfrak{B}. \tag{8}$$

$A(S, \mathfrak{B})$ will be called the space of *signed* (*or complex*) *measures* defined on $(S, \mathfrak{B})$.

Lemma 2. Let $\varphi \in A(S, \mathfrak{B})$ be real-valued. Then the *total variation* of φ on S defined by

$$V(\varphi; S) = \overline{V}(\varphi; S) + |\underline{V}(\varphi; S)| \tag{9}$$

is finite; here the *positive variation* and the *negative variation* of φ over $B \in \mathfrak{B}$ are given respectively by

$$\overline{V}(\varphi; B) = \sup_{B_1 \subseteq B} \varphi(B_1) \text{ and } \underline{V}(\varphi; B) = \inf_{B_1 \subseteq B} \varphi(B_1). \tag{10}$$

Proof. Since $\varphi(\emptyset) = 0$, we have $\overline{V}(\varphi; B) \geqq 0 \geqq \underline{V}(\varphi; B)$. Suppose that $V(\varphi; S) = \infty$. Then there exists a decreasing sequence $\{B_n\}$ of sets $\in \mathfrak{B}$ such that

$$V(\varphi; B_n) = \infty, \ |\varphi(B_n)| \geqq n - 1.$$

The proof is obtained by induction. Let us choose $B_1 = S$ and assume that the sets $B_2, B_3, \ldots, B_k$ have been defined so as to satisfy the above conditions. By the first condition with $n = k$, there exists a set $B \in \mathfrak{B}$

such that $B \subseteq B_k$, $|\varphi(B)| \geqq |\varphi(B_k)| + k$. We have only to set $B_{k+1} = B$ in the case $V(\varphi; B) = \infty$ and $B_{k+1} = B_k - B$ in the case $V(\varphi; B) < \infty$. For, in the latter case, we must have $V(\varphi; B_k - B) = \infty$ and $|\varphi(B_k - B)| \geqq |\varphi(B)| - |\varphi(B_k)| \geqq k$ which completes the induction.

By the decreasing property of the sequence $\{B_n\}$, we have

$$S - \bigcap_{n=1}^{\infty} B_n = \sum_{n=1}^{\infty} (S - B_n)$$
$$= (S - B_1) + (B_1 - B_2) + (B_2 - B_3) + \cdots + (B_n - B_{n+1}) + \cdots$$

so that, by the countable additivity of φ,

$$\varphi\left(S - \bigcap_{n=1}^{\infty} B_n\right) = \varphi(S - B_1) + \varphi(B_1 - B_2) + \varphi(B_2 - B_3) + \cdots$$
$$= [\varphi(S) - \varphi(B_1)] + [\varphi(B_1) - \varphi(B_2)] + [\varphi(B_2) - \varphi(B_3)] + \cdots$$
$$= \varphi(S) - \lim_{n\to\infty} \varphi(B_n) = \infty \text{ or } -\infty,$$

which is a contradiction of (7).

Theorem 2 (Jordan's decomposition). Let $\varphi \in A(S, \mathfrak{B})$ be real-valued. Then the positive variation $\overline{V}(\varphi; B)$, the negative variation $\underline{V}(\varphi; B)$ and the total variation $V(\varphi; B)$ are countably additive on B. Moreover, we have the Jordan decomposition

$$\varphi(B) = \overline{V}(\varphi; B) + \underline{V}(\varphi; B) \text{ for any } B \in \mathfrak{B}. \tag{11}$$

Proof. Let $\{B_n\}$ be a sequence of disjoint sets $\in \mathfrak{B}$. For any set $B \in \mathfrak{B}$ such that $B \subseteq \sum_{n=1}^{\infty} B_n$, we have $\varphi(B) = \sum_{n=1}^{\infty} \varphi(B \cap B_n) \leqq \sum_{n=1}^{\infty} \overline{V}(\varphi; B_n)$ and hence $\overline{V}\left(\varphi; \sum_{n=1}^{\infty} B_n\right) \leqq \sum_{n=1}^{\infty} \overline{V}(\varphi; B_n)$. On the other hand, if $C_n \in \mathfrak{B}$ is a subset of B_n $(n = 1, 2, \ldots)$, then we have $\overline{V}\left(\varphi; \sum_{n=1}^{\infty} B_n\right) \geqq \varphi\left(\sum_{n=1}^{\infty} C_n\right) = \sum_{n=1}^{\infty} \varphi(C_n)$ and so $\overline{V}\left(\varphi; \sum_{n=1}^{\infty} B_n\right) \geqq \sum_{n=1}^{\infty} \overline{V}(\varphi; B_n)$. Hence we have proved the countable additivity of $\overline{V}(\varphi; B)$ and those of $\underline{V}(\varphi; B)$ and of $V(\varphi; B)$ may be proved similarly.

To establish (11), we observe that, for every $C \in \mathfrak{B}$ with $C \subseteq B$, we have $\varphi(C) = \varphi(B) - \varphi(B - C) \leqq \varphi(B) - \underline{V}(\varphi; B)$ and so $\overline{V}(\varphi; B) \leqq \varphi(B) - \underline{V}(\varphi; B)$. Similarly we obtain $\underline{V}(\varphi; B) \geqq \varphi(B) - \overline{V}(\varphi; B)$. These inequalities together give (11).

Theorem 3 (Hahn's decomposition). Let $\varphi \in A(S, \mathfrak{B})$ be a signed measure. Then there exists a set $P \in \mathfrak{B}$ such that

$$\varphi(B) \geqq 0 \text{ for every } B \in \mathfrak{B} \text{ with } B \subseteq P,$$
$$\varphi(B) \leqq 0 \text{ for every } B \in \mathfrak{B} \text{ with } B \subseteq P^C = S - P.$$

The decomposition $S = P + (S - P)$ is called the *Hahn decomposition* of S pertaining to φ.

Proof. For each positive integer n we choose a set $B_n \in \mathfrak{B}$ such that $\varphi(B_n) \geqq \overline{V}(\varphi; S) - 2^{-n}$. Hence by (11), we have

$$\underline{V}(\varphi; B_n) \geqq -2^{-n} \quad \text{and} \quad \overline{V}(\varphi; S - B_n) \leqq 2^{-n}. \tag{12}$$

The latter inequality is obtained from $\overline{V}(\varphi; S - B_n) = \overline{V}(\varphi; S) - \overline{V}(\varphi; B_n)$ and $\overline{V}(\varphi; B_n) \geqq \varphi(B_n)$. We then put

$$P = \varliminf_{n\to\infty} B_n = \bigcup_{k=1}^{\infty} \bigcap_{n=k}^{\infty} B_n.$$

Then $S - P = \varlimsup_{n\to\infty} (S - B_n) = \bigcap_{k=1}^{\infty} \bigcup_{n=k}^{\infty} (S - B_n) \subseteqq \bigcup_{n=k}^{\infty} (S - B_n)$ for every k, and therefore, by the σ-additivity of $\overline{V}(\varphi; B)$,

$$\overline{V}(\varphi; S - P) \leqq \sum_{n=k}^{\infty} \overline{V}(\varphi; S - B_n) \leqq 2^{-(k-1)},$$

which gives $\overline{V}(\varphi; S - P) = 0$. On the other hand, the negative variation $\underline{V}(\varphi; B)$ is a non-positive measure and so, by (12) and similarly as above,

$$|\underline{V}(\varphi; P)| \leqq \varliminf_{n\to\infty} |\underline{V}(\varphi; B_n)| = 0,$$

which gives $\underline{V}(\varphi; P) = 0$. The proof is thus completed.

Corollary. The total variation $V(\varphi; S)$ of a signed measure φ is defined by

$$V(\varphi; S) = \sup_{\sup|x(s)| \leqq 1} \left| \int_S x(s)\, \varphi(ds) \right| \tag{13}$$

where $x(s)$ ranges through $\mathfrak{B}$-measurable functions defined on S such that $\sup_s |x(s)| \leqq 1$.

Proof. If we take $x(s) = 1$ or $= -1$ according as $s \in P$ or $s \in S - P$, then the right hand side of (13) gives $V(\varphi; S)$. On the other hand, it is easy to see that

$$\left| \int_S x(s)\, \varphi(ds) \right| \leqq \sup_s |x(s)| \cdot \int_S V(\varphi; ds) = \sup_s |x(s)| \cdot V(\varphi; S)$$

and hence (13) is proved.

Example 5. $A(S, \mathfrak{B})$. The space $A(S, \mathfrak{B})$ of signed measures φ on $\mathfrak{B}$ is a real linear space by

$$(\alpha_1 \varphi_1 + \alpha_2 \varphi_2)(B) = \alpha_1 \varphi_1(B) + \alpha_2 \varphi_2(B), \quad B \in \mathfrak{B}.$$

It is a normed linear space by the norm

$$\|\varphi\| = V(\varphi; S) = \sup_{\sup|x(s)| \leqq 1} \left| \int_S x(s)\, \varphi(ds) \right|. \tag{14}$$

Example 6. The space $A(S, \mathfrak{B})$ of complex measures φ is a complex linear space by

$$(\alpha_1 \varphi_1 + \alpha_2 \varphi_2)(B) = \alpha_1 \varphi_1(B) + \alpha_2 \varphi_2(B), \ B \in \mathfrak{B} \text{ with complex } \alpha_1, \alpha_2.$$

It is a normed linear space by the norm

$$\|\varphi\| = \sup_{\sup|x(s)| \leq 1} \left| \int_S x(s)\, \varphi(ds) \right|, \tag{15}$$

where complex-valued $\mathfrak{B}$-measurable functions $x(s)$ defined on S are taken into account. We shall call the right hand value of (15) the *total variation* of φ on S and denote it by $V(\varphi; S)$.

4. Examples of Quasi-normed Linear Spaces

Example 1. $\mathfrak{E}^k(\Omega)$. The linear space $\mathfrak{E}^k(\Omega)$, introduced in Chapter I, 1, is a quasi-normed linear space by the quasi-norm $\|x\| = d(x, 0)$, where the distance $d(x, y)$ is as defined there.

Example 2. $M(S, \mathfrak{B}, m)$. Let $m(S) < \infty$ and let $M(S, \mathfrak{B}, m)$ be the set of all complex-valued $\mathfrak{B}$-measurable functions $x(s)$ defined on S and such that $|x(s)| < \infty$ m-a.e. Then $M(S, \mathfrak{B}, m)$ is a quasi-normed linear space by the algebraic operations

$$(x + y)(s) = x(s) + y(s), \quad (\alpha x)(s) = \alpha x(s)$$

and (under the convention that $x = y$ iff $x(s) = y(s)$ m-a.e.)

$$\|x\| = \int_S |x(s)|\, (1 + |x(s)|)^{-1}\, m(ds). \tag{1}$$

The triangle inequality for the quasi-norm $\|x\|$ is clear from

$$\frac{|\alpha + \beta|}{1 + |\alpha + \beta|} \leq \frac{|\alpha| + |\beta|}{1 + |\alpha| + |\beta|} \leq \frac{|\alpha|}{1 + |\alpha|} + \frac{|\beta|}{1 + |\beta|}.$$

The mapping $\{\alpha, x\} \to \alpha x$ is continuous by the following

Proposition. The convergence $\underset{n\to\infty}{\text{s-lim}}\, x_n = x$ in $M(S, \mathfrak{B}, m)$ is equivalent to the *asymptotic convergence* (or the *convergence in measure*) in S of the sequence of functions $\{x_n(s)\}$ to $x(s)$:

$$\text{For any } \varepsilon > 0, \ \lim_{n\to\infty} m\{s \in S;\ |x(s) - x_n(s)| \geq \varepsilon\} = 0. \tag{2}$$

Proof. Clear from the inequality

$$\frac{\delta}{1+\delta} m(B_\delta) \leq \|x\| \leq m(B_\delta) + \frac{\delta}{1+\delta} m(S - B_\delta),\ B_\delta = \{s \in S;\ |x(s)| \geq \delta\}.$$

Remark. It is easy to see that the topology of $M(S, \mathfrak{B}, m)$ may also be defined by the quasi-norm

$$\|x\| = \inf_{\varepsilon > 0} \tan^{-1}[\varepsilon + m\{s \in S;\ |x(s)| \geq \varepsilon\}]. \tag{1'}$$

Example 3. $\mathfrak{D}_K(\Omega)$. The linear space $\mathfrak{D}_K(\Omega)$, introduced in Chapter I, 1, is a quasi-normed linear space by the quasi-norm $\|x\| = d(x, 0)$, where the distance $d(x, y)$ is defined in Chapter I, 1.

5. Pre-Hilbert Spaces

Definition 1. A real or complex normed linear space X is called a *pre-Hilbert space* if its norm satisfies the condition

$$\|x+y\|^2 + \|x-y\|^2 = 2(\|x\|^2 + \|y\|^2). \tag{1}$$

Theorem 1 (M. Fréchet-J. von Neumann-P. Jordan). We define, in a real pre-Hilbert space X,

$$(x, y) = 4^{-1}(\|x+y\|^2 - \|x-y\|^2). \tag{2}$$

Then we have the properties:

$$(\alpha x, y) = \alpha(x, y) \quad (\alpha \in R^1), \tag{3}$$

$$(x+y, z) = (x, z) + (y, z), \tag{4}$$

$$(x, y) = (y, x), \tag{5}$$

$$(x, x) = \|x\|^2. \tag{6}$$

Proof. (5) and (6) are clear. We have, from (1) and (2),

$$\begin{aligned}(x, z) + (y, z) &= 4^{-1}(\|x+z\|^2 - \|x-z\|^2 + \|y+z\|^2 - \|y-z\|^2)\\ &= 2^{-1}\left(\left\|\frac{x+y}{2} + z\right\|^2 - \left\|\frac{x+y}{2} - z\right\|^2\right) \qquad (7)\\ &= 2\left(\frac{x+y}{2}, z\right).\end{aligned}$$

If we take $y = 0$, we obtain $(x, z) = \left(2\frac{x}{2}, z\right)$, because $(0, z) = 0$ by (2). Hence, by (7), we obtain (4). Thus we see that (3) holds for rational numbers α of the form $\alpha = m/2^n$. In a normed linear space, $\|\alpha x + y\|$ and $\|\alpha x - y\|$ are continuous in α. Hence, by (2), $(\alpha x, y)$ is continuous in α. Therefore (3) is proved for every real number α.

Corollary (J. von Neumann-P. Jordan). We define, in a complex normed linear space X,

$$(x, y) = (x, y)_1 + i(x, iy)_1,$$

$$\text{where } i = \sqrt{-1},\ (x, y)_1 = 4^{-1}(\|x+y\|^2 - \|x-y\|^2). \tag{8}$$

Then, we have (4), (6) and

$$(\alpha x, y) = \alpha(x, y) \quad (\alpha \in C^1), \tag{3'}$$

$$(x, y) = \overline{(y, x)} \text{ (complex-conjugate number)}. \tag{5'}$$

Proof. X is also a real pre-Hilbert space and so (4) and (3') with real α hold good. We have, by (8), $(y, x)_1 = (x, y)_1$, $(ix, iy)_1 = (x, y)_1$ and hence $(y, ix)_1 = (-iiy, ix)_1 = -(iy, x)_1 = -(x, iy)_1$. Therefore

$$(y, x) = (y, x)_1 + i(y, ix)_1 = (x, y)_1 - i(x, iy)_1 = \overline{(x, y)}.$$

Similarly, we have

$$(ix, y) = (ix, y)_1 + i(ix, iy)_1 = -(x, iy)_1 + i(x, y)_1 = i(x, y),$$

and therefore we have proved (3'). Finally we have (6), because

$$(x, x)_1 = \|x\|^2 \text{ and } (x, ix)_1 = 4^{-1}(|1 + i|^2 - |1 - i|^2)\,\|x\|^2 = 0.$$

Theorem 2. A (real or) complex linear space X is a (real or) complex *pre-Hilbert space*, if to every pair of elements $x, y \in X$ there is associated a (real or) complex number (x, y) satisfying (3'), (4), (5') and

$$(x, x) \geqq 0, \text{ and } (x, x) = 0 \text{ iff } x = 0. \tag{9}$$

Proof. For any real number α, we have, by (3'), (4) and (5')

$$(x + \alpha(x, y)\, y, x + \alpha(x, y)\, y) = \|x\|^2 + 2\alpha\,|(x, y)|^2$$
$$+ \alpha^2\,|(x, y)|^2\,\|y\|^2 \geqq 0, \text{ where } \|x\| = (x, x)^{1/2}$$

so that we have $|(x, y)|^4 - \|x\|^2\,|(x, y)|^2\,\|y\|^2 \leqq 0$. Hence we obtain Schwarz' inequality

$$|(x, y)| \leqq \|x\| \cdot \|y\|, \tag{10}$$

where the equality is satisfied iff x and y are linearly dependent.

The latter part of (10) is clear from the latter part of (9).

We have, by (10), the triangle inequality for $\|x\|$:

$$\|x + y\|^2 = (x + y, x + y) = \|x\|^2 + (x, y) + (y, x) + \|y\|^2$$
$$\leqq (\|x\| + \|y\|)^2.$$

Finally, the equality (1) is verified easily.

Definition 2. The number (x, y) introduced above is called the *scalar product* (or *inner product*) of two vectors x and y of the pre-Hilbert space X.

Example 1. $L^2(S, \mathfrak{B}, m)$ is a pre-Hilbert space in which the scalar product is given by $(x, y) = \int_S x(s)\,\overline{y(s)}\,m(ds)$.

Example 2. The normed linear space (l^2) is a pre-Hilbert space in which the scalar product is given by $(\{\xi_n\}, \{\eta_n\}) = \sum_{n=1}^{\infty} \xi_n \overline{\eta_n}$.

Example 3. Let Ω be an open domain of R^n and $0 \leqq k < \infty$. Then the totality of functions $f \in C^k(\Omega)$ for which

$$\|f\|_k = \Big(\sum_{|j| \leqq k} \int_\Omega |D^j f(x)|^2\, dx\Big)^{1/2} < \infty, \text{ where } dx = dx_1 dx_2 \cdots dx_n \tag{11}$$

is the Lebesgue measure in R^n,

constitutes a pre-Hilbert space $\hat{H}^k(\Omega)$ by the scalar product

$$(f, g)_k = \sum_{|j| \leq k} \int_\Omega D^j f(x) \cdot \overline{D^j g(x)}\, dx. \tag{12}$$

Example 4. Let Ω be an open domain of R^n and $0 \leq k < \infty$. Then $C_0^k(\Omega)$ is a pre-Hilbert space by the scalar product (12) and the norm (11). We shall denote this pre-Hilbert space by $\hat{H}_0^k(\Omega)$.

Example 5. Let G be a bounded open domain of the complex z-plane. Let $A^2(G)$ be the set of all holomorphic functions $f(z)$ defined in G and such that

$$\|f\| = \Big(\iint_G |f(z)|^2\, dx\, dy\Big)^{1/2} < \infty, \quad (z = x + iy). \tag{13}$$

Then $A^2(G)$ is a pre-Hilbert space by the norm (13), the scalar product

$$(f, g) = \iint_G f(z)\, \overline{g(z)}\, dx\, dy \tag{14}$$

and the algebraic operations

$$(f + g)(z) = f(z) + g(z), \quad (\alpha f)(z) = \alpha f(z).$$

Example 6. Hardy-Lebesgue class H-L^2. Let H-L^2 be the set of all functions $f(z)$ which are holomorphic in the unit disk $\{z;\ |z| < 1\}$ of the complex z-plane and such that

$$\sup_{0<r<1} \left(\int_0^{2\pi} |f(re^{i\theta})|^2\, d\theta \right) < \infty. \tag{15}$$

Then, if $f(z) = \sum_{n=0}^{\infty} c_n z^n$ is the Taylor expansion of f,

$$F(r) = \frac{1}{2\pi} \int_0^{2\pi} |f(re^{i\theta})|^2\, d\theta = \frac{1}{2\pi} \sum_{n,m=0}^{\infty} \int_0^{2\pi} c_n \bar{c}_m r^{n+m} e^{i(n-m)\theta}\, d\theta$$
$$= \sum_{n=0}^{\infty} |c_n|^2 r^{2n}$$

is monotone increasing in r, $0 < r < 1$, and bounded from above. Thus it is easy to see that

$$\|f\| = \sup_{0<r<1} \left[\frac{1}{2\pi} \left(\int_0^{2\pi} |f(re^{i\theta})|^2\, d\theta \right) \right]^{1/2} = \left(\sum_{n=0}^{\infty} |c_n|^2 \right)^{1/2} \tag{16}$$

is a norm which satisfies condition (1), since (l^2) is a pre-Hilbert space.

Remark. Let a sequence $\{c_n\} \in (l^2)$ be given, and consider

$$f(z) = f(re^{i\theta}) = \sum_{n=0}^{\infty} c_n z^n = \sum_{n=0}^{\infty} c_n r^n e^{in\theta}, \quad |z| < 1.$$

By Schwarz' inequality, we have

$$\left| \sum_{n=k}^{\infty} c_n z^n \right| \leq \left(\sum_{n=k}^{\infty} |c_n|^2 \right)^{1/2} \left(\sum_{n=k}^{\infty} r^{2n} \right)^{1/2},$$

and so $\sum_{n=0}^{\infty} c_n z^n$ is uniformly convergent in any disk $|z| \leqq \varrho$ with $0 < \varrho < 1$. Thus $f(z)$ is a holomorphic function in the unit disk $|z| < 1$ such that (15) holds good, that is, $f(z)$ belongs to the class H-L^2.

Therefore we have proved

Theorem 3. The Hardy-Lebesgue class H-L^2 is in one-to-one correspondence with the pre-Hilbert space (l^2) as follows:

$$H\text{-}L^2 \ni f(z) = \sum_{n=0}^{\infty} c_n z^n \leftrightarrow \{c_n\} \in (l^2)$$

in such a way that

$$f(z) = \sum_{n=0}^{\infty} c_n z^n \leftrightarrow \{c_n\},\ \ g(z) = \sum_{n=0}^{\infty} d_n z^n \leftrightarrow \{d_n\} \text{ imply}$$

$$f(z) + g(z) \leftrightarrow \{c_n + d_n\},\ \alpha f(z) \leftrightarrow \{\alpha c_n\} \text{ and } \|f\| = \left(\sum_{n=0}^{\infty} |c_n|^2\right)^{1/2}.$$

Hence, as a pre-Hilbert space, H-L^2 is *isomorphic* with (l^2).

6. Continuity of Linear Operators

Proposition 1. Let X and Y be linear topological spaces over the same scalar field K. Then a linear operator Z on $D(T) \subseteqq X$ into Y is continuous everywhere on $D(T)$ iff it is continuous at the zero vector $x = 0$.

Proof. Clear from the linearity of the operator T and $T \cdot 0 = 0$.

Theorem 1. Let X, Y be locally convex spaces, and $\{p\}$, $\{q\}$ be the systems of semi-norms respectively defining the topologies of X and Y. Then a linear operator T on $D(T) \subseteqq X$ into Y is continuous iff, for every semi-norm $q \in \{q\}$, there exist a semi-norm $p \in \{p\}$ and a positive number β such that

$$q(Tx) \leqq \beta p(x) \quad \text{for all} \quad x \in D(T). \tag{1}$$

Proof. The condition is sufficient. For, by $T \cdot 0 = 0$, the condition implies that T is continuous at the point $x = 0 \in D(T)$ and so T is continuous everywhere on $D(T)$.

The condition is necessary. The continuity of T at $x = 0$ implies that, for every semi-norm $q \in \{q\}$ and every positive number ε, there exist a semi-norm $p \in \{p\}$ and a positive number δ such that

$$x \in D(T) \text{ and } p(x) \leqq \delta \text{ imply } q(Tx) \leqq \varepsilon.$$

Let x be an arbitrary point of $D(T)$, and let us take a positive number λ such that $\lambda p(x) \leqq \delta$. Then we have $p(\lambda x) \leqq \delta$, $\lambda x \in D(T)$ and so $q(T(\lambda x)) \leqq \varepsilon$. Thus $q(Tx) \leqq \varepsilon/\lambda$. Hence, if $p(x) = 0$, we can take λ arbitrarily large and so $q(Tx) = 0$; and if $p(x) \neq 0$, we can take $\lambda = \delta/p(x)$ and so, in any case, we have $q(Tx) \leqq \beta p(x)$ with $\beta = \varepsilon/\delta$.

Corollary 1. Let X be a locally convex space, and f a linear functional on $D(f) \subseteqq X$. Then f is continuous iff there exist a semi-norm p from the system $\{p\}$ of semi-norms defining the topology of X and a positive number β such that

$$|f(x)| \leqq \beta p(x) \text{ for all } x \in D(f). \tag{2}$$

Proof. For, the absolute value $|\alpha|$ itself constitutes a system of semi-norms defining the topology of the real or complex number field.

Corollary 2. Let X, Y be normed linear spaces. Then a linear operator T on $D(T) \subseteqq X$ into Y is continuous iff there exists a positive constant β such that

$$||Tx|| \leqq \beta ||x|| \text{ for all } x \in D(T). \tag{3}$$

Corollary 3. Let X, Y be normed linear spaces. Then a linear operator T on $D(T) \subseteqq X$ into Y admits a continuous inverse T^{-1} iff there exists a positive constant γ such that

$$||Tx|| \geqq \gamma ||x|| \text{ for every } x \in D(T). \tag{4}$$

Proof. By (4), $Tx = 0$ implies $x = 0$ and so the inverse T^{-1} exists. The continuity of T^{-1} is proved by (4) and the preceding Corollary 2.

Definition 1. Let T be a continuous linear operator on a normed linear space X into a normed linear space Y. We define

$$||T|| = \inf_{\beta \in B} \beta, \text{ where } B = \{\beta;\ ||Tx|| \leqq \beta ||x|| \text{ for all } x \in X\}. \tag{5}$$

By virtue of the preceding Corollary 2 and the linearity of T, it is easy to see that

$$||T|| = \sup_{||x|| \leqq 1} ||Tx|| = \sup_{||x||=1} ||Tx||. \tag{6}$$

$||T||$ is called the *norm* of T. A continuous linear operator on a normed linear space X into Y is called a *bounded linear operator* on X into Y, since, for such an operator T, the norm $||Tx||$ is bounded when x ranges over the *unit disk* or the *unit sphere* $\{x \in X;\ ||x|| \leqq 1\}$ of X.

Definition 2. Let T and S be linear operators such that

$$D(T) \text{ and } D(S) \subseteqq X, \text{ and } R(T) \text{ and } R(S) \subseteqq Y.$$

Then the *sum* $T + S$ and the *scalar multiple* αT are defined respectively by

$$(T + S)(x) = Tx + Sx \quad \text{for} \quad x \in D(T) \cap D(S),\ (\alpha T)(x) = \alpha (Tx).$$

Let T be a linear operator on $D(T) \subseteqq X$ into Y, and S a linear operator on $D(S) \subseteqq Y$ into Z. Then the *product* ST is defined by

$$(ST)x = S(Tx) \quad \text{for} \quad x \in \{x;\ x \in D(T) \quad \text{and} \quad Tx \in D(S)\}.$$

$T + S$, αT and ST are linear operators.

Remark. ST and TS do not necessarily coincide even if $X = Y = Z$. An example is given by $Tx = tx(t)$, $Sx(t) = \sqrt{(-1)}\frac{d}{dt}x(t)$ considered as linear operators from $L^2(R^1)$ into $L^2(R^1)$. In this example, we have the *commutation relation* $(ST - TS)\,x(t) = \sqrt{-1}\,x(t)$.

Proposition 2. If T and S are bounded linear operators on a normed linear space X into a normed linear space Y, then

$$\|T + S\| \leqq \|T\| + \|S\|, \quad \|\alpha T\| = |\alpha|\,\|T\|. \tag{7}$$

If T is a bounded linear operator on a normed linear space X into a normed linear space Y, and S a bounded linear operator on Y into a normed linear space Z, then

$$\|ST\| \leqq \|S\| \cdot \|T\|. \tag{8}$$

Proof. We prove the last inequality; (7) may be proved similarly. $\|STx\| \leqq \|S\|\,\|Tx\| \leqq \|S\|\,\|T\|\,\|x\|$ and so $\|ST\| \leqq \|S\|\,\|T\|$.

Corollary. If T is a bounded linear operator on a normed linear space X into X, then

$$\|T^n\| \leqq \|T\|^n, \tag{9}$$

where T^n is defined inductively by $T^n = TT^{n-1}$ $(n = 1, 2, \ldots;\ T^0 = I$ which maps every x onto x itself, i.e., $Ix = x$, and I is called the *identity operator*).

7. Bounded Sets and Bornologic Spaces

Definition 1. A subset B in a linear topological space X is said to be *bounded* if it is *absorbed* by any neighbourhood U of 0, i.e., if there exists a positive constant α such that $\alpha^{-1}B \subseteq U$. Here $\alpha^{-1}B = \{x \in X;\ x = \alpha^{-1}b,\ b \in B\}$.

Proposition. Let X, Y be linear topological spaces. Then a continuous linear operator on X into Y maps every bounded set of X onto a bounded set of Y.

Proof. Let B be a bounded set of X, and V a neighbourhood of 0 of Y. By the continuity of T, there exists a neighbourhood U of 0 of X such that $T \cdot U = \{Tu;\ u \in U\} \subseteq V$. Let $\alpha > 0$ be such that $B \subseteq \alpha U$. Then $T \cdot B \subseteq T(\alpha U) = \alpha(T \cdot U) \subseteq \alpha V$. This proves that $T \cdot B$ is a bounded set of Y.

Definition 2. A locally convex space X is called *bornologic* if it satisfies the condition:

If a balanced convex set M of X absorbs every bounded set of X, then M is a neighbourhood of 0 of X. (1)

Theorem 1. A locally convex space X is bornologic iff every semi-norm on X, which is bounded on every bounded set, is continuous.

Proof. We first remark that a semi-norm $p(x)$ on X is continuous iff it is continuous at $x = 0$. This we see from the subadditivity of the semi-norm: $p(x-y) \geqq |p(x) - p(y)|$ (Chapter I, 1, (4)).

Necessity. Let a semi-norm $p(x)$ on X be bounded on every bounded set of X. The set $M = \{x \in X; p(x) \leqq 1\}$ is convex and balanced. If B is a bounded set of X, then $\sup_{b \in B} p(b) = \alpha < \infty$ and therefore $B \subseteqq \alpha M$. Since, by the assumption, X is bornologic, M must be a neighbourhood of 0. Thus we see that p is continuous at $x = 0$.

Sufficiency. Let M be a convex, balanced set of X which absorbs every bounded set of X. Let p be the Minkowski functional of M. Then p is bounded on every bounded set, since M absorbs, by the assumption, every bounded set. Hence, by the hypothesis, $p(x)$ is continuous. Thus $M_1 = \{x \in X; p(x) < 1/2\}$ is an open set $\ni 0$ contained in M. This proves that M is a neighbourhood of 0.

Example 1. Normed linear spaces are bornologic.

Proof. Let X be a normed linear space. Then the unit disk $S = \{x \in X; \|x\| \leqq 1\}$ of X is a bounded set of X. Let a semi-norm $p(x)$ on X be bounded on S, i.e., $\sup_{x \in S} p(x) = \alpha < \infty$. Then, for any $y \neq 0$,

$$p(y) = p\left(\|y\| \cdot \frac{y}{\|y\|}\right) = \|y\| \, p\left(\frac{y}{\|y\|}\right) \leqq \alpha \, \|y\|.$$

Thus p is continuous at $y = 0$ and so continuous at every point of X.

Remark. As will be seen later, the quasi-normed linear space $M(S, \mathfrak{B})$ is not locally convex. Thus a quasi-normed linear space is not necessarily bornologic. However we can prove

Theorem 2. A linear operator T on one quasi-normed linear space into another such space is continuous iff T maps bounded sets into bounded sets.

Proof. As was proved in Chapter I, 2, Proposition 2, a quasi-normed linear space is a linear topological space. Hence the "only if" part is already proved above in the Proposition. We shall prove the "if" part.

Let T map bounded sets into bounded sets. Suppose that $s\text{-}\lim_{k \to \infty} x_k = 0$. Then $\lim_{k \to \infty} \|x_k\| = 0$ and so there exists a sequence of integers $\{n_k\}$ such that

$$\lim_{k \to \infty} n_k = \infty \quad \text{while} \quad \lim_{k \to \infty} n_k \|x_k\| = 0.$$

We may take, for instance, n_k as follows:

$$\begin{aligned} n_k &= \text{the largest integer} \leqq \|x_k\|^{-1/2} \quad \text{if} \quad x_k \neq 0, \\ &= k \quad \text{if} \quad x_k = 0. \end{aligned}$$

Now we have $||n_k x_k|| = ||x_k + x_k + \cdots + x_k|| \leqq n_k ||x_k||$ so that $\underset{k\to\infty}{s\text{-lim}}\, n_k x_k = 0$. But, in a quasi-normed linear space, the sequence $\{n_k x_k\}$, which converges to 0, is bounded. Thus, by the hypothesis, $\{T(n_k x_k)\} = \{n_k T x_k\}$ is a bounded sequence. Therefore

$$\underset{k\to\infty}{s\text{-lim}}\, T x_k = \underset{k\to\infty}{s\text{-lim}}\, n_k^{-1}(T(n_k x_k)) = 0,$$

and so T is continuous at $x = 0$ and hence is continuous everywhere.

Theorem 3. Let X be bornologic. If a linear operator T on X into a locally convex linear topological space Y maps every bounded set into a bounded set, then T is continuous.

Proof. Let V be a convex balanced neighbourhood of 0 of Y. Let p be the Minkowski functional of V. Consider $q(x) = p(Tx)$. q is a semi-norm on X which is bounded on every bounded set of X, because every bounded set of Y is absorbed by the neighbourhood V of 0. Since X is bornologic, q is continuous. Thus the set $\{x \in X;\, Tx \in V^a\} = \{x \in X;\, q(x) \leqq 1\}$ is a neighbourhood of 0 of X. This proves that T is continuous.

8. Generalized Functions and Generalized Derivatives

A continuous linear functional defined on the locally convex linear topological space $\mathfrak{D}(\Omega)$, introduced in Chapter I, 1, is the "distribution" or the "generalized function" of L. SCHWARTZ. To discuss the generalized functions, we shall begin with the proof of

Theorem 1. Let B be a bounded set of $\mathfrak{D}(\Omega)$. Then there exists a compact subset K of Ω such that

$$\operatorname{supp}(\varphi) \subseteqq K \text{ for every } \varphi \in B, \tag{1}$$

$$\sup_{x\in K,\varphi\in B} |D^j \varphi(x)| < \infty \text{ for every differential operator } D^j. \tag{2}$$

Proof. Suppose that there exist a sequence of functions $\{\varphi_i\} \subseteqq B$ and a sequence of points $\{p_i\}$ such that (i): $\{p_i\}$ has no accumulation point in Ω, and (ii): $\varphi_i(p_i) \neq 0$ $(i = 1, 2, \ldots)$. Then

$$p(\varphi) = \sum_{i=1}^{\infty} i\, |\varphi(p_i)|/|\varphi_i(p_i)|$$

is a continuous semi-norm on every $\mathfrak{D}_K(\varphi)$, defined in Chapter I, 1. Hence, for any $\varepsilon > 0$, the set $\{\varphi \in D_K(\Omega);\, p(\varphi) \leqq \varepsilon\}$ is a neighbourhood of 0 of $\mathfrak{D}_K(\Omega)$. Since $\mathfrak{D}(\Omega)$ is the inductive limit of $\mathfrak{D}_K(\Omega)$'s, we see that $\{\varphi \in \mathfrak{D}(\Omega);\, p(\varphi) \leqq \varepsilon\}$ is also a neighbourhood of 0 of $\mathfrak{D}(\Omega)$. Thus p is continuous at 0 of $\mathfrak{D}(\Omega)$ and so is continuous on $\mathfrak{D}(\Omega)$. Hence p must be bounded on the bounded set B of $\mathfrak{D}(\Omega)$. However, $p(\varphi_i) \geqq i$ $(i = 1, 2, \ldots)$. This proves that we must have (1).

We next assume that (1) is satisfied, and suppose (2) is not satisfied. Then there exist a differential operator D^{j_0} and a sequence of functions $\{\varphi_i\} \subseteqq B$ such that $\sup_{x \in K} |D^{j_0} \varphi_i(x)| > i$ $(i = 1, 2, \ldots)$. Thus, if we set

$$p(\varphi) = \sup_{x \in K} |D^{j_0} \varphi(x)| \quad \text{for} \quad \varphi \in \mathfrak{D}_K(\Omega),$$

$p(\varphi)$ is a continuous semi-norm on $\mathfrak{D}_K(\Omega)$ and $p(\varphi_i) > i$ $(i = 1, 2, \ldots)$. Hence $\{\varphi_i\} \subseteqq B$ cannot be bounded in $\mathfrak{D}_K(\Omega)$, and a fortiori in $\mathfrak{D}(\Omega)$. This contradiction proves that (2) must be true.

Theorem 2. The space $\mathfrak{D}(\Omega)$ is bornologic.

Proof. Let $q(\varphi)$ be a semi-norm on $\mathfrak{D}(\Omega)$ which is bounded on every bounded set of $\mathfrak{D}(\Omega)$. In view of Theorem 1 in Chapter I, 7, we have only to show that q is continuous on $\mathfrak{D}(\Omega)$. To this purpose, we show that q is continuous on the space $\mathfrak{D}_K(\Omega)$ where K is any compact subset of Ω. Since $\mathfrak{D}(\Omega)$ is the inductive limit of $\mathfrak{D}_K(\Omega)$'s, we then see that q is continuous on $\mathfrak{D}(\Omega)$.

But q is continuous on every $\mathfrak{D}_K(\Omega)$. For, by hypothesis, q is bounded on every bounded set of the quasi-normed linear space $\mathfrak{D}_K(\Omega)$, and so, by Theorem 2 of the preceding section, q is continuous on $\mathfrak{D}_K(\Omega)$. Hence q must be continuous on $\mathfrak{D}(\Omega)$.

We are now ready to define the *generalized functions*.

Definition 1. A linear functional T defined and continuous on $\mathfrak{D}(\Omega)$ is called a *generalized function*, or an *ideal function* or a *distribution* in Ω; and the value $T(\varphi)$ is called the value of the generalized function T at the *testing function* $\varphi \in \mathfrak{D}(\Omega)$.

By virtue of Theorem 1 in Chapter I, 7 and the preceding Theorem 2, we have

Proposition 1. A linear functional T defined on $\mathfrak{D}(\Omega)$ is a generalized function in Ω iff it is bounded on every bounded set of $\mathfrak{D}(\Omega)$, that is, iff T is bounded on every set $B \in \mathfrak{D}(\Omega)$ satisfying the two conditions (1) and (2).

Proof. Clear from the fact that $T(\varphi)$ is continuous iff the semi-norm $|T(\varphi)|$ is continuous.

Corollary. A linear functional T defined on $C_0^\infty(\Omega)$ is a generalized function in Ω iff it satisfies the condition:

To every compact subset K of Ω, there correspond a positive constant C and a positive integer k such that

$$|T(\varphi)| \leqq C \sup_{|j| \leqq k, x \in K} |D^j \varphi(x)| \quad \text{whenever} \quad \varphi \in \mathfrak{D}_K(\varphi). \tag{3}$$

Proof. By the continuity of T on the inductive limit $\mathfrak{D}(\Omega)$ of the $\mathfrak{D}_K(\Omega)$'s, we see that T must be continuous on every $\mathfrak{D}_K(\Omega)$. Hence the necessity of condition (3) is clear. The sufficiency of the condition

(3) is also clear, since it implies that T is bounded on every bounded set of $\mathfrak{D}(\Omega)$.

Remark. The above Corollary is very convenient for all applications, since it serves as a useful definition of the generalized functions.

Example 1. Let a complex-valued function $f(x)$ defined a.e. in Ω be *locally integrable* in Ω with respect to the Lebesgue measure $dx = dx\,dx \cdots dx$ in R^n, in the sense that, for any compact subset K of Ω, $\int_K |f(x)|\,dx < \infty$. Then

$$T_f(\varphi) = \int_\Omega f(x)\,\varphi(x)\,dx, \ \varphi \in \mathfrak{D}(\Omega), \tag{4}$$

defines a generalized function T_f in Ω.

Example 2. Let $m(B)$ be a σ-finite, σ-additive and complex-valued measure defined on Baire subsets B of an open set Ω of R^n. Then

$$T_m(\varphi) = \int_\Omega \varphi(x)\,m(dx), \ \varphi \in \mathfrak{D}(\Omega), \tag{5}$$

defines a generalized function T_m in Ω.

Example 3. As a special case of Example 2,

$$T_{\delta_p}(\varphi) = \varphi(p), \text{ where } p \text{ is a fixed point of } \Omega, \ \varphi \in \mathfrak{D}(\Omega), \tag{6}$$

defines a generalized function T_{δ_p} in Ω. It is called the *Dirac distribution* concentrated at the point $p \in \Omega$. In the particular case $p = 0$, the origin of R^n, we shall write T_δ or δ for T_{δ_0}.

Definition 2. The set of all generalized functions in Ω will be denoted by $\mathfrak{D}(\Omega)'$. It is a linear space by

$$(T + S)(\varphi) = T(\varphi) + S(\varphi), \ (\alpha T)(\varphi) = \alpha T(\varphi), \tag{7}$$

and we call $\mathfrak{D}(\Omega)'$ the *space of the generalized functions* in Ω or the *dual space* of $\mathfrak{D}(\Omega)$.

Remark. *Two distributions* T_{f_1} *and* T_{f_2} *are equal* as functionals ($T_{f_1}(\varphi) = T_{f_2}(\varphi)$ for every $\varphi \in \mathfrak{D}(\Omega)$) *iff* $f_1(x) = f_2(x)$ *a.e.* If this fact is proved, then the set of all locally integrable functions in Ω is, by $f \leftrightarrow T_f$, in a one-one correspondence with a subset of $\mathfrak{D}(\Omega)'$ in such a way that (f_1 and f_2 being considered equivalent iff $f_1(x) = f_2(x)$ a.e.)

$$T_{f_1} + T_{f_2} = T_{f_1+f_2}, \ \alpha T_f = T_{\alpha f}. \tag{7'}$$

In this sense, the notion of the generalized function is, in fact, a generalization of the notion of the locally integrable function. To prove the above assertion, we have only to prove that a locally integrable function f is $= 0$ a.e. in an open set Ω of R^n if $\int_\Omega f(x)\,\varphi(x)\,dx = 0$ for all $\varphi \in C_0^\infty(\Omega)$. By introducing the Baire measure $\mu(B) = \int_\Omega f(x)\,dx$, the latter condition implies that $\int_\Omega \varphi(x)\,\mu(dx) = 0$ for all $\varphi \in C_0^\infty(\Omega)$, which further implies that $\int_\Omega \varphi(x)\,\mu(dx) = 0$ for all $\varphi \in C_0^0(\Omega)$, by virtue of Proposition 8 in Chapter I, 1. Let B be a compact G_δ-set in Ω: $B = \bigcap_{n=1}^{\infty} G_n$, where G_n is an open relatively compact set in Ω. By applying Urysohn's theorem in Chapter 0, 2, there exists a continuous function $f_n(x)$ such that

$$0 \leqq f_n(x) \leqq 1 \text{ for } x \in \Omega,\ f_n(x) = 1 \text{ for } x \in G_{n+2}^a \text{ and } f_n(x) = 0$$
$$\text{for } x \in G_n^a - G_{n+1}\ (n = 1, 2, \ldots),$$

assuming that $\{G_n\}$ is a monotone decreasing sequence of open relatively compact sets of Ω such that $G_{n+2}^a \subseteqq G_{n+1}$. Setting $\varphi = f_n$ and letting $n \to \infty$, we see that $\mu(B) = 0$ for all compact G_δ-sets B of Ω. The Baire sets of Ω are the members of the smallest σ-ring containing compact G_δ-sets of Ω, we see, by the σ-additivity of the Baire measure μ, that μ vanishes for every Baire set of Ω. Hence the density f of this measure μ must vanish a.e. in Ω.

We can define the notion of differentiation of generalized functions through

Proposition 2. If T is a generalized function in Ω, then

$$S(\varphi) = -T\left(\frac{\partial \varphi}{\partial x_1}\right), \quad \varphi \in \mathfrak{D}(\Omega), \tag{8}$$

defines another generalized function S in Ω.

Proof. S is a linear functional on $\mathfrak{D}(\Omega)$ which is bounded on every bounded set of $\mathfrak{D}(\Omega)$.

Definition 3. The generalized functional S defined by (8) is called the *generalized derivative* or the *distributional derivative* of T (with respect to x_1), and we write

$$S = \frac{\partial}{\partial x_1} T, \tag{9}$$

so that we have

$$\frac{\partial}{\partial x_1} T(\varphi) = -T\left(\frac{\partial \varphi}{\partial x_1}\right). \tag{10}$$

Remark. The above notion is an extension of the usual notion of the derivative. For, if the function f is continuously differentiable with respect to x_1, then we have

$$\frac{\partial}{\partial x_1} T_f(\varphi) = -T_f\left(\frac{\partial \varphi}{\partial x_1}\right) = -\int \cdots \int_\Omega f(x) \frac{\partial \varphi}{\partial x_1} dx_1 \cdots dx_n$$
$$= \int \cdots \int_\Omega \frac{\partial}{\partial x_1} f(x) \cdot \varphi(x)\, dx_1 \cdots dx_n = T_{\partial f/\partial x_1}(\varphi),$$

as may be seen by partial integration observing that $\varphi(x)$ vanishes identically outside some compact subset of Ω.

Corollary. A generalized function T in Ω is infinitely differentiable in the sense of distributions defined above and

$$(D^j T)(\varphi) = (-1)^{|j|} T(D^j \varphi), \text{ where } |j| = \sum_{i=1}^{n} j_i, \; D^j = \frac{\partial^{|j|}}{\partial x_1^{j_1} \ldots \partial x_n^{j_n}}. \tag{11}$$

Example 1. The *Heaviside function* $H(x)$ is defined by

$$H(x) = 1 \text{ or } = 0 \text{ according as } x \geqq 0 \text{ or } x < 0. \tag{12}$$

Then we have

$$\frac{d}{dx} T_H = T_{\delta_0}, \tag{12'}$$

where T_{δ_0} is the *Dirac distribution* concentrated at the origin 0 of R^1. In fact, we have, for any $\varphi \in \mathfrak{D}(R^1)$,

$$\left(\frac{d}{dx} T_H\right)(\varphi) = -\int_{-\infty}^{\infty} H(x)\varphi'(x)\,dx = -\int_{0}^{\infty} \varphi'(x)\,dx = -[\varphi(x)]_0^\infty = \varphi(0).$$

Example 2. Let $f(x)$ have a bounded and continuous derivative in the open set $R^1 - \bigcup_{j=1}^{k} x_j$ of R^1. Let $s_j = f(x_j + 0) - f(x_j - 0)$ be the *saltus* or the *jump* of $f(x)$ at $x = x_j$. Since

$$\left(\frac{d}{dx} T_f\right)(\varphi) = -\int_{-\infty}^{\infty} f(x)\varphi'(x)\,dx = \sum_j \varphi(x_j)\, s_j + \int_{-\infty}^{\infty} f'(x)\varphi(x)\,dx,$$

we have

$$\frac{d}{dx} T_f = T_{f'} + \sum_j s_j\, \delta_{x_j}, \tag{12''}$$

where δ_{x_j} is defined by (6).

Example 3. Let $f(x) = f(x_1, x_2, \ldots, x_n)$ be a continuously differentiable function on a closed bounded domain $\Omega \subseteqq R^n$ having a smooth boundary S. Define f to be 0 outside Ω. By partial integration, we have

$$\begin{aligned}\left(\frac{\partial}{\partial x_j} T_f\right)(\varphi) &= -\int_\Omega f(x)\,\frac{\partial}{\partial x_j}\varphi(x)\,dx \\ &= \int_S f(x)\,\varphi(x)\cos(\nu, x_j)\,dS + \int_\Omega \frac{\partial f}{\partial x_j}\varphi(x)\,dx,\end{aligned}$$

where ν is the *inner normal* to S, $(\nu, x_j) = (x_j, \nu)$ is the angle between ν and the positive x_j-axis and dS is the surface element. We have thus

$$\frac{\partial}{\partial x_j} T_f = T_{\partial f/\partial x_j} + T_S, \text{ where } T_S(\varphi) = \int_S f(x)\cos(\nu, x_j)\,\varphi(x)\,dS. \tag{12''}$$

Corollary. If $f(x) = f(x_1, x_2, \ldots, x_n)$ is C^2 on Ω and is 0 outside, then, from (12'') and $\frac{\partial}{\partial \nu} = \sum_j \frac{\partial}{\partial x_j}\cos(x_j, \nu)$ we obtain *Green's* *integral theorem*

$$(\Delta T_f)(\varphi) = T_{\Delta f}(\varphi) + \int_S \frac{\partial f}{\partial \nu} \varphi(x)\, dS - \int_S f(x) \frac{\partial \varphi}{\partial \nu}\, dS, \qquad (12''')$$

where Δ is the Laplacian $\sum_{j=1}^{n} \partial^2/\partial x_j^2$.

Proposition 3. If T is a generalized function in Ω and $f \in C^\infty(\Omega)$, then

$$S(\varphi) = T(f\varphi),\ \varphi \in \mathfrak{D}(\Omega), \qquad (13)$$

defines another generalized function S in Ω.

Proof. S is a linear functional on $\mathfrak{D}(\Omega)$ which is bounded on every bounded set of $\mathfrak{D}(\Omega)$. This we see by applying Leibniz' formula to $f\varphi$.

Definition 4. The generalized function S defined by (13) is called the *product* of the function f and the generalized function T.

Leibniz' Formula. We have, denoting S in (13) by fT,

$$\frac{\partial}{\partial x_j}(fT) = \frac{\partial f}{\partial x_j} T + f \frac{\partial T}{\partial x_j} \qquad (14)$$

because we have

$$-T\left(f \frac{\partial \varphi}{\partial x_j}\right) = T\left(\frac{\partial f}{\partial x_j} \varphi\right) - T\left(\frac{\partial}{\partial x_j}(f\varphi)\right)$$

by Leibniz' formula for $\partial(f\varphi)/\partial x_j$. This formula is generalized as follows.

Let $P(\xi)$ be a polynomial in $\xi_1, \xi_2, \ldots, \xi_n$, and consider a linear partial differential operator $P(D)$ with constant coefficients, obtained by replacing ξ_j by $i^{-1}\, \partial/\partial x_j$. The introduction of the imaginary coefficient i^{-1} is suitable for the symbolism in the Fourier transform theory in Chapter VI.

Theorem 3 (Generalized Leibniz' Formula of L. Hörmander). We have

$$P(D)(fT) = \sum_{(\alpha) \geqq 0} \frac{1}{(\alpha)!} D_\alpha f \cdot P^{(\alpha)}(D)\, T, \qquad (15)$$

where, for $\alpha = (\alpha_1, \alpha_2, \ldots, \alpha_k)$ with $1 \leqq \alpha_j \leqq n$,

$$P^{(\alpha)}(\xi) = \frac{\partial^{\alpha_1 + \alpha_2 + \cdots + \alpha_k}}{\partial \xi_{\alpha_1} \partial \xi_{\alpha_2} \ldots \partial \xi_{\alpha_k}}, \quad D_\alpha = \prod_{j=1}^{k} \frac{1}{i} \frac{\partial}{\partial x_{\alpha_j}}, \qquad (16)$$

and

$$(\alpha) = k, \text{ and } P^{(0)}(\xi) = P(\xi), \text{ and } D_0 = I \text{ for } (\alpha) = 0. \qquad (17)$$

Proof. Repeated application of (14) gives an identity of the form

$$P(D)(fT) = \sum_{(\alpha) \geqq 0} D_\alpha f \cdot Q_\alpha(D)\, T, \qquad (18)$$

where $Q_\alpha(D)$ are differential operators to be determined. We may assume that they are so chosen that they are invariant under permutation of the indices α; the invariance may be attained by rearrangement and summation over the indices α with the same (α), divided by the number of summands. Since (17) is an identity, we may substitute in (17)

$$f(x) = e^{i\langle x,\xi\rangle} \text{ and } T = e^{i\langle x,\eta\rangle}, \text{ where } \langle x, \xi\rangle = \sum_{j=1}^{n} x_j \xi_j.$$

Then, by the symbolism

$$P(D)\, e^{i\langle x,\xi\rangle} = P(\xi)\, e^{i\langle x,\xi\rangle}, \tag{19}$$

we obtain

$$P(\xi+\eta) = \sum_{\alpha} \xi^{\alpha} Q_{\alpha}(\eta), \text{ where } \xi^{\alpha} = \prod_j \xi_{\alpha_j}.$$

On the other hand, we have, by Taylor's formula,

$$P(\xi+\eta) = \sum_{\alpha} \frac{1}{(\alpha)!} \xi^{\alpha} P^{(\alpha)}(\eta),$$

and so we get $Q_{\alpha}(\eta) = \frac{1}{(\alpha)!} P^{(\alpha)}(\eta)$.

9. *B*-spaces and *F*-spaces

In a quasi-normed linear space X, $\lim_{n\to\infty} \|x_n - x\| = 0$ implies, by the triangle inequality $\|x_n - x_m\| \leqq \|x_n - x\| + \|x - x_m\|$, that $\{x_n\}$ is a *Cauchy sequence*, i.e., $\{x_n\}$ satisfies *Cauchy's convergence condition*

$$\lim_{n,m\to\infty} \|x_n - x_m\| = 0. \tag{1}$$

Definition 1. A quasi-normed (or normed) linear space X is called an *F-space* (or a *B-space*) if it is *complete*, i.e., if every Cauchy sequence $\{x_n\}$ of X converges strongly to a point x_∞ of X:

$$\lim_{n\to\infty} \|x_n - x_\infty\| = 0. \tag{2}$$

Such a limit x_∞, if it exists, is uniquely determined because of the triangle inequality $\|x - x'\| \leqq \|x - x_n\| + \|x_n - x'\|$. A complete pre-Hilbert space is called a *Hilbert space*.

Remark. The names F-space and B-space are abbreviations of *Fréchet space* and *Banach space*, respectively. It is to be noted that Bourbaki uses the term Fréchet spaces for locally convex spaces which are quasi-normed and complete.

Proposition 1. Let Ω be an open set of R^n, and denote by $\mathfrak{E}(\Omega) = C^\infty(\Omega)$ the locally convex space, quasi-normed as in Proposition 6 in Chapter I, 1. This $\mathfrak{E}(\Omega)$ is an F-space.

Proof. The condition $\lim_{n,m\to\infty} \|f_n - f_m\| = 0$ in $\mathfrak{E}(\Omega)$ means that, for any compact subset K of Ω and for any differential operator D^α, the sequence $\{D^\alpha f_n(x)\}$ of functions converges, as $n \to \infty$, uniformly on K. Hence there exists a function $f(x) \in C^\infty(\Omega)$ such that $\lim_{n\to\infty} D^\alpha f_n(x) =$

$D^\alpha f(x)$ uniformly on K. D^α and K being arbitrary, this means that $\lim_{n\to\infty} \|f_n - f\| = 0$ in $\mathfrak{E}(\Omega)$.

Proposition 2. $L^p(S) = L^p(S, \mathfrak{B}, m)$ is a B-space. In particular, $L^2(S)$ and (l^2) are Hilbert spaces.

Proof. Let $\lim_{n,m\to\infty} \|x_n - x_m\| = 0$ in $L^p(S)$. Then we can choose a subsequence $\{x_{n_k}\}$ such that $\sum_k \|x_{n_{k+1}} - x_{n_k}\| < \infty$. Applying the triangle inequality and the Lebesgue-Fatou Lemma to the sequence of functions

$$y_t(s) = |x_{n_1}(s)| + \sum_{k=1}^{t} |x_{n_{k+1}}(s) - x_{n_k}(s)| \in L^p(S),$$

we see that

$$\int_\Omega \left(\lim_{t\to\infty} y_t(s)^p\right) m(ds) \leqq \lim_{t\to\infty} \|y_t\|^p \leqq \left(\|x_{n_1}\| + \sum_{k=1}^{\infty} \|x_{n_{k+1}} - x_{n_k}\|\right)^p.$$

Thus a finite $\lim_{t\to\infty} y_t(s)$ exists a.e. Hence a finite $\lim_{t\to\infty} x_{n_{t+1}}(s) = x_\infty(s)$ exists a.e. and $x_\infty(s) \in L^p(S)$, since $|x_{n_{t+1}}(s)| \leqq \lim_{t\to\infty} y_t(s) \in L^p(S)$. Applying again the Lebesgue-Fatou Lemma, we obtain

$$\|x_\infty - x_{n_k}\|^p = \int_\Omega \left(\lim_{t\to\infty} |x_{n_t}(s) - x_{n_k}(s)|^p\right) m(ds) \leqq \left(\sum_{l=k}^{\infty} \|x_{n_{l+1}} - x_{n_l}\|\right)^p.$$

Therefore $\lim_{k\to\infty} \|x_\infty - x_{n_k}\| = 0$, and hence, by the triangle inequality and Cauchy's convergence condition $\lim_{n,m\to\infty} \|x_n - x_m\| = 0$, we obtain

$$\overline{\lim_{n\to\infty}} \|x_\infty - x_n\| \leqq \overline{\lim_{k\to\infty}} \|x_\infty - x_{n_k}\| + \overline{\lim_{k,n\to\infty}} \|x_{n_k} - x_n\| = 0.$$

Incidentally we have proved the following important

Corollary. A sequence $\{x_n\} \in L^p(S)$ which satisfies Cauchy's convergence condition (1) contains a subsequence $\{x_{n_k}\}$ such that

$$\text{a finite } \lim_{k\to\infty} x_{k_n}(s) = x_\infty(s) \text{ exists a.e., } x_\infty(s) \in L^p(S) \quad \text{and}$$

$$s\text{-}\lim_{n\to\infty} x_n = x_\infty. \tag{3}$$

Remark. In the above Proposition and the Corollary, we have assumed in the proof that $1 \leqq p < \infty$. However the results are also valid for the case $p = \infty$, and the proof is somewhat simpler than for the case $1 \leqq p < \infty$. The reader should carry out the proof.

Proposition 3. The space $A^2(G)$ is a Hilbert space.

Proof. Let $\{f_n(z)\}$ be a Cauchy sequence of $A^2(G)$. Since $A^2(G)$ is a linear subspace of the Hilbert space $L^2(G)$, there exists a subsequence $\{f_{n_k}(z)\}$ such that

$$\text{a finite } \lim_{k\to\infty} f_{n_k}(z) = f_\infty(z) \text{ exists a.e., } f_\infty \in L^2(G) \quad \text{and}$$

$$\lim_{n\to\infty} \int_G |f_\infty(z) - f_n(z)|^2\, dx\, dy = 0.$$

We have to show that $f_\infty(z)$ is holomorphic in G. To do this, let the sphere $|z - z_0| \leqq \varrho$ be contained in G. The Taylor expansion $f_n(z) - f_m(z) = \sum_{j=0}^{\infty} c_j (z - z_0)^j$ implies

$$\begin{aligned} \| f_n - f_m \|^2 &\geqq \int_{|z - z_0| \leqq \varrho} | f_n(z) - f_m(z) |^2 \, dx \, dy \\ &= \int_0^\varrho \left(\int_0^{2\pi} \sum_j c_j r^j e^{ij\theta} \sum_k \bar{c}_k r^k e^{-ik\theta} \, d\theta \right) r \, dr = \sum_j 2\pi \int_0^\varrho |c_j|^2 r^{2j+1} \, dr \\ &= 2\pi \sum_{j=0}^{\infty} \varrho^{2j+2} |c_j|^2 (2j+2)^{-1} \geqq \pi |c_0|^2 \varrho^2 \qquad (4) \\ &= \pi \varrho^2 | f_n(z_0) - f_m(z_0) |^2. \end{aligned}$$

Thus the sequence $\{f_n(z)\}$ itself converges uniformly on any closed sphere contained in G. $f_n(z)$'s being holomorphic in G, we see that $f_\infty(z) = \lim_{n \to \infty} f_n(z)$ must be holomorphic in G.

Proposition 4. $M(S, \mathfrak{B}, m)$ with $m(S) < \infty$ is an F-space.

Proof. Let $\{x_n\}$ be a Cauchy sequence in $M(S, \mathfrak{B}, m)$. Since the convergence in $M(S, \mathfrak{B}, m)$ is the asymptotic convergence, we can choose a sub-sequence $\{x_{n_k}(s)\}$ of $\{x_n(s)\}$ such that

$$m(B_k) \leqq 2^{-k} \quad \text{for} \quad B_k = \{s \in S; \, 2^{-k} \leqq | x_{n_{k+1}}(s) - x_{n_k}(s) | \}.$$

The sequence $x_{n_k}(s) = x_{n_1}(s) + \sum_{j=1}^{k-1} (x_{n_{j+1}}(s) - x_{n_j}(s))$ $(k = 1, 2, \ldots)$ is s-convergent to a function $\in M(S, \mathfrak{B}, m)$, because, if $s \bar{\in} \bigcup_{j=t}^{\infty} B_j$, we have $\sum_{j=t}^{\infty} | x_{n_{j+1}}(s) - x_{n_j}(s) | \leqq \sum_{j=t}^{\infty} 2^{-j} \leqq 2^{1-t}$ and $m\left(\bigcup_{j=t}^{\infty} B_j\right) \leqq \sum_{j=t}^{\infty} m(B_j) \leqq \sum_{j=t}^{\infty} 2^{-j} \leqq 2^{1-t}$; consequently we see, by letting $t \to \infty$, that the sequence $\{x_{n_k}(s)\}$ converges m-a.e. to a function $x_\infty(s) \in M(S, \mathfrak{B}, m)$. Hence $\lim_{k \to \infty} \| x_{n_k} - x_\infty \| = 0$ and so, by $\lim_{n,m \to \infty} \| x_n - x_m \| = 0$, we obtain $\lim_{n \to \infty} \| x_n - x_\infty \| = 0$.

The Space (s). The set (s) of all sequences $\{\xi_n\}$ of numbers quasi-normed by

$$\| \{\xi_n\} \| = \sum_{j=1}^{\infty} 2^{-j} |\xi_j| / (1 + |\xi_j|)$$

constitutes an F-space by $\{\xi_n\} + \{\eta_n\} = \{\xi_n + \eta_n\}$, $\alpha\{\xi_n\} = \{\alpha \xi_n\}$. The proof of the completeness of (s) may be obtained as in the case of $M(S, \mathfrak{B}, m)$. The quasi-norm

$$\| \{\xi_n\} \| = \inf_{\varepsilon > 0} \tan^{-1} \{\varepsilon + \text{the number of } \xi_n\text{'s which satisfy } |\xi_n| > \varepsilon\}$$

also gives an equivalent topology of (s).

Remark. It is clear that $C(S)$, (c_0) and (c) are B-spaces. The completeness of the space (l^p) is a consequence of that of $L^p(S)$. Hence, by Theorem 3 in Chapter I, 5, the space H-L^2 is a Hilbert space with (l^2).

Sobolev Spaces $W^{k,p}(\Omega)$. Let Ω be an open set of R^n, and k a positive integer. For $1 \leqq p < \infty$, we denote by $W^{k,p}(\Omega)$ the set of all complex-valued functions $f(x) = f(x_1, x_2, \ldots, x_n)$ defined in Ω such that f and its distributional derivatives $D^s f$ of order $|s| = \sum_{j=1}^{n} |s_j| \leqq k$ all belong to $L^p(\Omega)$. $W^{k,p}(\Omega)$ is a normed linear space by

$$(f_1 + f_2)(x) = f_1(x) + f_2(x), \ (\alpha f)(x) = \alpha f(x) \ \text{ and}$$
$$\|f\|_{k,p} = \Big(\sum_{|s| \leqq k} \int_\Omega |D^s f(x)|^p \, dx\Big)^{1/p}, \ dx = dx_1 \, dx_2 \ldots dx_n$$

under the convention that we consider two functions f_1 and f_2 as the same vector of $W^{k,p}(\Omega)$ if $f_1(x) = f_2(x)$ a.e. in Ω. It is easy to see that $W^{k,2}(\Omega)$ is a pre-Hilbert space by the scalar product

$$(f, g)_{k,2} = \Big(\sum_{|s| \leqq k} \int_\Omega D^s f(x) \, \overline{D^s g(x)} \, dx\Big).$$

Proposition 5. The space $W^{k,p}(\Omega)$ is a B-space. In particular, $W^k(\Omega) = W^{k,2}(\Omega)$ is a Hilbert space by the norm $\|f\|_k = \|f\|_{k,2}$ and the scalar product $(f, g)_k = (f, g)_{k,2}$.

Proof. Let $\{f_h\}$ be a Cauchy sequence in $W^{k,p}(\Omega)$. Then, for any differential operator D^s with $|s| \leqq k$, the sequence $\{D^s f_h\}$ is a Cauchy sequence in $L^p(\Omega)$ and so, by the completeness of $L^p(\Omega)$, there exist functions $f^{(s)} \in L^p(\Omega)$ $(|s| \leqq k)$ such that $\lim_{h\to\infty} \int_\Omega |D^s f_h(x) - f^{(s)}(x)|^p \, dx = 0$. By virtue of Hölder's inequality in Chapter I, 3, applied to compact sets of Ω, we easily see that f_h is locally integrable in Ω. Hence, for any function $\varphi \in C_0^\infty(\Omega)$,

$$T_{D^s f_h}(\varphi) = \int_\Omega D^s f_h(x) \cdot \varphi(x) \, dx = (-1)^{|s|} \int_\Omega f_h(x) \, D^s \varphi(x) \, dx,$$

and so, again applying Hölder's inequality, we obtain, by $\lim_{h\to\infty} \int_\Omega |f_h(x) - f^{(0)}(x)|^p \, dx = 0$,

$$\lim_{h\to\infty} T_{D^s f_h}(\varphi) = (-1)^{|s|} \, T_{f^{(0)}}(D^s \varphi) = D^s T_{f^{(0)}}(\varphi).$$

Similarly we have, by $\lim_{h\to\infty} \int_\Omega |D^s f_h(x) - f^{(s)}(x)|^p \, dx = 0$,

$$\lim_{h\to\infty} T_{D^s f_h}(\varphi) = T_{f^{(s)}}(\varphi).$$

Hence we must have $D^s T_{f^{(0)}} = T_{f^{(s)}}$, that is, the distributional derivative $D^s f^{(0)}$ equals $f^{(s)}$. This proves that $\lim_{h\to\infty} \|f_h - f^{(0)}\|_{k,p} = 0$ and $W^{k,p}(\Omega)$ is complete.

10. The Completion

The completeness of an F-space (and a B-space) will play an important role in functional analysis in the sense that we can apply to such spaces Baire's category arguments given in Chapter 0, Preliminaries. The following theorem of *completion* will be of frequent use in this book.

Theorem (of completion). Let X be a quasi-normed linear space which is not complete. Then X is isomorphic and isometric to a dense linear subspace of an F-space $\tilde{X}$, i.e., there exists a one-to-one correspondence $x \leftrightarrow \tilde{x}$ of X onto a dense linear subspace of $\tilde{X}$ such that

$$\widetilde{(x+y)} = \tilde{x} + \tilde{y},\ \widetilde{(\alpha x)} = \alpha\,\tilde{x},\ \|\tilde{x}\| = \|x\|. \tag{1}$$

The space $\tilde{X}$ is uniquely determined up to isometric isomorphism. If X is itself a normed linear space, then $\tilde{X}$ is a B-space.

Proof. The proof proceeds as in Cantor's construction of real numbers from rational numbers.

The set of all Cauchy sequences $\{x_n\}$ of X can be classified according to the equivalence $\{x_n\} \sim \{y_n\}$ which means that $\lim\limits_{n\to\infty} \|x_n - y_n\| = 0$. We denote by $\{x_n\}'$ the class containing $\{x_n\}$. Then the set $\tilde{X}$ of all such classes $\tilde{x} = \{x_n\}'$ is a linear space by

$$\{x_n\}' + \{y_n\}' = \{x_n + y_n\}',\ \alpha\{x_n\}' = \{\alpha x_n\}'.$$

We have $\big|\|x_n\| - \|x_m\|\big| \leqq \|x_n - x_m\|$ and hence $\lim\limits_{n\to\infty} \|x_n\|$ exists. We put

$$\|\{x_n\}'\| = \lim_{n\to\infty} \|x_n\|.$$

It is easy to see that these definitions of the vector sum $\{x_n\}' + \{y_n\}'$, the scalar multiplication $\alpha\{x_n\}'$ and the norm $\|\{x_n\}'\|$ do not depend on the particular representations for the classes $\{x_n\}'$, $\{y_n\}'$, respectively. For example, if $\{x_n\} \sim \{x_n'\}$, then

$$\lim_{n\to\infty} \|x_n\| \leqq \lim_{n\to\infty} \|x_n'\| + \lim_{n\to\infty} \|x_n' - x_n\| \leqq \lim_{n\to\infty} \|x_n'\|$$

and similarly $\lim\limits_{n\to\infty} \|x_n'\| \leqq \lim\limits_{n\to\infty} \|x_n\|$, so that we have $\|\{x_n\}'\| = \|\{x_n'\}'\|$.

To prove that $\|\{x_n\}'\|$ is a quasi-norm, we have to show that

$$\lim_{\alpha\to 0} \|\alpha\{x_n\}'\| = 0 \quad \text{and} \quad \lim_{\|\{x_n\}'\|\to 0} \|\alpha\{x_n\}'\| = 0.$$

The former is equivalent to $\lim\limits_{\alpha\to 0}\lim\limits_{n\to\infty} \|\alpha x_n\| = 0$ and the latter is equivalent to $\lim\limits_{n\to\infty} \|\alpha x_n\| = 0$. And these are true because $\|\alpha x\|$ is continuous in both variables α and x.

To prove the completeness of $\tilde{X}$, let $\{\tilde{x}_k\} = \{\{x_n^{(k)}\}\}$ be a Cauchy sequence of $\tilde{X}$. For each k, we can choose n_k such that

$$||x_m^{(k)} - x_{n_k}^{(k)}|| < k^{-1} \quad \text{if} \quad m > n_k. \tag{2}$$

Then we can show that the sequence $\{\tilde{x}_k\}$ converges to the class containing the Cauchy sequence of X:

$$\{x_{n_1}^{(1)}, x_{n_2}^{(2)}, \ldots, x_{n_k}^{(k)}, \ldots\}. \tag{3}$$

To this purpose, we denote by $\bar{x}_{n_k}^{(k)}$ the class containing

$$\{x_{n_k}^{(k)}, x_{n_k}^{(k)}, \ldots, x_{n_k}^{(k)}, \ldots\}. \tag{4}$$

Then, by (2),

$$||\tilde{x}_k - \bar{x}_{n_k}^{(k)}|| = \lim_{m\to\infty} ||x_m^{(k)} - x_{n_k}^{(k)}|| \leqq k^{-1} \tag{5}$$

and hence

$$\begin{aligned} ||x_{n_k}^{(k)} - x_{n_m}^{(m)}|| &= ||\bar{x}_{n_k}^{(k)} - \bar{x}_{n_m}^{(m)}|| \leqq ||\bar{x}_{n_k}^{(k)} - \tilde{x}_k|| + ||\tilde{x}_k - \tilde{x}_m|| + ||\tilde{x}_m - \bar{x}_{n_m}^{(m)}|| \\ &\leqq ||\tilde{x}_k - \tilde{x}_m|| + k^{-1} + m^{-1}. \end{aligned}$$

Thus (3) is a Cauchy sequence of X. Let $\tilde{x}$ be the class containing (3). Then, by (5),

$$||\tilde{x} - \tilde{x}_k|| \leqq ||\tilde{x} - \bar{x}_{n_k}^{(k)}|| + ||\bar{x}_{n_k}^{(k)} - \tilde{x}_k|| \leqq ||\tilde{x} - \bar{x}_{n_k}^{(k)}|| + k^{-1}.$$

Since, as shown above,

$$||\tilde{x} - \bar{x}_{n_k}^{(k)}|| \leqq \lim_{p\to\infty} ||x_{n_p}^{(p)} - x_{n_k}^{(k)}|| \leqq \lim_{p\to\infty} ||\tilde{x}_p - \tilde{x}_k|| + k^{-1}$$

we prove that $\lim\limits_{k\to\infty} ||\tilde{x} - \bar{x}_{n_k}^{(k)}|| = 0$, and so $\lim\limits_{k\to\infty} ||\tilde{x} - \tilde{x}_k|| = 0$.

The above proof shows that the correspondence

$$X \ni x \leftrightarrow \tilde{x} = \{x, x, \ldots, x, \ldots\}' = \bar{x}$$

is surely isomorphic and isometric, and the image of X in $\tilde{X}$ by this correspondence is dense in $\tilde{X}$. The last part of the Theorem is clear.

Example of completion. Let Ω be an open set of R^n and $k < \infty$. The completion of the space $C_0^k(\Omega)$ normed by

$$||f||_k = \Big(\sum_{|s|\leqq k} \int_\Omega |D^s f(x)|^2 \, dx\Big)^{1/2}$$

will be denoted by $H_0^k(\Omega)$; thus $H_0^k(\Omega)$ is the completion of the pre-Hilbert space $\hat{H}_0^k(\Omega)$ defined in Chapter I, 5, Example 4. Therefore $H_0^k(\Omega)$ is a Hilbert space. The completion of the pre-Hilbert space $\hat{H}^k(\Omega)$ in Chapter I, 5, Example 3 will similarly be denoted by $H^k(\Omega)$.

The elements of $H_0^k(\Omega)$ are obtained concretely as follows: Let $\{f_h\}$ be a Cauchy sequence of $C_0^k(\Omega)$ with regard to the norm $||f||_k$. Then, by the completeness of the space $L^2(\Omega)$, we see that there exist functions $f^{(s)}(x) \in L^2(\Omega)$ with $|s| = \sum_{j=1}^{n} s_j \leqq k$ such that

$$\lim_{h\to\infty} \int_\Omega |f^{(s)}(x) - D^s f_h(x)|^2\, dx = 0 \quad (dx = dx_1\, dx_2 \ldots dx_n).$$

Since the scalar product is continuous in the norm of $L^2(\Omega)$, we see, for any test function $\varphi(x) \in C_0^\infty(\Omega)$, that

$$\begin{aligned} T_{f^{(s)}}(\varphi) &= \lim_{h\to\infty} \langle D^s f_h, \varphi\rangle = \lim_{h\to\infty} (-1)^{|s|}\, T_{f_h}(D^s\varphi) \\ &= (-1)^{|s|} \lim_{h\to\infty} \langle f_h, D^s\varphi\rangle = (-1)^{|s|} \langle f^{(0)}, D^s\varphi\rangle = (D^s T_{f^{(0)}})(\varphi). \end{aligned}$$

Therefore we see that $f^{(s)} \in L^2(\Omega)$ is, when considered as a generalized function, the distributional derivative of $f^{(0)}$: $f^{(s)} = D^s f^{(0)}$.

We have thus proved that the Hilbert space $H_0^k(\Omega)$ is a linear subspace of the Hilbert space $W^k(\Omega)$, the Sobolev space. In general $H_0^k(\Omega)$ is a proper subspace of $W^k(\Omega)$. However, we can prove

Proposition. $H_0^k(R^n) = W^k(R^n)$.

Proof. We know that the space $W^k(R^n)$ is the space of all functions $f(x) \in L^2(R^n)$ such that the distributional derivatives $D^s f(x)$ with $|s| = \sum_{j=1}^{n} s_j \leqq k$ all belong to $L^2(R^n)$, and the norm in $W^k(R^n)$ is given by

$$||f||_k = \Big(\sum_{|s|\leqq k} \int |D^s f(x)|^2\, dx\Big)^{1/2}.$$

Let $f \in W^k(R^n)$ and define f_N by

$$f_N(x) = \alpha_N(x)\, f(x),$$

where the functions $\alpha_N(x) \in C_0^\infty(R^n)$ $(N = 1, 2, \ldots)$ is such that

$$\alpha_N(x) = 1 \text{ for } |x| \leqq N \text{ and } \sup_{x\in R^n;\, |s|\leqq k;\, N=1,2,\ldots} |D^s \alpha_N(x)| < \infty.$$

Then by Leibniz' formula, we have

$$\begin{aligned} D^s f(x) - D^s f_N(x) &= 0 \text{ for } |x| \leqq N, \\ &= \text{a linear combination of terms} \\ &\qquad D^t \alpha_N(x) \cdot D^u f(x) \text{ with } |u| + |t| \leqq k \text{ for } |x| > N. \end{aligned}$$

Hence, by $D^s f \in L^2(R^n)$ for $|s| \leqq k$, we see that $\lim_{N\to\infty} ||D^s f_N - D^s f||_0 = 0$ and so $\lim_{N\to\infty} ||f_N - f||_k = 0$.

Therefore, it will be sufficient to show that, for any $f \in W^k(R^n)$ with compact support, there exists a sequence $\{f_a(x)\} \subseteq C_0^\infty(R^n)$ such that $\lim_{a\to\infty} ||f_a - f||_k = 0$. To this purpose, consider the regularization of f (see (16) in Chapter I, 1):

$$f_a(x) = \int_{R^n} f(y)\, \theta_a(x - y)\, dy, \quad a > 0.$$

By differentiation, we have

$$D^s f_a(x) = \int_{R^n} f(y)\, D_x^s \theta_a(x-y)\, dy = (-1)^{|s|} \int_{R^n} f(y)\, D_y^s \theta_a(x-y)\, dy$$
$$= (D^s T_f)\,(\theta_{a,x}) \quad (\text{where } \theta_{a,x}(y) = \theta_a(x-y))$$
$$= \int_{R^n} D^s f(y) \cdot \theta_a(x-y)\, dy \quad (\text{for } |s| \leqq k).$$

Hence, by Schwarz' inequality,

$$\int_{R^n} |D^s f_a(x) - D^s f(x)|^2\, dx$$
$$\leqq \Big(\int_{R^n} \theta_a(x-y)\, dy\Big) \int_{R^n} \Big[\int_{R^n} |D_y^s f(y) - D_x^s f(x)|^2\, \theta_a(x-y)\, dy\Big] dx$$
$$= \int_{|\varepsilon| \leqq a} \Big[\int_{R^n} |D_y^s f(y) - D_y^s f(y+\varepsilon)|^2\, dy\Big]\, \theta_a(\varepsilon)\, d\varepsilon, \text{ where}$$
$$y + \varepsilon = (y_1 + \varepsilon_1, y_2 + \varepsilon_2, \ldots, y_n + \varepsilon_n).$$

We know that the inner integral on the extreme right tends to 0 as $\varepsilon \to 0$ (see Theorem 1 in Chapter 0, 3), and hence $\lim_{a \to 0} \int_{R^n} |D^s f_a(x) - D^s f(x)|^2 dx = 0$. Thus $\lim_{a \to 0} \|f_a - f\|_k = 0$. Therefore, the completion $H_0^k(R^n)$ of $C_0^k(R^n)$ with regard to the norm $\|\ \|_k$ is identical with the space $W^k(R^n)$.

Corollary. $H_0^k(R^n) = H^k(R^n) = W^k(R^n)$.

11. Factor Spaces of a B-space

Suppose that X is a normed linear space and that M is a closed linear subspace in X. We consider the factor space X/M, i. e., the space whose elements are classes modulo M. In virtue of the fact that M is closed, all these classes ξ are closed in X.

Proposition. If we define

$$\|\xi\| = \inf_{x \in \xi} \|x\|, \tag{1}$$

then all the axioms concerning the norm are satisfied by $\|\xi\|$.

Proof. If $\xi = 0$, then ξ coincides with M and contains the zero vector of X; consequently, it follows from (1) that $\|\xi\| = 0$. Suppose, conversely, that $\|\xi\| = 0$. It follows then from (1) that the class contains a sequence $\{x_n\}$ for which we have $\lim_{n \to \infty} \|x_n\| = 0$, and hence the zero vector of X belongs to the closed set ξ of X. This proves that $\xi = M$ and hence is the zero vector in X/M.

Next suppose $\xi, \eta \in X/M$. By definition (1), there exists for any $\varepsilon > 0$, vectors $x \in \xi$, $y \in \eta$ such that

$$\|x\| \leqq \|\xi\| + \varepsilon, \ \|y\| \leqq \|\eta\| + \varepsilon.$$

Hence $\|x + y\| \leqq \|x\| + \|y\| \leqq \|\xi\| + \|\eta\| + 2\varepsilon$. On the other hand, $(x + y) \in (\xi + \eta)$, and therefore $\|\xi + \eta\| \leqq \|x + y\|$ by (1). Consequently, we have $\|\xi + \eta\| \leqq \|\xi\| + \|\eta\| + 2\varepsilon$ and so we obtain the triangle inequality $\|\xi + \eta\| \leqq \|\xi\| + \|\eta\|$.

Finally it is clear that the axiom $\|\alpha\xi\| = |\alpha| \, \|\xi\|$ holds good.

Definition. The space X/M, normed by (1), is called a *normed factor space*.

Theorem. If X is a B-space and M a closed linear subspace in X, then the normed factor space X/M is also a B-space.

Proof. Suppose $\{\xi_n\}$ is a Cauchy sequence in X/M. Then $\{\xi_n\}$ contains a subsequence $\{\xi_{n_k}\}$ such that $\|\xi_{n_{k+1}} - \xi_{n_k}\| < 2^{-k-2}$. Further, by definition (1) of the norm in X/M, one can choose in every class $(\xi_{n_{k+1}} - \xi_{n_k})$ a vector y_k such that

$$\|y_k\| < \|\xi_{n_{k+1}} - \xi_{n_k}\| + 2^{-k-2} < 2^{-k-1}.$$

Let $x_{n_1} \in \xi_{n_1}$. The series $x_{n_1} + y_1 + y_2 + \cdots$ converges in norm and consequently, in virtue of the completeness of X, it converges to an element x of X. Let ξ be the class containing x. We shall prove that $\xi = s\text{-}\lim_{n\to\infty} \xi_n$.

Denote by s_k the partial sum $x_{n_1} + y_1 + y_2 + \cdots + y_k$ of the above series. Then $\lim_{k\to\infty} \|x - s_k\| = 0$. On the other hand, it follows from the relations $x_{n_1} \in \xi_{n_1}$, $y_p \in (\xi_{n_{p+1}} - \xi_{n_p})$ that $s_k \in \xi_{n_k}$, and so, by (1),

$$\|\xi - \xi_{n_k}\| \leqq \|x - s_k\| \to 0 \quad \text{as} \quad k \to \infty.$$

Therefore, from the inequality $\|\xi - \xi_n\| \leqq \|\xi - \xi_{n_k}\| + \|\xi_{n_k} - \xi_n\|$ and the fact that $\{\xi_n\}$ is a Cauchy sequence, we conclude that $\lim_{n\to\infty} \|\xi - \xi_n\| = 0$.

12. The Partition of Unity

To discuss the support of a generalized function, in the next section, we shall prepare the notion and existence of the *partition of unity*.

Proposition. Let G be an open set of R^n. Let a family $\{U\}$ of open subsets U of G constitute an *open base* of G: any open subset of G is representable as the union of open sets belonging to the family $\{U\}$. Then there exists a system of open sets of the family $\{U\}$ with the properties:

> the union of open sets of this system equals G, (1)
>
> any compact subset of G *meets* (has a non-void intersection with) only a finite number of open sets of this system. (2)

Definition 1. The above system of open sets is said to constitute a *scattered open covering* of G *subordinate to* $\{U\}$.

Proof of the Proposition. G is representable as the union of a countable number of compact subsets. For example, we may take the system of all closed spheres contained in G such that the centres are of rational coordinates and the radii of rational numbers.

Hence we see that there exists a sequence of compact subsets K_r such that (i) $K_r \subseteqq K_{r+1}$ $(r = 1, 2, \ldots)$, (ii) G is the union of K_r's and (iii) each K_r is contained in the interior of K_{r+1}. Set

$U_r =$ (the interior of K_{r+1}) $- K_{r-2}$ and $V_r = K_r -$ (the interior of K_{r-1}), where for convention we set $K_0 = K_{-1} =$ the void set. Then U_r is open and V_r is compact such that $G = \bigcup_{r=1}^{\infty} V_r$. For any point $x \in V_r$, take an open set $U(x; r) \in \{U\}$ such that $x \in U(x; r) \subseteqq U_r$. Since V_r is compact, there exists a finite system of points $x^{(1)}, x^{(2)}, \ldots, x^{(h_r)}$ such that $V_r \subseteqq \bigcup_{i=1}^{h_r} U(x^{(i)}; r)$. Then, since any compact set of G meets only a finite number of U_r's, it is easy to see that the system of open sets $U(x^{(i)}; r)$ $(r = 1, 2, \ldots; 1 \leqq i \leqq h_r)$ is a scattered open covering of G subordinate to $\{U\}$.

Theorem (the partition of unity). Let G be an open set of R^n, and let a family of open sets $\{G_i; i \in I\}$ cover G, i.e., $G = \bigcup_{i \in I} G_i$. Then there exists a system of functions $\{\alpha_j(x); j \in J\}$ of $C_0^\infty(R^n)$ such that

$$\text{for each } j \in J,\ \operatorname{supp}(\alpha) \text{ is contained in some } G_i, \tag{3}$$

$$\text{for every } j \in J,\ 0 \leqq \alpha_j(x) \leqq 1, \tag{4}$$

$$\sum_{j \in J} \alpha_j(x) \equiv 1 \quad \text{for} \quad x \in G. \tag{5}$$

Proof. Let $x^{(0)} \in G$ and take a G_i which contains $x^{(0)}$. Let the closed sphere $S(x^{(0)}; r)$ of centre $x^{(0)}$ and radius r be contained in G_i. We construct, as in (14), Chapter I, 1, a function $\beta_{x^{(0)}}^{(r)}(x) \in C_0^\infty(R^n)$ such that

$$\beta_{x^{(0)}}^{(r)}(x) > 0 \quad \text{for} \quad |x - x^{(0)}| < r,\ \beta_{x^{(0)}}^{(r)}(x) = 0 \quad \text{for} \quad |x - x^{(0)}| \geqq r.$$

We put $U_{x^{(0)}}^{(r)} = \{x; \beta_{x^{(0)}}^{(r)}(x) \neq 0\}$. Then $U_{x^{(0)}}^{(r)} \subseteqq G_i$ and $\bigcup_{x^{(0)} \in G, r > 0} U_{x^{(0)}}^{(r)} = G$, and, moreover, $\operatorname{supp}(\beta_{x^{(0)}}^{(r)})$ is compact.

There exists, by the Proposition, a scattered open covering $\{U_j; j \in J\}$ subordinate to the open base $\{U_{x^{(0)}}^{(r)}; x^{(0)} \in G, r > 0\}$ of G. Let $\beta_j(x)$ be any function of the family $\{\beta_{x^{(0)}}^{(r)}(x)\}$ which is associated with U_j. Then, since $\{U_j; j \in J\}$ is a scattered open covering, only a finite number of $\beta_j(x)$'s do not vanish at a fixed point x of G. Thus the sum $s(x) = \sum_{j \in J} \beta_j(x)$ is convergent and is > 0 at every point x of G. Hence the functions

$$\alpha_j(x) = \beta_j(x)/s(x) \quad (j \in J)$$

satisfy the condition of our theorem.

Definition 2. The system $\{\alpha_j(x); j \in J\}$ is called a *partition of unity* subordinate to the covering $\{G_i; i \in I\}$.

13. Generalized Functions with Compact Support

Definition 1. We say that a distribution $T \in \mathfrak{D}(\Omega)'$ vanishes in an open set U of Ω if $T(\varphi) = 0$ for every $\varphi \in \mathfrak{D}(\Omega)$ with support contained in U. The *support* of T, denoted by $\operatorname{supp}(T)$, is defined as the smallest closed set F of Ω such that T vanishes in $\Omega - F$.

To justify the above definition, we have to prove the existence of the largest open set of Ω in which T vanishes. This is done by the following

Theorem 1. If a distribution $T \in \mathfrak{D}(\Omega)'$ vanishes in each U_i of a family $\{U_i; i \in I\}$ of open sets of Ω, then T vanishes in $U = \bigcup_{i \in I} U_i$.

Proof. Let $\varphi \in \mathfrak{D}(\Omega)$ be a function with $\operatorname{supp}(\varphi) \subseteqq U$. We construct a partition of unity $\{\alpha_j(x); j \in J\}$ subordinate to the covering of Ω consisting of $\{U_i; i \in I\}$ and $\Omega - \operatorname{supp}(\varphi)$. Then $\varphi = \sum_{j \in J} \alpha_j \varphi$ is a finite sum and so $T(\varphi) = \sum_{j \in J} T(\alpha_j \varphi)$. If the $\operatorname{supp}(\alpha_j)$ is contained in some U_i, $T(\alpha_j \varphi) = 0$ by the hypothesis; if the $\operatorname{supp}(\alpha_j)$ is contained in $\Omega - \operatorname{supp}(\varphi)$, then $\alpha_j \varphi = 0$ and so $T(\alpha_j \varphi) = 0$. Therefore we have $T(\varphi) = 0$.

Proposition 1. A subset B of the space $\mathfrak{E}(\Omega)$ is bounded iff, for any differential operator D^j and for any compact subset K of Ω, the set of functions $\{D^j f(x); f \in B\}$ is uniformly bounded on K.

Proof. Clear from the definition of the semi-norms defining the topology of $\mathfrak{E}(\Omega)$.

Proposition 2. A linear functional T on $\mathfrak{E}(\Omega)$ is continuous iff T is bounded on every bounded set of $\mathfrak{E}(\Omega)$.

Proof. Since $\mathfrak{E}(\Omega)$ is a quasi-normed linear space, the Proposition is a consequence of Theorem 2 of Chapter I, 7.

Proposition 3. A distribution $T \in \mathfrak{D}(\Omega)'$ with compact support can be extended in one and only one way to a continuous linear functional T_0 on $\mathfrak{E}(\Omega)$ such that $T_0(f) = 0$ if $f \in \mathfrak{E}(\Omega)$ vanishes in a neighbourhood of $\operatorname{supp}(T)$.

Proof. Let us put $\operatorname{supp}(T) = K$ where K is a compact subset of Ω. For any point $x^0 \in K$ and $\varepsilon > 0$, we take a sphere $S(x^0, \varepsilon)$ of centre x^0 and radius ε. For any $\varepsilon > 0$ sufficiently small, the compact set K is covered by a finite number of spheres $S(x^0, \varepsilon)$ with $x^0 \in K$. Let $\{\alpha_j(x); j \in J\}$ be the partition of the unity subordinate to this finite system of spheres. Then the function $\psi(x) = \sum_{\operatorname{supp}(\alpha_j) \cap K' \neq \text{void}} \alpha_j(x)$, where K' is a compact neighbourhood of K contained in the interior of the finite system of spheres above, satisfies:

$$\psi(x) \in C_0^\infty(\Omega) \quad \text{and} \quad \psi(x) = 1 \text{ in a neighbourhood of } K.$$

We define $T_0(f)$ for $f \in C^\infty(\Omega)$ by $T_0(f) = T(\psi f)$. This definition is independent of the choice of ψ. For, if $\psi_1 \in C_0^\infty(\Omega)$ equals 1 in a neighbourhood of K, then, for any $f \in C^\infty(\Omega)$, the function $(\psi - \psi_1) f \in \mathfrak{D}(\Omega)$ vanishes in a neighbourhood of K so that $T(\psi f) - T(\psi_1 f) = T((\psi - \psi_1) f) = 0$.

It is easy to see, by applying Leibniz' formula of differentiation to ψf, that $\{\psi f\}$ ranges over a bounded set of $\mathfrak{D}(\Omega)$ when $\{f\}$ ranges over a bounded set of $\mathfrak{E}(\Omega)$. Thus, since a distribution $T \in \mathfrak{D}(\Omega)'$ is bounded on bounded sets of $\mathfrak{D}(\Omega)$, the functional T_0 is bounded on bounded sets of $\mathfrak{E}(\Omega)$. Hence, by the Theorem 2 of Chapter I, 7 mentioned above, T_0 is a continuous linear functional on $\mathfrak{E}(\Omega)$. Let $f \in \mathfrak{E}(\Omega)$ vanish in a neighbourhood $U(K)$ of K. Then, by choosing a $\psi \in C_0^\infty(\Omega)$ that vanishes in $(\Omega - U(K))$, we see that $T_0(f) = T(\psi f) = 0$.

Proposition 4. Let K' be the support of ψ in the above definition of T_0. Then for some constants C and k

$$|T_0(f)| \leqq C \sup_{|j| \leqq k, x \in K'} |D^j f(x)| \quad \text{for all} \quad f \in C^\infty(\Omega).$$

Proof. Since T is a continuous linear functional on $\mathfrak{D}(\Omega)$, there exist, for any compact set K' of Ω, constants C' and k' such that

$$|T(\varphi)| \leqq C' \sup_{|j| \leqq k', x \in K'} |D^j \varphi(x)| \quad \text{for all} \quad \varphi \in \mathfrak{D}_{K'}(\Omega)$$

(the Corollary of Proposition 1 in Chapter I, 8). But, for any $g \in C^\infty(\Omega)$, we have $\varphi = \psi g \in D_{K'}(\Omega)$. Consequently, we see, by Leibniz' formula of differentiation, that

$$\sup_{|j| \leqq k', x \in K'} |D^j(\psi g)(x)| \leqq C'' \sup_{|j| \leqq k', x \in K'} |D^j g(x)|$$

with a constant C'' which is independent of g. Setting $g = f$ and $k = k'$, we obtain the Proposition.

Proposition 5. Let S_0 be a linear functional on $C^\infty(\Omega)$ such that, for some constant C and a positive integer k and compact subset K of Ω,

$$|S_0(f)| \leqq C \sup_{|j| \leqq k, x \in K} |D^j f(x)| \quad \text{for all} \quad f \in C^\infty(\Omega).$$

Then the restriction of S_0 to $C_0^\infty(\Omega)$ is a distribution T with support contained in K.

Proof. We observe that $S_0(f) = 0$ if f vanishes identically in a neighbourhood of K. Thus, if $\psi \in C_0^\infty(\Omega)$ equals 1 in a neighbourhood of K, then

$$S_0(f) = S_0(\psi f) \quad \text{for all} \quad f \in C^\infty(\Omega).$$

It is easy to see that if $\{f\}$ ranges over a bounded set of $\mathfrak{D}(\Omega)$, then, in virtue of Leibniz' formula, $\{\psi f\}$ ranges over a set which is contained in a set of the form

$$\{g \in C^\infty(\Omega);\ \sup_{|j| \leqq k, x \in K} |D^j g(x)| = C_k < \infty\}.$$

Thus $S_0(\psi f) = T(f)$ is bounded on bounded sets of $\mathfrak{D}(\Omega)$ so that T is a continuous linear functional on $\mathfrak{D}(\Omega)$.

We have thus proved the following

Theorem 2. The set of all distributions in Ω with compact support is identical with the space $\mathfrak{E}(\Omega)'$ of all continuous linear functionals on $\mathfrak{E}(\Omega)$, the *dual space* of $\mathfrak{E}(\Omega)$. A linear functional T on $C^\infty(\Omega)$ belongs to $\mathfrak{E}(\Omega)'$ iff, for some constants C and k and a compact subset K of Ω,

$$|T(f)| \leqq C \sup_{|j| \leqq k, x \in K} |D^j f(x)| \quad \text{for all} \quad f \in C^\infty(\Omega).$$

We next prove a theorem which gives the general expression of distributions whose supports reduce to a single point.

Theorem 3. Let an open set Ω of R^n contain the origin 0. Then the only distributions $T \in \mathfrak{D}(\Omega)'$ with supports reduced to the origin 0 are those which are expressible as finite linear combinations of Dirac's distribution and its derivatives at 0.

Proof. For such a distribution T, there exist, by the preceding Theorem 2, some constants C and k and a compact subset K of Ω which contains the origin 0 in such a way that

$$|T(f)| \leqq C \sup_{|j| \leqq k, x \in K} |D^j f(x)| \quad \text{for all} \quad f \in C^\infty(\Omega).$$

We shall prove that the condition

$$D^j f(0) = 0 \quad \text{for all} \quad j \quad \text{with } |j| \leqq k$$

implies $T(f) = 0$. To this purpose, we take a function $\psi \in C^\infty(\Omega)$ which is equal to 1 in a neighbourhood of 0 and put

$$f_\varepsilon(x) = f(x)\, \psi(x/\varepsilon).$$

We have $T(f) = T(f_\varepsilon)$ since $f = f_\varepsilon$ in a neighbourhood of the origin 0. By Leibniz' formula, the derivative of f_ε of order $\leqq k$ is a linear combination of terms of the form $|\varepsilon|^{-j} D^i \psi \cdot D^i f$ with $|i| + |j| \leqq k$. Since, by the assumption, $D^i f(0) = 0$ for $|i| \leqq k$, we see, by Taylor's formula, that a derivative of order $|s|$ of f_ε is $0(\varepsilon^{k+1-|s|})$ in the support of $\psi(x/\varepsilon)$. Thus, when $\varepsilon \downarrow 0$, the derivatives of f_ε of order $\leqq k$ converge to 0 uniformly in a neighbourhood of 0. Hence $T(f) = \lim_{\varepsilon \downarrow 0} T(f_\varepsilon) = 0$.

Now, for a general f, we denote by f_k the Taylor's expansion of f up to the order k at the origin. Then, by what we have proved above,

$$T(f) = T(f_k) + T(f - f_k) = T(f_k) + 0 = T(f_k).$$

This shows that T is a linear combination of linear functionals in the derivatives of f at the origin of order $\leqq k$.

14. The Direct Product of Generalized Functions

We first prove a theorem of approximation.

Theorem 1. Let $x = (x_1, x_2, \ldots, x_n) \in R^n$, $y = (y_1, y_2, \ldots, y_m) \in R^m$ and $z = x \times y = (x_1, x_2, \ldots, x_n, y_1, y_2, \ldots, y_m) \in R^{n+m}$. Then, for any function $\varphi(z) = \varphi(x, y) \in C_0^\infty(R^{n+m})$, we can choose functions $u_{ij}(x) \in C_0^\infty(R^n)$ and functions $v_{ij}(y) \in C_0^\infty(R^m)$ such that the sequence of functions

$$\varphi_i(z) = \varphi_i(x, y) = \sum_{j=1}^{k_i} u_{ij}(x)\, v_{ij}(y) \tag{1}$$

tends, as $i \to \infty$, to $\varphi(z) = \varphi(x, y)$ in the topology of $\mathfrak{D}(R^{n+m})$.

Proof. We shall prove Theorem 1 for the case $n = m = 1$. Consider

$$\Phi(x, y, t) = \left(2\sqrt{\pi t}\right)^{-2} \int_{-\infty}^{\infty} \int_{-\infty}^{\infty} \varphi(\xi, \eta) \exp\left(-((x-\xi)^2 + (y-\eta)^2)/4t\right) d\xi\, d\eta,$$

$$t > 0;\ \Phi(x, y, 0) = \varphi(x, y). \tag{2}$$

We have, by the change of variables $\xi_1 = (\xi - x)/2\sqrt{t}$, $\eta_1 = (\eta - y)/2\sqrt{t}$,

$$\Phi(x, y, t) = \left(\sqrt{\pi}\right)^{-2} \int_{-\infty}^{\infty} \int_{-\infty}^{\infty} \varphi\left(x + 2\xi_1\sqrt{t}, y + 2\eta_1\sqrt{t}\right) e^{-\xi_1^2 - \eta_1^2}\, d\xi_1\, d\eta_1.$$

Hence, by $\int_{-\infty}^{\infty} \int_{-\infty}^{\infty} e^{-\xi_1^2 - \eta_1^2}\, d\xi_1\, d\eta_1 = \pi$,

$$\begin{aligned} |\Phi(x, y, t) - \varphi(x, y)| &\leqq \left(\sqrt{\pi}\right)^{-2} \int_{-\infty}^{\infty} \int_{-\infty}^{\infty} \left|\varphi\left(x + 2\xi\sqrt{t}, y + 2\eta\sqrt{t}\right) - \varphi(x, y)\right| \\ &\quad \times e^{-\xi^2 - \eta^2}\, d\xi\, d\eta \\ &\leqq \pi^{-1} \left\{ \iint_{\xi^2 + \eta^2 \geqq T^2} + \iint_{\xi^2 + \eta^2 < T^2} \right\}. \end{aligned}$$

Since the function φ is bounded and $e^{-\xi^2 - \eta^2}$ is integrable in R^2, we see that the first term on the right tends to zero as $T \uparrow \infty$. The second term on the right tends, for fixed $T > 0$, to zero as $t \downarrow 0$. Hence we have proved that $\lim_{t \downarrow 0} \Phi(x, y, t) = \varphi(x, y)$ uniformly in (x, y).

Next, since $\mathrm{supp}(\varphi)$ is compact, we see, by partial integration,

$$\begin{aligned} \frac{\partial^{m+k}\Phi(x, y, t)}{\partial x^m\, \partial y^k} &= \iint_{-\infty}^{\infty} \left(2\sqrt{\pi t}\right)^{-2} \frac{\partial^{m+k}\varphi(\xi, \eta)}{\partial \xi^m\, \partial \eta^k}\, e^{-[(x-\xi)^2 + (y-\eta)^2]/4t}\, d\xi\, d\eta,\ t > 0, \\ &= \frac{\partial^{m+k}\varphi(x, y)}{\partial x^m\, \partial y^k},\ t = 0. \end{aligned}$$

Thus we see as above that

$$\lim_{t \downarrow 0} \frac{\partial^{m+k}\Phi(x, y, t)}{\partial x^m\, \partial y^k} = \frac{\partial^{m+k}\varphi(x, y)}{\partial x^m\, \partial y^k} \quad \text{uniformly in } (x, y). \tag{3}$$

It is easy to see that $\Phi(x, y, t)$ for $t > 0$ given by (2) may be extended as a holomorphic function of the complex variables x and y for $|x| < \infty$,

$|y| < \infty$. Hence, for any given $\gamma > 0$, the function $\Phi(x, y, t)$ for fixed $t > 0$ may be expanded into Taylor's series

$$\Phi(x, y, t) = \sum_{m=0}^{\infty} \sum_{s=0}^{m} c_s(t)\, x^s y^{m-s}$$

which is absolutely and uniformly convergent for $|x| \leqq \gamma$, $|y| \leqq \gamma$ and may be differentiated term by term:

$$\frac{\partial^{m+k}\Phi(x, y, t)}{\partial x^m\, \partial y^k} = \sum_{m_1=0}^{\infty} \sum_{s=0}^{m_1} c_s(t)\, \frac{\partial^{m+k} x^s y^{m_1-s}}{\partial x^m\, \partial y^k}.$$

Let $\{t_i\}$ be a sequence of positive numbers such that $t_i \downarrow 0$. By the above we can choose, for each t_i, a polynomial section $P_i(x, y)$ of the series $\sum_{m=0}^{\infty} \sum_{s=0}^{m} c_s(t)\, x^s y^{m-s}$ such that

$$\lim_{i\to\infty} P_i(x, y) = \varphi(x, y) \text{ in the topology of } \mathfrak{E}(R^2),$$

that is, for any compact subset K of R^2, $\lim_{i\to\infty} D^s P_i(x, y) = D^s \varphi(x, y)$ uniformly on K for every differential operator D^s. Let us take $\varrho(x) \in C_0^\infty(R^1)$ and $\sigma(y) \in C_0^\infty(R^1)$ such that $\varrho(x)\,\sigma(y) = 1$ on the $\operatorname{supp}(\varphi(x, y))$. Then we easily see that $\varphi_i(x, y) = \varrho(x)\,\sigma(y)\, P_i(x, y)$ satisfies the condition of Theorem 1.

Remark. We shall denote by $\mathfrak{D}(R^n) \times \mathfrak{D}(R^m)$ the totality of functions $\in \mathfrak{D}(R^{n+m})$ which are expressible as

$$\sum_{j=1}^{k} \varphi_j(x)\, \psi_j(y) \quad \text{with} \quad \varphi_j(x) \in \mathfrak{D}(R^n),\ \psi_j(y) \in \mathfrak{D}(R^m).$$

The above Theorem 1 says that $\mathfrak{D}(R^n) \times \mathfrak{D}(R^m)$ is dense in $\mathfrak{D}(R^{n+m})$ in the topology of $\mathfrak{D}(R^{n+m})$. The linear subspace $\mathfrak{D}(R^n) \times \mathfrak{D}(R^m)$ of $\mathfrak{D}(R^{n+m})$ equipped with the relative topology is called the *direct product* of $\mathfrak{D}(R^n)$ and $\mathfrak{D}(R^m)$.

We are now able to define the *direct product* of distributions. To indicate explicitly the independent variables $x = (x_1, x_2, \ldots, x_n)$ of the function $\varphi(x) \in \mathfrak{D}(R^n)$, we shall write $(\mathfrak{D}_x)$ for $\mathfrak{D}(R^n)$. We also write $(\mathfrak{D}_y)$ for $\mathfrak{D}(R^m)$ consisting of the functions $\psi(y)$, $y = (y_1, y_2, \ldots, y_m)$. Likewise we shall write $(\mathfrak{D}_{x\times y})$ for $\mathfrak{D}(R^{n+m})$ consisting of the functions $\chi(x, y)$. We shall accordingly write $T_{(x)}$ for the distribution $T \in \mathfrak{D}(R^n)' = (\mathfrak{D}_x)'$ in order to show that T is to be applied to functions $\varphi(x)$ of x.

Theorem 2. Let $T_{(x)} \in (\mathfrak{D}_x)'$, $S_{(y)} \in (\mathfrak{D}_y)'$. Then we can define in one and only one way a distribution $W = W_{(x\times y)} \in (\mathfrak{D}_{x\times y})'$ such that

$$W(u(x)\, v(y)) = T_{(x)}(u(x))\, S_{(y)}(v(y)) \quad \text{for} \quad u \in (\mathfrak{D}_x),\ v \in (\mathfrak{D}_y), \tag{4}$$

$$W(\varphi(x, y)) = S_{(y)}\big(T_{(x)}(\varphi(x, y))\big) = T_{(x)}\big(S_{(y)}(\varphi(x, y))\big) \quad \text{for} \quad \varphi \in (\mathfrak{D}_{x\times y}) \tag{5}$$

(Fubini's theorem).

Remark. The distribution W is called the *direct product* or the *tensor product* of $T_{(x)}$ and $S_{(y)}$, and we shall write

$$W = T_{(x)} \times S_{(y)} = S_{(y)} \times T_{(x)}. \tag{6}$$

Proof of Theorem 2. Let $\mathfrak{B} = \{\varphi(x, y)\}$ be a bounded set of the space $(\mathfrak{D}_{x\times y})$. For fixed $y^{(0)}$, the set $\{\varphi(x, y^{(0)}); \varphi \in \mathfrak{B}\}$ is a bounded set of $(\mathfrak{D}_x)$. We shall show that

$$\{\psi(y^{(0)}); \psi(y^{(0)}) = T_{(x)}(\varphi(x, y^{(0)})),\ \varphi \in \mathfrak{B}\} \tag{7}$$

is a bounded set of $(\mathfrak{D}_{y^{(0)}})$. The proof is given as follows.

Since $\mathfrak{B}$ is a bounded set of $(\mathfrak{D}_{x\times y})$, there exist a compact set $K_x \in R^n$ and a compact set $K_y \in R^m$ such that

$$\operatorname{supp}(\varphi) \subseteqq \{(x, y) \in R^{n+m}; x \in K_x, y \in K_y\} \quad \text{whenever} \quad \varphi \in \mathfrak{B}.$$

Hence $y^{(0)} \bar{\in} K_y$ implies $\varphi(x, y^{(0)}) = 0$ and $\psi(y^{(0)}) = T_{(x)}(\varphi(x, y^{(0)})) = 0$. Thus

$$\operatorname{supp}(\psi) \subseteqq K_y \quad \text{whenever} \quad \varphi \in \mathfrak{B}. \tag{8}$$

We have to show that, for any differential operator D_y in R^m,

$$\sup_{y,\psi} |D_y \psi(y)| < \infty \quad \text{where} \quad \psi(y) = T_{(x)}(\varphi(x, y)),\ \varphi \in \mathfrak{B}. \tag{9}$$

To prove this, we take, e.g., $D_{y_1} = \partial/\partial y_1$. Then, by the linearity of $T_{(x)}$,

$$\frac{\psi(y_1 + h, y_2, \ldots, y_m) - \psi(y_1, y_2, \ldots, y_m)}{h}$$
$$= T_{(x)}\left\{\frac{\varphi(x, y_1 + h, y_2, \ldots, y_m) - \varphi(x, y_1, y_2, \ldots, y_m)}{h}\right\}.$$

When φ ranges over $\mathfrak{B}$, the functions $\in \{\ \}$ of x, with parameters $y \in R^m$ and h such that $|h| \leqq 1$, constitute a bounded set of $(\mathfrak{D}_x)$. This we see from the fact that $\mathfrak{B}$ is a bounded set of $(\mathfrak{D}_{x\times y})$. Hence we see, by letting $h \to 0$ and remembering Proposition 1 in Chapter I, 8, that (9) is true.

Therefore, by the same Proposition 1, we see that the set of values

$$\{S_{(y)}(T_{(x)}(\varphi(x, y))); \varphi \in \mathfrak{B}\} \tag{10}$$

is bounded. Consequently, the same Proposition 1 shows that we have defined a distribution $W^{(1)} \in (\mathfrak{D}_{x\times y})'$ through

$$W^{(1)}(\varphi) = S_{(y)}(T_{(x)}(\varphi(x, y))). \tag{11}$$

Similarly we define a distribution $W^{(2)} \in (\mathfrak{D}_{x\times y})'$ through

$$W^{(2)}(\varphi) = T_{(x)}(S_{(y)}(\varphi(x, y))). \tag{12}$$

Clearly we have, for $u \in (\mathfrak{D}_x)$ and $v \in (\mathfrak{D}_y)$,

$$W^{(1)}(u(x)\, v(y)) = W^{(2)}(u(x)\, v(y)) = T_{(x)}(u(x)) \cdot S_{(y)}(v(y)). \tag{13}$$

Therefore, by the preceding Theorem 1 and the continuity of the distributions $W^{(1)}$ and $W^{(2)}$, we obtain $W^{(1)} = W^{(2)}$. This proves our Theorem 2 by setting $W = W^{(1)} = W^{(2)}$.

References for Chapter I

For locally convex linear topological spaces and Banach spaces, see N. Bourbaki [2], A. Grothendieck [1], G. Köthe [1], S. Banach [1], N. Dunford-J. Schwartz [1] and E. Hille-R. S. Phillips [1]. For generalized functions, see L. Schwartz [1], I. M. Gelfand-G. E. Šilov [1], L. Hörmander [6] and A. Friedman [1].

II. Applications of the Baire-Hausdorff Theorem

The completeness of a B-space (or an F-space) enables us to apply the Baire-Hausdorff theorem in Chapter 0, and we obtain such basic principles in functional analysis as the *uniform boundedness theorem*, the *resonance theorem*, the *open mapping theorem* and the *closed graph theorem*. These theorems are essentially due to S. Banach [1]. The *termwise differentiability of generalized functions* is a consequence of the uniform boundedness theorem.

1. The Uniform Boundedness Theorem and the Resonance Theorem

Theorem 1 (the uniform boundedness theorem). Let X be a linear topological space which is not expressible as a countable union of closed non-dense subsets. Let a family $\{T_a;\, a \in A\}$ of continuous mappings be defined on X into a quasi-normed linear space Y. We assume that, for any $a \in A$ and $x, y \in X$,

$$||T_a(x+y)|| \leqq ||T_a x|| + ||T_a y|| \text{ and } ||T_a(\alpha x)|| = ||\alpha T_a x|| \text{ for } \alpha \geqq 0.$$

If the set $\{T_a x;\, a \in A\}$ is bounded at each $x \in X$, then $\operatorname*{s-lim}_{x \to 0} T_a x = 0$ uniformly in $a \in A$.

Proof. For a given $\varepsilon > 0$ and for each positive integer n, consider $X_n = \left\{x \in X;\, \sup_{a \in A} \{||n^{-1} T_a x|| + ||n^{-1} T_a(-x)||\} \leqq \varepsilon\right\}$. Each set X_n is closed by the continuity of T_a. By the assumption of the boundedness of $\{||T_a x||;\, a \in A\}$, we have $X = \bigcup_{n=1}^{\infty} X_n$. Hence, by the hypothesis on X, some X_{n_0} must contain a neighbourhood $U = x_0 + V$ of some point $x_0 \in X$, where V is a neighbourhood of 0 of X such that $V = -V$. Thus $x \in V$ implies $\sup_{a \in A} ||n_0^{-1} T_a(x_0 + x)|| \leqq \varepsilon$. Therefore we have

$$\begin{aligned} ||T_a(n_0^{-1} x)|| = ||T_a(n_0^{-1}(x_0 + x - x_0))|| &\leqq ||n_0^{-1} T_a(x_0 + x)|| \\ &+ ||n_0^{-1} T_a(-x_0)|| \leqq 2\varepsilon \text{ for } x \in V,\ a \in A. \end{aligned}$$

Thus the Theorem is proved, because the scalar multiplication αx in a linear topological space is continuous in both variables α and x.

Corollary 1 (the resonance theorem). Let $\{T_a; a \in A\}$ be a family of bounded linear operators defined on a B-space X into a normed linear space Y. Then the boundedness of $\{||T_a x||; a \in A\}$ at each $x \in X$ implies the boundedness of $\{||T_a||; a \in A\}$.

Proof. By the uniform boundedness theorem, there exists, for any $\varepsilon > 0$, a $\delta > 0$ such that $||x|| \leqq \delta$ implies $\sup_{a \in A} ||T_a x|| \leqq \varepsilon$. Thus $\sup_{a \in A} ||T_a|| \leqq \varepsilon/\delta$.

Corollary 2. Let $\{T_n\}$ be a sequence of bounded linear operators defined on a B-space X into a normed linear space Y. Suppose that $s\text{-}\lim_{n\to\infty} T_n x = Tx$ exists for each $x \in X$. Then T is also a bounded linear operator on X into Y and we have

$$||T|| \leqq \varliminf_{n\to\infty} ||T_n||. \tag{1}$$

Proof. The boundedness of the sequence $\{||T_n x||\}$ for each $x \in X$ is implied by the continuity of the norm. Hence, by the preceding Corollary, $\sup_{n \geqq 1} ||T_n|| < \infty$, and so $||T_n x|| \leqq \sup_{n \geqq 1} ||T_n|| \cdot ||x||$ $(n = 1, 2, \ldots)$. Therefore, again by the continuity of the norm, we obtain

$$||Tx|| = \lim_{n\to\infty} ||T_n x|| \leqq \varliminf_{n\to\infty} ||T_n|| \cdot ||x||,$$

which is precisely the inequality (1). Finally it is clear that T is linear.

Definition. The operator T obtained above is called the *strong limit* of the sequence $\{T_n\}$ and we shall write $T = s\text{-}\lim_{n\to\infty} T_n$.

We next prove an existence theorem for the bounded inverse of a bounded linear operator.

Theorem 2 (C. Neumann). Let T be a bounded linear operator on a B-space X into X. Suppose that $||I - T|| < 1$, where I is the identity operator: $I \cdot x = x$. Then T has a unique bounded linear inverse T^{-1} which is given by *C. Neumann's series*

$$T^{-1}x = s\text{-}\lim_{n\to\infty} (I + (I - T) + (I - T)^2 + \cdots + (I - T)^n)x, \quad x \in X. \tag{2}$$

Proof. For any $x \in X$, we have

$$\left|\left| \sum_{n=0}^{k} (I - T)^n x \right|\right| \leqq \sum_{n=0}^{k} ||(I - T)^n x|| \leqq \sum_{n=0}^{k} ||(I - T)^n|| \, ||x||$$
$$\leqq \sum_{n=0}^{\infty} ||I - T||^n \, ||x||.$$

The right hand side is convergent by $||I - T|| < 1$. Hence, by the completeness of X, the bounded linear operator $s\text{-}\lim_{k\to\infty} \sum_{n=0}^{k} (I - T)^n$ is defined.

It is the inverse of T as may be seen from

$$T \cdot s\text{-}\lim_{k\to\infty} \sum_{n=0}^{k} (I-T)^n x = s\text{-}\lim_{k\to\infty} (I-(I-T))\left(\sum_{n=0}^{k} (I-T)^n x\right)$$
$$= x - s\text{-}\lim_{k\to\infty} (I-T)^{k+1} x = x,$$

and the similar equation $s\text{-}\lim_{k\to\infty} \left(\sum_{n=0}^{k} (I-T)^n\right) Tx = x$.

2. The Vitali-Hahn-Saks Theorem

This theorem is concerned with a convergent sequence of measures, and makes use of the following

Proposition. Let $(S, \mathfrak{B}, m)$ be a measure space. Let $\mathfrak{B}_0$ be the set of all $B \in \mathfrak{B}$ such that $m(B) < \infty$. Then by the distance

$$d(B_1, B_2) = m(B_1 \ominus B_2), \text{ where } B_1 \ominus B_2 = B_1 \cup B_2 - B_1 \cap B_2, \tag{1}$$

$\mathfrak{B}_0$ gives rise to a metric space $(\mathfrak{B}_0)$ by identifying two sets B_1 and B_2 of $\mathfrak{B}_0$ when $m(B_1 \ominus B_2) = 0$. Thus a point $\overline{B}$ of $(\mathfrak{B}_0)$ is the class of sets $B_1 \in \mathfrak{B}_0$ such that $m(B \ominus B_1) = 0$. Under the above metric (1), $(\mathfrak{B}_0)$ is a complete metric space.

Proof. If we denote by $C_B(s)$ the defining function of the set B:

$$C_B(s) = 1 \text{ or } 0 \quad \text{according as} \quad s \in B \text{ or } s \in B,$$

we have

$$d(B_1, B_2) = \int_S |C_{B_1}(s) - C_{B_2}(s)| \, m(ds). \tag{2}$$

Thus the metric space $(\mathfrak{B}_0)$ may be identified with a subset of the B-space $L^1(S, \mathfrak{B}, m)$. Let a sequence $\{C_{B_n}(s)\}$ with $B_n \in \mathfrak{B}_0$ satisfy the condition

$$\lim_{n,k\to\infty} d(B_n, B_k) = \lim_{n,k\to\infty} \int_S |C_{B_n}(s) - C_{B_k}(s)| \, m(ds) = 0.$$

Then, as in the proof of the completeness of the space $L^1(S, \mathfrak{B}, m)$, we can choose a subsequence $\{C_{B_{n'}}(s)\}$ such that $\lim_{n'\to\infty} C_{B_{n'}}(s) = C(s)$ exists m-a.e. and $\lim_{n'\to\infty} \int_S |C(s) - C_{B_{n'}}(s)| \, m(ds) = 0$. Clearly $C(s)$ is the defining function of a set $B_\infty \in \mathfrak{B}_0$, and hence $\lim_{n\to\infty} d(B_\infty, B_n) = 0$.

This proves that $(\mathfrak{B}_0)$ is a complete metric space.

Theorem (Vitali-Hahn-Saks). Let $(S, \mathfrak{B}, m)$ be a measure space, and $\{\lambda_n(B)\}$ a sequence of complex measures such that the total variations $|\lambda_n|(S)$ are finite for $n = 1, 2, \ldots$ Suppose that each $\lambda_n(B)$ is m-absolutely continuous and that a finite $\lim_{n\to\infty} \lambda_n(B) = \lambda(B)$ exists for every $B \in \mathfrak{B}$. Then the m-absolute continuity of $\lambda_n(B)$ is uniform in n, that is, $\lim m(B) = 0$ implies $\lim \lambda_n(B) = 0$ uniformly in n. If $m(S) < \infty$, then $\lambda(B)$ is σ-additive on $\mathfrak{B}$.

Proof. Each λ_n defines a one-valued function $\bar{\lambda}_n(\bar{B})$ on $(\mathfrak{B}_0)$ by $\bar{\lambda}_n(\bar{B}) = \lambda_n(B)$, since the value $\lambda_n(B)$ is, by the m-absolute continuity of $\lambda_n(B)$, independent of the choice of the set B from the class $\bar{B}$. The continuity of $\bar{\lambda}_n(\bar{B})$ is implied by and implies the m-absolute continuity of $\lambda_n(B)$.

Hence, for any $\varepsilon > 0$, the set

$$F_k(\varepsilon) = \left\{\bar{B};\ \sup_{n\geqq 1} |\bar{\lambda}_k(\bar{B}) - \bar{\lambda}_{k+n}(\bar{B})| \leqq \varepsilon\right\}$$

is closed in $(\mathfrak{B}_0)$ and, by the hypothesis $\lim_{n\to\infty} \lambda_n(B) = \lambda(B)$, we have $(\mathfrak{B}_0) = \bigcup_{k=1}^{\infty} F_k(\varepsilon)$. The complete metric space $(\mathfrak{B}_0)$ being of the second category, at least one $F_{k_0}(\varepsilon)$ must contain a non-void open set of $(\mathfrak{B}_0)$. This means that there exist a $\bar{B}_0 \in (\mathfrak{B}_0)$ and an $\eta > 0$ such that

$$d(B, B_0) < \eta \quad \text{implies} \quad \sup_{n\geqq 1} |\bar{\lambda}_{k_0}(\bar{B}) - \bar{\lambda}_{k_0+n}(\bar{B})| \leqq \varepsilon.$$

On the other hand, any $B \in \mathfrak{B}_0$ with $m(B) \leqq \eta$ can be represented as $B = B_1 - B_2$ with $d(B_1, B_0) \leqq \eta$, $d(B_2, B_0) \leqq \eta$. We may, for example, take $B_1 = B \cup B_0$, $B_2 = B_0 - B \cap B_0$. Thus, if $m(B) \leqq \eta$ and $k \geqq k_0$, we have

$$\begin{aligned} |\lambda_k(B)| &\leqq |\lambda_{k_0}(B)| + |\lambda_{k_0}(B) - \lambda_k(B)| \\ &\leqq |\lambda_{k_0}(B)| + |\lambda_{k_0}(B_1) - \lambda_k(B_1)| + |\lambda_{k_0}(B_2) - \lambda_k(B_2)| \\ &\leqq |\lambda_{k_0}(B)| + 2\varepsilon. \end{aligned}$$

Therefore, by the m-absolute continuity of λ_{k_0} and the arbitrariness of $\varepsilon > 0$, we see that $m(B) \to 0$ implies $\lambda_n(B) \to 0$ uniformly in n. Hence, in particular, $m(B) \to 0$ implies $\lambda(B) \to 0$. On the other hand, it is clear that λ is *finitely additive*, i.e., $\lambda\left(\sum_{j=1}^{n} B_j\right) = \sum_{j=1}^{n} \lambda(B_j)$. Thus, by $\lim_{m(B)\to 0} \lambda(B) = 0$ proved above, we easily see that λ is σ-additive if $m(S) < \infty$.

Corollary 1. Let $\{\lambda_n(B)\}$ be a sequence of complex measures on S such that the total variation $|\lambda_n|(S)$ is finite for every n. If a finite $\lim_{n\to\infty} \lambda_n(B)$ exists for every $B \in \mathfrak{B}$, then the σ-additivity of $\{\lambda_n(B)\}$ is uniform in n, in the sense that, $\lim_{k\to\infty} \lambda_n(B_k) = 0$ uniformly in n for any decreasing sequence $\{B_k\}$ of sets $\in \mathfrak{B}$ such that $\bigcap_{k=1}^{\infty} B_k = \emptyset$.

Proof. Let us consider

$$m(B) = \sum_{j=1}^{\infty} 2^{-j} \mu_j(B) \quad \text{where} \quad \mu_j(B) = \lambda_j(B)/|\lambda_j|(S).$$

The σ-additivity of m is a consequence of that of the λ_j's, and we have $0 \leqq m(B) \leqq 1$. Each μ_j and hence each λ_j is m-absolutely continuous. Thus, by the above Theorem, we have $\lim_{k\to\infty} \lambda_n(B_k) = 0$ uniformly in n, because $\lim_{k\to\infty} m(B_k) = 0$.

Corollary 2. $\lambda(B)$ in the Theorem is σ-additive and m-absolutely continuous even if $m(S) = \infty$.

3. The Termwise Differentiability of a Sequence of Generalized Functions

The discussion of the convergence of a sequence of generalized functions is very simple. We can in fact prove

Theorem. Let $\{T_n\}$ be a sequence of generalized functions $\in \mathfrak{D}(\Omega)'$. Let a finite $\lim_{n\to\infty} T_n(\varphi) = T(\varphi)$ exist for every $\varphi \in \mathfrak{D}(\Omega)$. Then T is also a generalized function $\in \mathfrak{D}(\Omega)'$. We say then that T is the limit in $\mathfrak{D}(\Omega)'$ of the sequence $\{T_n\}$ and write $T = \lim_{n\to\infty} T_n(\mathfrak{D}(\Omega)')$.

Proof. The linearity of the functional T is clear. Let K be any compact subset of Ω. Then each T_n defines a linear functional on the F-space $\mathfrak{D}_K(\Omega)$. Moreover, these functionals are continuous. For, they are bounded on every bounded set of $\mathfrak{D}_K(\Omega)$ as may be proved by Proposition 1 in Chapter I, 8. Thus, by the uniform boundedness theorem, T must be a linear functional on $\mathfrak{D}_K(\Omega)$ which is bounded on every bounded set of $\mathfrak{D}_K(\Omega)$. Hence T is a continuous linear functional on every $\mathfrak{D}_K(\Omega)$. Since $\mathfrak{D}(\Omega)$ is the inductive limit of $\mathfrak{D}_K(\Omega)$'s, T must be a continuous linear functional on $\mathfrak{D}(\Omega)$.

Corollary (termwise differentiability theorem). Let $T = \lim_{n\to\infty} T_n(\mathfrak{D}(\Omega)')$. Then, for any differential operator D^j, we have $D^j T = \lim_{n\to\infty} D^j T_n(\mathfrak{D}(\Omega)')$.

Proof. $\lim_{n\to\infty} T_n = T(\mathfrak{D}(\Omega)')$ implies that $\lim_{n\to\infty} T_n((-1)^{|j|} D^j\varphi) = T((-1)^{|j|} D^j\varphi)$ for every $\varphi \in \mathfrak{D}(\Omega)$. Thus we have

$$(D^j T)(\varphi) = \lim_{n\to\infty} (D^j T_n)(\varphi) \quad \text{for every} \quad \varphi \in \mathfrak{D}(\Omega).$$

4. The Principle of the Condensation of Singularities

The Baire-Hausdorff theorem may be applied to prove the existence of a function with various singularities. For instance, we shall prove the existence of a continuous function without a finite derivative.

Weierstrass' Theorem. There exists a real-valued continuous function $x(t)$ defined on the interval $[0, 1]$ such that it admits a finite differential quotient $x'(t_0)$ at no point t_0 of the interval $[0, 1/2]$.

Proof. A function $x(t)$ admits finite upper- and lower-derivatives on the right at $t = t_0$ iff there exists a positive integer n such that

$$\sup_{2^{-1}>h>0} h^{-1} |x(t_0+h) - x(t_0)| \leqq n.$$

Let us denote by M_n the set of all functions $x(t) \in C[0, 1]$ such that $x(t)$ satisfies the above inequality at a certain point t_0 of $[0, 1/2]$; here the point t_0 may vary with the function $x(t)$. We have only to show that each M_n is a non-dense subset of $C[0, 1]$. For, then, the set $C[0,1] - \bigcup_{n=1}^{\infty} M_n$ is non-void since, by the Baire-Hausdorff theorem, the complete metric space $C[0, 1]$ is not of the first category.

Now it is easy to see, by the compactness of the closed interval $[0, 1/2]$ and the norm of the space $C[0, 1]$, that M_n is closed in $C[0, 1]$. Next, for any polynomial $z(t)$ and $\varepsilon > 0$, we can find a function $y(t) \in C[0,1] - M_n$ such that $\sup_{0 \leqq t \leqq 1} |z(t) - y(t)| = ||z - y|| \leqq \varepsilon$. We may take, for example, a continuous function $y(t)$ represented graphically by a zig-zag line so that the above conditions are satisfied. Hence, by Weierstrass' polynomial approximation theorem, M_n is non-dense in $C[0, 1]$.

S. Banach [1] and H. Steinhaus have proved a *principle of condensation of singularities* given below which is based upon the following

Theorem (S. Banach). Given a sequence of bounded linear operators $\{T_n\}$ defined on a B-space X into a normed linear space Y_n. Then the set

$$B = \left\{x \in X;\ \overline{\lim_{n\to\infty}} ||T_n x|| < \infty\right\}$$

either coincides with X or is a set of the first category of X.

Proof. We shall show that $B = X$ under the hypothesis that B is of the second category. By the definition of B, we have $\lim_{k\to\infty} \sup_{n \geqq 1} ||k^{-1} T_n x|| = 0$ whenever $x \in B$. Thus, for any $\varepsilon > 0$,

$$B \subseteqq \bigcup_{k=1}^{\infty} B_k \quad \text{where} \quad B_k = \left\{x \in X;\ \sup_{n \geqq 1} ||k^{-1} T_n x|| \leqq \varepsilon\right\}.$$

Each B_k is a closed set by the continuity of the T_n's. Hence, if B is of the second category, then a certain B_{k_0} contains a sphere of positive radius. That is, there exist an $x_0 \in X$ and a $\delta > 0$ such that $||x - x_0|| \leqq \delta$ implies $\sup_{n \geqq 1} ||k_0^{-1} T_n x|| \leqq \varepsilon$. Hence, by putting $x - x_0 = y$, we obtain, for $||y|| \leqq \delta$, $||k_0^{-1} T_n y|| \leqq ||k_0^{-1} T_n x|| + ||k_0^{-1} T_n x_0|| \leqq 2\varepsilon$. We have thus

$$\sup_{n \geqq 1, ||z|| \leqq k_0^{-1}\delta} ||T_n z|| \leqq 2\varepsilon,$$

and so $B = X$.

Corollary (the principle of the condensation of singularities). Let $\{T_{p,q}\}$ $(q = 1, 2, \ldots)$ be a sequence of bounded linear operators defined on a B-space X into a normed linear space Y_p $(p = 1, 2, \ldots)$. Let, for each p, there exist an $x_p \in X$ such that $\overline{\lim\limits_{q\to\infty}} \|T_{p,q}x_p\| = \infty$. Then the set

$$B = \left\{x \in X;\ \overline{\lim_{q\to\infty}} \|T_{p,q}x\| = \infty \quad \text{for all} \quad p = 1, 2, \ldots\right\}$$

is of the second category.

Proof. For each p, the set $B_p = \left\{x \in X;\ \overline{\lim\limits_{q\to\infty}} \|T_{p,q}x\| < \infty\right\}$ is, by the preceding Theorem and the hypothesis, of the first category. Thus $B = X - \bigcup\limits_{p=1}^{\infty} B_p$ must be of the second category.

The above Corollary gives a general method of finding functions with many singularities.

Example. There exists a real-valued continuous function $x(t)$ of period 2π such that the *partial sum* of its *Fourier expansion*:

$$\begin{aligned} &f_q(x;t) = \sum_{k=0}^{q} (a_k \cos kt + b_k \sin kt) = \frac{1}{\pi} \int_{-\pi}^{\pi} x(s)\, K_q(s,t)\, ds, \\ &\text{where } K_q(s,t) = \sin((q + 2^{-1})(s-t))/2 \sin 2^{-1}(s-t), \end{aligned} \tag{1}$$

satisfies the condition that

$$\begin{aligned} &\overline{\lim_{q\to\infty}} |f_q(x;t)| = \infty \quad \text{on a set } P \subseteqq [0, 2\pi] \quad \text{which is of the} \\ &\text{power of the continuum.} \end{aligned} \tag{2}$$

Moreover, the set P may be taken so as to contain any countable sequence $\{t_j\} \subseteqq [0, 2\pi]$.

Proof. The totality of real-valued continuous functions $x(t)$ of period 2π constitutes a real B-space $C_{2\pi}$ by the norm $\|x\| = \sup\limits_{0\leqq t\leqq 2\pi} |x(t)|$. As may be seen from (1), $f_q(x;t)$ is, for a given $t_0 \in [0, 2\pi]$, a bounded linear functional on $C_{2\pi}$. Moreover, the norm of this functional $f_q(x;t_0)$ is given by

$$\begin{aligned} &\frac{1}{\pi} \int_{-\pi}^{\pi} |K_q(s,t_0)|\, ds = \text{the total variation of the function} \\ &\frac{1}{\pi} \int_{-\pi}^{t} K_q(s,t_0)\, ds. \end{aligned} \tag{3}$$

It is also easy to see that, for fixed t_0, (3) tends, as $q \to \infty$, to ∞.

Therefore, if we take a countable dense sequence $\{t_j\} \subseteqq [0, 2\pi]$, then, by the preceding Corollary, the set

$$B = \left\{x \in C_{2\pi};\ \overline{\lim_{q\to\infty}} |f_q(x;t)| = \infty \quad \text{for} \quad t = t_1, t_2, \ldots\right\}$$

is of the second category. Hence, by the completeness of the space $C_{2\pi}$, the set B is non-void. We shall show that, for any $x \in B$, the set

$$P = \left\{t \in [0, 2\pi];\ \overline{\lim_{q\to\infty}}\, |f_q(x;t)| = \infty\right\}$$

is of the power of the continuum. To this purpose, set

$$F_{m,q} = \{t \in [0, 2\pi];\ |f_q(x;t)| \leqq m\}, \quad F_m = \bigcap_{q=1}^{\infty} F_{m,q}.$$

By the continuity of $x(t)$ and the trigonometric functions, we see that the set $F_{m,q}$ and hence the set F_m are closed sets of $[0, 2\pi]$. If we can show that $\bigcup_{m=1}^{\infty} F_m$ is a set of the first category of $[0, 2\pi]$, then the set $P = \left([0, 2\pi] - \bigcup_{m=1}^{\infty} F_m\right) \ni \{t_j\}$ would be of the second category. Being a set of the second category of $[0, 2\pi]$, P cannot be countable and so, by the *continuum hypothesis*, P must have the power of the continuum.

Finally we will prove that each F_m is a set of the first category of $[0, 2\pi]$. Suppose some F_{m_0} be of the second category. Then the closed set F_{m_0} of $[0, 2\pi]$ must contain a closed interval $[\alpha, \beta]$ of $[0, 2\pi]$. This implies that $\sup_{q\geqq 1} |f_q(x;t)| \leqq m_0$ for all $t \in [\alpha, \beta]$, contradicting the fact that the set P contains a dense subset $\{t_j\}$ of $[0, 2\pi]$.

Remark. We can prove that the set P has the power of the continuum without appealing to the continuum hypothesis. See, for example, F. HAUSDORFF [1].

5. The Open Mapping Theorem

Theorem (the open mapping theorem of S. BANACH). Let T be a continuous linear operator on an F-space X onto an F-space Y. Then T maps every open set of X onto an open set of Y.

For the proof, we prepare

Proposition. Let X, Y be linear topological spaces. Let T be a continuous linear operator on X into Y, and suppose that the range $R(T)$ of T is a set of the second category in Y. Then, to each neighbourhood U of 0 of X, there corresponds some neighbourhood V of 0 of Y such that $V \subseteq (TU)^a$.

Proof. Let W be a neighbourhood of 0 of X such that $W = -W$, $W + W \subseteq U$. For every $x \in X$, $x/n \to 0$ as $n \to \infty$, and so $x \in nW$ for large n. Hence $X = \bigcup_{n=1}^{\infty} (nW)$, and therefore $R(T) = \bigcup_{n=1}^{\infty} T(nW)$. Since $R(T)$ is of the second category in Y, there is some positive integer n_0 such that $(T(n_0 W))^a$ contains a non-void open set. Since $(T(nW))^a = n(T(W))^a$

and since $n(T(W))^a$ and $(T(W))^a$ are homeomorphic[1] to each other, $(T(W))^a$ also contains a non-void open set. Let $y_0 = Tx_0$, $x_0 \in W$, be a point of this open set. Then, since the mapping $x \to -x_0 + x$ is a homeomorphic mapping, we see that there exists a neighbourhood V of 0 of Y such that $V \subseteq -y_0 + (T(W))^a$. The elements of $-y_0 + T(W)$ are expressible as $-y_0 + Tw = T(w - x_0)$ with $w \in W$. But $w - x_0 \in W + W \subseteq U$, since W is a neighbourhood of 0 of X such that $W = -W$ and $W + W \subseteq U$.

Therefore, $-y_0 + T(W) \subseteq T(U)$ and hence, by taking the closure, $-y_0 + (T(W))^a \subseteq (T(U))^a$ and so $V \subseteq -y_0 + (T(W))^a \subseteq (T(U))^a = (TU)^a$.

Proof of the Theorem. Since Y is a complete metric space, it is of the second category. Thus, by the preceding Proposition, the closure of the image by T of a neighbourhood of 0 of X contains a neighbourhood of 0 of Y.

Let X_ε, Y_ε denote the spheres in X, Y respectively, with centres at the origins and radii $\varepsilon > 0$. Let us set $\varepsilon_i = \varepsilon/2^i$ $(i = 0, 1, 2, \ldots)$. Then by what we have stated above, there exists a sequence of positive numbers $\{\eta_i\}$ such that

$$\lim_{i\to\infty} \eta_i = 0 \quad \text{and} \quad Y_{\eta_i} \subseteq (TX_{\varepsilon_i})^a \qquad (i = 0, 1, 2, \ldots). \tag{1}$$

Let $y \in Y_{\eta_0}$ be any point. We shall show that there is an $x \in X_{2\varepsilon_0}$ such that $Tx = y$. From (1) with $i = 0$, we see that there is an $x_0 \in X_{\varepsilon_0}$ such that $\|y - Tx_0\| < \eta_1$. Since $(y - Tx_0) \in Y_{\eta_1}$, we see again from (1) with $i = 1$ that there is an $x_1 \in X_{\varepsilon_1}$ with $\|y - Tx_0 - Tx_1\| < \eta_2$. Repeating the process, we find a sequence $\{x_i\}$ with $x_i \in X_{\varepsilon_i}$ such that

$$\left\| y - T\left(\sum_{i=0}^{n} x_i\right)\right\| < \eta_{n+1} \qquad (n = 0, 1, 2, \ldots).$$

We have $\left\|\sum_{k=m+1}^{n} x_k\right\| \leq \sum_{k=m+1}^{n} \|x_k\| \leq \sum_{k=m+1}^{n} \varepsilon_k \leq \left(\sum_{k=m+1}^{n} 2^{-k}\right)\varepsilon_0$, and so the sequence $\left\{\sum_{k=0}^{n} x_k\right\}$ is a Cauchy sequence. Hence, by the completeness of X, we see that $s\text{-}\lim_{n\to\infty} \sum_{k=0}^{n} x_k = x \in X$ exists. Moreover, we have

$$\|x\| = \lim_{n\to\infty} \left\|\sum_{k=0}^{n} x_k\right\| \leq \lim_{n\to\infty} \sum_{k=0}^{n} \|x_k\| \leq \left(\sum_{k=0}^{\infty} 2^{-k}\right)\varepsilon_0 = 2\varepsilon_0.$$

We must have $y = Tx$ since T is continuous. Thus we have proved that any sphere $X_{2\varepsilon_0}$ is mapped by T onto a set which contains a sphere Y_{η_0}.

After these preliminaries, let G be a non-void open set in X and $x \in G$. Let U be a neighbourhood of 0 of X such that $x + U \subseteq G$. Let V

[1] A one-to-one mapping M of a topological space S_1 onto a topological space S_2 is called a *homeomorphic mapping* if M and M^{-1} both map open sets onto open sets.

be a neighbourhood of 0 of Y such that $TU \supseteqq V$. Then $TG \supseteqq T(x+U) = Tx + TU \supseteqq Tx + V$. Hence TG contains a neighbourhood of each one of its point. This proves that T maps open set of X onto open sets of Y.

Corollary of the Theorem. If a continuous linear operator T on an F-space onto an F-space gives a one-to-one map, then the inverse operator T^{-1} is also a linear continuous operator.

6. The Closed Graph Theorem

Definition 1. Let X and Y be linear topological spaces on the same scalar field. Then the product space $X \times Y$ is a linear space by

$$\{x_1, y_1\} + \{x_2, y_2\} = \{x_1 + x_2, y_1 + y_2\}, \ \alpha\{x, y\} = \{\alpha x, \alpha y\}.$$

It is also a linear topological space if we call open the sets of the form

$$G_1 \times G_2 = \{\{x, y\};\ x \in G_1, y \in G_2\},$$

where G_1, G_2 are open sets of X, Y respectively. If, in particular, X and Y are quasi-normed linear spaces, then $X \times Y$ is also a quasi-normed linear space by the quasi-norm

$$\|\{x, y\}\| = (\|x\|^2 + \|y\|^2)^{1/2}. \tag{1}$$

Proposition 1. If X and Y are B-spaces (F-spaces), then $X \times Y$ is also a B-space (F-space).

Proof. Clear since $s\text{-}\lim \{x_n, y_n\} = \{x, y\}$ is equivalent to $s\text{-}\lim_{n\to\infty} x_n = x$, $s\text{-}\lim_{n\to\infty} y_n = y$.

Definition 2. The *graph* $G(T)$ of a linear operator T on $D(T) \subseteqq X$ into Y is the set $\{\{x, Tx\}; x \in D(T)\}$ in the product space $X \times Y$. Let X, Y be linear topological spaces. Then T is called a *closed linear operator* when its graph $G(T)$ constitutes a closed linear subspace of $X \times Y$. If X and Y are quasi-normed linear spaces, then a linear operator T on $D(T) \subseteqq X$ into Y is closed iff the following condition is satisfied:

$$\{x_n\} \subseteqq D(T),\ s\text{-}\lim_{n\to\infty} x_n = x \quad \text{and} \quad s\text{-}\lim_{n\to\infty} Tx_n = y \quad \text{imply that}$$
$$x \in D(T) \quad \text{and} \quad Tx = y. \tag{2}$$

Thus the notion of a closed linear operator is an extension of the notion of a bounded linear operator. A linear operator T on $D(T) \subseteqq X$ into Y is said to be *closable* or *pre-closed* if the closure in $X \times Y$ of the graph $G(T)$ is the graph of a linear operator, say S, on $D(S) \subseteqq X$ into Y.

Proposition 2. If X, Y are quasi-normed linear spaces, then T is closable iff the following condition is satisfied:

$$\{x_n\} \subseteqq D(T),\ s\text{-}\lim_{n\to\infty} x_n = 0 \ \text{and} \ s\text{-}\lim_{n\to\infty} Tx_n = y \ \text{imply that} \ y = 0. \tag{3}$$

Proof. The "only if" part is clear since the closure in $X \times Y$ of $G(T)$ is a graph $G(S)$ of a linear operator S and so $y = S \cdot 0 = 0$. The "if" part is proved as follows. Let us define a linear operator S by the following condition and call S the *smallest closed extension* of T:

$$\begin{aligned}&x \in D(S) \text{ iff there exists a sequence } \{x_n\} \subseteq D(T) \text{ such that}\\ &\underset{n\to\infty}{\text{s-lim}}\, x_n = x \text{ and } \underset{n\to\infty}{\text{s-lim}}\, Tx_n = y \text{ exist; and we define}\\ &Sx = y.\end{aligned} \tag{4}$$

That the value y is defined uniquely by x follows from condition (3). We have only to prove that S is closed. Let $w_n \in D(S)$, $\underset{n\to\infty}{\text{s-lim}}\, w_n = w$ and $\underset{n\to\infty}{\text{s-lim}}\, Sw_n = u$. Then there exists a sequence $\{x_n\} \subseteq D(T)$ such that $\|w_n - x_n\| \leqq n^{-1}$, $\|Sw_n - Tx_n\| \leqq n^{-1}$ $(n = 1, 2, \ldots)$. Therefore $\underset{n\to\infty}{\text{s-lim}}\, x_n = \underset{n\to\infty}{\text{s-lim}}\, w_n = w$, $\underset{n\to\infty}{\text{s-lim}}\, Tx_n = \underset{n\to\infty}{\text{s-lim}}\, Sw_n = u$, and so $w \in D(S)$, $S \cdot w = u$.

An example of a discontinuous but closed operator. Let $X = Y = C[0, 1]$. Let D be the set of all $x(t) \in X$ such that the derivative $x'(t) \in X$; and let T be the linear operator on $D(T) = D$ defined by $Tx = x'$. This T is not continuous, since, for $x_n(t) = t^n$,

$$\|x_n\| = 1, \|Tx_n\| = \sup_{0 \leqq t \leqq 1} |x_n'(t)| = \sup_{0 \leqq t \leqq 1} |nt^{n-1}| = n \qquad (n = 1, 2, \ldots).$$

But T is closed. In fact, let $\{x_n\} \subseteq D(T)$, $\underset{n\to\infty}{\text{s-lim}}\, x_n = x$ and $\underset{n\to\infty}{\text{s-lim}}\, Tx_n = y$. Then $x_n'(t)$ converges uniformly to $y(t)$, and $x_n(t)$ converges uniformly to $x(t)$. Hence $x(t)$ must be differentiable with continuous derivative $y(t)$. This proves that $x \in D(T)$ and $Tx = y$.

Examples of closable operators. Let D_x be a linear differential operator

$$D_x = \sum_{|j| \leqq k} c_j(x) D^j \tag{5}$$

with coefficients $c_j(x) \in C^k(\Omega)$, where Ω is an open domain of R^n. Consider the totality D of functions $f(x) \in L^2(\Omega) \cap C^k(\Omega)$ such that $D_x f(x) \in L^2(\Omega)$. We define a linear operator T on $D(T) = D \subseteq L^2(\Omega)$ into $L^2(\Omega)$ by $(Tf)(x) = D_x f(x)$. Then T is closable. For, let $\{f_h\} \subseteq D$ be such that $\underset{h\to\infty}{\text{s-lim}}\, f_h = 0$, $\underset{h\to\infty}{\text{s-lim}}\, D_x f_h = g$. Then, for any $\varphi(x) \in C_0^k(\Omega)$, we have, by partial integration,

$$\int_\Omega D_x f(x) \cdot \varphi(x)\, dx = \int_\Omega f(x) \cdot D_x' \varphi(x)\, dx, \tag{6}$$

where D_x' is the differential operator *formally adjoint* to D_x:

$$D_x' \varphi(x) = \sum_{|j| \leqq k} (-1)^{|j|} D^j (c_j(x) \varphi(x)). \tag{7}$$

The formula (6) is obtained, since the integrated term in the partial integration vanishes by $\varphi(x) \in C_0^k(\Omega)$. Hence, by the continuity of the scalar product in $L^2(\Omega)$, we obtain, by taking $f = f_h$ and letting $h \to \infty$ in (6),

$$\int_\Omega g(x) \cdot \varphi(x)\, dx = \int_\Omega 0 \cdot D_x' \varphi(x)\, dx = 0. \tag{8}$$

Since $\varphi(x) \in C_0^k(\Omega)$ was arbitrary, we must have $g(x) = 0$ a.e., that is, $g = 0$ in $L^2(\Omega)$.

Proposition 3. The inverse T^{-1} of a closed linear operator on $D(T) \subseteq X$ into Y, if it exists, is a closed linear operator.

Proof. The graph of T^{-1} is the set $\{\{Tx, x\};\, x \in D(T)\}$ in the product space $Y \times X$. Thus the Proposition is proved remembering the fact that the mapping $\{x, y\} \to \{y, x\}$ of $X \times Y$ onto $Y \times X$ is a homeomorphic one.

We next prove *Banach's closed graph theorem*:

Theorem 1. A closed linear operator defined on an F-space X into an F-space Y is continuous.

Proof. The graph $G(T)$ of T is a closed linear subspace of the F-space $X \times Y$. Hence, by the completeness of $X \times Y$, $G(T)$ is an F-space. The linear mapping U defined by $U\{x, Tx\} = x$ is a one-to-one, continuous linear transformation on the F-space $G(T)$ onto the F-space X. Hence, by virtue of the open mapping theorem, the inverse U^{-1} of U is continuous. The linear operator V defined by $V\{x, Tx\} = Tx$ is clearly continuous on $G(T)$ onto $R(T) \subseteq Y$. Therefore, $T = VU^{-1}$ is continuous on X into Y.

The following theorem of comparison of two linear operators is due to L. Hörmander:

Theorem 2. Let X_i ($i = 0, 1, 2$; $X_0 = X$) be B-spaces and T_i ($i = 1, 2$) be linear operators defined on $D(T_i) \subseteq X$ into X_i. Then, if T_1 is closed and T_2 closable in such a way that $D(T_1) \subseteq D(T_2)$, there exists a constand C such that

$$\|T_2 x\| \leq C(\|T_1 x\|^2 + \|x\|^2)^{1/2} \text{ for all } x \in D(T_1). \tag{9}$$

Proof. The graph $G(T_1)$ of T_1 is a closed subspace of $X \times X_1$. Hence the mapping

$$G(T_1) \ni \{x, T_1 x\} \to T_2 x \in X_2 \tag{10}$$

is a linear operator on the B-space $G(T_1)$ into the B-space X_2. We shall prove that this mapping is closed. Suppose that $\{x_n, T_1 x_n\}$ s-converges in $G(T_1)$ and that $T_2 x_n$ s-converges in X_2. Since T_1 is closed, there is an element $x \in D(T_1)$ such that $x = s\text{-}\lim_{n\to\infty} x_n$ and $T_1 x = s\text{-}\lim_{n\to\infty} T_1 x_n$. By the hypothesis, we have $x \in D(T_2)$, and, since T_2 is closable, the existing $s\text{-}\lim_{n\to\infty} T_2 x_n$ can only be $T_2 x$. Hence the mapping (10) is closed, and so, by the closed graph theorem, it must be continuous. This proves (9).

7. An Application of the Closed Graph Theorem (Hörmander's Theorem)

Any distribution solution $u \in L^2$ of the Laplace equation

$$\Delta u = f \in L^2$$

is defined by a function which is equal to a C^∞ function after correction on a set of measure zero in the domain where f is C^∞. This result is known as *Weyl's Lemma* and plays a fundamental role in the modern treatment of the theory of potentials. See H. Weyl [1]. There is an abundant literature on the extensions of Weyl's Lemma. Of these, the research due to L. Hörmander [1] seems to be the most far reaching. We shall begin with his definition of hypoellipticity.

Definition 1. Let Ω be an open domain of R^n. A function $u(x)$, $x \in \Omega$, is said to belong to $L^2_{\text{loc}}(\Omega)$ if $\int_{\Omega'} |u(x)|^2 dx < \infty$ for any open subdomain Ω' with compact closure in Ω. A linear partial differential operator $P(D)$ with constant coefficients:

$$\left.\begin{aligned} &P(D) = P\left(\frac{1}{i}\frac{\partial}{\partial x_1}, \frac{1}{i}\frac{\partial}{\partial x_2}, \ldots, \frac{1}{i}\frac{\partial}{\partial x_n}\right), \text{ where} \\ &P(\xi) = P(\xi_1, \xi_2, \ldots, \xi_n) \text{ is a polynomial in } \xi_1, \xi_2, \ldots, \xi_n, \end{aligned}\right\} \tag{1}$$

is said to be *hypoelliptic*, if every distribution solution $u \in L^2_{\text{loc}}(\Omega)$ of $P(D)u = f$ is C^∞ after correction on a set of measure zero in the subdomain where f is C^∞.

Theorem (Hörmander). If $P(D)$ is hypoelliptic, then there exists, for any large positive constant C_1, a positive constant C_2 such that, for any solution $\zeta = \xi + i\eta$ of the algebraic equation $P(\zeta) = 0$, we have

$$|\zeta| = \left(\sum_{j=1}^{n} |\zeta_j|^2\right)^{1/2} \leqq C_1 \quad \text{if} \quad |\eta| = \left(\sum_{j=1}^{n} |\eta_j|^2\right)^{1/2} \leqq C_2. \tag{2}$$

Proof. Let U be the totality of distribution solutions $u \in L^2(\Omega')$ of $P(D)u = 0$, that is, the totality of $u \in L^2(\Omega')$ such that

$$\int_{\Omega'} u \cdot P'(D)\varphi\, dx = 0 \quad \text{for all} \quad \varphi \in C_0^\infty(\Omega'), \tag{3}$$

where the *adjoint differential operator* $P'(D)$ of $P(D)$ is defined by

$$P'(\xi) = P(-\xi_1, -\xi_2, \ldots, -\xi_n). \tag{4}$$

We can prove that U is a closed linear subspace of $L^2(\Omega')$. The linearity of U is obvious from the linearity of the differential operator $P(D)$. Let a sequence $\{u_h\}$ of U be such that $s\text{-}\lim_{h\to\infty} u_h = u$ in $L^2(\Omega')$. Then, by the continuity of the scalar product in $L^2(\Omega')$, we have

$$0 = \lim_{h\to\infty} \int_{\Omega'} u_h \cdot P'(D)\varphi\, dx = \int_{\Omega'} u \cdot P'(D)\varphi\, dx = 0, \text{ i.e., } \quad u \in U.$$

Thus U is a closed linear subspace of $L^2(\Omega')$, and as such a B-space.

Since $P(D)$ is hypoelliptic, we may suppose that every $u \in U$ is C^∞ in Ω'. Let Ω'_1 be any open subdomain with compact closure in Ω'. Then, for any $u \in U$, the function $\partial u/\partial x_k$ is C^∞ in Ω' $(k = 1, 2, \ldots, n)$. By the argument used in the preceding section, the mapping

$$U \ni u \to \frac{\partial u}{\partial x_k} \in L^2(\Omega'_1) \qquad (k = 1, 2, \ldots, n)$$

is a closed linear operator. Hence, by the closed graph theorem, there exists a positive constant C such that

$$\int_{\Omega'_1} \sum_{k=1}^{n} \left|\frac{\partial u}{\partial x_k}\right|^2 dx \leqq C \int_{\Omega'} |u|^2 \, dx, \qquad \text{for all } u \in U.$$

If we apply this inequality to the function $u(x) = e^{i\langle x, \zeta\rangle}$, where $\zeta = \xi + i\eta = (\xi_1 + i\eta_1, \xi_2 + i\eta_2, \ldots, \xi_n + i\eta_n)$ is a solution of $P(\zeta) = 0$ and $\langle x, \zeta\rangle = \sum_{j=1}^{n} \langle x_j, \zeta_j\rangle$, we obtain

$$\sum_{k=1}^{n} |\zeta_k|^2 \cdot \int_{\Omega'_1} e^{-2\langle x, \eta\rangle} \, dx \leqq C \int_{\Omega'} e^{-2\langle x, \eta\rangle} \, dx.$$

Therefore, when $|\eta|$ is bounded, it follows that $|\zeta|$ must be bounded.

Remark. Later on, we shall prove that condition (2) implies the hypoellipticity of $P(D)$. This result is also due to Hörmander. Thus, in particular, we see that Weyl's Lemma is a trivial consequence of Hörmander's result. In fact, the root of the algebraic equation $-\sum_{j=1}^{n} \zeta_j^2 = 0$ satisfies (2).

References for Chapter II

S. Banach [1], N. Bourbaki [2], N. Dunford-J. Schwartz [1], E. Hille-R. S. Phillips [1] and L. Hörmander [6].

III. The Orthogonal Projection and F. Riesz' Representation Theorem

1. The Orthogonal Projection

In a pre-Hilbert space, we can introduce the notion of *orthogonality* of two vectors. Thanks to this fact, a Hilbert space may be identified with its dual space, i.e., the space of bounded linear functionals. This result is the *representation theorem of F. Riesz* [1], and the whole theory of Hilbert spaces is founded on this theorem.

Definition 1. Let x, y be vectors of a pre-Hilbert space X. We say that x is *orthogonal* to y and write $x \perp y$ if $(x, y) = 0$; if $x \perp y$ then $y \perp x$, and $x \perp x$ iff $x = 0$. Let M be a subset of a pre-Hilbert space X. We denote by $M^\perp$ the totality of vectors $\in X$ orthogonal to every vector m of M.

Theorem 1. Let M be a closed linear subspace of a Hilbert space X. Then $M^\perp$ is also a closed linear subspace of X, and $M^\perp$ is called the *orthogonal complement* of M. Any vector $x \in X$ can be decomposed uniquely in the form

$$x = m + n, \text{ where } m \in M \text{ and } n \in M^\perp. \tag{1}$$

The element m in (1) is called the *orthogonal projection* of x upon M and will be denoted by $P_M x$; P_M is called the *projection operator* or the *projector* upon M. We have thus, remembering that $M \subseteq (M^\perp)^\perp$,

$$x = P_M x + P_{M^\perp} x, \text{ that is, } I = P_M + P_{M^\perp}. \tag{1'}$$

Proof. The linearity of $M^\perp$ is a consequence of the linearity in x of the scalar product (x, y). $M^\perp$ is closed by virtue of the continuity of the scalar product. The uniqueness of the decomposition (1) is clear since a vector orthogonal to itself is the zero vector.

To prove the possibility of the decomposition (1), we may assume that $M \neq X$ and $x \bar{\in} M$, for, if $x \in M$, we have the trivial decomposition with $m = x$ and $n = 0$. Thus, since M is closed and $x \bar{\in} M$, we have

$$d = \inf_{m \in M} \|x - m\| > 0.$$

Let $\{m_n\} \subseteq M$ be a *minimizing sequence*, i.e., $\lim_{n\to\infty} \|x - m_n\| = d$. Then $\{m_n\}$ is a Cauchy sequence. For, by $\|a + b\|^2 + \|a - b\|^2 = 2(\|a\|^2 + \|b\|^2)$ valid in any pre-Hilbert space (see (1) in Chapter I, 5), we obtain

$$\begin{aligned}
\|m_k - m_n\|^2 &= \|(x - m_n) - (x - m_k)\|^2 = 2(\|x - m_n\|^2 + \|x - m_k\|^2) \\
&\quad - \|2x - m_n - m_k\|^2 \\
&= 2(\|x - m_n\|^2 + \|x - m_k\|^2) - 4\,\|x - (m_n + m_k)/2\|^2 \\
&\leqq 2(\|x - m_n\|^2 + \|x - m_k\|^2 - 4d^2) \\
&\quad (\text{since } (m_n + m_k)/2 \in M) \\
&\to 2(d^2 + d^2) - 4d^2 = 0 \text{ as } k, n \to \infty.
\end{aligned}$$

By the completeness of the Hilbert space X, there exists an element $m \in X$ such that $s\text{-}\lim_{n\to\infty} m_n = m$. We have $m \in M$ since M is closed. Also, by the continuity of the norm, we have $\|x - m\| = d$.

Write $x = m + (x - m)$. Putting $n = x - m$, we have to show that $n \in M^\perp$. For any $m' \in M$ and any real number α, we have $(m + \alpha m') \in M$

and so

$$d^2 \leqq \|x - m - \alpha m'\|^2 = (n - \alpha m', n - \alpha m')$$
$$= \|n\|^2 - \alpha(n, m') - \alpha(m', n) + \alpha^2 \|m'\|^2.$$

This gives, since $\|n\| = d$, $0 \leqq -2\alpha Re(n, m') + \alpha^2 \|m'\|^2$ for every real α. Hence $Re(n, m') = 0$ for every $m' \in M$. Replacing m' by im', we obtain $Im(n, m') = 0$ and so $(n, m') = 0$ for every $m' \in M$.

Corollary. For a closed linear subspace M of a Hilbert space X, we have $M = M^{\perp\perp} = (M^{\perp})^{\perp}$.

Theorem 2. The projector $P = P_M$ is a bounded linear operator such that

$$P = P^2 \text{ (\textit{idempotent} property of } P), \tag{2}$$

$$(Px, y) = (x, Py) \text{ (\textit{symmetric} property of } P). \tag{3}$$

Conversely, a bounded linear operator P on a Hilbert space X into X satisfying (2) and (3) is a projector upon $M = R(P)$.

Proof. (2) is clear from the definition of the orthogonal projection. We have, by (1') and $P_M x \perp P_{M^\perp} y$,

$$(P_M x, y) = (P_M x, P_{M^\perp} y + P_M y) = (P_M x, P_M y)$$
$$= (P_M x + P_{M^\perp} x, P_M y) = (x, P_M y).$$

Next let $y = x + z$, $x \in M$, $z \in M^\perp$ and $w = u + v$, $u \in M$, $v \in M^\perp$, then $y + w = (x + u) + (z + v)$ with $(x + u) \in M$, $(z + v) \in M^\perp$ and so, by the uniqueness of the decomposition (1), $P_M(y + w) = P_M y + P_M w$; similarly we obtain $P_M(\alpha y) = \alpha P_M y$. The boundedness of the operator P_M is proved by

$$\|x\|^2 = \|P_M x + P_{M^\perp} x\|^2 = (P_M x + P_{M^\perp} x, P_M x + P_{M^\perp} x)$$
$$= \|P_M x\|^2 + \|P_{M^\perp} x\|^2 \geqq \|P_M x\|^2.$$

Thus, in particular, we have

$$\|P_M\| \leqq 1. \tag{4}$$

The converse part of the Theorem is proved as follows. The set $M = R(P)$ is a linear subspace, since P is a linear operator. The condition $x \in M$ is equivalent to the existence of a certain $y \in X$ such that $x = P \cdot y$, and this in turn is equivalent, by (2), to $x = Py = P^2 y = Px$. Therefore $x \in M$ is equivalent to $x = Px$. M is a closed subspace; for, $x_n \in M$, $\text{s-}\lim_{n\to\infty} x_n = y$ imply, by the continuity of P and $x_n = Px_n$, $\text{s-}\lim_{n\to\infty} x_n = \text{s-}\lim_{n\to\infty} Px_n = Py$ so that $y = Py$.

We have to show that $P = P_M$. If $x \in M$, we have $Px = x = P_M \cdot x$; and if $y \in M^\perp$, we have $P_M y = 0$. Moreover, in the latter case, $(Py, Py) =$

$(y, P^2y) = (y, Py) = 0$ and so $Py = 0$. Therefore we obtain, for any $y \in X$,

$$Py = P(P_M y + P_{M^\perp} y) = PP_M y + PP_{M^\perp} y$$
$$= P_M y + 0, \text{ i.e., } Py = P_M y.$$

Another characterization of the projection operator is given by

Theorem 3. A bounded linear operator P on a Hilbert space X into X is a projector iff P satisfies $P = P^2$ and $\|P\| \leqq 1$.

Proof. We have only to prove the "if" part. Set $M = R(P)$ and $N = N(P) = \{y; Py = 0\}$. As in the proof of the preceding Theorem 2, M is a closed linear subspace and $x \in M$ is equivalent to $x = Px$. N is also a closed linear subspace in virtue of the continuity of P. In the decomposition $x = Px + (I - P)x$, we have $Px \in M$ and $(I - P)x \in N$. The latter assertion is clear from $P(I - P) = P - P^2 = 0$.

We thus have to prove that $N = M^\perp$. For every $x \in X$, $y = Px - x \in N$ by $P^2 = P$. Hence, if, in particular, $x \in N^\perp$, then $Px = x + y$ with $(x, y) = 0$. It follows then that $\|x\|^2 \geqq \|Px\|^2 = \|x\|^2 + \|y\|^2$ so that $y = 0$. Thus we have proved that $x \in N^\perp$ implies $x = Px$, that is, $N^\perp \subseteqq M = R(P)$. Let, conversely, $z \in M = R(P)$, so that $z = Pz$. Then we have the orthogonal decomposition $z = y + x$, $y \in N$, $x \in N^\perp$, and so $z = Pz = Py + Px = Px = x$, the last equality being already proved. This shows that $M = R(P) \subseteqq N^\perp$. We have thus obtained $M = N^\perp$, and so, by $N = (N^\perp)^\perp$, $N = M^\perp$.

2. "Nearly Orthogonal" Elements

In general, we cannot define the notion of orthogonality in a normed linear space. However, we can prove the

Theorem (F. Riesz [2]). Let X be a normed linear space, and M be its closed linear subspace. Suppose $M \neq X$. Then there exists, for any $\varepsilon > 0$ with $0 < \varepsilon < 1$, an $x_\varepsilon \in X$ such that

$$\|x_\varepsilon\| = 1 \quad \text{and} \quad \text{dis}(x_\varepsilon, M) = \inf_{m \in M} \|x_\varepsilon - m\| \geqq 1 - \varepsilon. \tag{1}$$

The element x_ε is thus "nearly orthogonal" to M.

Proof. Let $y \in X - M$. Since M is closed, $\text{dis}(y, M) = \inf_{m \in M} \|y - m\| = \alpha > 0$. Thus there exists an $m_\varepsilon \in M$ such that $\|y - m_\varepsilon\| \leqq \alpha \left(1 + \frac{\varepsilon}{1 - \varepsilon}\right)$. The vector $x_\varepsilon = (y - m_\varepsilon)/\|y - m_\varepsilon\|$ satisfies $\|x_\varepsilon\| = 1$ and

$$\|x_\varepsilon - m\| = \|y - m_\varepsilon\|^{-1} \|y - m_\varepsilon - \|y - m_\varepsilon\| \cdot m\| \geqq \|y - m_\varepsilon\|^{-1} \alpha$$
$$\geqq \left(\frac{1}{1 - \varepsilon}\right)^{-1} 1 - \varepsilon.$$

Corollary 1. Let there exist a sequence of closed linear subspaces M_n of a normed linear space X such that $M_n \subseteqq M_{n+1}$ and $M_n \neq M_{n+1}$ $(n = 1, 2, \ldots)$. Then there exists a sequence $\{y_n\}$ such that

$$y_n \in M_n,\ \|y_n\| = 1 \quad \text{and} \quad \operatorname{dis}(y_{n+1}, M_n) \geqq 1/2 \ (n = 1, 2, \ldots). \quad (2)$$

Corollary 2. The unit sphere $S = \{x \in X; \|x\| \leqq 1\}$ of a B-space X is compact iff X is of finite dimension.

Proof. The "if" part is a consequence of the Bolzano-Weierstrass theorem that a bounded closed set of R^n is compact. The "only if" part is proved as follows. Suppose X is not of finite dimension. Then there exists, by the preceding Corollary 1, a sequence $\{y_n\}$ satisfying the conditions: $\|y_n\| = 1$ and $\|y_m - y_n\| \geqq 1/2$ for $m > n$. This is a contradiction to the hypothesis that the unit sphere of X is compact.

3. The Ascoli-Arzelà Theorem

To give an example of a relatively compact infinite subset of a B-space of infinite dimension, we shall prove the

Theorem (Ascoli-Arzelà). Let S be a compact metric space, and $C(S)$ the B-space of (real- or) complex-valued continuous functions $x(s)$ normed by $\|x\| = \sup_{s \in S} |x(s)|$. Then a sequence $\{x_n(s)\} \subseteqq C(S)$ is relatively compact in $C(S)$ if the following two conditions are satisfied:

$$x_n(s) \text{ is } \textit{equi-bounded} \text{ (in } n\text{), i.e., } \sup_{n \geqq 1} \sup_{s \in S} |x_n(s)| < \infty, \quad (1)$$

$$x_n(s) \text{ is } \textit{equi-continuous} \text{ (in } n\text{), i.e.,} \quad (2)$$

$$\lim_{\delta \downarrow 0} \sup_{n \geqq 1, \operatorname{dis}(s', s'') \leqq \delta} |x_n(s') - x_n(s'')| = 0.$$

Proof. A bounded sequence of complex numbers contains a convergent subsequence (the Bolzano-Weierstrass theorem). Hence, for fixed s, the sequence $\{x_n(s)\}$ contains a convergent subsequence. On the other hand, since the metric space S is compact, there exists a countable dense subset $\{s_n\} \subseteqq S$ such that, for every $\varepsilon > 0$, there exists a finite subset $\{s_n; 1 \leqq n \leqq k(\varepsilon)\}$ of $\{s_n\}$ satisfying the condition

$$\sup_{s \in S} \inf_{1 \leqq j \leqq k(\varepsilon)} \operatorname{dis}(s, s_j) \leqq \varepsilon. \quad (3)$$

The proof of this fact is obtained as follows. Since S is compact, it is totally bounded (see Chapter 0, 2). Thus there exists, for any $\delta > 0$, a finite system of points $\in S$ such that any point of S has a distance $\leqq \delta$ from some point of the system. Letting $\delta = 1, 2^{-1}, 3^{-1}, \ldots$ and collecting the corresponding finite systems of points, we obtain a sequence $\{s_n\}$ with the stated properties.

We then apply the diagonal process of choice to the sequence $\{x_n(s)\}$, so that we obtain a subsequence $\{x_{n'}(s)\}$ of $\{x_n(s)\}$ which converges for

$s = s_1, s_2, \ldots, s_k, \ldots$ simultaneously. By the equi-continuity of $\{x_n(s)\}$, there exists, for every $\varepsilon > 0$, a $\delta = \delta(\varepsilon) > 0$ such that $\operatorname{dis}(s', s'') \leqq \delta$ implies $|x_n(s') - x_n(s'')| \leqq \varepsilon$ for $n = 1, 2, \ldots$ Hence, for every $s \in S$, there exists a j with $j \leqq k(\varepsilon)$ such that

$$|x_{n'}(s) - x_{m'}(s)| \leqq |x_{n'}(s) - x_{n'}(s_j)| + |x_{n'}(s_j) - x_{m'}(s_j)| + |x_{m'}(s_j) - x_{m'}(s)| \leqq 2\varepsilon + |x_{n'}(s_j) - x_{m'}(s_j)|.$$

Thus $\lim\limits_{n,m\to\infty} \max\limits_s |x_{n'}(s) - x_{m'}(s)| \leqq 2\varepsilon$, and so $\lim\limits_{n,m\to\infty} \|x_{n'} - x_{m'}\| = 0$.

4. The Orthogonal Base. Bessel's Inequality and Parseval's Relation

Definition 1. A set S of vectors in a pre-Hilbert space X is called an *orthogonal set*, if $x \perp y$ for each pair of distinct vectors x, y of S. If, in addition, $\|x\| = 1$ for each $x \in S$, then the set S is called an *orthonormal set.* An orthonormal set S of a Hilbert space X is called a *complete orthonormal system* or an *orthogonal base* of X, if no orthonormal set of X contains S as a proper subset.

Theorem 1. A Hilbert space X (having a non-zero vector) has at least one complete orthonormal system. Moreover, if S is any orthonormal set in X, there is a complete orthonormal system containing S as a subset.

Proof (by Zorn's Lemma). Let S be an orthonormal set in X. Such a set surely exists; for instance, if $x \neq 0$, the set consisting only of $x/\|x\|$ is orthonormal. We consider the totality $\{S\}$ of orthonormal sets which contain S as a subset. $\{S\}$ is partially ordered by writing $S_1 \prec S_2$ for the inclusion relation $S_1 \subseteqq S_2$. Let $\{S'\}$ be a linearly ordered subsystem of $\{S\}$, then $\bigcup\limits_{S' \in \{S'\}} S'$ is an orthonormal set and an upper bound of $\{S'\}$. Thus, by Zorn's Lemma, there exists a maximal element S_0 of $\{S\}$. This orthonormal set S_0 contains S and, by the maximality, it must be a complete orthonormal system.

Theorem 2. Let $S = \{x_\alpha; \alpha \in A\}$ be a complete orthonormal system of a Hilbert space X. For any $f \in X$, we define its *Fourier coefficients* (with respect to S)

$$f_\alpha = (f, x_\alpha). \tag{1}$$

Then we have *Parseval's relation*

$$\|f\|^2 = \sum_{\alpha \in A} |f_\alpha|^2. \tag{2}$$

Proof. We shall first prove *Bessel's inequality*

$$\sum_{\alpha \in A} |f_\alpha|^2 \leqq \|f\|^2. \tag{2'}$$

Let $\alpha_1, \alpha_2, \ldots, \alpha_n$ be any finite system of α's. For any finite system of complex numbers $c_{\alpha_1}, c_{\alpha_2}, \ldots, c_{\alpha_n}$, we have, by the orthonormality of $\{x_\alpha\}$,

$$\left\| f - \sum_{j=1}^{n} c_{\alpha_j} x_{\alpha_j} \right\|^2 = \left(f - \sum_{j=1}^{n} c_{\alpha_j} x_{\alpha_j}, f - \sum_{j=1}^{n} c_{\alpha_j} x_{\alpha_j} \right)$$
$$= \|f\|^2 - \sum_{j=1}^{n} c_{\alpha_j} \bar{f}_{\alpha_j} - \sum_{j=1}^{n} \bar{c}_{\alpha_j} f_{\alpha_j} + \sum_{j=1}^{n} |c_{\alpha_j}|^2 \quad (3)$$
$$= \|f\|^2 - \sum_{j=1}^{n} |f_{\alpha_j}|^2 + \sum_{j=1}^{n} |f_{\alpha_j} - c_{\alpha_j}|^2.$$

Hence the minimum of $\left\| f - \sum_{j=1}^{n} c_{\alpha_j} x_{\alpha_j} \right\|^2$, for fixed $\alpha_1, \alpha_2, \ldots, \alpha_n$, is attained when $c_{\alpha_j} = f_{\alpha_j}$ $(j = 1, 2, \ldots, n)$. We have thus

$$\left\| f - \sum_{j=1}^{n} f_{\alpha_j} x_{\alpha_j} \right\|^2 = \|f\|^2 - \sum_{j=1}^{n} |f_{\alpha_j}|^2, \text{ and hence } \sum_{j=1}^{n} |f_{\alpha_j}|^2 \leqq \|f\|^2. \quad (4)$$

By the arbitrariness of $\alpha_1, \alpha_2, \ldots, \alpha_n$, we see that Bessel's inequality (2') is true, and $f_\alpha \neq 0$ for at most a countable number of α's, say $\alpha_1, \alpha_2, \ldots, \alpha_n, \ldots$ We then prove that $f = \text{s-}\lim_{n\to\infty} \sum_{j=1}^{n} f_{\alpha_j} x_{\alpha_j}$. First, the sequence $\left\{ \sum_{j=1}^{n} f_{\alpha_j} x_{\alpha_j} \right\}$ is a Cauchy sequence, since, by the orthonormality of $\{x_\alpha\}$,

$$\left\| \sum_{j=k}^{n} f_{\alpha_j} x_{\alpha_j} \right\|^2 = \left(\sum_{j=k}^{n} f_{\alpha_j} x_{\alpha_j}, \sum_{j=k}^{n} f_{\alpha_j} x_{\alpha_j} \right) = \sum_{j=k}^{n} |f_{\alpha_j}|^2$$

which tends, by (4) proved above, to 0 as $k \to \infty$. We set $\text{s-}\lim_{n\to\infty} \sum_{j=1}^{n} f_{\alpha_j} x_{\alpha_j} = f'$, and shall prove that $(f - f')$ is orthogonal to every vector of S. By the continuity of the scalar product, we have

$$(f - f', x_{\alpha_j}) = \lim_{n\to\infty} \left(f - \sum_{k=1}^{n} f_{\alpha_k} x_{\alpha_k}, x_{\alpha_j} \right) = f_{\alpha_j} - f_{\alpha_j} = 0,$$

and, when $\alpha \neq \alpha_j$ $(j = 1, 2, \ldots)$,

$$(f - f', x_\alpha) = \lim_{n\to\infty} \left(f - \sum_{k=1}^{n} f_{\alpha_k} x_{\alpha_k}, x_\alpha \right) = 0 - 0 = 0.$$

Thus, by the completeness of the orthonormal system $S = \{x_\alpha\}$, we must have $(f - f') = 0$. Hence, by (4) and the continuity of the norm, we have

$$0 = \lim_{n\to\infty} \left\| f - \sum_{j=1}^{n} f_{\alpha_j} x_{\alpha_j} \right\|^2 = \|f\|^2 - \lim_{n\to\infty} \sum_{j=1}^{n} |f_{\alpha_j}|^2 = \|f\|^2 - \sum_{\alpha \in A} |f_\alpha|^2.$$

Corollary 1. We have the *Fourier expansion*

$$f = \sum_{j=1}^{\infty} f_{\alpha_j} x_{\alpha_j} = \text{s-}\lim_{n\to\infty} \sum_{j=1}^{n} f_{\alpha_j} x_{\alpha_j}. \quad (5)$$

Corollary 2. Let $l^2(A)$ be the space $L^2(A, \mathfrak{B}, m)$ where $m(\{\alpha\}) = 1$ for every point α of A. Then the Hilbert space X is *isometrically isomorphic* to the Hilbert space $l^2(A)$ by the correspondence

$$X \ni f \leftrightarrow \{f_\alpha\} \in l^2(A) \tag{6}$$

in the sense that

$$(f+g) \leftrightarrow \{f_\alpha + g_\alpha\},\ \beta f \leftrightarrow \{\beta f_\alpha\} \text{ and } \|f\|^2 = \|\{f_\alpha\}\|^2 = \sum_{\alpha\in A} |f_\alpha|^2. \tag{7}$$

Example. $\left\{\frac{1}{\sqrt{2\pi}}\, e^{int};\ n = 0, \pm 1, \pm 2, \ldots\right\}$ is a complete orthonormal system in the Hilbert space $L^2(0, 2\pi)$.

Proof. We have only to prove the completeness of this system. We have, by (3),

$$\left\| f - \sum_{j=-n}^{n} \frac{1}{2\pi} c_j\, e^{ijt} \right\|^2 \geqq \left\| f - \sum_{j=-n}^{n} \frac{1}{2\pi} f_j\, e^{ijt} \right\|^2 = \|f\|^2 - \sum_{j=-n}^{n} |f_j|^2,$$

where $f_j = (f, e^{ijt})$.

If $f \in L^2(0, 2\pi)$ is continuous and with period 2π, then the left hand side of the inequality above may be taken arbitrarily small by virtue of Weierstrass' trigonometric approximation theorem (see Chapter 0, 2). Thus the set of all the linear combinations $\sum_j c_j\, e^{ijt}$ is dense, in the sense of the norm, in the subspace of $L^2(0, 2\pi)$ consisting of all continuous functions with period 2π. Such a subspace is also dense, in the sense of the norm, in the space $L^2(0, 2\pi)$. Therefore, any function $f \in L^2(0, 2\pi)$, orthogonal to all the functions of $\left\{\frac{1}{\sqrt{2\pi}}\, e^{int}\right\}$ must be a zero vector of $L^2(0, 2\pi)$. This proves that our system of functions $\frac{1}{\sqrt{2\pi}}\, e^{int}$ is a complete orthonormal system of $L^2(0, 2\pi)$.

5. E. Schmidt's Orthogonalization

Theorem (E. Schmidt's orthogonalization). Given a finite or countably infinite sequence $\{x_j\}$ of linearly independent vectors of a pre-Hilbert space X. Then we can construct an orthonormal set having the same cardinal number as the set $\{x_j\}$ and spanning the same linear subspace as $\{x_j\}$.

Proof. Certainly $x_1 \neq 0$. We define $y_1, y_2, \ldots$ and $u_1, u_2, \ldots$ recurrently as follows:

$$\begin{aligned}
y_1 &= x_1, & u_1 &= y_1/\|y_1\|,\\
y_2 &= x_2 - (x_2, u_1)\, u_1, & u_2 &= y_2/\|y_2\|,\\
&\cdots & &\cdots\\
y_{n+1} &= x_{n+1} - \sum_{j=1}^{n} (x_{n+1}, u_j)\, u_j, & u_{n+1} &= y_{n+1}/\|y_{n+1}\|,\\
&\cdots & &\cdots
\end{aligned}$$

This process terminates if $\{x_j\}$ is a finite set. Otherwise it continues indefinitely. Observe that $y_n \neq 0$, because $x_1, x_2, \ldots, x_n$ are linearly independent. Thus u_n is well defined. It is clear, by induction, that each u_n is a linear combination of $x_1, x_2, \ldots, x_n$ and that each x_n is a linear combination of $u_1, u_2, \ldots, u_n$. Thus the closed linear subspace spanned by the u's is the same as that spanned by the x's.

We see, by $\|u_1\| = 1$, that $y_2 \perp u_1$ and hence $u_2 \perp u_1$. Thus, by $\|u_1\| = 1$, $y_3 \perp u_1$ and hence $u_3 \perp u_1$. Repeating the argument, we see that u_1 is orthogonal to $u_2, u_3, \ldots, u_n, \ldots$ Thus, by $\|u_2\| = 1$, we have $y_3 \perp u_2$ and so $u_3 \perp u_2$. Repeating the argument, we finally see that $u_k \perp u_m$ whenever $k > m$. Therefore $\{u_j\}$ constitutes an orthonormal set.

Corollary. Let a Hilbert space X be *separable*, i.e., let X have a dense subset which is at most countable. Then X has a complete orthonormal system consisting of an at most countable number of elements.

Proof. Suppose that an at most countable sequence $\{a_j\}$ of vectors $\in X$ be dense in X. Let x_1 be the first non-zero element in the sequence $\{a_j\}$, x_2 the first a_i which is not in the closed subspace spanned by x_1, and x_n the first a_i which is not in the closed subspace spanned by $x_1, x_2, \ldots, x_{n-1}$. It is clear that the a's and the x's span the same closed linear subspace. In fact, this closed linear subspace is the whole space X, because the set $\{a_j\}$ is dense in X. Applying Schmidt's orthogonalization to $\{x_j\}$, we obtain an orthonormal system $\{u_j\}$ which is countable and spans the whole space X.

This system $\{u_j\}$ is complete, since otherwise there would exist a non-zero vector orthogonal to every u_j and hence orthogonal to the space X spanned by u_j's.

Example of Orthogonalization. Let S be the interval (a, b), and consider the real Hilbert space $L^2(S, \mathfrak{B}, m)$, where $\mathfrak{B}$ is the set of all Baire subsets in (a, b). If we orthogonalize the set of monomials

$$1, s, s^2, s^3, \ldots, s^n, \ldots,$$

we get the so-called *Tchebyschev system of polynomials*

$$P_0(s) = \text{constant}, \; P_1(s), P_2(s), P_3(s), \ldots, P_n(s), \ldots$$

which satisfies

$$\int_a^b P_i(s)\, P_j(s)\, m(ds) = \delta_{ij} \;\; (= 0 \text{ or } 1 \text{ according as } i \neq j \text{ or } i = j).$$

In the particular case when $a = -1$, $b = 1$ and $m(ds) = ds$, we obtain the *Legendre polynomials*; when $a = -\infty$, $b = \infty$ and $m(ds) = e^{-s^2}\, ds$, we obtain the *Hermite polynomials* and finally when $a = 0$, $b = \infty$ and $m(ds) = e^{-s}\, ds$, we obtain the *Laguerre polynomials*.

It is easy to see that, when $-\infty < a < b < \infty$, the orthonormal system $\{P_j(s)\}$ is complete. For, we may follow the proof of the completeness of the trigonometric system (see the Example in the preceding section 4); we shall appeal to Weierstrass' polynomial approximation theorem, in place of Weierstrass' trigonometric approximation theorem. As to the completeness proof of the Hermite or Laguerre polynomials, we refer the reader to G. SZEGÖ [1] or to K. YOSIDA [1].

6. F. Riesz' Representation Theorem

Theorem (F. Riesz' representation theorem). Let X be a Hilbert space and f a bounded linear functional on X. Then there exists a uniquely determined vector y_f of X such that

$$f(x) = (x, y_f) \text{ for all } x \in X, \text{ and } \|f\| = \|y_f\|. \tag{1}$$

Conversely, any vector $y \in X$ defines a bounded linear functional f_y on X by

$$f_y(x) = (x, y) \text{ for all } x \in X, \text{ and } \|f_y\| = \|y\|. \tag{2}$$

Proof. The uniqueness of y_f is clear, since $(x, z) = 0$ for all $x \in X$ implies $z = 0$. To prove its existence, consider the null space $N = N(f) = \{x \in X; f(x) = 0\}$ of f. Since f is continuous and linear, N is a closed linear subspace. The Theorem is trivial in the case when $N = X$; we take in this case, $y_f = 0$. Suppose $N \neq X$. Then there exists a $y_0 \neq 0$ which belongs to $N^\perp$ (see Theorem 1 in Chapter III, 1). Define

$$y_f = (\overline{f(y_0)}/\|y_0\|^2)\, y_0. \tag{3}$$

We will show that this y_f meets the condition of the Theorem. First, if $x \in N$, then $f(x) = (x, y_f)$ since both sides vanish. Next, if x is of the form $x = \alpha y_0$, then we have

$$(x, y_f) = (\alpha y_0, y_f) = \left(\alpha y_0, \frac{\overline{f(y_0)}}{\|y_0\|^2} y_0\right) = \alpha f(y_0) = f(\alpha y_0) = f(x).$$

Since $f(x)$ and (x, y_f) are both linear in x, the equality $f(x) = (x, y_f)$, $x \in X$, is proved if we have proved that X is spanned by N and y_0. To show the last assertion, we write, remembering that $f(y_f) \neq 0$,

$$x = \left(x - \frac{f(x)}{f(y_f)} y_f\right) + \frac{f(x)}{f(y_f)} y_f.$$

The first term on the right is an element of N, since

$$f\left(x - \frac{f(x)}{f(y_f)} y_f\right) = f(x) - \frac{f(x)}{f(y_f)} f(y_f) = 0.$$

We have thus proved the representation $f(x) = (x, y_f)$.

Therefore, we have

$$\|f\| = \sup_{\|x\|\leq 1} |f(x)| = \sup_{\|x\|\leq 1} |(x, y_f)| \leqq \sup_{\|x\|\leq 1} \|x\| \cdot \|y_f\| = \|y_f\|,$$

and also $\|f\| = \sup_{\|x\|\leq 1} |f(x)| \geqq |f(y_f/\|y_f\|)| = \left(\frac{y_f}{\|y_f\|}, y_f\right) = \|y_f\|$.

Hence we have proved the equality $\|f\| = \|y_f\|$.

Finally, the converse part of the Theorem is clear from $|f_y(x)| = |(x, y)| \leqq \|x\| \cdot \|y\|$.

Corollary 1. Let X be a Hilbert space. Then the totality X' of bounded linear functionals on X constitutes also a Hilbert space, and there is a norm-preserving, one-to-one correspondence $f \leftrightarrow y_f$ between X' and X. By this correspondence, X' may be identified with X as an abstract set; but it is not allowed to identify, by this correspondence, X' with X as linear spaces, since the correspondence $f \leftrightarrow y_f$ is *conjugate linear*:

$$(\alpha_1 f_1 + \alpha_2 f_2) \leftrightarrow (\bar{\alpha}_1 y_{f_1} + \bar{\alpha}_2 y_{f_2}), \tag{4}$$

where α_1, α_2 are complex numbers.

Proof. It is easily verified that X' is made into a Hilbert space by defining its scalar product through $(f_1, f_2) = \overline{(y_{f_1}, y_{f_2})}$, so that the statement of Corollary 1 is clear.

Corollary 2. Any continuous linear functional T on the Hilbert space X' is thus identified with a uniquely determined element t of X as follows:

$$T(f) = f(t) \quad \text{for all} \quad f \in X'. \tag{5}$$

Proof. Clear from the fact that the product of two conjugate linear transformations is a linear transformation.

Definition. The space X' is called the *dual space* of X. We can thus identify a Hilbert space X with its *second dual* $X'' = (X')'$ in the above sense. This fact will be referred to as the *reflexivity* of Hilbert spaces.

Corollary 3. Let X be a Hilbert space, and X' its dual Hilbert space. Then, for any subset F of X' which is dense in the Hilbert space X', we have

$$\|x_0\| = \sup_{f\in F, \|f\|\leq 1} |f(x_0)|, \ x_0 \in X. \tag{6}$$

Proof. We may assume that $x_0 \neq 0$, otherwise the formula (6) is trivial. We have $(x_0, x_0/\|x_0\|) = \|x_0\|$, and so there exists a bounded linear functional f_0 on X such that $\|f_0\| = 1, f_0(x_0) = \|x_0\|$. Since $f(x_0) = (x_0, y_f)$ is continuous in y_f, and since the correspondence $f \leftrightarrow y_f$ is norm-preserving we see that (6) is true by virtue of the denseness of F in X'.

Remark. Hilbert's original definition of the "Hilbert space" is the space (l^2). See his paper [1]. It was J. von Neumann [1] who gave an axiomatic definition (see Chapter I, 9) of the Hilbert space assuming that the space is separable. F. RIESZ [1] proved the above representation theorem without assuming the separability of the Hilbert space. In this paper, he stressed that the whole theory of the Hilbert space may be founded upon this representation theorem.

7. The Lax-Milgram Theorem

Of recent years, it has been proved that a variant of F. Riesz' representation theorem, formulated by P. LAX and A. N. MILGRAM [1], is a useful tool for the discussion of the existence of solutions of linear partial differential equations of elliptic type.

Theorem (LAX-MILGRAM). Let X be a Hilbert space. Let $B(x, y)$ be a complex-valued functional defined on the product Hilbert space $X \times X$ which satisfies the conditions:

Sesqui-linearity, i.e.,

$$B(\alpha_1 x_1 + \alpha_2 x_2, y) = \alpha_1 B(x_1, y) + \alpha_2 B(x_2, y) \text{ and } \tag{1}$$

$$B(x, \beta_1 y_1 + \beta_2 y_2) = \bar{\beta}_1 B(x, y_1) + \bar{\beta}_2 B(x, y_2),$$

Boundedness, i.e., there exists a positive constant γ such that

$$|B(x, y)| \leqq \gamma \, \|x\| \cdot \|y\|, \tag{2}$$

Positivity, i.e., there exists a positive constant δ such that

$$B(x, x) \geqq \delta \, \|x\|^2. \tag{3}$$

Then there exists a uniquely determined bounded linear operator S with a bounded linear inverse S^{-1} such that

$$(x, y) = B(x, Sy) \text{ whenever } x \text{ and } y \in X, \text{ and } \|S\| \leqq \delta^{-1}, \; \|S^{-1}\| \leqq \gamma. \tag{4}$$

Proof. Let D be the totality of elements $y \in X$ for which there exists an element y^* such that $(x, y) = B(x, y^*)$ for all $x \in X$. D is not void, because $0 \in D$ with $0^* = 0 \cdot y^*$ is uniquely determined by y. For, if w be such that $B(x, w) = 0$ for all x, then $w = 0$ by $0 = B(w, w) \geqq \delta \, \|w\|^2$. By the sesqui-linearity of (x, y) and $B(x, y)$, we obtain a linear operator S with domain $D(S) = D$: $Sy = y^*$. S is continuous and $\|Sy\| \leqq \delta^{-1} \|y\|$, $y \in D(S)$, because

$$\delta \, \|Sy\|^2 \leqq B(Sy, Sy) = (Sy, y) \leqq \|Sy\| \cdot \|y\|.$$

Moreover, $D = D(S)$ is a closed linear subspace of X. Proof: if $y_n \in D(S)$ and $s\text{-}\lim_{n\to\infty} y_n = y_\infty$, then, by the continuity of S proved above, $\{Sy_n\}$ is a Cauchy sequence and so has a limit $z = s\text{-}\lim Sy_n$. By the continuity of the scalar product, we have $\lim_{n\to\infty} (x, y_n) = (x, y_\infty)$. We have also, by

(2), $\lim_{n\to\infty} B(x, S y_n) = B(x, z)$. Thus, by $(x, y_n) = B(x, S y_n)$, we must have $(x, y_\infty) = B(x, z)$ which proves that $y_\infty \in D$ and $S y_\infty = z$.

Therefore the first part of the Theorem, that is, the existence of the operator S is proved if we can show that $D(S) = X$. Suppose $D(S) \neq X$. Then there exists a $w_0 \in X$ such that $w_0 \neq 0$ and $w_0 \in D(S)^\perp$. Consider the linear functional $F(z) = B(z, w_0)$ defined on X. $F(z)$ is continuous, since $|F(z)| = |B(z, w_0)| \leqq \gamma \|z\| \cdot \|w_0\|$. Thus, by F. Riesz' representation theorem, there exists a $w_0' \in X$ such that $B(z, w_0) = F(z) = (z, w_0')$ for all $z \in X$. This proves that $w_0' \in D(S)$ and $S w_0' = w_0$. But, by $\delta \|w_0\|^2 \leqq B(w_0, w_0) = (w_0, w_0') = 0$, we obtain $w_0 = 0$ which is a contradiction.

The inverse S^{-1} exists. For, $Sy = 0$ implies $(x, y) = B(x, Sy) = 0$ for all $x \in X$ and so $y = 0$. As above, we show that, for every $y \in X$ there exists a y' such that $(z, y') = B(z, y)$ for all $z \in X$. Hence $y = Sy'$ and so S^{-1} is an everywhere defined operator, and, by $|(z, S^{-1}y)| = |B(z, y)| \leqq \gamma \|z\| \cdot \|y\|$, we see that $\|S^{-1}\| \leqq \gamma$.

Concrete applications of the Lax-Milgram theorem will be given in later chapters. In the following four sections, we shall give some examples of the direct application of F. Riesz' representation theorem.

8. A Proof of the Lebesgue-Nikodym Theorem

This theorem reads as follows.

Theorem (Lebesgue-Nikodym). Let $(S, \mathfrak{B}, m)$ be a measure space, and $\nu(B)$ be a σ-finite, σ-additive and non-negative measure defined on $\mathfrak{B}$. If ν is m-absolutely continuous, then there exists a non-negative, m-measurable function $p(s)$ such that

$$\nu(B) = \int_B p(s)\, m(ds) \quad \text{for all} \quad B \in \mathfrak{B} \text{ with } \nu(B) < \infty. \tag{1}$$

Moreover, the "density" $p(s)$ of $\nu(B)$ (with respect to $m(B)$) is uniquely determined in the sense that any two of them are equal m-a.e.

Proof (due to J. von Neumann [2]). It is easy to see that $\varrho(B) = m(B) + \nu(B)$ is a σ-finite, σ-additive and non-negative measure defined on $\mathfrak{B}$. Let $\{B_n\}$ be a sequence of sets $\in \mathfrak{B}$ such that $S = \bigcup_{n=1}^{\infty} B_n$, $B_n \subseteqq B_{n+1}$ and $\varrho(B_n) < \infty$ for $n = 1, 2, \ldots$ If we can prove the theorem for every $B \subseteqq B_n$ (for fixed n) and obtain the density $p_n(s)$, then the Theorem is true. For, we have only to take $p(s)$ as follows:

$$p(s) = p_1(s) \text{ for } s \in B_1, \text{ and } p(s) = p_{n+1}(s) \text{ for } s \in B_{n+1} - B_n$$
$$(n = 1, 2, \ldots).$$

Therefore we may assume that $\varrho(S) < \infty$ without losing the generality. Consider the Hilbert space $L^2(S, \mathfrak{B}, \varrho)$. Then

$$f(x) = \int_S x(s)\, \nu(ds), \quad x \in L^2(S, \mathfrak{B}, \varrho),$$

gives a bounded linear functional on $L^2(S, \mathfrak{B}, \varrho)$, because

$$|f(x)| \leqq \int_S |x(s)|\, \nu(ds) \leqq \Big(\int_S |x(s)|^2\, \nu(ds)\Big)^{1/2} \Big(\int_S 1 \cdot \nu(ds)\Big)^{1/2}$$
$$\leqq \|x\|_\varrho \cdot \nu(S)^{1/2},$$

where $\|x\|_\varrho = \Big(\int_S |x(s)|^2\, \varrho(ds)\Big)^{1/2}$. Thus, by F. Riesz' representation theorem, there exists a uniquely determined $y \in L^2(S, \mathfrak{B}, \varrho)$ such that

$$\int_S x(s)\, \nu(ds) = \int_S x(s)\, \overline{y(s)}\, \varrho(ds) = \int_S x(s)\, \overline{y(s)}\, m(ds) + \int_S x(s)\, \overline{y(s)}\, \nu(ds)$$

holds for all $x \in L^2(S, \mathfrak{B}, \varrho)$. Taking x as non-negative functions and considering the real part of both sides, we may assume that $y(s)$ is a real-valued function. Hence

$$\int_S x(s)\,(1 - y(s))\, \nu(ds) = \int_S x(s)\, y(s)\, m(ds) \quad \text{if } x(s) \in L^2(S, \mathfrak{B}, \varrho) \tag{2}$$

is non-negative.

We can prove $0 \leqq y(s) < 1$ ϱ-a.e. To this purpose, set $E_1 = \{s; y(s) < 0\}$ and $E_2 = \{s; y(s) \geqq 1\}$. If we take the defining function $C_{E_1}(s)$ of E_1 for $x(s)$ in (2), then the left hand side is $\geqq 0$ and hence $\int_{E_1} y(s)\, m(ds) \geqq 0$. Thus we must have $m(E_1) = 0$, and so, by the m-absolute continuity of ν, $\nu(E_1) = 0$, $\varrho(E_1) = 0$. We may also prove $\varrho(E_2) = 0$, by taking the defining function $C_{E_2}(s)$ for $x(s)$ in (2). Therefore $0 \leqq y(s) < 1$ ϱ-a.e. on S.

Let $x(s)$ be $\mathfrak{B}$-measurable and $\geqq 0$ ϱ-a.e. Then, by $\varrho(S) < \infty$, the "truncated functions" $x_n(s) = \min(x(s), n)$ belong to $L^2(S, \mathfrak{B}, \varrho)$ $(n = 1, 2, \ldots)$, and hence

$$\int_S x_n(s)\,(1 - y(s))\, \nu(ds) = \int_S x_n(s)\, y(s)\, m(ds) \quad (n = 1, 2, \ldots). \tag{3}$$

Since the integrals increase monotonely as n increases, we have

$$\lim_{n\to\infty} \int_S x_n(s)\,(1 - y(s))\, \nu(ds) = \lim_{n\to\infty} \int_S x_n(s)\, y(s)\, m(ds) = L \leqq \infty. \tag{4}$$

Since the integrands are $\geqq 0$ ϱ-a.e., we have, by the Lebesgue-Fatou Lemma,

$$\begin{aligned} L &\geqq \int_S \lim_{n\to\infty} (x_n(s)\,(1 - y(s))\, \nu(ds) = \int_S x(s)\,(1 - y(s))\, \nu(ds), \\ L &\geqq \int_S \lim_{n\to\infty} (x_n(s)\, y(s))\, m(ds) = \int_S x(s)\, y\, m(ds), \end{aligned} \tag{5}$$

under the convention that if $x(s)(1-y(s))$ is not ν-integrable, the right hand side is equal to ∞; and the same convention for $x(s)\,y(s)$. If $x(s)\,y(s)$ is m-integrable, then, by the Lebesgue-Fatou Lemma,

$$L \leqq \int_S \overline{\lim_{n\to\infty}}\,(x_n(s)\,y(s))\,m(ds) = \int_S x(s)\,y(s)\,m(ds). \tag{6}$$

This formula is true even if $x(s)\,y(s)$ is not m-integrable, under the convention that then $L=\infty$. Under a similar convention, we have

$$L \leqq \int_S x(s)\,(1-y(s))\,\nu(ds). \tag{7}$$

Therefore, we have

$$\int_S x(s)\,(1-y(s))\,\nu(ds) = \int_S x(s)\,y(s)\,m(ds) \text{ for every } x(s) \text{ which is } \mathfrak{B}\text{-measurable and } \geqq 0\ \varrho\text{-a.e.}, \tag{8}$$

under the convention that, if either side of the equality is $=\infty$ then the other side is also $=\infty$.

Now we put

$$x(s)\,(1-y(s)) = z(s),\quad y(s)\,(1-y(s))^{-1} = p(s).$$

Then, under the same convention as in (8), we have

$$\int_S z(s)\,\nu(ds) = \int_S z(s)\,p(s)\,m(ds) \text{ if } z(s) \text{ is } \mathfrak{B}\text{-measurable and } \geqq 0\ \varrho\text{-a.e.} \tag{9}$$

If we take the defining function $C_B(s)$ of $B\in\mathfrak{B}$ for $z(s)$, we obtain

$$\nu(B) = \int_B p(s)\,m(ds) \text{ for all } B\in\mathfrak{B}.$$

The last part of the Theorem is clear in view of definition (1).

Reference. For a straightforward proof of the Lebesgue-Nikodym theorem based upon Hahn's decomposition (Theorem 3 in Chapter I, 3) see K. Yosida [2]. This proof is reproduced in Halmos [1], p. 128. See also Saks [1] and Dunford-Schwartz [1].

9. The Aronszajn-Bergman Reproducing Kernel

Let A be an abstract set, and let a system X of complex-valued functions defined on A constitute a Hilbert space by the scalar product

$$(f, g) = (f(a), g(a))_a. \tag{1}$$

A complex-valued function $K(a, b)$ defined on $A\times A$ is called a *reproducing kernel* of X if it satisfies the condition:

$$\text{For any fixed } b,\ K(a,b)\in X \text{ as a function of } a, \tag{2}$$

$$f(b) = (f(a), K(a,b))_a \text{ and hence } \overline{f(b)} = (K(a,b), f(a))_a. \tag{3}$$

As for the existence of reproducing kernels, we have

Theorem 1 (N. ARONSZAJN [1], S. BERGMAN [1]). X has a reproducing kernel K iff there exists, for any $y_0 \in A$, a positive constant C_{y_0}, depending upon y_0, such that

$$|f(y_0)| \leqq C_{y_0} \|f\| \quad \text{for all} \quad f \in X. \tag{4}$$

Proof. The "only if" part is proved by applying Schwarz' inequality to $f(y_0) = (f(x), K(x, y_0))_x$:

$$|f(y_0)| \leqq \|f\| \cdot (K(x, y_0), K(x, y_0))_x^{1/2} = \|f\| \, K(y_0, y_0)^{1/2}. \tag{5}$$

The "if" part is proved by applying F. Riesz' representation theorem to the linear functional $F_{y_0}(f) = f(y_0)$ of $f \in X$. Thus, there exists a uniquely determined vector $g_{y_0}(x)$ of X such that, for every $f \in X$,

$$f(y_0) = F_{y_0}(f) = (f(x), g_{y_0}(x))_x,$$

and so $g_{y_0}(x) = K(x, y_0)$ is a reproducing kernel of X. The proof shows that the reproducing kernel is uniquely determined.

Corollary. We have

$$\sup_{\|f\| \leqq 1} |f(y_0)| = K(y_0, y_0)^{1/2}, \tag{6}$$

the supremum being attained by

$$f_0(x) = \varrho K(x, y_0)/K(y_0, y_0)^{1/2}, \quad |\varrho| = 1. \tag{7}$$

Proof. The equality in the Schwarz' inequality (5) holds iff $f(x)$ and $K(x, y_0)$ are linearly dependent. From the two conditions $f(x) = \alpha K(x, y_0)$ and $\|f\| = 1$, we obtain

$$1 = |\alpha| (K(x, y_0), K(x, y_0))_x^{1/2} = |\alpha| K(y_0, y_0)^{1/2}, \text{ that is, } |\alpha| = K(y_0, y_0)^{-1/2}.$$

Hence the equality sign in (5) is attained by $f_0(x)$.

Example. Consider the Hilbert space $A^2(G)$. For any $f \in A^2(G)$ and $z \in G$, we have (see (4) in Chapter I, 9)

$$|f(z_0)|^2 \leqq (\pi r^2)^{-1} \int_{|z - z_0| \leqq r} |f(z)|^2 \, dx \, dy \quad (z = x + iy).$$

Thus $A^2(G)$ has the reproducing kernel which will be denoted by $K_G(z, z')$. This $K_G(z, z')$ is called *Bergman's kernel* of the domain G of the complex plane. The following theorem of Bergman illustrates the meaning of $K_G(z, z')$ in the theory of conformal mapping.

Theorem 2. Let G be a simply connected bounded open domain of the complex plane, and z_0 be any point of G. By Riemann's theorem, there exists a uniquely determined regular function $w = f(z; z_0)$ of z which gives a one-to-one conformal map of the domain G onto the sphere $|w| < \varrho_G$ of the complex w-plane in such a way that

$$f_0(z_0; z_0) = 0, \quad (df_0(z; z_0)/dz)_{z = z_0} = 1.$$

Bergman's kernel $K_G(z; z_0)$ is connected with $f_0(z; z_0)$ by

$$f_0(z; z_0) = K_G(z_0; z_0)^{-1} \int_{z_0}^{z} K_G(t; z_0)\, dt, \tag{8}$$

where the integral is taken along any rectifiable curve lying in G and connecting z_0 with z.

Proof. We set

$A_1^2(G) = \{f(z); f(z)$ is holomorphic in $G, f'(z) \in A^2(G), f(z_0) = 0$ and $f'(z_0) = 1\}$,

and consider, for any $f \in A_1^2(G)$, the number

$$\|f'\|^2 = \int_G |f'(z)|^2\, dx\, dy, \quad z = x + iy. \tag{9}$$

If we denote by $z = \varphi(w)$ the inverse function of $w = f_0(z; z_0)$, then, for any $f \in A_1^2(G)$,

$$\|f'\|^2 = \iint_{|w|<\varrho_G} |f'(\varphi(w))|^2\, |\varphi'(w)|^2\, du\, dv, \quad w = u + iv.$$

For, by the Cauchy-Riemann partial differential equations

$$x_u = y_v, \quad x_v = -y_u,$$

we have

$$\begin{aligned} dx\, dy &= \frac{\partial(x, y)}{\partial(u, v)}\, du\, dv = (x_u y_v - y_u x_v)\, du\, dv = (x_u^2 + y_u^2)\, du\, dv \\ &= |\varphi'(w)|^2\, du\, dv. \end{aligned}$$

Let $f \in A_1^2(G)$, and let $F(w) = f(\varphi(w))$ be expanded into power series:

$$F(w) = f(\varphi(w)) = w + \sum_{n=2}^{\infty} c_n w^n \quad \text{for } |w| < \varrho_G.$$

Then $F'(w) = f'(\varphi(w))\, \varphi'(w) = 1 + \sum_{n=2}^{\infty} n\, c_n w^{n-1}$ and so

$$\begin{aligned} \|f'\|^2 &= \iint_{|w|<\varrho_G} \left|1 + \sum_{n=2}^{\infty} n\, c_n w^{n-1}\right|^2 du\, dv \\ &= \int_0^{\varrho_G} dr \left\{\int_0^{2\pi} \left(r + \sum_{n=2}^{\infty} n^2\, |c_n|^2\, r^{2n-1}\right) d\theta\right\} = \pi \varrho_G^2 + \sum_{n=2}^{\infty} \pi n\, |c_n|^2\, \varrho_G^{2n}. \end{aligned}$$

Therefore $\underset{f \in A_1^2(G)}{\text{minimum}} \|f'\| = \sqrt{\pi}\, \varrho_G$, and this minimum is attained iff $F(w) = f(\varphi(w)) = w$, that is, iff $f(z) = f_0(z; z_0)$.

We set, for any $f \in A_1^2(G)$, $g(z) = f(z)/\|f'\|$. Then $\|g'\| = 1$. If we put

$$\tilde{A}^2(G) = \{g(z);\ g(z) \text{ is holomorphic in } G,\ g(z_0) = 0,\ g'(z_0) > 0 \text{ and } \|g'\| = 1\},$$

then the above remark shows that

$$\underset{g \in \tilde{A}^2(G)}{\text{maximum}}\, g'(z_0) = 1/\|f_0'\| = (\sqrt{\pi}\, \varrho_G)^{-1},$$

and this maximum is attained iff $g(z)$ is equal to

$$g_0(z) = f_0(z; z_0)/\|f_0'\| = f_0(z; z_0) / \sqrt{\pi}\, \varrho_G.$$

Hence, by (7), we obtain

$$g_0'(z) = (\sqrt{\pi}\, \varrho_G)^{-1} \frac{d f_0(z; z_0)}{dz} = \lambda\, K_G(z; z_0)/K_G(z_0; z_0)^{1/2}, \quad |\lambda| = 1.$$

Hence, by putting $z = z_0$, we have

$$(\lambda \sqrt{\pi}\, \varrho_G)^{-1} = K_G(z_0; z_0)/K_G(z_0; z_0)^{1/2} = K_G(z_0; z_0)^{1/2},$$

and so we have proved the formula

$$\frac{d f_0(z; z_0)}{dz} = K_G(z; z_0)/K_G(z_0; z_0).$$

10. The Negative Norm of P. Lax

Let $H_0^s(\Omega)$ be the completion of the pre-Hilbert space $C_0^\infty(\Omega)$ endowed with the scalar product (φ, ψ) and the norm $\|\varphi\|_s$:

$$(\varphi, \psi)_s = \sum_{|j| \leqq s} \int_\Omega D^j \varphi(x) \overline{D^j \psi(x)}\, dx, \quad \|\varphi\|_s = (\varphi, \varphi)_s^{1/2}. \tag{1}$$

Any element $b \in H_0^0(\Omega) = L^2(\Omega)$ defines a continuous linear functional f_b on $H_0^s(\Omega)$ by

$$f_b(w) = (w, b)_0, \quad w \in H_0^s(\Omega). \tag{2}$$

For, by Schwarz' inequality, we have

$$|(w, b)_0| \leqq \|w\|_0 \cdot \|b\|_0 \leqq \|w\|_s \cdot \|b\|_0.$$

Therefore, if we define the *negative norm* of $b \in H_0^0(\Omega) = L^2(\Omega)$ by

$$\|b\|_{-s} = \sup_{w \in H_0^s(\Omega), \|w\|_s \leqq 1} |f_b(w)| = \sup_{w \in H_0^s(\Omega), \|w\|_s \leqq 1} |(w, b)_0|, \tag{3}$$

then we have

$$\|b\|_{-s} \leqq \|b\|_0, \tag{4}$$

and so, by $\|b\|_{-s} \geqq |(w/\|w\|_s, b)_0|$,

$$|(w, b)_0| \leqq \|w\|_s \cdot \|b\|_{-s}. \tag{5}$$

Hence we may write

$$\|b\|_{-s} = \|f_b\|_{-s} = \sup_{\|w\|_s \leqq 1} |(w, b)_0| \text{ for any } b \in H_0^0(\Omega). \tag{3'}$$

We shall prove

Theorem 1 (P. Lax [2]). The dual space $H_0^s(\Omega)'$ of the space $H_0^s(\Omega)$ may be identified with the completion of the space $H_0^0(\Omega) = L^2(\Omega)$ with respect to the negative norm.

For the proof we prepare

Proposition. The totality F of the continuous linear functionals on $H_0^s(\Omega)$ of the form f_b is dense in the Hilbert space $H_0^s(\Omega)'$, which is the dual space of $H_0^s(\Omega)$.

Proof. F is *total* on $H_0^s(\Omega)$ in the sense that, for a fixed $w \in H_0^s(\Omega)$, the simultaneous vanishing $f_b(w) = 0$, $b \in H_0^0(\Omega)$, takes place only if $w = 0$. This is clear since any element $w \in H_0^s(\Omega)$ is also an element of $H_0^0(\Omega)$.

Now if F is not dense in the Hilbert space $H_0^s(\Omega)'$, then there exists an element $T \neq 0$ of the second dual space $H_0^s(\Omega)'' = (H_0^s(\Omega)')'$ such that $T(f_b) = 0$ for all $f_b \in F$. By the reflexivity of the Hilbert space $H_0^s(\Omega)$, there exists an element $t \in H_0^s(\Omega)$ such that $T(f) = f(t)$ for all $f \in H_0^s(\Omega)'$. Thus $T(f_b) = f_b(t) = 0$ for all $b \in H_0^0(\Omega)$. This implies, by the totality of F proved above, that $t = 0$, contrary to $T \neq 0$.

Corollary. We have, dually to (3'),

$$||w||_s = \sup_{b \in H_0^0(\Omega), ||b||_{-s} \leq 1} |(w, b)_0| \text{ for any } w \in H_0^s(\Omega). \tag{6}$$

Proof. Clear from Corollary 3 in Chapter III, 6, because $F = \{f_b; b \in H_0^0(\Omega)\}$ is dense in $H_0^s(\Omega)'$.

Proof of Theorem 1. Clear from the facts that i) F is dense in the dual space $H_0^s(\Omega)'$ and ii) F is in one-to-one correspondence to the set $H_0^0(\Omega) = L^2(\Omega)$ preserving the negative norm, i.e.,

$$F \ni f_b \leftrightarrow b \in H_0^0(\Omega) \text{ and } ||f_b||_{-s} = ||b||_{-s}.$$

We shall denote by $H_0^{-s}(\Omega)$ the completion of $H_0^0(\Omega)$ with respect to the negative norm $||b||_{-s}$. Thus

$$H_0^s(\Omega)' = H_0^{-s}(\Omega). \tag{7}$$

For any continuous linear functional f on $H_0^s(\Omega)$, we shall denote by $\langle w, f \rangle$ the value of f at $w \in H_0^s(\Omega)$. Thus, for any $b \in H_0^0(\Omega)$, we may write

$$f_b(w) = (w, b)_0 = \langle w, f_b \rangle = \langle w, b \rangle, \quad w \in H_0^s(\Omega), \tag{8}$$

and have the generalized Schwarz' inequality

$$|\langle w, b \rangle| \leq ||w||_s \, ||b||_{-s}, \tag{9}$$

which is precisely (5).

Now we can prove

Theorem 2 (P. LAX [2]). Any continuous linear functional $g(b)$ on $H_0^{-s}(\Omega)$ can be represented, by a fixed element $w \in H_0^s(\Omega)$, as

$$g(b) = g_w(b) = \overline{\langle w, b \rangle}. \tag{10}$$

We have, in particular,

$$H_0^s(\Omega)' = H_0^{-s}(\Omega), \quad H_0^{-s}(\Omega)' = H_0^s(\Omega). \tag{11}$$

Proof. If $b \in H_0^0(\Omega)$, then $\langle w, b\rangle = f_b(w) = (w, b)_0$. Since $F = \{f_b; b \in H_0^0(\Omega)\}$ is dense in the Hilbert space $H_0^s(\Omega)'$, we know, by (9), that $\overline{\langle w, b\rangle} = (b, w)_0$ with a fixed $w \in H_0^s(\Omega)$, defines a linear functional g_w continuous on a dense subset F of $H_0^s(\Omega)'$. The norm of this functional g_w on F will be denoted by $\|g_w\|_s$. Then, by (6),

$$\|g_w\|_s = \sup_{\|b\|_{-s} \leq 1} |(b, w)_0| = \sup_{\|b\|_{-s} \leq 1} |(w, b)_0| = \|w\|_s. \tag{12}$$

We may thus extend, by continuity, the functional g_w on F to a continuous linear functional on the completion of F (with respect to the negative norm), that is, g_w can be extended to a continuous linear functional on $H_0^s(\Omega)' = H_0^{-s}(\Omega)$; we denote this extension by the same letter g_w. We have thus

$$\|g_w\| = \sup_{\|b\|_{-s} \leq 1} |g_w(b)| = \|w\|_s. \tag{13}$$

Hence, in view of the completeness of the space $H_0^s(\Omega)$, the totality G of the continuous linear functionals g_w on $H_0^{-s}(\Omega)$ may be considered as a closed linear subspace of $H_0^{-s}(\Omega)'$ by the correspondence $g_w \leftrightarrow w$. If this closed subspace G were not dense in $H_0^{-s}(\Omega)'$, then there exists a continuous linear functional $f \neq 0$ on $H_0^{-s}(\Omega)'$ such that $f(g_w) = 0$ for all $g_w \in G$. But, since the Hilbert space $H_0^{-s}(\Omega)$ is reflexive, such a functional f is given by $f(g_w) = g_w(f_0)$, $f_0 \in H_0^{-s}(\Omega)$, and so by **(13)** f_0 must be equal to 0, contrary to the fact $f \neq 0$. Therefore we have proved $H_0^{-s}(\Omega)' = H_0^s(\Omega)$.

Remark. The notion of the negative norm was introduced by P. Lax with the view of applying it to the genuine differentiability of distribution solutions of linear partial differential equations. We shall discuss such differentiability problems in later chapters. It is to be noted that the notion of the negative norm is also introduced naturally through the Fourier transform. This was done by J. Leray [1] earlier than Lax. We shall explain the point in a later chapter on the Fourier transform.

11. Local Structures of Generalized Functions

A generalized function is locally the distributional derivative of a function. More precisely, we can prove

Theorem (L. Schwartz [1]). Let T be a generalized function in $\Omega \subseteq R^n$. Then, for any compact subset K of Ω, there exist a positive integer $m_0 = m_0(T, K)$ and a function $f(x) = f(x; T, K, m_0) \in L^2(K)$ such that

$$T(\varphi) = \int_K f(x) \frac{\partial^{nm_0} \varphi(x)}{\partial x_1^{m_0} \partial x_2^{m_0} \cdots \partial x_n^{m_0}} dx \text{ whenever } \varphi \in \mathfrak{D}_K(\Omega). \tag{1}$$

Proof. By the Corollary in Chapter I, 8, there exist a positive constant C and a positive integer m such that

$$|T(\varphi)| \leqq C \sup_{|j| \leqq m, x \in K} |D^j \varphi(x)| \text{ whenever } \varphi \in \mathfrak{D}_K(\Omega). \tag{2}$$

Thus there exists a positive number δ such that

$$p_m(\varphi) = \sup_{|j| \leqq m, x \in K} |D^j \varphi(x)| \leqq \delta,\ \varphi \in \mathfrak{D}_K(\Omega), \text{ implies } |T(\varphi)| \leqq 1. \tag{3}$$

We introduce the notation

$$\frac{\partial^s}{\partial x^s} = \frac{\partial^{sn}}{\partial x_1^s \partial x_2^s \cdots \partial x_n^s}, \tag{4}$$

and prove that there exists a positive constant ε such that, for $m_0 = m + 1$,

$$\int_\Omega |\partial^{m_0} \varphi(x)/\partial x^{m_0}|^2 \, dx \leqq \varepsilon,\ \varphi \in \mathfrak{D}_K(\Omega), \text{ implies } p_m(\varphi) \leqq \delta. \tag{5}$$

This is proved by repeated application of the following inequality

$$|\psi(x)| \leqq \int_{K \frown (-\infty, x_i)} |\partial \psi(x_1, \ldots, x_{i-1}, y, x_{i+1}, \ldots, x_n)/\partial y| \, dy$$

$$\leqq \Big(\int_{K \frown (-\infty, x_i)} dy\Big)^{1/2} \Big(\int_{K \frown (-\infty, x_i)} |\partial \psi(x_1, \ldots, x_{i-1}, y, x_{i+1}, \ldots, x_n)/\partial y|^2 dy\Big)^{1/2}$$

$$= t^{1/2} \Big(\int_{K \frown (-\infty, x_i)} |\partial \psi(x_1, \ldots, x_{i-1}, y, x_{i+1}, \ldots, x_n)/\partial y|^2 \, dy\Big)^{1/2},$$

where t is the *diameter* of K, i.e., the maximum distance between two points of the compact set K.

Consider the mapping $\varphi(x) \to \psi(x) = \partial^{m_0} \varphi(x)/\partial x^{m_0}$ defined on $\mathfrak{D}_K(\Omega)$ into $\mathfrak{D}_K(\Omega)$. As may be seen by integration, $\psi(x) = 0$ implies $\varphi(x) = 0$. Hence the above mapping is one-to-one. Thus $T(\varphi)$, $\varphi \in \mathfrak{D}_K(\Omega)$, defines a linear functional $S(\psi)$, $\psi(x) = \partial^{m_0} \varphi(x)/\partial x^{m_0}$, by $S(\psi) = T(\varphi)$. By (3) and (5), S is a continuous linear functional on the pre-Hilbert space X consisting of such ψ's and topologized by the norm $\|\psi\| = \Big(\int_K |\psi(x)|^2 dx\Big)^{1/2}$.

Thus there exists, by F. Riesz' representation theorem, a uniquely determined function $f(x)$ from the completion of X such that

$$T(\varphi) = S(\psi) = \int_K (\partial^{m_0} \varphi(x)/\partial x^{m_0}) \cdot f(x) \, dx \text{ for all } \varphi \in \mathfrak{D}_K(\Omega).$$

Actually, the completion of X is contained in $L^2(K)$ as a closed linear subspace, and so the Theorem is proved.

References for Chapter III

For general account of Hilbert spaces, see N. I. ACHIESER-I. M. GLASMAN [1], N. DUNFORD-J. SCHWARTZ [2], B. SZ. NAGY [1], F. RIESZ-B. SZ. NAGY [3] and M. H. STONE [1].

IV. The Hahn-Banach Theorems

In a Hilbert space, we can introduce the notion of orthogonal coordinates through an orthogonal base, and these coordinates are the values of bounded linear functionals defined by the vectors of the base. This suggests that we consider continuous linear functionals, in a linear topological space, as generalized coordinates of the space. To ensure the existence of non-trivial continuous linear functionals in a general locally convex linear topological space, we must rely upon the Hahn-Banach extension theorems.

1. The Hahn-Banach Extension Theorem in Real Linear Spaces

Theorem (HAHN [2], BANACH [1]). Let X be a real linear space and let $p(x)$ be a real-valued function defined on X satisfying the conditions:

$$p(x+y) \leqq p(x) + p(y) \quad \text{(subadditivity)}, \tag{1}$$

$$p(\alpha x) = \alpha p(x) \quad \text{for} \quad \alpha \geqq 0. \tag{2}$$

Let M be a real linear subspace of X and f_0 a real-valued linear functional defined on M:

$$f_0(\alpha x + \beta y) = \alpha f_0(x) + \beta f_0(y) \quad \text{for} \quad x, y \in M \text{ and real } \alpha, \beta. \tag{3}$$

Let f_0 satisfy $f_0(x) \leqq p(x)$ on M. Then there exists a real-valued linear functional F defined on X such that i) F is an extension of f_0, i.e., $F(x) = f_0(x)$ for all $x \in M$, and ii) $F(x) \leqq p(x)$ on X.

Proof. Suppose first that X is spanned by M and an element $x_0 \in M$, that is, suppose that

$$X = \{x = m + \alpha x_0;\ m \in M,\ \alpha \text{ real}\}.$$

Since $x_0 \bar{\in} M$, the above representation of $x \in X$ in the form $x = m + \alpha x_0$ is unique. It follows that, if, for any real number c, we set

$$F(x) = F(m + \alpha x_0) = f_0(m) + \alpha c,$$

then F is a real linear functional on X which is an extension of f_0. We have to choose c such that $F(x) \leqq p(x)$, that is, $f_0(m) + \alpha c \leqq p(m + \alpha x_0)$. This condition is equivalent to the following two conditions:

$$f_0(m/\alpha) + c \leqq p(x_0 + m/\alpha) \quad \text{for} \quad \alpha > 0,$$

$$f_0(m/(-\alpha)) - c \leqq p(-x_0 + m/(-\alpha)) \quad \text{for} \quad \alpha < 0.$$

To satisfy these conditions, we shall choose c such that

$$f_0(m') - p(m' - x_0) \leqq c \leqq p(m'' + x_0) - f_0(m'') \quad \text{for all} \quad m', m'' \in M.$$

Such a choice of c is possible since

$$\begin{aligned} f_0(m') + f_0(m'') = f_0(m' + m'') &\leqq p(m' + m'') = p(m' - x_0 + m'' + x_0) \\ &\leqq p(m' - x_0) + p(m'' + x_0); \end{aligned}$$

we have only to choose c between the two numbers

$$\sup_{m' \in M} [f_0(m') - p(m' - x_0)] \quad \text{and} \quad \inf_{m'' \in M} [p(m'' + x_0) - f_0(m'')].$$

Consider now the family of all real linear extensions g of f_0 for which the inequality $g(x) \leqq p(x)$ holds for all x in the domain of g. We make this family into a partially ordered family by defining $h \succ g$ to mean that h is an extension of g. Then Zorn's Lemma ensures the existence of a maximal linear extension g of f_0 for which the inequality $g(x) \leqq p(x)$ holds for all x in the domain of g. We have to prove that the domain $D(g)$ of g coincides with X itself. If it does not, we obtain, taking $D(g)$ as M and g as f_0, a proper extension F of g which satisfies $F(x) \leqq p(x)$ for all x in the domain of F, contrary to the maximality of the linear extension g.

Corollary. Given a functional $p(x)$ defined on a real linear space X such that (1) and (2) are satisfied. Then there exists a linear functional f defined on X such that

$$-p(-x) \leqq f(x) \leqq p(x). \tag{4}$$

Proof. Take any point $x_0 \in X$ and define $M = \{x;\, x = \alpha x_0, \alpha \text{ real}\}$. Set $f_0(\alpha x_0) = \alpha p(x_0)$. Then f_0 is a real linear functional defined on M. We have $f_0(x) \leqq p(x)$ on M. In fact, $\alpha p(x_0) = p(\alpha x_0)$ if $\alpha > 0$, and if $\alpha < 0$, we have $\alpha p(x_0) \leqq -\alpha p(-x_0) = p(\alpha x_0)$ by $0 = p(0) \leqq p(x_0) + p(-x_0)$. Thus there exists a linear functional f defined on X such that $f(x) = f_0(x)$ on M and $f(x) \leqq p(x)$ on X. Since $-f(x) = f(-x) \leqq p(-x)$, we obtain $-p(-x) \leqq f(x) \leqq p(x)$.

2. The Generalized Limit

The notion of a sequence $\{x_n\}$ of a countable number of elements x_n is generalized to the notion of a *directed set* of elements depending on a parameter which runs through an uncountable set. The notion of the limit of a sequence of elements may be extended to the notion of the *generalized limit* of a directed set of elements.

Definition. A partially ordered set A of elements $\alpha, \beta, \ldots$ is called a *directed set* if it satisfies the condition:

For any pair α, β of elements of A, there exists a $\gamma \in A$ such that $\alpha \prec \gamma$, $\beta \prec \gamma$. (1)

Let, to each point α of a directed set A, there be associated a certain set of real numbers $f(\alpha)$. Thus $f(\alpha)$ is a, not necessarily one-valued, real function defined on the directed set A. We write

$$\lim_{\alpha \in A} f(\alpha) = a \quad (a \text{ is a real number})$$

if, for any $\varepsilon > 0$, there exists an $\alpha_0 \in A$ such that $\alpha_0 \prec \alpha$ implies $|f(\alpha) - a| \leqq \varepsilon$ for all possible values of f at α. We say, in such a case, that the value a is the *generalized limit* or *Moore-Smith limit* of $f(\alpha)$ through the directed set A.

Example. Consider a *partition* Δ of the real interval $[0, 1]$:

$$0 = t_0 < t_1 < \cdots < t_n = 1.$$

The totality P of the partition of $[0, 1]$ is a directed set A by defining the partial order $\Delta \prec \Delta'$ as follows: If the partition Δ' is given by $0 = t'_0 < t'_1 < \cdots < t'_m = 1$, then $\Delta \prec \Delta'$ means that $n \leqq m$ and that every t_i is equal to some t'_j. Let $x(t)$ be a real-valued continuous function defined on $[0, 1]$, and let $f(\Delta)$ be the totality of real numbers of the form

$$\sum_{j=0}^{n-1} (t_{j+1} - t_j)\, x(t'_j), \text{ where } t'_j \text{ is any point of } [t_j, t_{j+1}].$$

Thus $f(\Delta)$ is the totality of the *Riemann sum* of the function $x(t)$ pertaining to the partition Δ. The value of the Riemann integral $\int_0^1 x(t)\, dt$ is nothing but the generalized limit of $f(\Delta)$ through P.

As to the existence of a generalized limit, we have the

Theorem (S. Banach). Let $x(\alpha)$ be a real-valued bounded function defined on a directed set A. The totality of such functions constitutes a real linear space X by

$$(x + y)(\alpha) = x(\alpha) + y(\alpha), \ (\beta x)(\alpha) = \beta x(\alpha).$$

We can then define a linear functional, defined on X and which we shall denote by $\underset{\alpha \in A}{\mathrm{LIM}}\, x(\alpha)$, satisfying the condition

$$\underset{\alpha \in A}{\underline{\lim}}\, x(\alpha) \leqq \underset{\alpha \in A}{\mathrm{LIM}}\, x(\alpha) \leqq \underset{\alpha \in A}{\overline{\lim}}\, x(\alpha),$$

where

$$\underset{\alpha \in A}{\underline{\lim}}\, x(\alpha) = \sup_{\alpha} \inf_{\alpha > \beta} x(\beta), \ \underset{\alpha \in A}{\overline{\lim}}\, x(\beta) = \inf_{\alpha} \sup_{\alpha > \beta} x(\beta).$$

Therefore $\underset{\alpha \in A}{\mathrm{LIM}}\, x(\alpha) = \underset{\alpha \in A}{\lim}\, x(\alpha)$ if the latter generalized limit exists.

Proof. We put $p(x) = \underset{\alpha \in A}{\overline{\lim}}\, x(\alpha)$. It is easy to see that this $p(x)$ satisfies the condition of the Hahn-Banach extension theorem. Hence, there exists a linear functional f defined on X such that $-p(-x) \leqq f(x) \leqq p(x)$ on X. We can easily prove that $\underset{\alpha \in A}{\underline{\lim}}\, x(\alpha) = -p(-x)$ so that we obtain the Theorem, by putting $\underset{\alpha \in A}{\mathrm{LIM}}\, x(\alpha) = f(x)$.

3. Locally Convex, Complete Linear Topological Spaces

Definition. As in numerical case, we may define a directed set $\{x_\alpha\}$ in a linear topological space X. $\{x_\alpha\}$ is said to *converge to an element* x of

X, if, for every neighbourhood $U(x)$ of x, there exists an index α_0 such that $x_\alpha \in U(x)$ for all indices $\alpha \succ \alpha_0$. A directed set $\{x_\alpha\}$ of X is said to be *fundamental*, if every neighbourhood $U(0)$ of the zero vector 0 of X is assigned an index α_0 such that $(x_\alpha - x_\beta) \in U(0)$ for all indices $\alpha, \beta \succ \alpha_0$. A linear topological space X is said to be *complete* if every directed fundamental set of X converges to some element $x \in X$ in the sense above.

Remark. We can weaken the condition of the completeness, and require only that every sequence of X which is fundamental as a directed set converges to an element $x \in X$; a space X satisfying this condition is said to be *sequentially complete*. For normed linear spaces, the two definitions of completeness are equivalent. However, in the general case, not every sequentially complete space is complete.

Example of a locally convex, sequentially complete linear topological space. Let a sequence $\{f_h(x)\}$ of $\mathfrak{D}(\Omega)$ satisfy the condition $\lim_{h,k\to\infty} (f_h - f_k) = 0$ in $\mathfrak{D}(\Omega)$. That is, by the Corollary of Proposition 7 in Chapter I, 1, we assume that there exists a compact subset K of Ω for which $\operatorname{supp}(f_h) \subseteqq K$ $(h = 1, 2, \ldots)$ and $\lim_{h,k\to\infty} (D^s f_h(x) - D^s f_k(x)) = 0$ uniformly on K for any differential operator D^s. Then it is easy to see, by applying the Ascoli-Arzelà theorem, that there exists a function $f \in \mathfrak{D}(\Omega)$ for which $\lim_{h\to\infty} D^s f_h(x) = D f(x)$ uniformly on K for any differential operator D^s. Hence $\lim_{h\to\infty} f_h = f$ in $\mathfrak{D}(\Omega)$ and so $\mathfrak{D}(\Omega)$ is sequentially complete. Similarly, we can prove that $\mathfrak{E}(\Omega)$ is also sequentially complete.

As in the case of a normed linear space, we can prove the

Theorem. Every locally convex linear topological space X can be embedded in a locally convex, complete linear topological space, in which X forms a dense subset.

We omit the proof. The reader is referred to the literature listed in J. A. Dieudonné [1]. Cf. also G. Köthe [1].

4. The Hahn-Banach Extension Theorem in Complex Linear Spaces

Theorem (Bohnenblust-Sobczyk). Let X be a complex linear space and p a semi-norm defined on X. Let M be a complex linear subspace of X and f a complex linear functional defined on M such that $|f(x)| \leqq p(x)$ on M. Then there exists a complex linear functional F defined on X such that i) F is an extension of f, and ii) $|F(x)| \leqq p(x)$ on X.

Proof. We observe that a complex linear space is also a real linear space if the scalar multiplication is restricted to real numbers. If $f(x) = g(x) + ih(x)$, where $g(x)$ and $h(x)$ are the real and imaginary parts of $f(x)$ respectively, then g and h are real linear functionals defined on M.

Thus

$$|g(x)| \leqq |f(x)| \leqq p(x) \quad \text{and} \quad |h(x)| \leqq |f(x)| \leqq p(x) \text{ on } M.$$

Since, for each $x \in M$,

$$g(ix) + ih(ix) = f(ix) = if(x) = i(g(x) + ih(x)) = -h(x) + ig(x),$$

we have

$$h(x) = -g(ix) \quad \text{for all} \quad x \in M.$$

By the Theorem in Chapter IV, 1, we can extend g to a real linear functional G defined on X with the property that $G(x) \leqq p(x)$ on X. Hence $-G(x) = G(-x) \leqq p(-x) = p(x)$, and so $|G(x)| \leqq p(x)$. We define

$$F(x) = G(x) - iG(ix).$$

Then, by $F(ix) = G(ix) - iG(-x) = G(ix) + iG(x) = iF(x)$, we easily see that F is a complex linear functional defined on X. F is an extension of f. For, $x \in M$ implies that

$$F(x) = G(x) - iG(ix) = g(x) - ig(ix) = g(x) + ih(x) = f(x).$$

To prove $|F(x)| \leqq p(x)$, we write $F(x) = re^{-i\theta}$ so that $|F(x)| = e^{i\theta}F(x) = F(e^{i\theta}x)$ is real positive; consequently $|F(x)| = |G(e^{i\theta}x)| \leqq p(e^{i\theta}x) = |e^{i\theta}|\, p(x) = p(x)$.

5. The Hahn-Banach Extension Theorem in Normed Linear Spaces

Theorem 1. Let X be a normed linear space, M a linear subspace of X and f_1 a continuous linear functional defined on M. Then there exists a continuous linear functional f defined on X such that i) f is an extension of f_1, and ii) $||f_1|| = ||f||$.

Proof. Set $p(x) = ||f_1|| \cdot ||x||$. Then p is a continuous semi-norm defined on X such that $|f_1(x)| \leqq p(x)$ on M. There exists, by the Theorem in the preceding section 4, a linear extension f of f_1 defined on the whole space X and such that $|f(x)| \leqq p(x)$. Thus $||f|| \leqq \sup_{||x|| \leqq 1} p(x) = ||f_1||$. On the other hand, since f is an extension of f_1, we must have $||f|| \geqq ||f_1||$ and so we obtain $||f_1|| = ||f||$.

An Application to Moment Problems

Theorem 2. Let X be a normed linear space. Given a sequence of elements $\{x_n\} \subseteq X$, a sequence of complex numbers $\{\alpha_n\}$ and a positive number γ. Then a necessary and sufficient condition for the existence of a continuous linear functional f on X such that $f(x_i) = \alpha_i$ $(i = 1, 2, \ldots)$ and $||f|| \leqq \gamma$ is that the inequalities

$$\left|\sum_{i=1}^{n} \beta_i \alpha_i\right| \leqq \gamma \left|\left|\sum_{i=1}^{n} \beta_i x_i\right|\right|$$

hold for any choice of positive numbers n and complex numbers $\beta_1, \beta_2, \ldots, \beta_n$.

Proof. The necessity of this condition is clear from the definition of $||f||$. We shall prove the sufficiency. Consider the set

$$X_1 = \left\{z; z = \sum_{i=1}^{n} \beta_i x_i \quad \text{where } n \text{ and } \beta \text{ are arbitrary}\right\}.$$

For two representations $z = \sum_{i=1}^{n} \beta_i x_i = \sum_{i=1}^{m} \beta_{i'} x_{i'}$ of the same element $z \in X_1$, we have, by the condition of the Theorem,

$$\left|\sum_{i=1}^{n} \beta_i \alpha_i - \sum_{i=1}^{m} \beta_{i'} \alpha_{i'}\right| \leqq \gamma \left\|\sum_{i=1}^{n} \beta_i x_i - \sum_{i=1}^{m} \beta_{i'} x_{i'}\right\| = 0.$$

Thus a continuous linear functional f_1 is defined on X_1 by $f_1\left(\sum_{i=1}^{n} \beta_i x_i\right) = \sum_{i=1}^{n} \beta_i \alpha_i$. We have only to extend, by the preceding Theorem 1, f_1 to a continuous linear functional f on X with $||f|| = ||f_1||$.

Remark. As will be shown in section 9 of this chapter, any continuous linear functional f on $C[0, 1]$ is representable as

$$f(x) = \int_0^1 x(t)\, m(dt)$$

with a uniquely determined Baire measure m on the interval $[0, 1]$. Thus, if we take $x_j(t) = t^{j-1}$ $(j = 1, 2, \ldots)$, Theorem 2 gives the solvability condition of the *moment problem*:

$$\int_0^1 t^{j-1} m(dt) = \alpha_j \quad (j = 1, 2, \ldots).$$

6. The Existence of Non-trivial Continuous Linear Functionals

Theorem 1. Let X be a real or complex linear topological space, x_0 a point of X and $p(x)$ a continuous semi-norm on X. Then there exists a continuous linear functional F on X such that $F(x_0) = p(x_0)$ and $|F(x)| \leqq p(x)$ on X.

Proof. Let M be the set of all elements αx_0, and define f on M by $f(\alpha x_0) = \alpha p(x_0)$. Then f is linear on M and $|f(\alpha x_0)| = |\alpha p(x_0)| = p(\alpha x_0)$ there. Thus there exists, by the Theorem in Chapter IV, 4, an extension F of f such that $|F(x)| \leqq p(x)$ on X. Hence $F(x)$ is continuous at $x = 0$ with $p(x)$, and so, by the linearity of F, $F(x)$ is continuous at every point of X.

Corollary 1. Let X be a locally convex space and $x_0 \neq 0$ be an element of X. Then there exists a continuous semi-norm p on X such that

$p(x_0) \neq 0$. Thus, by Theorem 1, there exists a continuous linear functional f_0 on X such that

$$f_0(x_0) = p(x_0) \neq 0 \quad \text{and} \quad |f_0(x)| \leqq p(x) \text{ on } X.$$

Corollary 2. Let X be a normed linear space and $x_0 \neq 0$ be any element of X. Then there exists a continuous linear functional f_0 on X such that

$$f_0(x_0) = \|x_0\| \quad \text{and} \quad \|f_0\| = 1.$$

Proof. We take $\|x\|$ for $p(x)$ in Corollary 1. Thus $\|f_0\| \leqq 1$ from $|f_0(x)| \leqq \|x\|$. But, by $f_0(x_0) = \|x_0\|$, we must have the equality $\|f_0\| = 1$.

Remark. As above, we prove the following theorem by the Theorem in Chapter IV, 1.

Theorem 1′. Let X be a real linear topological space, x_0 a point of X and $p(x)$ a real continuous functional on X such that

$$p(x+y) \leqq p(x) + p(y) \quad \text{and} \quad p(\alpha x) = \alpha p(x) \quad \text{for} \quad \alpha \geqq 0.$$

Then there exists a continuous real linear functional F on X such that $F(x_0) = p(x_0)$ and $-p(-x) \leqq F(x) \leqq p(x)$ on X.

Theorem 2. Let X be a locally convex linear topological space. Let M be a linear subspace of X, and f a continuous linear functional on M. Then there exists a continuous linear functional F on X which is an extension of f.

Proof. Since f is continuous on M and X is locally convex, there exists an open, convex, balanced neighbourhood of 0, say U, of X such that $x \in M \cap U$ implies $|f(x)| \leqq 1$. Let p be the Minkowski functional of U. Then p is a continuous semi-norm on X and $U = \{x; p(x) < 1\}$. For any $x \in M$ choose $\alpha > 0$ so that $\alpha > p(x)$. Then $p(x/\alpha) < 1$ and so $|f(x/\alpha)| \leqq 1$, that is $|f(x)| \leqq \alpha$. We thus see, by letting $\alpha \downarrow p(x)$, that $|f(x)| \leqq p(x)$ on M. Hence, by the Theorem in Chapter IV, 4, we obtain a continuous linear functional F on X such that F is an extension of f and $|F(x)| \leqq p(x)$ on X.

Theorem 3 (S. Mazur). Let X be a real or complex, locally convex linear topological space, and M a closed convex balanced subset of X. Then, for any $x_0 \,\overline{\in}\, M$, there exists a continuous linear functional f on X such that $f_0(x_0) > 1$ and $|f_0(x)| \leqq 1$ on M.

Proof. Since M is closed, there exists a convex, balanced neighbourhood V of 0 such that $M \cap (x_0 + V) = \emptyset$. Since V is balanced and convex, we have $\left(M + \frac{V}{2}\right) \cap \left(x_0 + \frac{V}{2}\right) = \emptyset$. The set $\left(x_0 + \frac{V}{2}\right)$ being a neighbourhood of x_0, the closure U of $\left(M + \frac{V}{2}\right)$ does not contain x_0. Since $M \ni 0$, the closed convex balanced set U is a neighbourhood of 0, because U

contains $\frac{V}{2}$ as a subset. Let p be the Minkowski functional of U. Since U is closed, we have, for any $x_0 \,\bar{\in}\, U$, $p(x_0) > 1$ and $p(x) \leqq 1$, if $x \in U$.

Thus there exists, by Corollary 1 of Theorem 1, a continuous linear functional f_0 on X such that $f_0(x_0) = p(x_0) > 1$ and $|f_0(x)| \leqq p(x)$ on X. Hence, in particular, $|f_0(x)| \leqq 1$ on M.

Corollary. Let M be a closed linear subspace of a locally convex linear topological space X. Then, for any $x_0 \in X - M$, there exists a continuous linear functional f_0 on X such that $f_0(x_0) > 1$ and $f_0(x) = 0$ on M. Moreover, if X is a normed linear space and if dis $(x_0, M) > d$, then we may take $\|f_0\| \leqq 1/d$.

Proof. The first part is clear from the linearity of M. The second part is proved by taking $U = \{x;\ \text{dis}(x, M) \leqq d\}$ in the proof of Theorem 3.

Remark. As above, we prove the following theorem by virtue of Theorem 1'.

Theorem 3' (S. Mazur). Let X be a locally convex real linear topological space, and M a closed convex subset of X such that $M \ni 0$. Then, for any $x_0 \,\bar{\in}\, M$, there exists a continuous real linear functional f_0 on X such that $f_0(x_0) > 1$ and $f_0(x) \leqq 1$ on M.

Theorem 4 (S. Mazur). Let X be a locally convex linear topological space, and M a convex balanced neighbourhood of 0 of X. Then, for any $x_0 \,\bar{\in}\, M$, there exists a continuous linear functional f_0 on X such that

$$f_0(x_0) \geqq \sup_{x \in M} |f_0(x)|.$$

Proof. Let p be the Minkowski functional of M. Then $p(x_0) \geqq 1$ and $p(x) \leqq 1$ on M. p is continuous since M is a neighbourhood of 0 of X. Thus there exists, by Corollary 1 of Theorem 1, a continuous linear functional f_0 on X such that

$$f_0(x_0) = p(x_0) \geqq 1 \quad \text{and} \quad |f_0(x)| \leqq p(x) \leqq 1 \text{ on } M.$$

Theorem 5 (E. Helly). Let X be a B-space, and $f_1, f_2, \ldots, f_n$ be a finite system of bounded linear functionals on X. Given n numbers $\alpha_1, \alpha_2, \ldots, \alpha_n$. Then a necessary and sufficient condition that there exists, for each $\varepsilon > 0$, an element $x_\varepsilon \in X$ such that

$$f_i(x_\varepsilon) = \alpha_i \quad (i = 1, 2, \ldots, n) \quad \text{and} \quad \|x_\varepsilon\| \leqq \gamma + \varepsilon$$

is that the inequality

$$\left|\sum_{i=1}^{n} \beta_i \alpha_i\right| \leqq \gamma \left\|\sum_{i=1}^{n} \beta_i f_i\right\|$$

holds for any choice of n numbers $\beta_1, \beta_2, \ldots, \beta_n$.

Proof. The necessity is clear from the definition of the norm of a continuous linear functional. We shall prove the sufficiency. We may assume, without losing the generality, that f's are linearly independent; otherwise, we can work with a linearly independent subsystem of $\{f_i\}$ which spans the same subspace as the original system $\{f_i\}$.

We consider the mapping $x \to \varphi(x) = (f_1(x), f_2(x), \ldots, f_n(x))$ of X onto the Hilbert space $l^2(n)$ consisting of all vectors $x = (\xi_1, \xi_2, \ldots, \xi_n)$ normed by $||x|| = \left(\sum_{j=1}^{n} |\xi_j|^2\right)^{1/2}$. By the open mapping theorem in Chapter II, 5, the image $\varphi(S_\varepsilon)$ of the sphere $S_\varepsilon = \{x \in X; ||x|| \leqq \gamma + \varepsilon\}$ contains the vector 0 of $l^2(n)$ as an interior point for every $\varepsilon > 0$. Let us suppose that $(\alpha_1, \alpha_2, \ldots, \alpha_n)$ does not belong to $\varphi(S_\varepsilon)$. Then, by Mazur's theorem given above, there exists a continuous linear functional F on $l^2(n)$ such that

$$F((\alpha_1, \alpha_2, \ldots, \alpha_n)) \geqq \sup_{||x|| \leqq \gamma+\varepsilon} |F(\varphi(x))|.$$

Since $l^2(n)$ is a Hilbert space, the functional F is given by an element $(\beta_1, \beta_2, \ldots, \beta_n) \in l^2(n)$ in such a way that $F((\alpha_1, \alpha_2, \ldots, \alpha_n)) = \sum_{j=1}^{n} \alpha_j \beta_j$. Thus

$$\sum_{j=1}^{n} \alpha_j \beta_j \geqq \left|\sum_{j=1}^{n} f_j(x)\, \beta_j\right| \quad \text{for} \quad ||x|| \leqq \gamma + \varepsilon.$$

But the supremum of the right hand side for $||x|| \leqq \gamma + \varepsilon$ is $= (\gamma + \varepsilon) \times \left\|\sum_{j=1}^{n} f_j \beta_j\right\|$, and this contradicts the hypothesis of the Theorem.

7. Topologies of Linear Maps

Let X, Y be locally convex linear topological spaces on the same scalar field (real or complex number field). We denote by $L(X, Y)$ the totality of continuous linear operators on X into Y. $L(X, Y)$ is a linear space by

$$(\alpha T + \beta S)\, x = \alpha T x + \beta S x, \quad \text{where} \quad T, S \in L(X, Y) \quad \text{and} \quad x \in X.$$

We shall introduce various topologies for this linear space $L(X, Y)$.

i) Simple Convergence Topology. This is the topology of convergence at each point of X and thus it is defined by the family of semi-norms of the form

$$p(T) = p(T; x_1, x_2, \ldots, x_n; q) = \sup_{1 \leqq j \leqq n} q(T x_j),$$

where $x_1, x_2, \ldots, x_n$ are an arbitrary finite system of elements of X and q an arbitrary continuous semi-norm on Y. $L(X, Y)$ endowed with this topology will be denoted by $L_s(X, Y)$. It is clearly a locally convex linear topological space.

ii) Bounded Convergence Topology. This is the topology of uniform convergence on bounded sets of X. Thus it is defined by the family of semi-norms of the form

$$p(T) = p(T; B; q) = \sup_{x \in B} q(Tx)$$

where B is an arbitrary bounded set of X and q an arbitrary continuous semi-norm on Y. $L(X, Y)$ endowed with this topology will be denoted by $L_b(X, Y)$. It is clearly a locally convex linear topological space.

Since any finite set of X is bounded, the simple convergence topology is *weaker* than the bounded convergence topology, i.e., open sets of $L_s(X, Y)$ are also open sets of $L_b(X, Y)$, but not conversely.

Definition 1. If X, Y are normed linear spaces, then the topology of $L_s(X, Y)$ is usually called the *strong topology* (of operators); the one of $L_b(X, Y)$ is called the *uniform topology* (of operators).

Dual Spaces. Weak and Weak* Topologies

Definition 1'. In the special case when Y is the real or complex number field topologized in the usual way, $L(X, Y)$ is called the *dual space* of X and will be denoted by X'. Thus X' is the set of all continuous linear functionals on X. The simple convergence topology is then called the *weak* topology* of X'; provided with this topology, X' will sometimes be denoted by X'_{w^*} and we call it the *weak* dual* of X. The bounded convergence topology for X' is called the *strong topology* of X'; provided with this topology, X' is sometimes denoted by X'_s and we call it the *strong dual* of X.

Definition 2. For any $x \in X$ and $x' \in X'$, we shall denote by $\langle x, x' \rangle$ or $x'(x)$ the value of the functional x' at the point x. Thus the weak* topology of X', i.e., the topology of X'_{w^*} is defined by the family of semi-norms of the form

$$p(x') = p(x'; x_1, x_2, \ldots, x_n) = \sup_{1 \leq j \leq n} |\langle x_j, x' \rangle|,$$

where $x_1, x_2, \ldots, x_n$ are an arbitrary finite system of elements of X. The strong topology of X', i.e., the topology of X'_s is defined by the family of semi-norms of the form

$$p(x') = p(x'; B) = \sup_{x \in B} |\langle x, x' \rangle|$$

where B is an arbitrary bounded set of X.

Theorem 1. If X is a normed linear space, then the strong dual space X'_s is a B-space with the norm

$$\|f\| = \sup_{\|x\| \leq 1} |f(x)|.$$

Proof. Let B be any bounded set of X. Then $\sup_{b \in B} \|b\| = \beta < \infty$, and hence $\|f\| \leq \alpha$ implies $p(f; B) = \sup_{b \in B} |f(b)| \leq \sup_{\|x\| \leq \beta} |f(x)| \leq \alpha\beta$.

On the other hand, the unit sphere $S = \{x;\ ||x|| \leqq 1\}$ of X is a bounded set, and so $||f|| = p(f; S)$. This proves that the topology of X'_s is equivalent to the topology defined by the norm $||f||$.

The completeness of X'_s is proved as follows. Let a sequence $\{f_n\}$ of X'_s satisfy $\lim_{n,m\to\infty} ||f_n - f_m|| = 0$. Then, for any $x \in X$, $|f_n(x) - f_m(x)| \leqq ||f_n - f_m|| \cdot ||x|| \to 0$ (as $n, m \to \infty$), and hence a finite $\lim_{n\to\infty} f_n(x) = f(x)$ exists. The linearity of f is clear. The continuity of f is proved by observing that $\lim_{n\to\infty} f_n(x) = f(x)$ uniformly on the unit sphere S. Incidentally we have proved that $\lim_{n\to\infty} ||f_n - f|| = 0$.

Similarly we can prove

Theorem 2. If X, Y are normed linear spaces, then the uniform topology (of operators) $L_b(X, Y)$ is defined by the operator norm

$$||T|| = \sup_{||x|| \leqq 1} ||Tx||.$$

Definition 3. We define the *weak topology* of a locally convex linear topological space X by the family of semi-norms of the form

$$p(x) = p(x; x'_1, x'_2, \ldots, x'_n) = \sup_{1 \leqq j \leqq n} |\langle x, x'_j\rangle|,$$

where $x'_1, x'_2, \ldots, x'_n$ are an arbitrary finite system of elements of X'. Endowed with this topology, X is sometimes denoted by X_w.

8. The Embedding of X in its Bidual Space X''

We first prove

Theorem 1 (S. BANACH). Let X be a locally convex linear topological space, and X' its dual space. A linear functional $f(x')$ on X' is of the form

$$f(x') = \langle x_0, x'\rangle$$

with a certain $x_0 \in X$ iff $f(x')$ is continuous in the weak* topology of X'.

Proof. The "only if" part is clear since $|\langle x_0, x'\rangle|$ is one of the semi-norms defining the weak* topology of X'. The "if" part will be proved as follows. The continuity of $f(x')$ in the weak* topology of X' implies that there exists a finite system of points $x_1, x_2, \ldots, x_n$ such that $|f(x')| \leqq \sup_{1 \leqq j \leqq n} |\langle x_j, x'\rangle|$. Thus

$$f_i(x') = \langle x_i, x'\rangle = 0 \ \ (i = 1, 2, \ldots, n) \text{ implies } f(x') = 0.$$

Consider the linear map $L\colon X' \to l^2(n)$, defined by

$$L(x') = (f_1(x'), f_2(x'), \ldots, f_n(x')).$$

$L(x'_1) = L(x'_2)$ implies $L(x'_1 - x'_2) = 0$ so that $f_i(x'_1 - x'_2) = 0$ $(i = 1, 2, \ldots, n)$ and hence $f(x'_1 - x'_2) = 0$. Hence we may define a continuous linear map F

defined on the linear subspace $L(X')$ of $l^2(n)$ by

$$F(L(x')) = F(f_1(x'), \ldots, f_n(x')) = f(x').$$

This map can be extended to a continuous linear functional defined on the whole space $l^2(n)$; the extension is possible since $l^2(n)$ is of finite dimension (easier than using the Hahn-Banach extension theorem in infinite dimensional linear spaces). We denote this extension by the same letter F. Writing

$$(y_1, y_2, \ldots, y_n) = \sum_{j=1}^{n} y_j e_j, \quad \text{where} \quad e_j = (0, 0, \ldots, 0, 1, 0, 0, \ldots, 0)$$

with 1 at the j-th coordinate,

we easily see that

$$F(y_1, y_2, \ldots, y_n) = \sum_{j=1}^{n} y_j \alpha_j, \ \alpha_j = F(e_j).$$

Therefore

$$f(x') = \sum_{j=1}^{n} \alpha_j f_j(x') = \sum_{j=1}^{n} \alpha_j \langle x_j, x' \rangle = \left\langle \sum_{j=1}^{n} \alpha_j x_j, x' \right\rangle.$$

Corollary. Each $x_0 \in X$ defines a continuous linear functional $f_0(x')$ on X'_s by $f_0(x') = \langle x_0, x' \rangle$. The mapping

$$x_0 \to f_0 = J x_0$$

of X into $(X'_s)'_s$ satisfies the conditions

$$J(x_1 + x_2) = J x_1 + J x_2, \ J(\alpha x) = \alpha J(x).$$

Theorem 2. If X is a normed linear space, then the mapping J is *isometric*, i.e., $\|Jx\| = \|x\|$.

Proof. We have $|f_0(x')| = |\langle x_0, x' \rangle| \leqq \|x_0\| \cdot \|x'\|$ so that $\|f_0\| \leqq \|x_0\|$. On the other hand, if $x_0 \neq 0$, then there exists, by Corollary 2 of Theorem 1 in Chapter IV, 6, an element $x'_0 \in X'$ such that $\langle x_0, x'_0 \rangle = \|x_0\|$ and $\|x'_0\| = 1$. Hence $f_0(x'_0) = \langle x_0, x'_0 \rangle = \|x_0\|$ so that $\|f_0\| \geqq \|x_0\|$. We have thus proved $\|Jx\| = \|x\|$.

Remark. As the strong dual space of X'_s, the space $(X'_s)'_s$ is a B-space. Hence a normed linear space X may be considered as a linear subspace of the B-space $(X'_s)'_s$ by the embedding $x \to Jx$. Therefore the strong closure of JX in the B-space $(X'_s)'_s$ gives a concrete construction of the completion of X.

Definition 4. A normed linear space X is said to be *reflexive* if X may be identified with its *second dual* or the *bidual* $(X'_s)'_s$ by the correspondence $x \leftrightarrow Jx$ above. We know already (see Chapter III, 6) that a Hilbert space is reflexive. As remarked above, $(X'_s)'_s$ is a B-space and so any reflexive normed linear space must be a B-space.

Theorem 3. Let X be a B-space and x_0'' any bounded linear functional on X_s'. Then, for any $\varepsilon > 0$ and any finite system of elements $f_1, f_2, \ldots, f_n$ of X_s', there exists an $x_0 \in X$ such that

$$\|x_0\| \leqq \|x_0''\| + \varepsilon \quad \text{and} \quad f_j(x_0) = x_0''(f_j) \ (j = 1, 2, \ldots, n).$$

Proof. We apply Helly's theorem 5 in Chapter IV, 6. For any system of numbers $\beta_1, \beta_2, \ldots, \beta_n$, we have

$$\left|\sum_{j=1}^{n} \beta_j \alpha_j\right| = \left|\sum_{j=1}^{n} \beta_j x_0''(f_j)\right| = \left|x_0''\left(\sum_{j=1}^{n} \beta_j f_j\right)\right|$$

$$\leqq \gamma \cdot \left\|\sum_{j=1}^{n} \beta_j f_j\right\|, \text{ wehre } \gamma = \|x_0''\|, \alpha_j = x_0''(f_j),$$

and hence, again by Helly's theorem, we obtain an $x_0 \in X$ with the estimate $\|x_0\| \leqq \gamma + \varepsilon = \|x_0''\| + \varepsilon$ and $x_0(f_j) = \alpha_j$ $(j = 1, 2, \ldots, n)$.

Corollary. The unit sphere $S = \{x \in X; \|x\| \leqq 1\}$ of a B-space X is dense in the unit sphere of $(X_s')_s'$ in the weak* topology of $(X_s')'$.

9. Examples of Dual Spaces

Example 1. $(c_0)' = (l^1)$.

To any $f \in (c_0)'$, there corresponds a uniquely determined $y_f = \{\eta_n\} \in (l^1)$ such that, for all $x = \{\xi_n\} \in (c_0)$,

$$\langle x, f\rangle = \sum_{n=1}^{\infty} \xi_n \eta_n \text{ and } \|f\| = \|y_f\|. \tag{1}$$

And conversely, any $y = \{\eta_n\} \in (l^1)$ defines an $f_y \in (c_0)'$ such that, for any $x = \{\xi_n\} \in (c_0)$,

$$\langle x, f_y\rangle = \sum_{n=1}^{\infty} \xi_n \eta_n \text{ and } \|f_y\| = \|y\|. \tag{1'}$$

Proof. Let us define the unit vector e_k by

$$e_k = (\overbrace{0, 0, \ldots, 0}^{k-1}, 1, 0, 0, \ldots) \ (k = 1, 2, \ldots).$$

For any $x = \{\xi_n\} \in (c_0)$ and $f \in (c_0)'$, we have, by $s\text{-}\lim\limits_{k\to\infty} \sum\limits_{n=1}^{k} \xi_n e_n = x$,

$$\langle x, f\rangle = \lim_{k\to\infty} \left\langle \sum_{n=1}^{k} \xi_n e_n, f\right\rangle = \lim_{k\to\infty} \sum_{n=1}^{k} \xi_n \eta_n, \ \eta_n = f(e_n).$$

Let $\eta_n = \varepsilon_n |\eta_n|$ for $\eta_n \neq 0$, and $\varepsilon_n = \infty$ for $\eta_n = 0$. Take $x^{(n_0)} = \{\xi_n\} \in (c_0)$ in such a way that $\xi_n = \varepsilon_n^{-1}$ for $n \leqq n_0$, and $\xi_n = 0$ for $n > n_0$. Then $\|x^{(n_0)}\| \leqq 1$ and so $\|f\| = \sup\limits_{\|x\| \leqq 1} |\langle x, f\rangle| \geqq |\langle x^{(n_0)}, f\rangle| = \sum\limits_{n=1}^{n_0} |\eta_n|$. Thus, by letting $n_0 \to \infty$, we see that $y_f = \{\eta_n\} \in (l^1)$ and $\|y_f\| = \sum\limits_{n=1}^{\infty} |\eta_n| \leqq \|f\|$.

If, conversely, $y = \{\eta_n\} \in (l^1)$, then $\left|\sum_{n=1}^{\infty} \xi_n \eta_n\right| \leqq ||x|| \cdot ||y||$ for all $x = \{\xi_n\} \in (c_0)$, and so y defines an $f_y \in (c_0)'$ and $||f_y|| \leqq ||y||$.

Example 2. $(c)' = (l^1)$.

For any $x = \{\xi_n\} \in (c)$, we have the representation

$$x = \xi_0 e_0 + \text{s-}\lim_{k\to\infty} \sum_{n=1}^{k} (\xi_n - \xi_0)\, e_n, \text{ where } \xi_0 = \lim_{n\to\infty} \xi_n,\ e_0 = (1, 1, 1, \ldots).$$

Thus, for any $f \in (c)'$, we have

$$\langle x, f\rangle = \xi_0 \langle e_0, f\rangle + \lim_{k\to\infty} \left\langle \sum_{n=1}^{k} (\xi_n - \xi_0)\, e_n, f\right\rangle = \xi_0 \eta_0' + \sum_{n=1}^{\infty} (\xi_n - \xi_0) \eta_n, \tag{2}$$

where $\eta_0' = \langle e_0, f\rangle$ and $\eta_n = \langle e_n, f\rangle$ $(n = 1, 2, \ldots)$. As above, we may take $x^{(n_0)} = \{\xi_n\} \in (c_0) \subseteqq (c)$ which satisfies

$$||x^{(n_0)}|| \leqq 1,\ \xi_0 = \lim_{n\to\infty} \xi_n = 0 \text{ and } \langle x^{(n_0)}, f\rangle = \sum_{n=1}^{n_0} |\eta_n|.$$

Hence, by $|\langle x^{(n_0)}, f\rangle| \leqq ||x^{(n_0)}||\ ||f||$, we see that $\{\eta_n\}^{\infty} \in (l^1)$. We set $\eta_0' - \sum_{n=1}^{\infty} \eta_n = \eta_0$. Then, by (2), we have

$$\langle x, f\rangle = \xi_0 \eta_0 + \sum_{n=1}^{\infty} \xi_n \eta_n, \text{ where } x = \{\xi_n\} \in (c) \text{ and } \xi_0 = \lim_{n\to\infty} \xi_n. \tag{2'}$$

Let $\eta_n = \varepsilon_n |\eta_n|$ for $\eta_n \neq 0$, and $\varepsilon_n = \infty$ for $\eta_n = 0$ $(n = 0, 1, 2, \ldots)$. Take $x = \{\xi_n\} \in (c)$ such that

$$\xi_n = \varepsilon_n^{-1} \text{ if } n \leqq n_0, \text{ and } \xi_n = \varepsilon_0^{-1} \text{ if } n > n_0.$$

Then $||x|| \leqq 1$, $\xi_0 = \lim_{n\to\infty} \xi_n = \varepsilon_0^{-1}$, and $\langle x, f\rangle = |\eta_0| + \sum_{n=1}^{n_0} |\eta_n| + \varepsilon_0^{-1} \sum_{n=n_0+1}^{\infty} \eta_n$. Hence, we must have $|\eta_0| + \sum_{n=1}^{\infty} |\eta_n| \leqq ||f||$.

If, conversely, $y = \{\eta_n\}_0^{\infty}$ is such that $||y|| = |\eta_0| + \sum_{n=1}^{\infty} |\eta_n| < \infty$, then

$$\eta_0 \cdot \lim_{n\to\infty} \xi_n + \sum_{n=1}^{\infty} \xi_n \eta_n, \text{ where } x = \{\xi_n\}_1^{\infty} \in (c),$$

defines an $f_y \in (c)'$ such that $||f_y|| \leqq |\eta_0| + \sum_{n=1}^{\infty} |\eta_n|$.

Therefore, we have proved that $(c)' = (l^1)$ as explained above.

Example 3. $L^p(S, \mathfrak{B}, m)' = L^q(S, \mathfrak{B}, m)$ $(1 \leqq p < \infty$ and $p^{-1} + q^{-1} = 1)$. To any $f \in L^p(S)'$, there corresponds a $y_f \in L^q(S)$ such that

$$\langle x, f\rangle = \int_S x(s)\, y_f(s)\, m(ds) \text{ for all } x \in L^p(S) \text{ and } ||f|| = ||y_f||, \tag{3}$$

and conversely, any $y \in L^q(S)$ defines an $f_y \in L^p(S)'$ such that

$$\langle x, f_y\rangle = \int_S x(s)\, y(s)\, m(ds) \text{ for all } x \in L^p(S) \text{ and } ||f_y|| = ||y||. \tag{3'}$$

Proof. Let $S = \bigcup_{j=1}^{\infty} B_j$ with $0 < m(B_j) < \infty$ and set $B^{(n)} = \bigcup_{j=1}^{n} B_j$. For a fixed n, the defining function $C_B(s)$ of the set $B \subseteqq B^{(n)}$ is $\in L^p(S)$. Thus the set function $\psi(B) = \langle C_B, f\rangle$ is σ-additive and m-absolutely continuous in $B \subseteqq B^{(n)}$. By the differentiation theorem of Lebesgue-Nikodym (see Chapter III, 8), there exists a $y_n(s) \in L^1(B^{(n)}, \mathfrak{B}^{(n)}, m)$ such that

$$\psi(B) = \int_B y_n(s)\, m(ds) \text{ whenever } B \subseteqq B^{(n)},$$

the family $\mathfrak{B}^{(n)}$ of sets being defined by $\mathfrak{B}^{(n)} = \{B \cap B^{(n)}; B \in \mathfrak{B}\}$. Therefore, by setting $y(s) = y_n(s)$ for $s \in B^{(n)}$, we have

$$\langle C_B, f\rangle = \int_B y(s)\, m(ds) \text{ for } B \in \mathfrak{B}^{(n)} \quad (n = 1, 2, \ldots).$$

Hence, for any finitely-valued function x,

$$\langle x, f\rangle = \int_S x(s)\, y(s)\, m(ds). \tag{4}$$

Let $x \in L^p(S)$ and put

$$x_n(s) = x(s) \text{ if } |x(s)| \leqq n \text{ and } s \in B^{(n)},$$
$$= 0 \text{ otherwise.}$$

We decompose the set $\{z; |z| \leqq n\}$ of the complex plane into a finite number of disjoint Baire sets $M_{n,k,t}$ $(t = 1, 2, \ldots, d_{k,n})$ of diameters $\leqq 1/k$. Set, for the $x_n(s) \in L^\infty(S, \mathfrak{B}, m)$,

$x_{n,k}(s) =$ a constant z, such that $z \in$ (the closure $M^a_{n,k,t}$) and $|z| = \inf_{w \in M n,k,t} |w|$ whenever $x_n(s) \in M_{n,k,t}$.

Then $|x_{n,k}(s)| \leqq |x_n(s)|$ and $\lim_{k\to\infty} x_{n,k}(s) = x_n(s)$ and so, by the Lebesgue-Fatou Lemma, $s\text{-}\lim_{k\to\infty} x_{n,k} = x_n$ $(n = 1, 2, \ldots)$. Thus, again by the Lebesgue-Fatou Lemma,

$$\langle x_n, f\rangle = \lim_{k\to\infty} \langle x_{n,k}, f\rangle = \lim_{k\to\infty} \int_S x_{n,k}(s)\, y(s)\, m(ds) \tag{5}$$
$$= \int_S \lim_{k\to\infty} x_{n,k}(s) \cdot y(s)\, m(ds) = \int_S x_n(s)\, y(s)\, m(ds).$$

Since $s\text{-}\lim_{n\to\infty} x_n = x$, we see that $\langle x, f\rangle = \lim_{n\to\infty} \langle x_n, f\rangle$. We put, for any complex number z, $a(z) = e^{-i\theta}$ if $z = re^{i\theta}$ and $a(0) = 0$. Then $||x|| \geqq ||(|x_n| \cdot a(y))||$ and so

$$||f||\ ||x|| \geqq \langle |x_n|\, a(y), f\rangle = \int_S |x_n(s)| \cdot |y(s)|\, m(ds).$$

Thus, by the Lebesgue-Fatou Lemma, $||f|| \cdot ||x|| \geqq \int_S |x(s)||y(s)|\ m(ds)$ and so the function $x(s)\, y(s)$ belongs to $\in L^1(S)$. Therefore, letting

$n \to \infty$ in (5), we obtain

$$\langle x, f\rangle = \int_S x(s)\, y(s)\, m(ds) \quad \text{whenever } x \in L^p(S).$$

We shall show that $y \in L^q(S)$. To this purpose, set

$$\begin{aligned} y_n(s) &= y(s) \text{ if } |y(s)| \leqq n \text{ and } s \in B^{(n)}, \\ &= 0 \text{ otherwise}. \end{aligned}$$

Then $y_n \in L^q(S)$ and, as proved above,

$$\begin{aligned} \|f\| \cdot \|x\| \geqq \langle |x| \cdot a(y), f\rangle &= \int_S |x(s)|\, |y(s)|\, m(ds) \\ &\geqq \int_S |x(s)|\, |y_n(s)|\, m(ds). \end{aligned}$$

If we take $x(s) = |y_n(s)|^{q/p}$ and apply Hölder's equality, we obtain

$$\int_S |x(s)|\, |y_n(s)|\, m(ds) = \left(\int_S |x(s)|^p\, m(ds)\right)^{1/p} \left(\int_S |y_n(s)|^q\, m(ds)\right)^{1/q}.$$

Hence $\|f\| \geqq \|y_n\| = \left(\int_S |y_n(s)|^q\, m(ds)\right)^{1/q}$, with the understanding that, when $p = 1$ we have $\|f\| \geqq \|y_n\| = \operatorname{essential\ sup}_{s \in S} |y_n(s)|$.

Therefore, by letting $n \to \infty$ and applying the Lebesgue-Fatou Lemma, we see that $y \in L^q(S)$ and $\|f\| \geqq \|y\|$. On the other hand, any $y \in L^q(S)$ defines an $f \in L^p(S)'$ by $\langle x, f\rangle = \int_S x(s)\, y(s)\, m(ds)$ as may be seen by Hölder's inequality, and this inequality shows that $\|f\| \leqq \|y\|$.

Remark. We have incidentally proved that

$$(l^p)' = (l^q) \quad (1 \leqq p < \infty \text{ and } p^{-1} + q^{-1} = 1).$$

Example 4. Let the measure space $(S, \mathfrak{B}, m)$ with $m(S) < \infty$ have the property that, for any $B \in \mathfrak{B}$ with $0 < m(B) = \delta < \infty$ and positive integer n, there exists a subset B_n of B such that $\delta(n+1)^{-1} \leqq m(B_n) \leqq \delta n^{-1}$. Then no other continuous linear functionals $\in M(S, \mathfrak{B}, m)'$ than the zero functional can exist.

Proof. Any $x \in L^1(S, \mathfrak{B}, m)$ belongs to $M(S, \mathfrak{B}, m)$ and the topology of $L^1(S, \mathfrak{B}, m)$ is stronger than that of $M(S, \mathfrak{B}, m)$. Thus any $f \in M(S, \mathfrak{B}, m)'$, when restricted to the functions of $L^1(S, \mathfrak{B}, m)$, defines a continuous linear functional $f_0 \in L^1(S, \mathfrak{B}, m)'$. Thus there exists a $y \in L^\infty(S, \mathfrak{B}, m)$ such that

$$\langle x, f\rangle = \langle x, f_0\rangle = \int_S x(s)\, y(s)\, m(ds) \quad \text{whenever } x \in L^1(S, \mathfrak{B}, m).$$

Since $L^1(S, \mathfrak{B}, m)$ is dense in $M(S, \mathfrak{B}, m)$ in the topology of $M(S, \mathfrak{B}, m)$, the condition $f \neq 0$ implies that $f_0 \neq 0$. Thus there exists an $\varepsilon > 0$ such that $B = \{s;\, |y(s)| \geqq \varepsilon\}$ has its measure $m(B) = \delta > 0$. Let $B_n \subseteq B$

be as in the hypothesis, and let $y(s) = re^{i\theta}$ for $s \in B$. Set $x_n(s) = e^{-i\theta}$ for $s \in B_n$ and $x_n(s) = 0$ otherwise. Then $z_n(s) = n x_n(s)$ converges to 0 asymptotically, that is, $s\text{-}\lim_{n\to\infty} z_n = 0$ in $M(S, \mathfrak{B}, m)$. But

$$\lim_{n\to\infty} \langle z_n, f\rangle = \lim_{n\to\infty} \langle z_n, f_0\rangle = \lim_{n\to\infty} \int_S z_n(s)\, y(s)\, m(ds) \geqq \delta\varepsilon > 0,$$

contrary to the continuity of the functional f.

Example 5. $L^\infty(S, \mathfrak{B}, m)'$.

Let an $f \in L^\infty(S, \mathfrak{B}, m)'$ be given and set, for any $B \in \mathfrak{B}$, $f(C_B) = \psi(B)$ where $C_B(s)$ is the defining function of the set B. We have then:

$$B_1 \cap B_2 = \emptyset \ \text{ implies } \ \psi(B_1 + B_2) = \psi(B_1) + \psi(B_2), \tag{6}$$

that is ψ is *finitely additive*,

the real part $\psi_1(B)$ and the imaginary part $\psi_2(B)$ of $\psi(B)$ are of *bounded total variation*, that is, $\sup_B |\psi_i(B)| < \infty$ $(i = 1, 2)$, (7)

ψ is *m-absolutely continuous*, that is $m(B) = 0$ implies $\psi(B) = 0$. (8)

The condition (6) is a consequence of the linearity of f, and (7) and (8) are clear from $|\psi(B)| \leqq ||f|| \cdot ||C_B||$.

For any $x \in L^\infty(S, \mathfrak{B}, m)$, we consider a partition of the sphere $\{z; |z| \leqq ||x||\}$ of the complex plane into a finite system of disjoint Baire sets $A_1, A_2, \ldots, A_n$ of diameters $\leqq \varepsilon$. If we set $B_i = \{s \in S; x(s) \in A_i\}$, then, no matter what point α_i we choose from A_i $(i = 1, 2, \ldots, n)$, we have

$$\left\| x - \sum_{i=1}^n \alpha_i C_{B_i} \right\| \leqq \varepsilon,$$

and so

$$\left| f(x) - \sum_{i=1}^n \alpha_i \psi(B_i) \right| \leqq ||f|| \cdot \varepsilon.$$

Thus, if we let $n \to \infty$ in such a way that $\varepsilon \downarrow 0$, we obtain

$$f(x) = \lim \sum_{i=1}^n \alpha_i \psi(B_i), \tag{9}$$

independently of the manner of partition $\{z; |z| \leqq ||x||\} = \sum_{i=1}^n A_i$ and choice of points α's. The limit on the right of (9) is called *Radon's integral* of $x(s)$ with respect to the *finitely additive* measure m. Thus

$$f(x) = \int_S x(s)\, \psi(ds) \ \text{(Radon's integral) whenever } x \in L^\infty(S, \mathfrak{B}, m), \tag{10}$$

and so

$$||f|| = \sup_{\operatorname{ess.sup}|x(s)| \leqq 1} \left| \int_S x(s)\, \psi(ds) \right|. \tag{11}$$

Conversely, it is easy to see that any ψ satisfying (6), (7) and (8) defines an $f \in L^\infty(S, \mathfrak{B}, m)'$ through (10) and that (11) is true.

Therefore, we have proved that $L^\infty(S, \mathfrak{B}, m)'$ is the space of all set functions ψ satisfying (6), (7) and (8) and normed by the right hand side of (11), the so-called *total variation* of ψ.

Remark. We have so far proved that $L^p(S, \mathfrak{B}, m)$ is reflexive when $1 < p < \infty$. However, the space $L^1(S, \mathfrak{B}, m)$ is, in general, not reflexive.

Example 6. $C(S)'$.

Let S be a compact topological space. Then the dual space $C(S)'$ of the space $C(S)$ of complex-valued continuous functions on S is given as follows. To any $f \in C(S)'$, there corresponds a uniquely determined complex Baire measure μ on S such that

$$f(x) = \int_S x(s)\,\mu(ds) \quad \text{whenever } x \in C(S), \tag{12}$$

and hence

$$\|f\| = \sup_{\substack{\sup|x(s)| \leq 1 \\ s}} \left| \int_S x(s)\,\mu(ds) \right| = \text{the total variation of } \mu \text{ on } S. \tag{13}$$

Conversely, any Baire measure μ on S such that the right side of (13) is finite, defines a continuous linear functional $f \in C(S)'$ through (12) and we have (13). Moreover, if we are concerned with a real functional f on a real B-space $C(S)$, then the corresponding measure μ is real-valued; if, moreover f is *positive*, in the sense that $f(x) \geqq 0$ for non-negative functions $x(s)$, then the corresponding measure μ is *positive*, i.e., $\mu(B) \geqq 0$ for every $B \in \mathfrak{B}$.

Remark. The result stated above is known as the F. Riesz-A. Markov-S. Kakutani theorem, and is one of the fundamental theorems in topological measures. For the proof, the reader is referred to standard text books on measure theory, e.g., P. R. Halmos [1] and N. Dunford-J. Schwartz [1].

References for Chapter IV

For the Hahn-Banach theorems and related topics, see Banach [1], Bourbaki [2] and Köthe [1]. It was Mazur [2] who noticed the importance of convex sets in normed linear spaces. The proof of Helly's theorem given in this book is due to Y. Mimura (unpublished).

V. Strong Convergence and Weak Convergence

In this chapter, we shall be concerned with certain basic facts pertaining to strong-, weak- and weak* convergences, including the comparison of the strong notion with the weak notion, e.g., strong- and weak measurability, and strong- and weak analyticity. We also discuss the integration of B-space-valued functions, that is, the theory of Bochner's integrals. The general theory of weak topologies and duality in locally convex linear topological spaces will be given in the Appendix.

1. The Weak Convergence and The Weak* Convergence

Weak Convergence

Definition 1. A sequence $\{x_n\}$ in a normed linear space X is said to be *weakly convergent* if a finite $\lim_{n\to\infty} f(x_n)$ exists for each $f \in X'_s$; $\{x_n\}$ is said to *converge weakly to* an element $x_\infty \in X$ if $\lim_{n\to\infty} f(x_n) = f(x_\infty)$ for all $f \in X'_s$. In the latter case, x_∞ is uniquely determined, in virtue of the Hahn-Banach theorem (Corollary 2 of Theorem 1 in Chapter IV, 6); we shall write $w\text{-}\lim_{n\to\infty} x_n = x_\infty$ or, in short, $x_n \to x_\infty$ weakly. X is said to be *sequentially weakly complete* if every weakly convergent sequence of X converges weakly to an element of X.

Example. Let $\{x_n(s)\}$ be a sequence of equi-bounded continuous functions of $C[0, 1]$ which is convergent to a discontinuous function $z(s)$ on $[0, 1]$. Then, since $C[0, 1]'$ is the space of Baire measures on $[0, 1]$ of bounded total variation, we see easily that $\{x_n(s)\}$ gives an example of a weakly convergent sequence of $C[0,1]$ which does not converge weakly to an element of $C[0, 1]$.

Theorem 1. i) $s\text{-}\lim_{n\to\infty} x_n = x_\infty$ implies $w\text{-}\lim_{n\to\infty} x_n = x_\infty$, but not conversely. ii) A weakly convergent sequence $\{x_n\}$ is strongly bounded, and, in particular, if $w\text{-}\lim_{n\to\infty} x_n = x_\infty$, then $\{\|x_n\|\}$ is bounded and $\|x_\infty\| \leqq \varliminf_{n\to\infty} \|x_n\|$.

Proof. i) The first part is clear from $|f(x_n) - f(x_\infty)| \leqq \|f\| \cdot \|x_n - x_\infty\|$. The second part is proved by considering the sequence $\{x_n\}$ in the Hilbert space (l^2):

$$x_n = \{\xi_m^{(n)}\} \text{ where } \xi_m^{(n)} = \delta_{nm} \ (= 1 \text{ or } 0 \text{ according as } n = m \text{ or not}).$$

For, the value of a continuous linear functional $\in (l^2)'$ at $x = \{\xi_n\}$ is given by $\sum_{n=1}^{\infty} \xi_n \bar{\eta}_n$ with some $\{\eta_n\} \in (l^2)$; consequently, $w\text{-}\lim_{n\to\infty} x_n = 0$ but $\{x_n\}$ does not converge strongly to 0 because $\|x_n\| = 1$ $(n = 1, 2, \ldots)$. ii) Consider the sequence of continuous linear functionals X_n defined on the B-space X'_s by $X_n(f) = \langle x_n, f\rangle$, and apply the resonance theorem in Chapter II, 1.

Theorem 2 (Mazur). Let $w\text{-}\lim_{n\to\infty} x_n = x_\infty$ in a normed linear space X. Then there exists, for any $\varepsilon > 0$, a *convex combination* $\sum_{j=1}^{n} \alpha_j x_j$ $\left(\alpha_j \geqq 0, \ \sum_{j=1}^{n} \alpha_j = 1\right)$ of x_j's such that $\left\| x_\infty - \sum_{j=1}^{n} \alpha_j x_j \right\| \leqq \varepsilon$.

Proof. Consider the totality M_1 of elements of the form $\sum_{j=1}^{n} \alpha_j x_j$ with $\alpha_j \geqq 0$, $\sum_{j=1}^{n} \alpha_j = 1$. We may assume that $0 \in M_1$ by replacing x_∞ and x_j by $(x_\infty - x_1)$ and $(x_j - x_1)$, respectively. Suppose that $||x_\infty - u|| > \varepsilon$ for every $u \in M_1$. The set $M = \{v \in X;\ ||v - u|| \leqq \varepsilon/2$ for some $u \in M_1\}$ is a convex neighbourhood of 0 of X and $||x_\infty - v|| > \varepsilon/2$ for all $v \in M$. Let $p(y)$ be the Minkowski functional of M. Since $x_\infty = \beta^{-1} u_0$ with $p(u_0) = 1$ and $0 < \beta < 1$, we must have $p(x_\infty) = \beta^{-1} > 1$. Consider a real linear subspace $X_1 = \{x \in X;\ x = \gamma u_0, -\infty < \gamma < \infty\}$ and put $f_1(x) = \gamma$ for $x = \gamma u_0 \in X_1$. This real linear functional f_1 on X_1 satisfies $f(x) \leqq p(x)$ on X_1. Thus, by the Hahn-Banach extension theorem in Chapter IV, 1, there exists a real linear extension f of f_1 defined on the real linear space X and such that $f(x) \leqq p(x)$ on X. M being a neighbourhood of 0, the Minkowski functional $p(x)$ is continuous in x. Hence f is a continuous real linear functional defined on the real linear normed space X. We have, moreover,

$$\sup_{x \in M_1} f(x) \leqq \sup_{x \in M} f(x) \leqq \sup_{x \in M} p(x) = 1 < \beta^{-1} = f(\beta^{-1} u_0) = f(x_\infty).$$

Therefore, it is easy to see that x_∞ cannot be a weak accumulation point of M_1, contrary to $x_\infty = w\text{-}\lim_{n\to\infty} x_n$.

Theorem 3. A sequence $\{x_n\}$ of a normed linear space X converges weakly to an element $x_\infty \in X$ iff the following two conditions are satisfied: i) $\sup_{n \leqq 1} ||x_n|| < \infty$, and ii) $\lim_{n\to\infty} f(x_n) = f(x_\infty)$ for every f from any strongly dense subset D' of X'_s.

Proof. We have only to prove the sufficiency. For any $g \in X'_s$ and $\varepsilon > 0$, there exists an $f \in D'$ such that $||g - f|| < \varepsilon$. Thus

$$\begin{aligned} |g(x_n) - g(x_\infty)| &\leqq |g(x_n) - f(x_n)| + |f(x_n) - f(x_\infty)| + |f(x_\infty) - g(x_\infty)| \\ &\leqq \varepsilon\, ||x_n|| + |f(x_n) - f(x_\infty)| + \varepsilon\, ||x_\infty||, \end{aligned}$$

and hence $\lim |g(x_n) - g(x_\infty)| \leqq 2\varepsilon \sup_{\infty \geqq n \geqq 1} ||x_n||$. This proves that $\lim_{n\to\infty} g(x_n) = g(x_\infty)$.

Theorem 4. A sequence $\{x_n\}$ in $L^1(S, \mathfrak{B}, m)$ converges weakly to an element $x \in L^1(S, \mathfrak{B}, m)$ iff $\{||x_n||\}$ is bounded and a finite $\lim_{n\to\infty} \int_B x_n(s)\, m(ds)$ exists for every $B \in \mathfrak{B}$.

Proof. The "only if" part is clear since the defining function $C_B(s)$ of $B \in \mathfrak{B}$ belongs to $L^\infty(S, \mathfrak{B}, m) = L^1(S, \mathfrak{B}, m)'$.

The proof of the "if" part. The set function $\psi(B) = \lim_{n\to\infty} \int_B x_n(s)\, m(d,s)$ $B \in \mathfrak{B}$, is σ-additive and m-absolutely continuous by the Vitali-Hahn-Saks theorem. Hence, by the differentiation theorem of LEBESGUE-NIKODYM, there exists an $x_\infty \in L^1(S, \mathfrak{B}, m)$ such that

$$\lim_{n\to\infty} \int_B x_n(s)\, m(ds) = \int_B x_\infty(s)\, m(ds) \text{ for all } B \in \mathfrak{B}.$$

Thus, for any decomposition $S = \sum_{j=1}^{k} B_j$ with $B_j \in \mathfrak{B}$, we have

$$\lim_{n\to\infty} \int_S x_n(s)\, y(s)\, m(ds) = \int_S x_\infty(s)\, y(s)\, m(ds), \; y(s) = \sum_{j=1}^{k} \alpha_j C_{B_j}(s).$$

Since such functions as $y(s)$ constitute a strongly dense subset of the space $L^\infty(S, \mathfrak{B}, m) = L^1(S, \mathfrak{B}, m)'$, we see that the "if" part is true by Theorem 3.

Theorem 5. Let $\{x_n\}$ converge weakly to x_∞ in $L^1(S, \mathfrak{B}, m)$. Then $\{x_n\}$ converges strongly to x_∞ iff $\{x_n(s)\}$ converges to $x_\infty(s)$ in m-measure on every $\mathfrak{B}$-measurable set B such that $m(B) < \infty$.

Remark. $\{x_n(s)\}$ is said to converge to $x_\infty(s)$ in m-measure on B, if, for any $\varepsilon > 0$, the m-measure of the set $\{s \in B; |x_n(s) - x_\infty(s)| \geqq \varepsilon\}$ tends to zero as $n \to \infty$ (see the Proposition in Chapter I, 4). The space (l^1) is an example of $L^1(S, \mathfrak{B}, m)$ for which $S = \{1, 2, \ldots\}$ and $m(\{n\}) = 1$ for $n = 1, 2, \ldots$ In this case, we have $(l^1)' = (l^\infty)$ so that the weak convergence of $\{x_n\}$, $x_n = (\xi_1^{(n)}, \xi_2^{(n)}, \ldots, \xi_k^{(n)}, \ldots)$, to $x_\infty = (\xi_1^{(\infty)}, \xi_2^{(\infty)}, \ldots, \xi_k^{(\infty)}, \ldots)$ implies that $\lim_{n\to\infty} \xi_k^{(n)} = \xi_k^{(\infty)}$ $(k = 1, 2, \ldots)$, as may be seen by taking $f \in (l^1)'$ in such a way that $f(x) = \langle x, f\rangle = \xi_k$ for $x = \{\xi_j\} \in (l^1)$. Thus, in the present case, $\{x_n\}$ converges to x_∞ in m-measure on every $\mathfrak{B}$-measurable set B of finite m-measure. In this way we obtain the

Corollary (I. SCHUR). In the space (l^1), if a sequence $\{x_n\}$ converges weakly to an $x_\infty \in (l^1)$, then $s\text{-}\lim_{n\to\infty} x_n = x_\infty$.

Proof of Theorem 5. Since the strong convergence in $L^1(S, \mathfrak{B}, m)$ implies the convergence in m-measure, the "only if" part is clear. We shall prove the "if" part. The sequence $\{x_n - x_\infty\}$ converges weakly to 0, and so

$$\lim_{n\to\infty} \int_B (x_n(s) - x_\infty(s))\, m(ds) = 0 \text{ for every } B \in \mathfrak{B}. \tag{1}$$

Consider the sequence of non-negative measures

$$\psi_n(B) = \int_B |x_n(s) - x_\infty(s)|\, m(ds), \; B \in \mathfrak{B}.$$

Then we have

$$\lim_{k\to\infty} \psi_n(B_k) = 0 \text{ uniformly in } n, \text{ for any decreasing sequence } \{B_k\} \text{ of sets} \in \mathfrak{B} \text{ such that } \bigcap_{k=1}^{\infty} B_k = \emptyset. \tag{2}$$

If otherwise, there exists an $\varepsilon > 0$ such that for each k there exists some

n_k for which $\lim_{k\to\infty} n_k = \infty$ and $\psi_{n_k}(B_k) > \varepsilon$. Consequently, we must have $\int_{B_k} |Re(x_{n_k}(s) - x_\infty(s))| \, m(ds) > \varepsilon/\sqrt{2}$ or $\int_{B_k} |Im\,(x_{n_k}(s) - x_\infty(s))| \, m(ds) > \varepsilon/\sqrt{2}$, and so there must exist some $B'_k \subseteqq B_k$ such that

$$\left|\int_{B'_k} (x_{n_k}(s) - x_\infty(s)) \, m(ds)\right| > \varepsilon/2\sqrt{2} \quad (k = 1, 2, \ldots),$$

contrary to the fact that, in virtue of (1), the m-absolute continuity of the sequence of measures $\varphi_n(B) = \int_B (x_n(s) - x_\infty(s)) \, m(d)s$ is uniform in n (see the proof of the Vitali-Hahn-Saks Theorem in Chapter II, 2).

Next let B_0 be any set of $\mathfrak{B}$ such that $m(B_0) < \infty$. We shall show that

$$\lim_{n\to\infty} \psi_n(B_0) = 0. \tag{3}$$

Suppose there exist an $\varepsilon > 0$ and a subsequence $\{\psi_{n'}\}$ of $\{\psi_n\}$ such that

$$\psi_{n'}(B_0) > \varepsilon \quad (n = 1, 2, \ldots). \tag{4}$$

By the hypothesis that $\{(x_n(s) - x_\infty(s))\}$ converges to 0 in m-measure on B_0, there exist a subsequence $\{(x_{n''}(s) - x_\infty(s))\}$ of $\{(x_{n'}(s) - x_\infty(s))\}$ and some sets $B''_n \subseteqq B_0$ such that $m(B''_n) \leqq 2^{-n}$and $|x_{n''}(s) - x_\infty(s)| < \varepsilon/m(B_0)$ on $(B_0 - B''_n)$. We put $B_k = \bigcup_{n=k}^{\infty} B''_n$. Then $\{B_k\}$ is a decreasing sequence such that

$$m\left(\bigcap_{k=1}^{\infty} B_k\right) \leqq \sum_{n=k}^{\infty} m(B''_n) \leqq 2^{-k+1} \quad (k = 1, 2, \ldots) \text{ and so } m\left(\bigcap_{k=1}^{\infty} B_k\right) = 0.$$

Hence, by (1) and the Corollary of the Vitali-Hahn-Saks Theorem referred to above, $\lim_{k\to\infty} \psi_n(B_k) = 0$ uniformly in n. Therefore

$$\psi_{n''}(B_0) \leqq \psi_{n''}(B''_n) + \varepsilon\, m(B_0)^{-1} \cdot m(B_0 - B''_n) \to (\leqq \varepsilon) \text{ as } n \to \infty,$$

contrary to (4). This proves (3).

Now we take a sequence $\{B'_k\}$ of sets $\in \mathfrak{B}$ such that $m(B'_k) < \infty$ $(k = 1, 2, \ldots)$ and $S = \bigcup_{k=1}^{\infty} B'_k$. Then

$$\int_S |x_n(s) - x_\infty(s)| \, m(ds) = \int_{\bigcup_{k=1}^{t} B'_k} + \int_{S - \bigcup_{k=1}^{t} B'_k}$$

By (3) the first term on the right tends to zero as $n \to \infty$ for fixed t, and the second term on the right tends, by (2), to zero as $t \to \infty$ uniformly in n. Therefore we have proved that $s\text{-}\lim_{n\to\infty} x_n = x_\infty$ in $L^1(S, \mathfrak{B}, m)$.

A similar situation in the case of the space $\mathfrak{D}(\Omega)'$ is given by

Theorem 6. Let $\{T_n\}$ be a sequence of generalized functions $\in \mathfrak{D}(\Omega)'$. If $\lim_{n\to\infty} T_n = T$ in the weak* topology of $\mathfrak{D}(\Omega)'$, then $\lim_{n\to\infty} T_n = T$ in the strong topology of $\mathfrak{D}(\Omega)'$.

Proof. The strong topology of the space $\mathfrak{D}(\Omega)'$ is defined (see Definition 1 in Chapter IV, 7) through the family of semi-norms

$$p_B(T) = \sup_{\varphi \in \mathfrak{B}} |T(\varphi)|, \text{ where } \mathfrak{B} \text{ is any bounded set of } \mathfrak{D}(\Omega).$$

The weak* topology of the space $\mathfrak{D}(\Omega)'$ is defined through the family of semi-norms

$$p_F(T) = \sup_{\varphi \in \mathfrak{F}} |T(\varphi)|, \text{ where } \mathfrak{F} \text{ is any finite set of } \mathfrak{D}(\Omega).$$

Thus $\lim_{n\to\infty} T_n = T$ in the weak* topology of $\mathfrak{D}(\Omega)'$ is precisely $\lim_{n\to\infty} T_n = T(\mathfrak{D}(\Omega)')$ defined in Chapter II, 3.

Let $\mathfrak{B}$ be any bounded set of $\mathfrak{D}(\Omega)$. Then there exists a compact subset K in Ω such that $\operatorname{supp}(\varphi) \subseteq K$ for any $\varphi \in \mathfrak{B}$ and $\sup_{x\in K, \varphi \in \mathfrak{B}} |D^j\varphi(x)| < \infty$ for any differential operator D^j (Theorem 1 in Chapter I, 8). Thus, by the Ascoli-Arzelà Theorem, $\mathfrak{B}$ is compact in $\mathfrak{D}_K(\Omega)$. We apply the uniform boundedness theorem to the sequence $\{T_n - T\}$ to the effect that, for any $\varepsilon > 0$, there exists a neighbourhood U of 0 of $\mathfrak{D}_K(\Omega)$ such that

$$\sup_{n;\, \varphi \in U} |(T_n - T)(\varphi)| < \varepsilon.$$

The compact subset $\mathfrak{B}$ of $\mathfrak{D}_K(\Omega)$ is covered by a finite system of sets of the form $\varphi_i + U$, where $\varphi_i \in \mathfrak{B}$ $(i = 1, 2, \ldots, k)$. Hence

$$\begin{aligned} |(T_n - T)(\varphi_i + u)| &\leqq |(T_n - T)(\varphi_i)| + |(T_n - T)(u)| \\ &\leqq |(T_n - T)(\varphi_i)| + \varepsilon \text{ for any } u \in U. \end{aligned}$$

Since $\lim_{n\to\infty} (T_n - T)(\varphi_i) = 0$ for $i = 1, 2, \ldots, k$, we have

$$\lim_{n\to\infty} (T_n - T)(\varphi) = 0 \text{ uniformly in } \varphi \in \mathfrak{B}.$$

This proves our Theorem.

Theorem 7. A reflexive B-space X is sequentially weakly complete.

Proof. Let a sequence $\{x_n\}$ of X be weakly convergent. Each x_n defines a continuous linear functional X_n on X'_s by $X_n(x') = \langle x_n, x' \rangle$. Since X'_s is a B-space (Theorem 1 in Chapter IV, 8), we may apply the resonance theorem. Thus a continuous linear functional on X'_s is defined by a finite $\lim_{n\to\infty} X_n(x')$ which exists by hypothesis. Since X is reflexive, there exists an $x_\infty \in X$ such that $\langle x_\infty, x' \rangle = \lim_{n\to\infty} X_n(x') = \lim_{n\to\infty} \langle x_n, x' \rangle$, that is, $x_\infty = w\text{-}\lim_{n\to\infty} x_n$.

Theorem 8. Let X be a Hilbert space. If a sequence $\{x_n\}$ of X converges weakly to $x_\infty \in X$, then $s\text{-}\lim_{n\to\infty} x_n = x_\infty$ iff $\lim_{n\to\infty} ||x_n|| = ||x_\infty||$.

Proof. The "only if" part is clear from the continuity of the norm. The "if" part is clear from the equality

$$\begin{aligned} ||x_n - x_\infty||^2 &= (x_n - x_\infty, x_n - x_\infty) \\ &= ||x_n||^2 - (x_n, x_\infty) - (x_\infty, x_n) + ||x_\infty||^2. \end{aligned}$$

In fact, the limit, as $n \to \infty$, of the right hand side is $||x_\infty||^2 - ||x_\infty||^2 - ||x_\infty||^2 + ||x_\infty||^2 = 0$.

Weak* Convergence

Definition 2. A sequence $\{f_n\}$ in the dual space X'_s of a normed linear space X is said to be *weakly* convergent* if a finite $\lim\limits_{n\to\infty} f_n(x)$ exists for every $x \in X$; $\{f_n\}$ is said to *converge weakly* to* an element $f_\infty \in X'_s$ if $\lim\limits_{n\to\infty} f_n(x) = f_\infty(x)$ for all $x \in X$. In the latter case, we write $w^*\text{-}\lim\limits_{n\to\infty} f_n = f_\infty$ or, in short, $f_n \to f_\infty$ weakly*.

Theorem 9. i) $s\text{-}\lim\limits_{n\to\infty} f_n = f_\infty$ implies $w^*\text{-}\lim\limits_{n\to\infty} f_n = f_\infty$, but not conversely. ii) If X is a B-space, then a weakly* convergent sequence $\{f_n\} \subseteq X'_s$ converges weakly* to an element $f_\infty \in X'_s$ and $||f_\infty|| \leqq \varliminf\limits_{n\to\infty} ||f_n||$.

Proof. (i) The first part is clear from $|f_n(x) - f_\infty(x)| \leqq ||f_n - f_\infty|| \cdot ||x||$. The second part is proved by the counter example given in the proof of Theorem 1. (ii) By the resonance theorem, we see that $f_\infty(x) = \lim\limits_{n\to\infty} f_n(x)$ is a continuous linear functional on X and $||f_\infty|| \leqq \varliminf\limits_{n\to\infty} ||f_n||$.

Theorem 10. If X is a B-space, then a sequence $\{f_n\} \subseteq X'_s$ converges weakly* to an element $f_\infty \in X'_s$ iff (i) $\{||f_n||\}$ is bounded, and ii) $\lim\limits_{n\to\infty} f_n(x) = f_\infty(x)$ on a strongly dense subset of X.

Proof. The proof is similar to that of Theorem 3.

Strong and Weak Closure

Theorem 11. Let X be a locally convex linear topological space, and M a closed linear subspace of X. Then M is closed in the weak topology of X.

Proof. If otherwise, there exists a point $x_0 \in X - M$ such that x_0 is an accumulation point of the set M in the weak topology of X. Then, by the Corollary of Theorem 3 in Chapter IV, 6, there exists a continuous linear functional f_0 on X such that $f_0(x_0) = 1$ and $f_0(x) = 0$ on M. Hence x_0 cannot be an accumulation point of the set M in the weak topology of X.

2. The Local Sequential Weak Compactness of Reflexive B-spaces. The Uniform Convexity

Theorem 1. Let X be a reflexive B-space, and let $\{x_n\}$ be any sequence which is norm bounded. Then we can choose a subsequence $\{x_{n'}\}$ which converges weakly to an element of X.

We will prove this Theorem under the assumption that X is separable, since concrete function spaces appearing in applications of this Theorem are mostly separable. The general case of a non-separable space will be treated in the Appendix.

Lemma. If the strong dual X'_s of a normed linear space X is separable, then so is X.

Proof. Let $\{x'_n\}$ be a countable sequence which is strongly dense on the surface of the unit sphere $\{x' \in X'_s;\ \|x'\| = 1\}$ of X'_s. Choose $x_n \in X$ so that $\|x_n\| = 1$ and $|\langle x_n, x'_n\rangle| \geqq 1/2$. Let M be the closed linear subspace of X spanned by the sequence $\{x_n\}$. Suppose $M \neq X$ and $x_0 \in X - M$. By Corollary of Mazur's Theorem 3 in Chapter IV, 6, there exists an $x'_0 \in X'_s$ such that $\|x'_0\| = 1$, $\langle x_0, x'_0\rangle \neq 0$ and $\langle x, x'_0\rangle = 0$ whenever $x \in M$. Thus $\langle x_n, x'_0\rangle = 0$ $(n = 1, 2, \ldots)$, and so $1/2 \leqq |\langle x_n, x'_n\rangle| \leqq |\langle x_n, x'_n\rangle - \langle x_n, x'_0\rangle| + |\langle x_n, x'_0\rangle|$ which implies that $1/2 \leqq \|x_n\|\,\|x'_n - x'_0\| = \|x'_n - x'_0\|$. This is a contradiction to the fact that $\{x'_n\}$ is strongly dense on the surface of the unit sphere of X'_s. Thus $M = X$, and so linear combinations with rational coefficients of $\{x_n\}$ are dense in X. This proves our Lemma.

Proof of Theorem 1. As we have remarked above, we assume that X is separable and so $(X'_s)'_s = X$ is separable also. By the preceding Lemma, X'_s is separable. Let $\{x'_n\}$ be a countable sequence which is strongly dense in X'_s. Since $\{x_n\}$ is norm bounded, the sequence $\{\langle x_n, x'_1\rangle\}$ is bounded. Thus there exists a subsequence $\{x_{n_1}\}$ for which the sequence $\{\langle x_{n_1}, x'_1\rangle\}$ is convergent. Since the sequence $\{\langle x_{n_1}, x'_2\rangle\}$ is bounded, there exists a subsequence $\{x_{n_2}\}$ of $\{x_{n_1}\}$ such that $\{\langle x_{n_2}, x'_2\rangle\}$ is convergent. Proceeding in this way, we can choose a subsequence $\{x_{n_{i+1}}\}$ of the sequence $\{x_{n_i}\}$ such that the sequence of numbers $\{\langle x_{n_{i+1}}, x'_j\rangle\}$ converges for $j = 1, 2, \ldots, i+1$. Hence the diagonal subsequence $\{x_{n_n}\}$ of the original sequence $\{x_n\}$ satisfies the condition that the sequence $\{\langle x_{n_n}, x'_j\rangle\}$ converges for $j = 1, 2, \ldots$ Thus, by Theorem 3 in the preceding section, $\lim_{n\to\infty} \langle x_{n_n}, x'\rangle$ exists and is finite for every $x' \in X'$. Hence, by Theorem 7 of the preceding section, we see that $w\text{-}\lim_{n\to\infty} x_{n_n}$ exists.

Milman's Theorem

We owe to D. P. Milman a theorem that a B-space is reflexive when it is *uniformly convex* in the sense that, for any $\varepsilon > 0$, there exists a

$\delta = \delta(\varepsilon) > 0$ such that $\|x\| \leqq 1$, $\|y\| \leqq 1$ and $\|x - y\| \geqq \varepsilon$ implies $\|x + y\| \leqq 2(1 - \delta)$. A pre-Hilbert space is uniformly convex as may be seen from the formula

$$\|x + y\|^2 + \|x - y\|^2 = 2(\|x\|^2 + \|y\|^2)$$

valid in such a space. It is known that, for $1 < p < \infty$, the spaces L^p and (l^p) are uniformly convex (see J. A. CLARKSON [1]).

Theorem 2 (MILMAN [1]). A uniformly convex B-space X is reflexive.

Proof (due to S. KAKUTANI). Given an $x_0'' \in (X_s')_s'$ with $\|x_0''\| = 1$. Then there exists a sequence $\{f_n\} \subseteq X_s'$ with $\|f_n\| = 1$, $x_0''(f_n) \geqq 1 - n^{-1}$ $(n = 1, 2, \ldots)$. By Theorem 5 in Chapter IV, 6, there exists, for every n, an $x_n \in X$ such that

$$f_i(x_n) = x_0''(f_i) \ (i = 1, 2, \ldots, n) \quad \text{and} \quad \|x_n\| \leqq \|x_0''\| + n^{-1} = 1 + n^{-1}.$$

Since

$$1 - n^{-1} \leqq x_0''(f_n) = f_n(x_n) \leqq \|f_n\| \, \|x_n\| = \|x_n\| \leqq 1 + n^{-1},$$

we must have $\lim_{n\to\infty} \|x_n\| = 1$.

If the sequence $\{x_n\}$ does not converge strongly, there exists an $\varepsilon > 0$ and $n_1 < m_1 < n_2 < m_2 < \cdots < n_k < m_k < \cdots$ such that $\varepsilon \leqq \|x_{n_k} - x_{m_k}\|$ $(k = 1, 2, \ldots)$. Thus, by $\lim_{n\to\infty} \|x_n\| = 1$ and the uniform convexity of X, we obtain $\overline{\lim_{k\to\infty}} \|x_{n_k} + x_{m_k}\| \leqq 2(1 - \delta(\varepsilon)) < 2$. But, since $n_k < m_k$, $f_{n_k}(x_{n_k}) = f_{n_k}(x_{m_k}) = x_0''(f_{n_k})$ and so

$$2(1 - n_k^{-1}) \leqq 2x_0''(f_{n_k}) = f_{n_k}(x_{n_k} + x_{m_k}) \leqq \|f_{n_k}\| \cdot \|x_{n_k} + x_{m_k}\|.$$

Hence, by $\|f_{n_k}\| = 1$, we obtain a contradiction $\overline{\lim_{k\to\infty}} \|x_{n_k} + x_{m_k}\| \geqq 2$.

We have thus proved the existence of $s\text{-}\lim_{n\to\infty} x_n = x_0$, and x_0 satisfies

$$\|x_0\| = 1, \ f_i(x_0) = x_0''(f_i) \ (i = 1, 2, \ldots). \tag{1}$$

We show that the solution of the above equations in (1) is unique. Otherwise, there exists an $\hat{x}_0 \neq x_0$ which satisfies the same equations. By the uniform convexity, $\|\hat{x}_0 + x_0\| < 2$. We also have $f_i(\hat{x}_0 + x_0) = 2x_0''(f_i)$ $(i = 1, 2, \ldots)$. Thus

$$2(1 - i^{-1}) \leqq 2x_0''(f_i) = f_i(\hat{x}_0 + x_0) \leqq \|f_i\| \, \|\hat{x}_0 + x_0\| = \|\hat{x}_0 + x_0\|,$$

and so $\|\hat{x}_0 + x_0\| \geqq \lim_{i\to\infty} 2(1 - i^{-1}) = 2$ which is a contradiction.

Finally let f_0 be any point of X_s'. If we show that $f_0(x_0) = x_0''(f_0)$, then $(X_s')_s' \subseteq X$ and the reflexivity of X is proved. To prove that $f_0(x_0) = x_0''(f_0)$, we take $f_0, f_1, \ldots, f_n, \ldots$ in place of $f_1, f_2, \ldots, f_n, \ldots$ above, and hence we obtain $\hat{x}_0 \in X$ such that

$$\|\hat{x}_0\| = 1, \ f_i(\hat{x}_0) = x_0''(f_i) \ (i = 0, 1, \ldots, n, \ldots);$$

we must have $\hat{x}_0 = x_0$ by virtue of the uniqueness just proved above, and so the proof of Theorem 2 is completed.

3. Dunford's Theorem and The Gelfand-Mazur Theorem

Definition 1. Let Z be an open domain of the complex plane. A mapping $x(\zeta)$ defined in Z with values in a B-space X is called *weakly holomorphic* in ζ in the domain Z if, for each $f \in X'$, the numerical function

$$f(x(\zeta)) = \langle x(\zeta), f \rangle$$

of ζ is holomorphic in Z.

Theorem 1 (N. Dunford [2]). If $x(\zeta)$ is weakly holomorphic in Z, then there exists a mapping $x'(\zeta)$ defined in Z with values in X such that, for each $\zeta_0 \in Z$, we have

$$\operatorname*{s-lim}_{h \to 0} h^{-1}(x(\zeta_0 + h) - x(\zeta_0)) = x'(\zeta_0).$$

In other words, the weak holomorphic property implies the *strong holomorphic* property.

Proof. Let C be a rectifiable Jordan curve such that the closed bounded domain $\overline{C}$ enclosed by C lies entirely in Z and $\zeta_0 \in \overline{C} - C$. Let Z_0 be any open complex domain $\ni \zeta_0$ such that its closure lies in the interior of $\overline{C}$. Then, by Cauchy's integral representation, we have

$$f(x(\zeta_0)) = \frac{1}{2\pi i} \int_C \frac{f(x(\zeta))}{\zeta - \zeta_0} d\zeta.$$

Hence, if both $\zeta_0 + h$ and $\zeta_0 + g$ belong to Z_0,

$$(h-g)^{-1}\left\{\frac{f(x(\zeta_0+h)) - f(x(\zeta_0))}{h} - \frac{f(x(\zeta_0+g)) - f(x(\zeta_0))}{g}\right\}$$
$$= \frac{1}{2\pi i} \int_C f(x(\zeta)) \left\{\frac{1}{(\zeta-\zeta_0-h)\,(\zeta-\zeta_0-g)\,(\zeta-\zeta_0)}\right\} d\zeta.$$

By the assumption, the distance between Z_0 and C is positive. Hence, for fixed $f \in X'$, the absolute value of the right hand side is uniformly bounded when $\zeta_0, \zeta_0 + h$ and $\zeta_0 + g$ range over Z_0. Thus, by the resonance theorem, we have

$$\sup_{\zeta_0, \zeta_0+h, \zeta_0+g \in Z_0} \frac{1}{|h-g|} \left\| \left\{\frac{x(\zeta_0+h) - x(\zeta_0)}{h} - \frac{x(\zeta_0+g) - x(\zeta_0)}{g}\right\} \right\| < \infty.$$

Therefore, by the completeness of the space X, $x(\zeta)$ is strongly differentiable at every $\zeta_0 \in Z$.

Corollary 1 (Cauchy's integral theorem). The strong differentiability of $x(\zeta)$ implies its strong continuity in ζ. Thus we can define the curvilinear integral $\int_C x(\zeta)\, d\zeta$ with values in X. Actually we can prove that $\int_C x(\zeta)\, d\zeta = 0$, the zero vector of X.

Proof. We have, by the continuity and the linearity of $f \in X'$,

$$f\Big(\int_C x(\zeta)\, d\zeta\Big) = \int_C f(x(\zeta))\, d\zeta.$$

But the right hand side is zero, because of the ordinary Cauchy integral theorem. Since $f \in X'$ was arbitrary, we must have $\int_C x(\zeta)\, d\zeta = 0$ by Corollary 2 of Theorem 1 in Chapter IV, 6.

From the above Corollary, we can derive other Corollaries, as in the ordinary theory of functions of a complex variable.

Corollary 2 (Cauchy's integral representation).

$$x(\zeta_0) = \frac{1}{2\pi i} \int_C \frac{x(\zeta)}{\zeta - \zeta_0}\, d\zeta \quad \text{for any interior point } \zeta_0 \text{ of } \overline{C}.$$

Corollary 3 (Taylor's expansion). For any point ζ_0 which is in the interior of the closed domain $\overline{C}$, the Taylor expansion of $x(\zeta)$ at $\zeta = \zeta_0$ converges strongly in the interior of the circle with centre at ζ_0, if this circle does not extend outside of $\overline{C}$:

$$x(\zeta) = \sum_{n=0}^{\infty} (n!)^{-1} (\zeta - \zeta_0)^n\, x^{(n)}(\zeta_0), \quad \text{where}$$

$$x^{(n)}(\zeta_0) = \frac{n!}{2\pi i} \int_C \frac{x(\zeta)}{(\zeta - \zeta_0)^{n+1}}\, d\zeta.$$

Corollary 4 (Liouville's theorem). If $x(\zeta)$ is (strongly) holomorphic in the whole finite plane: $|\zeta| < \infty$, and $\sup_{|\zeta|<\infty} |x(\zeta)| < \infty$, then $x(\zeta)$ must reduce to a constant vector $x(0)$.

Proof. If we take $|\zeta| = r$ for the curve C, then, as $r \to \infty$,

$$\|x^{(n)}(0)\| = \frac{n!}{2\pi} \sup_{|\zeta|<\infty} \|x(\zeta)\| \int_C \frac{|d\zeta|}{r^{n+1}} \to 0 \quad (n = 1, 2, \ldots).$$

Hence the Taylor expansion of $x(\zeta)$ at $\zeta = 0$ reduces to the constant term $x(0)$ only.

We shall now apply Corollary 4 to the proof of the Gelfand-Mazur theorem. We first give

Definition 2. A commutative field X over the field of complex numbers is called a *normed field*, if it is also a B-space such that the following conditions are satisfied:

$$\begin{aligned} &\|e\| = 1, \text{ where } e \text{ is the } \textit{unit} \text{ of the multiplication in } X, \\ &\|xy\| \leq \|x\|\,\|y\|, \text{ where } xy \text{ is the multiplication in } X. \end{aligned} \tag{1}$$

Theorem 2 (Gelfand [2]-Mazur [1]). A normed field X is isometrically isomorphic to the complex number field. In other words, every element x of X is of the form $x = \xi e$ where ξ is a complex number.

Proof. Assume the contrary, and let there exist an $x \in X$ such that $(x - \xi e) \neq 0$ for any complex number ξ. Since X is a field, the non-zero element $(x - \xi e)$ has the inverse $(x - \xi e)^{-1} \in X$.

We shall prove that $(x - \lambda e)^{-1}$ is (strongly) holomorphic in λ for $|\lambda| < \infty$. We have, in fact,

$$\begin{aligned} &h^{-1}((x - (\lambda + h)e)^{-1} - (x - \lambda e)^{-1}) \\ &= h^{-1}(x - (\lambda + h)e)^{-1}\{e - (x - (\lambda + h)e)(x - \lambda e)^{-1}\} \\ &= h^{-1}(x - (\lambda + h)e)^{-1}\{e - e + h(x - \lambda e)^{-1}\} \\ &= (x - (\lambda + h)e)^{-1}(x - \lambda e)^{-1}. \end{aligned}$$

On the other hand, for sufficiently small $|h|$, the series $y^{-1}\left(e + \sum_{n=1}^{\infty}(hy^{-1})^n\right)$, where $y = (x - \lambda e)$, converges by (1), and it represents the inverse $(y - he)^{-1} = y^{-1}(e - hy^{-1})^{-1}$, as may be seen by multiplying the series by $(y - he)$. Hence, by the strong continuity in h of the series, we can prove that $(x - \lambda e)^{-1}$ is (strongly) holomorphic in λ with the strong derivative $(x - \lambda e)^{-2}$.

Now, if $|\lambda| \geqq 2\,||x||$, then, as above, $(x - \lambda e)^{-1} = -\lambda^{-1}(e - \lambda^{-1}x)^{-1}$ $= -\lambda^{-1}\left(e + \sum_{n=1}^{\infty}(\lambda^{-1}x)^n\right)$ and so

$$||(x - \lambda e)^{-1}|| \leqq |\lambda^{-1}|\left(1 + \sum_{n=1}^{\infty}(1/2)^n\right) \to 0 \quad \text{as} \quad |\lambda| \to \infty.$$

Moreover, the function $(x - \lambda e)^{-1}$, being continuous in λ, is bounded on the compact domain of λ: $|\lambda| \leqq 2\,||x||$. Hence, by Liouville's theorem, $(x - \lambda e)^{-1}$ must reduce to the constant vector $x^{-1} = (x - 0e)^{-1}$. But, since $\underset{|\lambda| \to \infty}{s\text{-}\lim}\,(x - \lambda e)^{-1} = 0$ as proved above, we have arrived at a contradiction $x^{-1} = 0$, $e = x^{-1}x = 0$.

4. The Weak and Strong Measurability. Pettis' Theorem

Definition 1. Let $(S, \mathfrak{B}, m)$ be a measure space, and $x(s)$ a mapping defined on S with values in a B-space X. $x(s)$ is called *weakly $\mathfrak{B}$-measurable* if, for any $f \in X'$, the numerical function $f(x(s)) = \langle x(s), f\rangle$ of s is $\mathfrak{B}$-measurable. $x(s)$ is said to be *finitely-valued* if it is constant $\neq 0$ on each of a finite number of disjoint $\mathfrak{B}$-measurable sets B_j with $m(B_j) < \infty$ and $x(s) = 0$ on $S - \bigcup_j B_j$. $x(s)$ is said to be *strongly $\mathfrak{B}$-measurable* if there exists a sequence of finitely-valued functions strongly convergent to $x(s)$ m-a.e. on S.

Definition 2. $x(s)$ is said to be *separably-valued* if its range $\{x(s);\, s \in S\}$ is separable. It is *m-almost separably-valued* if there exists a $\mathfrak{B}$-measurable set B_0 of m-measure zero such that $\{x(s);\, s \in S - B_0\}$ is separable.

Theorem (B. J. PETTIS [1]). $x(s)$ is strongly $\mathfrak{B}$-measurable iff it is weakly $\mathfrak{B}$-measurable and m-almost separably-valued.

Proof. The "only if" part is proved as follows. The strong $\mathfrak{B}$-measurability implies the weak $\mathfrak{B}$-measurability, because a finitely-valued function is weakly $\mathfrak{B}$-measurable, and, by the strong $\mathfrak{B}$-measurability of $x(s)$, there exists a sequence of finitely-valued functions $x_n(s)$ such that $\underset{n\to\infty}{\text{s-lim}}\, x_n(s) = x(s)$ except on a set $B_0 \in \mathfrak{B}$ of m-measure zero. Thus the union of the ranges of $x_n(s)$ $(n = 1, 2, \ldots)$ is a countable set, and the closure of this set is separable and contains the range $\{x(s); s \in S - B_0\}$.

The proof of the "if" part. Without losing the generality, we may assume that the range $\{x(s); s \in S\}$ is itself separable. So we may assume that the space X is itself separable; otherwise, we replace X by the smallest closed linear subspace containing the range of $x(s)$. We first prove that $\|x(s)\|$ is itself $\mathfrak{B}$-measurable. To this purpose, we shall make use of a lemma, to be proved later, which states that the dual space X' of a separable B-space satisfies the condition

$$\left.\begin{array}{l}\text{there exists a sequence } \{f_n\} \subseteq X' \text{ with } \|f_n\| \leqq 1 \text{ such} \\ \text{that, for any } f_0 \in X' \text{ with } \|f_0\| \leqq 1, \text{ we may choose a} \\ \text{subsequence } \{f_{n'}\} \text{ of } \{f_n\} \text{ for which we have } \lim\limits_{n\to\infty} f_{n'}(x) \\ = f_0(x) \text{ at every } x \in X.\end{array}\right\} \quad (1)$$

Now, for any real number a, put

$$A = \{s; \|x(s)\| \leqq a\} \quad \text{and} \quad A_f = \{s; |f(x(s))| \leqq a\}, \quad \text{where} \quad f \in X'.$$

If we can show that $A = \bigcap_{j=1}^{\infty} A_{f_j}$, then, by the weak $\mathfrak{B}$-measurability of $x(s)$, the function $\|x(s)\|$ is $\mathfrak{B}$-measurable. It is clear that $A \subseteq \bigcap_{\|f\|\leqq 1} A_f$. But, by Corollary 2 of Theorem 1 in Chapter IV, 6, there exists, for fixed s, an $f_0 \in X'$ with $\|f_0\| = 1$ and $f_0(x(s)) = \|x(s)\|$. Hence the reverse inclusion $A \supseteq \bigcap_{\|f\|\leqq 1} A_f$ is true and so we have $A = \bigcap_{\|f\|\leqq 1} A_f$. By the Lemma, we obtain $\bigcap_{\|f\|\leqq 1} A_f = \bigcap_{j=1}^{\infty} A_{f_j}$ and so $A = \bigcap_{j=1}^{\infty} A_{f_j}$.

Since the range $\{x(s); s \in S\}$ is separable, this range may, for any positive integer n, be covered by a countable number of open spheres $S_{j,n}$ $(j = 1, 2, \ldots)$ of radius $\leqq 1/n$. Let the centre of the sphere $S_{j,n}$ be $x_{j,n}$. As proved above, $\|x(s) - x_{j,n}\|$ is $\mathfrak{B}$-measurable in s. Hence the set $B_{j,n} = \{s \in S; x(s) \in S_{j,n}\}$ is $\mathfrak{B}$-measurable and $S = \bigcup_{j=1}^{\infty} B_{j,n}$. We set

$$x_n(s) = x_{i,n} \quad \text{if} \quad s \in B'_{i,n} = B_{i,n} - \bigcup_{j=1}^{i-1} B_{j,n}.$$

Then, by $S = \sum_{i=1}^{\infty} B'_{i,n}$, we have $|x(s) - x_n(s)| < 1/n$ for every $s \in S$.

Since B'_{in} is $\mathfrak{B}$-measurable, it is easy to see that each $x_n(s)$ is strongly $\mathfrak{B}$-measurable. Therefore $x(s)$, which is the strong limit of the sequence $\{x_n(s)\}$, is also strongly $\mathfrak{B}$-measurable.

Proof of the Lemma. Let a sequence $\{x_n\}$ be strongly dense in X. Consider a mapping $f \to \varphi_n(f) = \{f(x_1), f(x_2), \ldots, f(x_n)\}$ of the unit sphere $S' = \{f \in X'; \|f\| \leq 1\}$ of X' into an n-dimensional Hilbert space $l^2(n)$ of vectors $(\xi_1, \xi_2, \ldots, \xi_n)$ normed by $\|(\xi_1, \xi_2, \ldots, \xi_n)\| = \left(\sum_{j=1}^{n} |\xi_j|^2\right)^{1/2}$. The space $l^2(n)$ being separable, there exists, for fixed n, a sequence $\{f_{n,k}\}$ $(k = 1, 2, \ldots)$ of S' such that $\{\varphi_n(f_{n,k}); k = 1, 2, \ldots\}$ is dense in the image $\varphi_n(S')$ of S'.

We thus have proved that, for any $f_0 \in S'$, we can choose a subsequence $\{f_{n,m_n}\}$ $(n = 1, 2, \ldots)$ such that $|f_{n,m_n}(x_i) - f_0(x_i)| < 1/n$ $(i = 1, 2, \ldots, n)$. Hence $\lim_{n\to\infty} f_{n,m_n}(x_i) = f_0(x_i)$ $(i = 1, 2, \ldots)$, and so, by Theorem 10 in Chapter V, 1, we obtain that $\lim f_{n,m_n}(x) = f_0(x)$ for every $x \in X$.

5. Bochner's Integral

Let $x(s)$ be a finitely-valued function defined on a measure space $(S, \mathfrak{B}, m)$ with values in a B-space X; let $x(s)$ be equal to $x_i \neq 0$ on $B_i \in \mathfrak{B}$ $(i = 1, 2, \ldots, n)$ where B_i's are disjoint and $m(B_i) < \infty$ for $i = 1, 2, \ldots, n$, and moreover, $x(s) = 0$ on $\left(S - \sum_{i=1}^{n} B_i\right)$. Then we can define the m-integral $\int_S x(s)\, m(ds)$ of $x(s)$ over S by $\sum_{i=1}^{n} x_i\, m(B_i)$. By virtue of a limiting procedure, we can define the m-integral of more general functions. More precisely, we have the

Definition. A function $x(s)$ defined on a measure space $(S, \mathfrak{B}, m)$ with values in a B-space X is said to be *Bochner m-integrable*, if there exists a sequence of finitely-valued functions $\{x_n(s)\}$ which s-converges to $x(s)$ m-a.e. in such a way that

$$\lim_{n\to\infty} \int_S \|x(s) - x_n(s)\|\, m(ds) = 0. \tag{1}$$

For any set $B \in \mathfrak{B}$, the *Bochner m-integral* of $x(s)$ over B is defined by

$$\int_B x(s)\, m(ds) = s\text{-}\lim_{n\to\infty} \int_S C_B(s)\, x_n(s)\, m(ds), \text{ where } C_B \text{ is the defining function of the set } B. \tag{2}$$

To justify the above definition, we have to verify that the s-limit on the right of (2) exists and that the value of this s-limit is independent of the approximating sequence of functions $\{x_n(s)\}$.

Justification of the Definition. First, $x(s)$ is strongly $\mathfrak{B}$-measurable and consequently the condition (1) has a sense, since $\|x(s) - x_n(s)\|$ is

$\mathfrak{B}$-measurable as shown in the proof of Petti's Theorem. From the inequality

$$\left\| \int_B x_n(s)\, m(ds) - \int_B x_k(s)\, m(ds) \right\| = \left\| \int_B (x_n(s) - x_k(s))\, m(ds) \right\|$$
$$\leqq \int_B \|x_n(s) - x_k(s)\|\, m(ds) \leqq \int_S \|x_n(s) - x(s)\|\, m(ds)$$
$$+ \int_S \|x(s) - x_k(s)\|\, m(ds)$$

and the completeness of the space X, we see that $s\text{-}\lim_{n\to\infty} \int_B x_n(s)\, m(ds)$ exists. It is clear also that this s-lim is independent of the approximating sequence, since any two such sequences can be combined into a single approximating sequence.

Theorem 1 (S. Bochner [1]). A strongly $\mathfrak{B}$-measurable function $x(s)$ is Bochner m-integrable iff $\|x(s)\|$ is m-integrable.

Proof. The "only if" part. We have $\|x(s)\| \leqq \|x_n(s)\| + \|x(s) - x_n(s)\|$. By the m-integrability of $\|x_n(s)\|$ and the condition (1), it is clear that $\|x(s)\|$ is m-integrable and

$$\int_B \|x(s)\|\, m(ds) \leqq \int_B \|x_n(s)\|\, m(ds) + \int_B \|x(s) - x_n(s)\|\, m(ds).$$

Moreover, since

$$\int_B \big|\, \|x_n(s)\| - \|x_k(s)\| \,\big|\, m(ds) \leqq \int_B \|x_n(s) - x_k(s)\|\, m(ds),$$

we see from (1) that $\lim_{n\to\infty} \int_B \|x_n(s)\|\, m(ds)$ exists so that we have

$$\int_B \|x(s)\|\, m(ds) \leqq \varliminf_{n\to\infty} \int_B \|x_n(s)\|\, m(ds).$$

The "if" part. Let $\{x_n(s)\}$ be a sequence of finitely-valued functions strongly convergent to $x(s)$ m-a.e. Put

$$\begin{aligned} y_n(s) &= x_n(s) \quad \text{if} \quad \|x_n(s)\| \leqq \|x(s)\|\,(1 + n^{-1}), \\ &= 0 \quad \text{if} \quad \|x_n(s)\| > \|x(s)\|\,(1 + n^{-1}). \end{aligned}$$

Then the sequence of finitely-valued functions $\{y_n(s)\}$ satisfies $\|y_n(s)\| \leqq \|x(s)\| \cdot (1 + n^{-1})$ and $\lim_{n\to\infty} \|x(s) - y_n(s)\| = 0$ m-a.e. Thus, by the m-integrability of $\|x(s)\|$, we may apply the Lebesgue-Fatou Lemma to the functions $\|x(s) - y_n(s)\| \leqq 2\,\|x(s)\|\,(1 + n^{-1})$ and obtain

$$\lim_{n\to\infty} \int_S \|x(s) - y_n(s)\|\, m(ds) = 0,$$

that is, $x(s)$ is Bochner m-integrable.

Corollary 1. The above proof shows that

$$\int_B \|x(s)\|\, m(ds) \geqq \left\| \int_B x(s)\, m(ds) \right\|,$$

and hence $\int_B x(s)\, m(ds)$ is m-absolutely continuous in the sense that

$$\underset{m(B)\to 0}{\text{s-lim}} \int_B x(s)\, m(ds) = 0.$$

The finite additivity $\int_{\sum_{j=1}^{n} B_j} x(s)\, m(ds) = \sum_{j=1}^{n} \int_{B_j} x(s)\, m(ds)$ is clear and so, by virtue of the σ-additivity of $\int_B \|x(s)\|\, m(ds)$, we see that $\int_B x(s)\, m(ds)$ is σ-additive, i.e.,

$$B = \sum_{j=1}^{\infty} B_j \text{ with } m(B_j) < \infty \text{ implies } \int_{\sum_{j=1}^{\infty} B_j} x(s)\, m(ds) = \underset{n\to\infty}{\text{s-lim}} \sum_{j=1}^{n} \int_{B_j} x(s)\, m(ds).$$

Corollary 2. Let T be a bounded linear operator on a B-space X into a B-space Y. If $x(s)$ is an X-valued Bochner m-integrable function, then $Tx(s)$ is a Y-valued Bochner m-integrable function, and

$$\int_B Tx(s)\, m(ds) = T \int_B x(s)\, m(ds).$$

Proof. Let a sequence of finitely-valued functions $\{y_n(s)\}$ satisfy

$$\|y_n(s)\| \leqq \|x(s)\|\,(1 + n^{-1}) \quad \text{and} \quad \underset{n\to\infty}{\text{s-lim}}\, y_n(s) = x(s) \quad m\text{-a.e.}$$

Then, by the linearity and the continuity of T, we have $\int_B Ty_n(s)\, m(ds) = T \int_B y_n(s)\, m(ds)$. We have, moreover, by the continuity of T,

$$\|Ty_n(s)\| \leqq \|T\| \cdot \|y_n(s)\| \leqq \|T\| \cdot \|x(s)\| \cdot (1 + n^{-1}) \quad \text{and}$$
$$\underset{n\to\infty}{\text{s-lim}}\, Ty_n(s) = Tx(s) \quad m\text{-a.e.}$$

Hence $Tx(s)$ is also Bochner m-integrable and

$$\int_B Tx(s)\, m(ds) = \underset{n\to\infty}{\text{s-lim}} \int_B Ty_n(s)\, m(ds) = \underset{n\to\infty}{\text{s-lim}}\, T \int_B y_n(s)\, m(ds)$$
$$= T \int_B x(s)\, m(ds).$$

Theorem 2 (S. Bochner [1]). Let S be an n-dimensional euclidean space, $\mathfrak{B}$ the family of Baire sets of S, and $m(B)$ the Lebesgue measure of B. If $x(s)$ is Bochner m-integrable, and if $P(s_0; \alpha)$ is the parallelopiped with centre at $s_0 \in S$ and side length 2α, then we have the differentiation theorem

$$\underset{\alpha\downarrow 0}{\text{s-lim}}\, (2\alpha)^{-n} \int_{P(s_0;\alpha)} x(s)\, m(ds) = x(s_0) \text{ for } m\text{-a.e. } s_0.$$

Proof. Put

$$(2\alpha)^{-n} \int_{P(s_0;\alpha)} x(s)\, m(ds) = D(x; s_0, \alpha).$$

If $\{x_n(s)\}$ is a sequence of finitely-valued functions such that $\|x_n(s)\| \leqq \|x(s)\| \cdot (1 + n^{-1})$ and $\underset{n\to\infty}{\text{s-lim}}\, x_n(s) = x(s)$ m-a.e., then

$$D(x; s_0, \alpha) - x(s_0) = D(x - x_k; s_0, \alpha) + D(x_k; s_0, \alpha) - x(s_0),$$

and so

$$\overline{\lim_{\alpha\downarrow 0}} \|D(x; s_0, \alpha) - x(s_0)\| \leqq \overline{\lim_{\alpha\downarrow 0}} D(\|x - x_k\|; s_0, \alpha)$$
$$+ \overline{\lim_{\alpha\downarrow 0}} \|D(x_k; s_0, \alpha) - x_k(s_0)\| + \|x_k(s_0) - x(s_0)\|.$$

The first term on the right is, by Lebesgue's theorem of differentiation of numerical functions, equal to $\|x(s_0) - x_k(s_0)\|$ m-a.e. The second term on the right is $= 0$ m-a.e., since $x_k(s)$ is finitely-valued. Hence

$$\overline{\lim_{\alpha\to 0}} \|D(x; s_0, \alpha) - x(s_0)\| \leqq 2\,\|x_k(s_0) - x(s_0)\| \text{ for } m\text{-a.e. } s_0.$$

Therefore, by letting $k \to \infty$, we obtain Theorem 2.

Remark. Contrary to the case of numerical functions, a B-space-valued, σ-additive, m-absolutely continuous function need not necessarily be represented as a Bochner m-integral. This may be shown by a counter example.

A Counter Example. Let $S = [0, 1]$ and $\mathfrak{B}$ the family of Baire sets on $[0, 1]$, and $m(B)$ the Lebesgue measure of $B \in \mathfrak{B}$. Consider the totality $m[1/3, 2/3]$ of real-valued bounded functions $\xi = \xi(\theta)$ defined on the closed interval $[1/3, 2/3]$ and normed by $\|\xi\| = \sup_\theta |\xi(\theta)|$. We define an $m[1/3, 2/3]$-valued function $x(s) = \xi(\theta; s)$ defined on $[0, 1]$ as follows:

$$\left\{\begin{array}{l}\text{the graph in } s\text{-}y \text{ plane of the real-valued function } y = y_\theta(s),\\ \text{which is the } \theta\text{-coordinate } \xi(\theta; s) \text{ of } x(s), \text{ is the polygonal line}\\ \text{connecting the three points } (0, 0), (\theta, 1) \text{ and } (1, 0) \text{ in this order.}\end{array}\right.$$

Then, if $s \neq s'$, we have *Lipschitz' condition*:

$$\|(s - s')^{-1}(x(s) - x(s'))\| = \sup_\theta |(s - s')^{-1}(\xi(\theta; s) - \xi(\theta; s'))| \leqq 3.$$

Thus, starting with the interval function $(x(s) - x(s'))$ taking values in $m[1/3, 2/3]$, we can define a σ-additive, m-absolutely continuous set function $x(B)$ defined for Baire set B of $[0, 1]$.

If this function $x(B)$ is represented as a Bochner m-integral, then, by the preceding Theorem 2, the function $x(s)$ must be strongly differentiable with respect to s m-a.e. Let the corresponding strong derivative $x'(s)$ be denoted by $\eta(\theta; s)$ which takes values in $m[1/3, 2/3]$. Then, for every $\theta \in [1/3, 2/3]$ and m-a.e. s,

$$0 = \lim_{h\to 0} \|h^{-1}(x(s+h) - x(s)) - x'(s)\| \geqq \lim_{h\to 0} |h^{-1}(\xi(\theta; s+h) - \xi(\theta; s)) - \eta(\theta; s)|.$$

This proves that $\xi(\theta; s)$ must be differentiable in s m-a.e. for all $\theta \in [1/3, 2/3]$. This is contradictory to the construction of $\xi(\theta; s)$.

References for Chapter V

S. Banach [1], N. Dunford-J. Schwartz [1] and E. Hille-R. S. Phillips [1].

Appendix to Chapter V. Weak Topologies and Duality in Locally Convex Linear Topological Spaces

The present book is so designed that the reader may skip this appendix in the first reading and proceed directly to the following chapters.

1. Polar Sets

Definition. Let X be a locally convex linear topological space. For any set $M \subseteqq X$, we define its *(right) polar set* M^0 by

$$M^0 = \{x' \in X'; \sup_{x\in M} |\langle x, x'\rangle| \leqq 1\}. \tag{1}$$

Similarly, for any set $M' \subseteqq X'$, we define its *(left) polar set* ${}^0M'$ by

$${}^0M' = \{x \in X; \sup_{x'\in M'} |\langle x, x'\rangle| \leqq 1\} = X \cap (M')^0, \tag{2}$$

where we consider X to be embedded in its bidual $(X'_s)'$.

A fundamental system of neighbourhoods of 0 in the weak topology of X is given by the system of sets of the form ${}^0M'$ where M' ranges over arbitrary finite sets of X'. A fundamental system of neighbourhoods of 0 in the weak* topology of X' is given by the system of sets of the form M^0 where M ranges over arbitrary finite sets of X. A fundamental system of neighbourhoods of 0 in the strong topology of X' is given by the system of sets of the form M^0 where M ranges over arbitrary bounded sets of X.

Proposition. M^0 is a convex, balanced set closed in the weak* topology of X'.

Proof. For any fixed $x \in X$, the linear functional $f(x') = \langle x, x'\rangle$ is continuous in the weak* topology of X'. Thus $M^0 = \bigcap_{m\in M} \{m\}^0$ is closed in the weak* topology of X'. The balanced convexity of M^0 is clear.

An Application of Tychonov's Theorem

Theorem 1. Let X be a locally convex linear topological space, and A a convex, balanced neighbourhood of 0 of X. Then A^0 is compact in the weak* topology of X'.

Proof. Let $p(x)$ be the Minkowski functional of A. Consider, for each $x \in X$, a sphere $S_x = \{z \in C; |z| \leqq p(x)\}$ and the topological product $S = \prod_{x \in X} S_x$. S is compact by Tychonov's theorem. Any element $x' \in X'$ is determined by the set of values $x'(x) = \langle x, x' \rangle$, $x \in X$. Since $x \in (p(x) + \varepsilon) A$ for any $\varepsilon > 0$, we see that $x' \in X'$ implies $\langle x, x' \rangle = \langle (p(x) + \varepsilon) a, x' \rangle$ with a certain $a \in A$. Thus $x \in A^0$ implies that $|x'(x)| \leqq p(x) + \varepsilon$, that is, $x'(x) \in S_x$. Hence we may consider A^0 as a subset of S. Moreover, it is easy to verify that the topology induced on A^0 by the weak* topology of X' is the same as the topology induced on A^0 in the Cartesian product topology of $S = \prod_{x \in X} S_x$.

Hence it is sufficient to prove that A^0 is a closed subset of S. Suppose $y = \prod_{x \in X} y(x)$ is an element of the weak* closure of A^0 in S. Consider any $\varepsilon > 0$ and any $x_1, x_2 \in X$. The set of all $u = \prod_{x \in X} u(x) \in S$ such that

$$|u(x_1) - y(x_1)| < \varepsilon, \ |u(x_2) - y(x_2)| < \varepsilon \text{ and } |u(x_1 + x_2) - y(x_1 + x_2)| < \varepsilon$$

is a neighbourhood of y in S. This neighbourhood contains some point $x' \in A^0$ and, since x' is a continuous linear functional on X, we have

$$|y(x_1 + x_2) - y(x_1) - y(x_2)| \leqq |y(x_1 + x_2) - \langle x_1 + x_2, x' \rangle| \\ + |\langle x_1, x' \rangle - y(x_1)| + |\langle x_2, x' \rangle - y(x_2)| < 3\varepsilon.$$

This proves that $y(x_1 + x_2) = y(x_1) + y(x_2)$. Similarly, we prove that $y(\beta x) = \beta y(x)$, and so y defines a linear functional on X. By the fact that $y = \prod_{x \in X} y(x) \in S$, we know that $|y(x)| \leqq p(x)$. Since $p(x)$ is continuous, $y(x)$ is a continuous linear functional, i.e., $y \in X'$. On the other hand, since y is a weak* accumulation point of A^0, there exists, for any $\varepsilon > 0$ and $a \in A$, an $x' \in A^0$ such that $|y(a) - \langle a, x' \rangle| \leqq \varepsilon$. Hence $|y(a)| \leqq |\langle a, x' \rangle| + \varepsilon \leqq 1 + \varepsilon$, and so $|y(a)| \leqq 1$, that is, $y \in A^0$.

Corollary. The unit sphere $S^* = \{x' \in X'; \|x'\| \leqq 1\}$ of the dual space X'_s of a normed linear space X is compact in the weak* topology of X'.

An Application of Mazur's Theorem

Theorem 2. Let M be a convex, balanced closed set of a locally convex linear topological space X. Then $M = {}^0(M^0)$.

Proof. It is clear that $M \subseteq {}^0(M^0)$. If there exists an $x_0 \in {}^0(M^0) - M$, then, by Mazur's theorem 3 in Chapter IV, 6, there exists an $x'_0 \in X'$ such that $\langle x_0, x'_0 \rangle > 1$ and $|\langle x, x'_0 \rangle| \leqq 1$ for all $x \in M$. The last inequality shows that $x'_0 \in M^0$ and so x_0 cannot belong to ${}^0(M^0)$.

2. Barrel Spaces

Definition. In a locally convex linear topological space X, any convex, balanced and absorbing closed set is called a *barrel* (*tonneau* in Bourbaki's terminology). X is called a *barrel space* if each of its barrels is a neighbourhood of 0.

Theorem 1. A locally convex linear topological space X is a barrel space if X is not of the first category.

Proof. Let T be a barrel in X. Since T is absorbing, X is the union of closed sets $nT = \{nt; t \in T\}$, where n runs over positive integers. Since X is not of the first category, at least one of the $(nT)'$ s contains an interior point. Hence T itself contains an interior point x_0. If $x_0 = 0$, T is a neighbourhood of 0. If $x_0 \neq 0$, then $-x_0 \in T$ by the fact that T is balanced. Thus $-x_0$ is an interior point of T with x_0. This proves that the convex set T contains $0 = (x_0 - x_0)/2$ as an interior point.

Corollary 1. All locally convex F-spaces and, in particular, all B-spaces and $\mathfrak{E}(R^n)$ are barrel spaces.

Corollary 2. The metric linear space $\mathfrak{D}_K(R^n)$ is a barrel space.

Proof. Let $\{\varphi_k\}$ be a Cauchy sequence with respect to the distance

$$\operatorname{dis}(\varphi, \psi) = \sum_{m=0}^{\infty} 2^{-m} \frac{p_m(\varphi-\psi)}{1+p_m(\varphi-\psi)}, \text{ where } p_m(\varphi) = \sup_{|j|\leq m, x\in K} |D^j\varphi(x)|.$$

For any differential operator D^j, the sequence $\{D^j\varphi_k(x)\}$ is equi-continuous and equi-bounded, that is,

$$\lim_{|x^1-x^2|\downarrow 0} \sup_{k\geq 1} |D^j\varphi_k(x^1) - D^j\varphi_k(x^2)| = 0 \text{ and } \sup_{x\in K, k\geq 1} |D^j\varphi_k(x)| < \infty.$$

This we see from the fact that, for any coordinate x_s, $\sup_{x\in K, k\geq 1} \left|\frac{\partial}{\partial x_s} D^j\varphi_k(x)\right| < \infty$. Hence, by the Ascoli-Arzelà theorem, there exists a subsequence of $\{D^j\varphi_{k'}(x)\}$ which converges uniformly on K. By the diagonal method, we may choose a subsequence $\{\varphi_{k''}(x)\}$ of $\{\varphi_k(x)\}$ such that, for any differential operator D^j, the sequence $\{D^j\varphi_{k''}(x)\}$ converges uniformly on K. Thus

$$\lim_{k''\to\infty} D^j\varphi_{k''}(x) = D^j\varphi(x) \quad \text{where} \quad \varphi(x) = \lim_{k''\to\infty} \varphi_{k''}(x),$$

and these limit relations hold uniformly on K. Hence the metric space $\mathfrak{D}_K(R^n)$ is complete and so it is not of the first category.

Remark. (i) The above proof shows that a bounded set of $\mathfrak{D}(R^n)$ is relatively compact in the topology of $\mathfrak{D}(R^n)$. For, a bounded set B of $\mathfrak{D}(R^n)$ is contained in some $\mathfrak{D}_K(R^n)$ where K is a compact set of R^n, and, moreover, the boundedness condition of B implies the equi-boundedness and equi-continuity of $\{D^j\varphi; \varphi \in B\}$ for every D^j. (ii) Similarly, we see that any bounded set of $\mathfrak{E}(R^n)$ is a relatively compact set of $\mathfrak{E}(R^n)$.

Corollary 3. $\mathfrak{D}(R^n)$ is a barrel space.

Proof. $\mathfrak{D}(R^n)$ being an inductive limit of $\{\mathfrak{D}_K(R^n)\}$ when K ranges over compact subsets of R^n, the present corollary is a consequence of the following

Proposition. Let a locally convex linear topological space X be an inductive limit of its barrel subspaces X_α, $\alpha \in A$. Then X itself is a barrel space.

Proof. Let V be a barrel of X. By the continuity of the identical mapping $T_\alpha: x \to x$ of X_α into X, the inverse image $T_\alpha^{-1}(V) = V \cap X_\alpha$ is closed with V. Thus $V \cap X_\alpha$ is a barrel of X_α. X_α being a barrel space, $V \cap X_\alpha$ is a neighbourhood of 0 of X_α. X being an inductive limit of X_α's, V must be a neighbourhood of 0 of X.

Theorem 2. Let X be a barrel space. Then the mapping $x \to Jx$ of X into $(X_s')_s'$, defined in Chapter IV, 8, is a topological mapping of X onto JX, where the topology of JX is provided with the relative topology of JX as a subset of $(X_s')_s'$.

Proof. Let B' be a bounded set of X_s'. Then the polar set $(B')^0 = \{x'' \in (X_s')'; \sup_{x' \in B'} |\langle x', x''\rangle| \leqq 1\}$ of B' is a neighbourhood of 0 of $(X_s')_s'$, and it is a convex, balanced and absorbing set closed in $(X_s')_s'$. Thus $(B')^0 \cap X = {}^0(B')$ is a convex, balanced and absorbing set of X. As a (left) polar set, ${}^0(B')$ is closed in the weak topology of X, and hence ${}^0(B')$ is closed in the original topology of X. Thus ${}^0(B') = (B')^0 \cap X$ is a barrel of X, and so it is a neighbourhood of 0 of X. Therefore, the mapping $x \to Jx$ of X into $(X_s')_s'$ is continuous, because the topology of $(X_s')_s'$ is defined by a fundamental system of neighbourhoods of 0 of the form $(B')^0$, where B' ranges over bounded sets of X'.

Let, conversely, U be a convex, balanced and closed neighbourhood of 0 of X. Then, by the preceding section, $U = {}^0(U^0)$. Thus $JU = JX \cap (U^0)^0$. On the other hand, U^0 is a bounded set of X_s', since, for any bounded set B of X, there exists an $\alpha > 0$ such that $\alpha B \subseteqq U$ and so $(\alpha B)^0 \supseteqq U^0$. Hence $(U^0)^0$ is a neighbourhood of 0 of $(X_s')_s'$. Thus the image JU of the neighbourhood U of 0 of X is a neighbourhood of 0 of JX provided with the relative topology of JX as a subset of $(X_s')_s'$.

3. Semi-reflexivity and Reflexivity

Definition 1. A locally convex linear topological space X is called *semi-reflexive* if every continuous linear functional on X_s' is given by

$$\langle x, x'\rangle, \text{ with a certain } x \in X. \tag{1}$$

Thus X is semi-reflexive iff

$$X_w = (X_s')_{w^*}'. \tag{2}$$

Definition 2. A locally convex linear topological space X is called *reflexive* if

$$X = (X'_s)'_s. \tag{3}$$

By Theorem 2 in the preceding section, we have

Proposition 1. A semi-reflexive space X is reflexive if X is a barrel space.

It is also clear, from Definition 2, that we have

Proposition 2. The strong dual of a reflexive space is reflexive.

Theorem 1. A locally convex linear topological space X is semi-reflexive iff every closed, convex, balanced and bounded set of X is compact in the weak topology of X.

Proof. Let X be semi-reflexive, and T a closed, convex, balanced and bounded set of X. Then, by Theorem 2 in Section 1 of this Appendix, $T = {}^0(T^0)$. T being a bounded set of X, T^0 is a neighbourhood of 0 of X'_s. Thus, by Theorem 1 in Section 1 of this Appendix, $(T^0)^0$ is compact in the weak* topology of $(X'_s)'$. Hence, by the semi-reflexivity of X, $T = {}^0(T^0)$ is compact in the weak topology of X.

We next prove the sufficiency part of Theorem 1. Take any $x'' \in (X'_s)'$. The strong continuity of x'' on X'_s implies that there exists a bounded set B of X such that

$$|\langle x', x''\rangle| \leqq 1 \text{ whenever } x' \in B^0, \text{ that is, } x'' \in (B^0)^0.$$

We may assume that B is a convex, balanced and closed set of X. Thus, by the hypothesis of Theorem 1, B is a compact set in the weak topology of X. Hence $B = B^{wa}$ where B^{wa} denotes the closure of B in the weak topology of X. Since X_w is embedded in $(X'_s)'_{w^*}$ as a linear topological subspace, we must have $(B^0)^0 \supseteq B^{wa} = B$. Therefore we have to show that x'' is an accumulation point of B in $(X'_s)'_{w^*}$. Consider the mapping $x \to \varphi(x) = \{\langle x, x'_1\rangle, \ldots, \langle x, x'_n\rangle\}$ of X into $l^2(n)$, where $x'_1, \ldots, x'_n \in X'$. The image $\varphi(B)$ is convex, balanced and compact, since B is convex, balanced and weakly compact. If $\{\langle x'_1, x''\rangle, \ldots, \langle x'_n, x''\rangle\}$ does not belong to $\varphi(B)$, then, by Mazur's theorem, there would exists a point $\{c_1, \ldots, c_n\} \in l^2(n)$ such that $\sup_{b \in B} \left|\sum_i c_i \langle b, x'_i\rangle\right| \leqq 1$ and $\left(\sum_i c_i \langle x'_i, x''\rangle\right) > 1$, proving that $\sum_i c_i x'_i \in B^0$ and x'' cannot belong to B^{00}.

Theorem 2. A locally convex linear topological space X is reflexive iff it is a barrel space and every closed, convex, balanced and bounded set of X is compact in the weak topology of X. In particular, $\mathfrak{D}(R^n)$ and $\mathfrak{E}(R^n)$ are reflexive.

Proof. The sufficiency part is proved already. We shall prove that the first condition of Theorem 2 is necessary.

Let T be a barrel of X. We shall prove that T absorbs any bounded set B of X so that $B^0 \supseteq \alpha T^0, \alpha > 0$. B^0 being a neighbourhood of 0 of X'_s, we see that T^0 is a bounded set of X'_s. By the Proposition and Theorem 2 in Section 1 of this Appendix, we have $T = {}^0(T^0)$. By the hypothesis that X is reflexive, we have ${}^0(T^0) = (T^0)^0$ and so $T = (T^0)^0$. Therefore we have proved that the barrel T is a neighbourhood of 0 of $X = (X'_s)'_s$. Hence X is a barrel space.

By hypothesis, the closed, convex and balanced set $K = \operatorname{Conv}\left(\bigcup_{|\alpha| \leq 1} \alpha B\right)^a$ is compact in the weak topology of X. Here we denote by $\operatorname{Conv}(N)^a$ the closure in X of the convex closure (see p. 28) $\operatorname{Conv}(N)$ of N. Set $Y = \bigcup_{n=1}^{\infty} nK$ and let $p(x)$ be the Minkowski functional of K. Then since K is w-compact in Y, $p(x)$ defines a norm of Y. That is, the system $\{\alpha K\}$ with $\alpha > 0$ defines a fundamental system of neighbourhoods of the normed linear space Y, and Y is a B-space since K is w-compact. Hence Y is a barrel space. On the other hand, since K is a bounded set of X, the topology of Y defined by the norm $p(x)$ is stronger than the relative topology of Y as a subset of X. As a barrel of X, T is closed in X. Hence $T \cap Y$ is closed in Y with respect to the topology defined by the norm $p(x)$. Therefore $T \cap Y$ is a barrel of the B-space Y, and so $T \cap Y$ is a neighbourhood of 0 of the B-space Y. We have thus proved that $T \cap Y$ and, a fortiori, T both absorb $K \supseteq B$.

4. The Eberlein-Shmulyan Theorem

This theorem is very important in view of its applications.

Theorem (Eberlein-Shmulyan). A B-space X is reflexive iff it is locally sequentially compact; that is, X is reflexive iff every strongly bounded sequence of X contains a subsequence which converges weakly to an element of X.

For the proof we need two Lemmas:

Lemma 1. If the strong dual X'_s of a B-space X is separable, then X itself is separable.

Lemma 2 (S. Banach). A linear subspace M' of the dual space X' of a B-space X is weakly* closed iff M' is boundedly weakly* closed; that is, M' is weakly* closed iff M' contains all weak* accumulation points of every strongly bounded subset of M'.

Lemma 1 is already proved as a Lemma in Chapter V, 2. For Lemma 2, we have only to prove its "if" part. It reads as follows.

Proof. (E. Hille-R. S. Phillips [1]). We remark that M' is strongly closed by the hypothesis. Let $x'_0 \in M'$. Then we can prove that, for each constant C satisfying the condition $0 < C < \inf_{x' \in M'} \|x' - x_0'\|$, there exists an $x_0 \in X$ with $\|x_0\| \leqq 1/C$ such that

$$\langle x_0, x_0' \rangle = 1 \quad \text{and} \quad \langle x_0, x' \rangle = 0 \quad \text{for all} \quad x' \in M'. \tag{1}$$

Thus the strongly closed set M' must contain all of its weak* accumulation points.

To prove the existence of x_0, we choose an increasing sequence of numbers $\{C_n\}$ such that $C_1 = C$ and $\lim_{n\to\infty} C_n = \infty$. Then there exists a finite subset σ_1 of the unit sphere $S = \{x \in X;\ \|x\| \leqq 1\}$ such that

$$\|x' - x_0'\| \leqq C_2 \quad \text{and} \quad \sup_{x \in \sigma_1} |\langle x, x' \rangle - \langle x, x_0' \rangle| \leqq C_1 \quad \text{implies} \quad x' \,\overline{\in}\, M'.$$

If not, there wonld exists, corresponding to each finite subset σ of S, an $x_\sigma' \in M$ such that

$$\|x_\sigma' - x_0'\| \leqq C_2 \quad \text{and} \quad \sup_{x \in \sigma} |\langle x, x_\sigma' \rangle - \langle x, x_0' \rangle| \leqq C_1.$$

We order the sets σ by inclusion relation and denote the weak* closure of the set $\{x_{\sigma}';\ \sigma' \geqq \sigma\}$ by N_σ'. It is clear that N_σ' enjoys the finite intersection property. On the other hand, since M' is boundedly weakly* closed, the Corollary of Theorem 1 in this Appendix, 1, implies that the set

$$M_{C'}' = \{x' \in M';\ \|x'\| \leqq C'\}$$

is weakly* compact. Hence $N_\sigma' \subseteqq M_{C'}'$ for $C' = C_2 + \|x_0'\|$, and so there exists an $x_1' \in \bigcap_\sigma N_\sigma' \subseteqq M'$. Hence we have $\sup_{x \in S} |\langle x, x_1' \rangle - \langle x, x_0' \rangle| \leqq C_1$ and so $\|x_1' - x_0'\| \leqq C_1$, contrary to the hypothethis that $0 < C_1 < \inf_{x' \in M'} \|x' - x_0'\|$.

By a similar argument, we successively prove the existence of a sequence of finite subsets $\sigma_1, \sigma_2, \ldots$ of S such that

$$\begin{cases} \|x' - x_0'\| \leqq C_k \quad \text{and} \quad \sup_{x \in \sigma_i} |\langle x, x' \rangle - \langle x, x_0' \rangle| \leqq C_i \\ \qquad\qquad\qquad\qquad\qquad\qquad (i = 1, 2, \ldots, k-1) \\ \text{do not imply } x' \in M'. \end{cases}$$

Thus, since $\lim_{i\to\infty} C_i = \infty$, we see that $x' \,\overline{\in}\, M'$ if

$$|\langle x, x' \rangle - \langle x, x_0' \rangle| \leqq C \quad \text{for all} \quad x \in (C/C_j)\sigma_j \qquad (j = 1, 2, \ldots).$$

Let $\{x_n\}$ be a sequence which successively exhausts the sets $(C/C_j)\sigma_j$ $(j = 1, 2, \ldots)$. Then $\lim_{n\to\infty} x_n = 0$ and so $L(x') = \{\langle x_n, x' \rangle\}$ is a bounded linear transformation of X_s' into the B-space (c_0). We know that the point $\{\langle x_n, x_0' \rangle\} \in (c_0)$ lies at a distance $> C$ from the linear subspace $L(M')$. Thus, by Corollary of Theorem 3 in Capter IV, 6, there exists a continuous linear functional $\{\alpha_n\} \in (c_0)' = (l^1)$ such that

$$||\{\alpha_n\}|| = \sum_{n=1}^{\infty} |\alpha_n| \leqq 1/C, \sum_{n=1}^{\infty} \alpha_n \langle x_n, x_0' \rangle = 1 \quad \text{and}$$

$$\sum_{n=1}^{\infty} \alpha_n \langle x_n, x' \rangle = 0 \quad \text{for all} \quad x' \in M'.$$

The element $x_0 = \sum_{n=1}^{\infty} \alpha_n x_n$ clearly satisfies condition (1).

Corollary. Let $\langle x', x_0'' \rangle = F(x')$ be a linear functional defined on the dual space X' of a B-space X. If $N(F) = N(x_0'') = \{x' \in X'; F(x') = 0\}$ is weakly* closed, then there exists an element x_0 such that

$$F(x') = \langle x', x_0'' \rangle = \langle x_0, x' \rangle \quad \text{for all} \quad x' \in X'. \tag{2}$$

Proof. We may assume that $N(F) \neq X'$. Otherwise we can take $x_0 = 0$. Let $x_0' \in X'$ be such that $F(x_0') = 1$. By (1) of the preceding Lemma 2, there exists an $x_0 \in X$ such that

$$\langle x_0, x_0' \rangle = 1 \quad \text{and} \quad \langle x_0, x' \rangle = 0 \quad \text{for all} \quad x' \in N(F). \tag{3}$$

Hence, for any $x' \in X'$, the functional

$$x' - F(x') x_0' = y' \in X'$$

satisfies $F(y') = 0$, i.e. $y' \in N(F)$. Therefore, by (3), we obtain (2).

Proof of the Thorem. *"Only if" part.* Let $\{x_n\}$ be a sequence of X such that $||x_n|| = 1$. The strong closure X_0 of the subspace spanned by $\{x_n\}$ is a separable B-space. Being a B-space, X_0 is a barrel space. We shall show that X_0 is reflexive. Any strongly closed, bounded set B_0 of X_0 is also a strongly closed, bounded set of X and hence B_0 is compact in the weak topology of X by the reflexivity of X. But, as a strongly closed linear subspace of X, X_0 is closed in the weak topology of X (see Theorem 3 in Chapter IV, 6). Hence B_0 is compact in the weak topology of X_0. Thus X_0 is reflexive by Theorem 2 in the preceding section. We have thus $X_0 = ((X_0)_s')_s'$. By Lemma 1 above, $(X_0)_s'$ is thus separable. Let $\{x_n'\}$ be strongly dense in $(X_0)_s'$. Then the weak topology of X_0 is defined by an enumerable sequence of semi-norms $p_m(x) = |\langle x, x_m' \rangle|$ ($m = 1, 2, \ldots$). Hence it is easy to see that the sequence $\{x_n\}$, which is compact in the weak topology of X_0, is sequentially weakly compact in X_0 and in X as well. We have only to choose a subsequence $\{x_{n'}\}$ of $\{x_n\}$ such that finite $\lim_{n \to \infty} \langle x_{n'}, x_m' \rangle$ exists for $m = 1, 2, \ldots$

"If" part. Let M be a bounded set of X, and assume that every infinite sequence of M contains a subsequence which converges weakly to an element of X. We have to show that the closure $\overline{M}$ of M in X in the weak topology of X is weakly compact in X. For, then the barrel space X is reflexive by Theorem 2 in the preceding section.

Since $X_w \subseteqq (X'_s)'_{w^*}$, we have $\overline{M} = \overline{\overline{M}} \cap X_w$, where $\overline{\overline{M}}$ is the closure of $\overline{M}$ in the weak* topology of $(X'_s)'$. Let S'_r be the sphere of X'_s of radius $r > 0$ and centre 0. By the correspondence

$$\overline{\overline{M}} \ni m \leftrightarrow \{\langle x', m\rangle;\ \|x'\| \leqq 1\} \in \prod_{x' \in S'_1} I_{x'}, \text{ where}$$

$$I_{x'} = \{z;\ |z| \leqq \sup_{m \in \overline{M}} |\langle x', m\rangle|\},$$

$\overline{\overline{M}}$ may be identified with a closed subset of the topological product $\prod_{x' \in S'_1} I_{x'}$. By Tychonov's theorem, $\prod_{x' \in S'_1} I_{x'}$ is compact and hence $\overline{\overline{M}}$ is compact in the weak* topology of $(X'_s)'$. Hence we have only to show $\overline{\overline{M}} \subseteqq X_w$.

Let $x''_0 \in (X'_s)'$ be an accumulation point of the set $\overline{M}$ in the weak* topology of $(X'_s)'$. To prove that $x''_0 \in X_w$ we have only tc show that the set $N(x''_0) = \{x' \in X';\ \langle x', x''_0\rangle = 0\}$ is weakly* closed. For, then, by the above Corollary, there exists an $x_0 \in X$ such that $\langle x', x''_0\rangle = \langle x_0, x'\rangle$ for all $x' \in X'$. We shall first show that

for every finite set $x'_1, x'_2, \ldots, x'_n$ of X', there exists a $z \in \overline{M}$ such that $\langle x'_j, x''_0\rangle = \langle z, x'_j\rangle$ $(j = 1, 2, \ldots, n)$. (4)

The proof is as follows. Since x''_0 is in the weak* closure of $\overline{M}$, there is an element $z_m \in \overline{M}$ such that

$$|\langle z_m, x'_j\rangle - \langle x_j, x''_0\rangle| \leqq 1/m \ (j = 1, 2, \ldots, n).$$

By hypothesis there exists a subsequence of $\{z_m\}$ which converges weakly to an element $z \in X$ and so $z \in \overline{M}$, since the sequential weak closure of $\overline{M}$ is contained in $\overline{M}$. We have thus (4).

Now, by Lemma 2, $N(x''_0)$ is weakly* closed if, for every $r > 0$, the set $N(x''_0) \cap S'_r$ is weakly* closed. Let y'_0 be in the weak* closure of $N(x''_0) \cap S'_1$. We have to show that $y'_0 \in N(x''_0) \cap S'_1$. To this purpose, we choose an arbitrary $\varepsilon > 0$ and construct three sequences $\{z_n\} \subseteqq \overline{M}$, $\{x_n\} \subseteqq M$ and $\{y'_n\} \subseteqq N(x''_0) \cap S'_1$ as follows: By (4) we can choose a, $z_1 \in \overline{M}$ such that $\langle z_1, y'_0\rangle = \langle y'_0, x''_0\rangle$. z_1 being in the weak closure of M, there exists an $x_1 \in M$ such that $|\langle x_1, y'_0\rangle - \langle z_1, y'_0\rangle| \leqq \varepsilon/4$. y'_0 being in the weak* closure of $N(x''_0) \cap S'_1$, there exists a $y'_1 \in N(x''_0) \cap S'_1$ such that $|\langle x_1, y'_1\rangle - \langle x_1, y'_0\rangle| \leqq \varepsilon/4$. Repeating the argument and remembering (4), we obtain $\{z_n\} \subseteqq \overline{M}$, $\{x_n\} \subseteqq M$ and $\{y'_n\} \subseteqq N(x''_0) \cap S'_1$ such that

$$\left.\begin{array}{ll} \langle z_1, y'_0\rangle = \langle y'_0, x''_0\rangle, & \\ \langle z_n, y'_m\rangle = \langle y'_m, x''_0\rangle = 0 & (m = 1, 2, \ldots, n-1), \\ |\langle x_n, y'_m\rangle - \langle z_n, y'_m\rangle| \leqq \varepsilon/4 & (m = 0, 1, \ldots, n-1), \\ |\langle x_i, y'_n\rangle - \langle x_i, y'_0\rangle| \leqq \varepsilon/4 & (i = 1, 2, \ldots, n). \end{array}\right\} \tag{5}$$

Thus we have

$$|\langle y'_0, x''_0\rangle - \langle x_i, y'_n\rangle| \leqq \varepsilon/4 + \varepsilon/4 = \varepsilon/2 \ (i = 1, \quad 2, \ldots, n). \tag{6}$$

Since $\{x_n\} \subseteqq M$, there exists a subsequence of $\{x_n\}$ which converges weakly to an element $x \in \overline{M}$. Without losing the generality, we may assume that the sequence $\{x_n\}$ itself converges weakly to $x \in \overline{M}$. Hence, from (5), $|\langle x, y'_m\rangle| \leqq \varepsilon/4$. From $w\text{-}\lim_{n\to\infty} x_n = x$ and Mazur's theorem 2 in Chapter V, 1, there exists a convex combination $u = \sum_{j=1}^{n} \alpha_j x_j \left(\alpha_j \geqq 0, \sum_{j=1}^{n} \alpha_j = 1\right)$ such that $\|x - u\| \leqq \varepsilon/4$. Therefore, by (6),

$$|\langle y'_0, x''_0\rangle - \langle u, y'_n\rangle| \leqq \sum_{j=1}^{n} \alpha_j \, |\langle y'_0, x''_0\rangle - \langle x_j, y'_n\rangle| \leqq \varepsilon/2,$$

and hence

$$\begin{aligned} |\langle y'_0, x''_0\rangle| &\leqq |\langle y'_0, x''_0\rangle - \langle u, y'_n\rangle| + |\langle u, y'_n\rangle - \langle x, y'_n\rangle| + |\langle x, y'_n\rangle| \\ &\leqq \varepsilon/2 + \|u - x\| \, \|y'_n\| + \varepsilon/4 \leqq \varepsilon. \end{aligned}$$

As ε was arbitrary, we see that $\langle y'_0, x''_0\rangle = 0$ and so $y'_0 \in N(x''_0)$. Combined with the fact that S'_1 is weakly* closed, we finally obtain $y'_0 \in N(x''_0) \cap S'_1$.

Remark. As for the weak topologies and duality in B-spaces, there is an extensive literature. See, e.g., the references in N. Dunford-J. Schwartz [1]. Sections 1, 2 and 3 of this Appendix are adapted and modified from N. Bourbaki [1] and A. Grothendieck [1]. It is remarkable that necessary tools for proving this far reaching theorem of Eberlein [1]-Shmulyan [1] are found, in one form or other, in the book of S. Banach [1].

VI. Fourier Transform and Differential Equations

The Fourier transform is one of the most powerful tools in classical and modern analysis. Its scope has recently been strikingly extended thanks to the introduction of the notion of generalized functions of S. L. Sobolev [1] and L. Schwartz [1]. The extension has been applied successfully to the theory of linear partial differential equations by L. Ehrenpreis, B. Malgrange and especially by L. Hörmander [6].

1. The Fourier Transform of Rapidly Decreasing Functions

Definition 1. We denote by $\mathfrak{S}(R^n)$ the totality of functions $f \in C^\infty(R^n)$ such that

$$\sup_{x \in R^n} |x^\beta D^\alpha f(x)| < \infty \left(x^\beta = \prod_{j=1}^{n} x_j^{\beta_j} \right) \tag{1}$$

for every $\alpha = (\alpha_1, \alpha_2, \ldots, \alpha_n)$ and $\beta = (\beta_1, \beta_2, \ldots, \beta_n)$ with non-negative integers α_j and β_k. Such functions are called *rapidly decreasing* (at ∞).

Example. $\exp(-|x|^2)$ and functions $f \in C_0^\infty(R^n)$ are rapidly decreasing.

Proposition 1. $\mathfrak{S}(R^n)$ is a locally convex linear topological space by the algebraic operation of function sum and multiplication of functions by complex numbers, and by the topology defined by the system of semi-norms of the form

$$p(f) = \sup_{x \in R^n} |P(x) D^\alpha f(x)|, \text{ where } P(x) \text{ denotes a polynomial.} \tag{2}$$

Proposition 2. $\mathfrak{S}(R^n)$ is closed with respect to the application of linear partial differential operators with polynomial coefficients.

Proposition 3. With respect to the topology of $\mathfrak{S}(R^n)$, $C_0^\infty(R^n)$ is a dense subset of $\mathfrak{S}(R^n)$.

Proof. Let $f \in \mathfrak{S}(R^n)$ and take $\psi \in C_0^\infty(R^n)$ such that $\psi(x) = 1$ when $|x| \leqq 1$. Then, for any $\varepsilon > 0$, $f_\varepsilon(x) = f(x)\,\psi(\varepsilon x) \in C_0^\infty(R^n)$. By applying Leibniz' rule of differentiation of the product of functions, we see that

$$D^\alpha(f_\varepsilon(x) - f(x)) = D^\alpha\{f(x)\,(\psi(\varepsilon x) - 1)\}$$

is a finite linear combination of terms of the form

$$D^\beta f(x) \cdot (\varepsilon)^{|\gamma|} \{D^\gamma \psi(y)\}_{y=\varepsilon x}, \text{ where } |\beta| + |\gamma| = |\alpha| \text{ with } |\gamma| > 0,$$

and the term $D^\alpha f(x) \cdot (\psi(\varepsilon x) - 1)$. Thus it is easy to see that $f_\varepsilon(x)$ tends to $f(x)$ in the topology of $\mathfrak{S}(R^n)$ when $\varepsilon \downarrow 0$.

Definition 2. For any $f \in \mathfrak{S}(R^n)$, define its *Fourier transform* $\hat{f}$ by

$$\hat{f}(\xi) = (2\pi)^{-n/2} \int_{R^n} e^{-i\langle \xi, x \rangle} f(x)\, dx, \tag{3}$$

where $\xi = (\xi_1, \xi_2, \ldots, \xi_n)$, $x = (x_1, x_2, \ldots, x_n)$, $\langle \xi, x \rangle = \sum_{j=1}^{n} \xi_j x_j$ and $dx = dx_1\, dx_2 \ldots dx_n$. We also define the *inverse Fourier transform* $\tilde{g}$ of $g \in \mathfrak{S}(R^n)$ by

$$\tilde{g}(x) = (2\pi)^{-n/2} \int e^{i\langle x, \xi \rangle} g(\xi)\, d\xi. \tag{4}$$

Proposition 4. The Fourier transform: $f \to \hat{f}$ maps $\mathfrak{S}(R^n)$ linearly and continuously into $\mathfrak{S}(R^n)$. The inverse Fourier transform: $g \to \tilde{g}$ also maps $\mathfrak{S}(R^n)$ linearly and continuously into $\mathfrak{S}(R^n)$.

Proof. Differentiating formally under the integral sign, we have

$$D^\alpha \hat{f}(\xi) = (2\pi)^{-n/2} \int e^{-i\langle \xi, x\rangle} (-i)^{|\alpha|} x^\alpha f(x)\, dx. \tag{5}$$

The formal differentiation is permitted since the right hand side is, by (1), uniformly convergent in ξ. Thus $\hat{f} \in C^\infty(R^n)$. Similarly, we have, by integration by parts,

$$(i)^{|\beta|} \xi^\beta \hat{f}(\xi) = (2\pi)^{-n/2} \int e^{-i\langle \xi, x\rangle} D^\beta f(x)\, dx. \tag{6}$$

Thus we have

$$(i)^{|\beta|+|\alpha|} \xi^\beta D^\alpha \hat{f}(\xi) = (2\pi)^{-n/2} \int e^{-i\langle \xi, x\rangle} D^\beta (x^\alpha f(x))\, dx, \tag{7}$$

and (7) proves that the mapping $f \to \hat{f}$ is continuous in the topology of $\mathfrak{S}(R^n)$.

Theorem 1 (Fourier's integral theorem). *Fourier's inversion theorem holds:*

$$\tilde{\hat{f}}(x) = (2\pi)^{-n/2} \int e^{i\langle x, \xi\rangle} \hat{f}(\xi)\, d\xi = f(x), \text{ i.e., we have} \tag{8}$$

$$\tilde{\hat{f}} = f, \text{ and similarly } \hat{\tilde{f}} = f. \tag{8'}$$

Therefore it is easy to see that the Fourier transform maps $\mathfrak{S}(R^n)$ *onto* $\mathfrak{S}(R^n)$ linearly and continuously in both directions, and the inverse Fourier transform gives the inverse mapping of the Fourier transform.

Proof. We have

$$\int g(\xi) \hat{f}(\xi) e^{i\langle x, \xi\rangle} d\xi = \int \hat{g}(y) f(x+y)\, dy \quad (f \text{ and } g \in \mathfrak{S}(R^n)). \tag{9}$$

In fact, the left hand side is equal to

$$\begin{aligned} &\int g(\xi) \{(2\pi)^{-n/2} \int e^{-i\langle \xi, y\rangle} f(y)\, dy\} e^{i\langle x, \xi\rangle} d\xi \\ &= (2\pi)^{-n/2} \int \{\int g(\xi) e^{-i\langle \xi, y-x\rangle} d\xi\} f(y)\, dy \\ &= \int \hat{g}(y-x) f(y)\, dy = \int \hat{g}(y) f(x+y)\, dy. \end{aligned}$$

If we take $g(\varepsilon\xi)$ for $g(\xi)$, $\varepsilon > 0$, then

$$(2\pi)^{-n/2} \int e^{-i\langle y, \xi\rangle} g(\varepsilon\xi)\, d\xi = (2\pi)^{-n/2} \varepsilon^{-n} \int g(z) e^{-i\langle y, z/\varepsilon\rangle} dz = \varepsilon^{-n} \hat{g}(y/\varepsilon).$$

Hence, by (9),

$$\int g(\varepsilon\xi) \hat{f}(\xi) e^{i\langle x, \xi\rangle} d\xi = \int \hat{g}(y) f(x+\varepsilon y)\, dy.$$

We shall take, following F. Riesz, $g(x) = e^{-|x|^2/2}$ and let $\varepsilon \downarrow 0$. Then

$$g(0) \int \hat{f}(\xi) e^{i\langle x, \xi\rangle} d\xi = f(x) \int \hat{g}(y)\, dy.$$

This proves (8), since $g(0) = 1$ and $\int \hat{g}(y)\, dy = (2\pi)^{n/2}$ by the well-known facts:

$$(2\pi)^{-n/2} \int e^{-|x|^2/2} e^{-i\langle y, x\rangle} dx = e^{-|y|^2/2}, \tag{10}$$

$$(2\pi)^{-n/2} \int e^{-|x|^2/2} dx = 1. \tag{10'}$$

Remark. For the sake of completeness, we will give the proof of (10). We have

$$(2\pi)^{-1/2}\int_{-\lambda}^{\lambda} e^{-t^2/2}\, e^{-iut}\, dt = e^{-u^2/2}(2\pi)^{-1/2}\int_{-\lambda}^{\lambda} e^{-(t+iu)^2/2}\, dt.$$

Let $u > 0$, and integrate the function $e^{-z^2/2}$, which is holomorphic in $z = t + iu$, along the curve consisting of the oriented segments

$$\overrightarrow{-\lambda, \lambda},\ \overrightarrow{\lambda, \lambda + iu},\ \overrightarrow{\lambda + iu, -\lambda + iu} \text{ and } \overrightarrow{-\lambda + iu, -\lambda}$$

in this order. By Cauchy's integral theorem, the integral vanishes. Thus

$$\begin{aligned}(2\pi)^{-1/2}\int_{-\lambda}^{\lambda} e^{-(t+iu)^2/2}\, dt &= (2\pi)^{-1/2}\int_{-\lambda}^{\lambda} e^{-t^2/2}\, dt \\ &+ (2\pi)^{-1/2}\int_{u}^{0} e^{-(-\lambda+iu)^2/2}\, idu \\ &+ (2\pi)^{-1/2}\int_{0}^{u} e^{-(\lambda+iu)^2/2}\, idu.\end{aligned}$$

The second and third terms on the right tend to 0 as $\lambda\to\infty$, and so, by (10′),

$$(2\pi)^{-1/2}\int_{-\infty}^{\infty} e^{-t^2/2}\, e^{-itu}\, dt = e^{-u^2/2}\,(2\pi)^{-1/2}\int_{-\infty}^{\infty} e^{-t^2/2}\, dt = e^{-u^2/2}.$$

We have thus proved (10) for the case $n = 1$, and it is easy to prove the case of a general n by reducing it to the case $n = 1$.

Corollary (Parseval's relation). We have

$$\int \hat{f}(\xi)\, g(\xi)\, d\xi = \int f(x)\, \hat{g}(x)\, dx, \tag{11}$$

$$\int f(\xi)\, \overline{g}(\xi)\, d\xi = \int \hat{f}(x)\, \overline{\hat{g}}(x)\, dx, \tag{12}$$

$$\widehat{(f * g)} = (2\pi)^{n/2}\hat{f}\cdot\hat{g} \text{ and } (2\pi)^{n/2}\,\widehat{(f\cdot g)} = \hat{f} * \hat{g}, \tag{13}$$

where the *convolution* $f * g$ is defined through

$$(f * g)(x) = \int f(x-y)\, g(y)\, dy = \int g(x-y)\, f(y)\, dy. \tag{14}$$

Proof. (11) is obtained from (9) by putting $x = 0$. (12) is obtained from (11) by observing that the Fourier transform of $\overline{g}$ is equal to $\overline{\overline{g}}$. We next show that

$$\begin{aligned}&(2\pi)^{-n/2}\int (f * g)(x)\, e^{-i\langle\xi,x\rangle}\, dx \\ &= (2\pi)^{-n/2}\int g(y)\, e^{-i\langle\xi,y\rangle}\{\int f(x-y)\, e^{-i\langle\xi,x-y\rangle}\, dx\}\, dy \\ &= (2\pi)^{n/2}\,\hat{f}(\xi)\,\hat{g}(\xi).\end{aligned} \tag{15}$$

Since the product $\hat{f}\cdot\hat{g}$ of functions $\hat{f}$ and $\hat{g}\in\mathfrak{S}(R^n)$ is again a function of $\mathfrak{S}(R^n)$, we know that the right hand side of (15) belongs to $\mathfrak{S}(R^n)$.

It is easy to see that, the convolution $f * g$ of two functions of $\mathfrak{S}(R^n)$ belongs also to $\mathfrak{S}(R^n)$. Thus we have proved the first formula of (13). The second formula may be proved similarly to (9) by (15).

Theorem 2 (Poisson's Summation Formula). Let $\varphi \in \mathfrak{S}(R^1)$ and $\hat{\varphi} \in \mathfrak{S}(R^1)$ its Fourier transform. Then we have

$$\sum_{n=-\infty}^{\infty} \varphi(2\pi n) = \sum_{n=-\infty}^{\infty} \hat{\varphi}(n). \tag{16}$$

Proof. Set $f(x) = \sum_{n=-\infty}^{\infty} \varphi(x + 2\pi n)$. This series is absolutely convergent, $\in C^\infty$ and $f(x + 2\pi) = f(x)$, as may be proved by the fact that $\varphi(x)$ is rapidly decreasing at ∞. In particular, both sides of (16) are convergent. We have to prove the equality.

The Fourier coefficients c_k of $f(x)$ with respect to the complete orthonormal system $\{(2\pi)^{-1/2} e^{-ikx};\ k = 0, \pm 1, \pm 2, \ldots\}$ of $L^2(0, 2\pi)$ are given by

$$c_k = (2\pi)^{-1/2} \int_0^{2\pi} f(x)\, e^{-ikx}\, dx = \sum_{n=-\infty}^{\infty} (2\pi)^{-1/2} \int_0^{2\pi} \varphi(x + 2\pi n)\, e^{-ikx}\, dx$$

$$= \sum_{n=-\infty}^{\infty} (2\pi)^{-1/2} \int_{2\pi n}^{2\pi(n+1)} \varphi(x)\, e^{-ikx}\, dx = \hat{\varphi}(k).$$

Thus, by $f \in L^2(0, 2\pi)$, we have

$$f(x) = \sum_{n=-\infty}^{\infty} \varphi(x + 2\pi n) = \underset{s \uparrow \infty}{\text{l.i.m.}} \sum_{k=-s}^{s} \hat{\varphi}(k)\, e^{ikx}.$$

However, since $\hat{\varphi}(x) \in \mathfrak{S}(R^n)$, the series $\sum_{k=-\infty}^{\infty} \hat{\varphi}(k)\, e^{ikx}$ converges absolutely. Hence

$$\sum_{n=-\infty}^{\infty} \varphi(x + 2\pi n) \equiv \sum_{k=-\infty}^{\infty} \hat{\varphi}(k)\, e^{ikx},$$

and so we obtain (16) by setting $x = 0$.

Example. We have, by (10),

$$(2\pi)^{-1/2} \int_{-\infty}^{\infty} e^{-tx^2} e^{-ixy}\, dx = (2\pi)^{-1/2} \int_{-\infty}^{\infty} e^{-x^2/2} e^{-ixy/\sqrt{2t}} (2t)^{-1/2}\, dx$$

$$= (2t)^{-1/2} e^{-y^2/4t},\ t > 0.$$

Hence, by (16), we obtain the so-called *θ-formula*:

$$\sum_{n=-\infty}^{\infty} e^{-4t\pi^2 n^2} = \sum_{n=-\infty}^{\infty} (2t)^{-1/2} e^{-n^2/4t},\ t > 0. \tag{17}$$

2. The Fourier Transform of Tempered Distributions

Definition 1. A linear functional T defined and continuous on $\mathfrak{S}(R^n)$ is called a *tempered distribution (in R^n)*. The totality of tempered distributions is denoted by $\mathfrak{S}(R^n)'$. As a dual space of $\mathfrak{S}(R^n)$, $\mathfrak{S}(R^n)'$ is a locally convex linear topological space by the strong dual topology.

Proposition 1. Since $C_0^\infty(R^n)$ is contained in $\mathfrak{S}(R^n)$ as an abstract set, and since the topology in $\mathfrak{D}(R^n)$ is stronger than the topology in $\mathfrak{S}(R^n)$, the restriction of a tempered distribution to $C_0^\infty(R^n)$ is a distribution in R^n. Two different tempered distributions define, when restricted to $C_0^\infty(R^n)$, two different distributions in R^n, because $C_0^\infty(R^n)$ is dense in $\mathfrak{S}(R^n)$ with respect to the topology of $\mathfrak{S}(R^n)$, and hence a distribution $\in \mathfrak{S}(R^n)'$ which vanishes on $C_0^\infty(R^n)$ must vanish on $\mathfrak{S}(R^n)$. Therefore

$$\mathfrak{S}(R^n)' \subseteqq \mathfrak{D}(R^n)'. \tag{1}$$

Example 1. A distribution in R^n with a compact support surely belongs to $\mathfrak{S}(R^n)'$. Therefore

$$\mathfrak{E}(R^n)' \subseteqq \mathfrak{S}(R^n)'. \tag{2}$$

Example 2. A σ-finite, non-negative measure $\mu(dx)$ which is σ-additive on Baire sets of R^n is called a *slowly increasing measure*, if, for some non-negative k,

$$\int_{R^n} (1 + |x|^2)^{-k} \mu(dx) < \infty. \tag{3}$$

Such a measure μ defines a tempered distribution by

$$T_\mu(\varphi) = \int_{R^n} \varphi(x) \mu(dx), \ \varphi \in \mathfrak{S}(R^n). \tag{4}$$

For, by the condition $\varphi \in \mathfrak{S}(R^n)$, we have $\varphi(x) = 0((1 + |x|^2)^{-k})$ for large $|x|$.

Example 3. As a special case of Example 2, any function $f \in L^p(R^n)$, $p \geqq 1$, defines a tempered distribution

$$T_f(\varphi) = \int_{R^n} \varphi(x) f(x)\, dx, \ \varphi \in \mathfrak{S}(R^n). \tag{4'}$$

That an $f \in L^p(R^n)$ gives rise to a slowly increasing measure $\mu(dx) = |f(x)|\, dx$ may be proved by applying Hölder's inequality to

$$\int_{R^n} (1 + |x|^2)^{-k} |f(x)|\, dx.$$

Definition 2. A function $f \in C^\infty(R^n)$ is called *slowly increasing* (at ∞), if, for any differentiation D^j, there exists a non-negative integer N such that

$$\lim_{|x| \to \infty} |x|^{-N} |D^j f(x)| = 0. \tag{5}$$

The totality of slowly increasing functions will be denoted by $\mathfrak{O}_M(R^n)$. It is a locally convex linear topological space by the algebraic operations of function sum and multiplication of functions by complex numbers, and by the topology defined by the system of semi-norms of the form

$$p(f) = p_{h,D^j}(f) = \sup_{x \in R^n} |h(x)\, D^j f(x)|, \ f \in \mathfrak{O}_M(R^n), \tag{6}$$

where $h(x)$ is an arbitrary function $\in \mathfrak{S}(R^n)$ and D^j an arbitrary differentiation. As may be seen by applying Leibniz' formula of differentiation of the product of functions, $h(x)\, D^j f(x) \in \mathfrak{S}(R^n)$ and so $p_{h,Dj}(f)$ is finite for every $f \in \mathfrak{O}_M(R^n)$. Moreover, if $p_{h,Dj}(f) = 0$ for all $h \in \mathfrak{S}(R^n)$ and D^j, then $f(x) \equiv 0$ as may be seen by taking $D = I$ and $h \in \mathfrak{D}(R^n)$.

Proposition 2. $C_0^\infty(R^n)$ is dense in $\mathfrak{O}_M(R^n)$ with respect to the topology of $\mathfrak{O}_M(R^n)$.

Proof. Let $f \in \mathfrak{O}_M(R^n)$, and take $\psi \in C_0^\infty(R^n)$ such that $\psi(x) = 1$ for $|x| \leqq 1$. Then $f_\varepsilon(x) = f(x)\, \psi(\varepsilon x) \in C_0^\infty(R^n)$ for any $\varepsilon > 0$. As in Proposition 3 in Chapter VI, 1, we easily prove that $f_\varepsilon(x)$ tends to $f(x)$ in the topology of $\mathfrak{O}_M(R^n)$ when $\varepsilon \downarrow 0$.

Proposition 3. Any function $f \in \mathfrak{O}_M(R^n)$ defines a tempered distribution

$$T_f(\varphi) = \int_{R^n} f(x)\, \varphi(x)\, dx, \ \varphi \in \mathfrak{S}(R^n). \tag{7}$$

Definition 3. As in the case of a distribution in R^n, we can define the generalized derivative of a tempered distribution T by

$$D^j T(\varphi) = (-1)^{|j|} T(D^j \varphi), \ \varphi \in \mathfrak{S}(R^n), \tag{8}$$

since the mapping $\varphi(x) \to D^j \varphi(x)$ of $\mathfrak{S}(R^n)$ into $\mathfrak{S}(R^n)$ is linear and continuous in the topology of $\mathfrak{S}(R^n)$. We can also define a multiplication by a function $f \in \mathfrak{O}_M(R^n)$ to a distribution $T \in \mathfrak{S}(R^n)'$ through

$$(fT)(\varphi) = T(f\varphi), \ \varphi \in \mathfrak{S}(R)^n, \tag{9}$$

since the mapping $\varphi(x) \to f(x)\, \varphi(x)$ of $\mathfrak{S}(R^n)$ into $\mathfrak{S}(R^n)$ is linear and continuous in the topology of $\mathfrak{S}(R^n)$.

The Fourier Transform of Tempered Distributions

Definition 4. Since the mapping $\varphi(x) \to \hat{\varphi}(x)$ of $\mathfrak{S}(R^n)$ onto $\mathfrak{S}(R^n)$ is linear and continuous in the topology of $\mathfrak{S}(R^n)$, we can define the Fourier transform $\hat{T}$ of a tempered distribution T as the tempered distribution $\hat{T}$ defined through

$$\hat{T}(\varphi) = T(\hat{\varphi}), \ \varphi \in \mathfrak{S}(R^n). \tag{10}$$

Example 1. If $f \in L^1(R^n)$, then

$$\hat{T}_f = T_{\hat{f}}, \text{ where } \hat{f}(x) = (2\pi)^{-n/2} \int_{R^n} e^{-i\langle x,\xi\rangle} f(\xi)\, d\xi, \tag{11}$$

as may be seen by changing the order of integration in $\hat{T}_f(\varphi) = \int f(x) \cdot \hat{\varphi}(x)\, dx = (2\pi)^{-n/2} \int_{R^n} f(x) \left\{ \int_{R^n} e^{-i\langle x,\xi\rangle} \varphi(\xi)\, d\xi \right\} dx$.

Remark. In the above sense, the Fourier transform of a tempered distribution is a generalization of the ordinary Fourier transform of functions.

Proposition 4. If we define

$$\check{f}(x) = f(-x), \tag{12}$$

then Fourier's integral theorem in the preceding Section 1 is expressed by

$$\hat{\hat{f}} = \check{f}, \quad f \in \mathfrak{S}(R^n). \tag{13}$$

Corollary 1 (Fourier's integral theorem). Fourier's integral theorem is generalized to tempered distributions as follows:

$$\hat{\hat{T}} = \check{T}, \text{ where } \check{T}(\varphi) = T(\check{\varphi}). \tag{14}$$

In particular, the Fourier transform $T \to \hat{T}$ maps $\mathfrak{S}(R^n)'$ linearly onto $\mathfrak{S}(R^n)'$.

Proof. We have, by definition,

$$\hat{\hat{T}}(\varphi) = T(\hat{\hat{\varphi}}) = T(\check{\varphi}) = \check{T}(\varphi) \text{ for all } \varphi \in \mathfrak{S}(R^n).$$

Corollary 2. The Fourier transform $T \to \hat{T}$ and its inverse are linear and continuous on $\mathfrak{S}(R^n)$ onto $\mathfrak{S}(R^n)$ with respect to the weak* topology of $\mathfrak{S}(R^n)'$:

$$\lim T_h(\varphi) = T(\varphi) \text{ for all } \varphi \in \mathfrak{S}(R^n) \text{ implies that}$$

$$\lim \hat{T}_h(\varphi) = \hat{T}(\varphi) \text{ for all } \varphi \in \mathfrak{S}(R^n). \tag{15}$$

Here the inverse of the mapping $T \to \hat{T}$ is defined by the *inverse Fourier transform* $T \to \tilde{T}$ given by

$$\tilde{T}(\varphi) = T(\tilde{\varphi}), \quad \varphi \in \mathfrak{S}(R^n). \tag{10'}$$

Example 2.

$$\hat{T}_\delta = (2\pi)^{-n/2}\, T_1, \quad \hat{T}_1 = (2\pi)^{n/2} T_\delta. \tag{16}$$

Proof. $\hat{T}_\delta(\varphi) = T_\delta(\hat{\varphi}) = \hat{\varphi}(0) = (2\pi)^{-n/2} \int\limits_{R^n} 1 \cdot \varphi(y)\, dy =$ $(2\pi)^{-n/2}\, T_1(\varphi)$, and $T_\delta = \check{T}_\delta = \hat{\hat{T}}_\delta = (2\pi)^{-n/2}\, \hat{T}_1$.

Example 3.

$$\widehat{(\partial T/\partial x_j)} = i x_j \hat{T}, \tag{17}$$

$$\widehat{(i x_j T)} = -(\partial \hat{T}/\partial x_j). \tag{18}$$

Proof. We have, by (5) in Chapter VI, 1,

$$\widehat{(\partial T/\partial x_j)}(\varphi) = (\partial T/\partial x_j)(\hat{\varphi}) = -T(\partial\hat{\varphi}/\partial x_j) = -T(\widehat{-i x_j\,\varphi}(x))$$
$$= T(\widehat{i x_j\varphi}) = (i x_j \hat{T})(\varphi).$$

We also have, by (6) in Chapter VI, 1,

$$\widehat{(i x_j T)}(\varphi) = (i x_j T)(\hat{\varphi}) = T(i x_j\hat{\varphi}) = T(\widehat{\partial\varphi/\partial x_j}) = \hat{T}(\partial\varphi/\partial x_j)$$
$$= -(\partial\hat{T}/\partial x_j)(\varphi).$$

Plancherel's Theorem. If $f \in L^2(R^n)$, then the Fourier transform $\hat{T}_f$ of T_f is defined by a function $\hat{f} \in L^2(R^n)$, i.e.,

$$\hat{T}_f = T_{\hat{f}} \text{ with } \hat{f} \in L^2(R^n), \tag{19}$$

and

$$\|\hat{f}\| = \Big(\int_{R^n} |\hat{f}(x)|^2\,dx\Big)^{1/2} = \Big(\int_{R^n} |f(x)|^2\,dx\Big)^{1/2} = \|f\|. \tag{20}$$

Proof. We have, by Schwarz' inequality,

$$|\hat{T}_f(\varphi)| = |T_f(\hat{\varphi})| = \Big|\int_{R^n} f(x)\,\hat{\varphi}(x)\,dx\Big| \leqq \|f\|\,\|\hat{\varphi}\| = \|f\|\,\|\varphi\|. \tag{21}$$

The equality above $\|\hat{\varphi}\| = \|\varphi\|$ is proved in (12) of the preceding section 1. Hence, by F. Riesz' representation theorem in the Hilbert space $L^2(R^n)$, there exists a uniquely determined $\hat{f} \in L^2(R^n)$ such that

$$\begin{aligned} &\hat{T}_f(\varphi) = \int_{R^n} \varphi(x)\,\hat{f}(x)\,dx = T_{\hat{f}}(\varphi), \text{ that is,} \\ &\int_{R^n} \hat{f}(x)\,\varphi(x)\,dx = \int_{R^n} f(x)\,\hat{\varphi}(x)\,dx \text{ for all } \varphi \in \mathfrak{S}(R^n). \end{aligned} \tag{22}$$

Moreover, we have, from (21), $\|f\| \geqq \|\hat{f}\|$ since $\mathfrak{S}(R^n)$ is dense in $L^2(R^n)$ in the topology of $L^2(R^n)$. We have thus $\|\hat{\hat{f}}\| \leqq \|\hat{f}\| \leqq \|f\|$. On the other hand, we have, by (13) and (22),

$$\int_{Rn} \hat{\hat{f}}(x)\,\varphi(x)\,dx = \int_{R^n} \hat{f}(x)\,\hat{\varphi}(x)\,dx = \int_{R^n} f(-x)\,\varphi(x)\,dx \text{ for all } \varphi \in \mathfrak{S}(R^n),$$

that is,

$$\hat{\hat{f}}(x) = f(-x) = \check{f}(x) \text{ a.e.} \tag{23}$$

Thus $\|\hat{\hat{f}}\| = \|f\|$ and so, combined with $\|\hat{\hat{f}}\| \leqq \|\hat{f}\| \leqq \|f\|$, we obtain (20).

Definition 5. The above obtained $\hat{f}(x) \in L^2(R^n)$ is called the *Fourier transform* of the function $f(x) \in L^2(R^n)$.

Corollary 1. We have, for any $f \in L^2(R^n)$,

$$\hat{f}(x) = \underset{h \uparrow \infty}{\text{l.i.m.}}\ (2\pi)^{-n/2} \int_{|x| \leqq h} e^{-i\langle x,y\rangle} f(y)\, dy. \tag{24}$$

Proof. Put

$$f_h(x) = f(x) \text{ or } = 0 \text{ according as } |x| \leqq h \text{ or } |x| > h.$$

Then $\lim_{h\to\infty} ||f_h - f|| = 0$ and so, by (20), $\lim_{h\to\infty} ||\hat{f}_h - \hat{f}|| = 0$, that is, $\hat{f}(x) = \underset{h\to\infty}{\text{l.i.m.}}\ \hat{f}_h(x)$ a.e. But, by (22),

$$\begin{aligned}\int_{R^n} \hat{f}_h(x)\, \varphi(x)\, dx &= \int_{R^n} f_h(x)\, \hat{\varphi}(x)\, dx \\ &= \int_{|x| \leqq h} f(x) \left\{ (2\pi)^{-n/2} \int_{R^n} e^{-i\langle x,y\rangle} \varphi(y)\, dy \right\} dx,\end{aligned}$$

which is, by changing the order of integration, equal to

$$\int_{R^n} (2\pi)^{-n/2} \left\{ \int_{|x| \leqq h} e^{-i\langle x,y\rangle} f(x)\, dx \right\} \varphi(y)\, dy,$$

since $f_h(x)$ is integrable over $|x| \leqq h$ as may be seen by Schwarz' inequality. Thus $\hat{f}_h(x) = (2\pi)^{-n/2} \int_{|x| \leqq h} e^{-i\langle x,y\rangle} f(y)\, dy$ a.e., and so we obtain (24).

Corollary 2. The Fourier transform $f \to \hat{f}$ maps $L^2(R^n)$ onto $L^2(R^n)$ in a one-one-manner such that

$$(f, g) = (\hat{f}, \hat{g}) \text{ for all } f, g \in L^2(R^n). \tag{25}$$

Proof. Like the Fourier transform $f \to \hat{f}$, the inverse Fourier transform $f \to \tilde{f}$ defined by

$$\tilde{f}(x) = \underset{h\to\infty}{\text{l.i.m.}}\ (2\pi)^{-n/2} \int_{|y| \leqq h} e^{i\langle x,y\rangle} f(y)\, dy \tag{26}$$

maps $L^2(R^n)$ into $L^2(R^n)$ in such a way that $||\tilde{f}|| = ||f||$. Hence we see that the Fourier transform $f \to \hat{f}$ maps $L^2(R^n)$ onto $L^2(R^n)$ in a one-one way such that $||\hat{f}|| = ||f||$. Hence, by the linearity of the Fourier transform and

$$(x, y) = 4^{-1}(||x + y||^2 - ||x - y||^2) + 4^{-1} i(||x + iy||^2 - ||x - iy||^2),$$

we obtain (25).

Parseval's Theorem for the Fourier Transform. Let $f_1(x)$ and $f_2(x)$ both belong to $L^2(R^n)$, and let their Fourier transforms be $\hat{f}_2(u)$ and $\hat{f}_2(u)$, respectively. Then

$$\int_{R^n} \hat{f}_1(u) \hat{f}_2(u)\, du = \int_{R^n} f_1(x)\, f_2(-x)\, dx, \tag{27}$$

and so

$$\int_{R^n} \hat{f}_1(u)\, \hat{f}_2(u)\, e^{-i\langle u,x\rangle}\, du = \int_{R^n} f_1(y)\, f_2(x-y)\, dy. \tag{28}$$

Thus, if $\hat{f}_1(u)\,\hat{f}_2(u)$ as well as both its factors belong to $L^2(R^n)$, it is the Fourier transform of

$$(2\pi)^{-n/2} \int_{R^n} f_1(y)\, f_2(x-y)\, dy. \tag{29}$$

This will also be true if $f_1(x)$, $f_2(x)$ and (29) all belong to $L^2(R^n)$.

Proof. It is easy to see that

$$(2\pi)^{-n/2} \int_{R^n} \overline{f_2(-x)}\, e^{-i\langle u,x\rangle}\, dx = \overline{\hat{f}_2(u)},$$

and so, by (25), we obtain (27). Next, since the Fourier transform of the function of y, $\overline{f_2(x-y)}$, is $\overline{\hat{f}_2(u)}\, e^{i\langle u,x\rangle}$ containing a parameter x, we obtain (28) by (25). The rest of the Theorem is clear from (28) and $\tilde{\tilde{f}} = f$.

The Negative Norm. The Sobolev space $W^{k,2}(\Omega)$ was defined in Chapter I, 9. Let $f(x) \in W^{k,2}(R^n)$. Since $f(x) \in L^2(R^n)$, f gives rise to a slowly increasing measure $|f(x)|\, dx$ in R^n. Thus we can define the Fourier transform $\hat{T}_f$ of the tempered distribution T_f. We have, by (17),

$$\widehat{D^\alpha T_f} = (i)^{|\alpha|} \prod_{j=1}^{n} x_j^{\alpha_j}\, \hat{T}_f.$$

By the definition of the space $W^{k,2}(R^n)$, $D^\alpha T_f \in L^2(R^n)$ for $|\alpha| \leqq k$. Thus, by Plancherel's theorem for $L^2(R^n)$,

$$\|\widehat{D^\alpha T_f}\|_0 = \|D^\alpha T_f\|_0, \text{ where } \|\ \|_0 \text{ is the } L^2(R^n)\text{-norm}.$$

Hence we see that $(1+|x|^2)^{k/2}\,\hat{T}_f \in L^2(R^n)$, and so it is easy to show that the norm $\|f\|_k = \left(\sum_{|\alpha|\leqq k} \int_{R^n} |D^\alpha T_f|^2\, dx\right)^{1/2}$ is equivalent to the norm

$$\|(1+|x|^2)^{k/2}\,\hat{T}_f\|_0 = \|f\|'_k, \tag{30}$$

in the sense that there exist two positive constants c_1 and c_2 for which we have

$$c_1 \leqq \|f\|_k / \|f\|'_k \leqq c_2 \text{ whenever } f \in W^{k,2}(R^n).$$

We may thus renorm the space $W^{k,2}(R^n)$ by $\|f\|'_k$; $W^{k,2}(R^n)$ may thus be defined as the totality of $f \in L^2(R^n)$ such that $\|f\|'_k$ is finite. One advantage of the new formulation of $W^{k,2}(R^n)$ is that we may also consider the case of a negative exponent k. Then, as in the case of $L^2(R^n)$ pertaining to the ordinary Lebesgue measure dx, we see that the dual space of the renormed space $W^{k,2}(R^n)$ is the space $W^{-k,2}(R^n)$ normed by $\|f\|'_{-k}$. This observation due to L. Schwartz [5] is of an earlier date than the introduction of the negative by P. Lax [2].

3. Convolutions

We define the *convolution* (*Faltung*) of two functions f, g of $C(R^n)$, one of which has a compact support, by (cf. the case in which $f, g \in \mathfrak{S}(R^n)$ in Chapter VI, 1)

$$(f * g)(x) = \int_{R^n} f(x-y)\, g(y)\, dy = \int_{R^n} f(y)\, g(x-y)\, dy = (g * f)(x). \quad (1)$$

Suggested by this formula, we define the *convolution* of a $T \in \mathfrak{D}(R^n)'$ and a $\varphi \in \mathfrak{D}(R^n)$ (or a $T \in \mathfrak{E}(R^n)'$ and a $\varphi \in \mathfrak{E}(R^n)$) by

$$(T * \varphi)(x) = T_{[y]}(\varphi(x-y)), \quad (2)$$

where $T_{[y]}$ indicates that we apply the distribution T on test functions of y.

Proposition 1. $(T * \varphi)(x) \in C^\infty(R^n)$ and $\operatorname{supp}(T * \varphi) \subseteqq \operatorname{supp}(T) + \operatorname{supp}(\varphi)$, that is,

$$\operatorname{supp}(T * \varphi) \subseteqq \{w \in R^n;\, w = x + y,\, x \in \operatorname{supp}(T),\, y \in \operatorname{supp}(\varphi)\}.$$

Moreover, we have

$$D^\alpha(T * \varphi) = T * (D^\alpha \varphi) = (D^\alpha T) * \varphi. \quad (3)$$

Proof. Let $\varphi \in \mathfrak{D}(R^n)$ (or $\in \mathfrak{E}(R^n)$). If $\lim_{h\to 0} x^h = x$, then, as functions of y, $\lim_{h\to 0} \varphi(x^h - y) = \varphi(x-y)$ in $\mathfrak{D}(R^n)$ (or in $\mathfrak{E}(R^n)$). Hence $T_{[y]}(\varphi(x-y)) = (T * \varphi)(x)$ is continuous in x. The inclusion relation of the supports is proved by the fact that $T_{[y]}(\varphi(x-y)) = 0$ unless the support of T and that of $\varphi(x-y)$ as a function of y meet. Next let e_j be the unit vector of R^n along the x_j-axis and consider the expression

$$T_{[y]}((\varphi(x + h e_j - y) - \varphi(x-y))/h).$$

When $h \to 0$, the function enclosed by the outer parenthesis converges, as a function of y, to $\left(\frac{\partial \varphi}{\partial x_j}\right)(x-y)$ in $\mathfrak{D}(R^n)$ (or in $\mathfrak{E}(R^n)$). Thus we have proved

$$\frac{\partial}{\partial x_j}(T * \varphi)(x) = \left(T * \frac{\partial \varphi}{\partial x_j}\right)(x).$$

Moreover, we have

$$\left(T * \frac{\partial \varphi}{\partial x_j}\right)(x) = T_{[y]}\left(-\frac{\partial \varphi(x-y)}{\partial y_j}\right) = \frac{\partial T_{[y]}}{\partial y_j}(\varphi(x-y)) = \left(\frac{\partial T}{\partial x_j} * \varphi\right)(x).$$

Proposition 2. If φ and ψ are in $\mathfrak{D}(R^n)$ and $T \in \mathfrak{D}(R^n)'$ (or $\varphi \in \mathfrak{E}(R^n)$, $\psi \in \mathfrak{D}(R^n)$ and $T \in \mathfrak{E}(R^n)'$), then

$$(T * \varphi) * \psi = T * (\varphi * \psi). \quad (4)$$

Proof. We approximate the function $(\varphi * \psi)(x)$ by the Riemann sum

$$f_h(x) = h^{-n} \sum_h \varphi(x - kh)\, \psi(kh),$$

where $h > 0$ and k ranges over points of R^n with integral coordinates. Then, for every differentiation D^α and for every compact set of x,

$$D^\alpha f_h(x) = h^{-n} \sum_h D^\alpha \varphi(x - kh)\, \psi(kh)$$

converges, as $h \downarrow 0$, to $((D^\alpha \varphi) * \psi)(x) = (D^\alpha(\varphi * \psi))(x)$ uniformly in x. Hence we see that $\lim_{h \downarrow 0} f_h = \varphi * \psi$ in $\mathfrak{D}(R^n)$ (or in $\mathfrak{E}(R^n)$). Therefore, by the linearity and the continuity of T, we have

$$\begin{aligned}(T * (\varphi * \psi))(x) &= \lim_{h \downarrow 0} (T * f_h)(x) = \lim_{h \downarrow 0} h^{-n} \sum_k (T * \varphi)(x - kh)\, \psi(kh) \\ &= ((T * \varphi) * \psi)(x).\end{aligned}$$

Definition. Let $\varphi \in \mathfrak{D}(R^n)$ be non-negative, $\int_{R^n} \varphi\, dx = 1$ and such that $\operatorname{supp}(\varphi) \subseteq \{x \in R^n;\, |x| \leqq 1\}$. We may, for instance, take

$$\begin{aligned}\varphi(x) &= \exp(1/(|x|^2 - 1)) \Big/ \int_{|x|<1} \exp(1/(|x|^2 - 1))\, dx \text{ if } |x| < 1, \\ &= 0 \text{ if } |x| \geqq 1.\end{aligned}$$

We write $\varphi_\varepsilon(x)$ for $\varepsilon^{-n} \varphi(x/\varepsilon)$, $\varepsilon > 0$, and call $T * \varphi_\varepsilon$ the *regularization* of $T \in \mathfrak{D}(R^n)'$ (or $\in \mathfrak{E}(R^n)'$) through $\varphi_\varepsilon(x)$'s (Cf. Chapter I, 1).

Theorem 1. Let $T \in \mathfrak{D}(R^n)'$ (or $\in \mathfrak{E}(R^n)'$). Then $\lim_{\varepsilon \downarrow 0} (T * \varphi_\varepsilon) = T$ in the weak* topology of $\mathfrak{D}(R^n)'$ (or of $\mathfrak{E}(R^n)'$). In this sense, φ_ε is called an *approximate identity.*

For the proof we prepare a

Lemma. For any $\psi \in \mathfrak{D}(R^n)$ (or $\in \mathfrak{E}(R^n)$), we have $\lim_{\varepsilon \downarrow 0} \psi * \varphi_\varepsilon = \psi$ in $\mathfrak{D}(R^n)$ (or in $\mathfrak{E}(R^n)$).

Proof. We first observe that $\operatorname{supp}(\psi * \varphi_\varepsilon) \subseteq \operatorname{supp}(\psi) + \operatorname{supp}(\varphi_\varepsilon) = \operatorname{supp}(\psi) + \varepsilon$. We have, by (3), $D^\alpha(\psi * \varphi_\varepsilon) = (D^\alpha \psi) * \varphi_\varepsilon$. Hence we have to show that $\lim_{\varepsilon \downarrow 0} (\psi * \varphi_\varepsilon)(x) = \psi(x)$ uniformly on any compact set of x. But, by $\int \varphi_\varepsilon(y)\, dy = 1$,

$$(\psi * \varphi_\varepsilon)(x) - \psi(x) = \int_{R^n} \{\psi(x - y) - \psi(x)\}\, \varphi_\varepsilon(y)\, dy.$$

Therefore, by $\varphi_\varepsilon(x) \geqq 0$, $\int_{R^n} \varphi_\varepsilon(y)\, dy = 1$ and the uniform continuity of $\psi(x)$ on any compact interval of x, we obtain the Lemma.

Proof of Theorem 1. We have, by $\check{\psi}(x) = \psi(-x)$,

$$T(\psi) = (T * \check{\psi})(0). \tag{5}$$

Hence we have to prove that $\lim_{\varepsilon \downarrow 0} ((T * \varphi_\varepsilon) * \check{\psi})(0) = (T * \check{\psi})(0)$. But, as proved in (4), $(T * \varphi_\varepsilon) * \check{\psi} = T * (\varphi_\varepsilon * \check{\psi})$ and so, by (5), $(T * \varphi_\varepsilon)(\psi) = T((\varphi_\varepsilon * \check{\psi})\check{\ })$. Hence we obtain $\lim_{\varepsilon \downarrow 0} (T * \varphi_\varepsilon)(\psi) = T((\check{\psi})\check{\ }) = T(\psi)$ by the Lemma.

We next prove a theorem which concerns with a characterization of the operation of convolution.

Theorem 2 (L. Schwartz). Let L be a continuous linear mapping on $\mathfrak{D}(R^n)$ into $\mathfrak{E}(R^n)$ such that

$$L\tau_h\varphi = \tau_h L\varphi \text{ for any } h \in R^n \text{ and } \varphi \in \mathfrak{D}(R^n), \tag{6}$$

where the translation operator τ_h is defined by

$$\tau_h\varphi(x) = \varphi(x-h). \tag{7}$$

Then there exists a uniquely determined $T \in \mathfrak{D}(R^n)'$ such that $L\varphi = T * \varphi$. Conversely, any $T \in \mathfrak{D}(R^n)'$ defines a continuous linear map L on $\mathfrak{D}(R^n)$ into $\mathfrak{E}(R^n)$ by $L\varphi = T * \varphi$ such that L satisfies (6).

Proof. Since $\varphi \to \check{\varphi}$ is a continuous linear map of $\mathfrak{D}(R^n)$ onto $\mathfrak{D}(R^n)$, the linear map $T: \check{\varphi} \to (L\varphi)(0)$ defines a distribution $T \in \mathfrak{D}(R^n)'$. Hence, by (5), $(L\varphi)(0) = T(\check{\varphi}) = (T * \varphi)(0)$. If we replace φ by $\tau_h\varphi$ and make use of condition (6), then we obtain $(L\varphi)(h) = (T * \varphi)(h)$. The converse part of Theorem 2 is easily proved by (2), Proposition 1 and (5).

Corollary. Let $T_1 \in \mathfrak{D}(R^n)'$ and $T_2 \in \mathfrak{E}(R^n)'$. Then the convolution $T_1 * T_2$ may be defined, through the continuous linear map L on $\mathfrak{D}(R^n)$ into $\mathfrak{E}(R^n)$ as follows:

$$(T_1 * T_2) * \varphi = L(\varphi) = T_1 * (T_2 * \varphi), \varphi \in \mathfrak{D}(R^n). \tag{8}$$

Proof. The map $\varphi \to T_2 * \varphi$ is continuous linear on $\mathfrak{D}(R^n)$ into $\mathfrak{D}(R^n)$, since $\text{supp}(T_2)$ is compact. Hence the map $\varphi \to T_1 * (T_2 * \varphi)$ is continuous linear on $\mathfrak{D}(R^n)$ into $\mathfrak{E}(R^n)$. It is easily verified that condition (6) is satisfied for the present L.

Remark. We see, by (4), that the above definition of the convolution $T_1 * T_2$ agrees with the previous one in the case where T_2 is defined by a function $\in \mathfrak{D}(R^n)$. It is to be noted that we may also define $T_1 * T_2$ by

$$(T_1 * T_2)(\varphi) = (T_{1(x)} \times T_{2(y)})(\varphi(x+y)), \varphi \in \mathfrak{D}(R^n), \tag{8'}$$

where $T_{1(x)} \times T_{2(y)}$ is the tensor product of T_1 and T_2. See L. Schwartz [1].

Theorem 3. Let $T_1 \in \mathfrak{D}(R^n)'$ and $T_2 \in \mathfrak{E}(R^n)'$. Then we can define another "convolution" $T_2 \boxtimes T_1$ through the continuous linear map L:

$$\varphi \to T_2 * (T_1 * \varphi) \text{ on } \mathfrak{D}(R^n) \text{ into } \mathfrak{E}(R^n)$$

as follows. $(T_2 \boxtimes T_1)(*\varphi) = L(\varphi)$, $\varphi \in \mathfrak{D}(R^n)$. Then we can prove $T_2 \boxtimes T_1 = T_1 * T_2$ so that the convolution is commutative if it is defined either as $T_1 * T_2$ or as $T_2 \boxtimes T_1$.

Proof. The map $\varphi \to T_1 * \varphi$ is linear continuous on $\mathfrak{D}(R^n)$ into $\mathfrak{E}(R^n)$. Hence the map $\varphi \to T_2 * (T_1 * \varphi)$ is linear continuous on $\mathfrak{D}(R^n)$ into $\mathfrak{E}(R^n)$. Thus $T_2 \boxtimes T_1$ is well-defined. Next, we have for any $\varphi_1, \varphi_2 \in \mathfrak{D}(R^n)$,

$$\begin{aligned}(T_1 * T_2) * (\varphi_1 * \varphi_2) &= T_1 * (T_2 * (\varphi_1 * \varphi_2)) = T_1 * ((T_2 * \varphi_1) * \varphi_2) \\ &= T_1 * (\varphi_2 * (T_2 * \varphi_1)) = (T_1 * \varphi_2) * (T_2 * \varphi_1)\end{aligned}$$

by the commutativity of the convolution of functions and Proposition 2, observing that $\operatorname{supp}(T_2 * \varphi_1)$ is compact because of $T_2 \in \mathfrak{E}(R^n)'$. Similarly we obtain

$$\begin{aligned}(T_2 \boxtimes T_1) * (\varphi_1 * \varphi_2) &= T_2 * (T_1 * (\varphi_1 * \varphi_2)) = T_2 * ((T_1 * \varphi_2) * \varphi_1) \\ &= T_2 * (\varphi_1 * (T_1 * \varphi_2)) = (T_2 * \varphi_1) * (T_1 * \varphi_2).\end{aligned}$$

Thus $(T_1 * T_2) * (\varphi_1 * \varphi_2) = (T_2 \boxtimes T_1) * (\varphi_1 * \varphi_2)$, and so, by (5) and the Lemma above, we obtain $(T_1 * T_2)\ (\varphi) = (T_2 \boxtimes T_1)\ (\varphi)$ for all $\varphi \in \mathfrak{D}(R^n)$, that is, $(T_1 * T_2) = (T_2 \boxtimes T_1)$.

Corollary

$$T_1 * (T_2 * T_3) = (T_1 * T_2) * T_3 \quad \text{if all } T_j \text{ except one have compact support}, \tag{9}$$

$$D^\alpha (T_1 * T_2) = (D^\alpha T_1) * T_2 = T_1 * (D^\alpha T_2). \tag{10}$$

Proof. We have, by (5) and the definition of $T_1 * T_2$,

$$\begin{aligned}(T_1 * (T_2 * T_3))\ (\varphi) &= ((T_1 * (T_2 * T_3)) * \check{\varphi})\ (0) \\ &= (T_1 * ((T_2 * T_3) * \check{\varphi}))\ (0) \\ &= (T_1 * (T_2 * (T_3 * \check{\varphi})))\ (0)\end{aligned}$$

and similarly

$$((T_1 * T_2) * T_3)\ (\varphi) = (T_1 * (T_2 * (T_3 * \check{\varphi})))\ (0),$$

so that (9) holds.

The proof of (10) is as follows. We observe that, by (3),

$$(D^\alpha T_\delta) * \varphi = T_\delta * (D^\alpha \varphi) = D^\alpha (T_\delta * \varphi) = D^\alpha \varphi, \tag{11}$$

which implies

$$(D^\alpha T) * \varphi = T * (D^\alpha \varphi) = T * ((D^\alpha T_\delta) * \varphi) = (T * D^\alpha T_\delta) * \varphi,$$

that is, by (5),

$$D^\alpha T = (D^\alpha T_\delta) * T. \tag{12}$$

Hence, by the commutativity (Theorem 3) and the associativity (9),

$$\begin{aligned}D^\alpha (T_1 * T_2) &= (D^\alpha T_\delta) * (T_1 * T_2) = ((D^\alpha T_\delta) * T_1) * T_2 = (D^\alpha T_1) * T_2 \\ &= (D^\alpha T_\delta) * (T_2 * T_1) = ((D^\alpha T_\delta) * T_2) * T_1 = (D^\alpha T_2) * T_1.\end{aligned}$$

The Fourier Transform of the Convolution. We first prove a theorem which will be sharpened to the Paley-Wiener theorem in the next section.

Theorem 4. The Fourier transform of a distribution $T \in \mathfrak{E}(R^n)'$ is given by the function

$$\hat{T}(\xi) = (2\pi)^{-n/2}\, T_{[x]}(e^{-i\langle x,\xi\rangle}). \tag{13}$$

Proof. When $\varepsilon \downarrow 0$, the regularization $T_\varepsilon = T * \varphi_\varepsilon$ tends to T in the weak* topology of $\mathfrak{E}(R^n)'$ and a fortiori in the weak* topology of $\mathfrak{S}(R^n)'$. This we see from $(T * \varphi_\varepsilon)(\psi) = (T * (\varphi_\varepsilon * \check{\psi}))(0) = T_{[x]}((\varphi_\varepsilon * \check{\psi})(-x))$ and the Lemma. Thus, by the continuity of the Fourier transform in the weak* topology of $\mathfrak{S}(R^n)'$, $\lim_{\varepsilon\downarrow 0} \widehat{(T * \varphi_\varepsilon)} = \hat{T}$ in the weak* topology of $\mathfrak{S}(R^n)'$. Now formula (13) is clear for the distribution defined by the function $(T * \varphi_\varepsilon)(x)$. Hence

$$(2\pi)^{n/2}\, \widehat{(T * \varphi_\varepsilon)}(\xi) = (T * \varphi_\varepsilon)_{[x]}(e^{-i\langle x,\xi\rangle}),$$

which is, by (5), $= (T_{[x]} * (\varphi_\varepsilon * e^{-i\langle x,\xi\rangle}))(0) = T_{[x]}(\check{\varphi}_\varepsilon * e^{-i\langle x,\xi\rangle})$. The last expression tends, as $\varepsilon \downarrow 0$, to $T_{[x]}(e^{i\langle x,\xi\rangle})$ uniformly in ξ on any bounded set of ξ of the complex n-space. This proves Theorem 4.

Theorem 5. If we define the convolution of a distribution $T \in \mathfrak{S}(R^n)'$ and a function $\varphi \in \mathfrak{S}(R^n)$ by $(T * \varphi)(x) = T_{[y]}(\varphi(x-y))$, then the linear map L on $\mathfrak{S}(R^n)$ into $\mathfrak{E}(R^n): \varphi \to T * \varphi$ is characterized by the continuity and the translation invariance $\tau_h L = L\tau_h$.

Proof. Similar to the proof of Theorem 2.

Theorem 6. If $T \in \mathfrak{S}(R^n)'$ and $\varphi \in \mathfrak{S}(R^n)$, then

$$\widehat{(T * \varphi)} = (2\pi)^{n/2}\, \hat{\varphi}\,\hat{T} \tag{14}$$

If $T_1 \in \mathfrak{S}(R^n)'$ and $T_2 \in \mathfrak{E}(R^n)'$, then

$$\widehat{(T_1 * T_2)} = (2\pi)^{n/2}\, \hat{T}_2 \cdot \hat{T}_1, \tag{15}$$

which has a sense since, as proved above in Theorem 4, $\hat{T}_2$ is given by a function.

Proof. Let $\psi \in \mathfrak{S}(R^n)$. Then the Fourier transform of $\hat{\varphi} \cdot \psi$ is, by (13) of Chapter VI, 1, equal to $(2\pi)^{-n/2}\, \hat{\hat{\varphi}} * \hat{\psi} = (2\pi)^{-n/2}\, \check{\varphi} * \hat{\psi}$. Thus

$$\begin{aligned}
\widehat{(T * \varphi)}(\psi) &= (T * \varphi)(\hat{\psi}) = ((T * \varphi) * \check{\hat{\psi}})(0) = (T * (\varphi * \check{\hat{\psi}}))(0) \\
&= T((\varphi * \check{\hat{\psi}})\check{}\,) = T(\check{\varphi} * \hat{\psi}) = T((2\pi)^{n/2}(\hat{\varphi}\cdot\psi)\hat{}\,) \\
&= (2\pi)^{n/2}\, \hat{T}(\hat{\varphi}\psi) = (2\pi)^{n/2}\, \hat{\varphi}\,\hat{T}(\psi),
\end{aligned}$$

which proves (14).

Let ψ_ε be the regularization $T_2 * \varphi_\varepsilon$. Then the Fourier transform of $T_1 * \psi_\varepsilon = T_1 * (T_2 * \varphi_\varepsilon) = (T_1 * T_2) * \varphi_\varepsilon$ is, by (14), equal to

$$(2\pi)^{n/2}\, \hat{T}_1 \cdot \hat{\psi}_\varepsilon = (2\pi)^{n/2}\, \hat{T}_1 \cdot (2\pi)^{n/2}\, \hat{T}_2 \cdot \hat{\varphi}_\varepsilon = (2\pi)^{n/2}\, (\widehat{T_1 * T_2}) \cdot \hat{\varphi}_\varepsilon .$$

Hence we obtain (15), by letting $\varepsilon \downarrow 0$ and using $\lim_{\varepsilon\downarrow 0} \hat{\varphi}_\varepsilon(x) = 1$.

4. The Paley-Wiener Theorems. The One-sided Laplace Transform

The Fourier transform of a function $\in C_0^\infty(R^n)$ is characterized by the **Paley-Wiener Theorem for Functions.** An entire holomorphic function $F(\zeta) = F(\zeta_1, \zeta_2, \ldots, \zeta_n)$ of n complex variables $\zeta_j = \xi_j + i\,\eta_j$ $(j = 1, 2, \ldots, n)$ is the *Fourier-Laplace transform*

$$F(\zeta) = (2\pi)^{-n/2} \int_{R^n} e^{-i\langle\zeta,x\rangle} f(x)\, dx \tag{1}$$

of a function $f \in C_0^\infty(R^n)$ with $\mathrm{supp}(f)$ contained in the sphere $|x| \leqq B$ of R^n iff there exists for every integer N a positive constant C_N such that

$$|F(\zeta)| \leqq C_N (1 + |\zeta|)^{-N} e^{B|Im\zeta|}. \tag{2}$$

Proof. The necessity is clear from

$$\prod_{j=1}^n (i\zeta_j)^{\beta_j} F(\zeta) = (2\pi)^{-n/2} \int_{|x|\leqq B} e^{-i\langle\zeta,x\rangle} D^\beta f(x)\, dx$$

which is obtained by partial integration.

The sufficiency. Let us define

$$f(x) = (2\pi)^{-n/2} \int_{R^n} e^{i\langle x,\xi\rangle} F(\xi)\, d\xi. \tag{3}$$

Then, as for the case of functions $\in \mathfrak{S}(R^n)$, we can prove that the Fourier transform $\hat{f}(\xi)$ is equal to $F(\xi)$ and $f \in C^\infty(R^n)$. The last assertion is proved by differentiation:

$$D^\beta f(x) = (2\pi)^{-n/2} \int_{R^n} e^{i\langle x,\xi\rangle} \prod_{j=1}^n (i\xi_j)^{\beta_j} F(\xi)\, d\xi, \tag{4}$$

by making use of condition (2). The same condition (2) and Cauchy's integral theorem enable us to shift the real domain of integration in (3) into the complex domain so that we obtain

$$f(x) = (2\pi)^{-n/2} \int_{R^n} e^{i\langle x,\xi+i\eta\rangle} F(\xi + i\,\eta)\, d\xi \tag{3'}$$

for arbitrary real η of the form $\eta = \alpha x/|x|$ with $\alpha > 0$. We thus obtain, by taking $N = n + 1$,

$$|f(x)| \leqq C_N\, e^{B|\eta| - \langle x,\eta\rangle} (2\pi)^{-n/2} \int_{R^n} (1 + |\xi|)^{-N}\, d\xi.$$

If $|x| > B$, we obtain $f(x) = 0$ by letting $\alpha \uparrow +\infty$. Hence $\mathrm{supp}(f) \subseteq \{x \in R^n;\ |x| \leqq B\}$.

The above theorem may be generalized to distributions with compact support. Thus we have

The Paley-Wiener Theorem for Distributions $\in \mathfrak{E}(R^n)'$ (L. SCHWARTZ). An entire holomorphic function $F(\zeta) = F(\zeta_1, \ldots, \zeta_n)$ of n complex variables $\zeta_j = \xi_j + i\eta_j$ $(j = 1, 2, \ldots, n)$ is the Fourier-Laplace transform of a distribution $T \in \mathfrak{E}(R^n)'$ iff for some positive constants B, N and C

$$|F(\zeta)| \leqq C(1 + |\zeta|)^N e^{B|Im\zeta|}. \tag{5}$$

Proof. The necessity is clear from the fact (Theorem 2 in Chapter I, 13) that, if $T \in \mathfrak{E}(R^n)'$, there exist positive constants C, N and B such that

$$|T(\varphi)| \leqq C \sum_{|\beta| \leqq N} \sup_{|x| \leqq B} |D^\beta \varphi(x)| \quad \text{whenever} \quad \varphi \in \mathfrak{E}(R^n).$$

For, we have only to take $\varphi(x) = e^{-i\langle x, \xi\rangle}$ and apply (13) of the preceding section.

The sufficiency. $F(\xi)$ is in $\mathfrak{S}(R^n)'$ and so it is the Fourier transform of a distribution $T \in \mathfrak{S}(R^n)'$. The Fourier transform of the regularization $T_\varepsilon = T * \varphi_\varepsilon$ is $(2\pi)^{n/2}\,\hat{T} \cdot \hat{\varphi}_\varepsilon$ by (14) of the preceding section. Since the supp (φ_ε) is in the sphere $|x| \leqq \varepsilon$ of R^n, we have, by the preceding theorem,

$$|\hat{\varphi}_\varepsilon(\xi)| \leqq C' \cdot e^{\varepsilon|Im\zeta|}.$$

Moreover, since $\hat{T}$ is defined by the function $F(\xi)$, we see that $(2\pi)^{n/2}\,\hat{T} \cdot \hat{\varphi}_\varepsilon$ is defiend by a function $(2\pi)^{n/2} F(\xi) \cdot \hat{\varphi}_\varepsilon(\xi)$ which, when extended to the complex n-space analytically, satisfies the estimate of type (5) with B replaced by $B + \varepsilon$. Thus, by the preceding theorem, $T_\varepsilon = T * \varphi_\varepsilon$ has its support in the sphere $|x| \leqq B + \varepsilon$ of R^n. Thus, letting $\varepsilon \downarrow 0$ and making use of the Lemma in the preceding section, we see that the supp (T) is in the sphere $|x| \leqq B$ of R^n.

Remark. The formulation and the proofs of the Paley-Wiener theorems given above are adapted from L. HÖRMANDER [2].

The Fourier Transform and the One-sided Laplace Transform. Let $g(t) \in L^2(0, \infty)$. Then, for $x > 0$,

$$g(t)\, e^{-tx} \in L^1(0, \infty) \cap L^2(0, \infty)$$

as may be seen by Schwarz' inequality. Hence, by Plancherel's theorem, we have, for the Fourier transform

$$\begin{aligned} f(x + iy) &= (2\pi)^{-1/2} \int_0^\infty g(t)\, e^{-tx}\, e^{-ity}\, dt \\ &= (2\pi)^{-1/2} \int_0^\infty g(t)\, e^{-t(x+iy)}\, dt \qquad (x > 0), \end{aligned} \tag{6}$$

the inequality

$$\int_{-\infty}^{\infty} |f(x + iy)|^2\, dy = \int_0^\infty |g(t)|^2\, e^{-2tx}\, dt \leqq \int_0^\infty |g(t)|^2\, dt. \tag{7}$$

The function $f(x+iy)$ is holomorphic in $z = x + iy$ in the right half-plane $Re(z) = x > 0$, as may be seen by differentiating (6) under the integral sign observing that $g(t)\, te^{-tz}$ as a function of t belongs to $L^1(0,\infty)$ and to $L^2(0,\infty)$ when $Re(z) = x > 0$. We have thus proved

Theorem 1. Let $g(t) \in L^2(0,\infty)$. Then the *one-sided Laplace transform*

$$f(z) = (2\pi)^{-1/2} \int_0^\infty g(t)\, e^{-tz}\, dt \quad (Re(z) > 0) \tag{6'}$$

belongs to the so-called *Hardy-Lebesgue class* $H^2(0)$, that is, (i) $f(z)$ is holomorphic in the right half-plane $Re(z) > 0$, (ii) for each fixed $x > 0$, $f(x+iy)$ as a function of y belongs to $L^2(-\infty,\infty)$ in such a way that

$$\sup_{x>0} \left(\int_{-\infty}^{\infty} |f(x+iy)|^2\, dy \right) < \infty. \tag{7'}$$

This theorem admits a converse, that is, we have

Theorem 2 (Paley-Wiener). Let $f(z) \in H^2(0)$. Then the *boundary function* $f(iy) \in L^2(-\infty,\infty)$ of $f(x+iy)$ exists in the sense that

$$\lim_{x \downarrow 0} \int_{-\infty}^{\infty} |f(iy) - f(x+iy)|^2\, dy = 0 \tag{8}$$

in such a way that the inverse Fourier transform

$$g(t) = (2\pi)^{-1/2} \operatorname*{l.i.m.}_{N\to\infty} \int_{-N}^{N} f(iy)\, e^{ity}\, dy \tag{9}$$

vanishes for $t < 0$ and $f(z)$ may be obtained as the one-sided Laplace transform of $g(t)$.

Proof. By the local weak compactness of $L^2(-\infty,\infty)$, we see, that there exists a sequence $\{x_n\}$ and an $f(iy) \in L^2(-\infty,\infty)$ such that

$$x_n \downarrow 0 \quad \text{and} \quad \operatorname*{weak-lim}_{n\to\infty} f(x_n + iy) = f(iy).$$

For any $\delta > 0$, there exists a sequence $\{N_k\}$ such that

$$\lim_{k\to\infty} N_k = \infty \quad \text{and} \quad \lim_{k\to\infty} \int_{0+}^{\delta} |f(x \pm iN_k)|^2\, dx = 0,$$

as may be seen from

$$\int_{-N}^{N} \left\{ \int_{0+}^{\delta} |f(x+iy)|^2\, dx \right\} dy \leq \int_{0+}^{\delta} \left\{ \int_{-\infty}^{\infty} |f(x+iy)|^2\, dy \right\} dx < \infty \quad \text{(for all } N > 0).$$

Thus, by Schwarz' inequality,

$$\lim_{k\to\infty} \int_{0+}^{\delta} |f(x \pm iN_k)|\, dx = 0. \tag{10}$$

From this we obtain

$$f(z) = (2\pi)^{-1} \int_{-\infty}^{\infty} \frac{f(it)}{z - it} dt \quad (Re(z) > 0). \tag{11}$$

The proof may be obtained as follows. By Cauchy's integral representation, we obtain

$$f(z) = (2\pi i)^{-1} \int_C \frac{f(\zeta)}{\zeta - z} d\zeta \quad (Re(z) > 0), \tag{12}$$

where the path C of integration is composed of the segments

$$\overrightarrow{x_0 - iN_k, x_1 - iN_k}, \ \overrightarrow{x_1 - iN_k, x_1 + iN_k},$$

$$\overrightarrow{x_1 + iN_k, x_0 + iN_k}, \ \overrightarrow{x_0 + iN_k, x_0 - iN_k}$$

$$(x_0 < Re(z) < x_1, \ -N_k < Im(z) < N_k)$$

so that the closed contour C encloses the point z. Hence, by letting $k \to \infty$ and observing (10), we obtain

$$f(z) = (2\pi)^{-1} \int_{-\infty}^{\infty} \frac{f(x_0 + it)}{z - (x_0 + it)} dt + (2\pi)^{-1} \int_{-\infty}^{\infty} \frac{f(x_1 + it)}{(x_1 + it) - z} dt.$$

We see that the second term on the right tends to 0 as $x_1 \to \infty$, because of (7′) and Schwarz' inequality. We set, in the first term on the right, $x_0 = x_n$ and letting $n \to \infty$ we obtain (11). Similarly, we obtain

$$0 = (2\pi)^{-1} \int_{-\infty}^{\infty} \frac{f(it)}{it - z} dt \quad (Re(z) < 0). \tag{13}$$

Let us put, for $Re(z) > 0$,

$$h(x) = 0 \quad (\text{for } x < 0); \quad h(x) = e^{-zx} \quad (\text{for } x > 0).$$

We then have

$$\int_{-\infty}^{\infty} h(x)\, e^{ixt}\, dx = \int_0^{\infty} e^{ixt - zx}\, dx = (z - it)^{-1},$$

and so, by Plancherel's theorem,

$$\underset{N\to\infty}{\text{l.i.m.}}\, (2\pi)^{-1} \int_{-N}^{N} \frac{e^{-itx}}{z - it} dt = \begin{cases} 0 & (\text{for } x < 0) \\ e^{-zx} & (\text{for } x > 0) \end{cases} \quad \text{if } Re(z) > 0. \tag{14}$$

Similarly we have

$$\underset{N\to\infty}{\text{l.i.m.}}\, (2\pi)^{-1} \int_{-N}^{N} \frac{e^{-itx}}{z - it} dt = \begin{cases} -e^{-zx} & (\text{for } x < 0) \\ 0 & (\text{for } x > 0) \end{cases} \quad \text{if } Re(z) < 0. \tag{14′}$$

Therefore, by applying the Parseval theorem (27) of Chapter VI, 2 to (11), we obtain the result that $f(z)$ is the one-sided Laplace transform of $g(t)$ given by (9). By applying the Parseval theorem to (13), we see also that $g(t) = 0$ for $t < 0$.

We finally shall prove that (8) is true. By adding (11) and (13), we see that $f(z)$ admits *Poisson's integral representation*

$$f(z) = f(x + iy) = \frac{x}{\pi} \int_{-\infty}^{\infty} \frac{f(it)}{(t-y)^2 + x^2} dt \quad \text{whenever} \quad x > 0. \tag{15}$$

By virtue of

$$\frac{x}{\pi} \int_{-\infty}^{\infty} \frac{dt}{(t-y)^2 + x^2} = 1, \tag{16}$$

we have

$$|f(x+iy) - f(iy)| \leq \frac{x}{\pi} \int_{-\infty}^{\infty} \frac{|f^+(s+y) - f^+(y)|}{s^2 + x^2} ds, \text{ where } f^+(y) = f(iy),$$

and so

$$\int_{-\infty}^{\infty} |f(x+iy) - f(iy)|^2 dy \leq \left(\frac{x}{\pi}\right)^2 \int_{-\infty}^{\infty} \left\{ \int_{-\infty}^{\infty} \frac{|f^+(s+y) - f^+(y)|}{s^2 + x^2} ds \right\}^2 dy$$

$$\leq \frac{x^2}{\pi^2} \int_{-\infty}^{\infty} \left(\int_{-\infty}^{\infty} \frac{ds}{s^2 + x^2} \right) \left(\int_{-\infty}^{\infty} \frac{|f^+(s+y) - f^+(y)|^2}{s^2 + x^2} ds \right) dy$$

$$= \frac{x}{\pi} \int_{-\infty}^{\infty} \frac{ds}{s^2 + x^2} \left\{ \int_{-\infty}^{\infty} |f^+(s+y) - f^+(y)|^2 dy \right\}$$

$$= \frac{x}{\pi} \int_{-\infty}^{\infty} \frac{\mu(f^+; s)}{s^2 + x^2} ds,$$

where $0 \leq \mu(f^+; s) \leq 4 \, \|f^+\|^2$ and $\mu(f^+; s)$ is continuous in s and has value 0 at $s = 0$.

To prove that the right side tends to zero as $x \downarrow 0$, we take, for any $\varepsilon > 0$, a $\delta = \delta(\varepsilon) > 0$ such that $\mu(f^+; s) \leq \varepsilon$ whenever $|s| \leq \delta$. We decompose the integral

$$\frac{x}{\pi} \int_{-\infty}^{\infty} \frac{\mu(f^+; s)}{s^2 + x^2} ds = \frac{x}{\pi} \left\{ \int_{-\infty}^{-\delta} + \int_{-\delta}^{\delta} + \int_{\delta}^{\infty} \right\} = I_1 + I_2 + I_3.$$

We have $|I_2| \leq \varepsilon$ by (16), and $|I_j| \leq 4\pi^{-1} \, \|f^+\|^2 \cdot \cot^{-1}(\delta/x)$ $(j = 1, 2)$. Hence the left integral must tend to zero as $x \downarrow 0$. This proves the Theorem.

Remark 1. For the original Paley-Wiener theorem, see Paley-Wiener [1]. For the one-sided Laplace transform of tempered distributions, see L. Schwartz [2].

Remark 2. M. SATO [1] has introduced a happy idea of defining a "generalized function" as the "boundary value of an analytic function". His idea may be explained as follows. Let $\mathfrak{B}$ be the totality of functions $\varphi(z)$ which are defined and regular in the upper-half and the lower-half planes of the complex z-plane, and let $\mathfrak{R}$ be the totality of functions which are regular in the whole complex z-plane. Then $\mathfrak{B}$ is a ring with respect to the function sum and function multiplication, and $\mathfrak{R}$ is a subring of $\mathfrak{B}$. Sato calls the residue class (mod $\mathfrak{R}$) containing $\varphi(z)$ as the "generalized function" $\hat{\varphi}(x)$ on the real axis R^1 defined through $\varphi(z)$. The "generalized derivative $d\hat{\varphi}(x)/dx$ of the generalized function $\hat{\varphi}(x)$" is naturally defined as the residue class (mod $\mathfrak{R}$) containing $d\varphi(z)/dz$. Thus the "delta function $\delta(x)$" is the residue class (mod $\mathfrak{R}$) containing $-(2\pi i)^{-1}z^{-1}$. Sato's theory of "generalized function of many variables" admits the following very interesting topological interpretation. Let M be an n-dimensional real analytic manifold and let X be a complexification of M. Then the n-th relative cohomology group $H^n(X \bmod (X-M))$ with the germ of regular functions in X as coefficients gives rise to a notion of the "generalized function on M". That is, the relative cohomology class is a natural definition of the "generalized function".

Remark 3. For more detailed treatment of the Fourier transform of generalized functions, see L. SCHWARTZ [1] and GELFAND-ŠILOV [1]. In the latter book, many interesting classes of basic functions other than $\mathfrak{D}(R^n)$, $\mathfrak{S}(R^n)$ and $\mathfrak{O}_M(R^n)$ are introduced to define generalized functions; the Fourier transform of the corresponding generalized functions are also discussed in GELFAND-ŠILOV [1]. Cf. also A. FRIEDMAN [1] and L. HÖRMANDER [6].

5. Titchmarsh's Theorem

Theorem (E. C. TITCHMARSH). Let $f(x)$ and $g(x)$ be real- or complex-valued continuous functions, $0 \leqq x < \infty$, such that

$$(f * g)(x) = \int_0^x f(x-y)\, g(y)\, dy = \int_0^x g(x-y)\, f(y)\, dy = (g * f)(x) \tag{1}$$

vanishes identically. Then either one of $f(x)$ or $g(x)$ must vanish identically.

There is a variety of proofs of this important theorem, such as by TITCHMARSH [1] himself and also by CRUM and DUFRESNOY. The following proof is elementary in the sense that it does not appeal to the theory of functions of a complex variable. It is due to RYLL-NARDZEWSKI [1] and given in the book by J. MIKUSIŃSKI [1].

Lemma 1 (PHRAGMÉN). If $g(u)$ is continuous for $0 \leqq u \leqq T$, and $0 \leqq t \leqq T$, then

$$\lim_{x\to\infty} \sum_{k=1}^{\infty} \frac{(-1)^{k-1}}{k!} \int_0^T e^{kx(t-u)}\, g(u)\, du = \int_0^t g(u)\, du. \tag{2}$$

Proof. We have $\sum_{k=1}^{\infty}(-1)^{k-1}(k!)^{-1}e^{kx(t-u)}=1-\exp(-e^{x(t-u)})$, and, for fixed x and t, the series on the left converges uniformly in u for $0\leqq u\leqq T$. Thus the summation in (2) may be carried out under the integral sign and we obtain (2) by the Lebesgue-Fatou Lemma.

Lemma 2. If $f(t)$ is continuous for $0\leqq t\leqq T$ and $\left|\int_0^T e^{nt}f(t)\,dt\right|\leqq M$ for $n=1, 2, \ldots$, where M is a positive constant which is independent of n, then $f(t)$ must be $=0$ for $0\leqq t\leqq T$.

Proof. We have

$$\left|\sum_{k=1}^{\infty}\int_0^T\frac{(-1)^{k-1}}{k!}e^{kn(t-u)}f(T-u)\,du\right|\leqq\sum_{k=1}^{\infty}\frac{1}{k!}e^{kn(t-T)}\left|\int_0^T e^{kn(T-u)}f(T-u)\,du\right|$$
$$\leqq M\left(\exp\left(e^{-n(T-t)}\right)-1\right).$$

If $t<T$, the last expression tends to zero as $n\to\infty$. Hence, by Lemma 1, with $g(u)=f(T-u)$, we see that $\int_0^t f(T-u)\,du=0$ for $0\leqq t\leqq T$. Since f is continuous, it follows that $f(t)=0$ for $0\leqq t\leqq T$.

Corollary 1. If $g(x)$ is continuous for $1\leqq x\leqq X$ and if there exists a positive number N such that $\left|\int_1^X x^n g(x)\,dx\right|\leqq N$ $(n=1, 2, \ldots)$, then $g(x)=0$ for $1\leqq x\leqq X$.

Proof. Putting $x=e^t$, $X=e^T$ and $xg(x)=f(t)$, we obtain, by Lemma 2, that $f(t)=0$ for $0\leqq t\leqq T$. Thus $xg(x)=0$ for $1\leqq x\leqq X$, and so $g(x)=0$ for $1\leqq x\leqq X$.

Corollary 2 (Lerch's theorem). If $f(t)$ is continuous for $0\leqq t\leqq T$ and $\int_0^T t^n f(t)\,dt=0$ $(n=1, 2, \ldots)$, then $f(t)=0$ for $0\leqq t\leqq T$.

Proof. Let t_0 be any number from the open interval $(0, T)$, and put $t=t_0x$, $T=t_0X$, $f(t)=g(x)$. We then obtain

$$t_0^{n+1}\int_0^X x^n g(x)\,dx=0\quad(n=1, 2, \ldots),$$

and so $\left|\int_1^X x^n g(x)\,dx\right|=\left|\int_0^1 x^n g(x)\,dx\right|\leqq\int_0^1|g(x)|\,dx=N$ $(n=1, 2, \ldots)$. Hence, by Corollary 1, we obtain $g(x)=0$ for $1\leqq x\leqq X$, and so $f(t)=0$ for $t_0\leqq t\leqq T$. Since t_0 was an arbitrary point of $(0, T)$, we must have $f(t)=0$ for $0\leqq t\leqq T$.

The proof of Titchmarsh's theorem. We shall first prove the Theorem for the special case when $f=g$, that is, if $f(t)$ is continuous and $(f*f)(t)$ $=\int_0^t f(t-u)f(u)\,du=0$ for $0\leqq t\leqq 2T$, then $f(t)=0$ for $0\leqq t\leqq T$.

We have

$$\int_0^{2T} e^{n(2T-t)} \left(\int_0^t f(u)\, f(t-u)\, du \right) dt = 0,$$

and, by the change of variables $u = T - v$, $t = 2T - v - w$, we obtain

$$\iint_\Delta e^{n(v+w)}\, f(T-v)\, f(T-w)\, dv\, dw = 0$$

where Δ is the triangle $v + w \geqq 0$, $v \leqq T$, $w \leqq T$ in the $v - w$ plane. Let Δ' be the triangle $v + w \leqq 0$, $v \geqq -T$, $w \geqq -T$. Then the join $\Delta + \Delta'$ is the square $-T \leqq v, w \leqq T$. The above equality shows that the integral of $e^{n(v+w)} f(T-v) f(T-w)$ over $\Delta + \Delta'$ is equal to the integral over Δ'. The integral over $\Delta + \Delta'$ is the product of two single integrals, and in the integral over Δ' we have $e^{n(v+w)} \leqq 1$. Thus

$$\left| \int_{-T}^{T} e^{nu} f(T-u)\, du \right|^2 = \left| \iint_{\Delta+\Delta'} e^{n(v+w)}\, f(T-v)\, f(T-w)\, dv\, dw \right|$$
$$\leqq \iint_{\Delta'} |f(T-v)\, f(T-w)|\, dv\, dw \leqq 2T^2 \cdot A^2,$$

where A is the maximum of $|f(t)|$ for $0 \leqq t \leqq 2T$, and $2T^2$ is the area of Δ'. We thus have

$$\left| \int_{-T}^{T} e^{nu} f(T-u)\, du \right| \leqq \sqrt{2}\, T \cdot A,$$

and, moreover, $\left| \int_{-T}^{0} e^{nu} f(T-u)\, du \right| \leqq TA$. Therefore

$$\left| \int_0^T e^{nu} f(T-u)\, du \right| = \left| \int_{-T}^{T} - \int_{-T}^{0} \right| \leqq (1 + \sqrt{2})\, TA \quad (n = 1, 2, \ldots),$$

and so, by Lemma 2, $f(t) = 0$ for $0 \leqq t \leqq T$.

We are now ready to prove Titchmarsh's theorem for the general case. Let $\int_0^t f(t-u)\, g(u)\, du = 0$ for $0 \leqq t < \infty$. Then we have, for $0 \leqq t < \infty$,

$$\int_0^t (t-u) f(t-u)\, g(u)\, du + \int_0^t f(t-u)\, u g(u)\, du = t \int_0^t f(t-u)\, g(u)\, du = 0.$$

This may be written as

$$(f_1 * g)(t) + (f * g_1)(t) = 0 \quad (0 \leqq t < \infty),$$
where $f_1(t) = t f(t)$, $g_1(t) = t g(t)$.

Thus

$$[f * \{g_1 * (f_1 * g + f * g_1)\}](t) = 0$$

and so

$$[(f * g) * (f_1 * g_1)](t) + [(f * g_1) * (f * g_1)](t) = 0.$$

Thus, by $(f * g)(t) = 0$, we have $[(f * g_1) * (f * g_1)](t) = 0$, and so, by the special case proved above, $(f * g_1)(t) = 0$, that is,

$$\int_0^t f(t-u)\, u g(u)\, du = 0 \quad (0 \leqq t < \infty).$$

From this we obtain, similarly to above,

$$\int_0^t f(t-u)\, u^2 g(u)\, du = 0 \quad (0 \leqq t < \infty).$$

Repeating the argument, we find that

$$\int_0^t f(t-u)\, u^n g(u)\, du = 0 \quad (0 \leqq t < \infty;\ n = 1, 2, \ldots).$$

Hence, by Lerch's theorem proved above, we obtain

$$f(t-u)\, g(u) = 0 \quad \text{for} \quad 0 \leqq u \leqq t < \infty.$$

If a u_0 exists for which $g(u_0) \neq 0$, then $f(t - u_0) = 0$ for all $t \geqq u_0$, that is, $f(v) = 0$ for all $v \geqq 0$. Therefore, we have either $f(v) = 0$ for all $v \geqq 0$ or $g(v) = 0$ for all $v \geqq 0$.

6. Mikusiński's Operational Calculus

In his "Electromagnetic Theory", London (1899), the physicist O. Heaviside inaugurated an operational calculus which he successfully applied to ordinary linear differential equations connected with electrotechnical problems. In his calculus occured certain operators whose meaning is not at all obvious. The interpretation of such operators as given by Heaviside himself is difficult to justify. The interpretation given by his successors is unclear with regard to its range of validity, since it is based upon the theory of Laplace transforms. The theory of convolution quotients due to J. Mikusiński provides a clear and simple basis for an operational calculus applicable to ordinary differential equations with constant coefficients, as well as to certain partial differential equations with constant coefficients, difference equations and integral equations.

The convolution quotients. We denote by C the totality of complex-valued functions $f(t)$ defined for $0 \leqq t < \infty$. In this section, we write such functions by $\{f(t)\}$ or simply by f; whereas, $f(t)$ will mean the value at t of the function $f(t)$. We write $\{f(t)\} \cdot \{g(t)\}$ or simply $f \cdot g$ for the convolution function $\left\{\int_0^t f(t-s)\, g(s)\, ds\right\}$, i.e.,

$$\{f(t)\} \cdot \{g(t)\} = \{(f * g)(t)\} = \left\{\int_0^t f(t-s)\, g(s)\, ds\right\}. \tag{1}$$

We have, as proved in Chapter VI, 3,

$$f \cdot g = g \cdot f \text{ (commutativity)}, \tag{2}$$

$$f \cdot (g \cdot h) = (f \cdot g) \cdot h \text{ (associativity)}. \tag{3}$$

We can define, beside the convolution product $f \cdot g$, the sum through

$$\{f(t)\} + \{g(t)\} = \{f(t) + g(t)\}, \tag{4}$$

and we have the distributive law

$$h \cdot (f + g) = h \cdot f + h \cdot g. \tag{5}$$

Hence C is a *ring* with respect to the addition $f + g$ and the multiplication $f \cdot g$. The zero of this ring is represented by the function which is identically zero; we shall thus denote this function by 0. This ring C is *without zero factor*, that is, in C, $f \cdot g = 0$ implies either $f = 0$ or $g = 0$. This is a consequence of Titchmarsh's theorem. Hence, by introducing the *convolution quotient* $f/g = \dfrac{f}{g}$ of two functions $f, g \in C$ with $g \neq 0$, we obtain a commutative field Q:

$a/b = c/d$ is equivalent to $ad = bc$, and, in particular,

$$a/b = c \text{ is equivalent to } a = bc, \tag{6}$$

$$\frac{a}{b} \cdot \frac{c}{d} = \frac{a \cdot c}{b \cdot d}, \tag{7}$$

$$\frac{a}{b} + \frac{c}{d} = \frac{a \cdot d + b \cdot c}{b \cdot d}, \tag{8}$$

$$a/b = a \cdot c/b \cdot c \; (c \neq 0). \tag{9}$$

The operator. We shall call the quotient a/b an "operator". Any $a \in C$ gives an example of operators, since, by (9), we can identify a with $a \cdot b/b$ $(b \neq 0)$.

The unit- or δ-operator. The operator c/c $(c \neq 0)$ is the *unit* of the multiplication in the field Q. For, we have, by (7) and (9),

$$\frac{a}{b} \cdot \frac{c}{c} = \frac{a}{b}. \tag{10}$$

We have, by (6), $c/c = b/b$. We call c/c the *unit*- or *δ-operator*, and we shall henceforth denote this operator by 1:

$$1 \cdot \frac{a}{b} = \frac{a}{b} \cdot 1 = \frac{a}{b}. \tag{11}$$

The operator $1 = c/c$ does not belong to C, for, if we take for c the function $\{1\}$, then $\{1\}/\{1\} = \{f(t)\} \in C$ implies that $\{1\} \cdot \{f(t)\} = \left\{\int_0^t 1 \cdot f(s)\, ds\right\}$ $= \left\{\int_0^t f(s)\, ds\right\} = \{1\}$, which is surely a contradiction.

The operator of integration. We shall denote by h the operator defined by the function $\{1\}$:

$$h = \{1\}, \tag{12}$$

and call h the *operator of integration*, since, for any $f \in C$, we have, as above,

$$h \cdot \{f(t)\} = \{1\} \cdot \{f(t)\} = \left\{ \int_0^t f(s)\, ds \right\}. \tag{13}$$

Remark. In his book, referred to above, MIKUSIŃSKI uses the symbol l for $\{1\}$. We here use the symbol h in honour of Heaviside and for typographical convention. A locally integrable function $\{f(t)\}$, $t \geqq 0$, may be identified with the operator $\left\{ \int_0^t f(s)\, ds \right\} / h$. Thus the convolution quotient is another kind of "generalized function".

The scalar operator. Let α be any complex number, and $\{\alpha\}$ the function which is identically equal to α. Then the operator

$$[\alpha] = \{\alpha\}/\{1\} = \{\alpha\}/h \tag{14}$$

is called a *scalar operator*, because we have

$$[\alpha] + [\beta] = [\alpha + \beta],\ [\alpha] \cdot [\beta] = [\alpha\beta],\ [\alpha] \cdot \{f(t)\} = \{\alpha f(t)\}. \tag{15}$$

Proof.

$$[\alpha] + [\beta] = \frac{\{\alpha\}}{h} + \frac{\{\beta\}}{h} = \frac{\{\alpha + \beta\}}{h} = [\alpha + \beta],$$

$$[\alpha] \cdot [\beta] = \frac{\{\alpha\} \cdot \{\beta\}}{h^2} = \frac{\{\alpha\beta t\}}{h^2} = \frac{h \cdot \{\alpha\beta\}}{h^2} = \frac{\{\alpha\beta\}}{h} = [\alpha\beta],$$

$$[\alpha] \cdot \{f(t)\} = \frac{\{\alpha\} \cdot \{f(t)\}}{h} = \frac{\left\{ \int_0^t \alpha f(s)\, ds \right\}}{h} = \frac{h \cdot \{\alpha f(t)\}}{h} = \{\alpha f(t)\}.$$

Remark. As a corollary, we obtain

$$[\alpha] \cdot \frac{\{a(t)\}}{\{b(t)\}} = \frac{\{\alpha a(t)\}}{\{b(t)\}}, \tag{16}$$

so that the effect of the operator $[\alpha]$ is exactly the α-times multiplication. We can thus identify the scalar operator $\{1\}/\{1\}$ with the unit operator 1, and the operator $[\alpha]$ with the number α.

The operator of differentiation. We shall denote the operator $1/h$ by the symbol s:

$$s = 1/h = 1/\{1\}. \tag{17}$$

s is called the *operator of differentiation*, since, if $f = \{f(t)\} \in C$ has a continuous derivative $f' = \{f'(t)\}$, then

$$sf = f' + f(0), \quad \text{where} \quad f(0) = \{f(0)\}/h. \tag{18}$$

Proof. Multiply both sides of the equation

$$\{f(t)\} = \{f(0)\} + \left\{\int_0^t f'(s)\, ds\right\} = h \cdot f(0) + h \cdot \{f'(t)\}$$

by s and use the fact $s \cdot h = h \cdot s = 1$.

Corollary 1. If $f = \{f(t)\}$ has a continuous n-th derivative $f^{(n)} = \{f^{(n)}(t)\}$, then

$$f^{(n)} = s^n \cdot f - s^{n-1} \cdot f(0) - s^{n-2} \cdot f'(0) - \cdots - s \cdot f^{(n-2)}(0) - f^{(n-1)}(0),$$

where $f^{(j)}(0)$ is the operator of the $f^{(j)}(0)$-times multiplication, (19)

that is, $f^{(j)}(0) = \{f^{(j)}(0)\}/h$.

Proof. For $n = 2$, we have

$$s^2 \cdot f = s \cdot (s \cdot f) = s \cdot (f' + f(0)) = s \cdot f' + s \cdot f(0) = f'' + f'(0) + s \cdot f(0).$$

The general case may be proved by induction.

Corollary 2. We have

$$1/(s - \alpha) = \{e^{\alpha t}\}, \tag{20}$$

and more generally,

$$1/(s - \alpha)^n = \left\{\frac{t^{n-1}}{(n-1)!} e^{\alpha t}\right\} \quad (n = 1, 2, \ldots). \tag{21}$$

Proof. $s \cdot \{e^{\alpha t}\} = \{\alpha e^{\alpha t}\} + 1 = \alpha \{e^{\alpha t}\} + 1$ by (18). Thus (20) holds.

$$1/(s - \alpha)^2 = \{e^{\alpha t}\} \cdot \{e^{\alpha t}\} = \left\{\int_0^t e^{\alpha(t-s)}\, e^{\alpha s}\, ds\right\} = \left\{e^{\alpha t} \int_0^t ds\right\} = \left\{\frac{t}{1!} e^{\alpha t}\right\}.$$

Application to the integration of linear ordinary differential equations with constant coefficients. We shall show the applications by examples.

Example 1. Solve the equation

$$x''(t) - x'(t) - 6x(t) = 2, \; x(0) = 1, \; x'(0) = 0.$$

Solution. We write the equation in the operator form

$$x''(t) - x'(t) - 6x(t) = 2/s.$$

Then, by (19),

$$s^2 \cdot x - s \cdot x(0) - x'(0) - s \cdot x + x(0) - 6x = 2/s,$$

and so, by substituting the initial condition, we obtain

$$s^2 x - s - s \cdot x + 1 - 6x = 2/s, \text{ i.e., } (s^2 - s - 6) \cdot x = s - 1 + 2/s.$$

Hence, by (20),

$$x = \frac{s^2 - s + 2}{s \cdot (s-3) \cdot (s+2)} = -\frac{1}{3} \cdot \frac{1}{s} + \frac{8}{15} \cdot \frac{1}{s-3} + \frac{4}{5} \cdot \frac{1}{s+2}$$
$$= \left\{-\frac{1}{3} + \frac{8}{15} e^{3t} + \frac{4}{5} e^{-2t}\right\}.$$

Example 2. Let λ be a constant $\neq 0$. Solve the equation

$$x''(t) + \lambda^2 x(t) = 0, \ x(0) = \alpha, \ x'(0) = \beta.$$

Solution. The corresponding operator equation is

$$s^2 \cdot x - \alpha s - \beta + \lambda^2 x = 0, \text{ i.e., } (s^2 + \lambda^2) \cdot x = \alpha s + \beta.$$

Hence, expanding into partial fraction, we obtain

$$x = \frac{\alpha s + \beta}{s^2 + \lambda^2} = \frac{\gamma}{s + i\lambda} + \frac{\delta}{s - i\lambda}$$

with $\alpha s + \beta = \gamma(s - i\lambda) + \delta(s + i\lambda)$ so that

$$\gamma = 2^{-1}\left(\alpha + \frac{i\beta}{\lambda}\right), \ \delta = 2^{-1}\left(\alpha - \frac{i\beta}{\lambda}\right).$$

Hence

$$\begin{aligned} x &= \frac{1}{2}\left(\alpha + \frac{i\beta}{\lambda}\right)\frac{1}{s + i\lambda} + \frac{1}{2}\left(\alpha - \frac{i\beta}{\lambda}\right)\frac{1}{s - i\lambda} \\ &= \left\{\frac{1}{2}\left(\alpha + \frac{i\beta}{\lambda}\right)e^{-i\lambda t} + \frac{1}{2}\left(\alpha - \frac{i\beta}{\lambda}\right)e^{i\lambda t}\right\} \\ &= \left\{\alpha \cos \lambda t + \frac{\beta}{\lambda} \sin \lambda t\right\}. \end{aligned}$$

Example 3. Solve the system of equations

$$x'(t) - \alpha x(t) - \beta y(t) = \beta e^{\alpha t}, \ y'(t) + \beta x(t) - \alpha y(t) = 0,$$

under the initial condition $x(0) = 0$, $y(0) = 1$.

Solution. Solving the operator equations

$$s \cdot x - \alpha x - \beta y = \beta/(s - \alpha), \ s \cdot y - 1 + \beta x - \alpha y = 0,$$

we obtain,

$$x = \frac{2\beta}{(s - \alpha)^2 + \beta^2}, \ y = \frac{(s - \alpha)^2 - \beta^2}{(s - \alpha) \cdot ((s - \alpha)^2 + \beta^2)},$$

from which we obtain

$$x = \frac{1}{i}\left\{\frac{1}{s - \alpha - i\beta} - \frac{1}{s - \alpha + i\beta}\right\} = \frac{1}{i}\left\{e^{(\alpha + i\beta)t} - e^{(\alpha - i\beta)t}\right\} = \{2e^{\alpha t} \sin \beta t\},$$

$$\begin{aligned} y &= \frac{2(s - \alpha)}{(s - \alpha)^2 + \beta^2} - \frac{1}{s - \alpha} = \frac{1}{s - \alpha - i\beta} + \frac{1}{s - \alpha + i\beta} - \frac{1}{s - \alpha} \\ &= \{e^{(\alpha + i\beta)t} + e^{(\alpha - i\beta)t} - e^{\alpha t}\} = \{e^{\alpha t}(2 \cos \beta t - 1)\}. \end{aligned}$$

For further development and applications, see MIKUSIŃSKI's book, [1]. We also refer the reader to a book by A. ERDÉLYI [1].

7. Sobolev's Lemma

A generalized function is infinitely differentiable in the sense of the distribution (see Chapter I, 8). So the differentiability in the generalized sense has no bearing on the ordinary differentiability. However, we have the following result which is of fundamental importance in modern treatment of partial differential equations.

Theorem (Sobolev's Lemma). Let G be a bounded open domain of R^n. Let a function $u(x)$ belong to $W^k(G)$ for $k > 2^{-1}n + \sigma$, where σ is an integer $\geqq 0$. We thus assume that the distributional derivatives of $u(x)$ of order up to and including k all belong to $L^2(G)$. Then, for any open subset G_1 of G such that the closure G_1^a is a compact subset of G, there exists a function $u_1(x) \in C^\sigma(G_1)$ such that $u(x) = u_1(x)$ a.e. in G_1.

Proof. Let $\alpha(x)$ be a function $\in C_0^\infty(R^n)$ such that

$$G_1 \subseteqq \mathrm{supp}(\alpha) \subset G,\ 0 \leqq \alpha(x) \leqq 1 \quad \text{and} \quad \alpha(x) = 1 \text{ on } G_1.$$

We define a function $v(x)$ defined on R^n by

$$v(x) = \alpha(x)\, u(x) \text{ for } x \in G;\ v(x) = 0 \text{ for } x \in R^n - G.$$

Then $v(x) = u(x)$ whenever $x \in G_1$. Since $v(x)$ is locally integrable over R^n, it defines a distribution $\in \mathfrak{D}(R^n)'$. By the assumption that $u \in W^k(G)$, the distributional derivatives $D^s v(x) \in L^2(R^n)$ when $|s| \leqq k$. For example, the distributional derivative

$$\frac{\partial}{\partial x_j}(v) = \frac{\partial}{\partial x_j}(\alpha u) = \frac{\partial \alpha}{\partial x_j} \cdot u + \alpha \frac{\partial}{\partial x_j} u$$

belongs to $L^2(R^n)$ by the fact that u, $\partial u/\partial x_j$ both belong to $L^2(G)$ and that the infinitely differentiable function $\alpha(x)$ has its support contained in a compact subset of the open domain G. By the Fourier transformation $v(x) \to \hat{v}(y)$, we obtain

$$\widehat{(D^s v)}(y) = (i)^{|s|} y_1^{s_1} y_2^{s_2} \ldots y_n^{s_n} \cdot \hat{v}(y).$$

Since, by Plancherel's theorem, the L^2-norm is preserved by the Fourier transformation, we have $\widehat{(D^s v)}(y) \in L^2(R^n)$ for $|s| \leqq k$. Thus

$$\hat{v}(y)\, y_1^{s_1} y_2^{s_2} \ldots y_n^{s_n} \in L^2(R^n) \quad \text{for} \quad |s| \leqq k. \tag{1}$$

In particular we have

$$\hat{v}(y) \in L^2(R^n). \tag{1'}$$

Let $q = (q_1, q_2, \ldots, q_n)$ be a system of non-negative integers. Then by (1), we can prove that

$$\hat{v}(y)\, y_1^{q_1} y_2^{q_2} \ldots y_n^{q_n} \text{ is integrable over } R^n \text{ whenever } |q| + \frac{n}{2} < k. \tag{2}$$

For the proof, take any positive number C. We have, by Schwarz' inequality,

$$\int_{|y| \leqq C} |\hat{v}(y)\, y_1^{q_1} y_2^{q_2} \ldots y_n^{q_n}|\, dy$$
$$\leqq \Big(\int_{|y| \leqq C} |y_1^{q_1} y_2^{q_2} \ldots y_n^{q_n}|^2\, dy \cdot \int_{|y| \leqq C} |\hat{v}(y)|^2\, dy\Big)^{1/2} < \infty,$$
$$\int_{|y| > C} |\hat{v}(y)\, y_1^{q_1} y_2^{q_2} \ldots y_n^{q_n}|\, dy$$
$$\leqq \Big(\int_{|y| > C} |(1 + |y|^2)^{-k/2} y_1^{q_1} y_2^{q_2} \ldots y_n^{q_n}|^2\, dy \cdot \int_{|y| > C} |\hat{v}(y)\, (1 + |y|^2)^{k/2}|^2\, dy\Big)^{1/2}.$$

The second factor on the right of the last inequality is $< \infty$ by (1). The first factor is finite by

$$\left.\begin{array}{l} dy = dy_1\, dy_2 \ldots dy_n = r^{n-1}\, dr\, d\Omega_n, \\ \text{where } d\Omega_n \text{ is the hypersurface element of the surface} \\ \text{of the unit sphere of } R^n \text{ with the origin 0 as its centre,} \end{array}\right\} \tag{3}$$

provided

$$2\,|q| - 2k + n - 1 < -1, \quad \text{that is, if} \quad k > \frac{n}{2} + |q|.$$

Now, by Plancherel's theorem, we have

$$v(x) = \underset{h\to\infty}{\text{l.i.m.}}\ (2\pi)^{-n/2} \int_{|y|\leqq h} \hat{v}(y) \exp(i\langle y, x\rangle)\, dy$$

and so, as in the proof of the completeness of $L^2(R^n)$, we can choose a subsequence $\{h'\}$ of positive integers h such that

$$v(x) = \lim_{h'\to\infty} (2\pi)^{-n/2} \int_{|y|\leqq h'} \hat{v}(y) \exp(i\langle y, x\rangle)\, dy \quad \text{for a.e. } x \in R^n.$$

But, since $\hat{v}(y)$ is integrable over R^n as proved above, the right side is equal to

$$v_1(x) = (2\pi)^{-n/2} \int_{R^n} \hat{v}(y) \exp(i\langle x, y\rangle)\, dy,$$

that is, $v(x)$ is equal to $v_1(x)$ for a.e. $x \in R^n$. By (2), the differentiation of $v_1(x)$ under the integral sign is justified up to the order σ; and the result of the differentiation is continuous in x. By taking $u_1(x) = v_1(x)$ for $x \in G_1$, we have proved the Theorem.

Remark. For the original proof, see S. L. Sobolev [1], [2] and L. Kantorovitch-G. Akilov [1].

8. Gårding's Inequality

Consider a quadratic integral form defined for C^∞ function $u(x) = u(x_1, x_2, \ldots, x_n)$ with compact support in a bounded domain G of R^n:

$$B[u, u] = \sum_{|s|, |t| \leqq m} (c_{st} D^s u, D^t u)_0, \tag{1}$$

where the complex-valued coefficients c_{st} are continuous on the closure G^a of G, and $(u, v)_0$ denotes the scalar product in $L^2(G)$.

Then we have

Theorem (L. Gårding [1]). A sufficient condition for the existence of positive constants c, C so that the inequality

$$\|u\|_m^2 \leqq c\, Re\, B[u, u] + C\, \|u\|_0^2 \tag{2}$$

holds for all $u \in C_0^\infty(G)$, is that, for some positive constant c_0,

$$\left.\begin{array}{l} Re \sum_{|s|, |t| = m} c_{st}\, \xi^s \xi^t \geqq c_0 |\xi|^{2m} \quad \text{for all} \quad x \in G \quad \text{and all real vectors} \\ \qquad \xi = (\xi_1, \xi_2, \ldots, \xi_m). \end{array}\right. \tag{3}$$

Remark. The inequality (2) is called *Gårding's inequality*. If condition (3) is satisfied, then the differential operator

$$L = \sum_{|s|=|t|\leq m} (-1)^{|t|} D^t c_{st} D^s \tag{4}$$

is said to be *strongly elliptic* in G, assuming that c_{st} is C^m on G^a.

Proof. We first prove that, for every $\varepsilon > 0$, there is a constant $C(\varepsilon) > 0$ such that for every $C_0^\infty(G)$ function u,

$$\|u\|_{m-1}^2 \leqq \varepsilon \|u\|_m^2 + C(\varepsilon) \|u\|_0. \tag{5}$$

To this purpose, we consider u to belong to $C_0^\infty(R^n)$, defining its value as 0 outside G. By the Fourier transformation we have

$$\|D^s u\|_0^2 = \|\widehat{(D^s u)}\|_0^2 = \int_{R^n} \left| \prod_{j=1}^n y_j^{s_j} \hat{u}(y) \right|^2 dy,$$

in virtue of Plancherel's theorem. Thus (5) is a consequence from the fact that

$$\left(\sum_{|s|\leqq m-1} \prod_{j=1}^n y_j^{2s_j} \right) \Big/ \left(C + \sum_{|t|\leqq m} \prod_{j=1}^n y_j^{2t_j} \right) \left(|s| = \sum_{j=1}^n s_j,\ |t| = \sum_{j=1}^n t_j \right)$$

tends to zero uniformly with respect to $y = (y_1, y_2, \ldots, y_n)$ as $C \uparrow \infty$.

Suppose that the coefficients c_{st} are constant and vanish unless $|s| = |t| = m$. By the Fourier transformation $u(x) \to \hat{u}(\xi)$ and Plancherel's theorem, we have, by (3),

$$\begin{aligned} Re\, B[u, u] &= Re \int \sum_{s,t} c_{st} \xi^s \xi^t |\hat{u}(\xi)|^2 d\xi \\ &\geqq \int c_0 |\xi|^{2m} |\hat{u}(\xi)|^2 d\xi \geqq c_1 (\|u\|_m^2 - \|u\|_{m-1}^2), \end{aligned}$$

where $c_1 > 0$ is a constant which is independent of u. Hence, by (5), we see that (2) is true for our special case.

We next consider the case of variable coefficients c_{st}. Suppose, first, that the support of u is sufficiently small and contained, say, in a small sphere about the origin. By the preceding case, we have, with a constant $c_0' > 0$ which is independent of u,

$$\begin{aligned} c_0' \|u\|_m^2 \leqq \mathrm{Re}\, B[u, u] &+ \mathrm{Re} \sum_{|s|=|t|=m} \int (c_{st}(0) - c_{st}(x)) D^s u \cdot D^t \bar{u}\, dx \\ &- \mathrm{Re} \sum_{|s|+|t|<2m} \int c_{st}(x) D^s u \cdot D^t \bar{u}\, dx + C(\varepsilon) \|u\|_0^2. \end{aligned}$$

If the support of u is so small that c_{st} has very small oscillation there, we see that the second term on the right may be bounded by $2^{-1} c_0' \|u\|_m^2$. The third term on the right is bounded by constant times $\|u\|_m \cdot \|u\|_{m-1}$. Hence we find that, constants denoting positive constants,

$$2^{-1} c_0' \|u\|_m^2 \leqq Re\, B[u, u] + \text{constant } \|u\|_m \cdot \|u\|_{m-1} + C(\varepsilon) \|u\|_0^2.$$

Thus, by

$$2\,|\alpha|\cdot|\beta| \leqq \varepsilon\,|\alpha|^2 + \varepsilon^{-1}\,|\beta|^2 \text{ which is valid for every } \varepsilon > 0, \tag{6}$$

we obtain $||u||_m^2 \leqq \text{constant} \cdot Re\, B[u, u] + \text{constant} \cdot ||u||_{m-1}^2 + C(\varepsilon) \cdot ||u||_0^2$, and so, by (5), we obtain (2).

Next we consider the general case. Construct a partition of unity in G:

$$1 = \sum_{j=1}^{N} \omega_j^2,\ \omega_j \in C_0^\infty(G) \quad \text{and} \quad \omega_j(x) \geqq 0 \text{ in } G,$$

such that the support of each ω_j may be taken as small as we please. Then, by Leibniz' rule of differentiation of the product of functions, Schwarz' inequality and the estimate of the case obtained above, we have

$$\begin{aligned} Re\, B[u, u] &= Re \sum_{s,t} \int c_{st} D^s u D^t \bar{u}\, dx = Re \sum_{s,t} \sum_j \int \omega_j^2 c_{st} D^s u D^t \bar{u}\, dx \\ &= Re \sum_j \sum_{s,t} c_{st} D^s(\omega_j u)\, D^t\, \overline{(\omega_j u)}\, dx + 0(||u||_m\, ||u||_{m-1}) \\ &\geqq \sum_j \text{constant}\, (||\omega_j u||_m^2 - ||\omega_j u||_{m-1}^2) + 0(||u||_m \cdot ||u||_{m-1}) \\ &\geqq \text{constant}\, ||u||_m^2 + 0(||u||_m \cdot ||u||_{m-1}). \end{aligned}$$

Thus we obtain (2) by (5). We remark that the constants c, C in (2) depend upon c_0, c_{st} and the domain G.

9. Friedrichs' Theorem

Let

$$L = \sum_{|s|,|t| \leqq m} D^s c^{st}(x)\, D^t \tag{1}$$

be strongly elliptic with real C^∞ coefficients $c_{st}(x)$ in a bounded open domain of R^n. For a given locally square integrable function $f(x)$ in G, a locally square integrable function $u(x)$ in G is called a *weak solution* of

$$Lu = f, \tag{2}$$

if we have

$$(u, L^*\varphi)_0 = (f, \varphi)_0,\ L^* = \sum_{|s|,|t| \leqq m} (-1)^{|s|+|t|} D^t c_{st}(x)\, D^s, \tag{3}$$

for every $\varphi \in C_0^\infty(G)$. Here $(f, g)_0$ denotes the scalar product of the Hilbert space $L^2(G)$. Thus a weak solution u of (2) is a solution in the sense of the distribution. Concerning the differentiability of the weak solution u, we have the following fundamental result:

Theorem (K. Friedrichs [1]). Any weak solution u of (2) has square integrable (distributional) derivatives of order up to $(2m + p)$ in the domain $G_1 \subseteqq G$ where f has square integrable (distributional) derivatives of order up to p. In other words, any weak solution u of (2) belongs to $W^{p+2m}(G_1)$ whenever f belongs to $W^p(G_1)$.

Corollary. If $p = \infty$, then, by Sobolev's lemma, there exists a function $u_0(x) \in C^\infty(G_1)$ such that $u(x) = u_0(x)$ for a.e. $x \in G_1$. Thus, after a correction on a set of measure zero, any weak solution $u(x)$ of (2) is C^∞ in the domain $\subseteqq G$ where $f(x)$ is C^∞; the corrected solution is hence a *genuine solution* of the differential equation (2) in the domain where $f(x)$ is C^∞.

Remark. When $L = \Delta$, the Laplacian, the above Corollary is Weyl's Lemma (see Chapter II, 7). There is extensive literature concerning the extensions of Weyl's Lemma to general elliptic operator L; such extensions are sometimes called the Weyl-Schwartz theorem. Among an abundant literature, we refer to the papers by P. LAX [2], L. NIRENBERG [1] and L. NIRENBERG [2]. The proof below is due to the present author (unpublished). A similar proof was given by L. BERS [1]. It is to be noted that a non-differentiable, locally integrable function $f(x)$ is a distribution solution of the *hyperbolic equation*

$$\frac{\partial f}{\partial x\,\partial y} = 0,$$

as may be seen from

$$0 = \int_{-\infty}^{\infty} \left\{ \int_{-\infty}^{\infty} f(x) \frac{\partial \varphi(x, y)}{\partial y\,\partial x}\, dy \right\} dx \quad (\varphi(x, y) \in C_0^\infty(R^2)).$$

Proof of the Theorem. We shall be concerned only with real-valued functions. Replacing, if necessary, L by $I + \alpha L$ with a certain constant $\alpha \neq 0$, we may assume that the strongly elliptic operator L itself satisfies Gårding's inequality

$$\begin{aligned} &(\varphi, L^*\varphi)_0 \geqq \delta \,\|\varphi\|_m^2 \; (\delta > 0), \\ &|(\varphi, L^*\psi)_0| \leqq \gamma \,\|\varphi\|_m \,\|\psi\|_m \; (\gamma > 0), \text{ whenever } \varphi, \psi \in C_0^\infty(G). \end{aligned} \tag{4}$$

The latter inequality is proved by partial integration. We assume here that each of the derivatives of the coefficients $c_{st}(x)$ up to order m is bounded in G, so that the constants δ and γ are independent of the test functions $\psi, \varphi \in C_0^\infty(G)$.

Suppose that G_1 is a periodic parallelogram

$$0 \leqq x_j \leqq 2\pi \qquad (j = 1, 2, \ldots, n), \tag{5}$$

and that the coefficients of L and f periodic with the period 2π in each x_j. Under such assumptions, we are to deal with functions $\varphi(x)$ defined on a compact space without boundary, the n-dimensional torus G_1 given by (5), and the distribution $\in C^\infty(G_1)'$ associated with the space of test functions $\varphi \in C^\infty(G_1)$ consisting of C^∞ functions $\varphi(x) = \varphi(x_1, x_2, \ldots, x_n)$ periodic with period 2π in each of the variables x_j. It is to be noted that, since G_1 is without boundary, we need not restrict the supports of our test functions $\varphi(x)$.

The condition $v \in W^q(G_1)$ thus means that the Fourier coefficients v_k of $v(x)$ defined by the Fourier expansion

$$v(x) \sim \sum_k v_k \exp(ik \cdot x)$$
$$\left(k = (k_1, \ldots, k_n),\ x = (x_1, \ldots, x_n), \quad \text{and} \quad k \cdot x = \sum_{j=1}^{n} k_j x_j\right) \tag{6}$$

satisfy

$$\sum_k |v_k|^2 (1 + |k|^2)^q < \infty \quad \left(|k|^2 = \sum_{j=1}^{n} k_j^2\right). \tag{7}$$

For, by partial integration, the Fourier coefficients of the distributional derivative $D^s v$ satisfy

$$(D^s v(x), \exp(ik \cdot x))_0 = (-1)^{|s|} (v(x), D^s \exp(ik \cdot x))_0$$
$$= (-i)^{|s|} \prod_{j=1}^{n} k_j^{s_j} v_k,\ s = (s_1, s_2, \ldots, s_n),$$

and so, by Parseval's relation for the Fourier coefficients of $D^q v \in L^2(G_1)$, we obtain (7).

It is convenient to introduce the space $W^q(G_1)$ with integer $q \gtreqless 0$, by saying that a sequence $\{w_k;\, k = (k_1, k_2, \ldots, k_n)\}$ of complex numbers w_k with $w_k = \overline{w}_{-k}$ belongs to $W^q(G_1)$ if (7) holds. This space $W^q(G_1)$ is normed by $\|\{w_k\}\|_q = \left(\sum_k |w_k|^2 (1 + |k|^2)^q\right)^{1/2}$. By virtue of the Parseval relation with respect to the complete orthonormal system $\{(2\pi)^{-n/2} \exp(ik \cdot x)\}$ of $L^2(G_1)$, we see that, when $q \geqq 0$, the norm $\|v\|_q = \left(\sum_{|s| \leqq q} \int_{G_1} |D^s v(x)|^2 dx\right)^{1/2}$ is equivalent to the norm $\|\{v_k\}\|_q$, where $v(x) \sim \sum_k v_k \exp(ik \cdot x)$.

The above proof of (7) shows that, if $f \in W^p(G_1)$, then $D^s f \in W^{p-|s|}(G_1)$, and $\varphi f \in W^p(G_1)$ for $\varphi \in C^\infty(G_1)$. Hence

$$\text{if } f \in W^p(G_1), \text{ then, for any differential operator } N \text{ of order } q \text{ with } C^\infty(G_1) \text{ coefficients, } Nf \in W^{p-|q|}(G_1). \tag{8}$$

To prove the Theorem for our periodic case, we first show that we may assume that the weak solution $u \in L^2(G_1) = W^0(G_1)$ of (2) belongs to $W^m(G_1)$. This is justified as follows. Let

$$u(x) \sim \sum_k u_k \exp(ik \cdot x), \quad v(x) \sim \sum_k u_k (1 + |k|^2)^{-m} \exp(ik \cdot x).$$

Then it is easy to see that $v(x) \in W^{2m}(G_1)$ and that v is a weak solution of $(I - \Delta)^m v = u$, where Δ denotes the Laplacian $\sum_{j=1}^{n} \partial^2/\partial x_j^2$. Hence v is a weak solution of the strongly elliptic equation of order $4m$:

$$L(I - \Delta)^m v = f. \tag{2'}$$

If we can show that this weak solution $v \in W^{2m}(G_1)$ actually belongs to $W^{4m+p}(G_1)$, then, by (8), $u = (I - \Delta)^m v$ belongs to $W^{4m+p-2m}(G_1) =$

$W^{p+2m}(G_1)$. Therefore, without losing the generality, we may assume that the weak solution u of (2) belongs to $W^m(G_1)$, where m is half of the order $2m$ of L.

Next, by Gårding's inequality (4) for L and that for $(I-\Delta)^m$, we may apply the Lax-Milgram theorem (Chapter III, 7) to the following effect. The bilinear forms on $C^\infty(G_1)$:

$$(\varphi, \psi)' = (\varphi, L^*\psi)_0 \text{ and } (\varphi, \psi)'' = (\varphi, (I-\Delta)^m\psi)_0 \tag{9}$$

can both be extended to be continuous bilinear forms on $W^m(G_1)$ such that there exist one-one bicontinuous linear mappings T', T'' on $W^m(G_1)$ onto $W^m(G_1)$ satisfying the conditions

$$(T'\varphi, \psi)' = (\varphi, \psi)_m, \ (T''\varphi, \psi)'' = (\varphi, \psi)_m \quad \text{for} \quad \varphi, \psi \in W^m(G_1).$$

Therefore, there exists a one-one bicontinuous linear map $T_m = T''(T')^{-1}$ on $W^m(G_1)$ onto $W^m(G_1)$ such that

$$(\varphi, \psi)' = (T_m\varphi, \psi)'' \text{ whenever } \varphi, \psi \in W^m(G_1). \tag{10}$$

We can show that

$$\text{for any } j \geqq 1,\ T_m \text{ maps } W^{m+j}(G_1) \text{ onto } W^{m+j}(G_1) \text{ in a one-one and bicontinuous way.} \tag{11}$$

In fact, we have

$$(\varphi, L^*(I-\Delta)^j\psi)_0 = (T_m\varphi, (I-\Delta)^{m+j}\psi)_0 \quad \text{for} \quad \varphi, \psi \in C^\infty(G_1).$$

On the other hand, there exists, by the Lax-Milgram theorem as applied to the strongly elliptic operators $(I-\Delta)^j L$ and $(I-\Delta)^{m+j}$, a one-one bicontinuous linear map T_{m+j} of $W^{m+j}(G_1)$ onto $W^{m+j}(G_1)$ such that

$$(\varphi, L^*(I-\Delta)^j\psi)_0 = (T_{m+j}\varphi, (I-\Delta)^{m+j}\psi)_0 \quad \text{for} \quad \varphi, \psi \in C^\infty(G_1).$$

Therefore, the function $w = (T_{m+j} - T_m)\varphi$ is, for any $\varphi \in C^\infty(G_1)$, a weak solution of $(I-\Delta)^{m+j} w = 0$. But, such a $w(x)$ is identically zero. For, the Fourier coefficient w_k of $w(x)$ satisfies

$$\begin{aligned} 0 &= ((I-\Delta)^{m+j} w(x), \exp(ik\cdot x))_0 = (w(x), (I-\Delta)^{m+j} \exp(ik\cdot x))_0 \\ &= (1+|k|^2)^{m+j}(w(x), \exp(ik\cdot x))_0 = (1+|k|^2)^{m+j} w_k, \end{aligned}$$

and so $w_k = 0$ for all k. Thus $(T_{m+j} - T_m)$ is 0 on $C^\infty(G_1)$. The space $C^\infty(G_1)$ is dense in $W^{m+j}(G_1) \subseteqq W^m(G_1)$, since trigonometric polynomials $\sum_{|k|<\infty} w_k \exp(ik\cdot x)$ are dense in the space $W^{m+j}(G_1)$. Hence $T_{m+j} = T_m$ on $W^{m+j}(G_1)$.

We are now ready to prove the differentiability theorem for our periodic case. We have, for $\psi \in C^\infty(G_1)$,

$$(f, \psi)_0 = (u, L^*\psi)_0 = (u, \psi)' = (T_m u, \psi)'' = (T_m u, (I-\Delta)^m\psi)_0.$$

Hence, for

$$T_m u \sim \sum_k c_k \exp(ik \cdot x), \quad \psi(x) \sim \sum_k \psi_k \exp(ik \cdot x),$$

we obtain, by Parseval's relation,

$$(T_m u, (I - \Delta)^m \psi)_0 = \sum_k c_k (1 + |k|^2)^m \overline{\psi}_k = \sum_k f_k \overline{\psi}_k.$$

By the arbitrariness of $\psi \in C^\infty(G_1)$, we have $c_k(1 + |k|^2)^m = f_k$, and so, by $f \in W^p(G_1)$, we must have $T_m u \in W^{p+2m}(G_1)$. Hence, by (11), $u \in W^{p+2m}(G_1)$ .It is to be noted that the above conclusion $u \in W^{p+2m}(G_1)$ is true even in the case $0 \geqq p \geqq (1 - m)$, i.e., the case $\{f_k\} \in W^p(G)$ with $0 \geqq p \geqq (1 - m)$. For, by $p + 2m \geqq m + 1$, we can apply (11).

We finally shall prove, for the general non-periodic case, our differentiability theorem. The following argument is due to P. LAX [2].

We want to prove, for the general non-periodic case, the differentiability theorem in a vicinity of a point x^0 of G. Let $\beta(x) \in C_0^\infty(G)$ be identically one in a vicinity of x^0. Denote βu by u'. u' is a weak solution of

$$Lu' = \beta f + Nu, \tag{12}$$

where N is a differential operator of order at most equal to $(2m - 1)$ whose coefficients are, together with $\beta(x)$, zero outside some vicinity V of x^0, and the operator N is to be applied in the sense of the distribution. We denote the distribution $\beta f + Nu$ by f'.

Let the periodic parallelogram G_1 contain V, and imagine the coefficients of L so altered inside G_1 but outside V that they become periodic without losing their differentiability and ellipticity properties. Denote the so altered L by L'. Thus u' is a weak solution in G of

$$L'u' = f', \text{ where } f' = \beta f + Nu. \tag{13}$$

We can thus apply the result obtained above for the periodic case to our weak solution u'. We may assume that the weak solution u' belongs to $W^m(G_1)$. Thus, since N is of order $\leqq (2m - 1)$ and with coefficients vanishing ontside V, $f' = \beta f + Nu$ must satisfy, by (8),

$$f' \in W^{p'}(G_1) \text{ with } p' = \min(p, m - (2m - 1)) = \min(p, 1 - m) \geqq 1 - m.$$

Therefore, the weak solution u' of (13) must satisfy

$$u' \in W^{p''}(G_1) \text{ with } p'' = \min(p + 2m, 1 - m + 2m)$$
$$= \min(p + 2m, m + 1).$$

Hence, in a certain vicinity of x^0, u has square integrable distributional derivatives up to order p'' which is $\geqq$ $(m + 1)$.Thus $f' = \beta f + Nu$ has, in a vicinity of x^0, square integrable distributional derivatives of order up to

$$p''' = \min(p, p'' - (2m - 1)) \geqq \min(p, 2 - m).$$

Thus again applying the result already obtained, we see that u' has, in a certain vicinity of x^0, square integrable distributional derivatives up to order

$$p^{(4)} = \min(p + 2m, 2 - m + 2m) = \min(p + 2m, m + 2).$$

Repeating the process, we see that u has, in a vicinity of x^0, square integrable distributional derivatives up to order $p + 2m$.

10. The Malgrange-Ehrenpreis Theorem

There is a striking difference between ordinary differential equations and partial differential equations. A classical result of PEANO states that the ordinary differential equation $dy/dx = f(x, y)$ has a solution under a single condition of the continuity of the function f. This result has also been extended to equations of higher order or to systems of equations. However, for partial differential equations, the situation is entirely different. H. LEWY [1] constructed in 1957 the equation

$$-i\frac{\partial u}{\partial x_1} + \frac{\partial u}{\partial x_2} - 2(x_1 + i x_2)\frac{\partial u}{\partial x_3} = f(x_3),$$

which has no solution at all if f is not analytic, even if f is C^∞. Lewy's example led L. HÖRMANDER [3] to develop a systematic method of constructing linear partial differential equations without solutions. Thus it is important to single out classes of linear partial differential equations with solutions.

Let $P(\xi)$ be a polynomial in $\xi_1, \xi_2, \ldots, \xi_n$, and let $P(D)$ be the linear differential operator obtained by replacing ξ_j by $D_j = i^{-1}\partial/\partial x_j$. $P(D)$ may be written as

$$P(D) = \sum_{m \geqq (\alpha) \geqq 0} c_\alpha D_\alpha, \text{ where, for } \alpha = (\alpha_1, \alpha_2, \ldots, \alpha_k) \text{ with } 1 \leqq \alpha_j \leqq n,$$

$$D_\alpha = \prod_{j=1}^{k} D_{\alpha_j}, \text{ and } D_0 = I \text{ and, moreover, } (\alpha) = k.$$

Definition 1. By a *fundamental solution* of $P(D)$ we mean a distribution E in R^n such that

$$P(D)\,E = \delta = T_\delta.$$

The importance of the notion of the fundamental solution is due to the fact that

$$u = E * f, \text{ where } f \in C_0^\infty(R^n),$$

gives a solution of the equation

$$P(D)\,u = f.$$

In fact, the differentiation rule (10) in Chapter VI, 3 implies that $P(D)\,u = (P(D)\,E) * f = \delta * f = f$.

Example. Let $P(D)$ be the Laplacian $\Delta = \sum_{j=1}^{n} \frac{\partial^2}{\partial x_j^2}$ in R^n with $n \geqq 3$. Then the distribution

$E = T_g$, where $g(x) = \frac{1}{(2-n) S_n} |x|^{2-n}$ and $S_n =$ the area of the surface of the unit sphere of R^n,

is a fundamental solution of Δ.

Proof. We have, in the polar coordinates, $dx = |x|^{n-1} d|x| \, dS_n$, and so the function $g(x)$ is locally integrable in R^n. Hence

$$\Delta T_{|x|^{2-n}}(\varphi) = \lim_{\varepsilon \downarrow 0} \int_{|x| \geqq \varepsilon} |x|^{2-n} \cdot \Delta\varphi \, dx, \quad \varphi \in \mathfrak{D}(R^n).$$

Let us take two positive numbers ε and R ($> \varepsilon$) such that the supp(φ) is contained in the interior of the sphere $|x| \leqq R$. Consider the domain G: $\varepsilon \leqq |x| \leqq R$ of R^n and apply Green's integral theorem, obtaining

$$\int_G (|x|^{2-n} \cdot \Delta\varphi - \Delta |x|^{2-n} \cdot \varphi) \, dx = \int_{\partial G} \left(|x|^{2-n} \cdot \frac{\partial\varphi}{\partial\nu} - \frac{\partial |x|^{2-n}}{\partial\nu} \cdot \varphi \right) dS$$

where $S = \partial G$ is the boundary surfaces given by $|x| = \varepsilon$ and $|x| = R$, and ν denotes the outwards normal to S. Since φ vanishes around $|x| = R$, we have, remembering that $\Delta |x|^{2-n} = 0$ for $x \neq 0$ and that $-\frac{\partial}{\partial\nu} = \frac{\partial}{\partial |x|}$ at the points of the inner boundary surface $|x| = \varepsilon$,

$$\int_{R^n} |x|^{2-n} \Delta\varphi \, dx = - \int_{|x|=\varepsilon} \varepsilon^{2-n} \frac{\partial\varphi}{\partial |x|} dS + \int_{|x|=\varepsilon} (2-n) \, \varepsilon^{1-n} \varphi \, dS.$$

When $\varepsilon \downarrow 0$, the expression $\partial\varphi/\partial|x| = \sum_{j=1}^{n} (x_j/|x|) \cdot \partial\varphi/\partial x_j$ is bounded and the area of the boundary surface $|x| = \varepsilon$ is $S_n \varepsilon^{n-1}$. Consequently the first term on the right tends to zero as $\varepsilon \downarrow 0$. By the continuity of φ and a similar argument to above, the second term on the right tends, as $\varepsilon \downarrow 0$, to $(2-n) S_n \cdot \varphi(0)$. Thus T_g is a fundamental solution of Δ.

The existence of a fundamental solution for every linear partial differential equation with constant coefficients was proved independently by B. Malgrange [1] and L. Ehrenpreis [1] around 1954—55. The exposition of the result given below is due to L. Hörmander [4].

Definition 2. Set

$$\tilde{P}(\xi) = \left(\sum_{|\alpha| \geqq 0} |P^{(\alpha)}(\xi)|^2 \right)^{1/2} \quad \text{where} \quad P^{(\alpha)}(\xi) = D_\xi^\alpha P(\xi),$$
$$D_\xi^\alpha = \frac{\partial^{\alpha_1 + \alpha_2 + \cdots + \alpha_n}}{\partial\xi_1^{\alpha_1} \partial\xi_2^{\alpha_2} \dots \partial\xi_n^{\alpha_n}}. \tag{1}$$

We say that a differential operator $Q(D)$ with constant coefficients is *weaker* than $P(D)$ if

$$\tilde{Q}(\xi) \leqq C\tilde{P}(\xi),\ \xi \in R^n,\ C \text{ being a positive constant.} \tag{2}$$

Theorem 1. If Ω is a bounded domain of R^n and $f \in L^2(\Omega)$, then there exists a solution u of $P(D)\,u = f$ in Ω such that $Q(D)\,u \in L^2(\Omega)$ for all Q weaker than P. Here the differential operators $P(D)$ and $Q(D)$ are to be applied in the sense of the theory of distributions.

The proof is based on

Theorem 2. For every $\varepsilon > 0$, there is a fundamental solution E of $P(D)$ such that, with a constant C independent of u,

$$|(E * u)(0)| \leqq C \sup_{|\eta| \leqq \varepsilon} \int_{R_n} |\hat{u}(\xi + i\eta)|/\tilde{P}(\xi)\, d\xi,\quad u \in C_0^\infty(R^n). \tag{3}$$

Here $\hat{u}$ is the Fourier-Laplace transform of u:

$$\hat{u}(\zeta) = (2\pi)^{-n/2} \int_{R^n} e^{-i\langle x, \zeta\rangle} u(x)\, dx,\quad \zeta = \xi + i\eta,$$

and the finiteness of the right side of (3) is assured by the Paley-Wiener theorem in Chapter VI, 4.

Deduction of Theorem 1 from Theorem 2. Replace u in (3) by $Q(D)\,u * v$, where u and v are in $C_0^\infty(R^n)$. Then, by (10) in Chapter VI, 3,

$$|(Q(D)\,E * u * v)(0)| = |(E * Q(D)\,u * v)(0)| \leqq CN(Q(D)\,u * v),$$

$$\text{where}\quad N(u) = \sup_{|\eta| \leqq \varepsilon} \int_{R^n} |\hat{u}(\xi + i\eta)|/\tilde{P}(\xi) \cdot d\xi.$$

The Fourier-Laplace transform of $Q(D)\,u * v$ is, by (17) in Chapter VI, 2 and (15) in Chapter VI, 3, equal to $(2\pi)^{n/2}\, Q(\zeta)\,\hat{u}(\zeta)\,\hat{v}(\zeta)$. Since, by Taylor's formula,

$$Q(\xi + i\eta) = \sum_\alpha \frac{1}{(\alpha)!}(-\eta)^\alpha D_\alpha Q(\xi), \text{ where } (-\eta)^\alpha = \prod_j (-\eta_{\alpha_j}), \tag{4}$$

we have, by (2),

$$|Q(\xi + i\eta)|/\tilde{P}(\xi) \leqq C' \text{ for } |\eta| \leqq \varepsilon \text{ and } \xi \in R^n,$$

where the constant C' may depend on ε. Thus

$$N(Q(D)\,u * v) \leqq (2\pi)^{n/2}\, C' \cdot \sup_{|\eta| \leqq \varepsilon} \int_{R^n} |\hat{u}(\xi + i\eta)\,\hat{v}(\xi + i\eta)|\, d\xi.$$

By Parseval's theorem for the Fourier transform, we obtain, denoting by $\|\ \|$ the $L^2(R^n)$-norm,

$$\int_{R^n} |\hat{u}(\xi + i\eta)|^2\, d\xi = \int_{R^n} |u(x)|^2 e^{2\langle x, \eta\rangle}\, dx \leqq \|u(x)\, e^{\varepsilon|x|}\|^2 \text{ when } |\eta| \leqq \varepsilon,$$

and a similar estimate for $\hat{v}$. Thus, by Schwarz' inequality,

$$N(Q(D)\,u * v) \leqq C''\, \|u(x)\, e^{\varepsilon|x|}\|\, \|v(x)\, e^{\varepsilon|x|}\| \text{ whenever } u \text{ and } v \in C_0^\infty(R^n),$$

C'' denoting a constant which may depend on ε.

Therefore we have proved

$$\left|\int_{R^n} (Q(D)\, E * u)\,(x)\, v(-x)\, dx\right| \leqq (C C'') \|u e^{\varepsilon|x|}\|\, \|v e^{\varepsilon|x|}\| \quad (u, v \in C_0^\infty(R^n)). \tag{5}$$

We shall write $L_\varepsilon^2(R^n)$ for the Hilbert space of functions $w(x)$ normed by

$$\left(\int_{R^n} |w(x)|^2 e^{\varepsilon|x|}\, dx\right)^{1/2} = \|w(x)\, e^{\varepsilon|x|}\|.$$

Since $C_0^\infty(R^n)$ is dense in $L_\varepsilon^2(R^n)$ and, as may be proved easily, $L_{-\varepsilon}^2(R^n)$ is the conjugate space of $L_\varepsilon^2(R^n)$, we obtain, by dividing (5) by $\|v(x)\, e^{\varepsilon|x|}\|$ and taking the supremum over $v \in C_0^\infty(R^n)$,

$$\|(Q(D)\, E * u)\,(x)\, e^{-\varepsilon|x|}\| \leqq (C C'')\, \|u(x)\, e^{\varepsilon|x|}\|, \quad u \in C_0^\infty(R^n).$$

Hence the mapping

$$u \to Q(D)\, E * u \tag{6}$$

can be extended by continuity from $C_0^\infty(R^n)$ to $L_\varepsilon^2(R^n)$, so that it becomes a continuous linear mapping on $L_\varepsilon^2(R^n)$ into $L_{-\varepsilon}^2(R^n)$. Thus, to prove Theorem 1, we have to take $f_1 = f$ in Ω and $f_1 = 0$ in $R^n - \Omega$ and define u as equal to $u = E * f_1$.

For the proof of Theorem 2, we prepare three Lemmas.

Lemma 1 (Malgrange). If $f(z)$ is a holomorphic function of a complex variable z for $|z| \leqq 1$ and $p(z)$ is a polynomial in which the coefficient of the highest order term is A, then

$$|A\, f(0)| \leqq (2\pi)^{-1} \int_{-\pi}^{\pi} |f(e^{i\theta})\, p(e^{i\theta})|\, d\theta. \tag{7}$$

Proof. Let z_j's be the zeros of $p(z)$ in the unit circle $|z| < 1$ and put

$$p(z) = q(z) \prod_j \frac{z - z_j}{\bar{z}_j z - 1}.$$

Then $q(z)$ is regular in the unit circle and $|p(z)| = |q(z)|$ for $|z| = 1$. Hence we have

$$(2\pi)^{-1} \int_{-\pi}^{\pi} |f(e^{i\theta})\, p(e^{i\theta})|\, d\theta = (2\pi)^{-1} \int_{-\pi}^{\pi} |f(e^{i\theta})\, q(e^{i\theta})|\, d\theta$$

$$\geqq (2\pi)^{-1} \left| \int_{-\pi}^{\pi} f(e^{i\theta})\, q(e^{i\theta})\, d\theta \right| = |f(0)\, q(0)|.$$

Thus Lemma 1 is proved since $|q(0)/A|$ is equal to the product of the absolute values of zeros of $p(z)$ outside the unit circle.

Lemma 2. With the notations in Lemma 1, we have, if the degree of $p(z)$ is $\leqq m$,

$$|f(0)\, p^{(k)}(0)| \leqq \frac{m!}{(m-k)!} (2\pi)^{-1} \int_{-\pi}^{\pi} |f(e^{i\theta})\, p(e^{i\theta})|\, d\theta. \tag{8}$$

Proof. We may assume that the degree of $p(z)$ is m and that

$$p(z) = \prod_{j=1}^{m} (z - z_j).$$

Applying the preceding Lemma 1 to the polynomial $\prod_{j=1}^{k} (z - z_j)$ and the holomorphic function $f(z) \cdot \prod_{j=k+1}^{m} (z - z_j)$, we obtain

$$\left| f(0) \prod_{j=k+1}^{m} z_j \right| \leqq (2\pi)^{-1} \int_{-\pi}^{\pi} |f(e^{i\theta})\, p(e^{i\theta})|\, d\theta.$$

A similar inequality will hold for any product of $(m - k)$ of the numbers z_j on the left hand side. Since $p^{(k)}(0)$ is the sum of $m!/(m-k)!$ such terms, multiplied by $(-1)^{m-k}$, we have proved the inequality (8).

Lemma 3. Let $F(\zeta) = F(\zeta_1, \zeta_2, \ldots, \zeta_n)$ be holomorphic for $|\zeta| = \left(\sum_{j=1}^{n} |\zeta_j|^2\right)^{1/2} < \infty$ and $P(\zeta) = P(\zeta_1, \zeta_2, \ldots, \zeta_n)$ a polynomial of degree $\leqq m$. Let $\Phi(\zeta) = \Phi(\zeta_1, \zeta_2, \ldots, \zeta_n)$ be a non-negative integrable function with compact support depending only on $|\zeta_1|, |\zeta_2|, \ldots, |\zeta_n|$. Then we have

$$|F(0)\, D_\alpha P(0)| \int_{|\zeta|<\infty} |\zeta|^{(\alpha)} \Phi(\zeta)\, d\zeta \leqq \frac{m!}{(m-(\alpha))!} \int_{|\zeta|<\infty} |F(3)\, P(\zeta)|\, \Phi(\zeta)\, d\zeta, \tag{9}$$

where $d\zeta$ is the Lebesgue measure $d\xi_1\, d\eta_1 \cdots d\xi_n\, d\eta_n$ $(\zeta_k = \xi_k + i\,\eta_k)$.

Proof. Let $f(z)$ be an entire holomorphic function and apply (8) to the functions $f(rz)$ and $p(rz)$, where $r > 0$. Then we obtain

$$|f(0)\, p^{(k)}(0)| \cdot r^k \leqq \frac{m!}{(m-k)!} (2\pi)^{-1} \int_{-\pi}^{\pi} |f(r e^{i\theta})\, p(r e^{i\theta})|\, d\theta.$$

Let $\psi(r)$ be a non-negative, integrable function with compact support. Multiplying the above inequality by $2\pi r \psi(r)$ and integrating with respect to r, we obtain

$$|f(0)\, p^{(k)}(0)| \int |t|^k \psi(|t|)\, dt \leqq \frac{m!}{(m-k)!} \int |f(t)\, p(t)|\, \psi(|t|)\, dt \tag{10}$$

where $dt = r\,dr\,d\theta$ and the integration is extended over the whole complex t-plane. Lemma 3 is obtained by applying (10) successively to the variables $\zeta_1, \zeta_2, \ldots, \zeta_n$, one at a time.

Proof of Theorem 2. Put $P(D)\,u = v$, where $u \in C_0^\infty(R^n)$. Then $P(\zeta)\,\hat{u}(\zeta) = \hat{v}(\zeta)$. Apply Lemma 3 with $F(\zeta) = \hat{u}(\xi + \zeta)$, with $P(\zeta)$ replaced by $P(\xi + \zeta)$ and with $|\Phi(\zeta)| = 1$ when $|\zeta| \leqq \varepsilon$ and $= 0$ otherwise. Since $\tilde{P}(\xi) \leqq \sum_\alpha |D^\alpha P(\xi)|$, we obtain, from (9),

$$|\hat{u}(\xi)\, \tilde{P}(\xi)| \leqq C_1 \int_{|\zeta|\leqq\varepsilon} |\hat{u}(\xi+\zeta)\, P(\xi+\zeta)|\, d\zeta = C_1 \int_{|\zeta|\leqq\varepsilon} |\hat{v}(\xi+\xi)|\, d\zeta.$$

Hence, by Fourier's integral theorem,

$$|u(0)| \leqq (2\pi)^{-n/2} \int |\hat{u}(\xi)|\, d\xi \leqq C_1' \int_{|\zeta| \leqq \varepsilon} \left(\int |\hat{v}(\xi + \zeta)|/\tilde{P}(\xi)\, d\xi\right) d\zeta$$

$$\leqq C_1' \int \left(\int_{\xi'^2 + \eta'^2 \leqq \varepsilon} |\hat{v}(\xi + \xi' + i\eta')|/\tilde{P}(\xi)\, d\xi'\, d\eta'\right) d\xi.$$

On the other hand, we have

$$\tilde{P}(\xi + \xi')/\tilde{P}(\xi) \leqq C_2 \quad \text{when } |\xi'| \leqq \varepsilon,$$

because

$$D^\alpha P(\xi + \xi') = \sum_\beta \frac{(\xi')^\beta}{(\beta)!} D^{\alpha+\beta} P(\xi),$$

so that $|D^\alpha P(\xi + \xi')|/\tilde{P}(\xi)$ is bounded when $|\xi'| \leqq \varepsilon$. Therefore we have

$$|u(0)| \leqq C_1' C_2 \int \left(\int_{\xi'^2 + \eta'^2 \leqq \varepsilon^2} |\hat{v}(\xi + \xi' + i\eta')|/\tilde{P}(\xi + \xi')\, d\xi'\, d\eta'\right) d\xi$$

$$\leqq C_3 \|v\|', \text{ where} \tag{11}$$

$$\|v\|' = \int_{|\eta| \leqq \varepsilon} \left(\int |\hat{v}(\xi + i\eta)|/\tilde{P}(\xi)\, d\xi\right) d\eta \qquad (u \in C_0^\infty(R_n)),$$

C_3 denoting a constant depending only on ε.

By the way, the finiteness of $\|v\|'$ is a consequnece of the Paley-Wiener theorem in Chapter VI, 4. Consider the space $\tilde{C}_0^\infty(R^n)$ which is the completion of $C_0^\infty(R^n)$ with respect to the norm $\|v\|'$. Then, by the Hahn-Banach extension theorem, the linear functional L:

$$v = P(D)\, u \to u(0) \quad (\text{where } u \in C_0^\infty(R^n))$$

can be extended to a continuous linear functional L defined on $\tilde{C}_0^\infty(R^n)$. As in the case of the space $L^1(R^n)$, we see that there exists a Baire function $k(\xi + i\eta)$ bounded a.e. with respect to the measure $\tilde{P}(\xi)^{-1} d\xi\, d\eta$ such that the extended linear functional L is represented as

$$L(v) = \int_{|\eta| \leqq \varepsilon} \left(\int \hat{v}(\xi + i\eta)\, k(\xi + i\eta)/\tilde{P}(\xi) \cdot d\xi\right) d\eta. \tag{12}$$

When $v_h(x) \in C_0^\infty(R^n)$ tends, as $h \to \infty$, to 0 in the topology of $\mathfrak{D}(R^n)$, $v_h(x)\, e^{\langle x, \eta\rangle}$ also tends to zero in the topology of $\mathfrak{D}(R^n)$, uniformly with respect to η for $|\eta| \leqq \varepsilon$. Hence, as in Chapter VI, 1, we see easily that $\hat{v}_h(\xi + i\eta)$, as function of ξ, tends to zero in the topology of $\mathfrak{S}(R^n)$, uniformly in η for $|\eta| \leqq \varepsilon$. Therefore, by (12), L defines a distribution $T \in \mathfrak{D}(R^n)'$. Thus, by (5) in Chapter VI, 3.

$$L(v) = (T * \check{v})(0) = (\check{T} * v)(0). \tag{13}$$

We have thus proved Theorem 2 by taking $E = \check{T}$. (3) is clear from (11).

11. Differential Operators with Uniform Strength

The existence theorem of the preceding section may be extended to a linear differential operator

$$P(x, D) = \sum_{\alpha} a_{\alpha}(x) D_{\alpha} \tag{1}$$

whose coefficients $a_{\alpha}(x)$ are continuous in an open bounded domain Ω of R^n.

Definition. $P(x, D)$ is said to be of *uniform strength* in Ω, if

$$\sup_{x,y \in \Omega, \xi \in R^n} \tilde{P}(x, \xi)/\tilde{P}(y, \xi) < \infty, \tag{2}$$

where $\tilde{P}(x, \xi)$ is defined by $\left(\sum_{\alpha} |P^{(\alpha)}(x, \xi)|^2\right)^{1/2}$ considering x as parameters.

Examples. The differential operator $P(x, D) = \sum_{|s|, |t| \leqq m} D^s a_{s,t}(x) D^t$ with real, bounded C^{∞} coefficients $a_{s,t}(x) = a_{t,s}(x)$ in Ω is *strongly elliptic* in Ω (see Chapter VI, 9) if there exists a positive constant δ such that

$$\sum_{|s|, |t| = m} \xi^s a_{s,t}(x) \xi^t \geqq \delta \left(\sum_{j=1}^{n} \xi_j^2\right)^m \text{ in } \Omega. \tag{3}$$

In such a case, $P(x, D)$ satisfies the condition (2). Next let $P(x, D)$ be strongly elliptic in an open bounded domain Ω of R^{n-1}. Then

$$\frac{\partial}{\partial x_n} - P(x, D) \tag{4}$$

is said to be *parabolic* in the product space $\Omega \times \{x_n; 0 < x_n\}$. It is easy to see that the operator (4) is of uniform strength in the above product space.

Theorem (Hörmander [5]). Let $P(x, D)$ be of uniform strength in an open bounded domain Ω of R^n. For any point x^0, there exists an open subdomain Ω_1 of Ω such that $x^0 \in \Omega_1$ and the equation $P(x, D) u = f$ has, for every $f \in L^2(\Omega_1)$, a distribution solution $u \in L^2(\Omega_1)$ for which, moreover, $Q(D) u \in L^2(\Omega_1)$ for every $Q(D)$ weaker than $P(x, D)$ for any fixed $x \in \Omega_1$.

Proof. Write $P(x^0, D) = P_0(D)$. The set of all the differential operators with constant coefficients weaker than $P_0(D)$ is a finite dimensional linear space. For, the degree of such operators $Q(D)$ cannot exceed that of $P_0(D)$. Thus there exist $P_1(D), P_2(D), \ldots, P_N(D)$ which form a basis for the differential operators weaker than $P_0(D)$. Hence we can write

$$P(x, D) = P_0(D) + \sum_{j=1}^{N} b_j(x) P_j(D), \; b_j(x^0) = 0, \tag{5}$$

with uniquely determined $b_j(x)$ which are continuous in Ω.

By the result of the preceding section, there exists a bounded linear operator T on $L^2(\Omega_1)$ into $L^2(\Omega_1)$ such that

$$P_0(D)\,Tf = f \text{ for all } f \in L^2(\Omega_1), \tag{6}$$

and that the operators $P_j(D)\,T$ are all bounded as operators on $L^2(\Omega_1)$ into $L^2(\Omega_1)$. Here Ω_1 is any open subdomain of Ω. We have only to take Tf as the restriction to Ω_1 of $E * f_1$ where $f_1 = f$ in Ω_1 and $f_1 = 0$ in $R^n - \Omega_1$.

The equation $P(x, D)\,u = f$ is equivalent to

$$P_0(D)\,u + \sum_{j=1}^{N} b_j(x)\,P_j(D)\,u = f. \tag{7}$$

We shall seek a solution of the form $u = Tv$. Substituting this in (7), we obtain, by (6),

$$v + \sum_{j=1}^{N} b_j(x)\,P_j(D)\,Tv = f. \tag{8}$$

Let the sum of the norms of the bounded linear operators $P_j(D)\,T$ on $L^2(\Omega_1)$ into $L^2(\Omega_1)$ be denoted by C. Since $b_j(x)$ is continuous and $b_j(x^0) = 0$, we may choose $\Omega_1 \ni x^0$ so small that

$$C\,|b_j(x)| < 1/N \text{ whenever } x \in \Omega_1 \quad (j = 1, 2, \ldots, n).$$

We may assume that the above inequalities hold whenever x belongs to the compact closure of Ω_1. Thus the norm of the operator $\sum_{j=1}^{N} b_j(x)\,P_j(D)\,T$ is less than 1, and so the equation (8) may be solved by Neumann's series (Theorem 2 in Chapter II, 1):

$$v = \left(I + \sum_{j=1}^{N} b_j P_j(D)\,T\right)^{-1} f = Af,$$

where A is a bouned linear operator on $L^2(\Omega_1)$ into $L^2(\Omega_1)$. Hence $u = TAf$ is the required solution of $P(x, D)\,u = f$.

12. The Hypoellipticity (Hörmander's Theorem)

We have defined in Chapter II, 7 the notion of hypoellipticity of $P(D)$ and proved Hörmander's theorem to the effect that, if $P(D)$ is hypoelliptic, then there exists, for any large positive constant C_1, a positive constant C_2 such that, for all solutions $\zeta = \xi + i\eta$ of the algebraic equation $P(\zeta) = 0$,

$$|\xi| < C_2 \text{ if } |\eta| < C_1. \tag{1}$$

To prove conversely that (1) implies the hypoellipticity of $P(D)$, we prepare the

Lemma (Hörmander [1]), (1) implies that

$$\sum_{|\alpha|>0} |P^{(\alpha)}(\xi)|^2 \Big/ \sum_{|\alpha|\geq 0} |P^{(\alpha)}(\xi)|^2 \to 0 \text{ as } \xi \in R^n,\ |\xi| \to \infty. \tag{2}$$

Proof. We first show that, for any real vector $\Theta \in R^n$, we have

$$P(\xi+\Theta)/P(\xi) \to 1 \text{ as } \xi \in R^n, |\xi| \to \infty. \tag{3}$$

We may assume that the coordinates are so chosen that $\Theta = (1, 0, 0, \ldots, 0)$. We have, by (1),

$$P(\xi + i\eta) \neq 0 \text{ when } |\eta| < C_1 \text{ and } |\xi| > C_2.$$

Then the inequality $|\xi - \zeta'| \geqq C_1$ holds if $|\xi| \geqq C_1 + C_2$ and $P(\zeta') = 0$. For, setting $\zeta' = \xi' + i\eta'$, we have either $|\eta'| \geqq C_1$ or else $|\xi'| < C_2$ so that $|\xi - \xi'| \geqq C_1$. Giving fixed values to, $\xi_2, \xi_3, \ldots, \xi_n$ we can write

$$P(\xi) = C \prod_{k=1}^{m} (\xi_1 - t_k),\ C \neq 0,$$

where $(t_k, \xi_2, \ldots, \xi_n)$ is a zero of P. Hence we have $|t_k - \xi_1| \geqq C_1$ if $|\xi| \geqq C_1 + C_2$. Thus

$$\frac{P(\xi+\Theta)}{P(\xi)} = \prod_{k=1}^{m} \frac{\xi_1 + 1 - t_k}{\xi_1 - t_k} = \prod_{k=1}^{m}\left(1 + \frac{1}{\xi_1 - t_k}\right)$$

satisfies

$$\left|\frac{P(\xi+\theta)}{P(\xi)} - 1\right| \leqq m C_1^{-1}(1 + C_1^{-1})^{m-1} \quad \text{if} \quad |\xi| \geqq C_1 + C_2.$$

As we may take C_1 arbitrarily large by taking C_2 sufficiently large, we have proved (3).

We have, by Taylor's formula,

$$P(\xi+\eta) = \sum_{(\alpha)} \frac{1}{(\alpha)!} P^{(\alpha)}(\xi)\,\eta^\alpha,\ P^{(\alpha)}(\xi) = (i)^{(\alpha)} D_\alpha P(\xi),\ D_\alpha = \prod_{j=1}^{(\alpha)} D_{\alpha_j}$$

and so

$$\sum_{i=1}^{k} t_i P(\xi + \eta^{(i)}) = \sum_{(\alpha)} \frac{1}{(\alpha)!} P^{(\alpha)}(\xi) \sum_{i=1}^{k} t_i (\eta^{(i)})^\alpha, \tag{4}$$

where $\eta^{(i)}$ are arbitrary real vectors and t_i arbitrary complex numbers. The coefficients $\sum_{i=1}^{k} t_i(\eta^{(i)})^\alpha$, $(\alpha) \leqq m$, can be given arbitrary values, which are symmetric in α, by a convenient choice of k, t_i and $\eta^{(i)}$. If otherwise, there would exist constants C_α, $(\alpha) \leqq m$, symmetric in α and not all equal to zero and such that $\sum_{(\alpha)} C_\alpha \eta^\alpha = 0$ for every η. Thus

$$P^{(\alpha)}(\xi) = \sum_{i=1}^{k} t_i P(\xi + \eta^{(i)}) \text{ with real vectors } \eta^{(i)}.$$

Since the principal part on the right must cancel out when $(\alpha) \neq 0$, we must have $\sum_{i=1}^{k} t_i = 0$. Hence, by (3), we obtain (2).

Corollary. Suppose that $P_1(\xi)$ and $P_2(\xi)$ satisfy (2). Then $P(\xi) = P_1(\xi) \cdot P_2(\xi)$ also satisfies (2). Moreover, if $Q_j(D)$ is weaker than $P_j(D)$ $(j = 1, 2)$, it follows that $Q_1(D)\, Q_2(D)$ is weaker than $P(D)$.

Proof. By applying Leibniz' formula of differentiation of product of functions, we see that $P^{(\alpha)}(\xi)$ is a linear combination of products of derivatives of $P_1(\xi)$ and $P_2(\xi)$ of the order sum $\leqq (\alpha)$. Hence (2) holds for $P(\xi)$. The latter part of the Corollary may be proved similarly.

We are now ready to prove

Theorem (HÖRMANDER [1]). $P(D)$ is hypoelliptic iff the condition (2) is satisfied.

Proof. The "only if" part is already proved (Chapter II, 7 and what was proved above). We shall prove the "if" part.

Let Ω be an open subdomain of R^n. A distribution $u \in \mathfrak{D}(\Omega)'$ is said to belong to $H^k_{\mathrm{loc}}(\Omega)$ if, for any $\varphi_0 \in C_0^\infty(\Omega)$, the Fourier transform $\hat{u}_0$ of $u_0 = \varphi_0 u$ satisfies (see Chapter VI, 2)

$$\int_{R^n} (1 + |\xi|^2)^k \, |\hat{u}_0(\xi)|^2 \, d\xi < \infty, \text{ that is, if } u_0 = \varphi_0 u \in W^{k,2}(R^n). \quad (5)$$

By virtue of Sobolev's lemma in Chapter VI, 7, the "if" part is proved by the following proposition:

Let $P(\xi)$ satisfy (2). If a distribution $u \in \mathfrak{D}(\Omega)'$ satisfies $P(D)\,u \in H^s_{\mathrm{loc}}(\Omega)$ with a positive s, then u belongs to $H^s_{\mathrm{loc}}(\Omega)$. (6)

For, if $P(D)\,u \in C^\infty$ in Ω, then $P(D)\,u \in H^s_{\mathrm{loc}}(\Omega)$ for every positive s because of Leibniz' formula of differentiation.

The following proof of (6) is based upon two Lemmas:

Lemma 1. Let $f \in W^{s,2}(R^n)$ and $\psi \in C_0^\infty(R^n)$, $s \gtreqless 0$. Then $\psi f \in W^{s,2}(R^n)$.

Lemma 2. Let $P(\xi)$ satisfy (2). Then there exists a positive constant μ such that $|P^{(\alpha)}(\xi)\,\xi^\mu|/|P(\xi)| \to 0$ as $\xi \in R^n$, $|\xi| \to \infty$ for every $\alpha \neq 0$.

The proof of Lemma 1 will be given later, and the proof of Lemma 2 will not be given here (for the latter, we refer the reader to L. HÖRMANDER [6] or to A. FRIEDMAN [1]).

Next, let Ω_1 and Ω_0 be arbitrary open subdomains of Ω such that their closures Ω_1^a and Ω_0^a are compact and $\Omega_1^a \subseteqq \Omega_0$, $\Omega_0^a \subseteqq \Omega$. By Schwartz' Theorem in Chapter III, 11, the distribution $u \in \mathfrak{D}(\Omega)'$ is, when considered as a distribution $\in \mathfrak{D}(\Omega_0)'$, a distributional derivative of the form $D^t v$ of a function $v(x) \in L^2(\Omega_0)$. Let $\varphi \in C_0^\infty(\Omega_0)$ be such that $\varphi(x) = 1$ in Ω_1. Then $u = u_0 = D^t \varphi v$ as distributions $\in \mathfrak{D}(\Omega_1)'$. φv being $\in L^2(R^n)$, we see that there exists a (possibly) negative integer k such that

$$P^{(\alpha)}(D)\,u_0 = P^{(\alpha)}(D)\,D^t \varphi v \in W^{k,2}(R^n) \text{ for every } \alpha. \quad (7)$$

Hence, by Lemma 1 and the generalized Leibniz formula (see Chapter I, 8)

$$P(D)\,\varphi_1 u_0 = \varphi_1 P(D)\,u_0 + \sum_{(\alpha)>0} \frac{1}{(\alpha)!} D_\alpha \varphi_1 \cdot P^{(\alpha)}(D)\,u_0, \quad (8)$$

we see by $P(D)\,u_0 \in H^s_{\mathrm{loc}}(\Omega)$ that, whenever $\varphi_1 \in C_0^\infty(\Omega_1)$,

$$P(D)\,\varphi_1 u_0 \in W^{k_1,2}(R^n) \text{ with } k_1 = \min(s, k). \tag{9}$$

Thus the Fourier transform $\hat{u}_1(\xi)$ of $u_1(x) = \varphi_1(x)\,u_0(x)$ satisfies

$$\int_{R^n} |P(\xi)\,\hat{u}_1(\xi)|^2 (1 + |\xi|^2)^{k_1}\, d\xi < \infty \tag{10}$$

and hence, by Lemma 2,

$$\begin{aligned} &\int_{R^n} |P^{(\alpha)}(\xi)\,\hat{u}_1(\xi)|^2 (1 + |\xi|^2)^{k_1+\mu}\, d\xi < \infty, \text{ that is,} \\ &P^{(\alpha)}(D)\,u_1 \in W^{k_1+\mu,2}(R^n) \text{ for every } \alpha \neq 0. \end{aligned} \tag{11}$$

Let Ω_2 be any open subdomain of Ω_1 such that its closure Ω_2^a is compact and contained in Ω_1. Then, for any $\varphi_2 \in C_0^\infty(\Omega_2)$, we prove, by (8) and (11) as above, that

$$P(D)\,\varphi_2 u_1 \in W^{k_2,2}(R^n) \text{ with } k_2 = \min(s, k_1 + \mu) \text{ and hence}$$

$$P^{(\alpha)}(D)\,\varphi_2 u_1 \in W^{k_2+\mu,2}(R^n) \text{ for every } \alpha \neq 0.$$

Repeating the argument a finite number of times, we see that, for any open subdomain Ω' of Ω such that its closure is compact and contained in Ω,

$$P^{(\alpha)}(D)\,\varphi u \in W^{s,2}(R^n) \text{ for all } \alpha \neq 0 \text{ whenever } \varphi \in C_0^\infty(\Omega').$$

Thus, $P^{(\alpha)}(\xi) = \text{constant} \neq 0$ gives $\varphi u \in W^{s,2}(R^n)$.

Proof of Lemma 1. The Fourier transform of ψf is

$$(2\pi)^{-n/2} \int_{R^n} \hat{\psi}(\eta)\,\hat{f}(\xi - \eta)\, d\eta \text{ (see Theorem 6 in Chapter VI, 3)}$$

and thus we have to show that, for $s \gtreqless 0$,

$$\int_{R^n} (1 + |\xi|^2)^s \left| \int_{R^n} \hat{\psi}(\eta)\,\hat{f}(\xi - \eta)\, d\eta \right|^2 d\xi < \infty.$$

By Schwarz' inequality, this can be estimated from above by

$$\begin{aligned} &\int_{R^n} (1 + |\xi|^2)^s \left[\int_{R^n} |\hat{\psi}(\eta)|\, d\eta \cdot \int_{R^n} |\hat{\psi}(\eta)| \cdot |\hat{f}(\xi - \eta)|^2\, d\eta \right] d\xi \\ = &\int_{R^n} |\hat{\psi}(\eta)|\, d\eta \left[\int_{R^n} \int_{R^n} (1 + |\xi|^2)^s\, |\hat{\psi}(\eta)| \cdot |\hat{f}(\xi - \eta)|^2\, d\xi\, d\eta \right]. \end{aligned} \tag{12}$$

We then make use of the inequality

$$(1 + |\xi|^2)^s \leqq 4^{|s|} (1 + |\eta|^2)^{|s|} (1 + |\xi - \eta|^2)^s \tag{13}$$

which may be proved by

$$\frac{1 + |\xi|^2}{1 + |\xi - \eta|^2} \leqq 4(1 + |\eta|^2), \quad 4(1 + |\xi|^2) \geqq \frac{1 + |\xi - \eta|^2}{1 + |\eta|^2}.$$

By (13), the right side of (12) is estimated by $\int_{R^n} |\hat{\psi}(\eta)|\, d\eta$-times

$$4^{|s|} \int_{R^n} |\hat{\psi}(\eta)|\, (1 + |\eta|^2)^{|s|}\, d\eta \cdot \left(\int_{R^n} (1 + |\xi - \eta|^2)^s\, |\hat{f}(\xi - \eta)|^2\, d\xi \right).$$

This integral converges since $f \in W^{s,2}(R^n)$ and $\hat{\varphi}(\eta) \in \mathfrak{S}(R^n)$.

We have thus proved our Theorem.

Further Researches

1. A linear partial differential operator $P(x, D)$ with $C^\infty(\Omega)$ coefficients is said to be *formally hypoelliptic* in $\Omega \subseteqq R^n$ if the following two conditions are satisfied: i) $P(x^0, D)$ is hypoelliptic for every fixed $x^0 \in \Omega$ and ii) $P(x^0, \xi) = 0(P(x', \xi))$ as $\xi \in R^n$, $|\xi| \to \infty$ for every fixed x^0 and $x' \in \Omega$. L. Hörmander [5] and B. Malgrange [2] have proved that, for such an operator $P(x, D)$, any distribution solution $u \in \mathfrak{D}(\Omega)'$ of the equation $P(x, D)\, u = f$ is C^∞ after correction on a set of measure zero in the open subdomain $\subseteqq \Omega$ where f is C^∞. The proof above for the constant coefficients case may be modified so as to apply for the formally hypoelliptic case, see. e.g., J. Peetre [1].

2. It was proved essentially by I. G. Petrowsky [1] that all distribution solutions $u \in \mathfrak{D}(R^n)$ of $P(D)\, u = 0$ are analytic functions in R^n iff the homogeneous part $P_m(\xi)$ of $P(\xi)$ of the highest degree m does not vanish for $\xi \in R^n$. If this condition is satisfied then $P(D)$ is said to be (*analytically*) *elliptic*. It is proved that in such a case the degree m is even and $P(D)$ is hypoelliptic. It is to be noted that the hypoellipticity of an analytically elliptic operator $P(D)$ can also be proved by Friedrichs' Theorem in Chapter VI, 9. For, by the non-vanishing of $P_m(\xi)$, we easily see, by the Fourier transformation, that $P(D)$ or $-P(D)$ is strongly elliptic. For the proof of Petrowsky's Theorem, see, e.g., L. Hörmander [6], F. Trèves [1] and C. B. Morrey-L. Nirenberg [1].

VII. Dual Operators

1. Dual Operators

The notion of the *transposed matrix* may be extended to the notion of *dual operator* through

Theorem 1. Let X, Y be locally convex linear topological spaces and X'_s, Y'_s their strong dual spaces, respectively. Let T be a linear operator on $D(T) \subseteqq X$ into Y. Consider the points $\{x', y'\}$ of the product space $X'_s \times Y'_s$ satisfying the condition

$$\langle Tx, y' \rangle = \langle x, x' \rangle \quad \text{for all} \quad x \in D(T). \tag{1}$$

Then x' is determined uniquely by y' iff $D(T)$ is dense in X.

Proof. By the linearity of the problem, we have to consider the condition:

$$\langle x, x' \rangle = 0 \quad \text{for all} \quad x \in D(T) \quad \text{implies} \quad x' = 0.$$

Thus the "if" part is clear from the continuity of the linear functional x'. Assume that $D(T)^a \neq X$. Then there exists, by the Hahn-Banach theorem, an $x'_0 \neq 0$ such that $\langle x, x'_0 \rangle = 0$ for all $x \in D(T)$; consequently, the "only if" part is proved.

Definition 1. A linear operator T' such that $T'y' = x'$ is defined through (1) iff $D(T)^a = X$. T' is called the *dual* or *conjugate operator* of T; its domain $D(T')$ is the totality of those $y' \in Y'_s$ such that there exists $x' \in X'_s$ satisfying (1), and $T'y' = x'$. Hence T' is a linear operator defined on $D(T') \subseteq Y'_s$ into X'_s such that

$$\langle Tx, y' \rangle = \langle x, T'y' \rangle \quad \text{for all} \quad x \in D(T) \quad \text{and all} \quad y' \in D(T'). \tag{2}$$

Theorem 2. If $D(T) = X$ and T is continuous, then T' is a continuous linear operator defined on Y'_s into X'_s.

Proof. For any $y' \in Y'_s$, $\langle Tx, y' \rangle$ is a continuous linear functional of $x \in X$ and so there exists an $x' \in X'_s$ with $T'y' = x'$. Let B be a bounded set of X. Then, by the continuity of T, the image $T \cdot B = \{Tx; x \in B\}$ is a bounded set of Y. Thus, by the defining relation $\langle Tx, y' \rangle = \langle x, x' \rangle$, the convergence to 0 of y' in the bounded convergence topology of Y' (given in Chapter IV, 7) implies the convergence of x' in the bounded convergence topology of X'. Thus T' is a continuous linear operator on Y'_s into X'_s.

Example 1. Let $X = Y$ be n-dimensional euclidean spaces normed by the (l^2)-norm. For any continuous linear operator $T \in L(X, X)$, set

$$Tx = y, \quad \text{where} \quad x = (x_1, x_2, \ldots, x_n) \quad \text{and} \quad y = (y_1, y_2, \ldots, y_n).$$

Then $y_i = \sum_{j=1}^{n} t_{ij} x_j$ $(i = 1, 2, \ldots, n)$ and so, for $z = (z_1, z_2, \ldots, z_n)$,

$$\langle Tx, z \rangle = \langle y, z \rangle = \sum_j y_j z_j = \sum_i \Big(\sum_j t_{ij} x_j\Big) z_i = \sum_j x_j \Big(\sum_i t_{ij} z_i\Big).$$

Thus $T'z = w$ is given by $w_j = \sum_{i=1}^{n} t_{ij} z_i$ $(j = 1, 2, \ldots, n)$. This proves that the matrix corresponding to T' is the *transposed matrix* of the matrix corresponding to T.

Example 2. Let $X = Y$ be the real Hilbert space (l^2), and let $T_n \in L(X, X)$ be defined by

$$T_n(x_1, x_2, \ldots, x_k, \ldots) = (x_n, x_{n+1}, x_{n+2}, \ldots).$$

Then from

$$\langle T_n(x_1, x_2, \ldots), (z_1, z_2, \ldots) \rangle = x_n z_1 + x_{n+1} z_2 + x_{n+2} z_3 + \cdots,$$

we obtain

$$T'_n(z_1, z_2, \ldots) = (\overbrace{0, 0, \ldots, 0}^{n-1}, z_1, z_2, \ldots).$$

Since $\| T_n(x_1, x_2, \ldots) \| = \Big(\sum_{m=n}^{\infty} x_m^2\Big)^{1/2} \to 0$ as $n \to \infty$ and $\| T'_n(z_1, z_2, \ldots) \| = \| (z_1, z_2, \ldots) \|$, we have

Proposition 1. The mapping $T \to T'$ of $L(X, Y)$ into $L(Y'_s, X'_s)$ is not, in general, continuous in the simple convergence topology of operators, that is, $\lim_{n\to\infty} T_n x = Tx$ for all $x \in X$ does not necessarily imply $\lim_{n\to\infty} T'_n y' = T' y'$ for all $y' \in Y'$, in the strong dual topology of X'_s.

Theorem 2'. Let T be a bounded linear operator on a normed linear space X into a normed linear space Y. Then the dual operator T' is a bounded linear operator on Y'_s into X'_s such that

$$\|T\| = \|T'\|. \tag{3}$$

Proof. From the defining relation $\langle Tx, y'\rangle = \langle x, x'\rangle$, we obtain

$$\|T'y'\| = \|x'\| = \sup_{\|x\|\leq 1} |\langle x, x'\rangle|$$

$$= \sup_{\|x\|\leq 1} |\langle Tx, y'\rangle| \leqq \|y'\| \cdot \sup_{\|x\|\leq 1} \|Tx\| \leqq \|y'\| \cdot \|T\|,$$

and so $\|T'\| \leqq \|T\|$. The reverse inequality is proved as follows. For any $x_0 \in X$, there exists an $f_0 \in Y'$ such that $\|f_0\| = 1$ and $f_0(Tx_0) = \langle Tx_0, f_0\rangle = \|Tx_0\|$. Thus $f'_0 = T'f_0$ satisfies $\langle x_0, f'_0\rangle = \|Tx_0\|$ and so

$$\|Tx_0\| = \langle x_0, T'f_0\rangle \leqq \|T'\| \, \|f_0\| \cdot \|x_0\| = \|T'\| \cdot \|x_0\|, \text{ i.e.,}$$

$$\|T\| \leqq \|T'\|.$$

Theorem 3. i) If T and S are $\in L(X, Y)$, then $(\alpha T + \beta S)' = (\alpha T' + \beta S')$. ii) Let linear operators T, S be such that $D(T)$, $D(S)$, $R(T)$ and $R(S)$ are all contained in X. If S is $\in L(X, X)$ and $D(T)^a = X$, then

$$(ST)' = T'S'. \tag{4}$$

If, moreover, $D(TS)^a = X$, then

$$(TS)' \supseteqq S'T', \text{ i.e., } (TS)' \text{ is an extension of } S'T'. \tag{5}$$

Proof. i) is clear. ii) $D(ST) = D(T)$ is dense in X, and so $(ST)'$ exists. If $y \in D((ST)')$, then, for any $x \in D(T) = D(ST)$, $\langle Tx, S'y\rangle = \langle STx, y\rangle = \langle x, (ST)'y\rangle$. This shows that $S'y \in D(T')$ and $T'S'y = (ST)'y$, that is, $(ST)' \subseteqq T'S'$. Let, conversely, $y \in D(T'S')$, i.e., $S'y \in D(T')$. Then, for any $x \in D(T) = D(ST)$, $\langle STx, y\rangle = \langle Tx, S'y\rangle = \langle x, T'S'y\rangle$. This shows that $y \in D((ST)')$ and $(ST)'y = T'S'y$, that is, $T'S' \subseteqq (ST)'$. We have thus proved (4).

To prove (5), let $y \in D(S'T') = D(T')$. Then, for any $x \in D(TS)$, $\langle TSx, y\rangle = \langle Sx, T'y\rangle = \langle x, S'T'y\rangle$. This shows that $y \in D((TS)')$ and $(TS)'y = S'T'y$, that is, $S'T' \subseteqq (TS)'$.

2. Adjoint Operators

The notion of *transposed conjugate matrix* may be extended to the notion of *adjoint operator* through

Definition 1. Let X, Y be Hilbert spaces, and T a linear operator defined on $D(T) \subseteq X$ into Y. Let $D(T)^a = X$ and let T' be the dual operator of T. Thus $\langle Tx, y'\rangle = \langle x, T'y'\rangle$ for $x \in D(T)$, $y' \in D(T')$. If we denote by J_X the one-to-one norm-preserving conjugate linear correspondence $X'_s \ni f \leftrightarrow y_f \in X$ (defined in Corollary 1 in Chapter III, 6), then

$$\langle Tx, y'\rangle = y'(Tx) = (Tx, J_Y y'),\ \langle x, T'y'\rangle = (T'y')(x) = (x, J_X T'y').$$

We have thus

$$(Tx, J_Y y') = (x, J_X T'y'), \text{ that is, } (Tx, y) = (x, J_X T' J_Y^{-1} y).$$

In the special case when $Y = X$, we write

$$T^* = J_X T' J_X^{-1}$$

and call T^* the *adjoint operator* of T.

Remark. If X is the complex Hilbert space (l^2), we see, as in the Example in the preceding section, that the matrix corresponding to T^* is the transposed conjugate matrix of the matrix corresponding to T.

As in the case of dual operators, we can prove

Theorem 1. T^* exists iff $D(T)^a = X$. It is defined as follows: $y \in X$ is in the domain of $D(T^*)$ iff there exists a $y^* \in X$ such that

$$(Tx, y) = (x, y^*) \quad \text{holds for all} \quad x \in D(T), \tag{1}$$

and we define $T^* y = y^*$.

We can rewrite the above theorem in terms of the graph $G(A)$ of the linear operator A (the graph was introduced in Chapter II, 6):

Theorem 2. We introduce a continuous linear operator V on $X \times X$ into $X \times X$ by

$$V\{x, y\} = \{-y, x\}. \tag{2}$$

Then $(VG(T))^\perp$ is the graph of a linear operator iff $D(T)^a = X$, and, in fact, we have

$$G(T^*) = (VG(T))^\perp. \tag{3}$$

Proof. The condition $\{-Tx, x\} \perp \{y, y^*\}$ is equivalent to $(Tx, y) = (x, y^*)$. Thus Theorem 2 is proved by Theorem 1.

Corollary. T^* is a closed linear operator, since the orthogonal complement of a linear subspace is a closed linear subspace.

Theorem 3. Let T be a linear operator on $D(T) \subseteq X$ into X such that $D(T)^a = X$. Then T admits a closed linear extension iff $T^{**} = (T^*)^*$ exists, i.e., iff $D(T^*)^a = X$.

Proof. *Sufficiency*. We have $T^{**} \supseteq T$ by definition, and $T^{**} = (T^*)^*$ is closed by the above Corollary.

Necessity. Let S be a closed extension of T. Then $G(S)$ contains $G(T)^a$ as a closed linear subspace, and so $G(T)^a$ is the graph of a linear operator. But $G(T)^a = G(T)^{\perp\perp} = (G(T)^\perp)^\perp$ by the continuity of the scalar product, and, moreover, by $VG(T^*) = G(T)^\perp$, we obtain $(VG(T^*))^\perp = G(T)^{\perp\perp}$. Therefore, $(VG(T^*))^\perp$ is the graph of a linear operator. Thus by Theorem 2 $D(T^*)^a = X$ and T^{**} exists.

Corollary. Under the condition that $D(T)^a = X$, T is closed linear iff $T = T^{**}$.

Proof. The sufficiency is clear. Necessity is provėd by observing the formula $G(T)^a = G(T^{**})$ obtained above. For, $G(T) = G(T)^a$ implies that $T = T^{**}$.

Theorem 4. An everywhere defined closed linear operator T is a continuous linear operator.

Proof. Clear from the closed graph theorem.

Theorem 5. If T is a bounded linear operator, then T^* is also bounded linear and

$$\|T\| = \|T^*\|. \tag{4}$$

Proof. Similar to the case of dual operators.

3. Symmetric Operators and Self-adjoint Operators

A Hermitian matrix is a matrix which is equal to its transposed conjugate matrix. It is known that such a matrix can be transformed into a diagonal matrix by a suitable (complex) rotation of the vector space on which the matrix operates as a linear operator. The notion of the Hermitian matrices is extended to the notion of self-adjoint operators in a Hilbert space.

Definition 1. Let X be a Hilbert space. A linear operator on $D(T) \subseteq X$ into X is called *symmetric* if $T^* \supseteq T$, i.e., if T^* is an extension of T. Note that the condition of the existence of T^* implies that $D(T)^a = X$.

Proposition 1. If T is symmetric, then T^{**} is also symmetric.

Proof. Since T is symmetric, we have $D(T^*) \supseteq D(T)$ and $D(T)^a = X$. Hence $D(T^*)^a = X$ and so $T^{**} = (T^*)^*$ exists. T^{**} is surely an extension of T and so $D(T^{**}) \supseteq D(T)$. Thus, again by $D(T)^a = X$, we have $D(T^{**})^a = X$ and so $T^{***} = (T^{**})^*$ exists. We have, from $T^* \supseteq T$, $T^{**} \subseteq T^*$ and hence $T^{***} \supseteq T^{**}$ which proves that T^{**} is symmetric.

Corollary. A symmetric operator T has a closed symmetric extension $T^{**} = (T^*)^*$.

Definition 2. A linear operator T is called *self-adjoint* if $T = T^*$.

Proposition 2. A self-adjoint operator is closed. An everywhere defined symmetric operator is bounded and self-adjoint.

Proof. Being the adjoint of itself, a self-adjoint operator is closed. The last assertion is proved by the fact that an everywhere defined closed operator is bounded (closed graph theorem).

Example 1 (integral operator of the Hilbert-Schmidt type). Let $-\infty \leqq a < b \leqq \infty$ and consider $L^2(a, b)$. Let $K(s, t)$ be a complex-valued measurable function for $a \leqq s, t \leqq b$ such that

$$\int_a^b \int_a^b |K(s,t)|^2 \, ds \, dt < \infty.$$

For any $x(t) \in L^2(a, b)$, we define the operator K by

$$(K \cdot x)(s) = \int_a^b K(s,t)\, x(t)\, dt.$$

We have, by Schwarz' inequality and the Fubini-Tonelli theorem,

$$\int_a^b |(K \cdot x)(s)|^2 \, ds \leqq \int_a^b \int_a^b |K(s,t)|^2 \, dt \, ds \int_a^b |x(t)|^2 \, dt.$$

Hence K is a bounded linear operator on $L^2(a, b)$ into $L^2(a, b)$ such that $\|K\| \leqq \left(\int_a^b \int_a^b |K(s,t)|^2 \, ds \, dt\right)^{1/2}$. It is easy to see that the operator K^* is defined by $(K^* y)(t) = \int_a^b \overline{K(s,t)}\, y(s)\, ds$. Hence K is self-adjoint iff $K(s,t) = \overline{K(t,s)}$ for a.e. s, t.

Example 2 (the coordinate operator in quantum mechanics). Let $X = L^2(-\infty, \infty)$. Let $D = \{x(t);\, x(t) \text{ and } t \cdot x(t) \text{ both} \in L^2(-\infty, \infty)\}$. Then the operator T defined by $Tx(t) = t \cdot x(t)$ on D is self-adjoint.

Proof. It is clear that $D^a = X$, since the linear combinations of defining functions of finite intervals are strongly dense in $L^2(-\infty, \infty)$. Let $y \in D(T^*)$ and set $T^* y = y^*$. Then, for all $x \in D = D(T)$,

$$\int_{-\infty}^{\infty} t x(t) \overline{y(t)} \, dt = \int_{-\infty}^{\infty} x(t) \overline{y^*(t)} \, dt.$$

If we take for $x(t)$ the defining function of the interval $[\alpha, t_0]$, we have $\int_\alpha^{t_0} t \cdot \overline{y(t)} \, dt = \int_\alpha^{t_0} \overline{y^*(t)} \, dt$, and hence, by differentiation, $t_0 \cdot \overline{y(t_0)} = \overline{y^*(t_0)}$ for a.e. t_0. Thus $y \in D$ and $T^* y(t) = t \cdot y(t)$. Conversely, it is clear that $y \in D$ implies that $y \in D(T^*)$ and $T^* y(t) = t \cdot y(t)$.

Example 3 (the momentum operator in quantum mechanics). Let $X = L^2(-\infty, \infty)$. Let D be the totality of $x(t) \in L^2(-\infty, \infty)$ such that $x(t)$ is absolutely continuous on every finite interval with the derivative $x'(t) \in L^2(-\infty, \infty)$. Then the operator T defined by $Tx(t) = i^{-1} x'(t)$ on D is self-adjoint.

Proof. Let a continuous function $x_n(t)$ be defined by

$x_n(t) = 1$ for $t \in [\alpha, t_0]$,

$x_n(t) = 0$ for $t \leqq \alpha - n^{-1}$ and for $t \geqq t_0 + n^{-1}$,

$x_n(t)$ is a linear function on $[\alpha - n^{-1}, \alpha]$ and on $[t_0, t_0 + n^{-1}]$.

Then the linear combinations of functions of the form $x_n(t)$ with different values of α, t_0 and n are dense in $L^2(-\infty, \infty)$. Thus D is dense in X.

Let $y \in D(T^*)$ and $T^*y = y^*$. Then for any $x \in D$,

$$\int_{-\infty}^{\infty} i^{-1} x'(t) \overline{y(t)}\, dt = \int_{-\infty}^{\infty} x(t) \overline{y^*(t)}\, dt.$$

If we take $x_n(t)$ for $x(t)$, we obtain

$$n \int_{\alpha-n^{-1}}^{\alpha} i^{-1} \overline{y(t)}\, dt - n \int_{t_0}^{t_0+n^{-1}} i^{-1} \overline{y(t)}\, dt = \int_{-\infty}^{\infty} x_n(t)\, y^*(t)\, dt,$$

and so, by letting $n \to \infty$, we obtain $i^{-1}(\overline{y(\alpha)} - \overline{y(t_0)}) = \int_{\alpha}^{t_0} \overline{y^*(t)}\, dt$ for a.e. α and t_0. It is clear, by Schwarz' inequality, that $y^*(t)$ is integrable over any finite interval. Thus $y(t_0)$ is absolutely continuous in t_0 over any finite interval, and so we have $i^{-1}y'(t_0) = y^*(t_0)$ for a.e. t_0. Hence $y \in D$ and $T^*y(t) = i^{-1}y'(t)$. Let, conversely, $y \in D$. Then, by partial integration,

$$\int_a^b i^{-1} x'(t) \overline{y(t)}\, dt = i^{-1} [x(t) \overline{y(t)}]_a^b + \int_a^b x(t) \overline{(i^{-1}y'(t))}\, dt.$$

By the integrability of $x(t) \overline{y(t)}$ over $(-\infty, \infty)$, we see that $\lim_{a\downarrow-\infty, b\uparrow\infty} |[x(t)\, y(t)]_a^b| = 0$, and so $\int_{-\infty}^{\infty} i^{-1}x'(t) \overline{y(t)}\, dt = \int_{-\infty}^{\infty} x(t) \overline{(i^{-1}y'(t))}\, dt.$ Thus $y \in D(T^*)$ and $T^*y(t) = i^{-1}y'(t)$.

Theorem 1. If a self-adjoint operator T admits the inverse T^{-1}, then T^{-1} is also self-adjoint.

Proof. $T = T^*$ is equivalent to $(VG(T))^\perp = G(T)$. We have also $G(T^{-1}) = VG(-T)$. Hence, by $(-T)^* = -T^* = -T$, $(VG(-T))^\perp = G(-T)$ and so

$$(VG(T^{-1}))^\perp = G(-T)^\perp = (VG(-T))^{\perp\perp} = VG(-T) = G(T^{-1}),$$
$$\text{that is,} \quad (T^{-1})^* = T^{-1}.$$

We have used, in the above proof, the fact that $(VG(-T))^a = VG(-T)$ in virtue of the closedness of $(-T)$.

Corollary. A symmetric operator T in a Hilbert space X is self-adjoint if $D(T) = X$ or if $R(T) = X$.

Proof. The case $D(T) = X$ was proved already. We shall prove the case $R(T) = X$. $Tx = 0$ implies $0 = (Tx, y) = (x, Ty)$ for all $y \in D(T)$, and so, by $R(T) = X$, we must have $x = 0$. Therefore the inverse T^{-1} exists which is surely symmetric with T. $D(T^{-1}) = R(T) = X$, and so the everywhere defined symmetric operator T^{-1} must be self-adjoint. Hence $T = (T^{-1})^{-1}$ is self-adjoint by Theorem 1.

We can construct self-adjoint operators from a closed linear operator. More precisely, we have

Theorem 2 (J. VON NEUMANN [5]). For any closed linear operator T in a Hilbert space X such that $D(T)^a = X$, the operators T^*T and TT^* are self-adjoint, and $(I + T^*T)$ and $(I + TT^*)$ both admit bounded linear inverses.

Proof. We know that, in the product space $X\times X$, $G(T)$ and $VG(T^*)$ are closed linear subspaces orthogonal to each other and spanning the whole product space $X\times X$. Hence, for any $h\in X$, we have the uniquely determined decomposition

$$\{h, 0\} = \{x, Tx\} + \{-T^*y, y\} \quad \text{with} \quad x\in D(T),\ y\in D(T^*). \tag{1}$$

Thus $h = x - T^*y$, $0 = Tx + y$. Therefore

$$x\in D(T^*T) \quad \text{and} \quad x + T^*Tx = h. \tag{2}$$

Because of the uniqueness of decomposition (1), x is uniquely determined by h, and so the everywhere defined inverse $(I + T^*T)^{-1}$ exists.

For any $h, k\in X$, let

$$x = (I + T^*T)^{-1}h,\ y = (I + T^*T)^{-1}k.$$

Then x and $y\in D(T^*T)$ and, by the closedness of T, $(T^*)^* = T$. Hence

$$\begin{aligned}(h, (I + T^*T)^{-1}k) &= ((I + T^*T)\,x, y) = (x, y) + (T^*Tx, y)\\ &= (x, y) + (Tx, Ty) = (x, y) + (x, T^*Ty)\\ &= (x, (I + T^*T)\,y) = ((I + T^*T)^{-1}\,h, k),\end{aligned}$$

which proves that the operator $(I + T^*T)^{-1}$ is self-adjoint. As an everywhere defined self-adjoint operator, $(I + T^*T)^{-1}$ is a bounded operator. By Theorem 1, its inverse $(I + T^*T)$ and hence T^*T are self-adjoint.

Since T is closed, we have $(T^*)^* = T$, and so, by what was proved above, $TT^* = (T^*)^*T^*$ is self-adjoint and $(I + TT^*)$ has a bounded linear inverse.

We next give an example of a non-self-adjoint, symmetric operator:

Example 4. Let $X = L^2(0, 1)$. Let D be the totality of absolutely continuous functions $x(t)\in L^2(0, 1)$ such that $x(0) = x(1) = 0$ and $x'(t)\in L^2(0, 1)$. Then the operator T_1 defined by $T_1x(t) = i^{-1}x'(t)$ on $D = D(T_1)$ is symmetric but not self-adjoint.

Proof. We shall prove that $T_1^* = T_2$, where T_2 is defined by:

$T_2x(t) = i^{-1}x'(t)$ on $D(T_2) = \{x(t)\in L^2(0, 1);\ x(t)$ is absolutely continuous such that $x'(t)\in L^2(0, 1)\}$.

Since $D = D(T_1)$ is dense in $L^2(0, 1)$, the operator T_1^* is defined. Let $y\in D(T_1^*)$ and set $T_1^*y = y^*$. Then, for any $x\in D = D(T_1)$,

$$\int_0^1 i^{-1}x'(t)\,\overline{y(t)}\,dt = \int_0^1 x(t)\,\overline{y^*(t)}\,dt.$$

By partial integration, we obtain, remembering $x(0) = x(1) = 0$,

$$\int_0^1 x(t)\,\overline{y^*(t)}\,dt = -\int_0^1 x'(t)\,\overline{Y^*(t)}\,dt, \text{ where } Y^*(t) = \int_0^t y^*(s)\,ds.$$

Hence, by $x(1) = \int_0^1 x'(t)\,dt = 0$, we have, for any constant c,

$$\int_0^1 x'(t)\,\overline{(Y^*(t) - i^{-1}\,y(t) - c)}\,dt = 0 \quad \text{for all} \quad x \in D(T_1).$$

On the other hand, for any $z(t) \in L^2(0, 1)$, the function $Z(t) = \int_0^t z(t)\,dt - t\int_0^1 z(t)\,dt$ surely belongs to $D(T_1)$. Hence, taking $Z(t)$ for the above $x(t)$, we obtain

$$\int_0^1 \left\{z(t) - \int_0^1 z(t)\,dt\right\} \cdot \overline{(Y^*(t) - i^{-1}\,y(t) - c)}\,dt = 0.$$

If we take the constant c in such a way that $\int_0^1 (Y^*(t) - i^{-1}\,y(t) - c)\,dt = 0$, then

$$\int_0^1 z(t)\,\overline{(Y^*(t) - i^{-1}\,y(t) - c)}\,dt = 0,$$

and so, by the arbitrariness of $z \in L^2(0, 1)$, we must have $Y^*(t) = \int_0^t y^*(t)\,dt = i^{-1}\,y(t) + c$. Hence $y \in D(T_2)$ and $T_2 y = y^*$. This proves that $T_1^* \subseteq T_2$. It is also clear, by partial integration, that $T_2 \subseteq T_1^*$ and so $T_2 = T_1^*$.

Theorem 3. If H is a bounded self-adjoint operator, then

$$\|H\| = \sup_{\|x\| \leq 1} |(Hx, x)|. \tag{3}$$

Proof. Set $\sup_{\|x\| \leq 1} |(Hx, x)| = \gamma$. Then, by $|(Hx, x)| \leq \|Hx\|\,\|x\|$, $\gamma \leq \|H\|$. For any real number λ, we have

$$\begin{aligned} |(H(y \pm \lambda z), y \pm \lambda z)| &= |(Hy, y) \pm 2\lambda\, Re(Hy, z) + \lambda^2 (Hz, z)| \\ &\leq \gamma\,\|y \pm \lambda z\|^2. \end{aligned}$$

Hence

$$|4\lambda\, Re(Hy, z)| \leq \gamma(\|y + \lambda z\|^2 + \|y - \lambda z\|^2) = 2\gamma(\|y\|^2 + \lambda^2\,\|z\|^2).$$

By taking $\lambda = \|y\|/\|z\|$, we obtain $|Re(Hy, z)| \leq \gamma\,\|y\|\,\|z\|$. Hence, by substituting $ze^{i\theta}$ for z, we obtain $|(Hy, z)| \leq \gamma\,\|y\|\,\|z\|$, and so

$$(Hy, Hy) = \|Hy\|^2 \leq \gamma\,\|y\|\,\|Hy\|, \text{ i.e., } \|H\| \leq \gamma.$$

4. Unitary Operators. The Cayley Transform

A symmetric operator is not necessarily a bounded operator. Various investigations of a symmetric operator H may be made through the continuous operator $(H - iI)(H + iI)^{-1}$ called the *Cayley transform* of H. We shall begin with the notion of isometric operators.

Definition 1. A bounded linear operator T on a Hilbert space X into X is called (bounded) *isometric* if T leaves the scalar product invariant:

$$(Tx, Ty) = (x, y) \quad \text{for all} \quad x, y \in X. \tag{1}$$

If, in particular, $R(T) = X$, then a (bounded) isometric operator T is called a *unitary* operator.

Proposition 1. For a bounded linear operator T, condition (1) is equivalent to the condition of the isometry

$$\|Tx\| = \|x\| \quad \text{for all} \quad x \in X. \tag{2}$$

Proof. It is clear that (1) implies (2). We have, by (2),

$$\begin{aligned} 4\,Re(Tx, Ty) &= \|T(x+y)\|^2 - \|T(x-y)\|^2 \\ &= \|x+y\|^2 - \|x-y\|^2 = 4\,Re(x, y). \end{aligned}$$

By taking iy in place of y, we also obtain $4\,Im(Tx, Ty) = 4\,Im(x, y)$, and so (2) implies (1).

Proposition 2. A bounded linear operator on a Hilbert space X into X is unitary iff $T^* = T^{-1}$.

Proof. If T is unitary, then T^{-1} surely exists in virtue of condition (2), and $D(T^{-1}) = R(T) = X$. Moreover, by (1), $T^*T = I$ and so $T^* = T^{-1}$. Conversely, the condition $T^* = T^{-1}$ implies $T^*T = I$ which is the condition of the invariance of the scalar product. Moreover, $T^* = T^{-1}$ implies that $R(T) = D(T^{-1}) = D(T^*) = X$ and hence we see that T must be unitary.

Example 1. Let $X = L^2(-\infty, \infty)$. Then, for any real number a, the operator T defined by $Tx(t) = x(t + a)$ on $L^2(-\infty, \infty)$ is unitary.

Example 2. The Fourier transform on $L^2(R^n)$ onto $L^2(R^n)$ is unitary, since it leaves the scalar product $(f, g) = \int_{R^n} f(x)\,\overline{g(x)}\,dx$ invariant.

Definition 2. Let X be a Hilbert space. A linear operator T defined on $D(T) \subseteq X$ into X such that $D(T)^a = X$ is called *normal* if

$$TT^* = T^*T. \tag{3}$$

Self-adjoint operators and unitary operators are normal.

The Cayley Transform

Theorem 1 (J. von Neumann [1]). Let H be a closed symmetric operator in a Hilbert space X. Then the continuous (but not necessarily everywhere defined) inverse $(H + iI)^{-1}$ exists, and the operator

$$U_H = (H - iI)(H + iI)^{-1} \text{ with the domain } D(U_H) = D((H + iI)^{-1}) \quad (4)$$

is closed *isometric* ($\|U_H x\| = \|x\|$), and $(I - U_H)^{-1}$ exists.

We have, moreover,

$$H = i(I + U_H)(I - U_H)^{-1}. \quad (5)$$

Thus, in particular, $D(H) = R(I - U_H)$ is dense in X.

Definition 3. U_H is called the *Cayley transform* of H.

Proof of Theorem 1. We have, for any $x \in D(H)$,

$$((H \pm iI)x, (H \pm iI)x) = (Hx, Hx) \pm (Hx, ix) \pm (ix, Hx) + (x, x).$$

The symmetry condition for H implies $(Hx, ix) = -i(Hx, x) = -i(x, Hx) = -(ix, Hx)$ and so

$$\|(H \pm iI)x\|^2 = \|Hx\|^2 + \|x\|^2. \quad (6)$$

Hence $(H + iI)x = 0$ implies $x = 0$ and so the inverse $(H + iI)^{-1}$ exists. Since $\|(H + iI)x\| \geqq \|x\|$, the inverse $(H + iI)^{-1}$ is continuous. By (6), it is clear that $\|U_H y\| = \|y\|$, i.e., U_H is isometric.

U_H is closed. For, let $(H + iI)x_n = y_n \to y$ and $(H - iI)x_n = z_n \to z$ as $n \to \infty$. Then we have, by (6), $\|y_n - y_m\|^2 = \|H(x_n - x_m)\|^2 + \|x_n - x_m\|^2$, and so $(x_n - x_m) \to 0$, $H(x_n - x_m) \to 0$ as $n, m \to \infty$. Since H is closed, we must have $x = \underset{n\to\infty}{\text{s-lim}}\, x_n \in D(H)$ and $\underset{n\to\infty}{\text{s-lim}}\, Hx_n = Hx$. Thus $(H + iI)x_n \to y = (H + iI)x$, $(H - iI)x_n \to z = (H - iI)x$ and so $U_H y = z$. This proves that U_H is closed.

From $y = (H + iI)x$ and $U_H y = (H - iI)x$, we obtain $2^{-1}(I - U_H)y = ix$ and $2^{-1}(I + U_H)y = Hx$. Thus $(I - U_H)y = 0$ implies $x = 0$ and so $(I + U_H)y = 2Hx = 0$ which implies $y = 2^{-1}((I - U_H)y + (I + U_H)y) = 0$. Therefore the inverse $(I - U_H)^{-1}$ exists. By the same calculation as above, we obtain

$$Hx = 2^{-1}(I + U_H)y = i(I + U_H)(I - U_H)^{-1}x, \quad \text{that is,}$$
$$H = i(I + U_H)(I - U_H)^{-1}.$$

Theorem 2 (J. von Neumann [1]). Let U be a closed isometric operator such that $R(I - U)^a = X$. Then there exists a uniquely determined closed symmetric operator H whose Cayley transform is U.

Proof. We first show that the inverse $(I - U)^{-1}$ exists. Suppose that $(I - U)y = 0$. For any $z = (I - U)w \in R(I - U)$, we have, by the isometric property of U, $(y, w) = (Uy, Uw)$ as in Section 1. Hence

$$(y, z) = (y, w) - (y, Uw) = (Uy, Uw) - (y, Uw) = (Uy - y, Uw) = 0.$$

Hence, by the condition $R(I - U)^a = X$, y must be $= 0$. Thus $(I - U)^{-1}$ exists. Put $H = i(I + U)(I - U)^{-1}$. Then $D(H) = D((I - U)^{-1}) =$

$R(I-U)$ is dense in X. We first prove that H is symmetric. Let $x, y \in D(H) = R(I-U)$ and put $x = (I-U)u$, $y = (I-U)w$. Then $(Uu, Uw) = (u, w)$ implies that

$$(Hx, y) = (i(I+U)u, (I-U)w) = i((Uu, w) - (u, Uw))$$
$$= ((I-U)u, i(I+U)w) = (x, Hy).$$

The proof of $U_H = U$ is obtained as follows. For $x = (I-U)u$, we have $Hx = i(I+U)u$ and so $(H+iI)x = 2iu$, $(H-iI)x = 2iUu$. Thus $D(U_H) = \{2iu; u \in D(U)\} = D(U)$, and $U_H(2iu) = 2i \cdot Uu = U(2iu)$. Hence $U_H = U$.

To complete the proof of Theorem 2, we show that H is a closed operator. In fact, H is the operator which maps $(I-U)u$ onto $i(I+U)u$. If $(I-U)u_n$ and $i(I+U)u_n$ both converge as $n \to \infty$, then u_n and Uu_n both converge as $n \to \infty$. Hence by the closure property of U, we must have

$$u_n \to u,\ (I-U)u_n \to (I-U)u,\ i(I+U)u_n \to i(I+U)u.$$

This proves that H is a closed operator.

For the structure of the adjoint operator of a symmetric operator, we have

Theorem 3 (J. von Neumann [1]). *Let H be a closed symmetric operator in a Hilbert space X. For the Cayley transform $U_H = (H-iI)(H+iI)^{-1}$ of H, we set*

$$X_H^+ = D(U_H)^\perp,\ X_H^- = R(U_H)^\perp. \tag{7}$$

Then we have

$$X_H^+ = \{x \in X; H^*x = ix\},\ X_H^- = \{x \in X; H^*x = -ix\}, \tag{8}$$

and the element x of $D(H^*)$ is uniquely expressed as

$$x = x_0 + x_1 + x_2, \text{ where } x_0 \in D(H), x_1 \in X_H^+, x_2 \in X_H^- \text{ so that}$$
$$H^*x = Hx_0 + ix_1 + (-ix_2). \tag{9}$$

Proof. $x \in D(U_H)^\perp = D((H+iI)^{-1})^\perp$ implies $(x, (H+iI)y) = 0$ for all $y \in D(H)$. Hence $(x, Hy) = -(x, iy) = (ix, y)$ and so $x \in D(H^*)$, $H^*x = ix$. The last condition implies $(x, (H+iI)y) = 0$ for all $y \in D(H)$, i.e. $x \in D((H+iI)^{-1})^\perp = D(U_H)^\perp$. This proves the first half of (8); the latter half may be proved similarly.

Since U_H is a closed isometric operator, we see that $D(U_H)$ and $R(U_H)$ are closed linear subspaces of X. Hence any element $x \in X$ is uniquely decomposed as the sum of an element of $D(U_H)$ and an element of $D(U_H)^\perp$. If we apply this orthogonal decomposition to the element $(H^* + iI)x$, we obtain

$$(H^* + iI)x = (H+iI)x_0 + x' \quad \text{where} \quad x_0 \in D(H),\ x' \in D(U_H)^\perp.$$

But we have $(H + iI)\, x_0 = (H^* + iI)\, x_0$ by $x_0 \in D(H)$ and $H \subseteqq H^*$. We have also $H^* x' = ix'$ by $x' \in D(U_H)^{\perp}$ and (8). Thus

$$x' = (H^* + iI)\, x_1,\ x_1 = (2i)^{-1}\, x' \in D(U_H)^{\perp},$$

and so

$$(H^* + iI)\, x = (H^* + iI)\, x_0 + (H^* + iI)\, x_1 \quad \text{where}$$
$$x_0 \in D(H),\ x_1 \in D(U_H)^{\perp}$$

Therefore $(x - x_0 - x_1) \in R(U_H)^{\perp}$ by $H^*(x - x_0 - x_1) = -i(x - x_0 - x_1)$ and (8). This proves (9). The uniqueness of the representation (9) is proved as follows. Let $0 = x_0 + x_1 + x_2$ with $x_0 \in D(H)$, $x_1 \in D(U_H)^{\perp}$, $x_2 \in R(U_H)^{\perp}$. Then, by $H^* x_0 = H x_0$, $H^* x_1 = i x_1$, $H^* x_2 = -i x_2$,

$$0 = (H^* + iI)\, 0 = (H^* + iI)\, (x_0 + x_1 + x_2) = (H + iI)\, x_0 + 2i x_1.$$

But by the uniqueness of the orthogonal decomposition of X as the sum of $D(U_H)$ and $D(U_H)^{\perp}$, we obtain $(H + iI)\, x_0 = 0$, $2ix_1 = 0$. Since the inverse $(H + iI)^{-1}$ exists, we must have $x_0 = 0$ and so $x_2 = 0 - x_0 - x_1 = 0 - 0 - 0 = 0$.

Corollary. A closed symmetric operator H in a Hilbert space X is self-adjoint iff its Cayley transform U_H is unitary.

Proof. The condition $D(H) = D(H^*)$ is equivalent to the condition $D(U_H)^{\perp} = R(U_H)^{\perp} = \{0\}$. The last condition in turn is equivalent to the condition that U_H is unitary, i.e. the condition that U_H maps X onto X one-one and isometrically.

5. The Closed Range Theorem

The *closed range theorem* of S. Banach [1] reads as follows.

Theorem. Let X and Y be B-spaces, and T a closed linear operator defined in X into Y such that $D(T)^a = X$. Then the following propositions are all equivalent:

$$R(T) \text{ is closed in } Y, \tag{1}$$

$$R(T') \text{ is closed in } X', \tag{2}$$

$$R(T) = N(T')^{\perp} = \{y \in Y;\ \langle y, y^* \rangle = 0 \quad \text{for all} \quad y^* \in N(T')\}, \tag{3}$$

$$R(T') = N(T)^{\perp} = \{x^* \in X';\ \langle x, x^* \rangle = 0 \quad \text{for all} \quad x \in N(T)\}. \tag{4}$$

Proof. The proof of this theorem requires five steps.

The first step. The proof of the equivalence (1) $\leftrightarrow$ (2) is reduced to the equivalence (1) $\leftrightarrow$ (2) for the special case when T is a continuous linear operator such that $D(T) = X$.

The graph $G = G(T)$ of T is a closed linear subspace of $X \times Y$, and so G is a B-space by the norm $\|\{x, y\}\| = \|x\| + \|y\|$ of $X \times Y$. Consider a continuous linear operator S on G into Y:

$$S\{x, Tx\} = Tx.$$

Then the dual operator S' of S is a continuous linear operator on Y' into G', and we have

$$\begin{aligned}\langle\{x, Tx\}, S'y^*\rangle &= \langle S\{x, Tx\}, y^*\rangle = \langle Tx, y^*\rangle \\ &= \langle\{x, Tx\}, \{0, y^*\}\rangle,\ x \in D(T),\ y^* \in Y'.\end{aligned}$$

Thus the functional $S' \cdot y^* - \{0, y^*\} \in (X \times Y)' = X' \times Y'$ vanishes at every point of G. But, the condition $\langle\{x, Tx\}, \{x^*, y_1^*\}\rangle = 0$, $x \in D(T)$, is equivalent to the condition $\langle x, x^*\rangle = \langle -Tx, y_1^*\rangle$, $x \in D(T)$, that is, to the condition $-T'y_1^* = x^*$. Hence

$$S' \cdot y^* = \{0, y^*\} + \{-T'y_1^*, y_1^*\} = \{-T'y_1^*, y^* + y_1^*\},\ y^* \in Y'.$$

By the arbitrariness of y^*, we see that $R(S') = R(-T') \times Y' = R(T') \times Y'$.

Therefore $R(S')$ is closed in $X' \times Y'$ iff $R(T')$ is closed in X', and, since $R(S) = R(T)$, $R(S)$ is closed in Y iff $R(T)$ is closed in Y. Hence we have only to prove the equivalence (1) $\leftrightarrow$ (2) in the special case of a bounded linear operator S, instead of the original T.

The second step. Let X and Y be B-spaces, and T a bounded linear operator on X into Y. Then (1) $\rightarrow$ (2).

We consider T as a bounded linear operator T_1 on X into the B-space $Y_1 = R(T)^a = R(T)$. We have to prove that (2) is true. $T_1'y_1^*$, $y_1^* \in Y_1'$, is defined by

$$\langle T_1 x, y_1^*\rangle = \langle Tx, y_1^*\rangle = \langle x, T_1'y_1^*\rangle,\ x \in D(T_1) = D(T) = X.$$

By the Hahn-Banach theorem, the functional y_1^* can be extended to a $y^* \in Y'$ in such a way that $\langle Tx, y_1^*\rangle = \langle Tx, y^*\rangle$, $x \in D(T) = X$. Hence $T_1'y_1^* = T'y_1^*$ and so $R(T_1') = R(T')$. Thus it suffices to assume that $R(T) = Y$. Then, by the open mapping theorem in Chapter II, 5, there exists a $c > 0$ such that for each $y \in Y$, there exists an $x \in X$ with $Tx = y$, $\|x\| \leqq c\,\|y\|$. Thus, for each y^* in $D(T')$, we have

$$\begin{aligned}|\langle y, y^*\rangle| &= |\langle Tx, y^*\rangle| = |\langle x, T'y^*\rangle| \\ &\leqq \|x\| \cdot \|T'y^*\| \leqq c\,\|y\| \cdot \|T'y^*\|.\end{aligned}$$

Hence

$$\|y^*\| = \sup_{\|y\| \leqq 1} |\langle y, y^*\rangle| \leqq c\,\|T'y^*\|$$

and so $(T')^{-1}$ exists and is continuous. Moreover, $(T')^{-1}$ is a closed linear operator as the inverse of a bounded linear operator. Hence we see that the domain $D((T')^{-1}) = R(T')$ must be closed in X'.

The third step. Let X and Y be B-spaces, and T a bounded linear operator on X into Y. Then (2) $\rightarrow$ (1).

As in the second step, we consider T as a bounded linear operator T_1 on X into $Y_1 = R(T)^a$. Then T_1' has the inverse, since $T_1'y_1^* = 0$ implies

$$\langle T_1 x, y_1^*\rangle = \langle Tx, y_1^*\rangle = \langle x, T_1'y_1^*\rangle = 0,\ x \in D(T_1) = D(T) = X,$$

and so, since $R(T_1) = R(T)$ is dense in $Y_1 = R(T)^a$, y_1^* must be 0. Therefore, the condition that $R(T') = R(T_1')$ (proved above) is closed, implies that T_1' is a continuous linear operator on the B-space $(R(T)^a)' = Y_1'$ onto the B-space $R(T_1')$ in a one-one way. Hence, by the open mapping theorem, $(T_1')^{-1}$ is continuous.

We then prove that $R(T)$ is closed. To this purpose, it suffices to derive a contradiction from the condition

$$\begin{cases} \text{there exists a positive constant } \varepsilon \text{ such that the image} \\ \{T_1 x;\ \|x\| \leqq \varepsilon\} \text{ is not dense in all the spheres } \|y\| \leqq n^{-1} \\ (n = 1, 2, \ldots) \text{ of } Y_1 = R(T)^a = R(T_1)^a. \end{cases}$$

For, if otherwise, the proof of the open mapping theorem shows that $R(T_1) = R(T) = Y_1$. Thus we assume that there exists a sequence $\{y_n\} \subseteqq Y_1$ with

$$\operatorname*{s-lim}_{n\to\infty} y_n = 0,\ y_n \bar{\in} \{T_1 x;\ \|x\| \leqq \varepsilon\}^a \quad (n = 1, 2, \ldots).$$

Since $\{T_1 x;\ \|x\| \leqq \varepsilon\}^a$ is a closed convex, balanced set of the B-space Y_1 there exists, by Mazur's theorem in Chapter IV, 6, a continuous linear functional f_n on the B-space Y_1 such that

$$f_n(y_n) > \sup_{\|x\| \leqq \varepsilon} |f_n(T_1 x)| \quad (n = 1, 2, \ldots).$$

Hence $\|T_1' f_n\| < \varepsilon^{-1} \|f_n\| \, \|y_n\|$, and so, by $\operatorname*{s-lim}_{n\to\infty} y_n = 0$, T_1' does not have a continuous inverse. This is a contradiction, and so $R(T)$ must be closed.

The fourth step. We prove (1) → (3). First, it is clear, from

$$\langle Tx, y^*\rangle = \langle x, T'y^*\rangle,\ x \in D(T),\ y^* \in D(T'),$$

that $R(T) \subseteqq N(T')^\perp$. We show that (1) implies $N(T')^\perp \subseteqq R(T)$. Assume that there exists a $y_0 \in N(T')^\perp$ with $y_0 \bar{\in} R(T)$. Then, by the Hahn-Banach theorem, there exists a $y_0^* \in Y'$ such that $\langle y_0, y_0^*\rangle \neq 0$ and $\langle Tx, y_0^*\rangle = 0$ for all $x \in D(T)$. The latter condition implies $\langle x, T'y_0^*\rangle = 0$, $x \in D(T)$, and hence $T'y_0^* = 0$, i.e., $y_0 \bar{\in} N(T')^\perp$. This is a contradiction and so we must have $N(T')^\perp \subseteqq R(T)$.

The implication (3) → (1) is clear, since $N(T')^\perp$ is closed by virtue of the continuity in y of $\langle y, y^*\rangle$.

The fifth step. We prove (2) → (4). The inclusion $R(T') \subseteqq N(T)^\perp$ is clear as in the case of (3). We show that (2) implies that $N(T)^\perp \subseteqq R(T')$. To this purpose, let $x^* \in N(T)^\perp$, and define, for $y = Tx$, the functional $f_1(y)$ of y through $f_1(y) = \langle x, x^*\rangle$. It is a one-valued function of y, since $Tx = Tx'$ implies $(x - x') \in N(T)$ and so, by $x^* \in N(T)^\perp$, $\langle (x - x'), x^*\rangle = 0$. Thus $f_1(y)$ is a linear functional of y. (2) implies (1), and so, by the open mapping theorem applied to the operator S in the first step, we may choose the solution x of the equation $y = Tx$ in such a way that s-lim

$y = 0$ implies $s\text{-}\lim x = 0$. Hence $f_1(y) = \langle x, x^* \rangle$ is a continuous linear functional on $Y_1 = R(T)$. Let $f \in Y'$ be an extension of f_1. Then

$$f(Tx) = f_1(Tx) = \langle x, x^* \rangle .$$

This proves that $T'f = x^*$. Hence $N(T)^{\perp} \subseteq R(T')$.

That (4) implies (2) is clear, since $\langle x, x^* \rangle$ is a continuous linear functional of x.

Corollary 1. Let X and Y be B-spaces, and T a closed linear operator on $D(T) \subseteq X$ into Y such that $D(T)^a = X$. Then

$$R(T) = Y \quad \text{iff} \quad T' \quad \text{has a continuous inverse}, \tag{5}$$

$$R(T') = X' \quad \text{iff} \quad T \quad \text{has a continuous inverse}. \tag{6}$$

Proof. Suppose that $R(T) = Y$. Then, from $\langle Tx, y^* \rangle = \langle x, T'y^* \rangle$, $x \in D(T)$ and $T'y^* = 0$, we obtain $y^* = 0$; that is, T' must have the inverse $(T')^{-1}$. Since, by $R(T) = Y$ and (2), $R(T')$ is closed, the closed graph theorem implies that $(T')^{-1}$ is continuous. Next let T' admit a continuous inverse. Then $N(T') = \{0\}$ and also (2) holds since T' is closed. Thus, by (3), $R(T) = Y$.

Suppose that $R(T') = X'$. Then, from $\langle Tx, y^* \rangle = \langle x, T'y^* \rangle$, $y^* \in D(T')$ and $Tx = 0$, we obtain $x = 0$, i.e., T must have the inverse T^{-1}. Since, by $R(T') = X'$ and (1), $R(T)$ is closed, the closed graph theorem implies that T^{-1} must be continuous. Next let T admit a continuous inverse. Then $N(T) = \{0\}$ and also (1) holds since T is closed. Thus, by (4), $R(T') = X'$.

Corollary 2. Let X be a Hilbert space with a scalar product (u, v), and T a closed linear operator with dense domain $D(T) \subseteq X$ and range $R(T) \subseteq X$. Suppose that there exists a positive constant c such that

$$Re(Tu, u) \geqq c\, \|u\|^2 \quad \text{for all} \quad u \in D(T). \tag{7}$$

Then $R(T^*) = X$.

Proof. By Schwarz' inequality, we have

$$\|Tu\| \cdot \|u\| \geqq Re(Tu, u) \geqq c\, \|u\|^2 \quad \text{for all} \quad u \in D(T).$$

Hence $\|Tu\| \geqq c\, \|u\|$, $u \in D(T)$, and so T admits a continuous inverse. Thus, by the preceding Corollary, $R(T') = X$. Hence $R(T^*) = R(T') = X$.

Remark. A linear operator T on $D(T) \subseteq X$ into X is called *accretive* (the terminology is due to K. Friedrichs and T. Kato) if

$$Re(Tu, u) \geqq 0 \quad \text{for all} \quad u \in D(T). \tag{8}$$

T is called *dissipative* (the terminology is due to R. S. Phillips) if $-T$ is accretive.

References for Chapter VII

For a general account concerning Hilbert spaces, see M. H. Stone [1], N. I. Achieser-I. M. Glasman [1] and N. Dunford-J. Schwartz [5]. The closed range theorem is proved essentially in S. Banach [1].

VIII. Resolvent and Spectrum

Let T be a linear operator whose domain $D(T)$ and range $R(T)$ both lie in the same complex linear topological space X. We consider the linear operator

$$T_\lambda = \lambda I - T,$$

where λ is a complex number and I the identity operator. The distribution of the values of λ for which T_λ has an inverse and the properties of the inverse when it exists, are called the *spectral theory* for the operator T. We shall thus discuss the general theory of the inverse of T_λ.

1. The Resolvent and Spectrum

Definition. If λ_0 is such that the range $R(T_{\lambda_0})$ is dense in X and T_{λ_0} has a continuous inverse $(\lambda_0 I - T)^{-1}$, we say that λ_0 is in the *resolvent set* $\varrho(T)$ of T, and we denote this inverse $(\lambda_0 I - T)^{-1}$ by $R(\lambda_0; T)$ and call it the *resolvent* (at λ_0) of T. All complex numbers λ not in $\varrho(T)$ form a set $\sigma(T)$ called the *spectrum* of T. The spectrum $\sigma(T)$ is decomposed into disjoint sets $P_\sigma(T)$, $C_\sigma(T)$ and $R_\sigma(T)$ with the following properties:

$P_\sigma(T)$ is the totality of complex numbers λ for which T_λ does not have an inverse ;$P_\sigma(T)$ is called the *point spectrum* of T.

$C_\sigma(T)$ is the totality of complex numbers λ for which T_λ has a discontinuous inverse with domain dense in X; $C_\sigma(T)$ is called the *continuous spectrum* of T.

$R_\sigma(T)$ is the totality of complex numbers λ for which T_λ has an inverse whose domain is not dense in X; $R_\sigma(T)$ is called the *residual spectrum* of T.

From these definitions and the linearity of T we have the

Proposition. A necessary and sufficient condition for $\lambda_0 \in P_\sigma(T)$ is that the equation $Tx = \lambda_0 x$ has a solution $x \neq 0$. In this case λ_0 is called an *eigenvalue* of T, and x the corresponding *eigenvector*. The null space $N(\lambda_0 I - T)$ of T_{λ_0} is called the *eigenspace* of T corresponding to the eigenvalue λ_0 of T. It consists of the vector 0 and the totality of eigenvectors corresponding to λ_0. The dimension of the eigenspace corresponding to λ_0 is called the *multiplicity* of the eigenvalue λ_0.

Theorem. Let X be a complex B-space, and T a closed linear operator with its domain $D(T)$ and range $R(T)$ both in X. Then, for any $\lambda_0 \in \varrho(T)$, the resolvent $(\lambda_0 I - T)^{-1}$ is an everywhere defined continuous linear operator.

Proof. Since λ_0 is in the resolvent set $\varrho(T)$, $R(\lambda_0 I - T) = D((\lambda_0 I - T)^{-1})$ is dense in X in such a way that there exists a positive constant c for which

$$\|(\lambda_0 I - T)\, x\| \geqq c\, \|x\| \quad \text{whenever } x \in D(T).$$

We have to show that $R(\lambda_0 I - T) = X$. But, if $\underset{n\to\infty}{s\text{-}\lim}\, (\lambda_0 I - T)\, x_n = y$ exists, then, by the above inequality, $\underset{n\to\infty}{s\text{-}\lim}\, x_n = x$ exists, and so, by the closure property of T, we must have $(\lambda_0 I - T)\, x = y$. Hence, by the assumption that $R(\lambda_0 I - T)^a = X$, we must have $R(\lambda_0 I - T) = X$.

Example 1. If the space X is of finite dimension, then any bounded linear operator T is represented by a matrix (t_{ij}). It is known that the eigenvalues of T are obtained as the roots of the algebraic equation, the so-called *secular or characteristic equation* of the matrix (t_{ij}):

$$\det(\lambda\delta_{ij} - t_{ij}) = 0, \tag{1}$$

where $\det(A)$ denotes the determinant of the matrix A.

Example 2. Let $X = L_2(-\infty, \infty)$ and let T be defined by

$$T \cdot x(t) = t x(t),$$

that is, $D(T) = \{x(t);\, x(t) \text{ and } t x(t) \in L^2(-\infty, \infty)\}$ and $T x(t) = t x(t)$ for $x(t) \in D(T)$. Then every real number λ_0 is in $C_\sigma(T)$.

Proof. The condition $(\lambda_0 I - T)\, x = 0$ implies $(\lambda_0 - t)\, x(t) = 0$ a.e., and so $x(t) = 0$ a.e. Thus $(\lambda_0 I - T)^{-1}$ exists. The domain $D((\lambda_0 I - T)^{-1})$ comprises those $y(t) \in L^2(-\infty, \infty)$ which vanish identically in the neighbourhood of $t = \lambda_0$; the neighbourhood may vary with $y(t)$. Hence $D((\lambda_0 I - T)^{-1})$ is dense in $L^2(-\infty, \infty)$. It is easy to see that the operator $(\lambda_0 I - T)^{-1}$ is not bounded on the totality of such $y(t)$'s.

Example 3. Let X be the Hilbert space (l^2), and let T_0 be defined by

$$T_0(\xi_1, \xi_2, \ldots) = (0, \xi_1, \xi_2, \ldots).$$

Then 0 is in the residual spectrum of T, since $R(T_0)$ is not dense in X.

Example 4. Let H be a self-adjoint operator in a Hilbert space X. Then the resolvent set $\varrho(H)$ of H comprises all the complex numbers λ with $\operatorname{Im}(\lambda) \neq 0$, and the resolvent $R(\lambda; H)$ is a bounded linear operator with the estimate

$$\|R(\lambda; H)\| \leqq 1/|\operatorname{Im}(\lambda)|. \tag{2}$$

Moreover,

$$\operatorname{Im}((\lambda I - H)\, x, x) = \operatorname{Im}(\lambda)\, \|x\|^2, \quad x \in D(H). \tag{3}$$

Proof. If $x \in D(H)$, then (Hx, x) is real since $(Hx, x) = (x, Hx) = \overline{(Hx, x)}$. Therefore we have (3), and so, by Schwarz' inequality,

$$\|(\lambda I - H)\, x\| \cdot \|x\| \geqq |((\lambda I - H)\, x, x)| \geqq |\operatorname{Im}(\lambda)| \cdot \|x\|^2 \tag{4}$$

which implies that

$$\|(\lambda I - H)\, x\| \geqq |\operatorname{Im}(\lambda)| \cdot \|x\|, \quad x \in D(H). \tag{5}$$

Hence the inverse $(\lambda I - H)^{-1}$ exists if $\operatorname{Im}(\lambda) \neq 0$. Moreover, the range $R(\lambda I - H)$ is dense in X if $\operatorname{Im}(\lambda) \neq 0$. If otherwise, there would exist a $y \neq 0$ orthogonal to $R(\lambda I - H)$, i.e., $((\lambda I - H)\, x, y) = 0$ for all $x \in D(H)$ and so $(x, (\bar{\lambda} I - H)\, y) = 0$ for all $x \in D(H)$. Since the domain $D(H)$ of a self-adjoint operator H is dense in X, we must have $(\bar{\lambda}I - H)\, y = 0$, that is, $Hy = \bar{\lambda}y$, contrary to the reality of the value (Hy, y).

Therefore, by the above Theorem, we see that, for any complex number λ with $\mathrm{Im}(\lambda) \neq 0$, the resolvent $R(\lambda; H)$ is a bounded linear operator with the estimate (2).

2. The Resolvent Equation and Spectral Radius

Theorem 1. Let T be a closed linear operator with domain and range both in a complex B-space X. Then the resolvent set $\varrho(T)$ is an open set of the complex plane. In each component (the maximal connected sets) of $\varrho(T)$, $R(\lambda; T)$ is a holomorphic function of λ.

Proof. By the Theorem of the preceding section, $R(\lambda; T)$ for $\lambda \in \varrho(T)$ is an everywhere defined continuous operator. Let $\lambda_0 \in \varrho(T)$ and consider

$$S(\lambda) = R(\lambda_0; T)\left\{I + \sum_{n=1}^{\infty} (\lambda_0 - \lambda)^n R(\lambda_0; T)^n\right\}. \tag{1}$$

The series is convergent in the operator norm whenever $|\lambda_0 - \lambda| \cdot \|R(\lambda_0; T)\| < 1$, and within this circle of the complex plane, the series defines a holomorphic function of λ. Multiplication by $(\lambda I - T) = (\lambda - \lambda_0)\, I + (\lambda_0 I - T)$ on the left or right gives I so that the series $S(\lambda)$ actually represents the resolvent $R(\lambda; T)$. Thus we have proved that a circular neighbourhood of λ_0 belongs to $\varrho(T)$ and $R(\lambda; T)$ is holomorphic in this neighbourhood.

Theorem 2. If λ and μ both belong to $\varrho(T)$, and if $R(\lambda; T)$ and $R(\mu; T)$ are everywhere defined continuous operators, then the *resolvent equation* holds:

$$R(\lambda; T) - R(\mu; T) = (\mu - \lambda)\, R(\lambda; T)\, R(\mu; T). \tag{2}$$

Proof. We have

$$\begin{aligned} R(\lambda; T) &= R(\lambda; T)\,(\mu I - T)\, R(\mu; T) \\ &= R(\lambda; T)\,\{(\mu - \lambda)\, I + (\lambda I - T)\}\, R(\mu; T) \\ &= (\mu - \lambda)\, R(\lambda; T)\, R(\mu; T) + R(\mu; T). \end{aligned}$$

Theorem 3. If T is a bounded linear operator on a complex B-space X into X, then the following limit exists:

$$\lim_{n\to\infty} \|T^n\|^{1/n} = r_\sigma(T). \tag{3}$$

It is called the *spectral radius* of T, and we have

$$r_\sigma(T) \leqq \|T\|. \tag{4}$$

If $|\lambda| > r_\sigma(T)$, then the resolvent $R(\lambda; T)$ exists and is given by the series

$$R(\lambda; T) = \sum_{n=1}^{\infty} \lambda^{-n} T^{n-1} \tag{5}$$

which converges in the norm of operators.

Proof. Set $r = \inf_{n \geqq 1} \|T^n\|^{1/n}$. We have to show that $\overline{\lim}_{n \to \infty} \|T^n\|^{1/n} \leqq r$. For any $\varepsilon > 0$, choose m such that $\|T^m\|^{1/m} \leqq r + \varepsilon$. For arbitrary n, write $n = pm + q$ where $0 \leqq q \leqq (m-1)$. Then, by $\|AB\| \leqq \|A\| \cdot \|B\|$, we obtain

$$\|T^n\|^{1/n} \leqq \|T^m\|^{p/n} \cdot \|T\|^{q/n} \leqq (r+\varepsilon)^{mp/n} \|T\|^{q/n}.$$

Since $pm/n \to 1$ and $q/n \to 0$ as $n \to \infty$, we must have $\overline{\lim}_{n \to \infty} \|T^n\|^{1/n} \leqq r + \varepsilon$. Since ε was arbitrary, we have proved $\overline{\lim}_{n \to \infty} \|T^n\|^{1/n} \leqq r$.

Since $\|T^n\| \leqq \|T\|^n$, we have $\lim_{n \to \infty} \|T^n\|^{1/n} \leqq \|T\|$. The series (5) is convergent in the norm of operators when $\lambda > r_\sigma(T)$. For, if $|\lambda| \geqq r_\sigma(T) + \varepsilon$ with $\varepsilon > 0$, then, by (3), $\|\lambda^{-n} T^n\| \leqq (r_\sigma(T) + \varepsilon)^{-n} \cdot (r_\sigma(T) + 2^{-1}\varepsilon)^n$ for large n. Multiplication by $(\lambda I - T)$ on the left or right of this series gives I so that the series actually represents the resolvent $R(\lambda; T)$.

Corollary. The resolvent set $\varrho(T)$ is not empty when T is a bounded linear operator.

Theorem 4. For a bounded linear operator $T \in L(X, X)$, we have

$$r_\sigma(T) = \sup_{\lambda \in \sigma(T)} |\lambda|. \tag{6}$$

Proof. By Theorem 3, we know that $r_\sigma(T) \geqq \sup_{\lambda \in \sigma(T)} |\lambda|$. Hence we have only to show that $r_\sigma(T) \leqq \sup_{\lambda \in \sigma(T)} |\lambda|$.

By Theorem 1, $R(\lambda; T)$ is holomorphic in λ when $|\lambda| > \sup_{\lambda \in \sigma(T)} |\lambda|$. Thus it admits a uniquely determined Laurent expansion in positive and non-positive powers of λ convergent in the operator norm for $|\lambda| > \sup_{\lambda \in \sigma(T)} |\lambda|$. By Theorem 3, this Laurent series must coincide with $\sum_{n=1}^{\infty} \lambda^{-n} T^{n-1}$. Hence $\lim_{n \to \infty} \|\lambda^{-n} T^n\| = 0$ if $|\lambda| > \sup_{\lambda \in \sigma(T)} |\lambda|$, and so, for any $\varepsilon > 0$, we must have $\|T^n\| \leqq \left(\varepsilon + \sup_{\lambda \in \sigma(T)} |\lambda|\right)^n$ for large n. This proves that

$$r_\sigma(T) = \lim_{n \to \infty} \|T^n\|^{1/n} \leqq \sup_{\lambda \in \sigma(T)} |\lambda|.$$

Corollary. The series $\sum_{n=1}^{\infty} \lambda^{-n} T^{n-1}$ diverges if $|\lambda| < r_\sigma(T)$.

Proof. Let r be the smallest number $\geqq 0$ such that the series $\sum_{n=1}^{\infty} \lambda^{-n} T^{n-1}$ converges in the operator norm for $|\lambda| > r$. The existence of such an r is proved as for ordinary power series in λ^{-1}. Then, for $|\lambda| > r$, $\lim_{n \to \infty} \|\lambda^{-n} T^n\| = 0$ and so, as in the proof of $r_\sigma(T) \leqq \sup_{\lambda \in \sigma(T)} |\lambda|$, we must have $\lim_{n \to \infty} \|T^n\|^{1/n} \leqq r$. This proves that $r_\sigma(T) \leqq r$.

3. The Mean Ergodic Theorem

For a particular class of continuous linear operators, the *mean ergodic theorem* gives a method for obtaining the eigenspace corresponding to the eigenvalue 1. In this section, we shall state and prove the *mean ergodic theorem* from the view point of the spectral theory, as was formulated previously by the present author. The historical sketch of the ergodic theory in connection with statistical mechanics will be given in Chapter XIII.

Theorem 1. Let X be a locally convex linear topological space, and T a continuous linear operator on X into X. We assume that

the family of operators $\{T^n;\ n = 1, 2, \ldots\}$ is *equicontinuous* in the sense that, for any continuous semi-norm q on X, there exists a continuous semi-norm q' on X such that $\sup_{n \geqq 1} q(T^n x) \leqq q'(x)$ for all $x \in X$. (1)

Then the closure $R(I-T)^a$ of the range $R(I-T)$ satisfies

$$R(I-T)^a = \left\{x \in X;\ \lim_{n\to\infty} T_n x = 0,\ T_n = n^{-1} \sum_{m=1}^{n} T^m\right\}, \tag{2}$$

and so, in particular,

$$R(I-T)^a \cap N(I-T) = \{0\}. \tag{3}$$

Proof. We have $T_n(I-T) = n^{-1}(T - T^{n+1})$. Hence, by (1), $w \in R(I-T)$ implies that $\lim_{n\to\infty} T_n w = 0$. Next let $z \in R(I-T)^a$. Then, for any continuous semi-norm q' on X and $\varepsilon > 0$, there exists a $w \in R(I-T)$ such that $q'(z-w) < \varepsilon$. Thus, by (1), we have $q(T_n(z-w)) \leqq n^{-1} \sum_{m=1}^{n} q(T^m(z-w)) \leqq q'(z-w) < \varepsilon$. Hence $q(T_n z) \leqq q(T_n w) + q(T_n(z-w)) \leqq q(T_n w) + \varepsilon$, and so $\lim_{n\to\infty} T_n z = 0$. This proves that $R(I-T)^a \subseteq \{x \in X;\ \lim_{n\to\infty} T_n x = 0\}$.

Let, conversely, $\lim_{n\to\infty} T_n x = 0$. Then, for any continuous semi-norm q on X and $\varepsilon > 0$, there exists an n such that $q(x-(x-T_n x)) = q(T_n x) < \varepsilon$. But, by

$$\begin{aligned} x - T_n x &= n^{-1} \sum_{m=1}^{n} (I - T^m)\, x \\ &= n^{-1} \sum_{m=1}^{n} (I-T)\,(I + T + T^2 + \cdots + T^{m-1})\, x, \end{aligned}$$

$(x - T_n x) \in R(I-T)$. Hence x must belong to $R(I-T)^a$.

Theorem 2 (the mean ergodic theorem). Let condition (1) be satisfied. Let, for a given $x \in X$, there exist a subsequence $\{n'\}$ of $\{n\}$

such that

$$\underset{n'\to\infty}{\text{weak-lim}}\ T_{n'}x = x_0 \text{ exists.} \tag{4}$$

Then $Tx_0 = x_0$ and $\lim_{n\to\infty} T_n x = x_0$.

Proof. We have $TT_n - T_n = n^{-1}(T^{n+1} - T)$, and so, by (1), $\lim_{n\to\infty} (TT_n x - T_n x) = 0$. Thus, for any $f \in X'$, $\lim_{n\to\infty} \langle TT_{n'}x, f\rangle = \lim_{n\to\infty} \langle T_{n'}x, T'f\rangle$ exists and $= \lim_{n\to\infty} \langle T_{n'}x, f\rangle = \langle x_0, f\rangle$. Therefore $\langle x_0, f\rangle = \langle Tx_0, f\rangle$ and so, by the arbitrariness of $f \in X'$, we must have $Tx_0 = x_0$.

We have thus $T^m x = T^m x_0 + T^m(x - x_0) = x_0 + T^m(x - x_0)$ and so $T_n x = x_0 + T_n(x - x_0)$. But, $(x - x_0) = \underset{n\to\infty}{\text{weak-lim}}\ (x - T_{n'}x)$ and, as proved above, $(x - T_{n'}x) \in R(I - T)$. Therefore, by Theorem 11 in Chapter V, 1, $(x - x_0) \in R(I - T)^a$. Thus, by Theorem 1, $\lim_{n\to\infty} T_n(x - x_0) = 0$ and so we have proved that $\lim_{n\to\infty} T_n x = x_0$.

Corollary. Let condition (1) be satisfied, and X be locally sequentially weakly compact. Then, for any $x \in X$, $\lim_{n\to\infty} T_n x = x_0$ exists, and the operator T_0 defined by $T_0 x = x_0$ is a continuous linear operator such that

$$T_0 = T_0^2 = TT_0 = T_0 T, \tag{5}$$

$$R(T_0) = N(I - T), \tag{6}$$

$$N(T_0) = R(I - T)^a = R(I - T_0). \tag{7}$$

Moreover, we have the *direct sum decomposition*

$$X = R(I - T)^a \oplus N(I - T), \tag{8}$$

i.e., any $x \in X$ is represented uniquely as the sum of an element $\in R(I - T)^a$ and an element $\in N(I - T)$.

Proof. The linearity of T_0 is clear. The continuity of T_0 is proved by the equi-continuity of $\{T_n\}$ implied by (1). Next, since $Tx_0 = x_0$, we have $TT_0 = T_0$ and so $T^n T_0 = T_0$, $T_n T_0 = T_0$ which implies that $T_0^2 = T_0$. On the other hand, $T_n - T_n T = n^{-1}(T - T^{n+1})$ and (1) imply that $T_0 = T_0 T$. The equality (6) is proved as follows. Let $Tx = x$, then $T^n x = x$, $T_n x = x$ and so $T_0 x = x$, that is, $x \in R(T_0)$. Let conversely, $x \in R(T_0)$. Then, by $T_0^2 = T_0$, we have $T_0 x = x$ and so, by $TT_0 = T_0$, $Tx = TT_0 x = T_0 x = x$. Therefore, the eigenspace of T corresponding to the eigenvalue 1 of T is precisely the range $R(T_0)$. Hence (6) is proved. Moreover, we have, by Theorem 1, $N(T_0) = R(I - T)^a$. But, by $T_0^2 = T_0$, we have $R(I - T_0) \subseteq N(T_0)$, and if $x \in N(T_0)$, then $x = x - T_0 x \in R(I - T_0)$. Thus $N(T_0) = R(I - T_0)$.

Therefore, by $I = (I - T_0) + T_0$ and (6) and (7), we obtain (8).

Remark. The eigenspace $N(\lambda I - T)$ of T belonging to the eigenvalue λ with $|\lambda| = 1$ may be obtained as $R(T(\lambda))$, where $T(\lambda)\, x = \lim_{n\to\infty} n^{-1} \sum_{m=1}^{n} (T/\lambda)^m x$.

The Mean Ergodic Theorem of J. von Neumann. Let $(S, \mathfrak{B}, m)$ be a measure space, and P an *equi-measure transformation* of S, that is, P is a one-one mapping of S onto S such that $P \cdot B \in \mathfrak{B}$ iff B is $\in \mathfrak{B}$ and $m(P \cdot B) = m(B)$. Consider the linear operator T on $L^2(S, \mathfrak{B}, m)$ onto itself defined by

$$(Tx)(s) = x(Ps), \quad x \in L^2(S, \mathfrak{B}, m). \tag{9}$$

By the equi-measurable property of P, we easily see that the operator T is unitary and so the equi-continuity condition (1) is surely satisfied by $\|T^n\| = 1$ $(n = 1, 2, \ldots)$. Therefore, by the sequential weak compactness of the Hilbert space $L^2(S, \mathfrak{B}, m)$, we obtain the *mean ergodic theorem of J. von Neumann*:

$$\text{For any } x \in L^2(S, \mathfrak{B}, m), \ \text{s-}\lim_{n\to\infty} n^{-1} \sum_{m=1}^{n} T^m x = x_0 \in L^2(S, \mathfrak{B}, m) \text{ exists and } Tx_0 = x_0. \tag{10}$$

Remark. Theorem 1 and Theorem 2 are adapted from K. Yosida [3]. Cf. also S. Kakutani [1] and F. Riesz [4]. Neumann's mean ergodic theorem was published in J. von Neumann [3].

4. Ergodic Theorems of the Hille Type Concerning Pseudo-resolvents

The notion of resolvent is generalized to that of *pseudo-resolvent* by E. Hille. We can prove ergodic theorems for pseudo-resolvents by a similar idea to that used in the proof of the mean ergodic theorems in the preceding section. See K. Yosida [4]. Cf. T. Kato [1]. These ergodic theorems may be considered as extensions of the abelian ergodic theorems of E. Hille given in E. Hille-R. S. Phillips [1], p. 502.

We shall begin with the definition of the pseudo-resolvent.

Definition. Let X be a locally convex complex linear topological space, and $L(X, X)$ the algebra of all continuous linear operators defined on X into X. A pseudo-resolvent J_λ is a function defined on a subset $D(J)$ of the complex λ-plane with values in $L(X, X)$ such that

$$J_\lambda - J_\mu = (\mu - \lambda)\, J_\lambda J_\mu \ \text{(the resolvent equation)}. \tag{1}$$

Proposition. All J_λ, $\lambda \in D(J)$, have a common null space, which we denote by $N(J)$, and a common range which we denote by $R(J)$. Similarly, all $(I - \lambda J_\lambda)$, $\lambda \in D(J)$, have a common null space, which we denote by $N(I-J)$, and a common range which we denote by $R(I-J)$. Moreover, we have the commutativity:

$$J_\lambda J_\mu = J_\mu J_\lambda \quad (\lambda, \mu \in D(J)). \tag{2}$$

Proof. By interchanging λ and μ in (1), we obtain

$$J_\mu - J_\lambda = (\lambda - \mu) J_\mu J_\lambda = -(\mu - \lambda) J_\mu J_\lambda,$$

and hence (2) is true. The first part of the Proposition is clear from (1) and (2). The second part is also clear from

$$(I - \lambda J_\lambda) = (I - (\lambda - \mu) J_\lambda)(I - \mu J_\mu) \tag{1'}$$

which is a variant of (1).

Theorem 1. A pseudo-resolvent J_λ is a resolvent of a linear operator A iff $N(J) = \{0\}$; and then $R(J)$ coincides with the domain $D(A)$ of A.

Proof. The "only if" part is clear. Suppose $N(J) = \{0\}$. Then, for any $\lambda \in D(J)$, the inverse J_λ^{-1} exists. We have

$$\lambda I - J_\lambda^{-1} = \mu I - J_\mu^{-1} \quad (\lambda, \mu \in D(J)). \tag{3}$$

For, by (1) and (2),

$$\begin{aligned} J_\lambda J_\mu(\lambda I - J_\lambda^{-1} - \mu I + J_\mu^{-1}) &= (\lambda - \mu) J_\lambda J_\mu - J_\lambda J_\mu (J_\lambda^{-1} - J_\mu^{-1}) \\ &= (\lambda - \mu) J_\lambda J_\mu - (J_\mu - J_\lambda) = 0. \end{aligned}$$

We thus put

$$A = (\lambda I - J_\lambda^{-1}). \tag{4}$$

Then $J_\lambda = (\lambda I - A)^{-1}$ for $\lambda \in D(J)$.

Lemma 1. We assume that there exists a sequence $\{\lambda_n\}$ of numbers $\in D(J)$ such that

$$\lim_{n\to\infty} \lambda_n = 0 \text{ and the family of operators } \{\lambda_n J_{\lambda_n}\} \text{ is equi-continuous.} \tag{5}$$

Then we have

$$R(I-J)^a = \left\{x \in X; \lim_{n\to\infty} \lambda_n J_{\lambda_n} x = 0\right\}, \tag{6}$$

and hence

$$N(I-J) \cap R(I-J)^a = \{0\}. \tag{7}$$

Proof. We have, by (1),

$$\lambda J_\lambda (I - \mu J_\mu) = (1 - \mu(\mu - \lambda)^{-1}) \lambda J_\lambda - \lambda(\lambda - \mu)^{-1} \mu J_\mu. \tag{8}$$

Hence, by (5), the condition $x \in R(I - \mu J_\mu) = R(I-J)$ implies that $\lim_{n\to\infty} \lambda_n J_{\lambda_n} x = 0$. Let $y \in R(I-J)^a$. Then, for any continuous semi-norm q on X and $\varepsilon > 0$, there exists an $x \in R(I-J)$ such that $q(y-x) < \varepsilon$. By (5), we have, for any continuous semi-norm q' on X,

$$q'(\lambda_n J_{\lambda_n}(y-x)) \leqq q(y-x) \quad (n = 1, 2, \ldots)$$

with a suitable continuous semi-norm q on X. Therefore, by $\lambda_n J_{\lambda_n} y = \lambda_n J_{\lambda_n} x + \lambda_n J_{\lambda_n}(y-x)$, we must have $\lim_{n\to\infty} \lambda_n J_{\lambda_n} y = 0$.

Let, conversely, $\lim_{n\to\infty} \lambda_n J_{\lambda_n} x = 0$. Then, for any continuous semi-norm q on X and $\varepsilon > 0$, there exists a λ_n such that $q(x-(x-\lambda_n J_{\lambda_n} x)) < \varepsilon$. Hence x must belong to $R(I-\lambda_n J_{\lambda_n})^a = R(I-J)^a$.

Lemma 1′. We assume that there exists a sequence $\{\lambda_n\}$ of numbers $\in D(J)$ such that

$$\lim_{n\to\infty} |\lambda_n| = \infty \text{ and the family of operators } \{\lambda_n J_{\lambda_n}\} \text{ is equi-continuous.} \tag{5′}$$

Then we have

$$R(J)^a = \left\{x \in X;\ \lim_{n\to\infty} \lambda_n J_{\lambda_n} x = x\right\}, \tag{6′}$$

and hence

$$N(J) \cap R(J)^a = \{0\}. \tag{7′}$$

Proof. We have, by (1),

$$\lambda J_\lambda J_\mu = \frac{\mu}{\lambda} \lambda J_\lambda J_\mu - \frac{1}{\lambda} \lambda J_\lambda + J_\mu.$$

Hence, by (5′), the condition $x \in R(J_\mu) = R(J)$ implies that $\lim_{n\to\infty} \lambda_n J_{\lambda_n} x = x$. Let $y \in R(J)^a$. Then, for any continuous semi-norm q on X and $\varepsilon > 0$, there exists an $x \in R(J)$ such that $q(y-x) < \varepsilon$. By (5′), we have, for any continuous semi-norm q' on X,

$$q'(\lambda_n J_{\lambda_n}(y-x)) \leqq q(y-x) \quad (n = 1, 2, \ldots)$$

with a suitable continuous semi-norm q on X. Therefore, by (5′) and

$$\lambda_n J_{\lambda_n} y - y = (\lambda_n J_{\lambda_n} x - x) + (x-y) + \lambda_n J_{\lambda_n}(y-x),$$

we must have $\lim_{n\to\infty} \lambda_n J_{\lambda_n} y = y$.

Let, conversely, $\lim_{n\to\infty} \lambda_n J_{\lambda_n} x = x$. Then, for any continuous semi-norm q on X and $\varepsilon > 0$, there exists a λ_n such that $q(x - \lambda_n J_{\lambda_n} x) < \varepsilon$. Hence x must belong to $R(J_{\lambda_n})^a = R(J)^a$.

Theorem 2. Let (5) be satisfied. Let, for a given $x \in X$, there exist a subsequence $\{n'\}$ of $\{n\}$ such that

$$\text{weak-}\lim_{n\to\infty} \lambda_{n'} J_{\lambda_{n'}} x = x_h \text{ exists.} \tag{9}$$

Then $x_h = \lim_{n\to\infty} \lambda_n J_{\lambda_n} x$ and $x_h \in N(I-J)$, $x_p = (x - x_h) \in R(I-J)^a$.

Proof. Setting $\mu = \lambda_{n'}$ in (1′) and letting $n' \to \infty$, we see, by (5), that $(I-\lambda J_\lambda)\, x = (I-\lambda J_\lambda)\,(x-x_h)$, that is, $(I-\lambda J_\lambda)\, x_h = 0$. Hence $x_h \in N(I-J)$ and so

$$\lambda_n J_{\lambda_n} x = x_h + \lambda_n J_{\lambda_n}(x - x_h). \tag{10}$$

Therefore we have only to prove that $\lim_{n\to\infty} \lambda_n J_{\lambda_n}(x-x_h) = 0$, or, by Lemma 1, $(x-x_h) \in R(I-J)^a$. But $(x - \lambda_n J_{\lambda_n} x) \in R(I-J)$, and so, by Theorem 11 in Chapter V, 1, we must have $(x-x_h) \in R(I-J)^a$.

Corollary 1. Let (5) be satisfied, and let X be locally sequentially weakly compact. Then

$$X = N(I-J) \oplus R(I-J)^a. \tag{11}$$

Proof. For any $x \in X$, let $x_h = \lim_{n\to\infty} \lambda_n J_{\lambda_n} x$ and $x_p = (x - x_h)$ be the components of x in $N(I-J)$ and $R(I-J)^a$, respectively.

Theorem 2′. Let (5′) be satisfied. Let, for a given $x \in X$, there exist a subsequence $\{n'\}$ of $\{n\}$ such that

$$\text{weak-}\lim_{n\to\infty} \lambda_{n'} J_{\lambda_{n'}} x = x_{h'} \text{ exists.} \tag{9'}$$

Then $x_{h'} = \lim_{n\to\infty} \lambda_n J_{\lambda_n} x$ and $x_{h'} \in R(J)^a$, $x_{p'} = (x - x_{h'}) \in N(J)$.

Proof. Setting $\mu = \lambda_{n'}$ in (8) and letting $n' \to \infty$, we see, by (5′), that $\lambda J_\lambda (x - x_{h'}) = 0$, that is, $(x - x_{h'}) \in N(J)$. Hence

$$\lambda_n J_{\lambda_n} x = \lambda_n J_{\lambda_n} x_{h'}. \tag{10'}$$

Therefore we have only to prove that $\lim_{n\to\infty} \lambda_n J_{\lambda_n} x_{h'} = x_{h'}$, or, by Lemma 1′, $x_{h'} \in R(J)^a$. But $\lambda_{n'} J_{\lambda_{n'}} x_{h'} \in R(J)$, and so, by Theorem 11 in Chapter V, 1, we must have $x_{h'} \in R(J)^a$.

Corollary 1′. Let (5′) be satisfied, and let X be locally sequentially weakly compact. Then

$$X = N(J) \oplus R(J)^a. \tag{11'}$$

Proof. For any $x \in X$, let $x_{h'} = \lim_{n\to\infty} \lambda_n J_{\lambda_n} x$ and $x_{p'} = (x - x_{h'})$ be the componerts of X in $R(J)^a$ and $N(J)$, respectively.

Remark. As a Corollary we obtain: In a reflexive B-space X, a pseudo-resolvent J_λ satisfying (5′) is the resolvent of a closed linear operator A with dense domain iff $R(J)^a = X$. This result is due to *T.* Kato, loc. cit. The proof is easy, since, by Eberlein's theorem, a B-space X is locally sequentially weakly compact iff X is reflexive.

5. The Mean Value of an Almost Periodic Function

As an application of the mean ergodic theorem we shall give an existence proof of the mean value of an almost periodic function.

Definition 1. A set G of elements $g, h, \ldots$ is called a *group* if in G a product (in general non-commutative) gh of any pair (g, h) of elements $\in G$ is defined satisfying the following conditions:

$gh \in G$, (1)

$(gh)\,k = g(hk)$ (the associativity), (2)

there exists a unique element e in G such that $eg = ge = g$ for all $g \in G$; e is called the *identity element* of the group G, (3)

for every element $g \in G$, there exists a uniquely determined element in G, which is denoted by g^{-1}, such that $gg^{-1} = g^{-1}g = e$; the element g^{-1} is called the *inverse element* of g. (4)

Clearly, g is the inverse of g^{-1} so that $(g^{-1})^{-1} = g$. A group G is said to be *commutative* if $gh = hg$ for all $g, h \in G$.

Example. The totality of complex matrices of order n with determinants equal to unity is a group with respect to the matrix multiplication; it is called the *complex unimodular group* of order n. The identity of this group is the identity matrix and the inverse element of the matrix a is the inverse matrix a^{-1}. The *real unimodular group* is defined analogously. These groups are non-commutative when $n \geqq 2$.

Definition 2 (J. VON NEUMANN [4]). Given an abstract group G. A complex-valued function $f(g)$ defined on G is called *almost periodic* on G if the following condition is satisfied:

the set of functions $\{g_s(f, h); s \in G\}$, where $f_s(g, h) = f(gsh)$, defined on the direct product $G \times G$ is totally bounded with respect to the topology of uniform convergence on $G \times G$. (5)

Example. Let G be the set R^1 of all real numbers in which the group multiplication is defined as the addition of real numbers; this group R^1 is called the *additive group of real numbers*. The function $f(g) = e^{i\alpha g}$, where α is a real number and $i = \sqrt{-1}$, is almost periodic on R^1. This we easily see from the addition theorem $f(gsh) = e^{i\alpha g}\, e^{i\alpha s}\, e^{i\alpha h}$ and the fact that $\{e^{i\alpha t};\ t \in R^1\}$ is totally bounded as a set of complex numbers of absolute value 1.

Proposition 1. Suppose $f(g)$ is an almost periodic function on G. If we define, following A. WEIL,

$$\operatorname{dis}(s, u) = \sup_{g,h \in G} |f(gsh) - f(guh)|, \tag{6}$$

then

$$\operatorname{dis}(s, u) = \operatorname{dis}(asb,\ aub). \tag{7}$$

Proof. Clear from the definition of the group.

Corollary 1. The set E of all elements s which satisfies $\operatorname{dis}(s, e) = 0$ constitutes an *invariant subgroup* in G, that is, we have

$$\text{if } e_1, e_2 \in E, \text{ then } e_1 e_2 \in E \text{ and } a e_1 a^{-1} \in E \text{ for every } a \in G. \tag{8}$$

Proof. Let $\operatorname{dis}(e_1, e) = 0$, $\operatorname{dis}(e_2, e) = 0$. Then, by (7) and the triangle inequality, we obtain

$$\operatorname{dis}(e_1 e_2, e) \leqq \operatorname{dis}(e_1 e_2, e_1 e) + \operatorname{dis}(e_1 e, e) = 0 + 0 = 0.$$

Similarly we have $\operatorname{dis}(a e_1 a^{-1}, e) = \operatorname{dis}(a e_1 a^{-1}, a e a^{-1}) = 0$ from $\operatorname{dis}(e_1, e) = 0$.

Corollary 2. If we write $s \equiv u \pmod{E}$ when $su^{-1} \in E$, then $s \equiv u \pmod{E}$ is equivalent to $\operatorname{dis}(s, u) = 0$.

Proof. Clear from $\operatorname{dis}(su^{-1}, e) = \operatorname{dis}(s, eu) = \operatorname{dis}(s, u)$.

Corollary 3. The concept $s \equiv u \pmod{E}$ has all the general properties of equivalence, namely

$$s \equiv s \pmod{E}, \tag{9}$$

$$s \equiv u \pmod{E}, \text{ then } u \equiv s \pmod{E}, \tag{10}$$

$$\text{if } s_1 \equiv s_2 \pmod{E} \text{ and } s_2 \equiv s_3 \pmod{E}, \text{ then } s_1 \equiv s_3 \pmod{E}. \tag{11}$$

Proof. Clear from Corollary 2 and the triangle inequality for the $\operatorname{dis}(s, u)$.

Hence, as in the case of the factor space in a linear space, we can define the *factor group* or *residue class group* G/E as follows: we shall denote the set of all elements $\in G$ equivalent (mod E) to a fixed element $x \in G$ by ξ_x, the *residue class* (mod E) containing x; then the set of all residue classes ξ_x constitutes a group G/E by the notion of the product

$$\xi_x \xi_y = \xi_{xy}. \tag{12}$$

To justify this definition (12) of the product, we have to show that

$$\text{if } x_1 \equiv x_2 \pmod{E},\ y_1 \equiv y_2 \pmod{E}, \text{ then } x_1 y_1 \equiv x_2 y_2 \pmod{E}. \tag{13}$$

This is clear, since we have by (7) and Corollary 2,

$$\begin{aligned}\operatorname{dis}(x_1 y_1, x_2 y_2) &\leqq \operatorname{dis}(x_1 y_1, x_2 y_1) + \operatorname{dis}(x_2 y_1, x_2 y_2)\\ &= \operatorname{dis}(x_1, x_2) + \operatorname{dis}(y_1, y_2) = 0 + 0 = 0.\end{aligned}$$

Since the function $f(x)$ takes the same constant value on the residue class ξ_x, we may consider $f(x)$ as a function $F(\xi_x)$ defined on the residue class group G/E.

Corollary 4. The residue class group is a metric space by the distance

$$\operatorname{dis}(\xi_x, \xi_y) = \operatorname{dis}(x, y). \tag{14}$$

Proof. $x \equiv x_1 \pmod{E}$ and $y \equiv y_1 \pmod{E}$ imply

$$\operatorname{dis}(x, y) \leqq \operatorname{dis}(x, x_1) + \operatorname{dis}(x_1, y_1) + \operatorname{dis}(y_1, y) = 0 + \operatorname{dis}(x_1, y_1) + 0$$

and $\operatorname{dis}(x_1, y_1) \leqq \operatorname{dis}(x, y)$ to the effect that $\operatorname{dis}(x, y) = \operatorname{dis}(x_1, y_1)$. Thus (14) defines a distance in G/E.

Corollary 5. The group G/E is a *topological group* with respect to the distance $\operatorname{dis}(\xi_x, \xi_y)$, that is, the operation of multiplication $\xi_x \xi_y$ is continuous as a mapping from the product space $(G/E) \times (G/E)$ onto G/E, and the operation ξ_x^{-1} is continuous as a mapping from G/E onto G/E.

Proof. We have, by (7),

$$\operatorname{dis}(su, s'u') \leqq \operatorname{dis}(su, s'u) + \operatorname{dis}(s'u, s'u') = \operatorname{dis}(s, s') + \operatorname{dis}(u, u')$$

and

$$\operatorname{dis}(s^{-1}, u^{-1}) = \operatorname{dis}(ss^{-1}u, su^{-1}u) = \operatorname{dis}(u, s) = \operatorname{dis}(s, u).$$

We have thus proved the following

Theorem 1 (A. Weil). The topological group G/E, metrized by (14), is totally bounded, and the function $f(x)$ gives rise to a function $F(\xi_x)$ $(= f(x))$ which is uniformly continuous on this group G/E.

Proof. The uniform continuity of the function $F(\xi_x)$ is clear from

$$|F(\xi_x) - F(\xi_y)| = |f(x) - f(y)| \leqq \operatorname{dis}(x, y) = \operatorname{dis}(\xi_x, \xi_y).$$

The almost periodicity of the function $f(x)$ implies, by (7) and (14), that the metric space G/E is totally bounded.

By the above theorem, the theory of almost periodic functions is reduced to the study of a uniformly continuous function $f(g)$, defined on a totally bounded topological group G, metrized by a metric $\operatorname{dis}(g_1, g_2)$ satisfying condition (7). By virtue of this fact, we shall give a proof for the existence of the mean value of an almost periodic function.

Since G is totally bounded, there exists, for any $\varepsilon > 0$, a finite system of points $g_1, g_2, \ldots, g_n$ such that $\min\limits_{1\leqq i\leqq n} \operatorname{dis}(g, g_i) \leqq \varepsilon$ for any $g \in G$. Hence, collecting these finite systems of points corresponding to $\varepsilon = 1, 2^{-1}, 3^{-1}, \ldots$, respectively, we see that there exists a countable system $\{g_j\}$ of points $\in G$ such that $\{g_j\}$ is dense in G. We take a sequence of positive numbers α_j such that $\sum\limits_{j=1}^{\infty} \alpha_j = 1$. Let $C(G)$ be the set of all uniformly continuous complex-valued functions $h(g)$ defined on G. $C(G)$ is a B-space by the operation of the function sum and the norm $\|h\| = \sup\limits_{g\in G} |h(g)|$. We define an operator T defined on $C(G)$ into $C(G)$ by

$$(T \cdot h)(g) = \sum_{j=1}^{\infty} \alpha_j h(g_j g). \tag{15}$$

By the uniform continuity of $h(g)$ on G, there exists, for any $\varepsilon > 0$, a $\delta > 0$ such that $\operatorname{dis}(g, g') \leqq \delta$ implies $|h(g) - h(g')| \leqq \varepsilon$. Thus, by (7), $|h(g_j g) - h(g_j g')| \leqq \varepsilon$ $(j = 1, 2, \ldots)$ whenever $\operatorname{dis}(g, g') \leqq \delta$. Hence it is easy to see, by $\alpha_j > 0$ and $\sum\limits_{j=1}^{\infty} \alpha_j = 1$, that T is a bounded linear operator on $C(G)$ into $C(G)$. By the same reasoning we see that the set of functions $h_n(g)$ defined by

$$\begin{aligned} h_n(g) &= n^{-1} \sum_{m=1}^{n} (T^m h)(g), \text{ which is of the form} \\ h_n(g) &= \sum_{j=1}^{\infty} \beta_j h(g'_j g) \quad \text{with} \quad \beta_j > 0, \sum_{j=1}^{n} \beta_j = 1, \end{aligned} \tag{16}$$

is equi-bounded and equi-continuous with respect to n. Hence, by the Ascoli-Arzelà theorem, the sequence $\{h_n(g)\}$ contains a subsequence which is uniformly convergent on G.

Therefore, by the mean ergodic theorem, there exists an $h^*(g) \in C(G)$ such that

$$\lim_{n\to\infty} \sup_{g} |h_n(g) - h^*(g)| = 0 \text{ and } Th^* = h^*. \tag{17}$$

Proposition 2. $h^*(g)$ is identically equal to a constant.

Proof. We may assume, without losing the generality, that $h(g)$ and $h^*(g)$ are real-valued. Suppose that there exist a point $g_0 \in G$ and a positive constant δ such that

$$h^*(g_0) \leqq \beta - 2\delta, \text{ where } \beta = \sup_{g \in G} h^*(g).$$

By the continuity of $h^*(g)$, there exists a positive number ε such that $\operatorname{dis}(g', g'') \leqq \varepsilon$ implies $|h^*(g') - h^*(g'')| \leqq \delta$; in particular, we have $h^*(g'') \leqq \beta - \delta$ whenever $\operatorname{dis}(g_0, g'') \leqq \varepsilon$. Since the sequence $\{g_j\}$ is dense in G, there exists, for any $\varepsilon > 0$, an index n such that, for any $g \in G$, we have $\min_{1 \leqq j \leqq n} \operatorname{dis}(g, g_j) \leqq \varepsilon$. Hence, by (7), we have, for any $g \in G$,

$$\min_{1 \leqq j \leqq n} \operatorname{dis}(g_0, g_j g) \leqq \varepsilon.$$

Let the minimum be attained at $j = j_0$. Then

$$h^*(g) = (Th^*)(g) = \sum_{j=1}^{\infty} \alpha_j h^*(g_j g) \leqq \alpha_{j_0}(\beta - \delta) + (1 - \alpha_{j_0})\beta = \beta - \alpha_{j_0}\delta < \beta,$$

contrary to the assumption that g was an arbitrary point of G. Therefore $h^*(g)$ is identically equal to a constant.

Definition 3. We shall call the constant value $h^*(g)$ the *left mean value* of $h(g)$ and denote its value by $M_g^l(h(g))$:

$$M_g^l(h(g)) = \lim_{n \to \infty} n^{-1} \sum_{m=1}^{n} (T^m h)(g). \tag{18}$$

Theorem 2 (J. VON NEUMANN). We have

$$M_g^l(\alpha h(g)) = \alpha M_g^l(h(g)), \tag{i}$$

$$M_g^l(h_1(g) + h_2(g)) = M_g^l(h_1(g)) + M_g^l(h_2(g)), \tag{ii}$$

$$M_g^l(1) = 1, \tag{iii}$$

$$\text{if } h(g) \geqq 0 \text{ on } G, \text{ then } M_g^l(h(g)) \geqq 0;$$

$$\text{if, moreover, } h(g) \not\equiv 0, \text{ then } M_g^l(h(g)) > 0, \tag{iv}$$

$$|M_g^l(h(g))| \leqq M_g^l(|h(g)|), \tag{v}$$

$$M_g^l(\overline{h(g)}) = \overline{M_g^l(h(g))}, \tag{vi}$$

$$M_g^l(h(ga)) = M_g^l(h(g)), \tag{vii}$$

$$M_g^l(h(ag)) = M_g^l(h(g)), \tag{vii'}$$

$$M_g^l(h(g^{-1})) = M_g^l(h(g)). \tag{viii}$$

Proof. By definition (18), it is clear that (i), (ii), (iii), the first part of (iv), (v) and (vi) are true. The truth of (vii) is proved by Proposition 2. (vii') is proved as follows.

Starting with the linear operator T' defined by

$$(T'h)(g) = \sum_{j=1}^{n} \alpha_j h(g g_j),$$

we can also define a *right mean* $M_g^r(h(g))$ which, as a functional of $h(g)$, satisfies (i), (ii), (iii), the first part of (iv), (v), (vi) and (vii'). We thus have to prove that the left mean $M_g^l(h(g))$ coincides with the right mean $M_g^r(h(g))$. By its definition of the left mean, there exist, for any $\varepsilon > 0$, a sequence of elements $\{k_j\} \subseteqq G$ and a sequence of positive numbers β^j with $\sum_{j=1}^{\infty} \beta_j = 1$ such that

$$\sup_g \left| \sum_{j=1}^{\infty} \beta_j h(k_j g) - M_g^l(h(g)) \right| \leqq \varepsilon. \tag{19}$$

Similarly, there exist a sequence of elements $\{s_j\} \subseteqq G$ and a sequence of positive numbers γ_j with $\sum_{j=1}^{\infty} \gamma_j = 1$ such that

$$\sup_g \left| \sum_{j=1}^{\infty} \gamma_j h(g s_j) - M_g^r(h(g)) \right| \leqq \varepsilon. \tag{20}$$

We have, from (19) and (vii)

$$\sup_g \left| \sum_{i,j} \gamma_i \beta_j h(k_j g s_i) - M_g^l(h(g)) \right| \leqq \varepsilon,$$

and, similarly from (20) and (vii'),

$$\sup_g \left| \sum_{i,j} \gamma_i \beta_j h(k_j g s_i) - M_g^r(h(g)) \right| \leqq \varepsilon.$$

Hence we must have $M_g^l(h(g)) = M_g^r(h(g))$.

We next remark that a linear functional $M_g(h(g))$ defined on $C(G)$ is uniquely determined by the properties (i), (ii), (iii), the first part of (iv), (v), (vi) and (vii) (or (vii') as well). In fact, we have, by (20),

$$M_g^r(h(g)) - \varepsilon \leqq \sum_{i=1}^{\infty} \gamma_i h(g s_i) \leqq M_g^r(h(g)) + \varepsilon \quad \text{for real-valued } h(g).$$

Hence, for real-valued $h(g)$, $M_g(h(g))$ must coincide with the right mean $M_g^r(h(g))$ and hence with the left mean $M_g^l(h(g))$ as well. Therefore we see that we must have $M_g = M_g^r = M_g^l$. Being equal to the right mean, M_g must satisfy (vii'). Moreover, since $M_g^l(h(g^{-1}))$ satisfies, as a linear functional, (i), (ii), (iii), the first part of (iv), (v), (vi) and (vii'), we must have $M_g^l(h(g^{-1})) = M_g^r(h(g)) = M_g^l(h(g))$.

Finally we shall prove the last part of (iv). Suppose $h(g_0) > 0$. For any $\varepsilon > 0$, there exists, by the total boundedness of G, a finite system of elements $s_1, s_2, \ldots, s_n$ such that

$$\min_{1 \leqq i \leqq n} \sup_g |h(g s_i) - h(g s)| < \varepsilon$$

for all $s \in G$. This we see by the uniform continuity of $h(g)$ and the fact that $\text{dis}(gs_i, gs) = \text{dis}(s_i, s)$. Hence, for $\varepsilon = h(g_0)/2$, we obtain, for any $s \in G$, a suffix i, $1 \leqq i \leqq n$, such that

$$h(g_0 s_i^{-1} s) \geqq h(g_0)/2.$$

Thus, by the non-negativity of the function $h(g)$, we obtain

$$\sum_{i=1}^{n} h(g_0 s_i^{-1} s) \geqq h(g_0)/2 > 0 \quad \text{for all} \quad s \in G.$$

Therefore, by taking the right mean of both sides, we have

$$M_g^r\left(\sum_{i=1}^{n} h(g_0 s_i^{-1} s)\right) = n\, M_s^r(h(s)) \geqq h(g_0)/2 > 0.$$

Remark. The happy idea of introducing the distance (1) is shown in A. WEIL [1]. The application of the mean ergodic theorem to the existence proof of the mean value is due to the present author. See also W. MAAK [1].

6. The Resolvent of a Dual Operator

Lemma 1. Let X and Y be complex B-spaces. Let T be a linear operator with $D(T)^a = X$ and $R(T) \subseteq Y$. Then $(T')^{-1}$ exists iff $R(T)^a = Y$.

Proof. If $T' y_0^* = 0$, then

$$\langle x, T' y_0^* \rangle = \langle Tx, y_0^* \rangle = 0 \quad \text{for all} \quad x \in D(T),$$

and hence $y_0^*(R(T)^a) = 0$. Thus $R(T)^a = Y$ implies $y_0^* = 0$ and so T' has an inverse. On the other hand, if $y_0 \overline{\in} R(T)^a$, then, the Hahn-Banach theorem asserts that there exists a continuous linear functional $y_0^* \in Y'$ such that $y_0^*(y_0) = 1$ and $y_0^*(R(T)^a) = 0$. Hence $\langle Tx, y_0^* \rangle = 0$ for all $x \in D(T)$, and so $y_0^* \in D(T')$ and $T' y_0^* = 0$, whereas $y_0^*(y_0) \neq 0$, i.e., $y_0^* \neq 0$. Therefore, the condition $R(T)^a \neq Y$ implies that T' cannot have an inverse.

Theorem 1 (R. S. PHILLIPS [2]). Let T be a linear operator with an inverse and such that $D(T)^a = X$ and $R(T)^a = Y$, where X and Y are B-spaces. Then

$$(T')^{-1} = (T^{-1})'. \tag{1}$$

T^{-1} is bounded on Y iff T is closed and $(T')^{-1}$ is bounded on X'_s.

Proof. $(T^{-1})'$ exists because $D(T^{-1}) = R(T)$ is dense in Y. $(T')^{-1}$ exists by Lemma 1. We have to show the equality (1). If $y \in R(T)$ and $y^* \in D(T')$, then

$$\langle y, y^* \rangle = \langle T T^{-1} y, y^* \rangle = \langle T^{-1} y, T' y^* \rangle.$$

Hence $R(T') \subseteq D((T^{-1})')$ and $(T^{-1})'(T' y^*) = y^*$ for all $y^* \in D(T')$. Thus $(T^{-1})'$ is an extension of $(T')^{-1}$. Next, if $x \in D(T)$, then

$$\langle x, x^* \rangle = \langle T^{-1} T x, x^* \rangle = \langle Tx, (T^{-1})' x^* \rangle \quad \text{for all} \quad x^* \in D((T^{-1})').$$

Hence $R((T^{-1})') \subseteqq D(T')$ and $T'(T^{-1})' x^* = x^*$ for all $x^* \in D((T^{-1})')$. Thus $(T^{-1})'$ is a contraction of $(T')^{-1}$. Therefore we have proved (1).

If, in addition, T^{-1} is bounded on Y, then $(T^{-1})'$ is also bounded. Conversely, if $(T')^{-1}$ is bounded on X'_s, then, for all $x \in R(T)$ and $x^* \in X'$, we have, by (1),

$$|\langle T^{-1}x, x^*\rangle| = |\langle x, (T^{-1})' x^*\rangle| = |\langle x, (T')^{-1} x^*\rangle| \leqq \|(T')^{-1}\| \cdot \|x^*\| \cdot \|x\|.$$

Since T^{-1} is closed and $R(T)^a = Y$, T^{-1} must be bounded.

Lemma 2. Let T be a linear operator with $D(T)^a = X$ and $R(T) \subseteqq Y$, where X and Y are B-spaces. If $R(T')$ is weakly* dense in X', then T has an inverse.

Proof. Suppose that there exists an $x_0 \neq 0$ such that $Tx_0 = 0$. Then

$$\langle x_0, T'y^*\rangle = \langle Tx_0, y^*\rangle = 0 \quad \text{for all} \quad y^* \in D(T').$$

This shows that the weak* closure of $R(T')$ is a proper linear subspace of X', contrary to the hypothesis.

Theorem 2 (R. S. Phillips [2]). Let X be a complex B-space, and T a closed linear operator with $D(T)^a = X$ and $R(T) \subseteqq X$. Then

$$\varrho(T) = \varrho(T') \quad \text{and} \quad R(\lambda; T)' = R(\lambda; T') \quad \text{for} \quad \lambda \in \varrho(T). \tag{2}$$

Proof. If $\lambda \in \varrho(T)$, then, by Theorem 1, $\lambda \in \varrho(T')$ and $R(\lambda; T)' = R(\lambda; T')$. On the other hand, if $\lambda \in \varrho(T')$, then Lemma 2 shows that $(\lambda I - T)$ has an inverse $(\lambda I - T)^{-1}$ which is closed with $(\lambda I - T)$. Lemma 1 then shows that $D((\lambda I - T)^{-1}) = R(\lambda I - T)$ is strongly dense in Y. Hence, by Theorem 1, $\lambda \in \varrho(T)$.

7. Dunford's Integral

Let X be a complex B-space and T a bounded linear operator $\in L(X, X)$. We shall define a function $f(T)$ of T by Cauchy's type integral

$$f(T) = (2\pi i)^{-1} \int_C f(\lambda)\, R(\lambda; T)\, d\lambda.$$

To this purpose, we denote by $F(T)$ the family of all complex-valued functions $f(\lambda)$ which are holomorphic in some neighbourhood of the spectrum $\sigma(T)$ of T; the neighbourhood need not be connected, and can depend on the function $f(\lambda)$. Let $f \in F(T)$, and let an open set $U \supseteqq \sigma(T)$ of the complex plane be contained in the domain of holomorphy of f, and suppose further that the boundary ∂U of U consists of a finite number of rectifiable Jordan curves, oriented in positive sense. Then the bounded linear operator $f(T)$ will be defined by

$$f(T) = (2\pi i)^{-1} \int_{\partial U} f(\lambda)\, R(\lambda; T)\, d\lambda, \tag{1}$$

and the integral on the right may be called a *Dunford's integral.* By Cauchy's integral theorem, the value $f(T)$ depends only on the function f and the operator T, but not on the choice of the domain U.

The following *operational calculus* holds:

Theorem (N. DUNFORD). If f and g are in $F(T)$, and α and β are complex numbers, then

$$\alpha f + \beta g \text{ is in } F(T) \text{ and } \alpha f(T) + \beta g(T) = (\alpha f + \beta g)(T), \tag{2}$$

$$f \cdot g \text{ is in } F(T) \text{ and } f(T) \cdot g(T) = (f \cdot g)(T), \tag{3}$$

if f has the Taylor expansion $f(\lambda) = \sum_{n=0}^{\infty} \alpha_n \lambda^n$ convergent in a neighbourhood U of $\sigma(T)$, then $f(T) = \sum_{n=0}^{\infty} \alpha_n T^n$ (in the operator norm topology), (4)

let $f_n \in F(T)$ $(n = 1, 2, \ldots)$ be holomorphic in a fixed neighbourhood U of $\sigma(T)$. If $f_n(\lambda)$ converges to $f(\lambda)$ uniformly on U, then $f_n(T)$ converges to $f(T)$ in the operator norm topology, (5)

if $f \in F(T)$, then $f \in F(T')$ and $f(T') = f(T)'$. (6)

Proof. (2) is clear. *Proof of (3).* Let U_1 and U_2 be open neighbourhoods of $\sigma(T)$ whose boundaries ∂U_1 and ∂U_2 consist of a finite number of rectifiable Jordan curves, and assume that $U_1 + \partial U_1 \subseteq U_2$ and that $U_2 + \partial U_2$ is contained in the domain of holomorphy of f and g. Then, by virtue of the resolvent equation and Cauchy's integral theorem, we obtain

$$\begin{aligned}
f(T)\, g(T) &= -(4\pi^2)^{-1} \int_{\partial U_1} f(\lambda)\, R(\lambda; T)\, d\lambda \cdot \int_{\partial U_2} g(\mu)\, R(\mu; T)\, d\mu \\
&= -(4\pi^2)^{-1} \int_{\partial U_1} \int_{\partial U_2} f(\lambda)\, g(\mu)\, (\mu - \lambda)^{-1} (R(\lambda; T) - R(\mu; T))\, d\lambda\, d\mu \\
&= (2\pi i)^{-1} \int_{\partial U_1} f(\lambda)\, R(\lambda; T) \cdot \left\{(2\pi i)^{-1} \int_{\partial U_2} (\mu - \lambda)^{-1} g(\mu)\, d\mu\right\} d\lambda \\
&\quad - (2\pi i)^{-1} \int_{\partial U_2} g(\mu)\, R(\mu; T) \cdot \left\{(2\pi i)^{-1} \int_{\partial U_1} (\mu - \lambda)^{-1} f(\lambda)\, d\lambda\right\} d\mu \\
&= (2\pi i)^{-1} \int_{\partial U_1} f(\lambda)\, g(\lambda)\, R(\lambda; T)\, d\lambda = (f \cdot g)(T).
\end{aligned}$$

Proof of (4). By hypothesis, U must contain a circle $\{\lambda; |\lambda| \leqq r_\sigma(T) + \varepsilon\}$ $\varepsilon > 0$, in its interior, where $r_\sigma(T)$ is the spectral radius of T (Theorem 4 in Chapter VIII, 2). Hence the power series $f(\lambda) = \sum_{n=0}^{\infty} \alpha_n \lambda^n$ converges uniformly on the circle $C = \{\lambda; |\lambda| \leqq r_\sigma(T) + \varepsilon\}$, for some $\varepsilon > 0$. Hence, by Cauchy's integral theorem and Laurent's expansion $R(\lambda; T) = \sum_{n=1}^{\infty} \lambda^{-n} T^{n-1}$ of $R(\lambda; T)$ (Chapter VIII, 2, (5)),

$$f(T) = (2\pi i)^{-1} \int_{\partial C} \left(\sum_{k=0}^{\infty} \alpha_k \lambda^k\right) R(\lambda; T)\, d\lambda = (2\pi i)^{-1} \sum_{k=0}^{\infty} \alpha_k \int_{\partial C} \lambda^k R(\lambda; T)\, d\lambda$$

$$= (2\pi i)^{-1} \sum_{k=0}^{\infty} \alpha_k \sum_{n=1}^{\infty} \int_{\partial C} \lambda^{k-n} T^{n-1} d\lambda = \sum_{k=0}^{\infty} \alpha_k T^k.$$

(5) is proved by (1), and (6) is also proved by (1) and formula (2) of the preceding section.

Corollary 1 (Spectral Mapping Theorem). If f is in $F(T)$, then $f(\sigma(T)) = \sigma(f(T))$.

Proof. Let $\lambda \in \sigma(T)$, and define the function g by $g(\mu) = (f(\lambda) - f(\mu))/(\lambda - \mu)$. By the Theorem, $f(\lambda)\, I - f(T) = (\lambda I - T)\, g(T)$. Hence, if $(f(\lambda)\, I - f(T))$ had a bounded inverse B, then $g(T)\, B$ would be the bounded inverse of $(\lambda I - T)$. Thus $\lambda \in \sigma(T)$ implies that $f(\lambda) \in \sigma(f(T))$. Let, conversely, $\lambda \in \sigma(f(T))$, and assume that $\lambda \bar{\in} f(\sigma(T))$. Then the function $g(\mu) = (f(\mu) - \lambda)^{-1}$ must belong to $F(T)$, and so, by the preceding Theorem, $g(T)\,(f(T) - \lambda I) = I$ which contradicts the assumption $\lambda \in \sigma(f(T))$.

Corollary 2. If $f \in F(T)$, $g \in F(f(T))$ and $h(\lambda) = g(f(\lambda))$, then h is in $F(T)$ and $h(T) = g(f(T))$.

Proof. That $h \in F(T)$ follows from Corollary 1. Let U_1 be an open neighbourhood of $\sigma(f(T))$ whose boundary U_1 consists of a finite number of rectifiable Jordan curves such that $U_1 + \partial U_1$ is contained in the domain of holomorphy of g. Let U_2 be a neighbourhood of $\sigma(T)$ whose boundary ∂U_2 consists of a finite number of rectifiable Jordan curves such that $U_2 + \partial U_2$ is contained in the domain of holomorphy of f and $f(U_2 + \partial U_2) \subseteqq U_1$. Then we have, for $\lambda \in \partial U_1$,

$$R(\lambda; f(T)) = (2\pi i)^{-1} \int_{\partial U_2} (\lambda - f(\mu))^{-1} R(\mu; T)\, d\mu,$$

since the right hand operator S satisfies, by (3), the equation $(\lambda I - f(T))\, S = S(\lambda I - f(T)) = I$. Hence, by Cauchy's integral theorem,

$$g(f(T)) = (2\pi i)^{-1} \int_{\partial U_1} g(\lambda) R(\lambda; f(T))\, d\lambda$$

$$= (-4\pi^2)^{-1} \int_{\partial U_1} \int_{\partial U_2} g(\lambda)\, (\lambda - f(\mu))^{-1} R(\mu; T)\, d\mu\, d\lambda$$

$$= (2\pi i)^{-1} \int_{\partial U_2} R(\mu; T)\, g(f(\mu))\, d\mu = h(T).$$

Remark. The introduction of the operational calculus based on a formula like (1) goes back to the investigations by H. POINCARÉ on continuous groups (1899). The exposition of the operational calculus in this section is adapted from N. DUNFORD-J. SCHWARTZ [1]. In the next chapter on semi-groups, we shall frequently make use of Dunford's integral for a closed unbounded operator T.

8. The Isolated Singularities of a Resolvent

Let λ_0 be an isolated singular point of the resolvent $R(\lambda; T)$ of a closed linear operator T on a complex B-space X into X. Then $R(\lambda; T)$ can be expanded into Laurent series

$$R(\lambda; T) = \sum_{n=-\infty}^{\infty} (\lambda-\lambda_0)^n A_n, \quad A_n = (2\pi i)^{-1} \int_C (\lambda-\lambda_0)^{-n-1} R(\lambda; T)\, d\lambda, \tag{1}$$

where C is a circumference of sufficiently small radius: $|\lambda-\lambda_0| = \varepsilon$ such that the circle $|\lambda-\lambda_0| \leqq \varepsilon$ does not contain other singularities than $\lambda = \lambda_0$, and the integration is performed counter-clockwise. By virtue of the resolvent equation we obtain

Theorem 1. A's are mutually commutative bounded linear operators and

$$\begin{aligned} TA_k &= A_k T \quad (k = 0, \pm 1, \pm 2, \ldots), \\ A_k A_m &= 0 \quad \text{for} \quad k \geqq 0,\ m \leqq -1, \\ A_n &= (-1)^n A_0^{n+1} \quad (n \geqq 1), \\ A_{-p-q+1} &= A_{-p} A_{-q} \quad (p, q \geqq 1). \end{aligned} \tag{2}$$

Proof. The boundedness and the mutual commutativity of A's and the commutativity of A's with T are clear from the integral representation of A's.

We substitute the expansion of $R(\lambda; T)$ in the resolvent equation $R(\lambda; T) - R(\mu; T) = (\mu - \lambda) R(\lambda; T) R(\mu; T)$, and obtain

$$\sum_{k=-\infty}^{\infty} A_k \frac{(\lambda-\lambda_0)^k - (\mu-\lambda_0)^k}{(\lambda-\lambda_0) - (\mu-\lambda_0)} = - \sum_{k,m=-\infty}^{\infty} A_k A_m (\lambda-\lambda_0)^k (\mu-\lambda_0)^m.$$

The coefficient of A_n on the left is

$$(\lambda-\lambda_0)^{n-1} + (\lambda-\lambda_0)^{n-2}(\mu-\lambda_0) + \cdots + (\mu-\lambda_0)^{n-1}, \quad n \geqq 1,$$

$$-\{(\lambda-\lambda_0)^n (\mu-\lambda_0)^{-1} + (\lambda-\lambda_0)^{n+1}(\mu-\lambda_0)^{-2} + \cdots$$
$$+ (\lambda-\lambda_0)^{-1}(\mu-\lambda_0)^n\}, \quad n < 0.$$

Hence the terms containing $(\lambda-\lambda_0)^k (\mu-\lambda_0)^m$ with $k \geqq 0$ and $m \leqq -1$ are missing so that we must have the orthogonality $A_k A_m = 0$ ($k \geqq 0$, $m \leqq -1$). Hence

$$R^+(\lambda; T) = \sum_{n=0}^{\infty} A_n (\lambda-\lambda_0)^n \quad \text{and} \quad R^-(\lambda; T) = \sum_{n=-\infty}^{-1} A_n (\lambda-\lambda_0)^n$$

must both satisfy the resolvent equation. Substituting the expansion of $R^+(\lambda; T)$ in the resolvent equation

$$R^+(\lambda; T) - R^+(\lambda; T) = (\mu - \lambda)\, R^+(\lambda; T)\, R^+(\mu; T),$$

we obtain, setting $(\lambda - \lambda_0) = h$, $(\mu - \lambda_0) = k$,

$$\sum_{p=1}^{\infty} A_p (h^p - k^p) = (k - h)\left(\sum_{p=0}^{\infty} A_p h^p\right)\left(\sum_{q=0}^{\infty} A_q k^q\right).$$

Hence, dividing both sides by $(k - h)$, we obtain

$$-\sum_{p=1}^{\infty} A_p (h^{p-1} + h^{p-2} k + \cdots + k^{p-1}) = \sum_{p,q=0}^{\infty} h^p k^q A_p A_q$$

so that we have $-A_{p+q+1} = A_p A_q$ $(p, q \geqq 0)$. Thus, in particular, $A_1 = -A_0^2$, $A_2 = -A_1 A_0 = (-1)^2 A_0^3, \ldots, A_n = (-1)^n A_0^{n+1}$ $(n \geqq 1)$.

Similarly, from the resolvent equation for $R^-(\lambda; T)$, we obtain, setting $(\lambda - \lambda_0)^{-1} = h$, $(\mu - \lambda_0)^{-1} = k$,

$$\sum_{p=1}^{\infty} A_{-p} (h^{p-1} + h^{p-2} k + \cdots + k^{p-1}) = \sum_{p,q=1}^{\infty} h^{p-1} k^{q-1} A_{-p} A_{-q},$$

so that we have $A_{-p-q+1} = A_{-p} A_{-q}$ $(p, q \geqq 1)$. In particular, we have

$$A_{-1} = A_{-1}^2,\ A_{-2} = A_{-1} A_{-2}, \ldots, A_{-n} = A_{-1} A_{-n}\ (n \geqq 1).$$

Theorem 2. We have

$$\begin{aligned} &A_n = (T - \lambda_0 I)\, A_{n+1} \quad (n \geqq 0), \\ &(T - \lambda_0 I)\, A_{-n} = A_{-(n+1)} = (T - \lambda_0 I)^n A_{-1} \quad (n \geqq 1), \\ &(T - \lambda_0 I)\, A_0 = A_{-1} - I. \end{aligned} \tag{3}$$

Proof. By the integral representation of A_k, we see that the range $R(A_k)$ is in the domain of T, so that we can multiply A_k by T on the left. Thus our Theorem is proved by the identity

$$\begin{aligned} I &= (\lambda I - T) \sum_{k=-\infty}^{\infty} A_k (\lambda - \lambda_0)^k \\ &= \{(\lambda - \lambda_0)\, I + (\lambda_0 I - T)\} \sum_{k=-\infty}^{\infty} A_k (\lambda - \lambda_0)^k. \end{aligned}$$

Theorem 3. If λ_0 is a pole of $R(\lambda; T)$ of order m, then λ_0 is an eigenvalue of T. We have

$$R(A_{-1}) = N((\lambda_0 I - T)^n) \text{ and } R(I - A_{-1}) = R((\lambda_0 I - T)^n) \text{ for } n \geqq m, \tag{4}$$

so that, in particular,

$$X = N((\lambda_0 I - T)^n) \oplus R((\lambda_0 I - T)^n) \quad \text{for} \quad n \geqq m. \tag{5}$$

Proof. Since A_{-1} is a bounded linear operator satisfying $A_{-1}^2 = A_{-1}$, it is easy to see that

$$N(A_{-1}) = R(I - A_{-1}). \tag{6}$$

We put $X_1 = N(A_{-1}) = R(I - A_{-1})$, and put also

$$X_2 = R(A_{-1}),\ N_n = N((\lambda_0 I - T)^n) \text{ and } R_n = R_n((\lambda_0 I - T)^n). \tag{7}$$

Let $x \in N_n$ where $n \geqq 1$. Then we see, by $(T - \lambda_0 I)^n A_{n-1} = (T - \lambda_0 I) A_0 = A_{-1} - I$, that $0 = A_{n-1}(T - \lambda_0 I)^n x = (T - \lambda_0 I)^n A_{n-1} x = (T - \lambda_0 I) A_0 x = A_{-1}x - x$ so that $x = A_{-1}x \in X_2$. Thus N_n with $n \geqq 1$ belongs to X_2. Let, conversely, $x \in X_2$. Then we have $x = A_{-1}y$ and so $x = A_{-1}A_{-1}y = A_{-1}x$ by $A_{-1} = A_{-1}^2$; consequently, we have $(T - \lambda_0 I)^n x = A_{-(n+1)}\, x$ by $(T - \lambda_0 I)^n A_{-1} = A_{-(n+1)}$. Since $A_{-(n+1)} = 0$ for $n \geqq m$ by hypothesis, it follows that $X_2 \subseteqq N_n$ for $n \geqq m$ and so

$$N_n = X_2 \quad \text{if} \quad n \geqq m. \tag{8}$$

Because $(T - \lambda_0 I) A_{-m} = A_{-(m+1)} = 0$ and $A_{-m} \neq 0$, the number λ_0 is an eigenvalue of T.

We see that $X_1 = N(A_{-1}) = R(I - A_{-1}) \subseteqq R_n$ by $(T - \lambda_0 I)^n A_{n-1} = A_{-1} - I$. If $n \geqq m$, then $x \in R_n \cap N_n$ implies $x = 0$. For, if $x = (\lambda_0 I - T)^n y$ and $(\lambda_0 I - T)^n x = 0$, then, by (8), $y \in N_{2n} = N_n$ and therefore $x = 0$. Next suppose $x \in R_n$ with $n \geqq m$, and write $x = x_1 + x_2$ where $x_1 = (I - A_{-1})\, x \in X_1$, $x_2 = A_{-1}x \in X_2$. Then, since $X_1 \subseteqq R_n$, $x_2 = x - x_1 \in R_n$. But $x_2 \in X_2 = N_n$ by (8), and so $x_2 \in R_n \cap N_n$, $x_2 = 0$. This proves that $x = x_1 \in X_1$. Therefore we have proved that $R_n = X_1$ if $n \geqq m$.

Theorem 4. If, in particular, T is a bounded linear operator such that $X_2 = R(A_{-1})$ is a finite dimensional linear subspace of X, then λ_0 is a pole of $R(\lambda; T)$.

Proof. Let $x_1, x_2, \ldots, x_k$ be a base of the linear space X_2. Since $x_1, Tx_1, T^2x_1, \ldots, T^k x_1$ are linearly dependent vectors of X_2, there exists a non-zero polynomial $P_1(\lambda)$ such that $P_1(T)x_1 = 0$. Similarly there exist non-zero polynomials $P_2(\lambda), \ldots, P_k(\lambda)$ such that $P_j(T)x_j = 0$ $(j = 2, 3, \ldots, k)$. Then for the polynomial $P(\lambda) = \prod_{j=1}^{k} P_j(\lambda)$, we must have $P(T)x_j = 0$ $(j = 1, \ldots, k)$ and hence $P(T)x = 0$ for every $x \in X_2$.

Let

$$P(\lambda) = \alpha \prod_{j=0}^{s} (\lambda - \lambda_j)^{\nu_j} \quad (\alpha \neq 0)$$

be the factorization of $P(\lambda)$. Then we can prove that $(T - \lambda_0 I)^{\nu_0} x = 0$ for every $x \in X_2$. Assume the contrary, and let $x_0 \in X_2$ be such that $(T - \lambda_0 I)^{\nu_0} x_0 \neq 0$. Then, by $P(T)x_0 = 0$, we see that there exist at least one λ_j $(j \neq 0)$ and a polynomial $Q(\lambda)$ such that

$$(T - \lambda_j I)\, Q(T)\, (T - \lambda_0 I)^{\nu_0} x_0 = 0$$

and $y = Q(T)(T - \lambda_0 I)^{\nu_0} x_0 \neq 0$. Thus $y \in X_2$ is an eigenvector of T corresponding to the eigenvalue λ_j. Hence $(\lambda I - T)y = (\lambda - \lambda_j)y$ and so, multiplying both sides by $R(\lambda; T)$, we obtain $y = (\lambda - \lambda_j) R(\lambda; T)y$ which implies

$$y = A_{-1}y = (2\pi i)^{-1} \int_C R(\lambda; T)\, y\, d\lambda = (2\pi i)^{-1} \int_C (\lambda - \lambda_j)^{-1}\, y\, d\lambda = 0,$$

by taking the circumference C with λ_0 as centre sufficiently small. This is a contradiction, and so there must exist a positive integer m such that $(T - \lambda_0 I)^m X_2 = 0$. Thus, by $X_2 = R(A_{-1})$ and $(T - \lambda_0 I)^n A_{-1} = A_{-(n+1)}$, we see that $A_{-(n+1)} = 0$ for $n \geqq m$.

Comments and References

Section 6 is adapted from R. S. Phillips [2]. Section 8 is adapted from M. Nagumo [1] and A. Taylor [1]. Parts of these sections can easily be extended to the case of a locally convex linear topological space. See, e.g., section 13 of the following chapter.

IX. Analytical Theory of Semi-groups

The *analytical theory of semi-groups* of bounded linear operators in a B-space deals with the exponential functions in infinite dimensional function spaces. It is concerned with the problem of determining the most general bounded linear operator valued function $T(t)$, $t \geqq 0$, which satisfies the equations

$$T(t + s) = T(t) \cdot T(s), \quad T(0) = I.$$

The problem was investigated by E. Hille [2] and K. Yosida [5] independently of each other around 1948. They introduced the notion of the *infinitesimal generator* A of $T(t)$ defined by

$$A = s\text{-}\lim_{t \downarrow 0} t^{-1}(T(t) - I),$$

and discussed the generation of $T(t)$ in terms of A and obtained a characterization of the infinitesimal generator A in terms of the spectral property of A.

The basic result of the semi-group theory may be considered as a natural generalization of the theorem of M. H. Stone [2] on one-parameter group of unitary operators in a Hilbert space, which will be explained in a later section. Applications of the theory to *stochastic processes* and to the *integration of the equations of evolution*, which include diffusion equations, wave equations and Schrödinger equations, will be discussed in Chapter XIV.

In this chapter, we shall develop the theory of semi-groups of continuous linear operators in locally convex linear topological spaces rather than in Banach spaces.

1. The Semi-group of Class (C_0)

Proposition (E. HILLE). Let X be a B-space, and T_t, $t \geqq 0$, a one-parameter family of bounded linear operators $\in L(X, X)$ satisfying the *semi-group property*

$$T_t T_s = T_{t+s} \quad \text{for} \quad t, s > 0. \tag{1}$$

If $p(t) = \log \|T_t\|$ is bounded from above on the interval $(0, a)$ for each positive a, then

$$\lim_{t \to \infty} t^{-1} \log \|T_t\| = \inf_{t>0} t^{-1} \log \|T_t\|. \tag{2}$$

Proof. We have $p(t+s) \leqq p(t) + p(s)$ from $\|T_{t+s}\| = \|T_t T_s\| \leqq \|T_t\| \cdot \|T_s\|$. Let $\beta = \inf_{t>0} t^{-1} \cdot p(t)$. β is either finite or $-\infty$. Suppose that β is finite. We choose, for any $\varepsilon > 0$, a number $a > 0$ in such a way that $p(a) \leqq (\beta + \varepsilon)\, a$. Let $t > a$ and n be an integer such that $na \leqq t < (n+1)\, a$. Then

$$\begin{aligned} \beta &\leqq \frac{p(t)}{t} \leqq \frac{p(na)}{t} + \frac{p(t-na)}{t} \leqq \frac{na}{t} \frac{p(a)}{a} + \frac{p(t-na)}{t} \\ &\leqq \frac{na}{t}(\beta + \varepsilon) + \frac{p(t-na)}{t}. \end{aligned}$$

By hypothesis, $p(t - na)$ is bounded from above as $t \to \infty$. Thus, letting $t \to \infty$ in the above inequality, we obtain $\lim_{t \to \infty} t^{-1} p(t) = \beta$. The case $\beta = -\infty$ may be treated similarly.

Definition 1. If $\{T_t; t \geqq 0\} \subseteq L(X, X)$ satisfy the conditions

$$T_t T_s = T_{t+s} \ (\text{for } t, s \geqq 0), \tag{1'}$$

$$T_0 = I, \tag{3}$$

$$\text{s-}\lim_{t \to t_0} T_t x = T_{t_0} x \text{ for each } t_0 \geqq 0 \text{ and each } x \in X, \tag{4}$$

then $\{T_t\}$ is called a *semi-group of class* (C_0).

In virtue of the Proposition, we see that a semi-group $\{T_t\}$ of class (C_0) satisfies the condition

$$\|T_t\| \leqq M\, e^{\beta t} \quad (\text{for } 0 \leqq t < \infty), \tag{5}$$

with constants $M > 0$ and $\beta < \infty$.

The proof is easy. We have only to show that $\|T_t\|$ is, for any interval $(0, a)$ with $\infty > a > 0$, bounded on $(0, a)$. Assume the contrary and let there exist a sequence $\{t_n\} \subseteq (0, a)$ such that $\|T_{t_n}\| > n$ and $\lim_{n \to \infty} t_n = t_0 \leqq a$. By the resonance theorem, $\|T_{t_n} x\|$ must be unbounded at least for one $x \in X$, which surely contradicts the strong continuity condition (4).

Remark. By multiplying $e^{-\beta t}$, we may assume that a semi-group $\{T_t\}$ of class (C_0) is *equi-bounded*:

$$\|T_t\| \leqq M \quad (\text{for } 0 \leqq t < \infty). \tag{6}$$

If, in particular, M is $\leqq 1$, i.e., if

$$\|T_t\| \leqq 1 \quad (\text{for } 0 \leqq t < \infty), \tag{7}$$

then the semi-group $\{T_t\}$ is called a *contraction semi-group of class* (C_0).

As for the strong continuity condition (4), we have the following

Theorem. Let a family $\{T_t; t \geqq 0\}$ of operators $\in L(X, X)$ satisfy (1') and (3). Then condition (4) is equivalent to the condition

$$w\text{-}\lim_{t \downarrow 0} T_t x = x \quad \text{for every} \quad x \in X. \tag{8}$$

Proof. Suppose that (8) is satisfied. Let x_0 be any fixed element of X. We shall show that $s\text{-}\lim_{t \to t_0} T_t x_0 = T_{t_0} x_0$ for each $t_0 \geqq 0$. Consider the function $x(t) = T_t x_0$. For each $t_0 \geqq 0$, $x(t)$ is weakly right continuous at t_0, because $w\text{-}\lim_{t \downarrow t_0} T_t x = w\text{-}\lim_{h \downarrow 0} T_h T_{t_0} x_0 = T_{t_0} x_0$. We next prove that $\|T_t\|$ is bounded in a vicinity of $t = 0$. For, otherwise, there would exist a sequence $\{t_n\} \in$ such that $t_n \downarrow 0$ and $\lim_{n \to \infty} \|T_{t_n} x_0\| = \infty$, contrary to the resonance theorem implied by the weak right continuity of $x(t) = T_t x_0$. Thus by (1'), we see that $T_t x_0 = x(t)$ is bounded on any compact interval of t. Moreover, $x(t)$ is weakly measurable. For, a right continuous real-valued function $f(t)$ is Lebesgue measurable, as may be proved from the fact that, for any α, the set $\{t; f(t) < \alpha\}$ is representable as the union of intervals of positive length. Next let $\{t_n\}$ be the totality of positive rational numbers, and consider finite linear combinations $\sum_j \beta_j x(t_j)$ where β_j are rational numbers (if X is a complex linear space, we take $\beta_j = a_j + i b_j$ with rational coefficients a_j and b_j). These elements form a countable set $M = \{x_n\}$ such that $\{x(t); t \geqq 0\}$ is contained in the strong closure of M. If otherwise, there would exist a number t' such that $x(t')$ does not belong to M^a. But, being a closed linear subspace of X, M^a is weakly closed by Theorem 11 in Chapter V, 1; consequently, the condition $x(t') \overline{\in} M^a$ is contradictory to the weak right continuity of $x(t)$, i.e., to $x(t') = w\text{-}\lim_{t_n \downarrow t'} x(t_n)$.

We may thus apply Pettis' theorem in Chapter V, 4 to the effect that $x(t)$ is strongly measurable, and so by the boundedness of $\|x(t)\|$ on any compact interval of t, we may define the Bochner integral $\int_\alpha^\beta x(t)\,dt$ and we have $\left\| \int_\alpha^\beta x(t)\,dt \right\| \leqq \int_\alpha^\beta \|x(t)\|\,dt$ for $0 \leqq \alpha < \beta < \infty$. By virtue of the strong continuity in s of the integral $\int_\alpha^\beta x(t+s)\,dt = \int_{\alpha+s}^{\beta+s} x(t)\,dt$, which is implied by the boundedness of $x(t)$ on any compact interval of t, N. Dunford [3] proved that $x(t)$ is strongly continuous in $t > 0$. We shall follow his proof.

Let $0 \leqq \alpha < \eta < \beta < \xi - \varepsilon < \xi$ with $\varepsilon > 0$. Since $x(\xi) = T_\xi x_0 = T_\eta T_{\xi-\eta} x_0 = T_\eta x(\xi - \eta)$, we have

$$(\beta - \alpha)\, x(\xi) = \int_\alpha^\beta x(\xi)\, d\eta = \int_\alpha^\beta T_\eta\, x(\xi - \eta)\, d\eta,$$

and so, by $\sup_{\alpha \leqq \eta \leqq \beta} \|T_\eta\| < \infty$ which is implied by (1') and (3), combined with the boundedness of $\|T_t\|$ near $t = 0$, we obtain

$$(\beta - \alpha)\, \{x(\xi \pm \varepsilon) - x(\xi)\} = \int_\alpha^\beta T_\eta \{x(\xi \pm \varepsilon - \eta) - x(\xi - \eta)\}\, d\eta,$$

$$(\beta - \alpha)\, \|x(\xi \pm \varepsilon) - x(\xi)\| \leqq \sup_{\alpha \leqq \eta \leqq \beta} \|T_\eta\| \cdot \int_{\xi-\beta}^{\xi-\alpha} \|x(\tau \pm \varepsilon) - x(\tau)\|\, d\tau.$$

The right hand side tends to zero as $\varepsilon \downarrow 0$, as may be seen by approximating $x(\tau)$ by finitely-valued functions.

We have thus proved that $x(t)$ is strongly continuous in $t > 0$. To prove the strong continuity of $x(t)$ at $t = 0$, we proceed as follows. For any positive rational number t_n, we have $T_t x(t_n) = T_t T_{t_n} x_0 = T_{t+t_n} x_0 = x(t + t_n)$. Hence, by the strong continuity of $x(t)$ for $t > 0$ proved above, we have $s\text{-}\lim_{t \downarrow 0} T_t x(t_n) = x(t_n)$. Since each $x_m \in M$ is a finite linear combination of $x(t_n)$'s, we have $s\text{-}\lim_{t \downarrow 0} T_t x_m = x_m$ $(m = 1, 2, \ldots)$. On the other hand, we have, for any $t \in [0, 1]$,

$$\begin{aligned} \|x(t) - x_0\| &\leqq \|T_t x_m - x_m\| + \|x_m - x_0\| + \|T_t(x_0 - x_m)\| \\ &\leqq \|T_t x_m - x_m\| + \|x_m - x_0\| + \sup_{0 \leqq t \leqq 1} \|T_t\| \cdot \|x_0 - x_m\|. \end{aligned}$$

Hence $\overline{\lim_{t \downarrow 0}} \|x(t) - x_0\| \leqq (1 + \sup_{0 \leqq t \leqq 1} \|T_t\|)\, \|x_m - x_0\|$, and so $s\text{-}\lim_{t \downarrow 0} x(t) = x_0$ by $\inf_{x_m \in M} \|x_0 - x_m\| = 0$.

2. The Equi-continuous Semi-group of Class (C_0) in Locally Convex Spaces. Examples of Semi-groups

Suggested by the preceding section, we shall pass to a more general class of semi-groups.

Definition. Let X be a locally convex linear topological space, and $\{T_t;\ t \geqq 0\}$ be a one-parameter family of continuous linear operators $\in L(X, X)$ such that

$$T_t T_s = T_{t+s},\ T_0 = I, \tag{1}$$

$$\lim_{t \to t_0} T_t x = T_{t_0} x \quad \text{for any} \quad t_0 \geqq 0 \quad \text{and} \quad x \in X, \tag{2}$$

the family of mappings $\{T_t\}$ is *equi-continuous* in t, i.e., for any continuous semi-norm p on X, there exists a continuous semi-norm q on X such that $p(T_t x) \leqq q(x)$ for all $t \geqq 0$ and all $x \in X$. (3)

Such a family $\{T_t\}$ is called an *equi-continuous semi-group of class* (C_0).

The semi-groups $\{T_t\}$ satisfying conditions (1'), (3), (4) and (6) of Section 1 are example of such equi-continuous semi-groups of class (C_0). We shall give concrete examples.

Example 1. Let $C[0,\infty]$ be the space of bounded uniformly continuous real-valued (or complex-valued) functions on the interval $[0,\infty)$, and define T_t, $t \geqq 0$, on $C[0,\infty]$ into $C[0,\infty]$ by

$$(T_t x)(s) = x(t+s).$$

Condition (1) is trivially satisfied. (2) follows from the uniform continuity of $x(s)$. Finally $\|T_t\| \leqq 1$ and so $\{T_t\}$ is a contraction semi-group of class (C_0). In this example, we could replace $C[0,\infty]$ by $C[-\infty,\infty]$ or by $L^p(-\infty,\infty)$.

Example 2. Consider the space $C[-\infty,\infty]$. Let

$$N_t(u) = (2\pi t)^{-1/2} e^{-u^2/2t}, \quad -\infty < u < \infty, \ t > 0,$$

which is the Gaussian probability density. Define T_t, $t \geqq 0$, on $C[-\infty,\infty]$ into $C[-\infty,\infty]$ by

$$(T_t x)(s) = \int_{-\infty}^{\infty} N_t(s-u)\, x(u)\, du, \quad \text{for } t > 0,$$

$$= x(s), \quad \text{for } t = 0.$$

Each T_t is continuous, since, by $\int_{-\infty}^{\infty} N_t(s-u)\, du = 1$,

$$\|T_t x\| \leqq \|x\| \int_{-\infty}^{\infty} N_t(s-u)\, du = \|x\|.$$

$T_0 = I$ by the definition, and the semi-group property $T_t T_s = T_{t+s}$ is a consequence from the well-known formula concerning the Gaussian probability distribution:

$$\frac{1}{\sqrt{2\pi(t+t')}} e^{-u^2/2(t+t')} = \frac{1}{\sqrt{2\pi t}} \frac{1}{\sqrt{2\pi t'}} \int_{-\infty}^{\infty} e^{-(u-v)^2/2t} e^{-v^2/2t'}\, dv.$$

This formula may be proved by applying the Fourier transformation on both sides, remembering (10) and (13) in Chapter VI, 1. To prove the strong continuity in t of T_t, we observe that $x(s) = \int_{-\infty}^{\infty} N_t(s-u)\, x(s)\, du$. Thus

$$(T_t x)(s) - x(s) = \int_{-\infty}^{\infty} N_t(s-u)\,(x(u) - x(s))\, du,$$

which is, by the change of variable $(s-u)/\sqrt{t} = z$, equal to

$$\int_{-\infty}^{\infty} N_1(z)\,(x(s-\sqrt{t}\,z) - x(s))\, dz.$$

By the uniform continuity of $x(s)$ on $(-\infty, \infty)$, there exists, for any $\varepsilon > 0$, a number $\delta = \delta(\varepsilon) > 0$ such that $|x(s_1) - x(s_2)| \leqq \varepsilon$ whenever $|s_1 - s_2| \leqq \delta$. Splitting the last integral, we obtain

$$\begin{aligned} |(T_t x)(s) - x(s)| &\leqq \int_{|\sqrt{t}\,z| \leqq \delta} N_1(z)\, |x(s - \sqrt{t}\cdot z) - x(s)|\, dz + \int_{|\sqrt{t}\,z| > \delta} (\cdots)\, dz \\ &\leqq \varepsilon \int_{|\sqrt{t}\,z| \leqq \delta} N_1(z)\, dz + 2\,||x|| \int_{|\sqrt{t}\,z| > \delta} N_1(z)\, dz \\ &\leqq \varepsilon + 2\,||x|| \int_{|\sqrt{t}\,z| > \delta} N_1(z)\, dz. \end{aligned}$$

The second term on the right tends to 0 as $t \to 0$, since the integral $\int_{-\infty}^{\infty} N_1(z)\, dz$ converges. Thus $\varlimsup_{t \downarrow 0} \sup_s |(T_t x)(s) - x(s)| = 0$ and hence $s\text{-}\lim_{t \downarrow 0} T_t x = x$; consequently, by the Theorem in the preceding section, we have proved (2).

In the above example, we can replace $C[-\infty, \infty]$ by $L^p(-\infty, \infty)$. Consider, for example, $L^1(-\infty, \infty)$. In this case, we have

$$||T_t x|| \leqq \iint_{-\infty}^{\infty} N_t(s - u)\, |x(u)|\, ds\, du \leqq ||x||,$$

by Fubini's theorem. As for the strong continuity, we have, as above,

$$\begin{aligned} ||T_t x - x|| &= \int_{-\infty}^{\infty} \Big| \int N_1(z)\, (x(s - \sqrt{t}\cdot z) - x(s))\, dz \Big|\, ds \\ &\leqq \int_{-\infty}^{\infty} N_1(z) \left[\int_{-\infty}^{\infty} |x(s - \sqrt{t}\cdot z) - x(s)|\, ds \right] dz \\ &\leqq 2 \int_{-\infty}^{\infty} N_1(z)\, dz \cdot ||x||. \end{aligned}$$

Hence, by the Lebesgue-Fatou lemma, we obtain

$$\varlimsup_{t \downarrow 0} ||T_t x - x|| \leqq \int_{-\infty}^{\infty} N_1(z) \left(\varlimsup_{t \downarrow 0} \int_{-\infty}^{\infty} |x(s - \sqrt{t}\cdot z) - x(s)|\, ds \right) dz = 0,$$

because of the continuity in the mean of the Lebesgue integral. which may be proved by approximating $x(s)$ by finitely-valued functions.

Example 3. Consider $C[-\infty, \infty]$. Let $\lambda > 0$, $\mu > 0$. Define T_t, $t \geqq 0$, on $C[-\infty, \infty]$ into $C[-\infty, \infty]$ by

$$(T_t x)(s) = e^{-\lambda t} \sum_{k=0}^{\infty} \frac{(\lambda t)^k}{k!}\, x(s - k\mu).$$

We have

$$(T_w(T_t x))(s) = e^{-\lambda w} \sum_{m=0}^{\infty} \frac{(\lambda w)^m}{m!} \left[e^{-\lambda t} \sum_{k=0}^{\infty} \frac{(\lambda t)^k}{k!} x(s - k\mu - m\mu) \right].$$

$$= e^{-\lambda(w+t)} \sum_{p=0}^{\infty} \frac{1}{p!} \left[p! \sum_{m=0}^{p} \frac{(\lambda w)^m}{m!} \frac{(\lambda t)^{p-m}}{(p-m)!} x(s - p\mu) \right]$$

$$= e^{-\lambda(w+t)} \sum_{p=0}^{\infty} \frac{1}{p!} (\lambda w + \lambda t)^p x(s - p\mu) = (T_{w+t} x)(s).$$

Thus it is easy to verify that T_t is a contraction semi-group of class (C_0).

3. The Infinitesimal Generator of an Equi-continuous Semi-group of Class (C_0)

Let $\{T_t; t \geqq 0\}$ be an equi-continuous semi-group of class (C_0) defined on a locally convex linear topological space X which we assume to be *sequentially complete*. We define the infinitesimal generator A of T_t by

$$A x = \lim_{h \downarrow 0} h^{-1}(T_h - I) x, \tag{1}$$

i.e., A is the linear operator whose domain is the set $D(A) = \{x \in X; \lim_{h \downarrow 0} h^{-1}(T_h - I) x$ exists in $X\}$, and, for $x \in D(A)$, $A x = \lim_{h \downarrow 0} h^{-1}(T_h - I) x$. $D(A)$ is non-empty; it contains at least the vector 0. Actually $D(A)$ is larger. We can prove

Theorem 1. $D(A)$ is dense in X.

Proof. Let $\varphi_n(s) = n e^{-ns}$, $n > 0$. Consider the linear operator C_{φ_n} which is the *Laplace transform* of T_t multiplied by n:

$$C_{\varphi_n} x = \int_0^{\infty} \varphi_n(s) T_s x \, ds \quad \text{for} \quad x \in X, \tag{2}$$

the integral being defined in the sense of Riemann. The ordinary procedure of defining the Riemann integral of numerical functions can be extended to a function with values in a locally convex, sequentially complete space X, using continuous semi-norms p on X in place of the absolute value of a number. The convergence of the improper integral is a consequence of the equi-continuity of T_t, the inequality

$$p(\varphi_n(s) T_s x) = n e^{-ns} p(T_s x)$$

and the sequential completeness of X.

We see, by

$$p(C_{\varphi_n} x) \leqq \int_0^{\infty} n e^{-ns} p(T_s x) \, ds \leqq \sup_{s \geqq 0} p(T_s x),$$

that the operator C_{φ_n} is a continuous linear operator $\in L(X, X)$. We shall show that

$$R(C_{\varphi_n}) \subseteqq D(A) \quad \text{for each } n > 0, \tag{3}$$

and

$$\lim_{n\to\infty} C_{\varphi_n} x = x \quad \text{for each} \quad x \in X. \tag{4}$$

Then $\bigcup_{n=1}^{\infty} R(C_{\varphi_n})$ will be dense in X, and a fortiori $D(A)$ will be dense in X. To prove (3), we start with the formula

$$h^{-1}(T_h - I)\, C_{\varphi_n} x = h^{-1} \int_0^\infty \varphi_n(s)\, T_h T_s x\, ds - h^{-1} \int_0^\infty \varphi_n(s)\, T_s x\, ds.$$

The change of the order: $T_h \int_0^\infty \cdots = \int_0^\infty T_h \cdots$ is justified, using the linearity and the continuity of T_h. Thus

$$\begin{aligned} h^{-1}(T_h - I)\, C_{\varphi_n} x &= h^{-1} \int_0^\infty \varphi_n(s)\, T_{s+h} x\, ds - h^{-1} \int_0^\infty \varphi_n(s)\, T_s x\, ds \\ &= \frac{e^{nh}-1}{h}\, n \int_h^\infty e^{-n\sigma} T_\sigma x\, d\sigma - \frac{1}{h}\, n \int_0^h e^{-ns}\, T_s x\, ds \\ &= \frac{e^{nh}-1}{h} \left\{ C_{\varphi_n} x - \int_0^h n e^{-n\sigma} T_\sigma x\, d\sigma \right\} - \frac{1}{h} \int_0^h \varphi_n(s)\, T_s x\, ds. \end{aligned}$$

By the continuity of $\varphi_n(s)\, T_s x$ in s, the second term on the right tends to $-\varphi_n(0)\, T_0\, x = -nx$ as $h \downarrow 0$. Similarly, the first term on the right tends to $n C_{\varphi_n} x$ as $h \downarrow 0$. Hence we have

$$A\, C_{\varphi_n} x = n(C_{\varphi_n} - I)\, x, \; x \in X. \tag{5}$$

We next prove (4). We have, by $\int_0^\infty n e^{-ns}\, ds = 1$,

$$C_{\varphi_n} x - x = n \int_0^\infty e^{-ns}\, (T_s x - x)\, ds,$$

$$p(C_{\varphi_n} x - x) \leqq n \int_0^\infty e^{-ns}\, p(T_s x - x)\, ds = n \int_0^\delta + n \int_\delta^\infty = I_1 + I_2, \text{ say,}$$

where $\delta > 0$ is a positive number. For any $\varepsilon > 0$, we can choose, by the continuity in s of $T_s x$, a $\delta > 0$ such that $p(T_s x - x) \leqq \varepsilon$ for $0 \leqq s \leqq \delta$. Then

$$I_1 \leqq \varepsilon n \int_0^\delta e^{-ns}\, ds \leqq \varepsilon n \int_0^\infty e^{-ns}\, ds = \varepsilon.$$

For a fixed $\delta > 0$, we have, by the equi-boundedness of $\{T_s x\}$ in $s \geqq 0$,

$$I_2 \leqq n \int_\delta^\infty e^{-ns}\, (p(T_s x) + p(x))\, ds \to 0 \quad \text{as} \quad n \uparrow \infty.$$

Hence we have proved (4).

Definition. For $x \in X$, we define

$$D_t T_t x = \lim_{h \to 0} h^{-1} (T_{t+h} - T_t)\, x \tag{6}$$

if the right hand limit exists.

Theorem 2. If $x \in D(A)$, then $x \in D(D_t T_t)$ and

$$D_t T_t x = A T_t x = T_t A x, \ t \geqq 0. \tag{7}$$

Thus, in particular, the operator A is *commutative* with T_t.

Proof. If $x \in D(A)$, then we have, since T_t is continuous linear,

$$T_t A x = T_t \lim_{h \downarrow 0} h^{-1} (T_h - I) x = \lim_{h \downarrow 0} h^{-1} (T_t T_h - T_t) x = \lim_{h \downarrow 0} h^{-1} (T_{t+h} - T_t) x$$
$$= \lim_{h \downarrow 0} h^{-1} (T_h - I)\, T_t x = A T_t x.$$

Thus, if $x \in D(A)$, then $T_t x \in D(A)$ and $T_t A x = A T_t x = \lim_{h \downarrow 0} h^{-1} (T_{t+h} - T_t) x$. We have thus proved that the right derivative of $T_t x$ exists for each $x \in D(A)$. We shall show that the left derivative also exists and is equal to the right derivative.

For this purpose, take any $f_0 \in X'$. Then, for a fixed $x \in D(A)$, the numerical function $f_0(T_t x) = \langle T_t x, f_0 \rangle$ is continuous in $t \geqq 0$ and has right derivative $d^+ f_0(T_t x)/dt$ which is equal to $f_0(A T_t x) = f_0(T_t A x)$ by what we have proved above. Hence $d^+ f_0(T_t x)/dt$ is continuous in t. We shall prove below a well-known **Lemma**: if one of the Dini derivatives

$$\overline{D}^+ f(t),\ \underline{D}^+ f(t),\ \overline{D}^- f(t) \quad \text{and} \quad \underline{D}^- f(t)$$

of a continuous real-valued function $f(t)$ is finite and continuous, then $f(t)$ is differentiable and the derivative is, of course, continuous and equals $\overline{\underline{D}}^{\pm} f(t)$. Thus $f_0(T_t x)$ is differentiable in t and

$$f_0(T_t x - x) = f_0(T_t x) - f_0(T_0 x) = \int_0^t d^+ f_0(T_s x)/ds \cdot ds = \int_0^t f_0(T_s A x) ds$$
$$= f_0\left(\int_0^t T_s A x\, ds\right).$$

Since $f_0 \in X'$ was arbitrary, we must have

$$T_t x - x = \int_0^t T_s A x\, ds \quad \text{for each} \quad x \in D(A).$$

Since $T_s A x$ is continuous in s, it follows that $T_t x$ is differentiable in t in the topology of X and

$$D_t T_t x = \lim_{h \to 0} h^{-1} \int_t^{t+h} T_s A x\, ds = T_t A x.$$

Thus we have proved (7).

Proof of the Lemma. We first prove that the condition: $\overline{D}^+ f(t) \geqq 0$ for $a \leqq t \leqq b$ implies that $f(b) - f(a) \geqq 0$. Assume the contrary, and let $f(b) - f(a) < -\varepsilon(b-a)$ with some $\varepsilon > 0$. Then, for $g(t) = f(t) - f(a) + \varepsilon(t-a)$, we have $\overline{D}^+ g(a) = \overline{D}^+ f(a) + \varepsilon > 0$, and so, by $g(a) = 0$, we must have $g(t_0) > 0$ for some $t_0 > a$ near a. By the continuity of $g(t)$ and $g(b) < 0$, there must exist a t_1 with $a < t_0 < t_1 < b$ such that $g(t_1) = 0$ and $g(t) < 0$ for $t_1 < t < b$. We have then $\overline{D}^+ g(t_1) \leqq 0$ which surely contradicts the fact $\overline{D}^+ g(t_1) = \overline{D}^+ f(t_1) + \varepsilon > 0$.

By applying a similar argument to $f(t) - \alpha t$ and to $\beta t - f(t)$, we prove the following: if one of the Dini derivatives $Df(t)$ satisfies

$$\alpha \leqq Df(t) \leqq \beta \text{ on any interval } [t_1, t_2],$$

then $\alpha \leqq (f(t_2) - f(t_1))/(t_2 - t_1) \leqq \beta$. Hence, the suprema (and the infima) on $[t_1, t_2]$ of the four Dini derivatives of a continuous real-valued function $f(t)$ are the same. Thus, in particular, if one of the Dini derivatives of a continuous real-valued function $f(t)$ is continuous on $[t_1, t_2]$, then the four Dini derivatives of $f(t)$ must coincide on $[t_1, t_2]$.

4. The Resolvent of the Infinitesimal Generator A

Theorem 1. If $n > 0$, then the operator $(nI - A)$ admits an inverse $R(n; A) = (nI - A)^{-1} \in L(X, X)$, and

$$R(n; A)\, x = \int_0^\infty e^{-ns}\, T_s x\, ds \quad \text{for} \quad x \in X. \tag{1}$$

In other words, positive real numbers belong to the resolvent set $\varrho(A)$ of A.

Proof. We first show that $(nI - A)^{-1}$ exists. Suppose that there exists an $x_0 \neq 0$ such that $(nI - A)\, x_0 = 0$, that is, $Ax_0 = nx_0$. Let f_0 be a continuous linear functional $\in X'$ such that $f_0(x_0) = 1$, and set $\varphi(t) = f_0(T_t x_0)$. Since $x_0 \in D(A)$, $\varphi(t)$ is differentiable by Theorem 2 in the preceding section and

$$d\varphi(t)/dt = f_0(D_t T_t x_0) = f_0(T_t A x_0) = f_0(T_t n x_0) = n\varphi(t).$$

If we solve this differential equation under the initial condition $\varphi(0) = f_0(x_0) = 1$, we get $\varphi(t) = e^{nt}$. But, $\varphi(t) = f_0(T_t x_0)$ is bounded in t, because of the equi-boundedness of $T_t x_0$ in $t \geqq 0$ and the continuity of the functional f_0. This is a contradiction, and so the inverse $(nI - A)^{-1}$ must exist.

Since, by (5) of the preceding section, $A C_{\varphi_n} x = n(C_{\varphi_n} - I)\, x$, we have $(nI - A)\, C_{\varphi_n} x = nx$ for all $x \in X$. Thus $(nI - A)$ maps $R(C_{\varphi_n}) \subseteq D(A)$ onto X in a one-one way. Hence, a fortiori, $(nI - A)$ must map $D(A)$ onto X in a one-one way, in virtue of the existence of $(nI - A)^{-1}$. Therefore, we must have $R(C_{\varphi_n}) = D(A)$ and $(nI - A)^{-1} = n^{-1} C_{\varphi_n}$.

Corollary 1. The right half plane of the complex λ-plane is in the resolvent set $\varrho(A)$ of A, and we have

$$R(\lambda; A)\, x = (\lambda I - A)^{-1}\, x = \int_0^\infty e^{-\lambda t}\, T_t x\, dt \quad \text{for } Re(\lambda) > 0 \text{ and } x \in X. \tag{2}$$

Proof. For real fixed τ, $\{e^{-i\tau t}\, T_t;\, t \geqq 0\}$, constitutes an equi-continuous semi-group of class (C_0). It is easy to see that the infinitesimal generator of this semi-group is equal to $(A - i\tau I)$. Thus, for any $\sigma > 0$, the resolvent $R(\sigma + i\tau; A) = ((\sigma + i\tau)\, I - A)^{-1}$ exists and

$$R((\sigma + i\tau)\, I - A)\, x = \int_0^\infty e^{-(\sigma + i\tau)s}\, T_s x\, ds \quad \text{for all} \quad x \in X. \tag{2'}$$

Corollary 2.

$$D(A) = R((\lambda I - A)^{-1}) = R(R(\lambda; A)) \quad \text{when} \quad Re(\lambda) > 0, \tag{3}$$

$$A\, R(\lambda; A)\, x = R(\lambda; A)\, A\, x = (\lambda R(\lambda; A) - I)\, x \quad \text{for} \quad x \in D(A), \tag{4}$$

$$A\, R(\lambda; A)\, x = (\lambda R(\lambda; A) - I)\, x \quad \text{for} \quad x \in X, \tag{5}$$

$$\lim_{\lambda \uparrow \infty} \lambda R(\lambda; A)\, x = x \quad \text{for} \quad x \in X. \tag{6}$$

Proof. Clear from $R(\lambda; A) = (\lambda I - A)^{-1}$ and (4) of the preceding section.

Corollary 3. The infinitesimal generator A is a closed operator in the following sense (Cf. Chapter II, 6):

if $x_h \in D(A)$ and $\lim_{h\to\infty} x_h = x \in X$, $\lim_{h\to\infty} A\, x_h = y \in X$, then

$x \in D(A)$ and $A\, x = y$.

Proof. Put $(I - A)\, x_h = z_h$. Then $\lim_{h\to\infty} z_h = x - y$ and so, by the continuity of $(I - A)^{-1}$, $\lim_{h\to\infty} x_h = \lim_{h\to\infty} (I - A)^{-1}\, z_h = (I - A)^{-1}\, (x - y)$ that is, $x = (I - A)^{-1}\, (x - y)$, $(I - A)\, x = x - y$. This proves that $y = A\, x$.

Theorem 2. The family of operators

$$\{(\lambda R(\lambda; A))^n\} \tag{7}$$

is equi-continuous in $\lambda > 0$ and in $n = 0, 1, 2, \ldots$.

Proof. From the resolvent equation (Chapter VIII, 2, (2))

$$R(\mu; A) - R(\lambda; A) = (\lambda - \mu)\, R(\mu; A)\, R(\lambda; A),$$

we obtain

$$\lim_{\mu\to\lambda} (\mu - \lambda)^{-1} (R(\mu; A) - R(\lambda; a))\, x = dR(\lambda; A)\, x/d\lambda = -R(\lambda; A)^2\, x,\ x \in X.$$

To derive the above formula, we have to appeal to (2) in order to show that $\lim_{\mu\to\lambda} R(\mu; A)\, y = R(\lambda; A)\, y$, $y \in X$.

Therefore, $R(\lambda; A)\,x$ is infinitely differentiable with respect to λ when $Re(\lambda) > 0$ and

$$d^n R(\lambda; A)\,x/d\lambda^n = (-1)^n\, n!\, R(\lambda; A)^{n+1}\, x \qquad (n = 0, 1, 2, \ldots). \tag{8}$$

On the other hand, we have, by differentiating (2) n-times with respect to λ,

$$d^n R(\lambda; A)\,x/d\lambda^n = \int_0^\infty e^{-\lambda t}(-t)^n\, T_t x\, dt. \tag{9}$$

Here the differentiation under the integral sign is justified since $\{T_t x\}$ is equi-bounded in t and $\int_0^\infty e^{-\lambda t}\, t^n\, dt = (n!)/\lambda^{n+1}$ when $Re(\lambda) > 0$. Hence

$$(\lambda R(\lambda; A))^{n+1}\, x = \frac{\lambda^{n+1}}{n!} \int_0^\infty e^{-\lambda t}\, t^n\, T_t x\, dt \ \text{ for } x \in X \text{ and } Re(\lambda) > 0, \tag{10}$$

and so, for any continuous semi-norm p on X and $\lambda > 0$, $n > 0$,

$$p((\lambda R(\lambda; A))^{n+1}\, x) \leqq \frac{\lambda^{n+1}}{n!} \int_0^\infty e^{-\lambda t}\, t^n\, dt \cdot \sup_{t \geqq 0} p(T_t x) = \sup_{t \geqq 0} p(T_t x). \tag{11}$$

This proves Theorem 2 by the equi-continuity of $\{T_t\}$ in t.

5. Examples of Infinitesimal Generators

We first define, for $n > 0$,

$$J_n = (I - n^{-1}A)^{-1} = nR(n; A), \tag{1}$$

so that

$$AJ_n = n(J_n - I). \tag{2}$$

Example 1. $(T_t x)(s) = x(t + s)$ on $C[0, \infty]$.

Writing $y_n(s) = (J_n x)(s) = n \int_0^\infty e^{-nt}\, x(t+s)\, dt = n \int_s^\infty e^{-n(t-s)} x(t)\, dt$,

we obtain $y_n'(s) = -n\, e^{-n(s-s)} x(s) + n^2 \int_s^\infty e^{-n(t-s)}\, x(t)\, dt = -nx(s) + n\, y_n(s)$. Comparing this with the general formula (2):

$$(AJ_n x)(s) = n((J_n - I)\, x)(s),$$

we obtain $Ay_n(s) = y_n'(s)$. Since $R(J_n) = R(R(n; A)) = D(A)$, we have

$$Ay(s) = y'(s) \quad \text{for every} \quad y \in D(A).$$

Conversely, let $y(s)$ and $y'(s)$ both belong to $C[0, \infty]$. We will then show that $y \in D(A)$ and $Ay(s) = y'(s)$. To this purpose, define $x(s)$ by

$$y'(s) - ny(s) = -nx(s).$$

Setting $(J_n x)(s) = y_n(s)$, we obtain, as shown above,

$$y_n'(s) - ny_n(s) = -nx(s).$$

Hence $w(s) = y(s) - y_n(s)$ satisfies $w'(s) = n w(s)$ and so $w(s) = C e^{ns}$. But, w must belong to $C[0, \infty]$ and this is possible only if $C = 0$. Hence $y(s) = y_n(s) \in D(A)$ and $A y(s) = y'(s)$.

Therefore, the domain $D(A)$ of A is precisely the set of functions $y \in C[0, \infty]$ whose first derivative also $\in C[0, \infty]$, and for such a function y we have $Ay = y'$. We have thus characterized the differential operator d/ds as the infinitesimal generator of the semi-group associated with the operation of translation by t on the function space $C[0, \infty]$.

Example 2. We shall give a characterization of the second derivative d^2/ds^2 as the infinitesimal generator of the semi-group associated with the integral operator by Gaussian kernel. The space is $C[-\infty, \infty]$ and

$$(T_t x)(s) = (2\pi t)^{-1/2} \int_{-\infty}^{\infty} e^{-(s-v)^2/2t} x(v)\, dv \text{ if } t > 0, \; = x(s) \text{ if } t = 0.$$

We have

$$y_n(s) = (J_n x)(s) = \int_{-\infty}^{\infty} x(v) \left[\int_0^{\infty} n(2\pi t)^{-1/2} e^{-nt-(s-v)^2/2t}\, dt \right] dv$$

$$= \int_{-\infty}^{\infty} x(v) \left[2\sqrt{n} \int_0^{\infty} (2\pi)^{-1/2} e^{-\sigma^2 - n(s-v)^2/2\sigma^2}\, d\sigma \right] dv$$

(by substituting $t = \sigma^2/n$).

Assuming, for the moment, the formula

$$\int_0^{\infty} e^{-(\sigma^2 + c/\sigma^2)}\, d\sigma = \frac{\sqrt{\pi}}{2} e^{-2c}, \quad c = \sqrt{n}\,|s-v|/\sqrt{2} > 0, \tag{3}$$

we get

$$y_n(s) = \int_{-\infty}^{\infty} x(v)\, (n/2)^{1/2} e^{-\sqrt{2n}\,|s-v|}\, dv.$$

$x(v)$ being continuous and bounded, we can differentiate twice and obtain

$$y_n'(s) = n \int_s^{\infty} x(v)\, e^{-\sqrt{2n}\,(v-s)}\, dv - n \int_{-\infty}^{s} x(v)\, e^{-\sqrt{2n}\,(s-v)}\, dv,$$

$$y_n''(s) = n \left\{ -x(s) - x(s) + \sqrt{2}\sqrt{n} \int_s^{\infty} x(v)\, e^{-\sqrt{2n}\,(s-v)}\, dv \right.$$

$$\left. + \sqrt{2}\sqrt{n} \int_{-\infty}^{s} x(v)\, e^{-\sqrt{2n}\,(s-v)}\, dv \right\}$$

$$= -2n\, x(s) + 2n\, y_n(s).$$

Comparing this with the general formula (2):

$$(A y_n)(s) = (A J_n x)(s) = n((J_n - I)\, x)(s) = n(y_n(s) - x(s)),$$

we find that $A y_n(s) = y_n''(s)/2$. Since $R(J_n) = R(R(n; A)) = D(A)$, we have proved that

$$A y(s) = y''(s)/2 \quad \text{for every} \quad y \in D(A).$$

Conversely, let $y(s)$ and $y''(s)$ both belong to $C[-\infty, \infty]$. Define $x(s)$ by

$$y''(s) - 2n y(s) = -2n x(s).$$

Setting $y_n(s) = (J_n x)(s)$, we obtain, as shown above,

$$y_n''(s) - 2n y_n(s) = -2n x(s).$$

Hence $w(s) = y(s) - y_n(s)$ satisfies $w''(s) - 2n w(s) = 0$, and so $w(s) = C_1 e^{\sqrt{2n}\, s} + C_2 e^{-\sqrt{2n}\, s}$. This function cannot be bounded unless both C_1 and C_2 are zero. Thus $y(s) = y_n(s)$, and so $y \in D(A)$, $(Ay)(s) = y''(s)/2$.

Therefore, the *differential operator* $\frac{1}{2}\frac{d^2}{ds^2}$ is characterized as the infinitesimal generator of the semi-group associated with the integral transform by the Gaussian kernel on the function space $C[-\infty, \infty]$.

The proof of (3). We start with the well-known formula

$$\int_0^\infty e^{-x^2}\, dx = \sqrt{\pi}/2.$$

Setting $x = \sigma - c/\sigma$, we obtain

$$\frac{\sqrt{\pi}}{2} = \int_{\sqrt{c}}^\infty e^{-(\sigma - c/\sigma)^2}(1 + c/\sigma^2)\, d\sigma = e^{2c} \int_{\sqrt{c}}^\infty e^{-(\sigma^2 + c^2/\sigma^2)}\, (1 + c/\sigma^2)\, d\sigma$$

$$= e^{2c}\left\{\int_{\sqrt{c}}^\infty e^{-(\sigma^2 + c^2/\sigma^2)}\, d\sigma + \int_{\sqrt{c}}^\infty e^{-(\sigma^2 + c^2/\sigma^2)}\, \frac{c}{\sigma^2}\, d\sigma\right\}.$$

Setting $\sigma = c/t$ in the last integral, we obtain

$$\frac{\sqrt{\pi}}{2} = e^{2c}\left\{\int_{\sqrt{c}}^\infty e^{-(\sigma^2 + c^2/\sigma^2)}\, d\sigma - \int_{\sqrt{c}}^0 e^{-(c^2/t^2 + t^2)}\, dt\right\} = e^{2c}\int_0^\infty e^{-(t^2 + c^2/t^2)}\, dt.$$

Exercise. Show that the infinitesimal generator A of the semi-group $\{T_t\}$ on $C[-\infty, \infty]$ given by

$$(T_t x)(s) = e^{-\lambda t} \sum_{k=0}^\infty \frac{(\lambda t)^k}{k!} x(s - k\mu) \quad (\lambda, \mu > 0),$$

is the *difference operator* A:

$$(Ax)(s) = \lambda\{x(s - \mu) - x(s)\}.$$

6. The Exponential of a Continuous Linear Operator whose Powers are Equi-continuous

Proposition. Let X be a locally convex, sequentially complete, linear topological space. Let B be a continuous linear operator $\in L(X, X)$ such that $\{B^k;\ k=1,2,\ldots\}$ is equi-continuous. Then, for each $x \in X$, the series

$$\sum_{k=0}^{\infty} (k!)^{-1} (tB)^k x \quad (t \geqq 0) \tag{1}$$

converges.

Proof. For any continuous semi-norm p on X, there exists, by the equi-continuity of $\{B^k\}$, a continuous semi-norm q on X such that $p(B^k x) \leqq q(x)$ for all $k \geqq 0$ and $x \in X$. Hence

$$p\left(\sum_{k=n}^{m} (tB)^k x/k!\right) \leqq \sum_{k=n}^{m} t^k p(B^k x)/k! \leqq q(x) \cdot \sum_{k=n}^{m} t^k/k!.$$

Therefore $\left\{\sum_{k=0}^{n} (tB)^k x/k!\right\}$ is a Cauchy sequence in the sequentially complete space X. The limit of this sequence will be denoted by (1).

Corollary 1. The mapping $x \to \sum_{k=0}^{\infty} (tB)^k x/k!$ defines a continuous linear operator which we shall denote by $\exp(tB)$.

Proof. By the equi-continuity of $\{B^k\}$, we can prove that $B_n = \sum_{k=0}^{n} (tB)^k/k!$ $(n=0,1,2,\ldots)$ are equi-continuous when t ranges over any compact interval. In fact, we have

$$p(B_n x) \leqq \sum_{k=0}^{n} t^k p(B^k x)/k! \leqq q(x) \cdot \sum_{k=0}^{n} t^k/k! \leqq e^t \cdot q(x).$$

Hence the limit $\exp(tB)$ satisfies

$$p(\exp(tB)\, x) \leqq \exp(t) \cdot q(x) \quad (t \geqq 0). \tag{2}$$

Corollary 2. Let B and C be two continuous linear operators $\in L(X, X)$ such that $\{B^k\}$ and $\{C^k\}$ are equi-continuous. If, moreover, $BC = CB$, then we have

$$\exp(tB)\exp(tC) = \exp(t(B+C)). \tag{3}$$

Proof. We have

$$p((B+C)^k x) \leqq \sum_{s=0}^{k} {}_kC_s\, p(B^{k-s} C^s x) \leqq \sum_{s=0}^{k} {}_kC_s\, q(C^s x) \leqq 2^k \sup_{0 \leqq s} q(C^s x).$$

Hence $\{2^{-k}(B+C)^k\}$ is equi-continuous, and we can define $\exp(t(B+C))$. By making use of the commutativity $BC = CB$, we rearrange the series

$$\sum_{k=0}^{\infty} (t(B+C))^k x/k!$$

so that we obtain $\left(\sum_{k=0}^{\infty} (tB)^k/k!\right)\left(\sum_{k=0}^{\infty} (tC)^k x/k!\right)$ as in the case of numerical series $\sum_{k=0}^{\infty} (t(b+c))^k/k!$.

Corollary 3. For every $x \in X$,

$$\lim_{h \downarrow 0} h^{-1}(\exp(hB) - I)\, x = Bx, \tag{4}$$

and so, by making use of the semi-group property

$$\exp((t+h)B) = \exp(tB)\exp(hB) \tag{5}$$

proved above, we obtain

$$D_t \exp(tB)\,x = \exp(tB)\,Bx = B\exp(tB)\,x. \tag{6}$$

Proof. For any continuous semi-norm p on X, we obtain, as above

$$p(h^{-1}(\exp(hB)-I)\,x - Bx) \leqq \sum_{k=2}^{\infty} h^{k-1}\,p(B^k x)/k! \leqq q(x)\sum_{k=2}^{\infty} h^{k-1}/k!$$

which surely tends to 0 as $h \downarrow 0$.

7. The Representation and the Characterization of Equi-continuous Semi-groups of Class (C_0) in Terms of the Corresponding Infinitesimal Generators

We shall prove the following fundamental

Theorem. Let X be a locally convex, sequentially complete, linear topological space. Suppose A is a linear operator with domain $D(A)$ dense in X and range $R(A)$ in X such that the resolvent $R(n; A) = (nI - A)^{-1} \in L(X, X)$ exists for $n = 1, 2, \ldots$ Then A is the infinitesimal generator of a uniquely determined equi-continuous semi-group of class (C_0) iff the operators $\{(I - n^{-1}A)^{-m}\}$ are equi-continuous in $n = 1, 2, \ldots$ and $m = 0, 1, \ldots$

Proof. The "only if" part is already proved. We shall prove the "if" part.

Setting

$$J_n = (I - n^{-1}A)^{-1}, \tag{1}$$

we shall prove

$$\lim_{n\to\infty} J_n x = x \quad \text{for every} \quad x \in X. \tag{2}$$

In fact, we have, for $x \in D(A)$, $AJ_n x = J_n A x = n(J_n - I)\,x$ and so $J_n x - x = n^{-1} J_n(Ax)$ tends to 0 as $n \to \infty$, in virtue of the equi-boundedness of $\{J_n(Ax)\}$ in $n = 1, 2, \ldots$. Since $D(A)$ is dense in X and $\{J_n\}$ is equi-continuous in n, it follows that $\lim_{n\to\infty} J_n x = x$ for every $x \in X$.

Put

$$T_t^{(n)} = \exp(tAJ_n) = \exp(tn(J_n - I)) = \exp(-nt)\exp(ntJ_n),\ t \geqq 0. \tag{3}$$

Since $\{J_n^k\}$ is equi-bounded in n and k, the $\exp(tnJ_n)$ can be defined, and we have, as in (2) of the preceding section,

$$p(\exp(ntJ_n)\,x) \leqq \sum_{k=0}^{\infty} (nt)^k (k!)^{-1}\,p(J_n^k \cdot x) \leqq \exp(nt)\cdot q(x).$$

Consequently, the operators $\{T_t^{(n)}\}$ are equi-continuous in $t \geqq 0$ and $n = 1, 2, \ldots$ in such a way that

$$p(T_t^{(n)} x) \leqq q(x). \tag{4}$$

We next remark that $J_n J_m = J_m J_n$ for $n, m > 0$. Thus J_n is commutative with $T_t^{(m)}$. Hence, by $D_t T_t^{(n)} x = A J_n T_t^{(n)} x = T_t^{(n)} A J_n x$, proved in the preceding section,

$$\begin{aligned} p(T_t^{(n)} x - T_t^{(m)} x) &= p\left(\int_0^t D_s (T_{t-s}^{(m)} T_s^{(n)} x)\, ds\right) \\ &= p\left(\int_0^t T_{t-s}^{(m)} T_s^{(n)} (A J_n - A J_m)\, x\, ds\right). \end{aligned} \tag{5}$$

Hence, if $x \in D(A)$, there exists a continuous semi-norm $\tilde{q}$ on X such that

$$p(T_t^{(n)} x - T_t^{(m)} x) \leqq \int_0^t \tilde{q}((A J_n - A J_m)\, x)\, ds = t\tilde{q}((J_n A - J_m A)\, x).$$

Therefore, by (2), we have proved that $\lim_{n,m\to\infty} p(T_t^{(n)} x - T_t^{(m)} x) = 0$ uniformly in t when t varies on every compact interval. Since $D(A)$ is dense in the sequentially complete space X, and since the operators $\{T^{(n)}\}$ are equi-continuous in $t \geqq 0$ and in n, we see that $\lim_{n\to\infty} T_t^{(n)} x = T_t x$ exists for every $x \in X$ and $t \geqq 0$ uniformly in t on every compact interval of t. Thus the operators $\{T_t\}$ are equi-continuous in $t \geqq 0$, and from the uniform convergence in t, $T_t x$ is continuous in $t \geqq 0$.

We next prove that T_t satisfies the semi-group property $T_t T_s = T_{t+s}$. Since $T_{t+s}^{(n)} = T_t^{(n)} T_s^{(n)}$, we have

$$\begin{aligned} p(T_{t+s} x - T_t T_s x) &\leqq p(T_{t+s} x - T_{t+s}^{(n)} x) + p(T_{t+s}^{(n)} x - T_t^{(n)} T_s^{(n)} x) \\ &\quad + p(T_t^{(n)} T_s^{(n)} x - T_t^{(n)} T_s x) + p(T_t^{(n)} T_s x - T_t T_s x) \\ &\leqq p(T_{t+s} x - T_{t+s}^{(n)} x) + q(T_s^{(n)} x - T_s x) \\ &\quad + p((T_t^{(n)} - T_t)\, T_s x) \to 0 \text{ as } n \to \infty. \end{aligned}$$

Thus $p(T_{t+s} x - T_t T_s x) = 0$ for every continuous semi-norm p on X, which proves that $T_{t+s} = T_t T_s$.

Let $\hat{A}$ be the infinitesimal generator of this equi-continuous semi-group $\{T_t\}$ of class (C_0). We have to show that $\hat{A} = A$. Let $x \in D(A)$. Then $\lim_{n\to\infty} T_t^{(n)} A J_n x = T_t A x$ uniformly in t on every compact interval of t. For, we have, by (4),

$$\begin{aligned} p(T_t A x - T_t^{(n)} A J_n x) &\leqq p(T_t A x - T_t^{(n)} A x) + p(T_t^{(n)} A x - T_t^{(n)} A J_n x) \\ &\leqq p((T_t - T_t^{(n)})\, A x) + q(A x - J_n A x) \end{aligned}$$

which tends to 0 as $n \to \infty$, since $\lim_{n\to\infty} J_n A x = A x$. Hence

$$T_t x - x = \lim_{n\to\infty} (T_t^{(n)} x - x) = \lim_{n\to\infty} \int_0^t T_s^{(n)} A J_n x \, ds$$
$$= \int_0^t \left(\lim_{n\to\infty} T_s^{(n)} A J_n x\right) ds = \int_0^t T_s A x \, ds$$

so that $\lim_{t\downarrow 0} t^{-1}(T_t x - x) = \lim_{t\downarrow 0} t^{-1} \int_0^t T_s A x \, ds$ exists and equals $A x$. We have thus proved that $x \in D(A)$ implies $x \in D(\hat{A})$ and $A x = \hat{A} x$, i.e. $\hat{A}$ is an extension of A. $\hat{A}$ being the infinitesimal generator of the semi-group T_t, we know that, for $n > 0$, $(nI - \hat{A})$ maps $D(\hat{A})$ onto X in one-one way. But, by hypothesis, $(nI - A)$ also maps $D(A)$ onto X in one-one way. Therefore the extension $\hat{A}$ of A must coincide with A.

Finally, the uniqueness of the semi-group T_t is proved as follows. Let $\tilde{T}_t$ be an equi-continuous semi-group of class (C_0) whose infinitesimal generator is precisely A. We construct the semi-group $T_t^{(n)}$. Since A is commutative with $\tilde{T}_t$, we see that $A J_n$ and $T_s^{(n)}$ are commutative with $\tilde{T}_t$. Then, for $x \in D(A)$, we obtain, as in (5),

$$\begin{aligned} p(T_t^{(n)} x - \tilde{T}_t x) &= p\left(\int_0^t D_s(\tilde{T}_{t-s} T_s^{(n)} x)\, ds\right) \\ &= p\left(\int_0^t -\tilde{T}_{t-s} T_s^{(n)} (A - A J_n) x \, ds\right). \end{aligned} \tag{6}$$

Thus, by virtue of $\lim_{n\to\infty} A J_n x = A x$ for all $x \in D(A)$, we prove $\lim_{n\to\infty} T_t^{(n)} x = \tilde{T}_t x$ for all $x \in X$ similarly to the above proof of the existence of $\lim_{n\to\infty} T_t^{(n)} x$, $x \in X$.

Therefore, $T_t x = \tilde{T}_t x$ for all $x \in X$, that is, $T_t = \tilde{T}_t$.

Remark. The above proof shows that, if A is the infinitesimal generator of an equi-continuous semi-group T_t of class (C_0), then

$$T_t x = \lim_{n\to\infty} \exp(tA(I - n^{-1}A)^{-1})\, x, \quad x \in X, \tag{7}$$

and the convergence in (7) is uniform in t on every compact interval of t. This is the *representation theorem* for semi-groups.

Corollary 1. If X is a B-space, then the condition of the Theorem reads: $D(A)^a = X$ and the resolvent $(I - n^{-1}A)^{-1}$ exists such that

$$\|(I - n^{-1}A)^{-m}\| \leq C \qquad (n = 1, 2, \ldots; \; m = 1, 2, \ldots) \tag{8}$$

with a positive constant C which is independent of n and m. In particular, for the case of a contraction semi-group, the condition reads: $D(A)^a = X$

and the resolvent $(I - n^{-1}A)^{-1}$ exists such that

$$\|(I - n^{-1}A)^{-1}\| \leqq 1 \quad (n = 1, 2, \ldots). \tag{9}$$

Remark. The above result (9) was obtained independently by E. HILLE [2] and by K. YOSIDA [5]. The result was extended by W. FELLER [1], R. S. PHILLIPS [3] and I. MIYADERA [1] and the extension is given in the form (8). It is to be noted that in condition (8) and (9), we may replace $(n = 1, 2, \ldots)$ by (for all sufficiently large n). The extension of the semi-group theory to locally convex linear topological spaces, as given in the present book, is suggested by L. SCHWARTZ [3].

Corollary 2. Let X be a B-space, and $\{T_t;\ t \geqq 0\}$ be a family of bounded linear operators $L(X, X)$ such that

$$T_t T_s = T_{t+s}(t, s \geqq 0),\ T_0 = I, \tag{10}$$

$$\underset{t \downarrow 0}{\text{s-lim}}\ T_t x = x \ \text{ for all } x \in X, \tag{11}$$

$$\|T_t\| \leqq M e^{\beta t} \text{ for all } t \geqq 0, \text{ where } M > 0 \text{ and } \beta \geqq 0 \text{ are independent of } t. \tag{12}$$

Then $(A - \beta I)$ is the infinitesimal generator of the equi-continuous semi-group $S_t = e^{-\beta t} T_t$ of class (C_0), where A is the operator defined through $A x = \underset{t \downarrow 0}{\text{s-lim}}\ t^{-1}(T_t - I)\, x$. Thus, by Corollary 1, we see that a closed linear operator A with $D(A)^a = X$ and $R(A) \subseteq X$ is the infinitesimal generator of a semi-group T_t satisfying (10), (11) and (12) iff the resolvent $(I - n^{-1}(A - \beta I))^{-1}$ exists such that

$$\|(I - n^{-1}(A - \beta I))^{-m}\| \leqq M \quad (\text{for } m = 1, 2, \ldots \text{ and all large } n). \tag{13}$$

This condition may be rewritten as

$$\|(I - n^{-1}A)^{-m}\| \leqq M(1 - n^{-1}\beta)^{-m} \quad (\text{for } m = 1, 2, \ldots \text{ and all large } n. \tag{13'}$$

In particular, for those semi-groups T_t satisfying (10), (11) and

$$\|T_t\| \leqq e^{\beta t} \quad \text{for all} \quad t \geqq 0, \tag{14}$$

condition (13′) may be replaced by

$$\|(I - n^{-1}A)^{-1}\| \leqq (1 - n^{-1}\beta)^{-1} \quad (\text{for all large } n). \tag{13''}$$

An application of the representation theorem to the proof of Weierstrass' polynomial approximation theorem. Consider the semi-group T_t defined by $(T_t x)(s) = x(t + s)$ on $C[0, \infty]$. The representation theorem gives

$$(T_t x)(s) = x(t + s) = \underset{n \to \infty}{\text{s-lim}}\ \exp(tA J_n x)(s) = \underset{n \to \infty}{\text{s-lim}} \sum_{m=0}^{\infty} \frac{t^m}{m!}(A J_n)^m x(s),$$

and the above $\underset{n \to \infty}{\text{s-lim}}$ is uniform in t on any compact interval of t. From this result we can derive Weierstrass' polynomial approximation theorem. Let $z(s)$ be a continuous function on the closed interval $[0, 1]$. Let $x(s) \in C[0, \infty]$ be such that $x(s) = z(s)$ for $s \in [0, 1]$. Put $s = 0$ in the above representation of $x(t + s)$. Then we obtain

$$(T_t x)(0) = x(t) = \text{s-}\lim_{n\to\infty} \sum_{m=0}^{\infty} t^m (A J_n)^m x(0)/m! \quad \text{in} \quad C[0, 1],$$

uniformly in t on $[0, 1]$. Hence $z(t)$ is a uniform limit on $[0, 1]$ of a sequence of polynomials in t.

8. Contraction Semi-groups and Dissipative Operators

G. Lumer and R. S. Phillips have discussed contraction semi-groups by virtue of the notion of *semi-scalar product*. The infinitesimal generator of such a semi-group is *dissipative* in their terminology.

Proposition. (Lumer). To each pair $\{x, y\}$ of a complex (or real) normed linear space X, we can associate a complex (or real) number $[x, y]$ such that

$$[x + y, z] = [x, z] + [y, z], \quad [\lambda x, y] = \lambda[x, y], \quad [x, x] = \|x\|^2, \quad |[x, y]| \leqq \|x\| \cdot \|y\|. \tag{1}$$

$[x, y]$ is called a *semi-scalar product* of the vectors x and y.

Proof. According to Corollary 2 of Theorem 1 in Chapter IV, 6, there exists, for each $x_0 \in X$, at least one (and let us choose exactly one) bounded linear functional $f_{x_0} \in X'$ such that $\|f_{x_0}\| = \|x_0\|$ and $\langle x_0, f_{x_0}\rangle = \|x_0\|^2$. Then clearly

$$[x, y] = \langle x, f_y\rangle \tag{2}$$

defines a semi-scalar product.

Definition. Let a complex (or real) B-space X be endowed with a semi-scalar product $[x, y]$. A linear operator A with domain $D(A)$ and range $R(A)$ both in X is called *dissipative* (with respect to $[x, y]$) if

$$Re[Ax, x] \leqq 0 \quad \text{whenever} \quad x \in D(A). \tag{3}$$

Example. Let X be a Hilbert space. Then a symmetric operator A such that $(Ax, x) \leqq 0$ is surely dissipative with respect to the semi-scalar product $[x, y] = (x, y)$, where (x, y) is the ordinary scalar product.

Theorem (Phillips and Lumer). Let A be a linear operator with domain $D(A)$ and range $R(A)$ both in a complex (or real) B-space X such that $D(A)^a = X$. Then A generates a contraction semi-group of class (C_0) in X iff A is dissipative (with respect to any semi-scalar product $[x, y]$) and $R(I - A) = X$.

Proof. The "if" part. Let A be dissipative and $\lambda > 0$. Then the inverse $(\lambda I - A)^{-1}$ exists and $\|(\lambda I - A)^{-1} y\| \leqq \lambda^{-1} \|y\|$ when $y \in D((\lambda I - A)^{-1})$.

For, if $y = \lambda x - Ax$, then

$$\lambda \|x\|^2 = \lambda [x, x] \leqq Re(\lambda [x, x] - [Ax, x]) = Re[y, x] \leqq \|y\| \cdot \|x\|, \quad (4)$$

since A is dissipative. By hypothesis, $R(I - A) = X$ so that $\lambda = 1$ is in the resolvent set $\varrho(A)$ of A, and we have $\|R(1; A)\| \leqq 1$ by (4). If $|\lambda - 1| < 1$, then the resolvent $R(\lambda; A)$ exists and is given by

$$R(\lambda; A) = R(1; A)(I + (\lambda - 1) R(1; A))^{-1}$$

$$= R(1; A) \cdot \sum_{n=0}^{\infty} ((1 - \lambda) R(1; A))^n,$$

(see Theorem 1 in Chapter VIII, 2). Moreover, (4) implies that $\|R(\lambda; A)\| \leqq \lambda^{-1}$ for $\lambda > 0$ with $|\lambda - 1| < 1$. Hence, again by

$$R(\mu; A) = R(\lambda; A)(I + (\mu - \lambda) R(\lambda; A))^{-1},$$

which is valid for $\mu > 0$ with $|\mu - \lambda| \|R(\lambda; A)\| < 1$, we prove the existence of $R(\mu; A)$ and $\|R(\mu; A)\| \leqq \mu^{-1}$. Repeating the process, we see that $R(\lambda; A)$ exists for all $\lambda > 0$ and satisfies the estimate $\|R(\lambda; A)\| \leqq \lambda^{-1}$. As $D(A)$ is dense by hypothesis, it now follows, from Corollary 1 of the preceding section, that A generates a contraction semi-group of class (C_0).

The "only if" part. Suppose $\{T_t; t \geqq 0\}$ is a contraction semi-group of class (C_0). Then

$$Re[T_t x - x, x] = Re[T_t x, x] - \|x\|^2 \leqq \|T_t x\| \cdot \|x\| - \|x\|^2 \leqq 0.$$

Thus, for $x \in D(A)$, the domain of the infinitesimal generator A of T, we have $Re[Ax, x] = \lim_{t \downarrow 0} Re\{t^{-1}[T_t x - x, x]\} \leqq 0$. Hence A is dissipative. Moreover, we know that $R(I - A) = D(R(1; A)) = X$, since A is the infinitesimal generator of a contraction semi-group of class (C_0).

Corollary. If A is a densely defined closed linear operator such that $D(A)$ and $R(A)$ are both in a B-space X and if A and its dual operator A' are both dissipative, then A generates a contraction semi-group of class (C_0).

Proof. It sufficies to show that $R(I - A) = X$. But, since $(I - A)^{-1}$ is closed with A and continuous, $R(I - A)$ is a closed linear subspace of X. Thus $R(I - A) \neq X$ implies the existence of a non-trivial $x' \in X'$ such that

$$\langle (x - Ax), x' \rangle = 0 \quad \text{for all} \quad x \in D(A).$$

Hence $x' - A'x' = 0$, contrary to the dissipativity of A' and $x' \neq 0$.

Remark. For further details concerning dissipative operators, see G. Lumer-R. S. Phillips [1]. See also T. Kato [6].

9. Equi-continuous Groups of Class (C_0). Stone's Theorem

Definition. An equi-continuous semi-group $\{T_t\}$ of class (C_0) is called an *equi-continuous group of class* (C_0), if there exists an equi-continuous semi-group $\{\hat{T}_t\}$ of class (C_0) with the condition:

If we define S_t by $S_t = T_t$ for $t \geqq 0$ and $S_{-t} = \hat{T}_t$ for $t \geqq 0$, then the family of operators S_t, $-\infty < t < \infty$, has the group property

$$S_t S_s = S_{t+s} \ (-\infty < t, s < \infty), \ S_0 = I. \tag{1}$$

Theorem. Let X be a locally convex, sequentially complete linear topological space. Suppose A is a linear operator with domain $D(A)$ dense in X and range $R(A)$ in X. Then A is the infinitesimal generator of an equi-continuous group of class (C_0) of operators $S_t \in L(X, X)$ iff the operators $(I - n^{-1}A)^{-m}$ are everywhere defined and equi-continuous in $n = \pm 1, \pm 2, \ldots$ and in $m = 1, 2, \ldots$

Proof. The "only if" part. Let $T_t = S_t$ for $t \geqq 0$ and $\hat{T}_t = S_{-t}$ for $t \geqq 0$. Let A and $\hat{A}$ be the infinitesimal generator of $\hat{T}_t$ and T_t, respectively. We have to show that $\hat{A} = -A$. If $x \in D(\hat{A})$, then, by putting $x_h = h^{-1}(\hat{T}_h - I)\,x$ and making use of the equi-continuity of T_h,

$$\begin{aligned} p(T_h x_h - \hat{A}x) &\leqq p(T_h x_h - T_h \hat{A}x) + p((T_h - I)\,Ax) \\ &\leqq q(x_h - \hat{A}x) + p((T_h - I)\,\hat{A}x) \end{aligned}$$

where p and q are continuous semi-norms on X such that, for a given p, we can choose a q satisfying the above inequality for all $h \geqq 0$ and all $x \in D(\hat{A})$ simultaneously. Thus $\lim_{h \downarrow 0} T_h x_h = \hat{A}x$, and so $x \in D(\hat{A})$ implies $\hat{A}x = \lim_{h \downarrow 0} T_h(h^{-1}(\hat{T}_h - I))\,x = \lim_{h \downarrow 0} h^{-1}(I - T_h)\,x = -Ax$, Hence $-A$ is an extension of $\hat{A}$. In the same way, we can prove that $\hat{A}$ is an extension of $-A$. Therefore $\hat{A} = -A$.

The "if" part. Define, for $t \geqq 0$,

$$T_t x = \lim_{n\to\infty} T_t^{(n)} x = \lim_{n\to\infty} \exp(tA\,(I - n^{-1}A)^{-1})\,x,$$

$$\hat{T}_t x = \lim_{n\to\infty} \hat{T}_t^{(n)} x = \lim_{n\to\infty} \exp(t\hat{A}\,(I - n^{-1}\hat{A})^{-1})\,x, \text{ where } \hat{A} = -A.$$

Then T_t and $\hat{T}_t$ are both equi-continuous semi-groups of class (C_0). We have

$$\begin{aligned} p(T_t\hat{T}_t x - T_t^{(n)}\hat{T}_t^{(n)}x) &\leqq p(T_t\hat{T}_t x - T_t^{(n)}\hat{T}_t x) + p(T_t^{(n)}\hat{T}_t x - T_t^{(n)}\hat{T}_t^{(n)}x) \\ &\leqq p((T_t - T_t^{(n)})\,\hat{T}_t x) + q(\hat{T}_t x - \hat{T}_t^{(n)}x) \end{aligned}$$

by the equi-continuity of $\{T_t^{(n)}\}$ in n and $t \geqq 0$. Thus we have $\lim_{n\to\infty} T_t^{(n)}\hat{T}_t^{(n)}x = T_t\hat{T}_t x$. We have incidentally proved the equi-continuity of $T_t\hat{T}_t$ by the equi-continuity of $T_t^{(n)}\hat{T}_t^{(n)}$. On the other hand, we have $T_t^{(n)}\hat{T}_s^{(m)} = \hat{T}_s^{(m)}T_t^{(n)}$ by the commutativity of $(I - n^{-1}A)^{-1}$ and $(I - m^{-1}A)^{-1}$. Thus $(T_t^{(n)}\hat{T}_t^{(n)})\,(T_s^{(n)}\hat{T}_s^{(n)}) = T_{t+s}^{(n)}\hat{T}_{t+s}^{(n)}$, and so

$(T_t\hat{T}_t)$ with $t \geqq 0$ enjoys the semi-group property $(T_t\hat{T}_t)(T_s\hat{T}_s) = T_{t+s}\hat{T}_{t+s}$. Hence $\{T_t\hat{T}_t\}$ is an equi-continuous semi-group of class (C_0). If $x \in D(\hat{A}) = D(A)$, then

$$\lim_{h\downarrow 0} h^{-1}(T_h\hat{T}_h - I)\,x = \lim_{h\downarrow 0} T_h h^{-1}(\hat{T}_h - I)\,x + \lim_{h\downarrow 0} h^{-1}(T_h - I)\,x$$
$$= \hat{A}x + Ax = 0,$$

so that the infinitesimal generator A_1 of $\{T_t\hat{T}_t\}$ is 0 at every $x \in D(\hat{A})$. Since $(I - A_1)$ is the inverse of a continuous linear operator $(I - A_1)^{-1} \in L(X, X)$, we see that A_1 must be closed. Hence A_1 must be $= 0$, in virtue of the fact that A_1 vanishes on a dense subset $D(\hat{A}) = D(A)$ of X. Therefore $(T_t\hat{T}_t)\,x = \lim_{n\to\infty} \exp(t\cdot 0\cdot(I - n^{-1}\cdot 0)^{-1})\,x = x$, that is, $T_t\hat{T}_t = I$. We have thus proved the group property of $S_t, -\infty < t < \infty$, where $S_t = T_t$ and $S_{-t} = \hat{T}_t$ for $t \geqq 0$.

Corollary 1. For the case of a B-space X, the condition of the Theorem reads: $D(A)^a = X$ and the resolvent $(I - n^{-1}A)^{-1}$ exists such that

$$\|(I - n^{-1}A)^{-m}\| \leqq M \quad (\text{for } m = 1, 2, \ldots \text{ and all large } |n|,\ n \gtrless 0). \tag{2}$$

For the group S_t satisfying $\|S_t\| \leqq M e^{\beta|t|}$ $(\beta \geqq 0)$ for all $t, -\infty < t < \infty$, the condition reads: $D(A)^a = X$ and the resolvent $(I - n^{-1}A)^{-1}$ exists such that

$$\|(I - n^{-1}A)^{-m}\| \leqq M\,(1 - |n^{-1}|\,\beta)^{-m} \quad (\text{for } m = 1, 2, \ldots \text{ and all large } |n|,\ n \lessgtr 0). \tag{3}$$

For the particular case $\|S_t\| \leqq e^{\beta|t|}$ for all $t, -\infty < t < \infty$, the condition reads: $D(A)^a = X$ and the resolvent $(I - n^{-1}A)^{-1}$ exists such that

$$\|(I - n^{-1}A)^{-1}\| \leqq (1 - |n^{-1}|\beta)^{-1} \quad (\text{for all large } |n|,\ n \gtrless 0). \tag{4}$$

Proof. As the proof of Corollary 1 and Corollary 2 in Chapter IX, 7.

Corollary 2 (Stone's theorem). Let U_t, $-\infty < t < \infty$, be a group of unitary operators of class (C_0) in a Hilbert space X. Then the infinitesimal generator A of U_t is $\sqrt{-1}$ times a self-adjoint operator H.

Proof. We have $(U_t x, y) = (x, U_t^{-1}y) = (x, U_{-t}y)$ and so, by differentiation,

$$(Ax, y) = (x, -Ay) \quad \text{whenever } x \text{ and } y \in D(A).$$

Thus $-iA = H$ is symmetric. A being the infinitesimal generator of U_t, $(I - n^{-1}A)^{-1} = (I - n^{-1}iH)^{-1}$ must be a bounded linear operator such that $\|(I - n^{-1}iH)^{-1}\| \leqq 1$ for $n = \pm 1, \pm 2, \ldots$ Hence, taking the case $n = \pm 1$, we see that the Cayley transform of H is unitary. This proves that H is self-adjoint.

Remark. If A is of the form $A = \sqrt{-1} \cdot H$, where H is a self-adjoint operator in a Hilbert space X, then condition (4) of Corollary 1 is surely satisfied, as may be proved by the theory of Cayley's transform. Therefore, A is the infinitesimal generator of a group of contraction operators U_t in X. It is easy to see that such U_t is unitary. For, a contraction operator U_t on a Hilbert space X into X must be unitary, if $U_t^{-1} = U_{-t}$ is also a contraction operator on X into X.

10. Holomorphic Semi-groups

We shall introduce an important class of semi-groups, namely, semi-groups T_t which can, as functions of the parameter t, be continued holomorphically into a sector of the complex plane containing the positive t-axis. We first prove

Lemma. Let X be a locally convex, sequentially complete, linear topological space. Let $\{T_t; t \geqq 0\} \subseteq L(X, X)$ be an equi-continuous semi-group of class (C_0). Suppose that, for all $t > 0$, $T_t X \subseteq D(A)$, the domain of the infinitesimal generator A of T_t. Then, for any $x \in X$, $T_t x$ is infinitely differentiable in $t > 0$ and we have

$$T_t^{(n)} x = (T'_{t/n})^n x \quad \text{for all} \quad t > 0, \tag{1}$$

where $T'_t = D_t T_t$, $T''_t = D_t T'_t, \ldots, T_t^{(n)} = D_t T_t^{(n-1)}$.

Proof. If $t > t_0 > 0$, then $T'_t x = A T_t x = T_{t-t_0} A T_{t_0} x$ by the commutativity of A and T_s, $s \geqq 0$. Thus $T'_t X \subseteq T_{t-t_0} X \subseteq D(A)$ when $t > 0$, and so $T''_t x$ exists for all $t > 0$ and $x \in X$. Since A is a closed linear operator, we have

$$\begin{aligned} T''_t x = D_t (A T_t) x &= A \cdot \lim_{n\to\infty} \cdot n (T_{t+1/n} - T_t) x = A (A T_t) x \\ &= A T_{t/2} A T_{t/2} x = (T'_{t/2})^2 x. \end{aligned}$$

Repeating the same argument, we obtain (1).

Let X be a locally convex, sequentially complete, complex linear topological space. Let $\{T_t; t \geqq 0\} \subseteq L(X, X)$ be an equi-continuous semi-group of class (C_0). For such a semi-group, we consider the following three conditions:

(I) For all $t > 0$, $T_t x \in D(A)$, and there exists a positive constant C such that the family of operators $\{(C t T'_t)^n\}$ is equi-continuous in $n \geqq 0$ and t, $0 < t \leqq 1$.

(II) T_t admits a *weakly holomorphic extension* T_λ given by

$$T_\lambda x = \sum_{n=0}^{\infty} (\lambda - t)^n T_t^{(n)} x / n! \quad \text{for} \quad |\arg \lambda| < \tan^{-1}(C e^{-1}), \tag{2}$$

and

the family of operators $\{e^{-\lambda} T_\lambda\}$ is equi-continuous in λ for

$$|\arg \lambda| < \tan^{-1}(2^{-k} C e^{-1}) \text{ with some positive constant } k. \tag{3}$$

(III) Let A be the infinitesimal generator of T_t. Then there exists a positive constant C_1 such that the family of operators $\{(C_1\lambda R(\lambda;A))^n\}$ is equi-continuous in $n \geqq 0$ and in λ with $Re(\lambda) \geqq 1 + \varepsilon$, $\varepsilon > 0$.

We can prove

Theorem. The three conditions (I), (II) and (III) are mutually equivalent.

Proof. The implication (I) → (II). Let p be any continuous semi-norm on X. Then, by hypothesis, there exists a continuous semi-norm q on X such that $p((tT_t')^n x) \leqq C^{-n} q(x)$, $0 < C$ for $1 \geqq t > 0$, $n \geqq 0$ and $x \in X$. Hence, by (1), we obtain, for any $t > 0$,

$$p((\lambda - t)^n T_t^{(n)} x/n!) \leqq \frac{|\lambda - t|^n}{t^n} \frac{n^n}{n!} \frac{1}{C^n} p\left(\left(\frac{t}{n} C T'_{t/n}\right)^n x\right)$$
$$\leqq \left(\frac{|\lambda - t|}{t} C^{-1} e\right)^n \cdot q(x), \quad \text{whenever} \quad 0 \leqq t/n \leqq 1.$$

Thus the right side of (2) is convergent for $|\arg \lambda| < \tan^{-1}(Ce^{-1})$, and so, by the sequential completeness of the space X, $T_\lambda x$ is well defined and *weakly holomorphic* in λ for $|\arg \lambda| < \tan^{-1}(Ce^{-1})$. That is, for any $x \in X$ and $f \in X'$, the numerical function $f(T_t x)$ of t, $t > 0$, admits a holomorphic extension $f(T_\lambda x)$ for $|\arg \lambda| < \tan^{-1}(Ce^{-1})$; consequently, by the Hahn-Banach theorem, we see that $T_\lambda x$ is an extension of $T_t x$ for $|\arg \lambda| < \tan^{-1}(Ce^{-1})$. Next put $S_t = e^{-t} T_t$. Then $S_t' = e^{-t} T_t' - e^{-t} T_t$ and so, by $0 \leqq te^{-t} \leqq 1$ $(0 \leqq t)$ and (I), we easily see that the family of operators $\{(2^{-k} C t S_t')^n\}$ is equi-continuous in $t > 0$ and $n \geqq 0$, in virtue of the equi-continuity of $\{T_t\}$ in $t > 0$. The equi-continuous semi-group S_t of class (C_0) satisfies the condition that $t > 0$ implies $S_t X \subseteq D(A - I) = D(A)$, where $(A - I)$ is the infinitesimal generator of S_t. Therefore, by the same reasoning applied above to T_t, we can prove that the weakly holomorphic extension $e^{-\lambda} T_\lambda$ of $S_t = e^{-t} T_t$ satisfies the estimate (3).

By the way, we can prove the following

Corollary (due to E. Hille). If, in particular, X is a complex B-space and $\overline{\lim_{t \downarrow 0}} \|tT_t'\| < e^{-1}$, then $X = D(A)$.

Proof. For a fixed $t > 0$, we have $\overline{\lim_{n \to \infty}} \|(t/n)\, T'_{t/n}\| < e^{-1}$, and so the series

$$\sum_{n=0}^{\infty} (\lambda - t)^n T_t^{(n)} x/n! = \sum_{n=0}^{\infty} \frac{(\lambda - t)^n}{t^n} \frac{n^n}{n!} \left(\frac{t}{n} T'_{t/n}\right)^n x$$

strongly converges in some sector

$$\{\lambda;\ |\lambda - t|/t < 1 + \delta \quad \text{with some} \quad \delta > 0\}$$

of the complex λ-plane. This sector surely contains $\lambda = 0$ in its interior.

The implication (II) → (III). We have, by (10) in Chapter IX, 4,

$$(\lambda R(\lambda; A))^{n+1} x = \frac{\lambda^{n+1}}{n!} \int_0^\infty e^{-\lambda t} t^n T_t x \, dt \quad \text{for} \quad Re(\lambda) > 0, \ x \in X. \tag{4}$$

Hence, putting $S_t = e^{-t} T_t$, we obtain

$$((\sigma + 1 + i\tau) R(\sigma + 1 + i\tau; A))^{n+1} x$$
$$= \frac{(\sigma + 1 + i\tau)^{n+1}}{n!} \int_0^\infty e^{-(\sigma+i\tau)t} t^n S_t x \, dt, \ \sigma > 0.$$

Let $\tau < 0$. Since the integrand is weakly holomorphic, we can deform, by the estimate (3) and Cauchy's integral theorem, the path of integration: $0 \leqq t < \infty$ to the ray: $re^{i\theta}$ $(0 \leqq r < \infty)$ contained in the sector $0 < \arg \lambda < \tan^{-1}(2^{-k} C e^{-1})$ of the complex λ-plane. We thus obtain

$$((\sigma + 1 + i\tau) R(\sigma + 1 + i\tau; A))^{n+1} x$$
$$= \frac{(\sigma + 1 + i\tau)^{n+1}}{n!} \int_0^\infty e^{-(\sigma+i\tau)re^{i\theta}} r^n e^{in\theta} S_{re^{i\theta}} x e^{i\theta} dr,$$

and so, by (3),

$$p(((\sigma + 1 + i\tau) R(\sigma + 1 + i\tau; A))^{n+1} x)$$
$$\leqq \sup_{0<r<\infty} p(S_{re^{i\theta}} x) \frac{|(\sigma + 1 + i\tau)|^{n+1}}{n!} \int_0^\infty e^{(-\sigma\cos\theta + \tau\sin\theta)r} r^n dr$$
$$\leqq q'(x) \frac{|\sigma + 1 + i\tau|^{n+1}}{|\tau \sin \theta - \sigma \cos \theta|^{n+1}}, \quad \text{where } q' \text{ is a continuous semi-norm on } X.$$

A similar estimate is also obtained for the case $\tau > 0$. Hence, combined with (7) in Section 4, we have proved (III).

The implication (III) → (I). For any continuous semi-norm p on X, there exists a continuous semi-norm q on X such that

$$p\big((C_1 \lambda R(\lambda; A))^n x\big) \leqq q(x) \ \text{whenever} \ Re(\lambda) \geqq 1 + \varepsilon, \ \varepsilon > 0, \ \text{and} \ n \geqq 0.$$

Hence, if $Re(\lambda_0) \geqq 1 + \varepsilon$, we have

$$p((\lambda - \lambda_0)^n R(\lambda_0; A)^n x) \leqq \frac{|\lambda - \lambda_0|^n}{(C_1 |\lambda_0|)^n} q(x) \quad (n = 0, 1, 2, \ldots).$$

Thus, if $|\lambda - \lambda_0|/C_1 |\lambda_0| < 1$, the resolvent $R(\lambda, A)$ exists and is given by

$$R(\lambda; A) x = \sum_{n=0}^\infty (\lambda_0 - \lambda)^n R(\lambda_0; A)^{n+1} x \quad \text{such that}$$
$$p(R(\lambda; A) x) \leqq (1 - C_1^{-1} |\lambda_0|^{-1} |\lambda - \lambda_0|)^{-1} q(R(\lambda_0; A) x).$$

Therefore, by (III), there exists an angle θ_0 with $\pi/2 < \theta_0 < \pi$ such that $R(\lambda; A)$ exists and satisfies the estimate

$$p(R(\lambda; A) x) \leqq \frac{1}{|\lambda|} q'(x) \ \text{with a continuous semi-norm} \ q' \ \text{on} \ X \tag{5}$$

in the sectors $\pi/2 \leqq \arg \lambda \leqq \theta_0$ and $-\theta_0 \leqq \arg \lambda \leqq -\pi/2$ and also for $Re(\lambda) \geqq 0$, when $|\lambda|$ is sufficiently large.

Hence the integral

$$\hat{T}_t x = (2\pi i)^{-1} \int_{C_2} e^{\lambda t} R(\lambda; A)\, x\, d\lambda \quad (t > 0,\ x \in X) \tag{6}$$

converges if we take the path of integration $C_2 = \lambda(\sigma)$, $-\infty < \sigma < \infty$, in such a way that $\lim_{|\sigma|\uparrow\infty} |\lambda(\sigma)| = \infty$ and, for some $\varepsilon > 0$,

$$\pi/2 + \varepsilon \leqq \arg \lambda(\sigma) \leqq \theta_0 \quad \text{and} \quad -\theta_0 \leqq \arg \lambda(\sigma) \leqq -\pi/2 - \varepsilon$$

when $\sigma \uparrow +\infty$ and $\sigma \downarrow -\infty$, respectively; for $|\sigma|$ not large, $\lambda(\sigma)$ lies in the right half plane of the complex λ-plane.

We shall show that $\hat{T}_t$ coincides with the semi-group T_t itself. We first show that $\lim_{t\downarrow 0} \hat{T}_t x = x$ for all $x \in D(A)$. Let x_0 be any element $\in D(A)$, and choose any complex number λ_0 to the right of the contour C_2 of integration, and denote $(\lambda_0 I - A)\, x_0 = y_0$. Then, by the resolvent equation,

$$\begin{aligned}\hat{T}_t x_0 &= \hat{T}_t\, R(\lambda_0; A)\, y_0 = (2\pi i)^{-1} \int_{C_2} e^{\lambda t} R(\lambda; A)\, R(\lambda_0; A)\, y_0\, d\lambda \\ &= (2\pi i)^{-1} \int_{C_2} e^{\lambda t} (\lambda_0 - \lambda)^{-1} R(\lambda; A)\, y_0\, d\lambda \\ &\quad - (2\pi i)^{-1} \int_{C_2} e^{\lambda t} (\lambda_0 - \lambda)^{-1} R(\lambda_0; A)\, y_0\, d\lambda .\end{aligned}$$

The second integral on the right is equal to zero, as may be seen by shifting the path of integration to the left. Hence

$$\hat{T}_t x_0 = (2\pi i)^{-1} \int_{C_2} e^{\lambda t} (\lambda_0 - \lambda)^{-1} R(\lambda; A)\, y_0\, d\lambda, \quad y_0 = (\lambda_0 I - A)\, x_0 .$$

Because of the estimate (5), the passage to the limit $t \downarrow 0$ under the integral sign is justified, and so

$$\lim_{t\downarrow 0} \hat{T}_t x_0 = (2\pi i)^{-1} \int_{C_2} (\lambda_0 - \lambda)^{-1} R(\lambda; A)\, y_0\, d\lambda, \quad y_0 = (\lambda_0 I - A)\, x_0 .$$

To evaluate the right hand integral, we make a closed contour out of the original path of integration C_2 by adjoining the arc of the circle $|\lambda| = r$ which is to the right of the path C_2, and throwing away that portion of the original path C_2 which lies outside the circle $|\lambda| = r$. The value of the integral along the new arc and the discarded arc tends to zero as $r \uparrow \infty$, in virtue of (5). Hence the value of the integral is equal to the residue inside the new closed contour, that is, the value $R(\lambda_0; A)\, y_0 = x_0$. We have thus proved $\lim_{t\downarrow 0} \hat{T}_t x_0 = x_0$ when $x_0 \in D(A)$.

We next show that $\hat{T}' x = A \hat{T}_t x$ for $t > 0$ and $x \in X$. We have $R(\lambda; A)\, X = D(A)$ and $A R(\lambda; A) = \lambda R(\lambda; A) - I$, so that, by the convergence factor $e^{\lambda t}$, the integral $(2\pi i)^{-1} \int\limits_{C_2} e^{\lambda t} A R(\lambda; A)x \, d\lambda$ has a sense. This integral is equal to $A\hat{T}_t x$, as may be seen by approximating the integral (6) by Riemann sums and using the fact that A is *closed*: $\lim\limits_{n\to\infty} x_n = x$ and $\lim\limits_{n\to\infty} A x_n = y$ imply $x \in D(A)$ and $Ax = y$. Therefore

$$A\hat{T}_t x = (2\pi i)^{-1} \int\limits_{C_2} e^{\lambda t} A R(\lambda; A)\, x d\lambda, \ t > 0.$$

On the other hand, by differentiating (6) under the integral sign, we obtain

$$\hat{T}_t' x = (2\pi i)^{-1} \int\limits_{C_2} e^{\lambda t} \lambda R(\lambda; A)\, x d\lambda, \ t > 0. \tag{7}$$

The difference of these two integrals is $(2\pi i)^{-1} \int\limits_{C_2} e^{\lambda t} x d\lambda$, and the value of the last integral is zero, as may be seen by shifting the path of integration to the left.

Thus we have proved that $\hat{x}(t) = \hat{T}_t x_0$, $x_0 \in D(A)$, satisfies i) $\lim\limits_{t\downarrow 0} \hat{x}(t) = x_0$, ii) $d\hat{x}(t)/dt = A\hat{x}(t)$ for $t > 0$, and iii) $\{\hat{x}(t)\}$ is by (6) of exponential growth when $t \uparrow \infty$. On the other hand, since $x_0 \in D(A)$ and since $\{T_t\}$ is equi-continuous in $t \geqq 0$, we see that $x(t) = T_t x_0$ also satisfies $\lim\limits_{t\downarrow 0} x(t) = x_0$, $dx(t)/dt = Ax(t)$ for $t \geqq 0$, and $\{x(t)\}$ is bounded when $t \geqq 0$. Let us put $\hat{x}(t) - x(t) = y(t)$. Then $\lim\limits_{t\downarrow 0} y(t) = 0$, $dy(t)/dt = Ay(t)$ for $t > 0$ and $\{y(t)\}$ is of exponential growth when $t \uparrow \infty$. Hence we may consider the Laplace transform

$$L(\lambda; y) = \int\limits_0^\infty e^{-\lambda t} y(t)\, dt \quad \text{for large positive} \quad Re(\lambda).$$

We have

$$\int\limits_\alpha^\beta e^{-\lambda t} y'(t)\, dt = \int\limits_\alpha^\beta e^{-\lambda t} A y(t)\, dt = A \int\limits_\alpha^\beta e^{-\lambda t} y(t)\, dt, \ 0 \leqq \alpha < \beta < \infty,$$

by approximating the integral by Riemann sums and using the fact that A is closed. By partial integration, we obtain

$$\int\limits_\alpha^\beta e^{-\lambda t} y'(t)\, dt = e^{-\lambda\beta} y(\beta) - e^{-\lambda\alpha} y(\alpha) + \lambda \int\limits_\alpha^\beta e^{-\lambda t} y(t)\, dt,$$

which tends to $\lambda L(\lambda; y)$ as $\alpha \downarrow 0$, $\beta \uparrow \infty$. For, $y(0) = 0$ and $\{y(\beta)\}$ is of exponential growth as $\beta \uparrow \infty$. Thus again, by using the closure property of A, we obtain

$$AL(\lambda; y) = \lambda L(\lambda; y) \quad \text{for large positive} \quad Re(\lambda).$$

Since the inverse $(\lambda I - A)^{-1}$ exists for $Re(\lambda) > 0$, we must have $L(\lambda; y) = 0$ when $Re(\lambda)$ is large positive. Thus, for any continuous linear functional

$f \in X'$, we have

$$\int_0^\infty e^{-\lambda t} f(y(t))\, dt = 0 \quad \text{when} \quad Re(\lambda) \text{ is large positive.}$$

We set $\lambda = \sigma + i\tau$ and put

$$g_\sigma(t) = e^{-\sigma t} f(y(t)) \text{ or } = 0 \quad \text{according as} \quad t \geqq 0 \quad \text{or} \quad t < 0.$$

Then, the above equality shows that the Fourier transform $(2\pi)^{-1} \int_{-\infty}^{\infty} e^{-i\tau t} g_\sigma(t)\, dt$ vanishes identically in τ, $-\infty < \tau < \infty$, so that, by Fourier's integral theorem, $g_\sigma(t) = 0$ identically. Thus $f(y(t)) = 0$ and so we must have $y(t) = 0$ identically, in virtue of the Hahn-Banach theorem.

Therefore $\hat{T}_t x = T_t x$ for all $t > 0$ and $x \in D(A)$. $D(A)$ being dense in X and $\hat{T}_t$, T_t both belonging to $L(X, X)$, we easily conclude that $\hat{T}_t x = T_t x$ for all $x \in X$ and $t > 0$. Hence, by defining $\hat{T}_0 = I$, we have $\hat{T}_t = T_t$ for all $t \geqq 0$. Hence, by (7), $T'_t x = (2\pi i)^{-1} \int_{C_2} e^{\lambda t} \lambda R(\lambda; A)\, x\, d\lambda$, $t > 0$, and so, by (1) and (5), we obtain

$$(T'_{t/n})^n x = T_t^{(n)} x = (2\pi i)^{-1} \int_{C_2} e^{\lambda t} \lambda^n R(\lambda; A)\, x\, d\lambda, \quad t > 0.$$

Hence

$$(t T'_t)^n x = (2\pi i)^{-1} \int_{C_2} e^{n\lambda t} (t\lambda)^n R(\lambda; A)\, x\, d\lambda, \quad t > 0.$$

Therefore, by (III),

$$p((t T'_t)^n x) \leqq (2\pi)^{-1} q(x) \int_{C_2} |e^{n\lambda t}|\, t^n\, |\lambda|^{n-1}\, d|\lambda|.$$

If $0 < t \leqq 1$, then the last integral is majorized by C_3^n where C_3 is a certain positive constant. This we see, by splitting the integration path C_2 into the sum of that in the right half-plane $Re(\lambda) \geqq 0$ and those in the left half-plane $Re(\lambda) < 0$, and remembering the integral representation of the Γ-function.

References. The result of the present section is due to K. Yosida [6]. See also E. Hille [3] and E. Hille-R. S. Phillips [1].

11. Fractional Powers of Closed Operators

Let X be a B-space, and $\{T_t;\, t \geqq 0\} \subseteq L(X, X)$ an equi-continuous semi-group of class (C_0). We introduce

$$f_{t,\alpha}(\lambda) = \frac{1}{2\pi i} \int_{\sigma - i\infty}^{\sigma + i\infty} e^{z\lambda - t z^\alpha}\, dz \; (\sigma > 0,\, t > 0,\, \lambda \geqq 0,\, 0 < \alpha < 1),$$

$$= 0 \quad (\text{when } \lambda < 0), \tag{1}$$

17*

where the branch of z^α is so taken that $Re(z^\alpha) > 0$ for $Re(z) > 0$. This branch is a one-valued function in the z-plane cut along the negative real axis. The convergence of the integral (1) is apparent in virtue of the convergence factor e^{-tz^α}. Following S. BOCHNER [2] and R. S. PHILLIPS [5], we can show that the operators

$$\begin{aligned}\hat{T}_{t,\alpha}x = \hat{T}_t x &= \int_0^\infty f_{t,\alpha}(s)\, T_s x\, ds \quad (t > 0),\\ &= x \quad (t = 0),\end{aligned} \tag{2}$$

constitute an equi-continuous semi-group of class (C_0). Moreover, we can show that $\{\hat{T}_t\}$ is a holomorphic semi-group (K. YOSIDA [8] and V. BALAKRISHNAN [1]). The infinitesimal generator $\hat{A} = \hat{A}_\alpha$ of $\hat{T}_t$ is connected with the infinitesimal generator A of T_t by

$$\hat{A}_\alpha x = -(-A)^\alpha x \quad \text{for} \quad x \in D(A), \tag{3}$$

where the fractional power $(-A)^\alpha$ of $(-A)$ is given by

$$(-A)^\alpha x = \frac{\sin \alpha\pi}{\pi} \int_0^\infty \lambda^{\alpha-1} (\lambda I - A)^{-1} (-A x)\, d\lambda \quad \text{for} \quad x \in D(A) \tag{4}$$

and also by

$$(-A)^\alpha x = \Gamma(-\alpha)^{-1} \int_0^\infty \lambda^{-\alpha-1} (T_\lambda - I)\, x\, d\lambda \quad \text{for} \quad x \in D(A). \tag{5}$$

Formula (4) and (5) were obtained by V. BALAKRISHNAN. For the resolvent of $\hat{A}_\alpha$, we have the following formula due to T. KATO:

$$(\mu I - \hat{A}_\alpha)^{-1} = \frac{\sin \alpha\pi}{\pi} \int_0^\infty (r I - A)^{-1} \frac{r^\alpha}{\mu^2 - 2\mu r^\alpha \cos \alpha\pi + r^{2\alpha}}\, dr. \tag{6}$$

In this way, we see the abundance of holomorphic semi-groups among the class of equi-continuous semi-groups of class (C_0).

To prove the above result, we have to investigate the properties of the function $f_{t,\alpha}(\lambda)$ in a series of propositions.

Proposition 1. We have

$$e^{-ta^\alpha} = \int_0^\infty e^{-\lambda a} f_{t,\alpha}(\lambda)\, d\lambda \quad (t > 0,\ a > 0). \tag{7}$$

Proof. It is easy to see, by the convergence factor $e^{-z^\alpha t}$, that the function $f_{t,\alpha}(\lambda)$ is of exponential growth in λ. By Cauchy's integral theorem, integral (1) is independent of $\sigma > 0$. Let $a > \sigma = Re(z) > 0$. Then, by Cauchy's theorem of residue,

$$\begin{aligned}\int_0^\infty e^{-\lambda a} f_{t,\alpha}(\lambda)\, d\lambda &= \frac{1}{2\pi i} \int_{\sigma-i\infty}^{\sigma+i\infty} \left[\frac{e^{\lambda(z-a)}}{z-a}\right]_{\lambda=0}^{\lambda=\infty} e^{-z^\alpha t}\, dz\\ &= \frac{-1}{2\pi i} \int_{\sigma-i\infty}^{\sigma+i\infty} \frac{1}{z-a} e^{-z^\alpha t}\, dz = e^{-ta^\alpha}.\end{aligned}$$

Proposition 2. We have

$$f_{t,\alpha}(\lambda) \geqq 0 \quad \text{for all} \quad \lambda > 0. \tag{8}$$

Proof. If we set $a^{\alpha} = g(a)$, $e^{-tx} = h(x)$, then

$$(-1)^{n-1} \cdot g^{(n)}(a) \geqq 0 \quad (n = 1, 2, \ldots), \; g(a) \geqq 0 \quad \text{and}$$
$$(-1)^n \cdot h^{(n)}(x) \geqq 0 \quad (n = 0, 1, \ldots), \quad \text{when } a \geqq 0 \quad \text{and} \quad x \geqq 0.$$

Hence $k(a) = h(g(a)) = e^{-ta^{\alpha}}$ satisfies

$$\left.\begin{aligned}&(-1)^n k^{(n)}(a) = (-1)\, h'(x)\, (-1)^{n-1} g^{(n)}(a)\\&+ \sum_{(p)} C^{(n)}_{(p_0 p_1 \cdots p_\nu)} (-1)^{p_0} h^{(p_0)}(x) (-1)^{p_1} g^{(p_1+1)}(a) \ldots (-1)^{p_\nu} g^{(p_\nu+1)}(a)\\&\left(C^{(n)}_{(p)} \geqq 0, p_0 \geqq 2, p_1 \geqq 0, \ldots, p_\nu \geqq 0 \text{ with } p_0 \leqq \sum_{i=1}^{\nu} p_i = n, \text{ and } \nu \text{ arbitrary}\right)\\&\geqq 0, \; (n = 0, 1, \ldots).\end{aligned}\right\} \tag{9}$$

That is, the function $k(a) = e^{-ta^{\alpha}}$ is *completely monotone* in $a \geqq 0$.

We next prove the *Post-Widder inversion formula*

$$f_{t,\alpha}(\lambda) = \lim_{n \to \infty} \frac{(-1)^n}{n!} \left(\frac{n}{\lambda}\right)^{n+1} k^{(n)}\left(\frac{n}{\lambda}\right), \quad \lambda > 0 \tag{10}$$

so that, by (9), $f_{t,\alpha}(\lambda) \geqq 0$. The proof of (10) is obtained as follows. We find, differentiating (7) n-times,

$$k^{(n)}\left(\frac{n}{\lambda}\right) = (-1)^n \int_0^{\infty} s^n e^{-sn/\lambda} f_{t,\alpha}(s)\, ds.$$

Substituting this into the right side of (10), we get

$$\lim_{n \to \infty} \frac{n^{n+1}}{e^n n!} \frac{1}{\lambda} \int_0^{\infty} \left[\frac{s}{\lambda} \cdot \exp\left(1 - \frac{s}{\lambda}\right)\right]^n f_{t,\alpha}(s)\, ds.$$

Since

$$\lim_{n \to \infty} n^n / \sqrt{2\pi n}\; e^n n! = 1 \quad \text{(Stirling's formula)},$$

we have to prove that

$$f_{t,\alpha}(\lambda_0) = \lim_{n \to \infty} \frac{\sqrt{2\pi}}{\lambda_0} \int_0^{\infty} n^{3/2} \left[\frac{s}{\lambda_0} \cdot \exp\left(1 - \frac{s}{\lambda_0}\right)\right]^n f_{t,\alpha}(s)\, ds, \; \lambda_0 > 0. \tag{11}$$

Let η be a fixed positive number such that $\eta < \lambda_0$. We decompose the last integral into three parts,

$$\int_0^{\infty} = \int_0^{\lambda_0 - \eta} + \int_{\lambda_0 - \eta}^{\lambda_0 + \eta} + \int_{\lambda_0 + \eta}^{\infty} = J_1 + J_2 + J_3.$$

Since $x \cdot \exp(1 - x)$ increases monotonically in $[0, 1]$ from 0 to 1, we see, by the boundedness of $f_{t,\alpha}(s)$ in s, that $\lim_{n\to\infty} J_1 = 0$. Next, since $x \cdot \exp(1-x)$ decreases monotonically in $[1, \infty]$ from 1 to 0, we have

$$\frac{\lambda_0+\eta}{\lambda_0} \cdot \exp\left(1 - \frac{\lambda_0+\eta}{\lambda_0}\right) < \beta < 1,$$

and so, since $f_{t,\alpha}(s)$ is of exponential growth as $s \uparrow \infty$,

$$|J_3| \leqq n^{3/2} e^{n_0} \beta^{n-n_0} \int_0^\infty \left(\frac{s}{\lambda_0}\right)^{n_0} \exp\left(-\frac{n_0 s}{\lambda_0}\right) |f_{t,\alpha}(s)|\, ds \to 0 \quad \text{as} \quad n \to \infty.$$

By the continuity of $f_{t,\alpha}(s)$ in s, we have, for any positive number ε,

$$f_{t,\alpha}(\lambda_0) - \varepsilon \leqq f_{t,\alpha}(s) \leqq f_{t,\alpha}(\lambda_0) + \varepsilon \quad \text{whenever} \quad \lambda_0 - \eta \leqq s \leqq \lambda_0 + \eta,$$

if we take $\eta > 0$ sufficiently small. Thus

$$(f_{t,\alpha}(\lambda_0) - \varepsilon)\, J_0 \leqq J_2 \leqq (f_{t,\alpha}(\lambda_0) + \varepsilon)\, J_0, \tag{12}$$

where

$$J_0 = \int_{\lambda_0-\eta}^{\lambda_0+\eta} n^{3/2} \left[\frac{s}{\lambda_0} \cdot \exp\left(1 - \frac{s}{\lambda_0}\right)\right]^n ds. \tag{13}$$

The whole preceding argument is true for the particular case of the completely monotone function

$$k(a) = a^{-1} = \int_0^\infty e^{-\lambda a}\, d\lambda.$$

In this case, $k^{(n)}(n/\lambda_0) = (-1)^n\, n!\, (\lambda_0/n)^{n+1}$. Substituting this in (10), we find that (10) holds for $f_{t,\alpha}(\lambda) \equiv 1$. Since (10) and (11) are equivalent, (11) must also hold for $f_{t,\alpha}(\lambda) \equiv 1$. Thus, since $\lim_{n\to\infty} J_1 = 0$ and $\lim_{n\to\infty} J_3 = 0$ for a general $f_{t,\alpha}$, we obtain

$$1 = \lim_{n\to\infty} \sqrt{2\pi}\, \lambda_0^{-1} J_0.$$

Therefore, by (12), we get (11) and the equivalent formula (10) is proved.

Proposition 3.

$$\int_0^\infty f_{t,\alpha}(\lambda)\, d\lambda = 1, \tag{14}$$

$$f_{t+s,\alpha}(\lambda) = \int_0^\infty f_{t,\alpha}(\lambda - \mu)\, f_{s,\alpha}(\mu)\, d\mu. \tag{15}$$

Proof. Since the function $f_{t,\alpha}(\lambda)$ is non-negative, we have, by the Lebesgue-Fatou lemma and (7),

$$\int_0^\infty \lim_{a\downarrow 0} \left(e^{-\lambda a} f_{t,\alpha}(\lambda)\right) d\lambda \leqq \lim_{a\downarrow 0} e^{-t a^\alpha} = 1.$$

Thus $f_{t,\alpha}(\lambda)$ is integrable with respect to λ over $(0, \infty)$ and so, again by the Lebesgue-Fatou lemma and (7), we obtain (14). We next have, by (7),

$$\int e^{-\lambda a} \{\int f_{t,\alpha}(\lambda-\mu)\, f_{s,\alpha}(\mu)\, d\mu\}\, d\lambda$$
$$= \int e^{-(\lambda-\mu)a} f_{t,\alpha}(\lambda-\mu)\, d(\lambda-\mu) \cdot \int e^{-\mu a} f_{s,\alpha}(\mu)\, d\mu$$
$$= e^{-ta^\alpha} e^{-sa^\alpha} = e^{-(t+s)a^\alpha} = \int e^{-\lambda a} f_{t+s,\alpha}(\lambda)\, d\lambda, \quad a > 0.$$

Hence, by inverting the Laplace transform as in the preceding section, we get (15).

Proposition 4. We have

$$\int_0^\infty \partial f_{t,\alpha}(\lambda)/\partial t \cdot d\lambda = 0, \quad t > 0. \tag{16}$$

Proof. By deforming the path of integration in (1) to the union of two paths $re^{-i\theta}$ $(-\infty < -r < 0)$ and $re^{i\theta}$ $(0 < r < \infty)$, where $\pi/2 \leqq \theta \leqq \pi$, we obtain

$$f_{t,\alpha}(s) = \frac{1}{\pi} \int_0^\infty \exp(sr \cdot \cos\theta - tr^\alpha \cos\alpha\theta)$$
$$\times \sin(sr \cdot \sin\theta - tr^\alpha \sin\alpha\theta + \theta)\, dr. \tag{17}$$

Similarly, by deforming the path of integration in

$$\partial f_{t,\alpha}(\lambda)/\partial t = \frac{1}{2\pi i} \int_{\sigma-i\infty}^{\sigma+i\infty} e^{z\lambda - z^\alpha t} (-z^\alpha)\, dz$$

to the union of two paths, $re^{-i\theta}$ $(-\infty < -r < 0)$ and $re^{i\theta}$ $(0 < r < \infty)$, we obtain

$$f'_{t,\alpha}(s) = \partial f_{t,\alpha}(s)/\partial t = \frac{(-1)}{\pi} \int_0^\infty \exp(sr \cdot \cos\theta - tr^\alpha \cos\alpha\theta)$$
$$\times \sin(sr \cdot \sin\theta - tr^\alpha \sin\alpha\theta + \alpha\theta + \theta)\, r^\alpha dr. \tag{18}$$

If we take

$$\theta = \theta_\alpha = \pi/(1+\alpha),$$

then

$$f'_{t,\alpha}(s) = \frac{1}{\pi} \int_0^\infty \exp((sr + tr^\alpha)\cos\theta_\alpha) \cdot \sin((sr - tr^\alpha)\sin\theta_\alpha)\, r^\alpha dr. \tag{19}$$

Thus we see, by the factor r^α $(0 < \alpha < 1)$, that $f'_{t,\alpha}(s)$ is integrable with respect to s over $(0, \infty)$. Hence, by differentiating (14) with respect to t, we get (16).

We are now able to prove

Theorem 1. $\{\hat{T}_t\}$ is a holomorphic semi-group.

Proof. That $\{\hat{T}_t\}$ enjoys the semi-group property $\hat{T}_t \hat{T}_s = \hat{T}_{t+s}$ $(t, s > 0)$ is clear from (2) and (15). We have, by (2) and (17) with $\theta = \theta_\alpha$,

$$\hat{T}_t x = \frac{1}{\pi} \int_0^\infty T_s x\, ds \int_0^\infty \exp((sr + tr^\alpha)\cos\theta_\alpha)$$
$$\times \sin((sr - tr^\alpha)\sin\theta_\alpha + \theta_\alpha)\, dr, \tag{20}$$

which gives by the change of variables

$$s = vt^{1/\alpha}, \ r = ut^{-1/\alpha}, \tag{21}$$

$$\hat{T}_t x = \frac{1}{\pi} \int_0^\infty T_{vt^{1/\alpha}} \cdot x dv \int_0^\infty \exp((uv + u^\alpha) \cos \theta_\alpha)$$
$$\times \sin((uv - u^\alpha) \sin \theta_\alpha + \theta_\alpha) du. \tag{20'}$$

The second integral on the right is exactly $\pi \cdot f_{1,\alpha}(v)$, and so, by the equi-boundedness of $\{\|T_t x\|\}$ in $t \geqq 0$, we see, by (14), that

$$\|\hat{T}_t x\| \leqq \sup_{t \geqq 0} \|T_t x\| \int_0^\infty f_{1,\alpha}(v)\, dv = \sup_{t \geqq 0} \|T_t x\|. \tag{22}$$

By passing to the limit $t \downarrow 0$ in (20'), which is justified by the integrability of $f_{1,\alpha}(v)$ over $[0, \infty)$, we obtain, by (14),

$$s\text{-}\lim_{t \downarrow 0} \hat{T}_t x = \int_0^\infty f_{1,\alpha}(v)\, dv \cdot x = x.$$

Hence $\{\hat{T}_t\}$ is an equi-continuous semi-group of class (C_0) such that (22) holds.

By the integrability of $f'_{t,\alpha}(s) = \partial f_{t,\alpha}(s)/\partial t$ over $[0, \infty)$ and the equi-continuity of $\{T_t\}$, we obtain, by differentiating (2) with respect to t under the integral sign,

$$\hat{T}'_t x = \int_0^\infty f'_{t,\alpha}(s)\, T_s x ds$$
$$= \frac{1}{\pi} \int_0^\infty T_s x ds \int_0^\infty \exp((sr + tr^\alpha) \cos \theta_\alpha) \cdot \sin((sr - tr^\alpha) \sin \theta_\alpha)\, r^\alpha dr, \tag{23}$$

which is, by the change of variables (21),

$$= \int_0^\infty (T_{vt^{1/\alpha}} \cdot x) \cdot f'_{1,\alpha}(v)\, dv \cdot t^{-1}.$$

Thus, by the integrability of $f'_{1,\alpha}(v)$ over $[0, \infty)$ and the equi-continuity of $\{T_t\}$ in $t \geqq 0$, we see that

$$\overline{\lim_{t \downarrow 0}} \|t \hat{T}'_t\| < \infty,$$

that is, $\{\hat{T}_t\}$ is a holomorphic semi-group.

Theorem 2. The infinitesimal generator $\hat{A}_\alpha$ of $\hat{T}_t$ is connected with the infinitesimal generator A of T by (3), where $(-A)^\alpha$ is defined by (4) and also by (5). We have, moreover, (6).

Proof. By (16) and (23), we obtain

$$\hat{T}'_t x = \frac{1}{\pi}\int_0^\infty (T_s - I)\,x\,ds \cdot \int_0^\infty \exp((sr + tr^\alpha)\cos\theta_\alpha) \times \sin((sr - tr^\alpha)\sin\theta_\alpha)\, r^\alpha\, dr. \tag{24}$$

If $x \in D(A)$, then $s\text{-}\lim_{s\downarrow 0} s^{-1}(T_s - I)\,x = Ax$ and $\|(T_s - I)\cdot x\|$ is bounded in $s \geqq 0$. Thus, we obtain, letting $t \downarrow 0$ in (24),

$$\begin{aligned} s\text{-}\lim_{t\downarrow 0} \hat{T}'_t x &= \frac{1}{\pi}\int_0^\infty (T_s - I)\,x\,ds \int_0^\infty \exp(sr\cdot\cos\theta_\alpha)\cdot\sin(sr\cdot\sin\theta_\alpha)\, r^\alpha\, dr \\ &= (-\Gamma(-\alpha))^{-1}\int_0^\infty s^{-\alpha-1}(T_s - I)\,x\,ds, \end{aligned}$$

because, by the Γ-function formulae

$$\Gamma(z) = c^z \int_0^\infty e^{-cr}\, r^{z-1}\, dr \quad (Re(z) > 0,\ Re(c) > 0) \tag{25}$$

and

$$\Gamma(z)\,\Gamma(1-z) = \pi/\sin\pi z, \tag{26}$$

we obtain, by $(\alpha + 1)\,\theta_\alpha = \pi$,

$$\begin{aligned} &\frac{1}{\pi}\int_0^\infty \exp(sr\cdot\cos\theta_\alpha)\cdot\sin(sr\cdot\sin\theta_\alpha)\, r^\alpha\, dr = (\pi i)^{-1}\, Im\left\{\int_0^\infty e^{-r(-se^{i\theta_\alpha})}\, r^\alpha\, dr\right\} \\ &= (\pi i)^{-1}\,\mathrm{Im}((-s e^{i\theta_\alpha})^{-\alpha-1})\,\Gamma(1+\alpha) = s^{-\alpha-1}\,\pi^{-1}\sin(\alpha\pi)\,\Gamma(1+\alpha) \\ &= s^{-\alpha-1}\frac{-\Gamma(1+\alpha)}{\Gamma(-\alpha)\,\Gamma(1+\alpha)} = (-\Gamma(-\alpha))^{-1}\, s^{-\alpha-1}. \end{aligned}$$

Thus, by $\hat{T}'_t x = \hat{A}_\alpha \hat{T}_t x$ (when $t > 0$), the continuity at $t = 0$ of $\hat{T}_t x$ and the closure property of the infinitesimal generator $\hat{A}_\alpha$, we obtain

$$\hat{A}_\alpha\, x = (-\Gamma(-\alpha))^{-1}\int_0^\infty s^{-\alpha-1}(T_s - I)\,x\,ds \quad \text{when} \quad x \in D(A).$$

Hence, by (25), (26) and $(tI - A)^{-1} = \int_0^\infty e^{-ts}\, T_s\, ds$, we obtain

$$\begin{aligned} \hat{A}_\alpha x &= \Gamma(-\alpha)^{-1}\,\Gamma(1+\alpha)^{-1}\int_0^\infty \left\{\int_0^\infty e^{-st}\, t^\alpha\, dt\right\}(I - T_s)\,x\,ds \\ &= \frac{\sin\alpha\pi}{\pi}\int_0^\infty t^\alpha\,((tI - A)^{-1} - t^{-1}I)\,x\,dt \\ &= \frac{\sin\alpha\pi}{\pi}\int_0^\infty t^{\alpha-1}(tI - A)^{-1} A\,x\,dt \quad \text{for} \quad x \in D(A). \end{aligned}$$

Finally we have, by taking $\theta = \pi$ in (17) and (2),

$$(\mu I - \hat{A}_\alpha)^{-1} = \int_0^\infty e^{-\mu t} \hat{T}_t dt$$

$$= \pi^{-1} \int_0^\infty dr \int_0^\infty e^{-sr} T_s ds \int_0^\infty \exp(-\mu t - tr^\alpha \cos \alpha\pi) \cdot \sin(tr^\alpha \sin \alpha\pi)\, dt$$

$$= \pi^{-1} \int_0^\infty (rI - A)^{-1} \left\{ \int_0^\infty \exp(-\mu t - tr^\alpha \cos \alpha\pi) \cdot \sin(tr^\alpha \sin \alpha\pi)\, dt \right\} dr$$

$$= \frac{\sin \alpha\pi}{\pi} \int_0^\infty (rI - A)^{-1} \frac{r^\alpha}{\mu^2 - 2r^\alpha \mu \cos \alpha\pi + r^{2\alpha}} dr.$$

Remark. Formula (2) was devised by S. Bochner [2] without detailed proof. Cf. R. S. Phillips [5]. That $\hat{T}_t$ is a holomorphic semi-group was proved by K. Yosida [8], V. Balakrishnan [1] and T. Kato [2]. Formulae (4) and (5) are due to V. Balakrishnan [1], who, by virtue of (4), defined the fractional power $(-A)^\alpha$ of a closed linear operator A satisfying the condition:

$$\begin{aligned} &\text{the resolvent } R(\lambda; A) = (\lambda I - A)^{-1} \text{ exists for } Re(\lambda) > 0 \\ &\text{and } \sup_{Re(\lambda) > 0} |Re(\lambda)| \cdot \|R(\lambda; A)\| < \infty. \end{aligned} \tag{27}$$

He also proved that $(-A)^\alpha$ enjoys properties to be demanded for the fractional power. In fact we have

Theorem 3. Let a closed linear operator A satisfy condition (27). Then, by (4), a linear operator $(-A)^\alpha$ is defined, and we have

$$\begin{aligned} &(-A)^\alpha (-A)^\beta x = (-A)^{\alpha+\beta} x \quad \text{if} \quad x \in D(A^2) \quad \text{and} \quad 0 < \alpha, \beta \\ &\text{with} \quad \alpha + \beta < 1, \end{aligned} \tag{28}$$

$$\operatorname*{s-lim}_{\alpha \uparrow 1} (-A)^\alpha x = -Ax \quad \text{if} \quad x \in D(A), \tag{29}$$

$$\operatorname*{s-lim}_{\alpha \downarrow 0} (-A)^\alpha x = x \quad \text{if} \quad \operatorname*{s-lim}_{\lambda \downarrow 0} \lambda R(\lambda; A)\, x = 0, \tag{30}$$

and, if A is the infinitesimal generator of an equi-continuous semi-group T_t of class (C_0),

$$(A_\alpha)_\beta = A_{\alpha\beta}, \text{ where } A_\alpha \text{ is the operator } \hat{A}_\alpha \text{ defined through Kato's formula (6)}. \tag{31}$$

Remark. The last formula (31) is due to J. Watanabe [1].

Proof. $\|r^{\alpha-1}(rI - A)^{-1}(-Ax)\|$ is, by (27), of order $O(r^{\alpha-2})$ when $r \uparrow \infty$, and, by $(rI - A)^{-1}(-Ax) = x - r(rI - A)^{-1}x$ and (27), it is of order $O(r^{\alpha-1})$ when $r \downarrow 0$. Thus the right side of (4) is convergent.

It is clear that $x \in D(A^2)$ implies $(-A)^\beta x \in D(A)$. For, by approximating the integral by Riemann sums and making use of the closure property of A, we have $(-A)^\beta x \subseteq D(A)$. We can thus define $(-A)^\alpha(-A)^\beta x$.

$$(-A)^\alpha (-A)^\beta x = \frac{\sin \alpha\pi}{\pi} \frac{\sin \beta\pi}{\pi} \int_0^\infty \int_0^\infty \lambda^{\beta-1} \mu^{\alpha-1} R(\lambda; A) R(\mu; A) A^2 x \, d\lambda \, d\mu$$

can be rewritten, decomposing the domain of integration into those for which $\lambda \geqq \mu$ and $\lambda < \mu$, as follows.

$$\frac{\sin \alpha\pi}{\pi} \frac{\sin \beta\pi}{\pi} \int_0^1 (\sigma^{\beta-1} + \sigma^{\alpha-1}) \, d\sigma \int_0^\infty \lambda^{\alpha+\beta-1} R(\lambda\sigma; A) R(\lambda; A) A^2 x \, d\lambda.$$

By the resolvent equation $R(\lambda; A) - R(\mu; A) = (\mu - \lambda) R(\lambda; A) R(\mu; A)$ and $R(\lambda; A)(-A) = I - \lambda R(\lambda; A)$ valid on $D(A)$, we obtain

$$R(\lambda\sigma; A) R(\lambda; A) A^2 x = (1-\sigma)^{-1}\{-\sigma R(\lambda\sigma; A) + R(\lambda; A)\}(-A x),$$

and hence

$$\begin{aligned} &(-A)^\alpha (-A)^\beta x \\ &= \frac{\sin \alpha\pi}{\pi} \frac{\sin \beta\pi}{\pi} \operatorname*{s-lim}_{t \uparrow 1} \int_0^t (\sigma^{\beta-1} + \sigma^{\alpha-1})(1-\sigma)^{-1} \, d\sigma \\ &\quad \times \int_0^\infty \lambda^{\alpha+\beta-1} (-\sigma R(\lambda\sigma; A) + R(\lambda; A))(-A x) \, d\lambda \\ &= \int_0^\infty \left(\frac{\sin \alpha\pi}{\pi} \frac{\sin \beta\pi}{\pi} \int_0^1 \frac{\sigma^{\beta-1} + \sigma^{\alpha-1} - \sigma^{-\alpha} - \sigma^{-\beta}}{1-\sigma} \, d\sigma \right) \\ &\quad \times \lambda^{\alpha+\beta-1} R(\lambda; A)(-A x) \, d\lambda. \end{aligned}$$

The coefficient ( ) above is evaluated as $\pi^{-1} \sin \pi(\alpha + \beta)$, as may be seen by expanding $(1-\sigma)^{-1}$ into powers of σ. We have thus proved (28).

To prove (29), we shall make use of $\int_0^\infty \lambda^{\alpha-1}(1+\lambda)^{-1} d\lambda = \pi/\sin \alpha\pi$. Thus

$$(-A)^\alpha x - (-A) x = \frac{\sin \alpha\pi}{\pi} \int_0^\infty \lambda^{\alpha-1} \left(R(\lambda; A) - \frac{1}{\lambda+1} \right) (-A x) \, d\lambda.$$

We split the integral into two parts, one from 0 to C and the other from C to ∞. For a fixed C, the first part goes to zero as $\alpha \uparrow 1$, since $R(\lambda; A)(-A x) = x - \lambda R(\lambda; A) x$ is bounded in $\lambda > 0$. The second part is, in norm,

$$\leqq \frac{\sin \alpha\pi}{\pi(1-\alpha)} C^{\alpha-1} \sup_{\lambda \geqq C} \left\| \left(\lambda R(\lambda; A) - \frac{\lambda}{\lambda+1} \right) A x \right\|.$$

We have $\operatorname*{s-lim}_{\lambda \uparrow \infty} \lambda R(\lambda; A) x = x$ by $x - \lambda R(\lambda; A) x = R(\lambda; A)(-A x)$ and (27). Hence $\operatorname*{s-lim}_{\alpha \uparrow 1}$ of the second part is arbitrarily near 0 if we take C sufficiently large. This proves (29).

To prove (30), we split the integral into two parts, one from 0 to C and the other from C to ∞. Because of (27), the second part tends to zero as $\alpha \downarrow 0$. By $R(\lambda; A)(-Ax) = x - \lambda R(\lambda; A)x$ and the hypothesis $\underset{\lambda \downarrow 0}{s\text{-}\lim}\, \lambda R(\lambda; A)x = 0$, the first is, for sufficiently small C, arbitrarily near the value $(\alpha\pi)^{-1} \sin \alpha\pi \cdot C^\alpha x$, which tends to x when $\alpha \downarrow 0$. This proves (30).

We shall prove (31). By virtue of the representation (6), we obtain

$$(\mu I - (A_\alpha)_\beta)^{-1} = \int_0^\infty \int_0^\infty (2\pi i)^{-2} \left(\frac{1}{\mu - \lambda^\beta \cdot e^{-i\pi\beta}} - \frac{1}{\mu - \lambda^\beta \cdot e^{i\pi\beta}} \right)$$
$$\times \left(\frac{1}{\lambda - \zeta^\alpha \cdot e^{-i\pi\alpha}} - \frac{1}{\lambda - \zeta^\alpha \cdot e^{i\pi\alpha}} \right) (\zeta I - A)^{-1} \, d\lambda \, d\zeta .$$

This double integral is absolutely convergent in norm, and so we may interchange the order of integration. Hence we obtain (31), because the inner integral is

$$= (2\pi i)^{-2} \int_C \frac{-1}{\mu - z^\beta} \left(\frac{1}{z - \zeta^\alpha e^{-i\pi\alpha}} - \frac{1}{z - \zeta^\alpha e^{i\pi\alpha}} \right) dz$$
$$= (2\pi i)^{-1} \left(\frac{-1}{\mu - \zeta^{\alpha\beta} e^{-i\pi\alpha\beta}} - \frac{-1}{\mu - \zeta^{\alpha\beta} e^{i\pi\alpha\beta}} \right)$$

where the path of integration C runs from $\infty e^{i\pi}$ to 0 and from 0 to $\infty e^{-i\pi}$.

An example of the fractional power. If $\alpha = 1/2$, then we have, by taking $\theta = \pi$ in (17),

$$f_{t,1/2}(s) = \pi^{-1} \int_0^\infty e^{-sr} \sin(t r^{1/2}) \, dr = \pi^{-1} \sqrt{\pi}\, t \left(2^3 \sqrt{s}\right)^{-3} \cdot e^{-t^2/4s}. \quad (32)$$

Thus, if we take the semi-group $\{T_t\}$ associated with the *Gaussian kernel*:

$$(T_t x)(u) = \frac{1}{2\sqrt{\pi s}} \int_{-\infty}^\infty e^{-(u-v)^2/4s} x(v) \, dv, \quad x \in C[-\infty, \infty],$$

then

$$(\hat{T}_{t,1/2} x)(u) = \int_{-\infty}^\infty \left\{ \int_0^\infty x(v) \frac{t}{4\pi s^2} e^{-((u-v)^2 + t^2)/4s} \, ds \right\} dv$$
$$= \frac{t}{\pi} \int_{-\infty}^\infty \frac{1}{t^2 + (u-v)^2} x(v) \, dv,$$

that is, the semi-group $\{\hat{T}_t\}$ is associated with the *Poisson kernel*. In this case, the infinitesimal generator A of T_t is given by the differential operator d^2/ds^2, while the infinitesimal generator $\hat{A}$ of $\hat{T}_t$ is given by the singular integral operator

$$(\hat{A}_{1/2} x)(s) = \underset{h \downarrow 0}{s\text{-}\lim} \frac{1}{\pi} \int_{-\infty}^\infty \frac{x(s-v) - x(s)}{v^2 + h^2} \, dv,$$

and not by the differential operator d/ds.

12. The Convergence of Semi-groups. The Trotter-Kato Theorem

We shall denote by $\exp(tA)$ the semi-group of class (C_0) with the infinitesimal generator A. Concerning the convergence of semi-groups, we have

Theorem 1. Let X be a locally convex, sequentially complete complex linear space. Let $\{\exp(tA_n)\} \subseteqq L(X, X)$ be a sequence of equi-continuous semi-groups of class (C_0) such that the family of operators $\{\exp(tA_n)\}$ is equi-continuous in $t \geqq 0$ and in $n = 1, 2, \ldots$ Thus we assume that, for any continuous semi-norm $p(x)$ on X, there exists a continuous semi-norm $q(x)$ on X such that

$$p(\exp(tA_n)\,x) \leqq q(x) \quad \text{for all } t \geqq 0,\ x \in X \text{ and } n = 1, 2, \ldots \tag{1}$$

Suppose that, for some λ_0 with $Re(\lambda_0) > 0$,

$$\lim_{n\to\infty} R(\lambda_0; A_n)\,x = J(\lambda_0)\,x \text{ exists for all } x \in X \text{ in such a way that the range } R(J(\lambda_0)) \text{ is dense in } X. \tag{2}$$

Then $J(\lambda_0)$ is the resolvent of the infinitesimal generator A of an equi-continuous semi-group $\exp(tA)$ of class (C_0) in X and

$$\lim_{n\to\infty} \exp(tA_n)\,x = \exp(tA)\,x \quad \text{for every} \quad x \in X. \tag{3}$$

Moreover, the convergence in (3) is uniform in t on every compact interval of t.

For the proof, we prepare

Lemma. Let $T_t = \exp(tA)$ be an equi-continuous semi-group of class (C_0) in X. Then, for any continuous semi-norm $p(x)$ on X, there exists a continuous semi-norm $q(x)$ on X such that

$$p(T_t x - (I - tn^{-1}A)^{-n}\,x) \leqq (2n)^{-1}\,t^2\,q(A^2 x) \quad (n = 1, 2, \ldots) \text{ whenever } x \in D(A^2). \tag{4}$$

Proof. Set $T(t, n) = (I - n^{-1}tA)^{-n}$. Then we know (Chapter IX,7) that $\{T(t, n)\}$ is equi-continuous in $t \geqq 0$ and $n = 1, 2, \ldots$ Moreover, we have (Chapter IX,4), for any $x \in D(A)$,

$$D_t T(t, n) = (I - n^{-1}tA)^{-n-1}\,Ax = A\,(I - n^{-1}tA)^{-n-1}\,x,$$

$$D_t T x = T_t A x = A T_t x.$$

Thus, by the commutativity of T_t and $T(t, n)$,

$$\begin{aligned} T_t x - T(t, n)\,x &= \int_0^t [D_s T(t-s, n)\,T_s x]\,ds \\ &= \int_0^t T(t-s, n)\,T_s\Big(Ax - \Big(I - \frac{t-s}{n}A\Big)^{-1} Ax\Big)\,ds, \quad x \in D(A). \end{aligned} \tag{5}$$

Hence, if $x \in D(A^2)$, we have, by $(I - m^{-1}A)^{-1}Ax = -m(I - (I - m^{-1}A)^{-1})x$,

$$p(T_t x - T(t, n)x) \leqq \int_0^t p\left[T(t-s, n)\, T_s (I - n^{-1}(t-s)A)^{-1} \frac{s-t}{n} A^2 x\right] ds,$$

and so, by a continuous semi-norm $q(x)$ on X which is independent of x and n,

$$p(T_t x - T(t, n)x) \leqq (2n)^{-1} t^2 q(A^2 x).$$

Corollary. For any $x \in D(A^2)$, $s > 0$ and $t \geqq 0$,

$$\left.\begin{array}{l} p(T_t x - (I - sA)^{-[t/s]}x) \leqq s q_1(Ax) + \frac{ts}{2} q(A^2 x), \text{ where} \\ q_1(x) \text{ is a continuous semi-norm on } X \text{ which is indepen-} \\ \text{dent of } x, t \text{ and } s, \text{ and } [t/s] \text{ is the largest integer } \leqq t/s. \end{array}\right. \tag{6}$$

Proof. We have, for $t = ns$,

$$p(T_{ns}x - (I - sA)^{-n}x) \leqq 2^{-1} stq(A^2 x).$$

If $t = ns + u$ with $0 \leqq u < s$ and $n = [t/s]$,

$$p(T_t x - T_{ns}x) = p\left(\int_{ns}^t T'_\sigma x\, d\sigma\right) \leqq \int_{ns}^t p(T_\sigma Ax)\, d\sigma \leqq s q_1(Ax).$$

Proof of Theorem 1. By (1) and by (11) in Chapter IX, 4, we see that $\{(Re(\lambda) R(\lambda; A_n)^m)\}$ is equi-continuous in $Re(\lambda) > 0$, in $n = 1, 2, \ldots$ and in $m = 0, 1, 2, \ldots$ From this and by (2), we can prove that $J(\lambda_0) = (\lambda_0 I - A)^{-1}$ with some A and

$$\left.\begin{array}{l} \lim_{n\to\infty} R(\lambda; A_n)x = R(\lambda; A)x \text{ whenever } Re(\lambda) > 0, \text{ and} \\ \text{the convergence is uniform with respect to } \lambda \text{ on every com-} \\ \text{pact subset of the right half-plane } Re(\lambda) > 0. \end{array}\right\} \tag{7}$$

To this purpose, we observe that

$$R(\lambda; A_n)x = \sum_{m=0}^{\infty} (\lambda_0 - \lambda)^m R(\lambda_0; A_n)^{m+1} x \quad (\text{for } |\lambda - \lambda_0|/Re(\lambda_0) < 1)$$

and that the series is, by the equi-continuity of $\{(Re(\lambda_0) R(\lambda_0; A_n))^m\}$ in $n = 1, 2, \ldots$ and $m = 0, 1, 2, \ldots$, uniformly convergent for $|\lambda - \lambda_0|/Re(\lambda_0) \leqq 1 - \varepsilon$ and $n = 1, 2, \ldots$ Here ε is a fixed positive number. Therefore, for any $\delta > 0$, there exists an m_0 and a continuous semi-norm $q(x)$ on X such that, for $|\lambda - \lambda_0|/Re(\lambda_0) \leqq 1 - \varepsilon$,

$$p(R(\lambda; A_n)x - R(\lambda; A_{n'})x) \leqq \sum_{m=0}^{m_0} |\lambda_0 - \lambda|^m \cdot$$
$$p(R(\lambda_0; A_n)^{m_0+1}x - R(\lambda_0; A_{n'})^{m_0+1}x) + 2\delta\, q(x) \text{ for all } x \in X.$$

Hence, by (2), we see that $\lim_{n\to\infty} R(\lambda; A_n)\, x = J(\lambda)\, x$ exists uniformly for $|\lambda - \lambda_0|/Re(\lambda_0) \leqq 1 - \varepsilon$. In this way, extending the convergence domain of the sequence $\{R(\lambda; A_n)\}$, we see that

$$\lim_{n\to\infty} R(\lambda; A)\, x = J(\lambda)\, x \text{ exists and the convergence is uniform}$$

$$\text{on any compact set of } \lambda \text{ in the right half-plane } Re(\lambda) > 0.$$

Thus $J(\lambda)$ is a pseudo-resolvent, because $J(\lambda)$ satisfies the resolvent equation with $R(\lambda; A_n)$ $(n = 1, 2, \ldots)$. However, by $R(J(\lambda_0))^a = X$ and the ergodic theorem for pseudo-resolvents in Chapter VIII, 4, we see that $J(\lambda)$ is the resolvent of a closed linear operator A in such a way that $J(\lambda) = R(\lambda; A)$ and $D(A) = R(R(\lambda; A))$ is dense in X.

Thus we see that $\exp(tA)$ is an equi-continuous semi-group of class (C_0) in X. We have to show that (3) is true. But we have, by (6),

$$p((\exp(tA_n) - (I - sA_n)^{-[t/s]})(I - A_n)^{-2}x)$$
$$\leqq s q_1(A_n(I - A_n)^{-2}x) + 2^{-1} t s q(A_n^2(I - A_n)^{-2}x),$$

for any $x \in X$, $s > 0$ and $t \geqq 0$.

The operators

$$A_n(I - A_n)^{-1} = (I - A_n)^{-1} - I, \ A_n(I - A)^{-2} = A_n(I - A_n)^{-1}(I - A_n)^{-1}$$
$$\text{and} \quad A_n^2(I - A_n)^{-2} = (A_n(I - A_n)^{-1})^2$$

are equi-continuous in $n = 1, 2, \ldots$ On the other hand, by (7),

$$\lim_{n\to\infty} (I - sA_n)^{-[t/s]}(I - A_n)^{-2}x = (I - sA)^{-[t/s]}(I - A)^{-2}x$$

uniformly in s and t, if $s > 0$ is bounded away from 0 and ∞, and if t runs over a compact interval of $[0, \infty)$. Moreover, we have, by (6),

$$p\left((\exp(tA) - (I - sA)^{-[t/s]})(I - A)^{-2}x\right) \leqq s q_1(A(I - A)^{-2}x)$$
$$+ 2^{-1} t s q(A^2(I - A)^{-2}x)$$

for every $x \in X$, $s > 0$ and $t \geqq 0$. Thus, by taking $s > 0$ sufficiently small, we see that

$$\lim_{n\to\infty} \exp(tA_n)\, y = \exp(tA)\, y \quad \text{for any } y \in R(1; A)^2 \cdot X,$$

and the convergence is uniform in t on every compact interval of t. $R(1; A)^2 \cdot X$ being dense in X, we see, by the equi-continuity of exp (tA) and $\exp(tA_n)$ in $t \geqq 0$ and in $n = 1, 2, \ldots$, that (3) holds.

Theorem 2. Let a sequence $\{\exp(tA_n)\}$ of equi-continuous semi-groups of class (C_0) in X be such that $\{\exp(tA_n)\}$ is equi-continuous in $t \geqq 0$ and in $n = 1, 2, \ldots$ If, for each $x \in X$,

$$\lim_{n\to\infty} \exp(tA_n)\, x = \exp(tA)\, x$$

uniformly with respect to t on every compact interval of t, then

$\lim_{n\to\infty} R(\lambda; A_n)\, x = R(\lambda; A)\, x$ for each $x \in X$ and $Re(\lambda) > 0$

and the convergence is uniform on every compact set of λ in the right half-plane $Re(\lambda) > 0$.

Proof. We have

$$R(\lambda; A)\, x - R(\lambda; A_n)\, x = \int_0^\infty e^{-\lambda t} (\exp(tA) - \exp(tA_n))\, x\, dt.$$

Hence, splitting the integral into two parts, one from 0 to C and the other from C to ∞, we obtain the result.

Remark. For the case of a Banach space X, Theorem 1 was first proved by H. F. TROTTER [1]. In his paper, the proof that $J(\lambda)$ is the resolvent $R(\lambda; A)$ is somewhat unclear. This was pointed out by T. KATO. The proof given above is adapted from Kato's modification of Trotter's proof of the Banach space case (unpublished).

13. Dual Semi-groups. Phillips' Theorem

Let X be a locally convex sequentially complete linear topological space, and $\{T_t; t \geqq 0\} \subseteq L(X, X)$ an equi-continuous semi-group of class (C_0). Then the family $\{T_t^*; t \geqq 0\}$ of operators $\in L(X', X')$ where (*) denotes the dual operators in this section, satisfies the semi-group property: $T_t^* T_s^* = T_{t+s}^*$, $T_0^* = I^* =$ the identity in X_s' (see Theorem 3 in Chapter VII, 1). However, it is not of class (C_0) in general. For, the mapping $T_t \to T_t^*$ does not necessarily conserve the continuity in t (see Proposition 1 in Chapter VII, 1). But we can show that $\{T^*\}$ is equi-continuous in $t \geqq 0$. For, we can prove

Proposition 1. If $\{S_t; t \geqq 0\} \subseteq L(X, X)$ is equi-continuous in $t \geqq 0$, then $\{S_t^*; t \geqq 0\} \subseteq L(X', X')$ is also equi-continuous in $t \geqq 0$.

Proof. For any bounded set B of X, the set $\bigcup_{t \geqq 0} S_t \cdot B$ is by hypothesis a bounded set of X. Let U' and V' be the polar sets of B and $\bigcup_{t \geqq 0} S_t \cdot B$:

$$U' = \{x' \in X';\ \sup_{b \in B} |\langle b, x'\rangle| \leqq 1\},\quad V' = \{x' \in X';\ \sup_{b \in B, t \geqq 0} |\langle S_t \cdot b, x'\rangle| \leqq 1\}.$$

Then (see Chapter IV, 7), U' and V' are neighbourhoods of 0 of X_s'. From

$$|\langle S_t \cdot b, x'\rangle| = |\langle b, S_t^* x'\rangle| \leqq 1 \quad (\text{when } b \in B,\ x' \in V')$$

we see that $S_t^* \cdot V' \subseteq U'$ for all $t \geqq 0$. This proves that $\{S_t^*\}$ is equi-continuous in $t \geqq 0$.

Let A be the infinitesimal generator of the semi-group T_t. Then $D(A)^a = X$, $R(A) \subseteq X$, and, for $\lambda > 0$, the resolvent $(\lambda I - A)^{-1} \in L(X, X)$ exists in such a way that

$$\{\lambda^m (\lambda I - A)^{-m}\} \text{ is equi-continuous in } \lambda > 0 \text{ and in } m = 0, 1 \ldots \tag{1}$$

We can prove (cf. Theorem 2 in Chapter VIII, 6)

Proposition 2. For $\lambda > 0$, the resolvent $(\lambda I^* - A^*)^{-1}$ exists and

$$(\lambda I^* - A^*)^{-1} = ((\lambda I - A)^{-1})^*. \tag{2}$$

Proof. We have $(\lambda I - A)^* = \lambda I^* - A^*$. Since $(\lambda I - A)^{-1} \in L(X, X)$, the operator $((\lambda I - A)^{-1})^* \in L(X', X')$ exists. We shall prove that $(\lambda I^* - A^*)^{-1}$ exists and equals $((\lambda I - A)^{-1})^*$. Suppose there exists an $x' \in X'$ such that $(\lambda I^* - A^*)\, x' = 0$. Then $0 = \langle x, (\lambda I^* - A^*)\, x'\rangle = \langle(\lambda I - A)\, x, x'\rangle$ for all $x \in D(A)$. But, since $R(\lambda I - A) = X$, we must have $x' = 0$. Hence the inverse $(\lambda I^* - A^*)^{-1}$ must exist. We have, for $x \in X$, $x' \in D(A^*)$,

$$\langle x, x'\rangle = \langle(\lambda I - A)\,(\lambda I - A)^{-1}\, x, x'\rangle = \langle(\lambda I - A)^{-1}\, x, (\lambda I^* - A^*)\, x'\rangle.$$

Thus $D(((\lambda I - A)^{-1})^*) \supseteq R(\lambda I^* - A^*)$ and $((\lambda I - A)^{-1})^* \cdot (\lambda I^* - A^*)\, x' = x'$ for every $x' \in D(A^*)$. This proves that $((\lambda I - A)^{-1})^* \supseteq (\lambda I^* - A^*)^{-1}$. On the other hand, if $x \in D(A)$ and $x' \in D\big(((\lambda I - A)^{-1})^*\big)$, then

$$\langle x, x'\rangle = \langle(\lambda I - A)^{-1}\,(\lambda I - A)\, x, x'\rangle = \langle(\lambda I - A)\, x, ((\lambda I - A)^{-1})^*\, x'\rangle.$$

This proves that $D(A^*) = D((\lambda I - A)^*) \supseteq R(((\lambda I - A)^{-1})^*)$ and $(\lambda I - A)^* \cdot ((\lambda I - A)^{-1})^*\, x' = x'$ for every $x' \in D(((\lambda I - A)^{-1})^*)$, that is, $((\lambda I - A)^{-1})^* \subseteq (\lambda I^* - A^*)^{-1}$. We have thus proved (2).

We are now ready to prove

Theorem. Let X be a locally convex sequentially complete linear topological space such that its strong dual space X' is also sequentially complete. Let $\{T_t\} \subseteq L(X, X)$, $t \geqq 0$, be an equi-continuous semi-group of class (C_0) with the infinitesimal generator A. Let us denote by X^+ the closure $D(A^*)^a$ of the domain $D(A^*)$ in the strong topology of X'. Let T_t^+ be the restriction of T_t^* to X^+. Then $T_t^+ \in L(X^+, X^+)$ and $\{T_t^+\,;\, t \geqq 0\}$ is an equi-continuous semi-group of class (C_0) such that its infinitesimal generator A^+ is the largest restriction of A^* with domain and range in X^+.

Remark. The above Theorem was proved by R. S. Phillips [2] for the spepial case of a B-space X. The extension given above is due to H. Komatsu [4].

Proof of the Theorem. We have the resolvent equation $R(\lambda; A) - R(\mu; A) = (\mu - \lambda)\, R(\lambda; A)\, R(\mu; A)$ and the equi-continuity of $\{\lambda^m R(\lambda; A)^m\}$ in $\lambda > 0$ and in $m = 0, 1, 2, \ldots$ Thus, by Proposition 1 and 2,

$$(\lambda I^* - A^*)^{-1} - (\mu I^* - A^*)^{-1} = (\mu - \lambda)\,(\lambda I^* - A^*)^{-1}\,(\mu I^* - A^*)^{-1} \tag{3}$$

$$\{\lambda^m\,(\lambda I^* - A^*)^{-m}\} \text{ is equi-continuous in } \lambda > 0 \text{ and in } m = 0, 1, 2, \ldots \tag{4}$$

Thus, if we denote by $J(\lambda)$ the restriction to X^+ of $(\lambda I^* - A^*)^{-1}$, we have

$$J(\lambda) - J(\mu) = (\mu - \lambda)\, J(\lambda)\, J(\mu), \tag{3'}$$

$$\{\lambda^m\, J(\lambda)^m\} \text{ is equi-continuous in } \lambda > 0 \text{ and in } m = 0, 1, \ldots \tag{4'}$$

Since $D(A^*)$ is dense in X^+ and since (4′) holds, we see, as in Chapter IX, 7, that $\lim_{\lambda\to\infty} \lambda J(\lambda)\, x = x$ for $x \in X^+$. Thus we have $R(J(\lambda))^a = X^+$ and so, by (7′) in Chapter VIII, 4, that $N(J(\lambda)) = \{0\}$. Thus the pseudo-resolvent $J(\lambda)$ must be the resolvent of a closed linear operator A^+ in X^+. Hence, by the sequential completeness of X^+ and (4′), A^+ is the infinitesimal generator of an equi-continuous semi-group of class (C_0) of operators $T_t^+ \in L(X^+, X^+)$. For any $x \in X$ and $y' \in X^+$, we have

$$\langle (I - m^{-1}tA)^{-m}\, x, y'\rangle = \langle x, (I^* - m^{-1}tA^+)^{-m}\, y'\rangle,$$

and so, by the result of the preceding section, we obtain by letting $m \to \infty$ the equality $\langle T_t x, y'\rangle = \langle x, T_t^+ y'\rangle$. Hence $T_t^*\, y' = T_t^+\, y'$, that is, T_t^+ is the restriction to X^+ of T_t^*.

We finally show that A^+ is the largest restriction of A^* with domain and range in X^+. It is clear, by the above derivation of the operator A^+, that A^+ is a restriction of A^*. Suppose that $x' \in D(A^*)$ and that $x' \in X^+$, $A^*\, x' \in X^+$. Then $(\lambda I^* - A^*)\, x' \in X^+$ and hence $(\lambda I^* - A^+)^{-1}(\lambda I^* - A^*)\, x' = x'$. Thus, applying $(\lambda I^* - A^+)$ from the left on both sides, we obtain $A^* x' = A^+ x'$. This proves that A^+ is the largest restriction of A^* with domain as well as range in X^+.

X. Compact Operators

Let X and Y be complex B-spaces, and let S be the unit sphere in X. An operator $T \in L(X, Y)$ is said to be *compact* or *completely continuous* if the image $T \cdot S$ is relatively compact in Y. For a compact operator $T \in L(X, X)$, the eigenvalue problem can be treated fairly completely, in the sense that the classical theory of Fredholm concerning linear integral equations may be extended to the linear functional equation $Tx - \lambda x = y$ with a complex parameter λ. This result is known as the *Riesz-Schauder theory*. F. Riesz [2] and J. Schauder [1].

1. Compact Sets in *B*-spaces

A compact set in a linear topological space must be bounded. The converse is, however, not true in general; we know (Chapter III, 2) that the closed unit sphere of a normed linear space X is strongly compact iff X **is** of finite dimension. Let S be a compact metric space and $C(S)$ the B-space of real- or complex-valued continuous functions $x(s)$ on S, normed by $\|x\| = \sup_{s\in S} |x(s)|$. We know (Chapter III, 3) that a subset $\{x_\alpha(s)\}$ of $C(S)$ is strongly relatively compact in $C(S)$ iff $\{x_\alpha(s)\}$ is equi-bounded and equi-continuous in α. For the case of the space $L^p(S, \mathfrak{B}, m)$, $1 \leq p < \infty$, we have

Theorem (Fréchet-Kolmogorov). Let S be the real line, $\mathfrak{B}$ the σ-ring of Baire subsets B of S and $m(B) = \int_B dx$ the ordinary Lebesgue measure of B. Then a subset K of $L^p(S, \mathfrak{B}, m)$, $1 \leqq p < \infty$, is strongly pre-compact iff it satisfies the conditions:

$$\sup_{x\in K} \|x\| = \sup_{x\in K} \left(\int_S |x(s)|^p\, ds \right)^{1/p} < \infty, \tag{1}$$

$$\lim_{t\to 0} \int_S |x(t+s) - x(s)|^p\, ds = 0 \text{ uniformly in } x \in K, \tag{2}$$

$$\lim_{\alpha\uparrow\infty} \int_{|s|>\alpha} |x(s)|^p\, ds = 0 \text{ uniformly in } x \in K. \tag{3}$$

Proof. Let K be strongly relatively compact. Then K is bounded and so (1) is true. Let $\varepsilon > 0$ be given. Then there exists a finite number of functions $\in L^p: f_1, f_2, \ldots, f_n$ such that, for each $f \in K$, there is a j with $\|f - f_j\| \leqq \varepsilon$. Otherwise, we would have an infinite sequence $\{f_j\} \subseteq K$ with $\|f_j - f_i\| > \varepsilon$ for $j \neq i$, contrary to the relative compactness of K. We then find, by the definition of the Lebesgue integral, finitely-valued functions $g_1, g_2, \ldots, g_n$ such that $\|f_j - g_j\| \leqq \varepsilon$ $(j = 1, 2, \ldots, n)$. Since each finitely-valued function $g_j(x)$ vanishes outside some sufficiently large interval, we have, for large α,

$$\left(\int_\alpha^\infty + \int_{-\infty}^{-\alpha} |f(s)|^p\, ds \right)^{1/p} \leqq \left(\int_\alpha^\infty + \int_{-\infty}^{-\alpha} |f(s) - g_j(s)|^p\, ds \right)^{1/p}$$
$$+ \left(\int_\alpha^\infty + \int_{-\infty}^{-\alpha} |g_j(s)|^p\, ds \right)^{1/p} \leqq \|f - g_j\| + \left(\int_\alpha^\infty + \int_{-\infty}^{-\alpha} |g_j(s)|^p\, ds \right)^{1/p}.$$

This proves (3) by $\|f - g_j\| \leqq \|f - f_j\| + \|f_j - g_j\| \leqq 2\varepsilon$.

The proof of (2) is based on the fact that, for the defining function $C_I(s)$ of a finite interval I, $\lim_{t\to 0} \int_{-\infty}^{\infty} |C_I(s+t) - C_I(s)|^p\, ds = 0$ (see Chapter 0, 3). Thus (2) holds for finitely-valued functions $g_j(s)$ $(j = 1, 2, \ldots, n)$. Hence we have, for any $f \in K$,

$$\overline{\lim_{t\to 0}} \left(\int_{-\infty}^{\infty} |f(s+t) - f(s)|^p\, ds \right)^{1/p} \leqq \overline{\lim_{t\to 0}} \left(\int_{-\infty}^{\infty} |f(s+t) - f_j(s+t)|^p\, ds \right)^{1/p}$$
$$+ \overline{\lim_{t\to 0}} \left(\int_{-\infty}^{\infty} |f_j(s+t) - g_j(s+t)|^p\, ds \right)^{1/p} + \overline{\lim_{t\to 0}} \left(\int_{-\infty}^{\infty} |g_j(s+t) - g_j(s)|^p\, ds \right)^{1/p}$$
$$+ \left(\int_{-\infty}^{\infty} |g_j(s) - f_j(s)|^p\, ds \right)^{1/p} + \left(\int_{-\infty}^{\infty} |f_j(s) - f(s)|^p\, ds \right)^{1/p} \leqq \varepsilon + \varepsilon + 0 + \varepsilon + \varepsilon,$$

by taking f_j in such a way that $\|f - f_j\| \leqq \varepsilon$. This proves (2).

We next prove the converse part of the Theorem. We define the translation operator T_t by $(T_t f)(s) = f(t + s)$. Condition (2) says that

$\text{s-}\lim_{t\to 0} T_t f = f$ uniformly in $f \in K$. We next define the mean value $(M_a f)(s) = (2a)^{-1} \int_{-a}^{a} (T_t f)(s)\, dt$. Then, by Hölder's inequality and the Fubini-Tonelli theorem,

$$\|M_a f - f\| \leqq \left(\int_{-\infty}^{\infty} \left\{ \int_{-a}^{a} (2a)^{-1} |f(s+t) - f(s)|\, dt \right\}^p ds \right)^{1/p}$$

$$\leqq (2a)^{-1} \left(\int_{-\infty}^{\infty} \int_{-a}^{a} |f(s+t) - f(s)|^p\, dt \cdot (2a)^{p/p'}\, ds \right)^{1/p}$$

$$\leqq \left((2a)^{-1} \int_{-a}^{a} dt \int_{-\infty}^{\infty} |f(s+t) - f(s)|^p\, ds \right)^{1/p} \quad \text{if } 1 \leqq p < \infty.$$

Thus we have

$$\|M_a f - f\| \leqq \sup_{|t| \leqq a} \|T_t f - f\|,$$

so that $\text{s-}\lim_{a\downarrow 0} M_a f = f$ uniformly in $f \in K$. Therefore, we have to prove the relative compactness of the set $\{M_a f;\, f \in K\}$ for a sufficiently small fixed $a > 0$.

We will show that, for a fixed $a > 0$, the set of functions $\{(M_a f)(s);\, f \in K\}$ is equi-bounded and equi-continuous. In fact, we have, as above,

$$|(M_a f)(s_1) - (M_a f)(s_2)| \leqq (2a)^{-1} \int_{-a}^{a} |f(s_1 + t) - f(s_2 + t)|\, dt$$

$$\leqq \left((2a)^{-1} \int_{-a}^{a} |f(s_1 + t) - f(s_2 + t)|^p\, dt \right)^{1/p}.$$

Thus, by (2), we have proved the equi-continuity of the set of functions $\{(M_a f)(s);\, f \in K\}$ for a fixed $a > 0$. The equi-boundedness of the set may be proved similarly. Thus, by the Ascoli-Arzelà theorem, there exists, for any positive $a > 0$, a finite number of functions $M_a f_1, M_a f_2, \ldots, M_a f_n$ with $f_j \in K$ $(j = 1, 2, \ldots, n)$ such that, for any $f \in K$, there exists some j for which $\sup_{|s| \leqq \alpha} |(M_a f)(s) - (M_a f_j)(s)| \leqq \varepsilon$. Therefore

$$\|M_a f - M_a f_j\|^p \leqq \int_{-\alpha}^{\alpha} |(M_a f)(s) - (M_a f_j)(s)|^p\, ds + \int_{|s|>\alpha} |(M_a f)(s) - (M_a f_j)(s)|^p\, ds. \tag{4}$$

The second term on the right is, by Minkowski's inequality, smaller than $\left(\|M_a f - f\| + \left(\int_{|s|>\alpha} |f(s) - f_j(s)|^p\, ds \right)^{1/p} + \left(\int_{|s|>\alpha} |f_j(s) - (M_a f_j)(s)|^p\, ds \right)^{1/p} \right)^p.$

The term $\|M_a f - f\|$ is small for sufficiently small $a > 0$, and, by virtue of (3), $\int_{|s|>\alpha} |f(s) - f_j(s)|^p\, ds$ and $\int_{|s|>\alpha} |f_j(s) - (M_a f_j)(s)|^p\, ds$ are both small for sufficiently large $\alpha > 0$, if $a > 0$ is bounded. Also the first term on the right of (4) is $\leqq 2\alpha\varepsilon^p$ for an appropriate choice of j. These estimates are valid uniformly with respect to $f \in K$. Thus we have proved the relative compactness in L^p of the set $\{M_a f; f \in K\}$ for sufficiently small $a > 0$.

2. Compact Operators and Nuclear Operators

Definition 1. Let X and Y be B-spaces, and let S be the unit sphere of X. An operator $T \in L(X, Y)$ is said to be *compact* or *completely continuous* if the image $T \cdot S$ is relatively compact in Y.

Example 1. Let $K(x, y)$ be a real- or complex-valued continuous function defined for $-\infty < a \leqq x, y \leqq b < \infty$. Then the *integral operator* K defined by

$$(Kf)(x) = \int_a^b K(x, y) f(y)\, dy \tag{1}$$

is compact as an operator $\in L(C[a, b], C[a, b])$.

Proof. Clearly K maps $C[a, b]$ into $C[a, b]$. Set $\sup_{x,y} |K(x, y)| = M$. Then $\|K \cdot f\| \leqq (b - a) M \|f\|$ so that $K \cdot S$ is equi-bounded. By Schwarz' inequality, we have

$$|(Kf)(x_1) - (Kf)(x_2)|^2 \leqq \int_a^b |K(x_1, y) - K(x_2, y)|^2 dy \cdot \int_a^b |f(y)|^2 dy,$$

and hence $K \cdot S$ is equi-continuous, that is,

$$\lim_{\delta \downarrow 0} \sup_{|x_1 - x_2| \leqq \delta} |(Kf)(x_1) - (Kf)(x_2)| = 0 \text{ uniformly in } f \in S.$$

Therefore, by the Ascoli-Arzelà theorem (Chapter III, 3), the set $K \cdot S$ is relatively compact in $C[a, b]$.

Example 2. Let $K(x, y)$ be a real- or complex-valued $\mathfrak{B}$-measurable function on a measure space $(S, \mathfrak{B}, m)$ such that

$$\int_S \int_S |K(x, y)|^2\, m(dx)\, m(dy) < \infty. \tag{2}$$

Then the integral operator K defined by the *kernel* $K(x, y)$:

$$(Kf)(x) = \int_S K(x, y) f(y)\, m(dy), \quad f \in L^2(S) = L^2(S, \mathfrak{B}, m), \tag{3}$$

is compact as an operator $\in L(L^2(S), L^2(S))$. The kernel $K(x, y)$ satisfying (2) is said to be of the *Hilbert-Schmidt type*.

Proof. Take any sequence $\{f_n\}$ from the unit sphere of $L^2(S)$. We have to show that the sequence $\{K \cdot f_n\}$ is relatively compact in $L^2(S)$. Since a Hilbert space $L^2(S)$ is locally sequentially weakly compact, we may assume that $\{f_n\}$ converges weakly to an element $f \in L^2(S)$; otherwise, we choose a suitable subsequence. By (2) and the Fubini-Tonelli theorem,

$\int\limits_S |K(x, y)|^2 m(dy) < \infty$ for m-a.e. x. Hence, for such an x, $\lim\limits_{n\to\infty} (Kf_n)(x) = \lim\limits_{n\to\infty} \int\limits_S K(x, y) f_n(y) m(dy) = \lim\limits_{n\to\infty} (f_n(.), \overline{K(x,.)}) = (f(.), \overline{K(x,.)}) = \int\limits_S K(x, y) f(y) m(dy)$. On the other hand, we have, by Schwarz' inequality,

$$|(Kf_n)(x)|^2 \leqq \int\limits_S |K(x,y)|^2 m(dy) \cdot \int\limits_S |f_n(y)|^2 m(dy) \leqq \int\limits_S |K(x, y)|^2 m(dy)$$
$$\text{for } m\text{-a.e. } x. \tag{4}$$

Hence, by the Lebesgue-Fatou theorem, $\lim\limits_{n\to\infty} \int\limits_S |(Kf_n)(x)|^2 m(dx) = \int\limits_S |(Kf)(x)|^2 m(dx)$. This result, if combined with $w\text{-}\lim\limits_{n\to\infty} K \cdot f_n = K \cdot f$, implies $s\text{-}\lim\limits_{n\to\infty} K \cdot f_n = K \cdot f$ by Theorem 8 in Chapter V, 1. But, as proved above in (4), we have

$$\int\limits_S |(Kh)(x)|^2 m(dx) \leqq \int\limits_S\int\limits_S |K(x, y)|^2 m(dy)\, m(dx) \cdot \int\limits_S |h(y)|^2 m(dy),$$

and so

$$||K|| \leqq \Big(\int\limits_S\int\limits_S |K(x,y)|^2 m(dx)\, m(dy)\Big)^{1/2}. \tag{5}$$

Hence, from $w\text{-}\lim\limits_{n\to\infty} f_n = f$, we obtain $w\text{-}\lim\limits_{n\to\infty} K \cdot f_n = K \cdot f$, because, for any $g \in L^2(S)$, $\lim\limits_{n\to\infty} (K \cdot f_n, g) = \lim\limits_{n\to\infty} (f_n, K^* g) = (f, K^* g) = (K \cdot f, g)$.

Theorem. (i) A linear combination of compact operators is compact. (ii) The product of a compact operator with a bounded linear operator is compact; thus the set of compact operators $\in L(X, X)$ constitutes a closed two-sided *ideal* of the algebra $L(X, X)$ of operators. (iii) Let a sequence $\{T_n\}$ of compact operators $\in L(X, Y)$ converge to an operator T in the sense of the uniform operator topology, i.e., $\lim\limits_{n\to\infty} ||T - T_n|| = 0$. Then T is also compact.

Proof. (i) and (ii) are clear from the definition of compact operators. The closedness, in the sense of the uniform operator topology, of the ideal of compact operators in the algebra $L(X, X)$ is implied by (iii).

We shall prove (iii). Let $\{x_h\}$ be a sequence from the closed unit sphere S of X. By the compact property of each T_n, we can choose, by the diagonal method, a subsequence $\{x_{h'}\}$ such that $s\text{-}\lim\limits_{h\to\infty} T_n x_{h'}$ exists for every fixed n. We have

$$\begin{aligned} ||Tx_{h'} - Tx_{k'}|| &\leqq ||Tx_{h'} - T_n x_{h'}|| + ||T_n x_{h'} - T_n x_{k'}|| + ||T_n x_{k'} - T x_{k'}|| \\ &\leqq ||T - T_n|| + ||T_n x_{h'} - T_n x_{k'}|| + ||T_n - T||, \end{aligned}$$

and so $\overline{\lim\limits_{h,k\to\infty}} ||T \cdot x_{h'} - T \cdot x_{k'}|| \leqq 2\, ||T - T_n||$. Hence $\{Tx_{k'}\}$ is a Cauchy sequence in the B-space Y.

Nuclear Operator. As an application of the Theorem, we shall consider the nuclear operator introduced by A. GROTHENDIECK [2].

Definition 2. Let X, Y be B-spaces and $T \in L(X, Y)$. If there exist a sequence $\{f'_n\} \subseteq X'$, a sequence $\{y_n\} \subseteq Y$ and a sequence $\{c_n\}$ of numbers such that

$$\begin{aligned} &\sup_n \|f'_n\| < \infty, \sup_n \|y_n\| < \infty, \sum_n |c_n| < \infty \text{ and} \\ &T \cdot x = \operatorname*{s-lim}_{m\to\infty} \sum_{n=1}^{m} c_n \langle x, f'_n\rangle y_n \text{ in } Y \text{ for every } x \in X, \end{aligned} \tag{6}$$

then T is called a *nuclear operator* on X into Y.

Remark. The existence of the s-lim in (6) is clear, since

$$\left\| \sum_{j=n}^{m} c_j \langle x, f'_j\rangle y_j \right\| \leq \sum_{j=n}^{m} |c_j| \cdot \|x\| \cdot \|f'_j\| \cdot \|y_j\| \leq \text{constant} \cdot \sum_{j=n}^{m} |c_j| \cdot \|x\|.$$

The nuclear condition says that the s-lim is equal to $T \cdot x$ for every $x \in X$.

Proposition. A nuclear operator T is compact.

Proof. Define the operator T_n by

$$T_n x = \sum_{j=1}^{n} c_j \langle x, f'_j\rangle y_j. \tag{7}$$

Since the range $R(T_n)$ is of finite dimension, T_n is compact as may be proved by the Bolzano-Weierstrass theorem. Moreover, by (6) and

$$\|Tx - T_n x\| = \left\| \sum_{j=n+1}^{\infty} c_j \langle x, f'_j\rangle y_j \right\| \leq \text{constant} \sum_{j=n+1}^{\infty} |c_j| \cdot \|x\|,$$

we have $\lim_{n\to\infty} \|T - T_n\| = 0$ and so T must be compact.

An Example of the Nuclear Operator. Let G be a bounded open domain of R^n, and consider the Hilbert space $H_0^k(G)$. Suppose $(k - j) > n$. Then the mapping T

$$H_0^k(G) \ni \varphi \to \varphi \in H_0^j(G) \tag{8}$$

is a nuclear operator $\in L(H_0^k(G), H_0^j(G))$.

Proof. We may assume that the bounded domain G is contained in the interior of the parallelogram P:

$$0 \leq x_j \leq 2\pi \quad (j = 1, 2, \ldots, n).$$

We recall that $H_0^k(G)$ is the completion of $\hat{H}_0^k(G) = C_0^k(G)$ with respect to the norm $\|\varphi\|_k = \Big(\sum_{|s|\leq k} \int_G |D^s\varphi(x)|^2\, dx\Big)^{1/2}$ (see Chapter I, 10). We extend the functions $\in \hat{H}_0^k(G)$ to be periodic with period 2π in each variable x_s by defining the function values as 0 in $P - G$. The functions

$$\begin{aligned} &f_\beta(x) = (2\pi)^{-n/2} \exp(i\beta \cdot x), \text{ where } \beta = (\beta_1, \beta_2, \ldots, \beta_n) \\ &\text{is an } n\text{-tuple of integers and } \beta \cdot x = \sum_{s=1}^{n} \beta_s x_s, \end{aligned} \tag{9}$$

form a complete orthonormal system of $L^2(P) = H_0^0(P)$. Thus, denoting by D^s the distributional derivative, we have, for $|s| \leqq k$, the Fourier expansion in $L^2(P)$ of functions $D^s\varphi(x)$ where $\varphi \in \hat{H}_0^k(G)$:

$$D^s\varphi(x) = \sum_\beta (D^s\varphi, f_\beta)_0 f_\beta, \text{ where } (\psi, f_\beta)_0 = \int_P \psi(x)\overline{f_\beta(x)}\,dx. \quad (10)$$

We have, by

$$(D^s\varphi, f_\beta)_0 = (-1)^{|s|}(\varphi, D^s f_\beta)_0 = \prod_{m=1}^n (i\beta_m)^{s_m}(\varphi, f_\beta)_0$$

and the Parseval relation

$$\sum_\beta |(D^s\varphi, f_\beta)_0|^2 = \int_P |D^s\varphi(x)|^2\,dx \leqq \|\varphi\|_k^2 \quad (|s| \leqq k),$$

the inequality

$$|(\varphi, (1+|\beta|^2)^{k/2} f_\beta)_0|^2 \leqq \text{constant} \sum_{|s|\leqq k} |(D^s\varphi, f_\beta)_0|^2 \leqq \text{constant } \|\varphi\|_k^2.$$

Therefore the functional $f'_\beta \in H_0^k(G)'$ defined by

$$\langle \varphi, f'_\beta \rangle = (\varphi, (1+|\beta|^2)^{k/2} f_\beta)_0$$

satisfies $\sup_\beta \|f'_\beta\| < \infty$. Moreover,

$$y_\beta = (1+|\beta|^2)^{-j/2} f_\beta$$

satisfies $\sup_\beta \|y_\beta\|_j < \infty$ by $D^s f_\beta = \prod_{t=1}^n (i\beta_t)^{s_t} f_\beta$. We also have

$$\sum_\beta |c_\beta| < \infty, \text{ where } c_\beta = (1+|\beta|^2)^{(j-k)/2},$$

because, for positive integers β_s,

$$\begin{aligned}\sum_\beta \frac{1}{(\beta_1+\beta_2+\cdots+\beta_n)^{k-j}} &= \sum_\beta \left(\frac{1}{(\beta_1+\beta_2+\cdots+\beta_n)^n}\right)^{(k-j)/n}\\ &\leqq \sum_\beta \left(\frac{1}{\beta_1\beta_2\cdots\beta_n}\right)^{(k-j)/n} = \sum_{\beta_1}\left(\frac{1}{\beta_1}\right)^{(k-j)/n}\cdot\sum_{\beta_2}\left(\frac{1}{\beta_2}\right)^{(k-j)/n}\\ &\cdots\sum_{\beta_n}\left(\frac{1}{\beta_n}\right)^{(k-j)/n} < \infty \text{ by } \frac{(k-j)}{n} > 1.\end{aligned}$$

Therefore, we have proved the (Fourier) expansion

$$\varphi = \sum_\beta c_\beta \langle \varphi, f'_\beta \rangle y_\beta.$$

Remark. If there exists a complete orthonormal system $\{\varphi_j\}$ of eigenfunctions of a given bounded linear operator $K \in L(L^2(S), L^2(S))$ such that $K\varphi_j = \lambda_j\varphi_j$ $(j = 1, 2, \ldots)$, then, from the Fourier expansion

$$f = \sum_{j=1}^\infty (f, \varphi_j)\varphi_j, \quad f \in L^2(S),$$

we obtain

$$Kf = \sum_{j=1}^\infty \lambda_j (f, \varphi_j)\varphi_j.$$

We have $\lambda_j = (K\varphi_j, \varphi_j)$, and so, if the eigenvalues λ_j are all > 0 and $\sum_{j=1}^{\infty} \lambda_j < \infty$, then the operator K is nuclear. The condition $\sum_{j=1}^{\infty} |(K\varphi_j, \varphi_j)| < \infty$ is surely satisfied, if the operator K is defined by a kernel

$$\begin{cases} K(x, y) = \int_S \overline{K_2(z, x)}\, K_1(z, y)\, m(dz), \text{ where the kernels} \\ K_1(x, y) \text{ and } K_2(x, y) \text{ are of the Hilbert-Schmidt type.} \end{cases}$$

For,

$$\sum_{j=1}^{\infty} |(K_2^* K_1\varphi_j, \varphi_j)| = \sum_{j=1}^{\infty} |(K_1\varphi_j, K_2\varphi_j)| \leqq \left(\sum_{j=1}^{\infty} \|K_1\varphi_j\|^2 \cdot \sum_{j=1}^{\infty} \|K_2\varphi_j\|^2\right)^{1/2}$$

and, by Parseval's relation, we have

$$\begin{aligned} \sum_{j=1}^{\infty} \|K_1\varphi_j\|^2 &= \sum_{j=1}^{\infty} \int_S \left| \int_S K_1(z, y)\, \varphi_j(y)\, m(dy) \right|^2 m(dz) \\ &= \int_S \sum_{j=1}^{\infty} \left| \int_S K_1(z, y)\, \varphi_j(y)\, m(dy) \right|^2 m(dz) \\ &= \int_S \left\{ \int_S |K_1(z, y)|^2\, m(dy) \right\} m(dz) < \infty, \end{aligned}$$

and similary for $\sum_{j=1}^{\infty} \|K_2\varphi_j\|^2$. A bounded linear operator K in a separable Hilbert space X is said to be of the *trace class* if $\sum_{j=1}^{\infty} |(K\varphi_j, \psi_j)| < \infty$ for arbitrary complete orthonormal systems $\{\varphi_j\}$ and $\{\psi_j\}$ of X. For a general account concerning the trace class and the nuclear operator, see R. Schatten [1], and I. M. Gelfand-N. Y. Vilenkin [3].

3. The Rellich-Gårding Theorem

Theorem (Gårding [1]). Let G be a bounded open domain of R^n. If an operator $T \in L(H_0^k(G), H_0^k(G))$ satisfies, for $j < k$,

$$\|T\varphi\|_k \leqq C\, \|\varphi\|_j \text{ for all } \varphi \in H_0^k(G), \text{ where } C \text{ is a constant,} \tag{1}$$

then T is compact as an operator $\in L(H_0^k(G), H_0^k(G))$.

Proof. By the definition of the space $H_0^k(G)$ (see Chapter I, 10), it would be sufficient to show the following: Let a sequence $\{\varphi_\nu\} \subseteq \hat{H}_0^k(G) = C_0^k(G)$ be such that $\|\varphi_\nu\|_k \leqq 1$ ($\nu = 1, 2, \ldots$). Then the sequence $\{T\varphi_\nu\}$ contains a subsequence strongly convergent in $H_0^k(G)$. The Fourier transform $\hat{\varphi}_\nu(\xi) = (2\pi)^{-n/2} \int_G \varphi_\nu(x) \exp(-ix\xi)\, dx$ satisfies, by Schwarz' inequality,

$$|\hat{\varphi}_\nu(\xi)|^2 \leqq (2\pi)^{-n} \int_G dx \int_G |\varphi_\nu(x)|^2\, dx \leqq (2\pi)^{-n} \int_G dx,$$

and hence $\{\hat{\varphi}_\nu(\xi)\}$ is equi-bounded in $\xi \in R^n$ and in ν. We may assume, by the boundedness of $\|\varphi_\nu\|_0$, that a subsequence $\{\varphi_{\nu'}\}$ is weakly convergent in $L^2(G) = H_0^0(G)$. Since, for each ξ, the function $\exp(-ix\xi)$ belongs to $L^2(G)$, we know that the sequence of bounded functions $\hat{\varphi}_{\nu'}(\xi) = (\varphi_{\nu'}, (2\pi)^{-n/2}\exp(-ix\xi))_0$ converges at every ξ. Thus, by (1) and the Parseval relation for the Fourier transform (Chapter VI, 2),

$$\begin{aligned}
||T\varphi_{\nu'} - T\varphi_{\mu'}||_k^2 &= ||T(\varphi_{\nu'} - \varphi_{\mu'})||_k^2 \leqq C^2 ||\varphi_{\nu'} - \varphi_{\mu'}||_j^2 \\
&= C^2 \sum_{|s| \leqq j} ||D^s(\varphi_{\nu'} - \varphi_{\mu'})||_0^2 = C^2 \sum_{|s| \leqq j} ||\widehat{(D^s(\varphi_{\nu'} - \varphi_{\mu'}))}||_0^2 \\
&= C^2 \sum_{|s| \leqq j} \left|\left| \prod_{t=1}^{n} (i\xi_t)^{s_t} (\hat{\varphi}_{\nu'} - \hat{\varphi}_{\mu'})(\xi) \right|\right|_0^2 \\
&\leqq C^2 \sum_{|s| \leqq j} \int_{|\xi| \leqq r} \left| \prod_{t=1}^{n} \xi_t^{s_t} (\hat{\varphi}_{\nu'}(\xi) - \hat{\varphi}_{\mu'}(\xi)) \right|^2 d\xi \\
&\quad + C^2 C_1 \int_{|\xi| > r} |\xi|^{2j} |\hat{\varphi}_{\nu'}(\xi) - \hat{\varphi}_{\mu'}(\xi)|^2 d\xi,
\end{aligned}$$

where C_1 is a positive constant.

The first term on the right converges, for fixed r, to 0 as ν' and $\mu' \to \infty$. This we see by the Lebesgue-Fatou Lemma. The second term on the right is, for $r > 1$,

$$\begin{aligned}
&\leqq C^2 C_1 r^{2j-2k} \int_{|\xi| > r} |\xi|^{2k} |\hat{\varphi}_{\nu'}(\xi) - \hat{\varphi}_{\mu'}(\xi)|^2 d\xi \\
&\leqq C^2 C_1 r^{2j-2k} \int_{R^n} |\xi|^{2k} |\hat{\varphi}_{\nu'}(\xi) - \hat{\varphi}_{\mu'}(\xi)|^2 d\xi \\
&\leqq C^2 C_1 C_2 r^{2j-2k} \sum_{|s| \leqq k} ||\widehat{(D^s\varphi_{\nu'} - D^s\varphi_{\mu'})}||_0^2 \\
&= C^2 C_1 C_2 r^{2j-2k} \sum_{|s| \leqq k} ||D^s(\varphi_{\nu'} - \varphi_{\mu'})||_0^2 \\
&= C^2 C_1 C_2 r^{2j-2k} ||\varphi_{\nu'} - \varphi_{\mu'}||_k^2 \leqq 4C^2 C_1 C_2 r^{2j-2k}
\end{aligned}$$

with a constant C_2.

The last term converges, by $j < k$, to 0 as $r \to \infty$. Therefore, $\lim_{\nu,\mu\to\infty} ||T\varphi_{\nu'} - T\varphi_{\mu'}||_k = 0$.

4. Schauder's Theorem

Theorem (SCHAUDER). An operator $T \in L(X, Y)$ is compact iff its dual operator T' is compact.

Proof. Let S, S' be the closed unit sphere in X, Y', respectively. Let $T \in L(X, Y)$ be compact. Let $\{y_j'\}$ be an arbitrary sequence in S'. The functions $F_j(y) = \langle y, y_j' \rangle$ are equi-continuous in the sense that

$$|F_j(y) - F_j(z)| = |\langle y - z, y_j' \rangle| \leqq ||y - z||.$$

Moreover, $\{F_j(y)\}$ is equi-bounded in j on any bounded set of y, since $|F_j(y)| \leqq ||y||$. Therefore, by the Ascoli-Arzelà theorem, as applied to the functions $\{F_j(y)\}$ defined on the compact set $(T \cdot S)^a$, we see that some subsequence $\{F_{j'}(y)\}$ converges uniformly in $y \in (T \cdot S)^a$. Hence $\langle Tx, y_j' \rangle = \langle x, T' y_j' \rangle$ converges uniformly in $x \in S$, and so $\{T' \cdot y_j'\}$ converges in the strong topology of X'. This proves that T' is compact.

Conversely, let T' be compact. Then, by what we have proved above, T'' is compact. Hence, if S'' is the closed unit sphere in X'', $(T'' \cdot S'')$ is relatively compact. We know that Y is isometrically embedded in Y'' (see Theorem 2 in Chapter IV, 8). Hence $T \cdot S \subseteq T'' \cdot S''$ and so $T \cdot S$ is relatively compact in the strong topology of Y'' and so in the strong topology of Y. Therefore, T is compact.

5. The Riesz-Schauder Theory

We prepare

Lemma (F. RIESZ [2]). Let V be a compact operator $\in L(X, X)$, where X is a B-space. Then, for any complex number $\lambda_0 \neq 0$, the range $R(\lambda_0 I - V)$ is strongly closed.

Proof. We may assume that $\lambda_0 = 1$. Let $\{x_n\}$ be a sequence of X such that $y_n = (I - V)\,x_n$ converges strongly to y. If $\{x_n\}$ is bounded, then by, the compactness of the operator V, there exists a subsequence $\{x_{n'}\}$ such that $\{V x_n\}$ converges strongly. Since $x_{n'} = y_{n'} + V x_{n'}$, $\{x_{n'}\}$ converges to some x, and so $y = (I - V)\,x$.

We next assume that $\{\|x_n\|\}$ is unbounded. Set $T = (I - V)$ and put $\alpha_n = \operatorname{dis}(x_n, N(T))$, where $N(T) = \{x;\, Tx = 0\}$. Take a $w_n \in N(T)$ such that $\alpha_n \leqq \|x_n - w_n\| \leqq (1 + n^{-1})\,\alpha_n$. Then $T(x_n - w_n) = Tx_n$, and so in the case when $\{\alpha_n\}$ is bounded, we can prove, as above, that $y \in R(T) = R(I - V)$. Suppose that $\lim_{n\to\infty} \alpha_n = \infty$. Since $z_n = (x_n - w_n)/\|x_n - w_n\|$ satisfies $\|z_n\| = 1$ and $s\text{-}\lim_{n\to\infty} Tz_n = 0$, we can prove, as above, that there exists a subsequence $\{z_{n'}\}$ such that $s\text{-}\lim_{n\to\infty} z_{n'} = w_0$, $s\text{-}\lim_{n\to\infty} Tz_{n'} = 0$. Hence $w_0 \in N(T)$. But, if we put $z_n - w_0 = u_n$, then, in

$$x_n - w_n - w_0 \, \|x_n - w_n\| = u_n \, \|x_n - w_n\|,$$

the second and the third terms on the left belong to $N(T)$ so that we must have $\|u_n\| \cdot \|x_n - w_n\| \geqq \alpha_n$. This is a contradiction, since $s\text{-}\lim_{n\to\infty} u_n = 0$, $\|x_n - w_n\| \leqq (1 + n^{-1})\,\alpha_n$ and $\lim_{n\to\infty} \alpha_n = \infty$.

We are now able to prove the *Riesz-Schauder theory*; for convenience sake, we shall state the theory in a series of three theorems.

Theorem 1. Let V be a compact operator $\in L(X, X)$. If $\lambda_0 \neq 0$ is not an eigenvalue of V, then λ_0 is in the resolvent set of V.

Proof. By the preceding Lemma and the hypothesis, the operator $T_{\lambda_0} = (\lambda_0 I - V)$ gives a one-one map of X onto the set $R(T_{\lambda_0})$ which is strongly closed in X. Hence, by the Corollary of the open mapping theorem in Chapter II, 5, T_{λ_0} has a continuous inverse. We have to show that

$R(T_{\lambda_0}) = X$. If not, the topological image $X_1 = T_{\lambda_0}X$ of X is a proper closed subspace of X. Hence, if we set $X_2 = T_{\lambda_0}X_1$, $X_3 = T_{\lambda_0}X_2, \ldots$, then X_{n+1} is a proper closed subspace of X_n ($n = 0, 1, 2, \ldots; X_0 = X$). By F. Riesz' theorem in Chapter III, 2, there exists a sequence $\{y_n\}$ such that $y_n \in X_n$, $\|y_n\| = 1$ and $\operatorname{dis}(y_n, X_{n+1}) \geqq 1/2$. Thus, if $n > m$,

$$\lambda_0^{-1}(Vy_m - Vy_n) = y_m + \{-y_n - (T_{\lambda_0}y_m - T_{\lambda_0}y_n)/\lambda_0\} = y_m - y$$
$$\text{with some } y \in X_{m+1}.$$

Hence $\|Vy_n - Vy_m\| \geqq |\lambda_0|/2$, contrary to the compactness of the operator V.

Theorem 2. Let V be a compact operator $\in L(X, X)$. Then, (i) its spectrum consists of an at most countable set of points of the complex plane which has no point of accumulation except possibly $\lambda = 0$; (ii) every non-zero number in the spectrum of V is an eigenvalue of V of finite multiplicity; (iii) a non-zero number is an eigenvalue of V iff it is an eigenvalue of V'.

Proof. By Theorem 1, a non-zero number in the spectrum of V is an eigenvalue of V. The same is also true of V', since, by Schauder's theorem, V' is compact when V is. But the resolvent sets are the same for V and V' (see Chapter VIII, 6). Hence (iii) is proved. Since the eigenvectors belonging respectively to different eigenvalues of V are linearly independent, the proofs of (i) and (ii) are completed if we derive a contradiction from the following situation:

$$\begin{cases} \text{There exists a sequence } \{x_n\} \text{ of linearly independent vectors} \\ \text{such that } Vx_n = \lambda_n x \ (n = 1, 2, \ldots) \text{ and } \lim\limits_{n\to\infty} \lambda_n = \lambda \neq 0. \end{cases}$$

To derive a contradiction, we consider the closed subspace X_n spanned by $x_1, x_2, \ldots, x_n$. By F. Riesz' theorem in Chapter III, 2, there exists a sequence $\{y_n\}$ such that $y_n \in X_n$, $\|y_n\| = 1$ and $\operatorname{dis}(y_n, X_{n-1}) \geqq 1/2$ ($n = 2, 3, \ldots$). If $n > m$, then

$$\lambda_n^{-1}Vy_n - \lambda_m^{-1}Vy_m = y_n + (-y_m - \lambda_n^{-1}T_{\lambda_n}y_n + \lambda_m^{-1}T_{\lambda_m}y_m) = y_n - z,$$
$$\text{where } z \in X_{n-1}.$$

For, if $y_n = \sum_{j=1}^{n} \beta_j x_j$, then we have $y_n - \lambda_n^{-1}Vy_n = \sum_{j=1}^{n} \beta_j x_j - \sum_{j=1}^{n} \beta_j \lambda_n^{-1}\lambda_j x_j \in X_{n-1}$ and similarly $T_{\lambda_m}y_m \in X_m$. Therefore $\|\lambda_n^{-1}Vy_n - \lambda_m^{-1}Vy_m\| \geqq 1/2$. This contradicts the compactness of V combined with the hypothesis $\lim\limits_{n\to\infty} \lambda_n \neq 0$.

Theorem 3. Let $\lambda_0 \neq 0$ be an eigenvalue of a compact operator $V \in L(X, X)$. Then λ_0 is also an eigenvalue of V' by the preceding theorem. We can prove: (i) the multiplicities for the eigenvalue λ_0 are the same for V

and V'. (ii) The equation $(\lambda_0 I - V)x = y$ admits a solution x iff $y \in N(\lambda_0 I' - V')^{\perp}$, that is, iff $V'f = \lambda_0 f$ implies $\langle y, f\rangle = 0$. (iii) The equation $(\lambda_0 I' - V'f) = g$ admits a solution f iff $g \in N(\lambda_0 I - V)^{\perp}$, that is, iff $Vx = \lambda_0 x$ implies $\langle x, g\rangle = 0$.

Proof. Since the eigenvalue $\lambda_0 \neq 0$ is an isolated singularity of the resolvent $R(\lambda; V) = (\lambda I - V)^{-1}$, we can expand $R(\lambda; V)$ in Laurent series

$$R(\lambda; V) = \sum_{n=-\infty}^{\infty} (\lambda - \lambda_0)^n A_n.$$

We are particularly interested in the residue $A_{-1} = (2\pi i)^{-1} \int_{|\lambda-\lambda_0|=\varepsilon} R(\lambda; V)\, d\lambda$. As was proved in Chapter VIII, 8, A_{-1} is an idempotent, i.e., $A_{-1}^2 = A_{-1}$. If we set $(\lambda I - V)^{-1} = \lambda^{-1} I + V_\lambda$, then from $(\lambda I - V)(\lambda^{-1} I + V_\lambda) = I$ we obtain $V_\lambda = V(\lambda^{-1} V_\lambda + \lambda^{-2} I)$, and so V_λ is compact when V is. Hence, by

$$A_{-1} = (2\pi i)^{-1} \int_{|\lambda-\lambda_0|=\varepsilon} R(\lambda; V)\, d\lambda = (2\pi i)^{-1} \int_{|\lambda-\lambda_0|=\varepsilon} \lambda^{-1} d\lambda \cdot I$$
$$+ (2\pi i)^{-1} \int_{|\lambda-\lambda_0|=\varepsilon} V_\lambda\, d\lambda = (2\pi i)^{-1} \int_{|\lambda-\lambda_0|=\varepsilon} V_\lambda\, d\lambda.$$

Thus, by the Theorem in Chapter X, 2, A_{-1} is a compact operator.

Therefore, by $A_{-1}X = A_{-1}(A_{-1}X)$ and the compactness of A_{-1}, the unit sphere of the normed linear space $A_{-1}X$ is relatively compact. Hence, by F. Riesz' theorem in Chapter III, 2, the range $R(A_{-1})$ is of finite dimension. On the other hand, $Vx = \lambda_0 x$, $x \neq 0$, implies that $(\lambda I - V)^{-1} x = (\lambda - \lambda_0)^{-1} x$ by $(\lambda I - V)x = (\lambda - \lambda_0)x$, and so $A_{-1}x = (2\pi i)^{-1} \int_{|\lambda-\lambda_0|=\varepsilon} (\lambda - \lambda_0)^{-1} d\lambda \cdot x = x$. Therefore, the eigenvalue equation $Vx = \lambda_0 x$ is equivalent to $Vx = \lambda_0 x$, $x \in R(A_{-1})$. In the same way, we can prove that the eigenvalue equation $V'f = \lambda_0 f$ is equivalent to $V'f = \lambda_0 f$, $f \in R(A'_{-1})$. But $R(A_{-1})$ and $R(A'_{-1})$ are of the same dimension. For, $A'_{-1}f = g$ satisfies $A'_{-1}g = A'_{-1}(A'_{-1}f) = g$, and this is equivalent to $\langle x, g\rangle = \langle A_{-1}x, g\rangle$ for all $x \in X$ and so the functional g may be considered as a functional defined on the finite dimensional space $R(A_{-1})$.

Now, by the well-known theorem in matrix theory, the eigenvalue equation $Vx = \lambda_0 x$ (in $R(A_{-1})$) and its transposed equation $V'f = \lambda_0 f$ (in $R(A'_{-1})$) both have the same number of linearly independent solutions. We have thus proved (i). The propositions (ii) and (iii) are already proved by the Lemma and the closed range theorem (Chapter VII, 5).

Extension of the Riesz-Schauder Theory. Let a power V^n of $V \in L(X, X)$ be compact for some positive integer n. Then, by the spectral mapping theorem in Chapter VIII, 7, $\sigma(V^n) = \sigma(V)^n$ and, by the compactness of V^n, $\sigma(V^n)$ is either a finite set or else a countable set accumulating only at 0. Therefore, $\sigma(V)$ is either a finite set or a countable set accumulating only at 0. V^n being compact,

$$(2\pi i)^{-1} \int_{|\lambda-\lambda_0|=\varepsilon} R(\lambda; V^n)\, d\lambda$$

is, for any $\lambda_0 \neq 0$ of $\sigma(V^n)$ and sufficiently small $\varepsilon > 0$, of finite-dimensional range. Hence λ_0 is a pole of $R(\lambda; V^n)$ (see Chapter VIII, 8). But $(\lambda^n I - V^n) = (\lambda - V)(\lambda^{n-1} I + \lambda^{n-2} V + \cdots + V^{n-1})$ and hence

$$(\lambda^n I - V^n)^{-1} (\lambda^{n-1} I + \cdots + V^{n-1}) = (\lambda I - V)^{-1},$$

which proves that any $\lambda_0 \neq 0$ of $\sigma(V^n)$ is a pole of $R(\lambda; V)$ and so is an eigenvalue of V. These facts enable us to extend the Riesz-Schauder theory to operators V for which some power V^n is compact. This extension is highly important in view of its application to concrete problems of integral equations, such as the Dirichlet problem pertaining to potentials. See, e.g., O. D. KELLOGG [1]. It can be proved that the Riesz-Schauder theory for $\lambda_0 = 1$ is valid also for an operator $V \in L(X, X)$ if there exist a positive integer m and a compact operator $K \in L(X, X)$ such that $\|K - V^m\| < 1$. See K. YOSIDA [9]. It is to be noted here that, if $K_1(s, t)$ and $K_2(s, t)$ are bounded measurable for $0 \leqq s, t \leqq 1$, then the integral operator T defined by

$$x(s) \to (Tx)(s) = (K_1 K_2 x)(s), \text{ where } (K_j x)(s) = \int_0^1 K_j(s, t)\, x(t)\, dt,$$

is compact as an operator $L(L^1(0, 1), L^1(0, 1))$. See K. YOSIDA-Y. MIMURA-S. KAKUTANI [10].

6. Dirichlet's Problem

Let G be a bounded open domain of R^n, and

$$L = \sum_{|s|, |t| \leqq m} D^s c_{st}(x) D^t$$

a strongly elliptic differential operator with real $C^\infty(G^a)$ coefficients $c_{st}(x) = c_{ts}(x)$. We shall deal only with real-valued functions. Let $f \in L^2(G)$ and $u_1 \in H^m(G)$ be given. Consider a distribution solution $u_0 \in L^2(G)$ of

$$Lu = f \text{ such that } (u_0 - u_1) \in H_0^m(G). \tag{1}$$

The latter condition $(u_0 - u_1) \in H_0^m(G)$ means that each of the distributional derivatives

$$(D^j u_0 - D^j u_1) \text{ for } |j| \leqq m \tag{2}$$

is the $L^2(G)$-limit of a sequence $\{D^j \varphi_{h,j}\}$, where $\varphi_{h,j} \in C_0^\infty(G)$ (see Chapter I, 10). Thus it gives roughly the *boundary conditions*:

$$D^j u_0 = D^j u_1 \text{ on the boundary } \partial G \text{ of } G \text{ for } |j| < m. \tag{3}$$

In such a sense, (1) will be called a *Dirichlet's problem* for the operator

L. We shall follow the treatment of the problem as formulated and solved by L. GÅRDING [1].

We first solve

$$u+\alpha L u=f,\ (u-u_1)\in H_0^m(G), \tag{4}$$

where the positive constant α is so chosen that Gårding's inequality

$$(\varphi+\alpha L^*\varphi,\varphi)_0 \geqq \delta\,\|\varphi\|_m^2 \text{ holds whenever } \varphi\in C_0^\infty(G). \tag{5}$$

Here $L^*=\sum_{|s|,|t|\leqq m}(-1)^{|s|+|t|}D^t c_{st}(x)D^s$ and δ is a positive constant. The existence of such an α is guaranteed if the coefficients $c_{st}(x)$ are continuous on the closure G^a of G. We also have, by m-times partial differentiation, the inequality

$$|(\varphi+\alpha L^*\varphi,\psi)_0|\leqq\gamma\,\|\varphi\|_m\cdot\|\psi\|_m \text{ whenever } \varphi,\psi\in C_0^\infty(G), \tag{6}$$

where γ is another positive constant independent of φ and ψ.

We have, for $u_1\in H^m(G)$ and $\varphi\in C_0^\infty(G)$,

$$(L^*\varphi,u_1)_0=\sum_{s,t}((-1)^{|s|+|t|}D^t c_{st}D^s\varphi,u_1)_0=\sum_{s,t}(-1)^{|s|}(c_{st}D^s\varphi,D^t u_1)_0,$$

by partial differentiation. By Schwarz' inequality, we obtain, remembering that the coefficients c_{st} are bounded on G^a,

$$|(L^*\varphi,u_1)_0|\leqq\eta\sum_{|s|,|t|\leqq m}\|D^s\varphi\|_0\,\|D^t u_1\|_0\quad\Big(\sup_{s,t;x}|c_{st}(x)|=\eta\Big).$$

The right hand side is smaller than constant times $\|\varphi\|_m$.

Thus the linear functional

$$F(\varphi)=(\varphi+\alpha L^*\varphi,u_1)_0,\ \varphi\in C_0^\infty(G),$$

can be extended to a bounded linear functional defined on $H_0^m(G)$ which is the completion of $C_0^\infty(G)$ with respect to the norm $\|\varphi\|_m$. Similarly, we see, from

$$|(\varphi,f)_0|\leqq\|\varphi\|_0\cdot\|f\|_0\leqq\|\varphi\|_m\cdot\|f\|_0,$$

that the linear functional $(\varphi,f)_0$ of $\varphi\in C_0^\infty(G)$ can be extended to a bounded linear functional of $\varphi\in H_0^m(G)$. Hence, by F. Riesz' representation theorem as applied to the Hilbert space $H_0^m(G)$, there exists an $f'=f'(f,u_1)\in H_0^m(G)$ such that

$$(\varphi,f)_0-(\varphi+\alpha L^*\varphi,u_1)_0=(\varphi,f')_m \text{ whenever } \varphi\in C_0^\infty(G).$$

Hence, by the Milgram-Lax theorem in Chapter III, 7, applied to the Hilbert space $H_0^m(G)$, we have

$$(\varphi,f)_0-(\varphi+\alpha L^*\varphi,u_1)_0=(\varphi,f')_m=B(\varphi,Sf'),\ Sf'\in H_0^m(G), \tag{7}$$

where

$$B(\varphi,\psi)=(\varphi+\alpha L^*\varphi,\psi)_0 \quad\text{for}\quad \varphi\in C_0^\infty(G),\ \psi\in H_0^m(G). \tag{8}$$

Thus

$$(\varphi,f)_0=(\varphi+\alpha L^*\varphi,u_1+Sf')_0 \quad\text{whenever}\quad \varphi\in C_0^\infty(G),$$

and so $u_0=u_1+Sf'$ is the desired solution $\in L^2(G)$ of (4).

We shall next discuss the original equation (1). If $u_0 \in L^2(G)$ satisfies (1), then $u_2 = u_0 - u_1 \in H_0^m(G)$ satisfies

$$(u_0, L^*\varphi)_0 = (u_1, L^*\varphi)_0 + (u_2, L^*\varphi)_0 = (f, \varphi)_0, \ \varphi \in C_0^\infty(G).$$

We obtain, by partial integration as above,

$$|(u_1, L^*\varphi)_0| \leqq a \text{ constant times } \|\varphi\|_m,$$

$$|(f, \varphi)_0| \leqq \|f\|_0 \, \|\varphi\|_0 \leqq \|f\|_0 \cdot \|\varphi\|_m.$$

We may thus apply F. Riesz' representation theorem in $H_0^m(G)$ to the linear functional $(f, \varphi)_0 - (u_1, L^*\varphi)_0$ of $\varphi \in C_0^\infty(G)$. Hence there exists a uniquely determined $v \in H_0^m(G)$ such that

$$(f, \varphi)_0 - (u_1, L^*\varphi)_0 = (v, \varphi)_m \quad \text{whenever} \quad \varphi \in C_0^\infty(G).$$

By the Milgram-Lax theorem, applied to $(v, \varphi)_m$, we obtain an $S_1 v \in H_0^m(G)$ such that

$$(v, \varphi)_m = B(S_1 v, \varphi) \quad \text{whenever} \quad \varphi \in C_0^\infty(G), \ v \in H_0^m(G).$$

Thus the Dirichlet problem (1) is equivalent to the problem: For a given $S_1 v \in H_0^m(G)$, find a solution $u_2 \in H_0^m(G)$ of

$$(u_2, L^*\varphi)_0 = B(S_1 v, \varphi), \ \varphi \in C_0^\infty(G). \tag{1'}$$

Now, for a given $u \in L^2(G) = H_0^0(G)$,

$$|(u, \varphi)_0| \leqq \|u\|_0 \cdot \|\varphi\|_0 \leqq \|u\|_0 \cdot \|\varphi\|_m$$

so that, by F. Riesz' representation theorem in the Hilbert space $H_0^m(G)$, there exists a uniquely determined $u' = Tu \in H_0^m(G)$ such that, whenever $\varphi \in C_0^\infty(G)$,

$$(u, \varphi)_0 = (u', \varphi)_m \quad \text{and} \quad \|u'\|_m \leqq \|u\|_0.$$

Hence, by the Milgram-Lax theorem, we obtain

$$(u, \varphi)_0 = (u', \varphi)_m = B(S_1 u', \varphi) = B(S_1 Tu, \varphi), \ \|S_1 Tu\|_m \leqq \delta^{-1} \|u\|_0. \tag{9}$$

Therefore, by (1'), we have, whenever $\varphi \in C_0^\infty(G)$,

$$B(u_2, \varphi) = (u_2, \varphi + \alpha L^*\varphi)_0 = (u_2, \varphi)_0 + \alpha (u_2, L^*\varphi)_0$$
$$= B(S_1 T u_2, \varphi) + \alpha B(S_1 v, \varphi),$$

that is,

$$B(u_2 - S_1 T u_2 - \alpha S_1 v, \varphi) = 0.$$

Because of the positivity $B(\varphi, \varphi) > 0$ of B, we must have

$$u_2 - S_1 T u_2 = \alpha S_1 v. \tag{1''}$$

The right hand term $\alpha S_1 v \in H_0^m(G)$ is a known function. By $\|S_1 Tu\|_m \leqq \delta^{-1} \|u\|_0$, we see that the operator $S_1 T$ defined on $H_0^m(G)$ into $H_0^m(G)$ is compact (the Rellich-Gårding theorem in Chapter X, 3). Therefore, we may apply the Riesz-Schauder theory to the effect that one of the following alternatives holds:

$$\begin{cases}\text{Either the homogeneous equation } u - S_1 T u = 0 \text{ has a non-trivial} \\ \text{solution } u \in H_0^m(G)\text{, or the inhomogeneous equation } u - S_1 T u = w \text{ has,} \\ \text{for every given } w \in H_0^m(G)\text{, a uniquely determined solution } u \in H_0^m(G).\end{cases}$$

The first alternative corresponds to the case $(u, \varphi + \alpha L^*\varphi)_0 = (u, \varphi)_0$, that is, to the case $Lu = 0$. Hence, returning to the original equation (1), we have

Theorem. One of the following alternatives holds: Either i) the homogeneous equation $Lu = 0$ has a non-trivial solution $u \in H_0^m(G)$, or ii) for any $f \in L^2(G)$ and any $u_1 \in H^m(G)$, there exists a uniquely determined solution $u_0 \in L^2(G)$ of $Lu = f$, $u - u_1 \in H_0^m(G)$.

Appendix to Chapter X. The Nuclear Space of A. Grothendieck

The nuclear operator defined in Chapter X, 2 may be extended to locally convex spaces as follows.

Proposition 1. Let X be a locally convex linear topological space, and Y a B-space. Suppose that there exist an equi-continuous sequence $\{f_j'\}$ of continuous linear functionals on X, a bounded sequence $\{y_j\}$ of elements $\in Y$ and a sequence of non-negative numbers $\{c_j\}$ with $\sum_{j=1}^{n} c_j < \infty$. Then

$$T \cdot x = \text{s-}\lim_{n\to\infty} \sum_{j=1}^{n} c_j \langle x, f_j' \rangle y_j \tag{1}$$

defines a continuous linear operator on X into Y.

Proof. By the equi-continuity of $\{f_j'\}$, there exists a continuous semi-norm p on X such that $\sup_j |\langle x, f_j'\rangle| \leqq p(x)$ for $x \in X$. Hence, for $m > n$,

$$\left\| \sum_{j=n}^{m} c_j \langle x, f_j'\rangle y_j \right\| \leqq p(x) \sup_{j\geqq 1} \|y_j\| \cdot \sum_{j=n}^{m} c_j .$$

This proves that the right hand side of (1) exists and defines a continuous linear operator T on X into the B-space Y.

Definition 1. An operator T of the form (1) is said to be a *nuclear operator* on X into the B-space Y.

Corollary. A nuclear operator T is a compact operator in the sense that it maps a neighbourhood of O of X into a relatively compact set of Y.

Proof. We define

$$T_n \cdot x = \sum_{j=1}^{n} c_j \langle x, f_j' \rangle y_j.$$

T_n is compact, since the image by T_n of the set $V = \{x; p(x) \leqq 1\}$ of X is relatively compact in Y. On the other hand, we have

$$\|Tx - T_n x\| = \left\| \sum_{j=n+1}^{\infty} c_j \langle x, f_j' \rangle y_j \right\| \leqq p(x) \sup_{j \geqq 1} \|y_j\| \sum_{j=n+1}^{\infty} c_j,$$

and so $T_n x$ converges to Tx strongly and uniformly on V. Hence the operator T is compact.

As was proved in Chapter X, 2, we have a typical example of a nuclear operator:

Example. Let K be a compact subset of R. Then, for $(k - j) > n$, the identity mapping T of $H_0^k(K)$ into $H_0^j(K)$ is a nuclear operator.

Proposition 2. Let X be a locally convex linear topological space, and V a convex balanced neighbourhood of 0 of X. Let $p_V(x) = \inf_{x/\lambda \in V, \lambda > 0} \lambda$ be the Minkowski functional of V. p_V is a continuous semi-norm on X. Set

$$N_V = \{x \in X;\ p_V(x) = 0\} = \{x \in X; \lambda x \in V \quad \text{for all} \quad \lambda > 0\}.$$

Then N_V is a closed linear subspace of X, and the quotient space $X_V = X/N_V$ is a normed linear space by the norm

$$\|\tilde{x}\|_V = p_V(x), \text{ where } \tilde{x} \text{ is the residue class mod } N_V \text{ containing the element } x. \tag{2}$$

Proof. Let $(x - x_1) \in N_V$. Then $p_V(x_1) \leqq p_V(x) + p_V(x_1 - x) = p_V(x)$, and similarly $p_V(x) \leqq p_V(x_1)$. Thus $p_V(x) = p_V(x_1)$ if x and x_1 are in the same residue class mod N_V. We have $\|\tilde{x}\|_V \geqq 0$ and $\|\tilde{0}\|_V = 0$. If $\|\tilde{x}\|_V = 0$, then $x \in \tilde{x}$ implies $x \in N_V$ and so $\tilde{x} = \tilde{0}$. The triangle inequality is proved by $\|\tilde{x} + \tilde{y}\|_V = p_V(x + y) \leqq p_V(x) + p_V(y) = \|\tilde{x}\|_V + \|\tilde{y}\|_V$. We have also $\|\alpha\tilde{x}\|_V = p_V(\alpha x) = |\alpha| p_V(x) = |\alpha| \|\tilde{x}\|_V$.

Corollary. By the equivalence

$$(p_{V_1} \leqq p_{V_2}) \leftrightarrow (V_2 \subseteq V_1), \tag{3}$$

we can define the canonical mapping

$$X_{V_2} \to X_{V_1} \text{ (when } V_2 \subseteq V_1)$$

by associating the residue class $\tilde{x}_{V_2}$ (mod N_{V_2}) containing x to the residue class $\tilde{x}_{V_1}$ (mod N_{V_1}) containing $\tilde{x}$. The mapping thus obtained is continuous, since

$$\|\tilde{x}_{V_2}\|_{V_2} = p_{V_2}(x) \geqq p_{V_1}(x) = \|\tilde{x}_{V_1}\|_{V_1}.$$

We are now ready to give the notion of a nuclear space, introduced in analysis by A. GROTHENDIECK [2].

Definition 2. A locally convex linear topological space X is said to be a *nuclear space*, if, for any convex balanced neighbourhood V of 0, there exists another convex balanced neighbourhood $U \subseteqq V$ of 0 such that the canonical mapping

$$T : X_U \to \hat{X}_V \tag{4}$$

is nuclear. Here $\hat{X}_V$ is the completion of the normed linear space X_V.

Example 1. Let R^A be the topological product of real number field R in such a way that R^A is the totality of real-valued finite functions $x(a)$ defined on A and topologized by the system of semi-norms

$$p_a(x) = |x(a)|, \ a \in A. \tag{5}$$

Then R^A is a nuclear space.

Proof. N_V is the totality of functions $x(a) \in R^A$ such that, for some finite set $\{a_j \in A;\ j = 1, 2, \ldots, n\}$, $x(a_j) = 0$ $(j = 1, 2, \ldots, n)$. Hence $X_V = R^A/N_V$ is equivalent to the space of functions $x_V(a)$ such that $x_V(a) = 0$ for $a \neq a_j$ $(j = 1, 2, \ldots, n)$ and normed by

$$\|x_V(a)\|_V = \sup_{1 \leqq j \leqq n} |x(a_j)|.$$

We take N_U the totality of functions $x(a) \in R^A$ such that $x(a_\alpha) = 0$ for $\alpha \in A'$ where A' is any finite set of integers containing $1, 2, \ldots, n$. Thus, for $U \subseteqq V$, the canonical mapping $X_U = R^A/N_U \to R^A/N_V = X_V$ is nuclear. For the mapping is a continuous linear mapping with a finite-dimensional range.

Example 2. A nuclear B-space X must be of finite dimension.

Proof. Since $X = X_V$ for any convex balanced neighbourhood V of 0 of a B-space, the compactness of the identity mapping $X \to X$ implies that X is of finite dimension by F. Riesz' theorem in Chapter III, 2.

Example 3. Let K be a compact subset of R^n. Then the space $\mathfrak{D}_K(R^n)$ introduced in Chapter I, 1 is a nuclear space.

Proof. As in Chapter I, 1, let

$$p_{K,k}(f) = \sup_{x \in K, |s| \leqq k} |D^s f(x)|$$

be one of the semi-norms which define the topology of $\mathfrak{D}_K(R^n)$. Let $V_k = \{f \in \mathfrak{D}_K(R^n);\ p_{K,k}(f) \leqq 1\}$. Then N_{V_k} is $\{0\}$, and $X_{V_k} = X/N_{V_k} = \mathfrak{D}_K(R^n)/N_{V_k}$ is precisely the space $\mathfrak{D}_K(R^n)$ normed by $p_{K,k}$. If $(k-j) > n$, then it is easy to prove, as in the example following the Corollary of Definition 1 above, that the canonical mapping of X_{V_k} into X_{V_j} is a nuclear transformation. Hence $\mathfrak{D}_K(R^n)$ is a nuclear space.

Theorem 1. A locally convex linear topological space X is nuclear, iff, for any convex balanced neighbourhood V of 0, the canonical mapping $X \to \hat{X}_V$ is nuclear.

Proof. *Necessity.* Let $U \subseteqq V$ be a convex balanced neighbourhood of 0 of X such that the canonical mapping $X_U \to \hat{X}_V$ is nuclear. The canonical mapping $T: X \to \hat{X}_V$ is the product of the canonical mapping $X \to \hat{X}_U$ and the canonical nuclear transformation $X_U \to \hat{X}_V$. Hence T must be a nuclear transformation.

Sufficiency. Let the canonical mapping $T: X \to \hat{X}_V$ be given by a nuclear transformation

$$T x = \sum_{j=1}^{\infty} c_j \langle x, f_j' \rangle y_j .$$

For any $\alpha > 0$, the set $\{x \in X; |\langle x, f_j' \rangle| \leqq \alpha \text{ for } j = 1, 2, \ldots\}$ is a convex balanced neighbourhood U_α of 0 of X, because of the equi-continuity of $\{f_j'\} \subseteqq X'$. Moreover,

$$\|Tx\|_V = \Big\|\sum_j c_j \langle x, f_j' \rangle y_j\Big\|_V \leqq \alpha \sup_j \|y_j\|_V \sum_j c_j \quad \text{whenever } x \in U_\alpha .$$

Let α be so small that the right hand side is < 1. Then $\|Tx\|_V < 1$ and $U_\alpha \subseteqq V$. Each f_j' may be considered as belonging to the dual space X'_{U_α}, and so

$$Tx = Tz = \sum_j c_j \langle x, f_j' \rangle y_j \quad \text{whenever} \quad (x - z) \in N_{U_\alpha} .$$

Thus the canonical mapping $X_{U_\alpha} \to \hat{X}_V$ is given by a nuclear transformation

$$\tilde{x}_{U_\alpha} \to \sum_j c_j \langle \tilde{x}_{U_\alpha}, f_j' \rangle y_j .$$

Theorem 2. Let a locally convex linear topological space X be nuclear. Then, for any convex balanced neighbourhood V of 0 of X, there exists a convex balanced neighbourhood $W \subseteqq V$ of 0 of X such that $\hat{X}_W$ is a Hilbert space.

Proof. The nuclear canonical mapping $X_U \to \hat{X}_V$ $(U \subseteqq V)$ defined by

$$T\tilde{x}_U = \sum_j c_j \langle \tilde{x}_U, f_j' \rangle y_j'$$

is factored as the product of the two mappings

$\alpha: X_U \to (l^2)$ and $\beta: (l^2) \to \hat{X}_V$, where α is given by $\tilde{x}_U \to \{c_j^{1/2} \langle \tilde{x}_U, f_j' \rangle\}$ and β is given by $\{\xi_j\} \to \sum_j c_j^{1/2} \xi_j y_j$.

The continuity of α is clear from

$$\sum_j |c_j^{1/2} \langle \tilde{x}_U, f_j' \rangle|^2 \leqq \Big(\sup_j \|f_j'\| \cdot \|\tilde{x}_U\|_U\Big)^2 \cdot \sum_j c_j ,$$

and that of β is proved by

$$\Big\|\sum_j c_j^{1/2} \xi_j y_j\Big\|_V^2 \leqq \sum_j c_j \|y_j\|_V^2 \cdot \sum_j |\xi_j|^2 \leqq \sup_j \|y_j\|_V^2 \cdot \|\{\xi_j\}\|_{l^2}^2 \cdot \sum_j c_j .$$

Let U_2 be the inverse image in (l^2) by β of the unit sphere of $\hat{X}_V$. Then U_2 is a neighbourhood of 0 of (l^2) and so contains a sphere S of centre 0 of (l^2).

Let W be the inverse image in X of S by the continuous mapping α defined as the product of continuous canonical mapping $X \to X_U$ and the continuous mapping $\alpha: X_U \to (l^2)$. Then clearly $W \subseteqq V$ and, for any $\tilde{x}_W \in X_W$,

$$\|\tilde{x}_W\|_W = \inf_{x/\lambda \in W, \lambda > 0} \lambda = \inf_{\tilde{\alpha} x/\lambda \in S, \lambda > 0} \lambda = \|\tilde{\alpha} x\|_{l^2} \text{ (the radius of } S).$$

Since $\| \ \|_{l^2}$ is the norm in the Hilbert space (l^2), X_W is a pre-Hilbert space.

Corollary. Let X be a locally convex nuclear space. Then, for any convex balanced neighbourhood V of 0 of X, there exist convex balanced neighbourhoods W_1 and W_2 of 0 of X with the properties:

$W_2 \subseteqq W_1 \subseteqq V$, $\hat{X}_{W_1}$ and $\hat{X}_{W_2}$ are Hilbert spaces and the canonical mappings $X \to \hat{X}_{W_2}$, $\hat{X}_{W_2} \to \hat{X}_{W_1}$, $\hat{X}_{W_1} \to \hat{X}_V$ are all nuclear.

Therefore, a nuclear space X has a fundamental system $\{V_\alpha\}$ of neighbourhoods of 0 such that the spaces $\hat{X}_{V_\alpha}$ are Hilbert spaces.

Further Properties of Nuclear Spaces. It can be proved that:

1. A linear subspace and a factor space of a nuclear space are also nuclear.

2. The topological vector product of a family of nuclear spaces and the inductive limit of a sequence of nuclear spaces are also nuclear.

3. The strong dual of the inductive limit of a sequence of nuclear spaces, each of which is an F-space, is also nuclear.

For the proof, see the book by GROTHENDIECK [1] referred to above, p. 47. As a consequence of 2., the space $\mathfrak{D}(R^n)$, which is the inductive limit of the sequence $\{\mathfrak{D}_{K_r}(R^n); r = 1, 2, \ldots\}$ (here K_r is the sphere $|x| \leqq r$ of R^n), is nuclear. Hence, by 3., the space $\mathfrak{D}(R^n)'$ is also nuclear. The spaces $\mathfrak{E}(R^n)$, $\mathfrak{E}(R^n)'$, $\mathfrak{S}(R^n)$ and $\mathfrak{S}(R^n)'$ are also nuclear spaces.

The importance of the notion of the nuclear space has recently been stressed by R. A. MINLOS [1]. He has proved the following generalization of *Kolmogorov's extension theorem of measures*:

Let X be a nuclear space whose topology is defined through a countable system of convex balanced neighbourhoods of 0. Let X' be the strong dual space of X. A *cylinder set* of X' is defined as a set of the form

$$Z' = \{f' \in X';\ a_i < \langle x_i, f' \rangle < b_i \quad (i = 1, 2, \ldots, n)\}.$$

Suppose there is given a set function μ_0, defined and $\geqq 0$ for all cylinder sets. Let μ_0 be σ-additive for those cylinder sets Z' with fixed $x_1, x_2, \ldots, x_n$. Then, under a *compatibility condition* and a *continuity condition*, there exists a uniquely determined extension of μ_0 which is σ-additive and $\geqq 0$ for all sets of the smallest σ-additive family of sets of X' containing all the cylinder sets of X'.

For a detailed proof and applications of this result, see I. M. GELFAND-N. Y. VILENKIN [3].

XI. Normed Rings and Spectral Representation

A linear space A over a scalar field (F) is said to be an *algebra* or a *ring* over (F), if to each pair of elements $x, y \in A$ a unique product $xy \in A$ is defined with the properties:

$$\left.\begin{aligned} &(xy)\,z = x(yz) \text{ (associativity)},\\ &x(y+z) = xy + xz \text{ (distributivity)},\\ &\alpha\beta(xy) = (\alpha x)(\beta y). \end{aligned}\right\} \tag{1}$$

If there exists a *unit element* e such that $ex = xe = x$ for every $x \in A$, then A is said to be an *algebra with a unit*. A unit e of A, if it exists, is uniquely determined. For, if e' be another unit of A, then we must have $ee' = e = e'$. If the multiplication is commutative, i.e., $xy = yx$ for every pair $x, y \in A$, then A is called a *commutative algebra*. Let A be an algebra with a unit e. If, for an $x \in A$, there exists an $x' \in A$ such that $xx' = x'x = e$, then x' is called an *inverse* of x. An inverse x' of x, if it exists, is uniquely determined. For, if x'' be another inverse of x, then we must have

$$x''(xx') = x''e = x'' = (x''x)\,x' = ex' = x'.$$

Thus we shall denote by x^{-1} the inverse of x if x has an inverse.

An algebra is called a *Banach algebra*, or in short a *B-algebra* if it is a B-space and satisfies

$$\|xy\| \leqq \|x\|\,\|y\|. \tag{2}$$

The inequality

$$\begin{aligned} \|x_n y_n - xy\| &\leqq \|x_n(y_n - y)\| + \|(x_n - x)\,y\| \\ &\leqq \|x_n\|\,\|(y_n - y)\| + \|(x_n - x)\|\,\|y\| \end{aligned}$$

shows that xy is a continuous function of both variables together.

Example 1. Let X be a B-space. Then $L(X, X)$ is a B-algebra with a unit by the operator sum $T + S$ and operator product TS; the identity operator I is the unit of this algebra $L(X, X)$, and the operator norm $\|T\|$ is the norm of the element T of this algebra $L(X, X)$.

Example 2. Let S be a compact topological space. Then $C(S)$ is a B-algebra by $(x_1 + x_2)(s) = x_1(s) + x_2(s)$, $(\alpha x)(s) = \alpha x(s)$, $(x_1 x_2)(s) = x_1(s)\,x_2(s)$ and $\|x\| = \sup_{s \in S} |x(s)|$.

Example 3. Let B be the totality of continuous functions $x(s)$, $0 \leqq s \leqq 1$, which are representable as absolutely convergent Fourier series:

$$x(s) = \sum_{n=-\infty}^{\infty} c_n e^{2\pi i n s} \quad \text{with} \quad \sum_{n=-\infty}^{\infty} |c_n| < \infty. \tag{3}$$

Then it is easy to see that B is a commutative B-algebra with a unit by the ordinary function sum and function multiplication, normed by

$$||x|| = \sum_{n=-\infty}^{\infty} |c_n|. \tag{4}$$

In the last two examples, the unit is given by the function $e(s) \equiv 1$ and $||e|| = 1$. In the following sections, we shall be concerned with the commutative B-algebra with the unit e such that

$$||e|| = 1. \tag{5}$$

Such an algebra is called a *normed ring*.

A Historical Sketch. The notion of Banach algebras was introduced in analysis by M. NAGUMO [1]. He proved that Cauchy's complex function theory can be extended to functions with values in such an algebra, and applied it to the investigation of the resolvent of a bounded linear operator around an isolated singular point. The result is an abstract treatment of those given in our Chapter VIII, 8. K. YOSIDA [11] proved that a connected group embedded in a B-algebra is a Lie group iff the group is locally compact. This result is an extension of a result due to J. VON NEUMANN [6] concerning matrix groups. Cf. E. HILLE-R. S. PHILLIPS [1], in which the result of K. YOSIDA [11] is reproduced.

The *ideal theory* of normed rings was initiated by I. M. GELFAND [2]. He has shown that such a ring can be represented as the ring of continuous functions defined on the space of maximal ideals of the ring. By virtue of this representation, we can give an integration free treatment of the spectral resolution of bounded normal operators in a Hilbert space; see K. YOSIDA [12]. This result will be exposed in the following sections. The *Gelfand representation* may also be applied to a new proof of the Tauberian theorem of N. WIENER [2]. We shall expose this application in the last section of this chapter. For further details about B-algebras, see N. A. NAIMARK [1], C. E. RICKART [1] and I. M. GELFAND-D. A. RAIKOV-G. E. ŠILOV [5].

1. Maximal Ideals of a Normed Ring

We shall be concerned with a commutative B-algebra B with a unit e such that $||e|| = 1$.

Definition 1. A subset J of B is called an *ideal* of B if $x, y \in J$ implies that $(\alpha x + \beta y) \in J$ and $zx \in J$ for every $z \in B$. B itself and $\{0\}$ are ideals of B. Ideals other than B and $\{0\}$ are called *non-trivial ideals*. A non-trivial ideal J is said to be a *maximal ideal* if there exists no non-trivial ideal containing J as a proper subset.

Proposition 1. Each non-trivial ideal J_0 of B is contained in a maximal ideal J.

Proof. Let $[J_0]$ be the set of all non-trivial ideals containing J_0. We order the ideals of $[J_0]$ by inclusion relation, that is, we denote $J_1 \prec J_2$ if J_1 is a subset of J_2. Suppose that $\{J_\alpha\}$ is a linearly ordered subset of $[J_0]$ and put $J_\beta = \bigcup_{J_\alpha \in \{J_\alpha\}} J_\alpha$. We shall show that J_β is an upper bound of $\{J_\alpha\}$. For, if $x, y \in J_\beta$, then there exist ideals J_{α_1} and J_{α_2} such that $x \in J_{\alpha_1}$ and $y \in J_{\alpha_2}$. Since $\{J_\alpha\}$ is linearly ordered, $J_{\alpha_1} \prec J_{\alpha_2}$ (or $J_{\alpha_1} \succ J_{\alpha_2}$) and so x and y both belong to J_{α_2}; consequently $(x - y) \in J_{\alpha_2} \subseteqq J_\beta$ and $zx \in J_{\alpha_2} \subseteqq J_\beta$ for any $z \in B$. This proves that J_β is an ideal. Since the unit element e is not contained in any J_α, e is not contained in $J_\beta = \bigcup_{J_\alpha \in \{J_\alpha\}} J_\alpha$. Thus J_β is a non-trivial ideal containing every J_α. Therefore, by Zorn's Lemma, there exists at least one maximal ideal which contains J_0.

Corollary. An element x of B has the inverse $x^{-1} \in B$ such that $x^{-1}x = xx^{-1} = e$ iff x is contained in no maximal ideal.

Proof. If $x^{-1} \in B$ exists, then any ideal $J \ni x$ must contain $e = xx^{-1}$ so that J must coincide with B itself. Let conversely, x be contained in no maximal ideal. Then the ideal $xB = \{xb; b \in B\} \neq \{0\}$ must coincide with B itself, since, otherwise, there exists at least one maximal ideal containing $xB \ni x = xe$. It follows that $xB = B$, and so there must exist an element $b \in B$ such that $xb = e$. By the commutativity of B, we have $xb = bx = e$, that is, $b = x^{-1}$.

Proposition 2. A maximal ideal J is a closed linear subspace of B.

Proof. By the continuity of the algebraic operations (addition, multiplication and scalar multiplication) in B, the strong closure J^a is also an ideal containing J. Suppose $J^a \neq J$. Then $J^a = B$, because of the maximality of the ideal J. Thus $e \in J^a$, and so there exists an $x \in J$ such that $\|e - x\| < 1$. x has the inverse $x^{-1} \in B$ which is given by Neumann's series

$$e + (e - x) + (e - x)^2 + \cdots$$

For, by $\|(e - x)^n\| \leqq \|e - x\|^n$, the series converges to an element $\in B$ which is the inverse of x, as may be seen by multiplying the series by $x = e - (e - x)$. Hence $e = x^{-1}x \in J$ and so J cannot be a maximal ideal.

Proposition 3. For any ideal J of B, we write

$$x \equiv y \ (\text{mod } J) \text{ or } x \sim y \ (\text{mod } J) \text{ or in short } x \sim y, \text{ if } (x - y) \in J. \tag{1}$$

Then $x \sim y$ is an equivalence relation, that is, we have

$$\begin{cases} x \sim x \ \text{(reflexivity)}, \\ x \sim y \text{ implies } y \sim x \ \text{(symmetry)}, \\ x \sim y \text{ and } y \sim z \text{ implies } x \sim z \ \text{(transitivity)}. \end{cases}$$

We denote by $\bar{x}$ the set $\{y; (y - x) \in J\}$; it is called the *class* (mod J) containing x. Then the classes $\overline{(x + y)}$, $\overline{\alpha x}$ and $\overline{(xy)}$ are determined independently of the choice of elements x and y from the classes $\bar{x}$ and $\bar{y}$, respectively.

Proof We have to show that $x \sim x'$, $y \sim y'$ implies that $(x + y) \sim (x' + y')$, $\alpha x \sim \alpha x'$ and $xy \sim x'y'$. These are clear from the condition that J is an ideal. For instance, we have $xy - x'y' = (x - x')\, y + x'(y - y') \in J$ by $(x - x') \in J$ and $(y - y') \in J$.

Corollary. The set of classes $\bar{x}$ (mod J) thus constitutes an algebra by

$$\bar{x} + \bar{y} = \overline{x + y},\ \alpha \bar{x} = \overline{\alpha x},\ \bar{x}\,\bar{y} = \overline{xy}. \tag{2}$$

Definition 2. The above obtained algebra is called the residue class algebra of B (mod J) and is denoted by B/J. Thus the mapping $x \to \bar{x}$ of the algebra B onto $\bar{B} = B/J$ is a homomorphism, that is, relation (2) holds.

Proposition 4. Let J be a maximal ideal of B. Then $\bar{B} = B/J$ is a field, that is, each non-zero element $\bar{x} \in \bar{B}$ has an inverse $\bar{x}^{-1} \in \bar{B}$ such that $\bar{x}^{-1}\,\bar{x} = \bar{x}\,\bar{x}^{-1} = \bar{e}$.

Proof. Suppose the inverse $\bar{x}^{-1}$ does not exist. Then the set $\bar{x}\bar{B} = \{\bar{x}\,\bar{b};\, \bar{b} \in \bar{B}\}$ is an ideal of $\bar{B}$. It is non-trivial since it does not contain $\bar{e}$, but does contain $\bar{x} \neq 0$. The inverse image of an ideal by the homomorphism is an ideal. Therefore, B contains a non-trivial ideal containing J as a proper subset, contrary to the maximality of the ideal J.

We are now able to prove

Theorem. Let B be a normed ring over the field of complex numbers, and J a maximal ideal of B. Then the residue class algebra $\bar{B} = B/J$ is isomorphic to the complex number field, in the sense that each $\bar{x} \in \bar{B}$ is represented uniquely as $\bar{x} = \xi \bar{e}$, where ξ is a complex number.

Proof. We shall prove that $\bar{B} = B/J$ is a normed ring by the norm

$$\|\bar{x}\| = \inf_{x \in \bar{x}} \|x\|. \tag{3}$$

If this is proved, then B/J is a normed field and so, by the Gelfand-Mazur theorem in Chapter V, 3, $\bar{B} = B/J$ is isomorphic to the complex number field.

Now we have $\|\alpha \bar{x}\| = |\alpha|\, \|\bar{x}\|$, and $\|\bar{x} + \bar{y}\| = \inf_{x \in \bar{x}, y \in \bar{y}} \|x + y\| \leqq \inf_{x \in \bar{x}} \|x\| + \inf_{y \in \bar{y}} \|y\| = \|\bar{x}\| + \|\bar{y}\|$; $\|\bar{x}\,\bar{y}\| \leqq \|\bar{x}\|\,\|\bar{y}\|$ is proved similarly. If $\|\bar{x}\| = 0$, then there exists a sequence $\{x_n\} \subseteq \bar{x}$ such that $\text{s-}\lim_{n\to\infty} x_n = 0$. Hence, for any $x \in \bar{x}$, $(x - x_n) \in J$ and so $\text{s-}\lim_{n\to\infty} (x - x_n) = x$ which proves that $x \in J^a = J$, that is, $\bar{x} = \bar{0}$. Hence $\|\bar{x}\| = 0$ is equivalent to $\bar{x} = \bar{0}$. We have $\|\bar{e}\| \leqq \|e\| = 1$. If $\|\bar{e}\| < 1$, then there exists an element $x \in J$ such that $\|e - x\| < 1$. As in the proof of Proposition 2, the inverse x^{-1} exists, which is contradictory to the Corollary of Proposition 1. Thus we must have $\|\bar{e}\| = 1$. Finally, since B is a B-space and J is a closed linear subspace by Proposition 2, the factor space $\bar{B} = B/J$ is complete with respect to the norm (3) (see Chapter I, 11). We have thus proved the Theorem.

Corollary. We shall denote by $x(J)$ the number ξ in the representation $\bar{x} = \xi\bar{e}$. Thus, for each $x \in B$, we obtain a complex-valued function $x(J)$ defined on the set $\{J\}$ of all the maximal ideals of B. Then we have

$$\begin{aligned}(x+y)(J) &= x(J) + y(J),\ (\alpha x)(J) = \alpha x(J),\\ (xy)(J) &= x(J)\,y(J), \quad \text{and} \quad e(J) \equiv 1.\end{aligned} \tag{4}$$

We have, moreover,

$$\sup_{J\in\{J\}} |x(J)| \leqq \|x\|, \tag{5}$$

and

$$\sup_{J\in\{J\}} |x(J)| = 0 \text{ implies } x = 0 \text{ iff } \bigcap_{J\in\{J\}} J = \{0\}. \tag{6}$$

Proof. The mapping $x \to \bar{x} = x(J)\,\bar{e}$ of the algebra B onto the residue class algebra $\bar{B} = B/J$ is a homomorphism, that is, relation (2) holds. Hence we have (4). Inequality (5) is proved by

$$|\xi| = |\xi|\,\|\bar{e}\| = \|\bar{x}\| = \inf_{x\in\bar{x}} \|x\| \leqq \|x\|.$$

Property (6) is clear, since $x(J) = 0$ identically on $\{J\}$ iff $x \in \bigcap_{J\in\{J\}} J$.

Definition 3. The representation

$$x \to x(J) \tag{7}$$

of the normed ring B, by the ring of functions $x(J)$ defined on the set $\{J\}$ of all the maximal ideals J of B, is called the *Gelfand-representation* of B.

2. The Radical. The Semi-simplicity

Definition 1. Let B be a normed ring over the complex number field, and $\{J\}$ the totality of the maximal ideals J of B. Then the ideal $\bigcap_{J\in\{J\}} J$ is called the *radical* of the ring B. B is said to be *semi-simple* if its radical $R = \bigcap_{J\in\{J\}} J$ reduces to the zero ideal $\{0\}$.

Theorem 1. For any $x \in B$, $\lim_{n\to\infty} \|x^n\|^{1/n}$ exists and we have

$$\lim_{n\to\infty} \|x^n\|^{1/n} = \sup_{J\in\{J\}} |x(J)|. \tag{1}$$

Proof. Set $\alpha = \sup_{J\in\{J\}} |x(J)|$. Then, by $\|x^n\| \geqq |x^n(J)| = |x(J)|^n$, we have $\|x^n\| \geqq \alpha^n$, and so $\varliminf_{n\to\infty} \|x^n\|^{1/n} \geqq \alpha$. We have thus to prove $\varlimsup_{n\to\infty} \|x^n\|^{1/n} \leqq \alpha$.

Let $|\beta| > \alpha$. Then, for any $J \in \{J\}$, $x(J) - \beta \neq 0$, i.e., $(x - \beta e) \bar{\in} J$. Hence the inverse $(\beta e - x)^{-1}$ exists. Setting $\beta^{-1} = \lambda$, we see that the inverse $(\beta e - x)^{-1} = \lambda(e - \lambda x)^{-1}$ exists whenever $|\lambda| < \alpha^{-1}$. Moreover, as in Theorem 1 in Chapter VIII, 2, we see that $\lambda(e - \lambda x)^{-1}$ is, for

$|\lambda| < \alpha^{-1}$, holomorphic in λ. Hence we have the Taylor expansion

$$\lambda(e-\lambda x)^{-1} = \lambda(e + \lambda x_1 + \lambda^2 x_2 + \cdots + \lambda^n x_n + \cdots).$$

That $x_n = x^n$ may be seen by Neumann's series $(e-\lambda x)^{-1} = \sum_{n=0}^{\infty} \lambda^n x^n$ which is valid for $||\lambda x|| < 1$. By the convergence of the above Taylor series, we see that

$$\lim_{n\to\infty} ||\lambda^n x^n|| = 0 \quad \text{if} \quad |\lambda| < \alpha^{-1}.$$

Thus $||x^n|| = |\lambda|^{-n} ||\lambda^n x^n|| < |\lambda|^{-n}$ for large n when $|\lambda| < \alpha^{-1}$, and so

$$\overline{\lim_{n\to\infty}} ||x^n||^{1/n} \leqq |\lambda|^{-1} \text{ when } |\lambda|^{-1} > \alpha, \text{ that is, } \overline{\lim_{n\to\infty}} ||x^n||^{1/n} \leqq \alpha.$$

Corollary. The radical $R = \bigcap_{J\in\{J\}} J$ of B coincides with the totality of the *generalized nilpotent elements* $x \in B$ which are defined by

$$\lim_{n\to\infty} ||x^n||^{1/n} = 0. \tag{2}$$

Definition 2. A complex number λ is said to belong to the *spectrum* of $x \in B$, if the inverse $(x - \lambda e)^{-1}$ does not exist in B.

If λ belongs to the spectrum of x, then there exists a maximal ideal J such that $(x - \lambda e) \in J$. Conversely, if $(x - \lambda e)$ belongs to a maximal ideal J, then the inverse $(x - \lambda e)^{-1}$ does not exist. Hence we obtain

Theorem 2. The spectrum of $x \in B$ coincides with the totality of values taken by the function $x(J)$ on the space $\{J\}$ consisting of all the maximal ideals J of B.

Application of Tychonov's theorem. We define, for any $J_0 \in \{J\}$, a fundamental system of neighbourhoods of J_0 by

$$\{J \in \{J\};\ |x_i(J) - x_i(J_0)| < \varepsilon_i \ \ (i = 1, 2, \ldots, n)\}, \tag{3}$$

where $\varepsilon_i > 0$, n and $x_i \in B$ are arbitrary. Then $\{J\}$ becomes a topological space and each $x(J)$, $x \in B$, becomes a continuous function on $\{J\}$. We have only to verify that if $J_0 \neq J_1$, then there exist a neighbourhood V_0 of J_0 and a neighbourhood V_1 of J_1 with empty intersection. This may be done as follows. Let $x_0 \in J_0$ and $x_0 \,\overline{\in}\, J_1$ so that $x_0(J_0) = 0$ and $x_0(J_1) = \alpha \neq 0$. Then $V_0 = \{J \in \{J\};\ |x_0(J)| < |\alpha|/2\}$ and $V_1 = \{J \in \{J\};\ |x_0(J) - x_0(J_1)| < |\alpha|/2\}$ have an empty intersection.

Theorem 3. The space $\{J\}$, topologized as above, is a compact space.

Proof. We attach, to each $x \in B$, the compact set

$$K_x = \{z;\ |z| \leqq ||x||\}$$

of the complex z-plane. Then the topological product

$$S = \prod_{x\in B} K_x$$

is a compact space, by virtue of Tychonov's theorem. See Chapter 0. To any maximal ideal $J_0 \in \{J\}$, we assign the point

$$\prod_{x \in B} x(J_0) = s(J_0) \in S.$$

$\{J\}$ is in one-one correspondence with a subset S_1 of S by the above correspondence $J_0 \to s(J_0)$. Moreover, the topology of $\{J\}$ is the same as the relative topology of S_1 as a subset of S. Hence, if we can show that S_1 is a closed subset of the compact space S, then its topological image $\{J\}$ is compact.

To prove that S_1 is a closed set, we consider an accumulation point $\omega = \prod_{x \in B} \lambda_x \in S$ of the set S_1 in S. We shall show that the mapping $x \to \lambda_x$ is a homomorphism of the algebra B into the complex number field (K). Then $J_0 = \{x; \lambda_x = 0\}$ becomes, as may be seen from the isomorphism of B/J_0 with (K), a maximal ideal of B and $(x - \lambda_x e) \in J_0$, that is, $x(J_0) = \lambda_x$. This proves that the point $\omega = \prod_{x \in B} \lambda_x = \prod_{x \in B} x(J_0)$ belongs to S_1.

We thus have to show that

$$\lambda_{x+y} = \lambda_x + \lambda_y, \ \lambda_{\alpha x} = \alpha \lambda_x, \ \lambda_{xy} = \lambda_x \lambda_y, \ \lambda_e = 1.$$

We shall, for instance, prove that $\lambda_{x+y} = \lambda_x + \lambda_y$. Since $\omega = \prod_{x \in B} \lambda_x$ is an accumulation point of S_1, there exists, for any $\varepsilon > 0$, a maximal ideal J such that

$$|\lambda_x - x(J)| < \varepsilon, \ |\lambda_y - y(J)| < \varepsilon, \ |\lambda_{x+y} - (x+y)(J)| < \varepsilon.$$

By $(x + y)(J) = x(J) + y(J)$ and the arbitrariness of $\varepsilon > 0$, we easily see that $\lambda_{x+y} = \lambda_x + \lambda_y$ is true.

We can now state the fundamental facts about the Gelfand representation $x \to x(J)$ of the normed ring B in the form of

Theorem 4. A normed ring B over the complex number field is represented homomorphically by the ring of functions $x(J)$ on the compact space $\{J\}$ of all the maximal ideals J of B. The radical R of B consists of those and only those elements which are represented by functions identically equal zero on $\{J\}$. The representation $x \to x(J)$ is isomorphic iff the ring B is semi-simple.

Application of the Stone-Weierstrass Theorem (of Chapter 0). The above obtained ring of functions is dense in the space of all complex-valued continuous functions on $\{J\}$ with uniform convergence topology if the ring B is *symmetric* (or *involutive*) in the following sense:

For any $x \in B$, there exists an $x^* \in B$ such that $x^*(J) = \overline{x(J)}$ on $\{J\}$. (4)

Examples of Gelfand Representations

Example 1. Let $B = C(S)$ where S is a compact topological space, and J_0 a maximal ideal of $C(S)$. Then there exists a point $s_0 \in S$ such that

$x(s_0) = 0$ for all $x \in J_0$. Otherwise, for any $s_\alpha \in S$, there exists an $x_\alpha \in J_0$ such that $x_\alpha(s_\alpha) \neq 0$. $x_\alpha(s)$ being a continuous function, there exists a neighbourhood V_α of s_α such that $x_\alpha(s) \neq 0$ in V_α. Since S is compact, there exists a finite system, say, $V_{\alpha_1}, V_{\alpha_2}, \ldots, V_{\alpha_n}$ such that $\bigcup_{j=1}^{n} V_{\alpha_j} = S$. Hence the function

$$x(s) = \sum_{i=1}^{n} \overline{x_{\alpha_i}(s)}\, x_{\alpha_i}(s) \in J_0$$

does not vanish over S, and the inverse x^{-1}, $x^{-1}(s) = x(s)^{-1}$, of $x \in J_0$ exists, contrary to the maximality of the ideal J_0. Thus we see that J_0 is contained in the maximal ideal $J' = \{x \in B;\, x(s_0) = 0\}$. By the maximality of J_0, we must have $J_0 = J'$. In this way we see that the space $\{J\}$ of the maximal ideals J of B is in one-one correspondence with the points s of S.

Example 2. Let B be the totality of functions $x(s)$, $0 \leqq s \leqq 1$, which can be represented by absolutely convergent Fourier series:

$$x(s) = \sum_{n=-\infty}^{\infty} c_n e^{2\pi i s n}, \quad \sum_{n=-\infty}^{\infty} |c_n| < \infty.$$

B is a normed ring by $(x + y)(s) = x(s) + y(s)$, $(xy)(s) = x(s)\, y(s)$ and $\|x\| = \sum_j |c_j|$. Let J_0 be a maximal ideal of B. Set $e^{2\pi i s} = x_1$. Then $x_1^{-1} = e^{-2\pi i s}$, and so, by $|x_1(J_0)| \leqq \|x_1\| = 1$, $|x_1^{-1}(J)| = |x_1(J)^{-1}| \leqq \|x_1^{-1}\| = 1$, we see that $|x_1(J_0)| = 1$. Hence there exists a point s_0, $0 \leqq s_0 \leqq 1$, such that $x_1(J_0) = e^{2\pi i s_0}$. Thus $x_n = e^{2\pi i s n} = x_1^n$ satisfies $x_n(J_0) = e^{2\pi i s_0 n}$, and so $x(J_0) = \sum_{n=-\infty}^{\infty} c_n e^{2\pi i s_0 n} = x(s_0)$. In this way, we see that for any maximal ideal J_0 of B, there exists a point s_0, $0 \leqq s_0 \leqq 1$, such that the homomorphism $x \to x(J_0)$ is given by $x(J_0) = x(s_0)$, for all $x \in B$. It is also clear that the mapping $x \to x(s_0)$ gives a homomorphism of the algebra B into the complex number field. Therefore we see that the maximal ideal space of B coincides with $\{e^{2\pi i s};\, 0 \leqq s \leqq 1\}$.

Corollary (N. Wiener's theorem). If an absolutely convergent Fourier series $x(s) = \sum_{n=-\infty}^{\infty} c_n e^{2\pi i s n}$ does not vanish on $[0, 1]$, then the function $1/x(s)$ is also representable as an absolutely convergent Fourier series. For, x does not belong to any of the maximal ideals of the normed ring of the above Example 2.

Example 3. We take $B_1 = C[0, 1]$, and define, for $x, y \in B_1$,

$$(x + y)(s) = x(s) + y(s), \quad (\alpha x)(s) = \alpha x(s), \quad (xy)(s) = \int_0^s x(s - t)\, y(t)\, dt \text{ and } \|x\| = \sup_{s \in [0,1]} |x(s)|.$$

B_1 is a commutative B-algebra without unit. Adjoining formally a unit e by the rule $ex = xe = x$, $\|e\| = 1$, the set $B = \{z = \lambda e + x; x \in B_1\}$ becomes a normed ring by the operations

$$\begin{cases} (\lambda_1 e + x_1) + (\lambda_2 e + x_2) = (\lambda_1 + \lambda_2)\, e + (x_1 + x_2),\ \alpha(\lambda e + x) \\ = \alpha\lambda e + \alpha x,\ (\lambda_1 e + x_1)(\lambda_2 e + x_2) = \lambda_1\lambda_2 e + \lambda_1 x_2 + \lambda_2 x_1 + x_1 x_2 \\ \text{and } \|\lambda e + x\| = |\lambda| + \|x\|. \end{cases}$$

We have, by induction,

$$|x^2(s)| \leqq M^2 s,\ |x^3(s)| \leqq M^3 \frac{s^2}{2}, \ldots, |x^n(s)| \leqq M^n \frac{s^{n-1}}{(n-1)!}, \ldots$$

where $M = \sup_{s \in S} |x(s)| = \|x\|$. Thus every $x \in B_1$ is a generalized nilpotent element of B, due to the fact that $\lim_{n=\infty} (n!)^{1/n} = \infty$.

3. The Spectral Resolution of Bounded Normal Operators

Let X be a Hilbert space, and let a system M of bounded normal operators $\in L(X, X)$ satisfy the conditions:

$$T, S \in M \text{ implies } TS = ST \text{ (commutativity)}, \tag{1}$$

$$T \in M \text{ implies } T^* \in M. \tag{2}$$

A system M consisting of a bounded normal operator $T \in L(X, X)$ and its adjoint T^* surely satisfies (1) and (2).

Let M' be the totality of operators $\in L(X, X)$ which commute with every $T \in M$, and let $B = M'' = (M')'$ be the totality of operators $\in L(X, X)$ commutative with every operator $S \in M'$.

Proposition 1. Every element of B is a normal operator. B is a normed ring over the complex number field by the operator sum, the operator product, the unit I (the identity operator) and the operator norm $\|T\|$.

Proof. $M \subseteq M'$ by (1), and so $M' \supseteq M''$. Hence $M''' = (M'')' \supseteq M''$ and so $B = M''$ is a commutative ring. The identity operator I belongs to B and is the unit of this algebra B. By (2), we easily see that every operator $\in B$ is normal. Since the multiplication TS and the adjoint formation $T \to T^*$ in the algebra B are continuous with respect to the norm of the operator, it is easy to see that the ring B is complete with respect to the operator norm.

Theorem 1. By the Gelfand representation

$$B \ni T \to T(J), \tag{3}$$

the ring B is represented isomorphically by the algebra $C(\{J\})$ of all continuous functions $T(J)$ on the compact space $\{J\}$ of all the maximal ideals J of B in such a way that

$$||T|| = \sup_{J \in \{J\}} |T(J)|, \tag{4}$$

$$T(J) \text{ is real-valued on } \{J\} \text{ iff } T \text{ is self-adjoint}, \tag{5}$$

$$T(J) \geqq 0 \text{ on } \{J\} \text{ iff } T \text{ is self-adjoint and positive, that is, } (Tx, x) \geqq 0 \text{ for all } x \in X. \tag{6}$$

Proof. We first show that, for any bounded normal operator T,

$$||T^2|| = ||T||^2. \tag{7}$$

By the normality of T, we see that $H = TT^* = T^*T$ is self-adjoint. Hence, by Theorem 3 in Chapter VII, 3,

$$||T||^2 = \sup_{||x|| \leq 1} (Tx, Tx) = \sup_{||x|| \leq 1} |(T^*Tx, x)| = ||H|| = ||T^*T|| = ||TT^*||.$$

Since $(T^*)^2 = (T^2)^*$, T^2 is normal with T. Thus, as above, we obtain $||T^2||^2 = ||T^{*2}T^2||$, which is, by the commutativity $TT^* = T^*T$, equal to $||(T^*T)^2|| = ||H^2||$. Since H^2 is self-adjoint, we obtain, again by Theorem 3 in Chapter VII, 3,

$$||H||^2 = \sup_{||x|| \leq 1} (Hx, Hx) = \sup_{||x|| \leq 1} |(H^2x, x)| = ||H^2||.$$

Therefore, $||T^2||^2 = ||H^2|| = ||H||^2 = (||T||^2)^2$, that is, $||T^2|| = ||T||^2$.

We have, by (7), $||T|| = \lim_{n \to \infty} ||T^n||^{1/n}$, because we know already that the right hand limit exists (see (3) in Chapter VIII, 2). Hence, by Theorem 4 of the preceding section, the representation (3) is isomorphic and (4) is true.

Proof of (5). Let a self-adjoint $T \in B$ satisfy, for a certain $J_0 \in \{J\}$, $T(J_0) = a + ib$ with $b \neq 0$. Then the self-adjoint operator $S = (T - aI)/b \in B$ satisfies $(I + S^2)(J_0) = 1 + i^2 = 0$, and so $(I + S^2)$ does not have an inverse in B. But, by Theorem 2 in Chapter VII, 3, $(I + S^2)$ has an inverse which surely belongs to B. Thus, if $T \in B$ is self-adjoint, $T(J)$ must be real-valued. Let $T \in B$ be not self-adjoint, and put

$$T = \frac{T + T^*}{2} + i\frac{T - T^*}{2i}.$$

Then, since the first term on the right is self-adjoint, the self-adjoint operator $(T - T^*)/2i$ must be $\neq 0$. Thus, by the isomorphism of representation (3), there must exist a $J_0 \in \{J\}$ such that $\frac{T - T^*}{2i}(J_0) \neq 0$. Hence $T(J_0) = \frac{T + T^*}{2}(J_0) + i\frac{T - T^*}{2i}(J_0)$ is not real. For, as proved above, self-adjoint operators are represented by real-valued functions.

Proof of (6). We first show that

$$T^*(J) = \overline{T(J)} \text{ on } \{J\}. \tag{8}$$

This is clear, since self-adjoint operators $(T + T^*)/2$ and $(T - T^*)/2i$ are represented by real-valued functions. Therefore, by (4) and the result of the preceding section, the ring B is represented by the ring of all continuous complex-valued functions on $\{J\}$ satisfying (5) and (8). Let $T(J) \geqq 0$ over $\{J\}$. Then $S(J) = T(J)^{1/2}$ is a continuous function on $\{J\}$. Hence, by the isomorphism of the representation (3), $S^2 = T$. By (5), we have $S = S^*$. Hence $(Tx, x) = (S^2x, x) = (Sx, Sx) \geqq 0$. To prove, conversely, that the condition $(Tx, x) \geqq 0$ for all $x \in X$ implies $T(J) \geqq 0$ over $\{J\}$, we set $T_1(J) = \max(T(J), 0)$ and $T_2(J) = T_1(J) - T(J)$. Then, by what we have proved above, T_1 and T_2 are both $\in B$, self-adjoint and positive: $(T_jx, x) \geqq 0$ for all $x \in X$ $(j = 1, 2)$. Moreover, we have $T_1T_2 = 0$ and $T_2 = T_1 - T$. The former equality is implied by $T_1(J)\,T_2(J) = 0$.

Therefore, we have

$$0 \leqq (TT_2x, T_2x) = (-T_2^2x, T_2x) = -(T_2^3x, x) = -(T_2T_2x, T_2x) \leqq 0.$$

Thus $(T_2^3x, x) = 0$ and so, by Theorem 3 in Chapter VII, 3, we must have $T_2^3 = 0$. Hence, by $\|T_2\| = \lim_{n\to\infty} \|T_2^n\|^{1/n}$, we obtain $T_2 = 0$. We have thus proved $T = T_1$ and hence $T(J) \geqq 0$ on $\{J\}$.

We shall thus write $T \geqq 0$ if T is self-adjoint and positive. We also write $S \geqq T$ if $(S - T) \geqq 0$.

Theorem 2. Let $\{T_n\} \subseteqq B$ be a sequence of self-adjoint operators such that

$$0 \leqq T_1 \leqq T_2 \leqq \cdots \leqq T_n \leqq \cdots \leqq S \in B. \tag{9}$$

Then, for any $x \in X$, $s\text{-}\lim_{n\to\infty} T_nx = Tx$ exists, i.e., $s\text{-}\lim_{n\to\infty} T_n = T$ exists and $T \in B$, $S \geqq T \geqq T_n$ $(n = 1, 2, \ldots)$.

Proof. We first remark that, by (6),

$$E, F \in B \text{ and } E \geqq 0,\ F \geqq 0 \text{ imply } E + F \geqq 0 \text{ and } EF \geqq 0. \tag{10}$$

Thus $0 \leqq T_1^2 \leqq T_2^2 \leqq \cdots \leqq T_n^2 \leqq \cdots \leqq S^2$. Hence, for any $x \in X$, a finite $\lim_{n\to\infty} (T_n^2x, x)$ exists. Since, by (6), $T_{n+k}^2 \geqq T_{n+k}T_n \geqq T_n^2$, we also have

$$\lim_{n,k\to\infty} (T_{n+k}^2x, x) = \lim_{n,k\to\infty} (T_{n+k}T_nx, x) = \lim_{n\to\infty} (T_n^2x, x).$$

Therefore, $\lim_{n,m\to\infty} ((T_n - T_m)^2 x, x) = \lim_{n,m\to\infty} \|T_nx - T_mx\|^2 = 0$ so that $s\text{-}\lim_{n\to\infty} T_nx = Tx$ exists. That $T \in B$ and $S \geqq T \geqq T_n$ is clear from the process of the proof.

Theorem 3. Let a sequence of real-valued functions $\{T_n(J)\}$, where $T_n \in B$, satisfy the condition

$$0 \leqq T_1(J) \leqq T_2(J) \leqq \cdots \leqq T_n(J) \leqq \cdots \leqq a \text{ finite constant on } \{J\}. \tag{11}$$

Then, by (6) and Theorem 2, $s\text{-}\lim_{n\to\infty} T_n = T$ exists. In such a case, we can prove that

$$D = \{J \in \{J\};\, T(J) \neq \lim_{n\to\infty} T_n(J)\}$$

is a set of the first category, and so $D^C = \{J\} - D$ is dense in $\{J\}$.

Proof. By Theorem 2, $T \geqq T_n$ and so $T(J) \geqq \lim_{n\to\infty} T_n(J)$ on $\{J\}$. By Baire's theorem in Chapter 0, 2, the set of points of discontinuity of the function $\lim_{n\to\infty} T_n(J)$ is of the first category. Hence, if the set D is not of the first category, then there exists at least one point $J_0 \in D$ at which $\lim_{n\to\infty} T_n(J)$ is continuous. In other words, there exists a positive number δ and an open set $V(J_0) \ni J_0$ of $\{J\}$ such that

$$T(J) \geqq \delta + \lim_{n\to\infty} T_n(J) \text{ whenever } J \in V(J_0).$$

Since the compact space $\{J\}$ is normal, and since $T(J) \geqq \lim_{n\to\infty} T_n(J)$ on $\{J\}$, we may construct, by Urysohn's theorem an open set $V_1(J_0) \ni J_0$ and a function $W(J) \in C(\{J\})$ such that $0 \leqq W(J) \leqq \delta$ on $\{J\}$, $V_1(J_0)^a \subseteqq V(J_0)$, $W(J) = \delta/2$ on $V_1(J_0)$ and $W(J) = 0$ on $V(J_0)^C$. Hence $T(J) - W(J) \geqq \lim_{n\to\infty} T_n(J)$ on $\{J\}$, and so, by (6), $T - W \geqq T_n$ $(n = 1, 2, \ldots)$. $W(J) \not\equiv 0$ implies, by the isomorphism (3), that $W \neq 0$, $W \geqq 0$. Thus, again by (6), $T - W \geqq s\text{-}\lim_{n\to\infty} T_n$, contrary to $T = s\text{-}\lim_{n\to\infty} T_n$.

Finally, since $\{J\}$ is a compact space, the complement $D^C = \{J\} - D$ of the set D of the first category must be dense in $\{J\}$.

We are now able to prove (K. Yosida [12])

The spectral resolution or the spectral representation of operators $\in B$.

Consider the set $C'(\{J\})$ of all complex-valued bounded functions $T'(J)$ on $\{J\}$ such that $T'(J)$ is different from a continuous function $T(J)$ only on a set of the first category. We identify two functions from $C'(\{J\})$ if they differ only on a set of the first category. Then $C'(\{J\})$ is divided into classes. Since the complement of a set of the first category is dense in the compact space $\{J\}$, each class T' contains exactly one continuous function $T(J)$ which corresponds, by the isomorphism $B \leftrightarrow C(\{J\})$, to an element $T \in B$.

For any $T \in B$ and for any complex number $z = \lambda + i\mu$, we put E_z = the element $\in B$ which corresponds to the class E'_z containing the defining function $E'(J)$ of the set $\{J \in \{J\};\, Re\, T(J) < \lambda, Im\, T(J) < \mu\}$. It is clear that there exists a monotone increasing sequence of continuous functions f_n of complex argument such that $E'_z(J) = \lim_{n\to\infty} f_n(T(J))$ and so $E'_z(J) \in C'(\{J\})$. We have then

$$|T(J) - \sum_{j=2}^{n} (\lambda_j + i\mu_j)(E'_{\lambda_j+i\mu_j}(J) + E'_{\lambda_{j-1}+i\mu_{j-1}}(J) - E'_{\lambda_{j-1}+i\mu_j}(J) - E'_{\lambda_j+i\mu_{j-1}}(J))| \leqq \varepsilon \text{ on } \{J\} \text{ if}$$

$$\lambda_1 = -\alpha - \frac{\varepsilon}{\sqrt{2}} \leqq \lambda_2 \leqq \cdots \leqq \lambda_n = \alpha = \sup_{J\in\{J\}} |Re\,T(J)|,$$

$$\mu_1 = -\beta - \frac{\varepsilon}{\sqrt{2}} \leqq \mu_2 \leqq \cdots \leqq \mu_n = \beta = \sup_{J\in\{J\}} |Im\,T(J)|,$$

$$\left(\sup_j (\lambda_j - \lambda_{j-1})^2 + \sup_j (\mu_j - \mu_{j-1})^2\right)^{1/2} \leqq \varepsilon.$$

Thus, by the definition of $E'_z(J)$, we have

$$|T(J) - \sum_{j=2}^{n} (\lambda_j + i\mu_j)(E_{\lambda_j+i\mu_j}(J) + E_{\lambda_{j-1}+i\mu_{j-1}}(J) - E_{\lambda_{j-1}+i\mu_j}(J) - E_{\lambda_j+i\mu_{j-1}}(J))| \leqq \varepsilon$$

on $\{J\}$, since the complement of a set of the first category is dense in the compact space $\{J\}$. Therefore, by (4), we have

$$||T - \sum_{j=2}^{n} (\lambda_j + i\mu_j)(E_{\lambda_j+i\mu_j} + E_{\lambda_{j-1}+i\mu_{j-1}} - E_{\lambda_{j-1}+i\mu_j} - E_{\lambda_j+i\mu_{j-1}})|| \leqq \varepsilon,$$

which we shall write as

$$T = \iint z\, dE_z, \tag{12}$$

and it is called the *spectral resolution* of the normal operator T.

4. The Spectral Resolution of a Unitary Operator

If T is a unitary operator $\in B$, then, by

$$T(J)\, T^*(J) = T(J)\, \overline{T(J)} = 1, \tag{1}$$

we see that the values taken by the function $T(J)$ on $\{J\}$ are complex numbers of absolute value 1. From this fact, we can simplify the spectral resolution $\iint z\, dE_z$ of T.

The defining function $E'_\theta(J)$ of the set $\{J \in \{J\}; \arg(T(J)) \in (0, \theta)\}$, $0 < \theta < 2\pi$, belongs to $C'(\{J\})$, and we have, by setting $E'_0(J) = 0$, $E'_{2\pi}(J) = I$,

$$|T(J) - \sum_{j=1}^{n} e^{i\theta_j}(E'_{\theta_j}(J) - E'_{\theta_{j-1}}(J))| \leqq \max_j |e^{i\theta_j} - e^{i\theta_{j-1}}|$$

$$(0 = \theta_0 < \theta_1 < \cdots < \theta_n = 2\pi).$$

Let $E_\theta(J)$ be the continuous function on $\{J\}$ which is different from $E'_\theta(J)$ only on a set of the first category, and let E_θ be the operator $\in B$ which corresponds to $E_\theta(J)$ by the isomorphic representation $B \ni T \leftrightarrow T(J)$. Then, as in the preceding section,

$$||T - \sum_{j=1}^{n} e^{i\theta_j}(E_{\theta_j} - E_{\theta_{j-1}})|| \leqq \max_j |e^{i\theta_j} - e^{i\theta_{j-1}}|.$$

This we write, in view of the fact $e^{2\pi i} = 1$,

$$T = \int_0^{2\pi} e^{i\theta}\, dF(\theta), \text{ where}$$

$$F(\theta) = E_{\theta+0} - E_{+0} \text{ for } 0 < \theta < 2\pi,\ F(0) = 0,\ F(2\pi) = I. \tag{2}$$

Here $E_{\theta+0}$ is defined by $E_{\theta+0}x = \underset{\theta\downarrow 0}{s\text{-lim}}\, E_\theta x$, the existence of this limit will be proved below.

Theorem 1. The system of operators $F(\theta)$, $0 \leqq \theta \leqq 2\pi$, satisfies the conditions:

each $F(\theta)$ is a projection operator, commutative with every bounded linear operator commutative with T, (3)

$$F(\theta)\, F(\theta') = F(\min(\theta, \theta')), \tag{4}$$

$$F(0) = 0,\ F(2\pi) = I, \tag{5}$$

$F(\theta + 0) = F(\theta)$, $0 \leqq \theta < 2\pi$, in the sense that $\underset{\theta'\downarrow\theta}{s\text{-lim}}\, F(\theta')\, x = F(\theta)\, x$ for every $x \in X$. (6)

Proof. It is sufficient to prove that E_θ, $0 \leqq \theta \leqq 2\pi$, satisfies the conditions:

$$\text{each } E_\theta \text{ is a projection operator} \in B, \tag{3'}$$

$$E_\theta E_{\theta'} = E_{\min(\theta,\theta')}, \tag{4'}$$

$$E_0 = 0,\ E_{2\pi} = I, \tag{5'}$$

$$E_\theta x = \underset{\theta'\downarrow\theta}{s\text{-lim}}\, E_{\theta'} x \text{ for every } x \in X \text{ and } 0 \leqq \theta < 2\pi. \tag{6'}$$

We have $E'_\theta(J) = \overline{E'_\theta(J)}$ and $E'_\theta(J)^2 = E'_\theta(J)$. Hence, by the result of the preceding section, we obtain $E_\theta = E_\theta^*$ and $E_\theta^2 = E_\theta$. This proves (3'). (4') is proved similarly from $E'_\theta(J)\, E'_{\theta'}(J) = E'_{\min(\theta,\theta')}(J)$, and (5') is proved similarly. Next let $\theta_n \downarrow \theta$. Then $E'_{\theta_n}(J) \geqq E'_{\theta_{n+1}}(J) \geqq E'_\theta(J)$, and so, by the result in the preceding section, $\underset{n\to\infty}{s\text{-lim}}\, E_{\theta_n} = E$ exists and $E(J) = E'_\theta(J) = \underset{\theta_n\downarrow\theta}{\lim}\, E'_{\theta_n}(J)$ on $\{J\}$ except possibly on a set of the first category. Thus $E = E_\theta$.

Example 1. Let a linear operator T defined by

$$Ty(s) = e^{i\theta} y(s), \quad \text{where} \quad y(s) \in L^2(-\infty, \infty).$$

T is unitary. We define, when $2n\pi < s \leqq 2(n+1)\pi$,

$$F(\theta)\, y(s) = y(s) \quad \text{for} \quad s \leqq \theta + 2n\pi \leqq 2(n+1)\pi,$$

$$F(\theta)\, y(s) = 0 \quad \text{for} \quad \theta + 2n\pi < s.$$

It is easy to see that $T = \int_0^{2\pi} e^{i\theta}\, dF(\theta)$.

Example 2. Let a linear operator T_1 be defined by

$$T_1 x(t) = x(t+1) \quad \text{in} \quad L^2(-\infty, \infty).$$

T_1 is unitary. By the Fourier transformation

$$y(s) = Ux(t) = \underset{n\to\infty}{\text{l.i.m.}}\ (2\pi)^{-1/2} \int_{-n}^{n} e^{-ist} x(t)\, dt,$$

we obtain $Ux(t+1) = e^{is} Ux(t) = e^{is} y(s)$. Thus

$$T_1 x(t) = x(t+1) = U^{-1} e^{is} y(s) = U^{-1} T y(s) = U^{-1} T U x(t),$$

that is, $T_1 = U^{-1} T U$. Therefore, we have

$$T_1 = \int_0^{2\pi} e^{i\theta}\, dF_1(\theta), \text{ where } F_1(\theta) = U^{-1} F(\theta)\, U.$$

The uniqueness of the spectral resolution. Since $T^{-1} = T^*$ and $T^{-1}(J) = T^*(J) = T(J)^{-1}$, we easily see that

$$T^{-1} = \int_0^{2\pi} e^{-i\theta}\, dF(\theta). \tag{7}$$

Let $\max_j |e^{i\theta_j} - e^{i\theta_{j-1}}| < \varepsilon$. Then, from

$$T = \sum_j e^{i\theta_j}(F(\theta_j) - F(\theta_{j-1})) + \delta, \ \ ||\delta|| < \varepsilon,$$

we obtain, by (4),

$$T^2 = \sum_j e^{2i\theta_j}(F(\theta_j) - F(\theta_{j-1})) + \delta', \text{ where}$$

$$||\delta'|| \leqq ||(T-\delta)\,\delta|| + ||\delta(T-\delta)|| + ||\delta^2||$$

$$\leqq (||T|| + \varepsilon)\,\varepsilon + \varepsilon(||T|| + \varepsilon) + \varepsilon^2.$$

Hence we have $T^2 = \int_0^{2\pi} e^{2i\theta}\, dF(\theta)$, and, more generally,

$$T^n = \int_0^{2\pi} e^{in\theta}\, dF(\theta) \quad (n = 0, \pm 1, \pm 2, \ldots). \tag{8}$$

Therefore, if there exists another spectral resolution $T = \int_0^{2\pi} e^{i\theta}\, dF_1(\theta)$ satisfying (3) to (6), then, for any polynomial $p(\theta)$ in $e^{i\theta}$ and $e^{-i\theta}$,

$$\int_0^{2\pi} p(\theta)\, d((F(\theta)\,x, y) - (F_1(\theta)\,x, y)) = 0, \ (x, y \in X).$$

Thus, by continuity, the above equality holds for any continuous function $p(\theta)$ with $p(0) = p(2\pi)$. Let $0 < \theta_0 < \theta_1 < 2\pi$, and take

$$\begin{aligned} p_n(\theta) &= 0 \text{ for } 0 \leqq \theta \leqq \theta_0 \text{ and for } \theta_1 + \frac{1}{n} \leqq \theta \leqq 2\pi, \\ &= 1 \text{ for } \theta_0 + \frac{1}{n} \leqq \theta \leqq \theta_1, \\ &= \text{linear for } \theta_0 \leqq \theta \leqq \theta_0 + \frac{1}{n} \text{ and for } \theta_1 \leqq \theta \leqq \theta_1 + \frac{1}{n}. \end{aligned}$$

Then letting $n \to \infty$, we obtain, by (6),

$$\lim_{n\to\infty} \int_0^{2\pi} p_n(\theta)\, d[(F(\theta)x,y)-(F_1(\theta)x,y)] = [(F(\theta)x,y)-(F_1(\theta)x,y)]_{\theta_0}^{\theta_1} = 0,$$

which is valid for all $x, y \in X$. Hence, letting $\theta_0 \downarrow 0$ and making use of conditions (5) and (6), we see that $F(\theta_1) = F_1(\theta)$. Therefore, the spectral resolution for a unitary operator is uniquely determined.

5. The Resolution of the Identity

Definition 1. A family of projections $E(\lambda)$, $-\infty < \lambda < \infty$, in a Hilbert space X is called a (real) *resolution of the identity* if it satisfies the conditions:

$$E(\lambda)\,E(\mu) = E(\min(\lambda,\mu)), \tag{1}$$

$$E(-\infty) = 0,\ E(+\infty) = I, \text{ where } E(-\infty)\,x = \underset{\lambda\downarrow-\infty}{\text{s-lim}}\, E(\lambda)\,x \text{ and}$$
$$E(+\infty)\,x = \underset{\lambda\uparrow\infty}{\text{s-lim}}\, E(\lambda)\,x, \tag{2}$$

$$E(\lambda+0) = E(\lambda), \text{ where } E(\lambda+0)\,x = \underset{\mu\downarrow\lambda}{\text{s-lim}}\, E(\mu)\,x. \tag{3}$$

Proposition 1. For any $x, y \in X$, the function $(E(\lambda)\,x, y)$ is, as a function of λ, of bounded variation.

Proof. Let $\lambda_1 < \lambda_2 < \cdots < \lambda_n$. Then, by (1), $E(\alpha,\beta] = E(\beta) - E(\alpha)$ is a projection. Thus we have, by Schwarz' inequality,

$$\begin{aligned}\sum_j |(E(\lambda_{j-1},\lambda_j]\,x, y)| &= \sum_j |(E(\lambda_{j-1},\lambda_j]\,x, E(\lambda_{j-1},\lambda_j]\,y)|\\ &\leq \sum_j \|E(\lambda_{j-1},\lambda_j]\,x\| \cdot \|E(\lambda_{j-1},\lambda_j]\,y\|\\ &\leq \Big(\sum_j \|E(\lambda_{j-1},\lambda_j]\,x\|^2\Big)^{1/2} \cdot \Big(\sum_j \|E(\lambda_{j-1},\lambda_j]\,y\|^2\Big)^{1/2}\\ &= (\|E(\lambda_1,\lambda_n]\,x\|^2)^{1/2} \cdot (\|E(\lambda_1,\lambda_n]\,y\|^2)^{1/2} \leq \|x\| \cdot \|y\|.\end{aligned}$$

For, by the orthogonality

$$E(\lambda_{j-1},\lambda_j] \cdot E(\lambda_{i-1},\lambda_i] = 0 \qquad (i \neq j) \tag{4}$$

implied by (1), we have, for $m > n$,

$$\|x\|^2 \geq \|E(\lambda_n,\lambda_m]\,x\|^2 = \sum_{i=n}^{m-1} \|E(\lambda_i,\lambda_{i+1}]\,x\|^2. \tag{5}$$

Corollary. For any λ, $-\infty < \lambda < \infty$, the operators $E(\lambda+0) = \underset{\lambda'\downarrow\lambda}{\text{s-lim}}\, E(\lambda')$ and $E(\lambda-0) = \underset{\lambda'\uparrow\lambda}{\text{s-lim}}\, E(\lambda')$ do exist.

Proof. From (5), we see that, if $\lambda_n \uparrow \lambda$, then

$$\lim_{j,k\to\infty} \|E(\lambda_j,\lambda_k]\,x\|^2 = 0,$$

and the same is true for the case $\lambda_n \downarrow \lambda$.

Proposition 2. Let $f(\lambda)$ be a complex-valued continuous function on $(-\infty, \infty)$, and let $x \in X$. Then we can define, for $-\infty < \alpha < \beta < \infty$,

$$\int_{\alpha}^{\beta} f(\lambda)\, dE(\lambda)\, x$$

as the s-lim of the Riemann sums

$\sum_j f(\lambda_j')\, E(\lambda_j, \lambda_{j+1}]\, x$, where $\alpha = \lambda_1 < \lambda_2 < \cdots < \lambda_n = \beta$, $x_j \in (\lambda_j, \lambda_{j+1}]$, when the $\max_j |\lambda_{j+1} - \lambda_j|$ tends to zero.

Proof. $f(\lambda)$ is uniformly continuous on the compact interval $[\alpha, \beta]$. Let $|f(\lambda) - f(\lambda')| \leqq \varepsilon$ whenever $|\lambda - \lambda'| \leqq \delta$. We consider two partitions of $[\alpha, \beta]$:

$$\alpha = \lambda_1 < \cdots < \lambda_n = \beta, \max_j |\lambda_{j+1} - \lambda_j| \leqq \delta,$$

$$\alpha = \mu_1 < \cdots < \mu_m = \beta, \max_j |\mu_{j+1} - \mu_j| \leqq \delta,$$

and let

$$\alpha = \nu_1 < \cdots < \nu_p = \beta,\ p \leqq m + n,$$

be the superposition of these two partitions. Then, if $\mu_k' \in (\mu_k, \mu_{k+1}]$, we have

$$\sum_j f(\lambda_j')\, E(\lambda_j, \lambda_{j+1}]\, x - \sum_k f(\mu_k')\, E(\mu_k, \mu_{k+1}]\, x$$

$$= \sum_s \varepsilon_s\, E(\nu_s, \nu_{s+1}]\, x, \text{ with } |\varepsilon_s| \leqq 2\varepsilon,$$

and so the square of the norm of the left side is, as in (5),

$$\leqq \varepsilon^2 \Big\| \sum_s E(\nu_s, \nu_{s+1}]\, x \Big\|^2 = \varepsilon^2\, \|E(\alpha, \beta]\, x\|^2 \leqq \varepsilon^2\, \|x\|^2.$$

Corollary. We may define $\int_{\infty}^{\infty} f(\lambda)\, dE(\lambda)\, x$ as the $\underset{\alpha \downarrow -\infty, \beta \uparrow \infty}{s\text{-lim}} \int_{\alpha}^{\beta} f(\lambda)\, dE(\lambda)\, x$, when the right side limit exists.

Theorem 1. For a given $x \in X$, the following three conditions are mutually equivalent:

$$\int_{-\infty}^{\infty} f(\lambda)\, dE(\lambda)\, x \text{ exists}, \tag{6}$$

$$\int_{-\infty}^{\infty} |f(\lambda)|^2\, d\, \|E(\lambda)\, x\|^2 < \infty, \tag{7}$$

$$F(y) = \int_{-\infty}^{\infty} f(\lambda)\, d(E(\lambda)\, y, x) \text{ defines a bounded linear functional.} \tag{8}$$

Proof. We shall prove the implications (6) → (8) → (7) → (6).

(6) → (8). The scalar product of y with the approximate Riemann

sum of $\int_{-\infty}^{\infty} f(\lambda)\, dE(\lambda)\, x$ is a bounded linear functional of y. Hence, by $(y, E(\lambda)\, x) = (E(\lambda)\, y, x)$ and the resonance theorem, we obtain (8).

(8) $\to$ (7). We apply the operator $E(\alpha, \beta]$ to the approximate Riemann sum of $y = \int_{\alpha}^{\beta} \overline{f(\lambda)}\, dE(\lambda)\, x$. We then see, by (1), that $y = E(\alpha, \beta] y$. Thus, again by (1),

$$\overline{F(y)} = \int_{-\infty}^{\infty} \overline{f(\lambda)}\, d(E(\lambda)\, x, y) = \lim_{\alpha' \downarrow -\infty, \beta' \uparrow \infty} \int_{\alpha'}^{\beta'} \overline{f(\lambda)}\, d(E(\lambda)\, x, y)$$

$$= \lim_{\alpha' \downarrow -\infty, \beta' \uparrow \infty} \int_{\alpha'}^{\beta'} \overline{f(\lambda)}\, d(E(\lambda)\, x, E(\alpha, \beta]\, y)$$

$$= \lim_{\alpha' \downarrow -\infty, \beta' \uparrow \infty} \int_{\alpha'}^{\beta'} \overline{f(\lambda)}\, d(E(\alpha, \beta]\, E(\lambda)\, x, y)$$

$$= \int_{\alpha}^{\beta} \overline{f(\lambda)}\, d(E(\lambda)\, x, y) = \|y\|^2.$$

Hence $\|y\|^2 \leqq \|F\| \cdot \|y\|$, i.e., $\|y\| \leqq \|F\|$. On the other hand, by approximating $y = \int_{\alpha}^{\beta} \overline{f(\lambda)}\, dE(\lambda)\, x$ by Riemann sums, we obtain, by (1),

$$\|y\|^2 = \left\| \int_{\alpha}^{\beta} \overline{f(\lambda)}\, dE(\lambda)\, x \right\|^2 = \int_{\alpha}^{\beta} |f(\lambda)|^2\, d\, \|E(\lambda)\, x\|^2,$$

so that $\int_{\alpha}^{\beta} |f(\lambda)|^2\, d\, \|E(\lambda)\, x\|^2 \leqq \|F\|^2$. Therefore, by letting $\alpha \downarrow -\infty$, $\beta \downarrow \infty$, we see that (7) is true.

(7) $\to$ (6). We have, for $\alpha' < \alpha < \beta < \beta'$,

$$\left\| \int_{\alpha'}^{\beta'} f(\lambda)\, dE(\lambda)\, x - \int_{\alpha}^{\beta} f(\lambda)\, dE(\lambda)\, x \right\|^2$$

$$= \int_{\alpha'}^{\alpha} |f(\lambda)|^2\, d\, \|E(\lambda)\, x\|^2 + \int_{\beta}^{\beta'} |f(\lambda)|^2\, d\, \|E(\lambda)\, x\|^2$$

as above. Hence (7) implies (6).

Theorem 2. Let $f(\lambda)$ be a real-valued continuous function. Then, a self-adjoint operator H with $D(H) = D$ is defined through

$$(Hx, y) = \int_{-\infty}^{\infty} f(\lambda)\, d(E(\lambda)\, x, y), \text{ where} \tag{9}$$

$$x \in D = \left\{x;\ \int_{-\infty}^{\infty} |f(\lambda)|^2\, d\, \|E(\lambda)\, x\|^2 < \infty\right\} \text{and any } y \in X,$$

and we have $HE(\lambda) \supseteq E(\lambda)\, H$, that is, $HE(\lambda)$ is an extension of $E(\lambda)\, H$.

Proof. For any $y \in X$ and for any $\varepsilon > 0$, there exist α and β with $-\infty < \alpha < \beta < \infty$ such that $\|y - E(\alpha, \beta] y\| < \varepsilon$. Moreover, we have

$$\int_{-\infty}^{\infty} |f(\lambda)|^2 d\, \|E(\lambda) E(\alpha, \beta] y\|^2 = \int_{\alpha}^{\beta} |f(\lambda)|^2 d\, \|E(\lambda) y\|^2.$$

Hence $E(\alpha, \beta] y \in D$ and so, by (2), $D^a = X$. H is symmetric by

$$f(\lambda) = \overline{f(\lambda)}, \ (E(\lambda) x, y) = \overline{(E(\lambda) y, x)}.$$

If $y \in D(H^*)$ and $H^* y = y^*$, then, by $E(\alpha, \beta] z \in D$ and (1),

$$(z, E(\alpha, \beta] y^*) = (E(\alpha, \beta] z, H^* y) = (H E(\alpha, \beta] z, y) = \int_{\alpha}^{\beta} f(\lambda) d(E(\lambda) z, y).$$

Thus, by the resonance theorem,

$$\lim_{\alpha \downarrow -\infty, \beta \uparrow \infty} (z, E(\alpha, \beta] y^*) = \int_{-\infty}^{\infty} f(\lambda)\, d(E(\lambda) z, y) = F(z)$$

is a bounded linear functional. Hence, by the preceding theorem,

$$\int_{-\infty}^{\infty} |f(\lambda)|^2 d\, \|E(\lambda) y\|^2 < \infty, \text{ that is, } y \in D.$$

Therefore, $D = D(H) \supseteq D(H^*)$. Since H is a symmetric operator, we have $H \subseteq H^*$ and so H must be self-adjoint, i.e., $H = H^*$.

Finally, let $x \in D(H)$. Then, by applying $E(\mu)$ to the approximate Riemann sums of $Hx = \int_{-\infty}^{\infty} f(\lambda)\, dE(\lambda)\, x$, we obtain, by (1),

$$\begin{aligned} E(\mu) Hx &= \int_{-\infty}^{\infty} f(\lambda)\, d(E(\mu) E(\lambda) x) \\ &= \int_{-\infty}^{\infty} f(\lambda)\, d(E(\lambda) E(\mu) x) = HE(\mu) x. \end{aligned}$$

Corollary 1. In the particular case $f(\lambda) = \lambda$, we have

$$(Hx, y) = \int_{-\infty}^{\infty} \lambda\, d(E(\lambda) x, y), \quad \text{for} \quad x \in D(H),\ y \in X. \tag{10}$$

We shall write it symbolically

$$H = \int_{-\infty}^{\infty} \lambda\, dE(\lambda),$$

and call it the *spectral resolution* or the *spectral representation* of the self-adjoint operator H.

Corollary 2. We have, for $H = \int_{-\infty}^{\infty} f(\lambda)\, dE(\lambda)$ given by (9),

$$\|Hx\|^2 = \int_{-\infty}^{\infty} |f(\lambda)|^2 d\, \|E(\lambda) x\|^2 \quad \text{whenever} \quad x \in D(H). \tag{11}$$

In particular, if H is a bounded self-adjoint operator, then

$$(H^n x, y) = \int_{-\infty}^{\infty} f(\lambda)^n \, d(E(\lambda)\, x, y) \quad \text{for} \quad x, y \in X \ (n = 0, 1, 2, \ldots). \tag{12}$$

Proof. Since $E(\lambda)\, Hx = HE(\lambda)\, x$ for $x \in D(H)$, we have, by (1),

$$\begin{aligned}(Hx, Hx) &= \int f(\lambda)\, d(E(\lambda)\, x, Hx) = \int f(\lambda)\, d(HE(\lambda)\, x, x) \\ &= \int f(\lambda)\, d_\lambda \{\int f(\mu)\, d_\mu (E(\mu)\, E(\lambda)\, x, x)\} \\ &= \int f(\lambda)\, d_\lambda \left\{\int_{-\infty}^{\lambda} f(\mu)\, d(E(\mu)\, x, x)\right\} = \int f(\lambda)^2\, d\, \|E(\lambda)\, x\|^2.\end{aligned}$$

The last part of the Corollary may be proved similarly.

Example. It is easy to see that the multiplication operator

$$Hx(t) = tx(t) \quad \text{in} \quad L^2(-\infty, \infty)$$

admits the spectral resolution $H = \int_{-\infty}^{\infty} \lambda \, dE(\lambda)$, where

$$\begin{aligned} E(\lambda)\, x(t) &= x(t) \quad \text{for} \quad t \leqq \lambda, \\ &= 0 \quad \text{for} \quad t > \lambda. \end{aligned} \tag{13}$$

For,

$$\int_{-\infty}^{\infty} \lambda^2\, d\, \|E(\lambda)\, x\|^2 = \int_{-\infty}^{\infty} \lambda^2\, d_\lambda \int_{-\infty}^{\lambda} |x(t)|^2\, dt = \int_{-\infty}^{\infty} t^2\, |x(t)|^2\, dt = \|Hx\|^2,$$

$$\int_{-\infty}^{\infty} \lambda\, d(E(\lambda)\, x, y) = \int_{-\infty}^{\infty} \lambda\, d_\lambda \int_{-\infty}^{\lambda} x(t)\, \overline{y(t)}\, dt = \int_{-\infty}^{\infty} t \cdot x(t)\, \overline{y(t)}\, dt = (Hx, y).$$

6. The Spectral Resolution of a Self-adjoint Operator

Theorem 1. A self-adjoint operator H in a Hilbert space X admits a uniquely determined spectral resolution.

Proof. The Cayley transform $U = U_H = (H - iI)\,(H + iI)^{-1}$ of the self-adjoint operator H is unitary (see Chapter VII, 4). Let $U = \int_0^{2\pi} e^{i\theta}\, dF(\theta)$ be the spectral resolution of U. Then we have

$$F(2\pi - 0) = \text{s-}\lim_{\theta \downarrow 0} F(2\pi - \theta) = F(2\pi) = I.$$

If otherwise, the projection $F(2\pi) - F(2\pi - 0)$ would not be equal to the zero operator. Thus there exists an element $y \neq 0$ such that

$$(F(2\pi) - F(2\pi - 0))\, y = y.$$

Hence, by $F(\theta)\, F(\theta') = F(\min(\theta, \theta'))$,

$$U y = \int_0^{2\pi} e^{i\theta} d\left(F(\theta)\,(F(2\pi) - F(2\pi - 0))\right) y$$

$$= (F(2\pi) - F(2\pi - 0))\, y = y.$$

Thus $(y, z) = (Uy, Uz) = (y, Uz)$ and so $(y, z - Uz) = 0$ for every $z \in X$. U being the Cayley transform of a self-adjoint operator H, we know (see Chapter VII, 4) that the range $R(I - U)$ is dense in X. Hence we must have $y = 0$, which is a contradiction.

Thus, if we set

$$\lambda = -\cot\theta, \quad E(\lambda) = F(\theta),$$

then $0 < \theta < 2\pi$ and $-\infty < \lambda < \infty$ are in a topological correspondence. Hence $E(\lambda)$ is a resolution of the identity with $F(\theta)$. We shall show that the self-adjoint operator

$$H' = \int_{-\infty}^{\infty} \lambda\, dE(\lambda)$$

is equal to H. Since $H = i(I + U)(I - U)^{-1}$, we have only to show that

$$(H'(y - Uy), x) = (i(y + Uy), x) \quad \text{for all} \quad x, y \in X.$$

But, since $D(H')^a = X$, we may restrict x to be in the domain $D(H')$. Now, by $F(\theta) \cdot F(\theta') = F(\min(\theta, \theta'))$,

$$(y - Uy, F(\theta)\, x) = \int_0^{2\pi} (1 - e^{i\theta'})\, d_{\theta'}(F(\theta')\, y, F(\theta)\, x)$$

$$= \int_0^{2\pi} (1 - e^{i\theta'})\, d_{\theta'}(F(\theta)\, F(\theta')\, y, x)$$

$$= \int_0^{\theta} (1 - e^{i\theta'})\, d(F(\theta')\, y, x).$$

Hence

$$(y - Uy, H'x) = \int_{-\infty}^{\infty} \lambda\, d(y - Uy, E(\lambda)\, x)$$

$$= \int_0^{2\pi} -\cot\theta\, d\left\{\int_0^{\theta} (1 - e^{i\theta'})\, d(F(\theta')\, y, x)\right\}$$

$$= \int_0^{2\pi} i(1 + e^{i\theta})\, d(F(\theta)\, y, x) = (i(y + Uy), x).$$

The uniqueness of the spectral representation. Suppose $H = \int_{-\infty}^{\infty} \lambda\, dE(\lambda)$ admits another spectral representation $H = \int_{-\infty}^{\infty} \lambda\, dE'(\lambda)$ such that

$E'(\lambda_0) \neq E(\lambda_0)$ for some λ_0. Then, by setting

$$\lambda = -\cot\theta, \; E'(\lambda) = F'(\theta),$$

we have $F'(\theta_0) \neq F(\theta_0)$, where $\lambda_0 = -\cot\theta_0$. By a similar calculation to the above, we can prove that the Cayley transform of $\int_{-\infty}^{\infty} \lambda \, dE'(\lambda)$ is equal to $\int_0^{2\pi} e^{i\theta} \, dF'(\theta)$. Hence the unitary operator U admits two different spectral representations $U = \int_0^{2\pi} e^{i\theta} \, dF(\theta)$ and $U = \int_0^{2\pi} e^{i\theta} \, dF'(\theta)$, contrary to what we have proved in Chapter XI, 4.

We have thus proved (see Chapter VII, 3 and 4) the fundamental result due to J. VON NEUMANN [1]:

Theorem 2. A symmetric operator H has a closed symmetric extension H^{**}. A closed symmetric operator H admits a uniquely determined spectral representation iff H is self-adjoint. H is self-adjoint iff its Cayley transform is unitary.

Remark. It sometimes happens, in applications, that H is not self-adjoint but H^* is self-adjoint. In such a case, H is said to be essentially self-adjoint. In this connection, see T. KATO [7] concerning Schrödinger's operators in quantum mechanics.

The spectral representation of the momentum operator H_1:

$$H_1 x(t) = \frac{1}{i}\frac{d}{dt} x(t) \quad \text{in} \quad L^2(-\infty, \infty).$$

The Fourier transform U defined by

$$x(t) = U \cdot y(s) = \underset{n\to\infty}{\text{l.i.m.}} \, (2\pi)^{-1/2} \int_{-n}^{n} e^{ist} y(s) \, ds$$

is unitary and $U^{-1}x(t) = U^* x(t) = Ux(-t)$. Hence, denoting by $E(\lambda)$ the resolution of the identity given by (13) in Chapter XI, 5, we obtain a resolution of the identity $\{E'(\lambda)\}$, $E'(\lambda) = UE(\lambda)\,U^{-1}$. If both $y(s)$ and $sy(s)$ belong to $L^2(-\infty, \infty) \cap L^1(-\infty, \infty)$, then

$$\frac{1}{i}\frac{d}{dt} x(t) = \frac{1}{i}\frac{d}{dt}\left((2\pi)^{-1/2} \int_{-\infty}^{\infty} e^{ist} y(s) \, ds\right)$$

$$= (2\pi)^{-1/2} \int_{-\infty}^{\infty} e^{ist} s y(s) \, ds = U(sy(s)) = UsU^{-1}x(t),$$

or, symbolically,

$$\frac{1}{i}\frac{d}{dt} = UsU^{-1}. \tag{1}$$

Hence, for the self-adjoint operator $H = s \cdot = \int_{-\infty}^{\infty} \lambda \, dE(\lambda)$, we have

$$\begin{cases} U^{-1}H_1 U \cdot y(s) = s \cdot y(s) = Hy(s) & \text{whenever } y(s), \\ sy(s) \text{ both belong to } L^2(-\infty,\infty) \cap L^1(-\infty,\infty). \end{cases}$$

For any $y(s) \in D(H) = D(s \cdot)$, set

$$y_n(s) = y(s) \text{ or } = 0 \text{ according as } |s| \leqq n \text{ or } |s| > n.$$

Then surely $y_n(s)$, $sy_n(s)$ both $\in L^2(-\infty,\infty) \cap L^1(-\infty,\infty)$ and, moreover, $\text{s-}\lim_{n\to\infty} y_n = y$, $\text{s-}\lim_{n\to\infty} Hy_n = Hy$. Thus, since the self-adjoint operators $U^{-1}H_1U$ and H are closed, we have, by $(U^{-1}H_1U)\, y_n = Hy_n$,

$$(U^{-1}H_1U)\, y = Hy \quad \text{whenever} \quad y \in D(H),$$

that is, $U^{-1}H_1U$ is a self-adjoint extension of the self-adjoint operator H. Hence, by taking the adjoint, we see that $H^* = H$ is also an extension of $(U^{-1}H_1U)^* = U^{-1}H_1U$; consequently $U^{-1}H_1U = H$ and so

$$H_1 = UHU^{-1} = \int_{-\infty}^{\infty} \lambda d(UE(\lambda)\, U^{-1}) = \int_{-\infty}^{\infty} \lambda \, dE'(\lambda).$$

7. Real Operators and Semi-bounded Operators. Friedrichs' Theorem

Real operators and *semi-bounded operators*, defined below, have self-adjoint extensions. Thus we can apply von Neumann's theorem to these extensions to the effect that they admit spectral resolutions.

Definition 1. Let $X = L^2(S, \mathfrak{B}, m)$ and let H be a symmetric operator defined in X into X. H is said to be a *real operator*, if i) $x(s) \in D(H)$ then $\overline{x(s)} \in D(H)$, and ii) H maps real-valued functions into real-valued functions.

Example. Let $f(s)$ be a real-valued continuous function in $(-\infty,\infty)$. Then, for $X = L^2(-\infty,\infty)$, the operator of multiplication by $f(s)$ is a real operator.

Theorem 1 (J. von Neumann [1]). A real operator H admits a self-adjoint extension.

Proof. Let $U = U_H$ be the Cayley transform of H. Then $D(U) = \{(H + iI)\, x;\, x \in D(H)\}$ consists of the functions obtained by taking the complex-conjugate of the functions of $R(U) = \{(H - iI)\, x;\, x \in D(H)\}$. Thus, if we define an extension U_1 of U through

$$U_1 = U \text{ in } D(U),$$

$$U_1\Big(\sum_\alpha c_\alpha \varphi_\alpha\Big) = \sum_\alpha c_\alpha \bar{\varphi}_\alpha, \text{ where } \{\varphi_\alpha\} \text{ is a complete orthonormal system of the Hilbert space } D(U)^\perp,$$

then U_1 is a unitary extension of U. Therefore the self-adjoint extension H_1 of H exists such that $U_1 = U_{H_1}$ (see Chapter VII, 4).

Definition 2. A symmetric operator H is said to be *upper semi-bounded* (or *lower semi-bounded*) if there exists a real constant α such that

$$(Hx, x) \leqq \alpha \|x\|^2 \text{ (or } (Hx, x) \geqq \alpha \|x\|^2) \text{ for all } x \in D(H).$$

If $(H x, x) \geqq 0$ for all $x \in D(H)$, then H is said to be a *positive operator*.

Example. Let $q(s)$ be continuous and non-negative in $(-\infty, \infty)$. Consider the operator H defined for C^2 functions $x(s)$ with compact support by

$$(Hx)(s) = -x''(s) + q(s)\, x(s).$$

Then H is a positive operator in the Hilbert space $L^2(-\infty, \infty)$, as may be verified by partial integration.

Theorem 2 (K. Friedrichs [3]). A semi-bounded operator H admits a self-adjoint extension.

Proof (due to H. Freudenthal [1]). —H is lower semi-bounded if H is upper semi-bounded. If H is lower semi-bounded as above, then $H_1 = H + (1-\alpha) I$ satisfies the condition that $(H_1 x, x) \geqq \|x\|^2$ for all $x \in D(H_1)$. Therefore, since αI is self-adjoint, we may assume that the symmetric operator H satisfies the condition

$$(Hx, x) \geqq \|x\|^2 \text{ for all } x \in D(H). \tag{1}$$

We introduce a new norm $\|x\|'$ and the associated new scalar product $(x, y)'$ in $D(H)$ through

$$\|x\|' = (Hx, x), \quad (x, y)' = (Hx, y). \tag{2}$$

Since H is symmetric and satisfies (1), it is easy to see that $D(H)$ becomes a pre-Hilbert space with respect to $\|x\|'$ and $(x, y)'$. We denote by $D(H)'$ the completion of this pre-Hilbert space.

We shall show that $D(H)'$ is, as an abstract set without topology, a subset of the set X which is the original Hilbert space. Proof: A Cauchy sequence $\{x_n\}'$ of the pre-Hilbert space $D(H)$ satisfies $\|x_n - x_m\|' \geqq \|x_n - x_m\|$ and $\lim_{n,m\to\infty} \|x_n - x_m\|' = 0$; consequently $\{x_n\}$ is also a Cauchy sequence of the original Hilbert space X. If we can show, for a Cauchy sequence $\{y_n\}$ of $D(H)'$, that

$$\lim_{n\to\infty} \|y_n\|' \neq 0 \text{ does not imply } \lim_{n\to\infty} \|y_n\| = 0, \tag{3}$$

then the correspondence

$$\{x_n\}' \to \{x_n\} \tag{4}$$

is one-one from the Cauchy sequences of $D(H)$ into the Cauchy sequences of X. Two Cauchy sequences $\{x_n\}'$, $\{z_n\}'$ of $D(H)$ (of X) being identified if $\lim_{n\to\infty} \|x_n - z_n\|' = 0$ $\left(\text{if } \lim_{n\to\infty} \|x_n - z_n\| = 0\right)$. Since X is complete, we may thus identify its Cauchy sequence $\{x_n\}$ with the element $x \in X$ such that $\lim_{n\to\infty} ||x_n - x|| = 0$. Therefore, $D(H)'$ may, as an abstract set without topology, be identified with a subset of X by the correspondence (4). The proof of (3) is obtained, remembering the continuity of the scalar product in $D(H)'$ and in X, as follows. $\lim_{n,m\to\infty} ||x_n - x_m||' = 0$, $\lim_{n\to\infty} ||x_n||' = \alpha > 0$ and $\lim_{n\to\infty} ||x_n|| = 0$ imply a contradiction

$$\alpha^2 = \lim_{n,m\to\infty} (x_n, x_m)' = \lim_{n,m\to\infty} (Hx_n, x_m) = \lim_{n\to\infty} (Hx_n, 0) = 0.$$

We next set

$$\tilde{D} = D(H^*) \cap D(H)'. \tag{5}$$

Since $D(H) \subseteqq D(H^*)$, we must have $D(H) \subseteqq \tilde{D} \subseteqq D(H^*)$. Hence we can define an extension $\tilde{H}$ of H by restricting H^* to the domain $\tilde{D} = D(\tilde{H})$. We have to show that $\tilde{H}$ is self-adjoint.

We first show that $\tilde{H}$ is symmetric. Suppose $x, y \in \tilde{D}$; there exist two sequences $\{x_n\}'$, $\{y_n\}'$ of $D(H)$ such that $||x - x_n||' \to 0$, $||y - y_n||' \to 0$ as $n \to \infty$. Hence, by the continuity of the scalar product in $D(H)'$, we see that a finite $\lim_{n,m\to\infty} (x_n, y_m)' = \lim_{n,m\to\infty} (Hx_n\, y_m), = \lim_{n,m\to\infty} (x_n, \tilde{H}y_m)$ exists. This limit is equal to

$$\lim_{n\to\infty,}\lim_{m\to\infty} (Hx_n, y_m) = \lim_{n\to\infty} (Hx_n, y) = \lim_{n\to\infty} (x_n, \tilde{H}y) = (x, \tilde{H}y)$$

and also to

$$\lim_{m\to\infty}\lim_{n\to\infty} (Hx_n, y_m) = \lim_{m\to\infty} (x, Hy_m) = \lim_{m\to\infty} (\tilde{H}x, y_m) = (\tilde{H}x, y).$$

Hence $\tilde{H}$ is symmetric, that is, $\tilde{H} \subseteqq (\tilde{H})^*$.

Next let $x \in D(H)$, $y \in X$. Then, by

$$|(x, y)| \leqq ||x||\, ||y|| \leqq ||x||' \cdot ||y||,$$

we see that $f(x) = (x, y)$ is a bounded linear functional on the pre-Hilbert space $D(H)$. Hence $f(x)$ can, by continuity, be extended to a bounded linear functional on the Hilbert space $D(H)'$. Hence, by F. Riesz' representation theorem as applied to the Hilbert space $D(H)'$, there exists a uniquely determined $y' \in D(H)'$ such that

$$f(x) = (x, y) = (x, y')' = (Hx, y') \quad \text{for all} \quad x \in D(H).$$

This proves that $y' \in D(H^*)$ and $H^*y' = y$. Hence $y' \in \tilde{D}$ and $\tilde{H}y' = y$. We have thus proved that $R(\tilde{H}) = X$, and so, by the Corollary of Theorem 1 in Chapter VII, 3, $\tilde{H}$ must be self-adjoint.

8. The Spectrum of a Self-adjoint Operator. Rayleigh's Principle and the Krylov-Weinstein Theorem. The Multiplicity of the Spectrum

Theorem 1. Let $H = \int \lambda \, dE(\lambda)$ be a self-adjoint operator in a Hilbert space X. Let $\sigma(H)$, $P_\sigma(H)$, $C_\sigma(H)$ and $R_\sigma(H)$ be the spectrum, the point spectrum, the continuous spectrum and the residual spectrum of H, respectively. Then (i) $\sigma(H)$ is a set on the real line; (ii) $\lambda_0 \in P_\sigma(H)$ is equivalent to the condition $E(\lambda_0) \neq E(\lambda_0 - 0)$ and the eigenspace of H corresponding to the eigenvalue λ_0 is $R(E(\lambda_0) - E(\lambda_0 - 0))$; (iii) $\lambda_0 \in C_\sigma(H)$ is equivalent to the condition $E(\lambda_0) = E(\lambda_0 - 0)$ in such a way that $E(\lambda_1) \neq E(\lambda_2)$ whenever $\lambda_1 < \lambda_0 < \lambda_2$; (iv) $R_\sigma(H)$ is void.

Proof. We know already that, for a self-adjoint operator H, the resolvent set $\varrho(H)$ of H comprises all the complex numbers λ with $\mathrm{Im}(\lambda) \neq 0$ (see Chapter VIII, 1). Hence (i) is clear. We have $\lambda_0 I = \lambda_0 \int_{-\infty}^{\infty} dE(\lambda)$ by the definition of the resolution of the identity $\{E(\lambda)\}$ and so $(H - \lambda_0 I) = \int_{-\infty}^{\infty} (\lambda - \lambda_0) \, dE(\lambda)$. Hence, as in Corollary 2 of Theorem 2 in Chapter XI, 5, we obtain

$$\|(H - \lambda_0 I)\, x\|^2 = \int_{-\infty}^{\infty} (\lambda - \lambda_0)^2 \, d\, \|E(\lambda)\, x\|^2, \quad x \in D(H). \tag{1}$$

Thus, by $E(-\infty) = 0$ and the right continuity of $\|E(\lambda)\, x\|^2$ in λ, we see that $Hx = \lambda_0 x$ iff

$$\begin{cases} E(\lambda)\, x = E(\lambda_0 + 0)\, x = E(\lambda_0)\, x & \text{for } \lambda \geqq \lambda_0 \text{ and} \\ E(\lambda)\, x = E(\lambda_0 - 0)\, x = 0 & \text{for } \lambda < \lambda_0, \end{cases}$$

that is, $Hx = \lambda_0 x$ iff $(E(\lambda_0) - E(\lambda_0 - 0))\, x = x$. This proves (ii). Next we shall prove (iv). If $\lambda_0 \in R_\sigma(H)$, then, by (i), λ_0 is a real number. By the condition $R(H - \lambda_0 I)^a = D((H - \lambda_0 I)^{-1})^a \neq X$, we see that there exists a $y \neq 0$ which is orthogonal to $R(H - \lambda_0 I)$, i.e., $((H - \lambda_0 I)\, x, y) = 0$ for all $x \in D(H)$. Hence $(Hx, y) = (\lambda_0 x, y) = (x, \lambda_0 y)$ and so $y \in D(H^*)$, $H^* y = \lambda_0 y$. This proves that $Hy = \lambda_0 y$, i.e., λ_0 is an eigenvalue of H. Hence we have obtained a contradiction $\lambda_0 \in R_\sigma(H) \cap P_\sigma(H)$, and so $R_\sigma(H)$ must be a void set.

Let λ_0 be a real number not belonging to $\sigma(H)$. Then the resolvent $(\lambda_0 I - H)^{-1}$ exists. Hence $H_{\lambda_0} = (H - \lambda_0 I)$ has a continuous inverse $(H - \lambda_0 I)^{-1}$. The last condition is, by (iv), equivalent to $\lambda_0 \in \varrho(H)$ and to the condition that there exists a positive number α such that

$$\|(H - \lambda_0 I)\, x\| \geqq \alpha \|x\| \quad \text{for all} \quad x \in D(H).$$

This condition is by (1) equivalent to

$$\int_{-\infty}^{\infty} (\lambda - \lambda_0)^2 \, d \, ||E(\lambda)\, x||^2 \geqq \alpha^2 \, ||x||^2 \quad \text{for all} \quad x \in D(H). \tag{2}$$

Suppose that we have, for $\lambda_1 < \lambda_0 < \lambda_2$ with $\lambda_0 - \lambda_1 = \lambda_2 - \lambda_0 < \alpha$, $E(\lambda_1) = E(\lambda_2)$. Then, contrary to (2), we obtain

$$\int_{-\infty}^{\infty} (\lambda - \lambda_0)^2 \, d \, ||E(\lambda)\, x||^2 < \alpha^2 \int_{-\infty}^{\infty} d \, ||E(\lambda)\, x||^2 = \alpha^2 \, ||x||^2.$$

This proves (iii) by (i), (ii), (iv) and (2).

Remark. The Example in Chapter XI, 5 gives a self-adjoint operator H for which all real numbers are in the continuous spectrum of H.

Theorem 2. Let H be a bounded self-adjoint operator. Then

$$\sup_{\lambda \in \sigma(H)} \lambda = \sup_{||x|| \leqq 1} (Hx, x), \quad \inf_{\lambda \in \sigma(H)} \lambda = \inf_{||x|| \leqq 1} (Hx, x). \tag{3}$$

Proof. Since $(Hx, x) = (x, Hx) = \overline{(Hx, x)} =$ real number, we can consider

$$\alpha_1 = \inf_{||x|| \leqq 1} (Hx, x) \quad \text{and} \quad \alpha_2 = \sup_{||x|| \leqq 1} (Hx, x).$$

Let $\lambda_0 \in \sigma(H)$. Then, by Theorem 1, there exists, for every pair (λ_1, λ_2) of real numbers with $\lambda_1 < \lambda_0 < \lambda_2$, a $y = y_{\lambda_1, \lambda_2} \neq 0$ such that $(E(\lambda_2) - E(\lambda_1))\, y = y$. We may assume that $||y|| = 1$. Hence

$$\begin{aligned}(Hy, y) &= \int \lambda \, d\,(E(\lambda)\, y, y) = \int \lambda \, d\, ||E(\lambda)\, y||^2 \\ &= \int \lambda \, d\, ||E(\lambda)\, (E(\lambda_2) - E(\lambda_1))\, y||^2 \\ &= \int_{\lambda_1}^{\lambda_2} \lambda \, d\, ||(E(\lambda) - E(\lambda_1))\, y||^2.\end{aligned}$$

Thus, by letting $\lambda_1 \uparrow \lambda_0$, $\lambda_2 \downarrow \lambda_0$, we obtain $\lim (Hy_{\lambda_1,\lambda_2}, y_{\lambda_1,\lambda_2}) = \lambda_0$. This proves that $\sup_{\lambda \in \sigma(H)} \lambda = \sup \lambda_0 \leqq \alpha_2$.

Let us assume that $\alpha_2 \overline{\in} \sigma(H)$. Then, by Theorem 1, there exists a pair (λ_1, λ_2) of real numbers such that $\lambda_1 < \alpha_2 < \lambda_2$ and $E(\lambda_2) = E(\lambda_1)$. Hence $I = I - E(\lambda_2) + E(\lambda_1)$, $(I - E(\lambda_2))\, E(\lambda_1) = E(\lambda_1)\, (I - E(\lambda_2)) = 0$ and so either $(I - E(\lambda_2))$ or $E(\lambda_1)$ is not equal to the zero operator. If $(I - E(\lambda_2)) \neq 0$, then there exists a y with $||y|| = 1$, $(I - E(\lambda_2))\, y = y$. In this case, we have

$$\begin{aligned}(Hy, y) &= \int \lambda \, d\, ||E(\lambda)\, y||^2 = \int \lambda \, d\, ||E(\lambda)\, (I - E(\lambda_2))\, y||^2 \\ &= \int_{\lambda_2}^{\infty} \lambda \, d\, ||E(\lambda)\, y||^2 \geqq \lambda_2 > \alpha_2;\end{aligned}$$

and in the case $E(\lambda_1) \neq 0$, we obtain

$$(Hz, z) \leqq \lambda_1 < \alpha_2 \quad \text{for a } z \text{ with } ||z|| = 1, \; E(\lambda_1)\, z = z.$$

Therefore the assumption $\alpha_2 \bar{\in} \sigma(H)$ is absurd and so we have proved that $\sup_{\lambda\in\sigma(H)} \lambda = \sup_{||x||\leq 1} (Hx, x)$. Similarly, we can prove that $\inf_{\lambda\in\sigma(H)} \lambda = \inf_{||x||\leq 1} (Hx, x)$.

Theorem (KRYLOV-WEINSTEIN). Let H be self-adjoint, and define, for any $x \in D(H)$ with $||x|| = 1$,

$$\alpha_x = (Hx, x), \ \beta_x = ||Hx||. \tag{4}$$

Then, for any $\varepsilon > 0$, we can find a $\lambda_0 \in \sigma(H)$ satisfying the inequality

$$\alpha_x - (\beta_x^2 - \alpha_x^2)^{1/2} - \varepsilon \leqq \lambda_0 \leqq \alpha_x + (\beta_x^2 - \alpha_x^2)^{1/2} + \varepsilon. \tag{5}$$

Proof. We have

$$\beta_x^2 = (Hx, Hx) = (H^2 x, x) = \int \lambda^2 \, d\, ||E(\lambda)\, x||^2,$$

$$\alpha_x = (Hx, x) = \int \lambda \, d\, ||E(\lambda)\, x||^2,$$

$$||x||^2 = \int d\, ||E(\lambda)\, x||^2,$$

and so

$$\beta_x^2 - \alpha_x^2 = \int \lambda^2 \, d\, ||E(\lambda)\, x||^2 - 2\alpha_x \int \lambda \, d\, ||E(\lambda)\, x||^2 + \alpha_x^2 \int d\, ||E(\lambda)\, x||^2$$
$$= \int (\lambda - \alpha_x)^2 \, d\, ||E(\lambda)\, x||^2.$$

Therefore, if $||E(\lambda)\, x||^2$ does not vary in the interval given by (5), we would obtain a contradiction

$$\beta_x^2 - \alpha_x^2 \geqq ((\beta_x^2 - \alpha_x^2)^{1/2} + \varepsilon)^2 > \beta_x^2 - \alpha_x^2.$$

Remark. The so-called *Rayleigh principle* consists in taking α_x as an approximation to the spectrum of the operator H. If we calculate β_x, then Theorem 3 gives an upper bound of the error when we take α_x as an approximation of the spectrum of H. For a concrete application of such error estimate, we refer the reader to K. YOSIDA [1].

The Multiplicity of the Spectrum. We shall begin with the cace of a self-adjoint matrix $H = \int \lambda \, dE(\lambda)$ in an n-dimensional Hilbert space X_n. Let $\lambda_1, \lambda_2, \ldots, \lambda_p$ ($p \leqq n$) be the eigenvalues of H with the multiplicity $m_1, m_2, \ldots, m_p$, respectively $\left(\sum_{j=1}^{p} m_j = n\right)$. Let $x_{j_1}, x_{j_2}, \ldots, x_{j m_j}$ be orthonormal eigenvectors of H belonging to the eigenvalue λ_j ($H x_{j_s} = \lambda_j x_{j_s}$) so that $\{x_{j_s};\, s = 1, 2, \ldots, m_j\}$ spans the eigenspace $E_{\lambda_j} = R(E(\lambda_j) - E(\lambda_j - 0))$ of H belonging to the eigenvalue λ_j. Then the set $\{x_{j_s};\, j = 1, 2, \ldots, p$ and $s = 1, 2, \ldots, m_j\}$ is a complete orthonormal system of vectors of the space X_n, and hence every vector y of X_n is represented uniquely as the linear combination of x_{j_s}'s:

$$y = \sum_{j=1}^{p} \sum_{s=1}^{m_j} \alpha_{j_s} x_{j_s}. \tag{6}$$

Thus, denoting by P_{λ_j} the projection $(E(\lambda_j) - E(\lambda_j - 0))$ upon the eigenspace E_{λ_j}, we have, for any $\alpha < \beta$,

$$(E(\beta) - E(\alpha))\, y = \sum_{\alpha < \lambda_j \leqq \beta} \left(\sum_{s=1}^{m_j} \alpha_{j_s} x_{j_s} \right) = \sum_{\alpha < \lambda_j \leqq \beta} P_{\lambda_j} y, \tag{7}$$

$$P_{\lambda_j}(E(\beta) - E(\alpha))\, y = \sum_{s=1}^{m_j} \alpha_{j_s} x_{j_s} \text{ or } = 0 \text{ according as } \alpha < \lambda_j \leqq \beta \text{ or not.} \tag{8}$$

Hence, for fixed $\alpha < \beta$ and a fixed linear subspace M of X, the set

$$\{(E(\beta) - E(\alpha))\, y;\, y \in M\}$$

does not contain E_{λ_j} if the dimension of M, $\dim(M)$, is $< m_j$. Moreover, for a suitable M with $\dim(M) = m_j$, the set $\{(E(\beta) - E(\alpha))\, y;\ y \in M\}$ with $\alpha < \lambda_j \leqq \beta$ contains E_{λ_j}. In fact, the statement is true for M containing $x_{j_1}, x_{j_2}, \ldots, x_{j_{m_j}}$. In particular, $m_1 = m_2 = \cdots = m_p = 1$ with $p = n$ iff there exists a fixed vector $y \in X_n$ such that the set of vectors

$$\{(E(\beta) - E(\alpha))\, y;\, \alpha < \beta\}$$

spans the whole space X_n.

These considerations lead to the following definitions:

Definition 1. The spectrum of a self-adjoint operator $H = \int \lambda\, dE(\lambda)$ in a Hilbert space X is said to be *simple*, if there exists a fixed vector $y \in X$ such that the set of vectors $\{(E(\beta) - E(\alpha))\, y;\, \alpha < \beta\}$ spans a linear subspace strongly dense in X.

Definition 2. Let $H = \int \lambda\, dE(\lambda)$ be a self-adjoint operator in a Hilbert space X. For fixed $\alpha < \beta$, consider the linear subspace M of $(E(\beta) - E(\alpha)) \cdot X$ such that

$$(E(\beta) - E(\alpha)) \cdot M = (E(\beta) - E(\alpha)) \cdot X. \tag{9}$$

We may take, for example, $M = (E(\beta) - E(\alpha)) \cdot X$ to meet condition (9). The minimum of the set of values $\dim(M)$, where M satisfies condition (9), will be called the *total multiplicity of the spectrum* of H contained in the interval $(\alpha, \beta]$.

Definition 3. The *multiplicity of the spectrum of a self-adjoint operator* $H = \int \lambda\, dE(\lambda)$ *at* $\lambda = \lambda_0$ is defined as the limit, as $n \to \infty$, of the total multiplicity of the spectrum of H contained in the interval $(\lambda_0 - n^{-1}, \lambda_0 + n^{-1}]$.

Example. The coordinate operator H, i.e., the operator H defined by $H \cdot x(t) = t \cdot x(t)$ in $L^2(-\infty, \infty)$, is of the simple spectrum.

Proof. We know that the spectral resolution $H = \int \lambda\, dE(\lambda)$ is given by

$$E(\lambda)\, x(t) = x(t) \text{ or } = 0 \text{ according as } t \leqq \lambda \text{ or } t > \lambda.$$

Let $y(t)$ be defined by

$$y(t) = c_k > 0 \text{ for } k - 1 < t \leqq k \quad (k = 0, \pm 1, \pm 2, \ldots)$$
$$\text{with } \sum_k c_k^2 < \infty \text{ so that } y(t) \in L^2(-\infty, \infty).$$

Then it is easy to see that the linear combinations of vectors of the form $(E(\beta) - E(\alpha))\, y, \alpha < \beta$, are strongly dense in the totality of step functions with compact support and consequently strongly dense in $L^2(-\infty, \infty)$.

The Problem of the Unitary Equivalence. Two self-adjoint operators H_1 and H_2 in an n-dimensional Hilbert space X_n are said to be *unitarily equivalent* to each other if there exists a unitary matrix U in X_n such that $H_1 = UH_2U^{-1}$. It is well-known that H_1 and H_2 are unitarily equivalent to each other iff they have the same system of eigenvalues with respectively the same multiplicities. Thus the eigenvalues together with the respective multiplicities are the *unitary invariants* of a self-adjoint matrix.

The investigation of the unitary invariants for a self-adjoint operator in an infinite dimensional Hilbert space goes back to a paper by E. Hellinger [1] published in 1909. See, e.g., M. H. Stone [1]. There the Hilbert space is assumed to be separable. For the non-separable Hilbert space case, see F. Wecken [1] and H. Nakano [1] and also P. R. Halmos [2]. K. Yosida [13] proved the following theorem:

Let H be a self-adjoint operator in a Hilbert space X, and let us denote by $(H)'$ the totality of bounded linear operators $\in L(X, X)$ which are commutative with H. Then two self-adjoint operators H_1 and H_2 in X are unitarily equivalent to each other iff there exists a one-to-one mapping T of $(H_1)'$ onto $(H_2)'$ such that T defines a ring-isomorphism of the ring $(H_1)'$ with the ring $(H_2)'$ in such a way that $(T \cdot B)^* = T \cdot B^*$ for every $B \in (H_1)'$.

Thus the algebraic structure of the ring $(H_1)'$ is the unitary invariant of H_1.

9. The General Expansion Theorem. A Condition for the Absence of the Continuous Spectrum

Let $H = \int \lambda\, dE(\lambda)$ be a self-adjoint operator in a Hilbert space X. Then, by $E(+\infty) = I$ and $E(-\infty) = 0$, we have the representation

$$x = \underset{\lambda_1 \downarrow -\infty, \lambda_2 \uparrow \infty}{s\text{-lim}} \int_{\lambda_1}^{\lambda_2} dE(\lambda)\, x = \underset{\lambda_1 \downarrow -\infty, \lambda_2 \uparrow \infty}{s\text{-lim}} (E(\lambda_2) - E(\lambda_1))\, x \text{ for every } x \in X. \tag{1}$$

We shall call (1) the *general expansion theorem* associated with the self-adjoint operator H. In concrete cases, it sometimes happens that the resolvent $(\lambda I - H)^{-1}$ is obtainable more easily than the spectral resolution $H = \int \lambda\, dE(\lambda)$. In such a case, the general expansion theorem (1) may be replaced by

$$x = \underset{\alpha \downarrow -\infty, \beta \uparrow \infty}{s\text{-lim}} \; \underset{v \downarrow 0}{s\text{-lim}} \frac{1}{2\pi i} \left[\int_\alpha^\beta ((u - iv) I - H)^{-1} x \, du \right.$$

$$\left. + \int_\beta^\alpha ((u + iv) I - H)^{-1} x \, du \right], \quad x \in X. \tag{1'}$$

Proof. If $v \neq 0$, then

$$((u + iv) I - H)^{-1} x = \int_{-\infty}^{\infty} \frac{1}{u + iv - \lambda} dE(\lambda) \, x, \text{ for every } x \in X.$$

For, by approximating the integral by Riemann sums and remembering the relation $E(\lambda) E(\lambda') = E(\min(\lambda, \lambda'))$, we obtain, for $Im(\mu) \neq 0$,

$$\int_{-\infty}^{\infty} (\lambda - \mu) \, dE(\lambda) \left\{ \int_{-\infty}^{\infty} \frac{1}{\lambda' - \mu} dE(\lambda') \right\}$$

$$= \int_{-\infty}^{\infty} (\lambda - \mu) \, d_\lambda \left\{ \int_{-\infty}^{\infty} \frac{1}{(\lambda' - \mu)} d_{\lambda'} (E(\lambda) E(\lambda')) \right\}$$

$$= \int_{-\infty}^{\infty} (\lambda - \mu) \, d_\lambda \left\{ \int_{-\infty}^{\lambda} \frac{1}{\lambda' - \mu} dE(\lambda') \right\} = \int_{-\infty}^{\infty} dE(\lambda) = I,$$

$$\int_{-\infty}^{\infty} \frac{1}{\lambda - \mu} d_\lambda \left\{ E(\lambda) \int_{-\infty}^{\infty} (\lambda' - \mu) \, dE(\lambda') \right\}$$

$$= \int_{-\infty}^{\infty} \frac{1}{\lambda - \mu} d_\lambda \left\{ \int_{-\infty}^{\infty} (\lambda' - \mu) \, d_{\lambda'} (E(\lambda) E(\lambda')) \right\}$$

$$= \int_{-\infty}^{\infty} \frac{1}{\lambda - \mu} d_\lambda \left\{ \int_{-\infty}^{\lambda} (\lambda' - \mu) \, dE(\lambda') \right\} = \int_{-\infty}^{\infty} dE(\lambda) = I.$$

Therefore, we have

$$\int_\alpha^\beta ((u - iv) I - H)^{-1} x \, du + \int_\beta^\alpha ((u + iv) I - H)^{-1} x \, du$$

$$= \int_{-\infty}^{\infty} dE(\lambda) \, x \left\{ \int_\alpha^\beta \frac{du}{u - iv - \lambda} + \int_\beta^\alpha \frac{du}{u + iv - \lambda} \right\}$$

$$= \int_{-\infty}^{\infty} dE(\lambda) \, x \left\{ \int_\alpha^\beta d_u \log(u - iv - \lambda) + \int_\beta^\alpha d_u \log(u + iv - \lambda) \right\}$$

which tends, when $v \downarrow 0$, strongly to

$$\int_{\alpha+0}^{\beta-0} 2\pi i dE(\lambda)\, x + \pi i(E(\beta) - E(\beta-0))\, x + \pi i(E(\alpha) - E(\alpha-0))\, x$$

$$= \pi i(E(\beta) + E(\beta-0))\, x - \pi i(E(\alpha) + E(\alpha-0))\, x.$$

This proves formula (1').

Remark. The eigenfunction expansion associated with the second order differential operator

$$-\frac{d^2}{dx^2} + q(x)$$

with real continuous $q(x)$ in an open interval (a, b) was inaugurated by H. Weyl [2], further developed by M. H. Stone [1], and completed by E. C. Titchmarsh [2] and K. Kodaira [1] who gave a formula which determines the expansion explicitly. The expansion is exactly a concrete application of (1). The crucial point in their theory is to give the possible boundary conditions at $x = a$ and $x = b$ so that the operator

$$-\frac{d^2}{dx^2} + q(x)$$

becomes a self-adjoint operator H in the Hilbert space $L^2(a, b)$. Their theory is very important in that it gives a unified treatment of the classical expansions in terms of special functions, such as the Fourier series expansion, the Fourier integral representation, the Hermite polynomials expansion, the Laguerre polynomials expansion and the Bessel functions expansion. We do not go into details, and refer the reader to the above cited book by Titchmarsh and the paper by Kodaira. Cf. also N. A. Naimark [2], N. Dunford-J. Schwartz [5] and K. Yosida [1]. The last cited book gives an elementary treatment of the theory.

If the continuous spectrum $C_\sigma(H)$ is absent, then the expansion (1) will be replaced by a series rather than the integral. We have, for instance, the following

Theorem 1. Let $H = \int \lambda\, dE(\lambda)$ be a self-adjoint compact operator in a Hilbert space X. Then (i) $C_\sigma(H)$ contains no real number except possibly 0; (ii) the eigenvalues of H constitute at most a countable system of real numbers accumulating only at 0; (iii) for any eigenvalue $\lambda_0 \neq 0$ of H, the corresponding eigenspace E_{λ_0} is of finite dimension.

Proof. Suppose a closed interval $[\lambda', \lambda'']$ on the real line does not contain the number 0. Then the range $R(E(\lambda'') - E(\lambda'))$ is of finite dimension. If otherwise, there exists, by E. Schmidt's orthogonalization in Chapter III, 5, a countable orthonormal system $\{x_j\}$ contained in $R(E(\lambda'') - E(\lambda'))$. We have $w\text{-}\lim_{j\to\infty} x_j = 0$, since, by Bessel's inequality,

$$\sum_{j=1}^{\infty} |(f, x_j)|^2 \leq \|f\|^2 \quad \text{for any} \quad f \in X.$$

Hence, by the compactness of the operator H, there exists a subsequence $\{x_{j'}\}$ such that $s\text{-}\lim_{j\to\infty} Hx_{j'} = w\text{-}\lim_{j\to\infty} Hx_{j'} = 0$. On the other hand, we have

$$\|Hx_j\|^2 = \int \lambda^2 d\,\|E(\lambda)\,x_j\|^2 = \int \lambda^2 d\,\|E(\lambda)\,(E(\lambda'') - E(\lambda'))\,x_j\|^2$$

$$= \int_{\lambda'}^{\lambda''} \lambda^2 d\,\|(E(\lambda) - E(\lambda'))\,x_j\|^2 \geqq \|x_j\|^2 \cdot \min(|\lambda'|^2, |\lambda''|^2)$$

$$= \min(|\lambda'|^2, |\lambda''|^2),$$

which is a contradiction.

If $C_\sigma(H)$ contains a number $\lambda_0 \neq 0$, then, by Theorem 1 in the preceding section, $s\text{-}\lim_{\varepsilon\downarrow 0} (E(\lambda_0 - \varepsilon) - E(\lambda_0 - \varepsilon))\,x = 0$ for any $x \in X$. As proved above, the range $R(E(\lambda_0 + \varepsilon) - E(\lambda_0 - \varepsilon))$ is of finite dimension and this dimension number is monotone decreasing as $\varepsilon \downarrow 0$. Hence we see that $(E(\lambda_0 + \varepsilon) - E(\lambda_0 - \varepsilon)) = 0$ for sufficiently small $\varepsilon > 0$, and so, by Theorem 1 in the preceding section, λ_0 cannot be contained in $C_\sigma(H)$.

This proves our Theorem.

Corollary 1. Let $\{\lambda_j\}$ be the system of all eigenvalues of H different from 0. Then, for any $x \in X$, we have

$$x = (E(0) - E(0-0))\,x + s\text{-}\lim_{n\to\infty} \sum_{j=1}^{n} (E(\lambda_j) - E(\lambda_j - 0))\,x. \tag{2}$$

Proof. Clear from (1).

Corollary 2 (The Hilbert-Schmidt Expansion Theorem). For any $x \in X$, we have

$$Hx = s\text{-}\lim_{n\to\infty} \sum_{j=1}^{n} \lambda_j (E(\lambda_j) - E(\lambda_j - 0))\,x. \tag{3}$$

Proof. Clear from the continuity of the operator H and the fact that $H(E(0) - E(0-0)) = 0$ and $H\big(E(\lambda_j) - E(\lambda_j - 0)\big)\,x = \lambda_j\,(E(\lambda_j) - E(\lambda_j - 0))\,x$, the latter being implied by $R(E(\lambda_j) - E(\lambda_j - 0)) = E_{\lambda_j}$, the eigenspace of H belonging to the eigenvalue λ_j.

Remark. The strong convergence in (3) is uniform in the unit circle $\{x;\ \|x\| \leqq 1\}$. For, we have

$$\left\|Hx - \int_{|\lambda|>\varepsilon} \lambda\,dE(\lambda)\,x\right\|^2 = \left\|\int_{|\lambda|\leqq\varepsilon} \lambda\,dE(\lambda)\,x\right\|^2 = \int_{-\varepsilon}^{\varepsilon} \lambda^2 d\,\|E(\lambda)\,x\|^2$$

$$\leqq \varepsilon^2 \int_{-\varepsilon}^{\varepsilon} d\|E(\lambda)\,x\|^2 \leqq \varepsilon^2 \int d\,\|E(\lambda)\,x\|^2 = \varepsilon^2\,\|x\|^2 \leqq \varepsilon^2.$$

10. The Peter-Weyl-Neumann Theory

Let G be a totally bounded topological group, metrized by a distance satisfying the condition (see Chapter VIII, 5)

$$\operatorname{dis}(x, y) = \operatorname{dis}(axb, ayb) \text{ for every } x, y, a \text{ and } b \in G. \tag{1}$$

Let $f(g)$ be a complex-valued bounded uniformly continuous function defined on G. For any $\varepsilon > 0$, set

$$V = \left\{y \in G; \sup_{x \in G} |f(x) - f(y^{-1}x)| < \varepsilon\right\}. \tag{2}$$

Then, by the continuity of f, the set V is an open set containing the unit e of the group G. Hence the set $U = V \cap V^{-1}$, where $V^{-1} = \{y^{-1}; y \in V\}$, is also open and $\ni e$. If we put

$$\begin{aligned} &k(x) = 2^{-1}(k_1(x) + k_1(x^{-1})), \text{ where} \\ &k_1(x) = \frac{\operatorname{dis}(x, U^C)}{\operatorname{dis}(x, e) + \operatorname{dis}(x, U^C)} \quad \left(\operatorname{dis}(x, U^C) = \inf_{y \in U^C} \operatorname{dis}(x, y)\right), \end{aligned} \tag{3}$$

then we have the results:

$$\begin{aligned} &k(x) \text{ is bounded uniformly continuous on } G, \text{ and} \\ &k(x) = k(x^{-1}), \quad 0 \leqq k(x) \leqq 1 \text{ on } G, \quad k(e) = 1 \text{ and} \\ &k(x) = 0 \text{ whenever } x \in U^C. \end{aligned} \tag{4}$$

Hence, for all $x, y \in G$, we obtain

$$|k(y)\,(f(x) - f(y^{-1}x))| \leqq \varepsilon\, k(y).$$

By taking the mean value (see Chapter VIII, 5) of both sides, we obtain

$$|M_y(k(y))\, f(x) - M_y(k(y)\, f(y^{-1}x))| \leqq \varepsilon\, M_y(k(y)).$$

We have $M_y(k(y)) > 0$ by $k(y) \geqq 0$ and $k(y) \not\equiv 0$. Hence

$$|f(x) - M_y(k_0(y)\, f(y^{-1}x))| \leqq \varepsilon, \quad \text{where} \quad k_0(x) = k(x)/M_x(k(x)). \tag{5}$$

Thus, by virtue of the invariance $M_y(g(y^{-1})) = M_y(g(y)) = M_y(g(ay)) = M_y(g(ya))$ of the mean value, we obtain, from (5),

$$|f(x) - M_y(k_0(xy^{-1})\, f(y))| \leqq \varepsilon \quad \text{for all} \quad x \in G. \tag{6}$$

Proposition 1. We shall denote by $C(G)$ the set of all complex-valued bounded uniformly continuous functions $h(g)$ defined on G. Then $C(G)$ is a B-space by the norm $\|h\|_0 = \sup |h(g)|$. Then, for any b and $h \in C(G)$,

$$(b \times h)\,(x) = M_y(b(xy^{-1})\, h(y)) \tag{7}$$

also belongs to $C(G)$.

Proof. By $\operatorname{dis}(x, z) = \operatorname{dis}(axc, azc)$ and the uniform continuity of the function $b(g)$, there exists, for any $\delta > 0$, an $\eta = \eta(\delta) > 0$ such that

$$\sup_y |b(xy^{-1}) - b(x'y^{-1})| \leqq \delta \quad \text{whenever} \quad \operatorname{dis}(x, x') \leqq \eta.$$

Hence, as in the case of Schwarz' inequality, we obtain

$$\begin{aligned} &|M_y(b(xy^{-1})\, h(y)) - M_y(b(x'y^{-1})\, h(y))|^2 \\ &\leqq M_y((b(xy^{-1}) - b(x'y^{-1}))^2) \cdot M_y(|h(y)|^2) \leqq \delta^2 M_y(|h(y)|^2) \end{aligned} \tag{8}$$

whenever $\operatorname{dis}(x, x') \leqq \eta$. This proves our Proposition 1.

Proposition 2. $C(G)$ is a pre-Hilbert space by the operation of the function sum and the scalar product

$$(b, h) = (b\times h^*)(e) = M_y(b(y^{-1})\overline{h(y^{-1})}) = M_y(b(y)\overline{h(y)}), \text{ where } h^*(y) = \overline{h(y^{-1})}. \text{ We shall denote this pre-Hilbert space by } \hat{C}(G). \tag{9}$$

Proof. Easy.

Proposition 3. We shall denote the completion of the pre-Hilbert space $\hat{C}(G)$ by $\tilde{C}(G)$, and the norm in the Hilbert space $\tilde{C}(G)$ by $\|h\| = (h, h)^{1/2}$. Then the linear mapping T on $\hat{C}(G)$ into $\hat{C}(G)$ defined by

$$(Th)(x) = (k_0\times h)(x), \ x \in G, \tag{10}$$

can, by the continuity in $\hat{C}(G)$, be extended to a compact linear operator $\tilde{T}$ on $\tilde{C}(G)$ into $\tilde{C}(G)$.

Proof. By virtue of $M_y(1) = 1$, we obtain

$$\|h\| = (h, h)^{1/2} = M_y(h(y)\overline{h(y)})^{1/2} \leqq \sup_y |h(y)| = \|h\|_0. \tag{11}$$

The continuity of the operator T in $\hat{C}(G)$ is clear from the Schwarz inequality for (7), and so, by the denseness of $\hat{C}(G)$ in $\tilde{C}(G)$, we can extend T to a bounded linear operator $\tilde{T}$ in $\tilde{C}(G)$. On the other hand, we see, by (8), that T is a compact operator on $\hat{C}(G)$ into $C(G)$. We prove this by the Ascoli-Arzelà theorem. Thus, by (11), we easily see that $\tilde{T}$ is a compact operator on $\hat{C}(G)$ into $\hat{C}(G)$.

Therefore, by the denseness of $\hat{C}(G)$ in $\tilde{C}(G)$, we see that the extended operator $\tilde{T}$ is also compact as an operator on $\tilde{C}(G)$ into $\tilde{C}(G)$.

We are now ready to prove the *Peter-Weyl-Neumann theory* on the representation of almost periodic functions.

It is easy to see that, by $k_0(xy^{-1}) = k_0(yx^{-1})$, the compact operator $\tilde{T}$ is self-adjoint in the Hilbert space $\tilde{C}(G)$. Hence, by the Hilbert-Schmidt expansion theorem in the preceding section, we obtain

$$\tilde{T}h = \underset{n\to\infty}{s\text{-lim}} \sum_{m=1}^{n} \lambda_m P_{\lambda_m} h \text{ uniformly in } h \text{ satisfying } \|h\| \leqq 1, \tag{12}$$

if we denote by $\{\lambda_m\}$ the system of all eigenvalues of $\tilde{T}$ different from 0, and by P_{λ_m} the projection upon the eigenspace of $\tilde{T}$ belonging to the eigenvalue λ_m.

Since $f(g)$, introduced at the beginning of this section, belongs to $C(G)$, we have $\tilde{T}f = Tf \in C(G)$. Since the eigenspace $R(P_{\lambda_m}) = P_{\lambda_m}\cdot\tilde{C}(G)$ is of finite dimension, there exists, for every eigenvalue λ_m, a finite system $\{h_{mj}\}_{j=1,\ldots,n_m}$ of elements $\in \tilde{C}(G)$ such that each $h \in R(P_{\lambda_m}) =$

$P_{\lambda_m} \cdot \tilde{C}(G)$ can be represented as a uniquely determined linear combination of h_{mj}'s ($j = 1, 2, \ldots, n_m$). Let

$$P_{\lambda_m} h = \sum_{j=1}^{n_m} c_j h_{mj}, \text{ where } c\text{'s are complex numbers.} \tag{13}$$

Then, since $h_{mj} \in R(P_{\lambda_m})$, we have $\tilde{T} h_{mj} = \lambda_m h_{mj}$, and so, by applying (8) to the operator T given by (10), we see that $h_{mj} = \lambda_m^{-1} (\tilde{T} h_{mj})$ must belong to $C(G)$. Hence, by (13), we see that, for each eigenvalue λ_m of $\tilde{T}$, the eigenspace $R(P_{\lambda_m}) = P_{\lambda_m} \cdot \tilde{C}(G)$ is spanned by the functions $h_{mj} \in C(G)$.

Therefore, we have, by (12),

$$\left.\begin{aligned} &(Tf)(x) = s\text{-}\lim_{n\to\infty} \sum_{m=1}^{n} \lambda_m f_m(x) \text{ in the strong topology of } \tilde{C}(G), \\ &\text{where } f_m = P_{\lambda_m} \cdot f \in C(G) \text{ for each } m. \end{aligned}\right\}$$

By applying (8) to the operator T given by (10), we see that

$$(T^2 f)(x) = \lim_{n\to\infty} M_y\left(k_0(xy^{-1}) \cdot \sum_{m=1}^{n} \lambda_m f_m(y)\right) \text{ uniformly in } x. \tag{14}$$

On the other hand, by (6) and $M_y(k_0(y)) = 1$, we obtain

$$|M_z(k_0(xz^{-1}) f(z)) - M_z(k_0(xz^{-1}) M_y(k_0(zy^{-1}) f(y))| \leqq \varepsilon,$$

and so, combined with (6), we have

$$|f(x) - M_z(k_0(xz^{-1}) M_y(k_0(zy^{-1}) f(y)))| \leqq 2\varepsilon. \tag{15}$$

The left hand side is precisely $|f(x) - (T^2 f)(x)| \leqq 2\varepsilon$.

Since $T \cdot R(P_{\lambda m}) \subseteqq R(P_{\lambda m})$, we have proved

Theorem 1. The function $f(x)$ can be approximated uniformly on G by linear combinations of the eigenfunctions of $\tilde{T}$ belonging to the eigenvalues which are different from 0.

We shall take a fixed eigenvalue $\lambda \neq 0$ of $\tilde{T}$, and shall denote the base $\{h_j\} \subseteqq C(G)$ of the corresponding eigenspace $P_\lambda \cdot \tilde{C}(G)$ by $e_1(x)$, $e_2(x), \ldots, e_k(x)$. Then, by the invariance of the mean value, we obtain, for any $a \in G$,

$$\begin{aligned} M_y(k_0(xy^{-1}) e_j(ya)) &= M_y(k_0(xa \cdot a^{-1}y^{-1}) e_j(ya)) = M_z(k_0(xa \cdot z^{-1}) e_j(z)) \\ &= (Te_j)(xa) = \lambda e_j(xa). \end{aligned}$$

Since the left hand side is equal to the result of applying the operator T upon the function $e_j(ya)$ of y, we see that, for any given $a \in G$, the function $e_j(xa)$ of x must be uniquely represented as a linear combination of the functions $e_1(x), e_2(x), \ldots, e_k(x)$. We have thus

$$e_j(xa) = \sum_{i=1}^{k} d_{ji}(a)\, e_i(x) \quad (j = 1, 2, \ldots, k), \tag{16}$$

or, in vectorial notation,

$$e(xa) = D(a)\, e(x). \tag{16'}$$

By $e(x \cdot ab) = D(ab)\, e(x)$, $e(xa \cdot b) = D(b)\, e(xa) = D(b)\, D(a)\, e(x)$, we see, remembering the linear independence of $e_1(x), e_2(x), \ldots, e_n(x)$, that

$$D(ab) = D(b)\, D(a),\; D(e) = \text{the unit matrix of degree } k. \tag{17}$$

By applying E. Schmidt's orthogonalization, we may assume that $\{e_j(x)\}$ constitutes an orthonormal system in $\tilde{C}(G)$. Then, by (16),

$$M_x(e_j(xa)\, e_i^*(x)) = d_{ji}(a) \tag{18}$$

and so the elements $d_{ji}(a)$ of the matrix $D(a)$ belong to $C(G)$. By the invariance of the mean value, we see that

$$M_y(e_j(ya)\, \overline{e_i(ya)}) = M_y(e_j(y)\, \overline{e_i(y)}) = \delta_{ij}.$$

Hence the matrix $D(a)$ gives a linear mapping transforming the orthonormal system $\{e_j(x)\}$ onto the orthonormal system $\{e_j(xa)\}$. Therefore $D(a)$ must be a unitary matrix. Hence the tranposed matrix $D(a)'$ of $D(a)$ is also unitary and we have

$$D(ab)' = D(a)'\, D(b)',\; D(e)' = \text{the unit matrix of degree } k. \tag{17'}$$

Hence $D(a)'$ gives a *unitary matrix representation* of the group G such that its matrix elements are continuous functions of a. Letting $x = e$ in (16), we see that each $e_j(a)$ is a linear combination of the matrix elements of the representation $D(a)'$.

We have thus proved

Theorem 2 (Peter-Weyl-Neumann). Let G be a totally bounded topological group, metrized by a distance satisfying $\operatorname{dis}(x, y) = \operatorname{dis}(axb, ayb)$. Let $f(g)$ be any complex-valued bounded uniformly continuous function defined on G. Then, $f(g)$ can be approximated uniformly on G by linear combinations of the matrix elements of unitary uniformly continuous matrix representations $D(g)'$ of the group G.

Referring to A. Weil's reduction given in Chapter VIII, 5, we obtain the following

Corollary. Let G be an abstract group, and $f(g)$ an almost periodic function on G. Then $f(g)$ can be approximated uniformly on G by linear combinations of the matrix elements of unitary matrix representations $D(g)'$ of the group G.

Remark 1. Let the *degree* of a unitary matrix representation $D(g)'$ of the group G be d, i.e., the degree of the matrix $D(g)'$ be d. Then each $D(g)'$ gives a linear mapping of a fixed d-dimensional complex Hilbert space X_d onto itself. The representation $D(g)'$ is said to be *irreducible* if there is no proper linear subspace $\neq \{0\}$ of X_d which is invariant by applying the mappings $D(g)'$, $g \in G$. Otherwise, the representation

$D(g)'$ is called *reducible*, and there exists a proper linear subspace $X_{d,1} \neq \{0\}$ invariant by every $D(g)'$, $g \in G$. Then the orthogonal complement $X_{d,1}^{\perp}$ of $X_{d,1}$ in X_d is, by the unitarity of the representation $D(g)'$, also invariant by every $D(g)$, $g \in G$. If we take for the base of the linear space X_d the orthonormal system of vectors composed of one orthonormal system of vectors of $X_{d,1}$ and one of $X_{d,1}^{\perp}$, then, by this choice of an orthonormal base of X_d, the representation $D(g)'$ will be transformed into the form

$$U D(g)' U^{-1} = \begin{pmatrix} D_1(g)' & 0 \\ 0 & D_2(g)' \end{pmatrix}, \quad \text{where } U \text{ is a fixed unitary matrix.}$$

Hence a reducible unitary representation $D(g)'$ is *completely reducible* into the *sum* of two unitary representations $D_1(g)'$ and $D_2(g)'$ of the group G acting respectively upon $X_{d,1}$ and $X_{d,1}^{\perp}$. In this way, we finally can choose a fixed unitary matrix U_d such that the representation $U_d D(g)' U_d^{-1}$ is the *sum* of irreducible unitary representations of the group G. Therefore, in the statements of Theorem 2 and its Corollary, we can impose the condition that the matrix representations $D(g)'$ are all irreducible.

Remark 2. In the particular case when G is the additive group of real numbers, a unitary irreducible representation $D(g)'$ is given by

$$D(g)' = e^{i\alpha g}, \quad \text{where } \alpha \text{ is a real number and } i = \sqrt{-1}. \tag{19}$$

For, by the commutativity of the unitary matrices $D(g)'$, $g \in G$, the representation $D(g)'$ is completely reducible into the sum of one-dimensional unitary representations $\chi(g)$, that is, complex-valued solutions $\chi(g)$ of

$$\chi(g_1 + g_2) = \chi(g_1) \cdot \chi(g_2), \ |\chi(g_1)| = 1 \ (g_1, g_2 \in G), \chi(0) = 1. \tag{20}$$

It is well known that any continuous solution of (20) is of the form $\chi(g) = e^{i\alpha g}$. Hence, any continuous almost periodic function $f(g)$ on the additive group G of real numbers can be approximated uniformly on G by linear combinations of $e^{i\alpha g}$ with real α's. This constitutes the fundamental theorem in *H. Bohr's theory of almost periodic functions*. According to the original definition by Bohr, a continuous function $f(x)$ $(-\infty < x < \infty)$ is called *almost periodic* if for each $\varepsilon > 0$ there exists a positive number $p = p(\varepsilon)$ such that any interval of the form $(t, t + p)$ contains at least one τ such that

$$|f(x + \tau) - f(x)| \leqq \varepsilon \quad \text{for} \quad -\infty < x < \infty.$$

See H. Bohr [1]. S. Bochner [4] has shown that a continuous function $f(x)$ $(-\infty < x < \infty)$ is almost periodic in Bohr's sense iff the following condition is satisfied: For any sequence of real numbers $\{a_n\}$, the system of functions $\{f_{a_n}(x); f_{a_n}(x) = f(x + a_n)\}$ is totally bounded in the topology of the uniform convergence on $(-\infty, \infty)$. It was extended by J. VON NEUMANN [4] to *almost periodic functions in a group*. Neumann's result contains as a special case, the *Peter-Weyl theory of continuous representation of a compact Lie group* (PETER-WEYL [1]). According to our treatment, Bohr's result is easily proved by obsering that $\lim |s-t| = 0$ implies $\lim \left[\sup_{a,b} |f(asb) - f(atb)|\right] = 0$.

11. Tannaka's Duality Theorem for Non-commutative Compact Groups

Let G be a *compact (topological) group*. This means that G is a compact topological space as well as a group in such a way that the mapping

$$(x, y) \to xy^{-1}$$

of the product space $G \times G$ onto G is continuous.

Proposition 1. A complex-valued continuous function $f(g)$ defined on a compact group G is *uniformly continuous* in the following sense:

$$\left.\begin{array}{l}\text{for any } \varepsilon > 0, \text{ there exists a neighbourhood } U(e) \text{ of} \\ \text{the unit element } e \text{ of } G \text{ such that } |f(x) - f(y)| < \varepsilon \\ \text{whenever } xy^{-1} \in U(e) \text{ and also whenever } x^{-1}y \in U(e).\end{array}\right\} \quad (1)$$

Proof. Since $f(x)$ is continuous at every point $a \in G$, there exists a neighbourhood V_a of a such that $x \in V_a$ implies $|f(x) - f(a)| < \varepsilon/2$. If we denote by U_a the neighbourhood of e defined by $U_a = V_a a^{-1} = \{va^{-1}; v \in V\}$, then $xa^{-1} \in U_a$ implies $|f(x) - f(a)| < \varepsilon/2$. Let us denote by W_a the neighbourhood of e such that $W_a^2 \subseteq U_a$, where $W_a^2 = \{w_1 w_2; w_i \in W_a \ (i = 1, 2)\}$. Obviously, the system of all open sets of the form $W_a \cdot a$, where a is an arbitrary element of G, covers the whole space G. G being compact, there exists a finite set $\{a_i; i = 1, 2, \ldots, n\}$ such that the system of open sets $W_{a_i} \cdot a_i$ $(i = 1, 2, \ldots, n)$ covers G. We denote by $U(e)$ the intersection of all open sets of the system $\{W_{a_i}\}$. Then $U(e)$ is a neighbourhood of e. We shall show that if $xy^{-1} \in U(e)$, then $|f(x) - f(y)| \leqq \varepsilon$. Since the system $W_{a_i} \cdot a_i$ covers G, there exists a number k such that $ya_k^{-1} \in W_{a_k} \subseteq U_{a_k}$ and therefore $|f(y) - f(a_k)| < \varepsilon/2$. Furthermore we have $xa_k^{-1} = xy^{-1} ya_k^{-1} \in U(e)\, W_{a_k} \subseteq W_{a_k}^2 \subseteq U_{a_k}$ so that $|f(x) - f(a_k)| < \varepsilon/2$. Combining these two inequalities we get $|f(x) - f(y)| < \varepsilon$.

If we start with a neighbourhood U_a of e such that $x^{-1}a \in U_a$ implies $|f(x) - f(a)| < \varepsilon/2$, we would obtain a neighbourhood $U(e)$ of e such that $|f(x) - f(y)| < \varepsilon$ whenever $x^{-1}y \in U(e)$. Thus taking the intersection of these two $U(e)$'s as the $U(e)$ in the statement of the Proposition, we complete the proof.

Corollary. A complex-valued continuous function $f(x)$ on a compact group G is almost periodic on G.

Proof. Let $U(e)$ be the neighbourhood of e given in Proposition 1. For any $a \in G$, $U(e)\, a$ is a neighbourhood of a. The compact space G is covered by the system of open sets $U(e)\, a$, $a \in G$, and therefore some finite subsystem $\{U(e)\, a_i;\ i = 1, 2, \ldots, n\}$ covers G. That is, for any $a \in G$, there exists some a_k with $k \leqq n$ such that $a a_k^{-1} \in U(e)$. Hence, by $(ax)(a_k x)^{-1} = a a_k^{-1}$, we have $\sup_x |f(ax) - f(a_k x)| < \varepsilon$. Similarly, we can find a finite system $\{b_j,\ j = 1, 2, \ldots, m\}$ such that, for any $b \in G$, there exists some b_j with $j \leqq m$ satisfying the inequality

$$\sup_{i,x} |f(a_i x b) - f(a_i x b_j)| < \varepsilon.$$

Therefore, for any pair a, b of elements $\in G$, we can find a_k and b_j $(k \leqq n, j \leqq m)$ such that

$$\sup_x |f(axb) - f(a_k x b_j)| < 2\varepsilon.$$

This proves that the system of functions $\{f_{a,b}(x);\ f_{a,b}(x) = f(axb),\ a \text{ and } b \in G\}$ is totally bounded by the maximum norm $\|h\| = \sup_x |h(x)|$. Hence $f(x)$ is almost periodic on G.

We are now able to extend the Peter-Weyl-Neumann Theory of the preceding section to complex-valued continuous functions $f(x)$ on a compact group G. For such a function $f(x)$ and $\varepsilon > 0$, we set

$$V = \left\{y \in G;\ \sup_x |f(x) - f(y^{-1}x)| < \varepsilon\right\}.$$

Then, by the continuity of f, the set V is an open set containing e. By Urysohn's theorem as applied to the normal space G, we see that there exists a continuous function $k_1(x)$ defined on G such that

$$0 \leqq k(x) \leqq 1 \text{ on } G,\ k_1(e) = 1 \quad \text{and} \quad k_1(x) = 0 \text{ whenever } x \in V^C.$$

Then the continuous function

$$k(x) = 2^{-1}(k_1(x) + k_1(x^{-1})) \tag{2}$$

satisfies the condition $k(x^{-1}) = k(x)$ and

$$\left.\begin{array}{l} 0 \leqq k(x) \leqq 1 \text{ on } G,\ k(e) = 1 \quad \text{and} \quad k(x) = 0 \\ \text{whenever } x \in U^C, \text{ where } U = V \cup V^{-1}. \end{array}\right\}$$

Therefore, if we denote by $C(G)$ the B-space of all complex-valued continuous functions $h(x)$ defined on G normed by the maximum norm, we can define a linear operator T on $\hat{C}(G)$ into $\hat{C}(G)$ by

$$(Th)(x) = (k_0 \times h)(x),\ x \in G. \tag{3}$$

Here, as in the preceding section,

$$k_0(x) = k(x)/M_x(k(x)) \tag{4}$$

and $\hat{C}(G)$ is the the space $C(G)$ endowed witht the scalar product

$$(b, h) = M_y(b(y)\,\overline{h(y)}) = (b \times h^*)(e), \; h^*(y) = \overline{h(y^{-1})}. \tag{5}$$

Thus, as in the preceding section, we obtain

Theorem 1. Any complex-valued continuous function $f(g)$ defined on a compact group G is almost periodic, and $f(g)$ can be approximated uniformly on G by linear combinations of the matrix elements of unitary, continuous, irreducible matrix representations of G.

We shall say that two matrix representations $A_1(g)$ and $A_2(g)$ of a group G are *equivalent* if there exists a fixed non-singular matrix B such that $B^{-1}A_1(g)B = A_2(g)$ for all $g \in G$.

Proposition 2 (I. Schur's lemma). If the representations $A_1(g)$, $A_2(g)$ are irreducible and inequivalent, then there is no matrix B such that

$$A_1(g)\,B = B\,A_2(g) \tag{6}$$

holds identically in g, except $B = 0$. In (6), the matrix B is assumed to be of n_1 rows and n_2 columns, where n_1, n_2 are the degrees of $A_1(g)$, $A_2(g)$, respectively.

Proof. Let X_1 and X_2 be the linear spaces subject to the linear transformations $A_1(g)$ and $A_2(g)$, respectively. B in (6) can be interpreted as a linear mapping $x_2 \to x_1 = Bx_2$ of X_2 onto X_1. The linear subspace of X_1 consisting of all vectors x_1 of the form Bx_2 is *invariant*, for $A_1(g)\,x_1 = Bx_2'$ with $x_2' = A_2(g)\,x_2$. By virtue of the irreducibility of $A_1(g)$, there are only two possibilities: either $Bx_2 = 0$ for all x_2 in X_2, i.e. $B = 0$, or $X_1 = BX_2$. On the other hand, the set of all vectors x_2 in X_2 such that $Bx_2 = 0$ is an invariant subspace of X_2, for $BA_2(g)\,x_2 = A_1(g)\,Bx_2 = 0$. From the irreducibility of $A_2(g)$ we conclude: either $Bx_2 = 0$ for all x_2 in X_2, i.e. $B = 0$, or $x_2 = 0$ is the only vector in X_2 such that $Bx_2 = 0$, so that distinct vectors in X_2 go into distinct vectors in X_1 under the linear mapping B. Hence if $B \neq 0$ we conclude that B defines a one-one linear mapping of X_2 onto X_1. But this means that B is a non-singular matrix ($n_1 = n_2$) and so $A_1(g)$ and $A_2(g)$ would be equivalent.

Proposition 3 (the orthogonality relations). Let $A_1(g) = (a^1_{ij}(g))$ and $A_2(g) = (a^2_{kl}(g))$ be unitary, continuous, irreducible matrix representations of a compact group G. Then we have the *orthogonality relations*:

$$\begin{aligned} &M_g(a^1_{ij}(g)\,\overline{a^2_{kl}(g)}) = 0, \text{ if } A_1(g) \text{ is inequivalent to } A_2(g), \\ &M_g(a^1_{ij}(g)\,\overline{a^1_{kl}(g)}) = n_1^{-1}\,\delta_{ik}\,\delta_{jl} \text{ where } n_1 \text{ is the degree of } A_1(G). \end{aligned} \tag{7}$$

Proof. Let n_1, n_2 be the degrees of $A_1(g)$, $A_2(g)$ respectively. Take any matrix B of n_1 rows and n_2 columns, and set $A(g) = A_1(g)\,B\,A_2(g^{-1})$.

Then the matrix $A = M_g(A(g))$ satisfies $A_1(g)\,A = A\,A_2(g)$. For, we have, by the invariance of the mean value,

$$A_1(y)\,A\,A_2(y^{-1}) = M_g(A_1(y)\,A_1(g)\,B\,A_2(g^{-1})\,A_2(y^{-1}))$$
$$= M_g(A_1(yg)\,B\,A_2((yg)^{-1}) = A\,.$$

By Schur's lemma, A must be equal to a zero matrix. If we take $B = (b_{jl})$ in such a way that only the element b_{jl} is not zero, then, by the unitarity condition $A_2(g^{-1}) = \overline{A_2(g)}'$, we obtain

$$M_g(a^1_{ij}(g)\,\overline{a^2_{kl}(g)}) = 0\,.$$

Next we have, as above, $A_1(g)\,A = A\,A_1(g)$ for $A = M_g(A_1(g)\,B\,A_1(g^{-1}))$. Let α be any one of the eigenvalues of the matrix A. Then the matrix $(A - \alpha I_{n_1})$, where I_{n_1} denotes the unit matrix of degree n_1, satisfies

$$A_1(g)\,(A - \alpha I_{n_1}) = (A - \alpha I_{n_1})\,A_1(g)\,.$$

Therefore, by Schur's lemma, the matrix $(A - \alpha I_{n_1})$ is either non-singular, or $(A - \alpha I_{n_1}) = 0$. The first possibility is excluded since α is an eigenvalue of A. Thus $A = \alpha I_{n_1}$. By taking the *trace* (the sum of the diagonal elements) of both sides of $A = M_g(A_1(g)\,B\,A_1(g^{-1}))$, we obtain

$$n_1\alpha = \text{trace}(A) = M_g(\text{trace}(A_1(g)\,B\,A_1(g)^{-1}) = M_g(\text{trace}(B))$$
$$= \text{trace}(B)\,.$$

Therefore, if we take $B = (b_{jl})$ in such a way that $b_{jl} = 1$ while the other elements are all zero, then, from $M_g(A_1(g)\,B\,A_1(g^{-1})) = n_1^{-1}$ trace$(B)\,I_{n_1}$, we obtain

$$M_g(a^1_{ij}(g)\,\overline{a^1_{kl}(g)}) = n_1^{-1}\,\delta_{ik}\,\delta_{jl}\,.$$

Corollary. There exists a set $\mathfrak{U}$ of mutually inequivalent, continuous, unitary, irreducible matrix representations $U(g) = (u_{ij}(g))$ of G satisfying the following three conditions:

i) for any pair of distinct points g_1, g_2 of G, there exists an $U(g) \in \mathfrak{U}$ such that $U(g_1) \neq U(g_2)$,

ii) if $U(g) \in \mathfrak{U}$, then the *complex conjugate representation* $\overline{U}(g)$ of $U(g)$ also belongs to $\mathfrak{U}$,

iii) if $U_1(g)$, $U_2(g)$ are in $\mathfrak{U}$, then the *product representation* $U_1(g) \times U_2(g)$, explained below, is completely reducible to a sum of a finite number of representations $\in \mathfrak{U}$.

Proof. The definition of the product representation $U_1(g) \times U_2(g)$ is given as follows. Let $(e^1_1, e^1_2, \ldots, e^1_n)$ and $(e^2_1, e^2_2, \ldots, e^2_m)$ be orthonormal bases respectively of the finite dimensionai complex Hilbert spaces subject to the linear mappings $U_1(g)$ and $U_2(g)$, respectively. The product space of these two Hilbert spaces is spanned by the *product base* consisting of nm vectors $e_i^1 \times e_j^2$ ($i = 1, 2, \ldots, n$; $j = 1, 2, \ldots, m$). Referring to this base, the product representation $U_1(g) \times U_2(g)$ of $U_1(g) = (u_{ij}^1(g))$ and $U_2(g) = (u_{kl}^2(g))$ is given by

$$(U_1(g) \times U_2(g))\,(e_i^1 \times e_j^2) = \sum_{s,t} u_{si}^1(g)\, u_{ij}^2(g)\,(e_i^1 \times e_j^2).$$

Let us take a maximal set $\mathfrak{U}$ of mutually inequivalent, continuous, unitary, irreducible matrix representations $U(g)$ satisfying the condition ii). Then, by Theorem 1, condition i) is satisfied. Condition iii) is also satisfied since the product representation of two unitary representations is also unitary and hence is completely reducible.

We are now ready to formulate *T. Tannaka's duality theorem*. Let $\mathfrak{R}$ be the set of all *Fourier polynomials*:

$$x(g) = \sum \gamma_{ij}^{(\alpha)} u_{ij}^{(\alpha)}(g),$$

that is, finite linear combinations of $u_{ij}^{(\alpha)}(g)$, where $(u_{ij}^{(\alpha)}(g)) \in \mathfrak{U}$ and $\gamma_{ij}^{(\alpha)}$ denote complex numbers. Then $\mathfrak{R}$ is a *ring* with the *multiplicative unit* u ($u(g) \equiv 1$ on G) and with *complex multipliers*; the sum and the multiplication in the ring $\mathfrak{R}$ being understood as the function sum and the function multiplication, respectively. Let $\mathfrak{T}$ be the set of all *linear homomorphisms* T of the ring $\mathfrak{R}$ onto the field of complex numbers such that

$$Tu = 1,\ T\bar{x} = \overline{Tx}, \text{ the bar indicating the complex-conjugate.} \tag{8}$$

$\mathfrak{T}$ is not void since each $g \in G$ induces such a homomorphism T_g:

$$T_g x = x(g). \tag{9}$$

By the condition i) for $\mathfrak{U}$, we see that

$$g_1 \neq g_2 \quad \text{implies} \quad T_{g_1} \neq T_{g_2}. \tag{10}$$

Proposition 4. $\mathfrak{T}$ may be considered as a group which contains G as a subgroup.

Proof. We shall define a product $T_1 \cdot T_2 = T$ in $\mathfrak{T}$ as follows. Let

$$U^{(\alpha)}(g) = (u_{ij}^{(\alpha)}(g)) \qquad (i, j = 1, 2, \ldots, n)$$

be members of $\mathfrak{U}$. We put, for $u_{ij}^{(\alpha)}(gh) = \sum_k u_{ik}^{(\alpha)}(g)\, u_{kj}^{(\alpha)}(h)$,

$$T u_{ij}^{(\alpha)} = \sum_k T_1 u_{ik}^{(\alpha)} \cdot T_2 u_{kj}^{(\alpha)}. \tag{11}$$

By the orthogonal relations (7), we see that the functions $u_{ij}^{(\alpha)}(g)$'s, where $(u_{ij}^{(\alpha)}(g)) \in \mathfrak{U}$, are linearly independent on G. Hence, we see that T may be extended linearly on the whole $\mathfrak{R}$. It is easy to see that the extension T is also a member of $\mathfrak{T}$ and that $\mathfrak{T}$ is a group with the unit T_e (e = the unit of G) and $T_g^{-1} = T_{g^{-1}}$. G is isomorphically embedded in this group $\mathfrak{T}$ by the correspondence $g \leftrightarrow T_g$, as will be seen from (11) and (10).

In truth, we have

Theorem 2 (T. Tannaka). $\mathfrak{T} = G$, viz. every $T \in \mathfrak{T}$ is equal to a certain T_g:

$$T x = x(g) \quad \text{for all} \quad x \in \mathfrak{R}. \tag{12}$$

Proof. We introduce a *weak topology* in the group $\mathfrak{T}$ by taking the sets of the form

$$\{T \in \mathfrak{T};\, |T x_i - T_e x_i| < \varepsilon_i \ (i = 1, 2, \ldots, n)\} \tag{13}$$

as neighbourhoods of the unit T_e of the group $\mathfrak{T}$. Then $\mathfrak{T}$ is a compact space. For, by $\sum_s |T u_{si}^{(\alpha)}(g)|^2 = (T \cdot T)(1) = 1$ implied by (8) and (11), $\mathfrak{T}$ is a closed subset of the topological product of compact spaces:

$$\prod_{x \in \mathfrak{R}} \left\{z;\ |z| \leqq \sup_{T \in \mathfrak{T}} |T x|\right\},$$

and so we can apply Tychonov's theorem. It is easily seen that the isomorphic embedding $g \leftrightarrow T_g$ is also a topological one, because a one-one continuous map of a compact space onto a compact space is a topological map. Therefore G may be considered as a closed subgroup of the compact group $\mathfrak{T}$.

By the above weak topology of $\mathfrak{T}$, each $x(g) \in \mathfrak{R}$ gives rise to a continuous function $x(T)$ on the compact group $\mathfrak{T}$ such that $x(T_g) = x(g)$. We have only to set $x(T) = T \cdot x$. The set of all these continuous functions $x(T)$ constitutes a ring $\mathfrak{R}(\mathfrak{T})$ of complex-valued continuous functions on $\mathfrak{T}$ satisfying the conditions:

1) $1 = u(T) \in \mathfrak{R}(\mathfrak{T})$,

2) for any pair of distinct points T_1, T_2 of $\mathfrak{T}$, there exists an $x(T) \in \mathfrak{R}(\mathfrak{T})$ such that $x(T_1) \neq x(T_2)$,

3) for any $x(T) \in \mathfrak{R}(\mathfrak{T})$, there exists the complex-conjugate function $\bar{x}(T) = \overline{x(T)}$ which belongs to $\mathfrak{R}(\mathfrak{T})$.

Now let us suppose that $\mathfrak{T} - G$ is not void. Since a compact space $\mathfrak{T}$ is normal, we may apply Urysohn's theorem to the effect that there exist a point $T_0 \in (\mathfrak{T} - G)$ and a continuous function $y(T)$ on $\mathfrak{R}$ such that

$$y(T) \geqq 0 \text{ on } \mathfrak{T},\ y(g) = 0 \text{ on } G \text{ and } y(T_0) = 1. \tag{14}$$

By the Stone-Weierstrass theorem in Chapter 0, as applied to the ring $\mathfrak{R}(\mathfrak{T})$ satisfying 1), 2) and 3), we see that there exists, for any $\varepsilon > 0$, a function $x(g) = \sum \gamma_{ij}^{(\alpha)} u_{ji}^{(\alpha)}(g) \in \mathfrak{R}$ such that $|y(T) - \sum \gamma_{ij}^{(\alpha)} u_{ij}^{(\alpha)}(T)| < \varepsilon$ on $\mathfrak{T}$, and, in particular, $|y(g) - \sum \gamma_{ij}^{(\alpha)} u_{ij}^{(\alpha)}(g)| < \varepsilon$ on G. Let $u_{11}^{(\alpha_0)}(g) = u(g) = 1$. Then, by taking the mean value and remembering the orthogonality relation (7), as applied to the compact groups $\mathfrak{T}$ and G, we obtain

$$|M_T(y(T)) - \gamma_{11}^{(\alpha_0)}| < \varepsilon,\ |M_g(y(g)) - \gamma_{11}^{(\alpha_0)}| < \varepsilon. \tag{15}$$

We have thus arrived at a contradiction, since $M_T(y(T)) > 0$ and $M_g(y(g)) = 0$ by (14).

Remark 1. The above proof of Tannaka's theorem is adapted from K. Yosida [14]. The original proof is given in T. Tannaka [1]. Since a continuous, unitary and irreducible matrix representation of a compact abelian group G is precisely a continuous function $\chi(g)$ on G satisfying

$$\chi(g_1)\,\chi(g_2) = \chi(g_1\,g_2) \quad \text{and} \quad |\chi(g)| = 1,$$

Tannaka's theorem contains, as a special case, the duality theorem of L. Pontrjagin [1]. For further references, see M. A. Naimark [1].

Remark 2. To define mean values of continuous functions $f_k(g)$ $(k = 1, 2, \ldots, m)$ on G by the method given in Chapter VIII, 5, we have only to replace the dis (g_1, g_2) there by dis $(g_1, g_2) = \sup_{g,h \in G; k=1,2,\ldots,m} |f_k(g\,g_1\,h) - f_k(g\,g_2\,h)|$.

12. Functions of a Self-adjoint Operator

Let $H = \int \lambda dE(\lambda)$ be the spectral resolution of a self-adjoint operator H in a Hilbert space X. For a complex-valued Baire function $f(\lambda)$, we consider the set

$$D(f(H)) = \left\{x \in X;\ \int_{-\infty}^{\infty} |f(\lambda)|^2 d\,||E(\lambda)\,x||^2 < \infty\right\}, \tag{1}$$

where the integral is taken with respect to the Baire measure m determined by $m((\lambda_1, \lambda_2]) = ||E(\lambda_2)\,x||^2 - ||E(\lambda_1)\,x||^2$. As in the case of a continuous function $f(\lambda)$, treated in Chapter XI, 5, we see that the integral

$$\int_{-\infty}^{\infty} f(\lambda)\, d(E(\lambda)\,x, y),\ x \in D(f(H)),\ y \in X, \tag{2}$$

exists and is finite with respect to the Baire measure m determined by $m((\lambda_1, \lambda_2]) = (E(\lambda_2)\,x, y) - (E(\lambda_1)\,x, y)$. Further we see that (2) gives the complex conjugate of a bounded linear functional of y. Hence we may write (2) as $(f(H)\,x, y)$ by virtue of F. Riesz' representation theorem in Chapter III, 6. In this way, we can define functions of H:

$$f(H) = \int_{-\infty}^{\infty} f(\lambda)\, dE(\lambda) \tag{3}$$

through (1) and (2).

Example 1. If H is bounded self-adjoint and $f(\lambda) = \sum_{j=1}^{n} \alpha_j \lambda^j$, then, as in Chapter XI, 5,

$$f(H) = \int_{-\infty}^{\infty} f(\lambda)\, dE(\lambda) = \sum_{j=1}^{n} \alpha_j H^j.$$

Example 2. If $f(\lambda) = (\lambda - i)(\lambda + i)^{-1}$, then $f(H)$ is equal to the Cayley transform U_H of H. We have, in this case, $D(f(H)) = X$ since $|f(\lambda)| = 1$ for real λ. It is easy to see, by $E(\lambda_1)E(\lambda_2) = E(\min(\lambda_1, \lambda_2))$, that if we apply the bounded self-adjoint operator $(H - iI)^{-1} = \int (\lambda - i)^{-1} dE(\lambda)$ to the operator $f(H) = \int f(\lambda) dE(\lambda)$ then the result is equal to $(H + iI)^{-1} = \int (\lambda + i)^{-1} dE(\lambda)$. Hence $f(H) = U_H$.

Example 3. As in Chapter XI, 5, we have

$$\|f(H)x\|^2 = \int_{-\infty}^{\infty} |f(\lambda)|^2 d\|E(\lambda)x\|^2 \quad \text{whenever} \quad x \in D(f(H)). \tag{4}$$

Definition. Let A be a not necessarily bounded linear operator, and B a bounded linear operator in a Hilbert space. If

$$x \in D(A) \text{ implies } Bx \in D(A) \text{ and } ABx = BAx, \tag{5}$$

that is, if $AB \supseteq BA$, then we shall write $B \in (A)'$ and say that B is *commutative* with A. We shall thus write the totality of bounded linear operators B commutative with A by $(A)'$.

Theorem 1. For a function $f(H)$ of a self-adjoint operator $H = \int \lambda dE(\lambda)$ in a Hilbert space X, we have

$$(f(H))' \supseteq (H)', \tag{6}$$

that is, $f(H)$ is commutative with every bounded linear operator which commutes with H. (Since $E(\lambda) \in (H)'$ by Theorem 2 in Chapter XI, 5, we see, in particular, that $f(H)$ is commutative with every $E(\lambda)$.)

Proof. Suppose $S \in (H)'$. Then we can show that S is commutative with every $E(\lambda)$. We first show that S is commutative with the Cayley transform U_H of H. For, if $x \in D(H)$, then we have, by $S \in (H)'$,

$$S(H + iI)x = (H + iI)Sx, \quad (H - iI)Sx = S(H - iI)x.$$

By putting $(H + iI)x = y$, we see, from the first of the above relations, that

$$(H + iI)^{-1}Sy = S(H + iI)^{-1}y \quad \text{for all} \quad y \in X = R(H + iI).$$

Hence

$$S(H - iI)(H + iI)^{-1} = (H - iI)(H + iI)^{-1}S, \text{ that is, } SU_H = U_H S.$$

Therefore, S is commutative with $(U_H)^n = \int_0^{2\pi} e^{ni\theta} dF(\theta)$ $(n = 0, \pm 1, \ldots)$ and so

$$\int_0^{2\pi} e^{ni\theta} d(SF(\theta)x, y) = \left(S\int_0^{2\pi} e^{ni\theta} dF(\theta)x, y\right) = \int_0^{2\pi} e^{ni\theta} d(F(\theta)Sx, y).$$

Hence, as in the proof of the uniqueness of the spectral resolution of a unitary operator, given in Chapter XI, 4, we obtain $SF(\theta) = F(\theta)S$. This proves $SE(\lambda) = E(\lambda)S$ since $E(-\cot\theta) = F(\theta)$. Thus, for $x \in D(f(H))$, we obtain

$$(S \int f(\lambda)\, dE(\lambda)\, x, y) = \int f(\lambda)\, d(SE(\lambda)\, x, y) = \int f(\lambda)\, d(E(\lambda)\, S\, x, y),$$

that is, $Sf(H) \subseteq f(H)\, S$.

The above Theorem 1 admits a converse in the form of

Theorem 2 (NEUMANN-RIESZ-MIMURA). Let H be a self-adjoint operator in a separable Hilbert space X. Let T be a closed linear operator in X such that $D(T)^a = X$. Then a necessary and sufficient condition in order that T be a function $f(H)$ of H with an everywhere finite Baire function $f(\lambda)$ is that

$$(T)' \supseteq (H)'. \tag{7}$$

Proof. We have only to show that condition (7) is sufficient. We may assume that H is a bounded self-adjoint operator. If H is not bounded, then we consider $H_1 = \tan^{-1} H$. Since $|\tan^{-1}\lambda| \leq \pi/2$, we easily see that H_1 is bounded self-adjoint. By Theorem 1, the operator $H = \tan H_1$ is commutative with every operator $\in (H_1)'$. Thus, by hypothesis, we have

$$(T)' \supseteq (H)' \supseteq (H_1)'.$$

Therefore, if Theorem 2 is proved for bounded H_1, then $T = f_1(H_1) = f_1(\tan^{-1} H) = f_2(H)$, where $f_2(\lambda) = f_1(\tan^{-1}\lambda)$.

We may thus assume that H is bounded and self-adjoint.

The first step. For any fixed $x_0 \in D(T)$, we can find a Baire function $F(\lambda)$ such that $Tx_0 = F(H)\, x_0$. This may be proved as follows. Let $M(x_0)$ be the smallest closed linear subspace of X spanned by x_0, Hx_0, $H^2 x_0, \ldots, H^n x_0, \ldots$ We denote by L the projection upon $M(x_0)$. Then $(T)' \ni L$. For, by $HM(x_0) \subseteq M(x_0)$, we obtain $HL = LHL$ and so $LH = (HL)^* = (LHL)^* = LHL = HL$, that is, $L \in (H)'$. Hence, by hypothesis, $(T)' \ni L$.

Therefore $Tx_0 = TLx_0 = LTx_0 \in M(x_0)$, and so there exists a sequence $\{p_n(\lambda)\}$ of polynomials such that

$$Tx_0 = \underset{n\to\infty}{s\text{-}\lim}\, p_n(H)\, x_0. \tag{8}$$

Hence, by (4), we obtain

$$\|p_n(H)\, x_0 - p_m(H)\, x_0\|^2 = \int_{-\infty}^{\infty} |p_n(\lambda) - p_m(\lambda)|^2\, d\, \|E(\lambda)\, x_0\|^2.$$

As in the proof of the completeness of $L^2(S, \mathfrak{B}, m)$, we see that there exists a Baire function $F(\lambda)$ which is square-integrable with respect to the measure m determined by $m((\lambda_1, \lambda_2]) = \|E(\lambda_2)\, x_0\|^2 - \|E(\lambda_1)\, x_0\|^2$ such that

$$\lim_{n\to\infty} \int_{-\infty}^{\infty} |F(\lambda) - p_n(\lambda)|^2\, d\, \|E(\lambda)\, x_0\|^2 = 0.$$

Hence, for $F(H)$, we have

$$\lim_{n\to\infty} \|(F(H) - p_n(H))\, x_0\|^2 = \lim_{n\to\infty} \int_{-\infty}^{\infty} |F(\lambda) - p_n(\lambda)|^2 \, d\, \|E(\lambda)\, x_0\|^2 = 0.$$

This proves that $T x_0 = F(H)\, x_0$. Since $F(\lambda)$ is finite almost everywhere with respect to the measure m determined by $m((\lambda_1, \lambda_2]) = \|E(\lambda_2)\, x_0\|^2 - \|E(\lambda_1)\, x_0\|^2$, we may assume that $F(\lambda)$ is a Baire function which takes a finite value at every λ; we may put $F(\lambda) = 0$ for those λ for which $|F(\lambda)| = \infty$.

The second step. Since X is separable and $D(T)^a = X$, we may choose a countable sequence $\{g_n\} \subseteq D(T)$ such that $\{g_n\}$ is strongly dense in X. We set

$$f_1 = g_1,\ f_2 = g_2 - L_1\, g_2, \ldots, f_n = g_n - \sum_{k=1}^{n-1} L_k g_n, \text{ where } L_k \text{ is the projection on the closed linear subspace } M(f_k). \tag{9}$$

By the first step, we have $(T)' \ni L_k$ and so

$$L_k g_n \in D(T) \text{ which implies } f_n \in D(T) \quad (n = 1, 2, \ldots). \tag{10}$$

We may show that

$$L_i L_k = 0 \quad (\text{for } i \neq k) \tag{11}$$

and

$$I = \sum_{k=1}^{\infty} L_k. \tag{12}$$

Suppose that (11) is proved for $i, k < n$. Then, for $i < n$,

$$L_i f_n = L_i g_n - L_i \left(\sum_{k=1}^{n-1} L_k g_n \right) = L_i g_n - L_i^2 g_n = L_i g_n - L_i g_n = 0,$$

$$L_i H^{k'} f_n = H^{k'} L_i f_n = 0.$$

Thus $M(f_n)$ is orthogonal to $M(f_i)$. This proves $L_i L_n = L_n L_i = 0$.

Next we put $\sum_{k=1}^{\infty} L_k = P$ and show that $P g_n = g_n$ $(n = 1, 2, \ldots)$. Then, since $\{g_n\}$ is dense in X, we obtain $P = I$. But, by (9), we have

$$P g_n = P f_n + \sum_{k=1}^{n-1} P L_k g_n.$$

We also have $P f_n = f_n$ by $f_n \in M(f_n)$. Hence, by $P L_k = L_k$ implied by (11), we obtain $P g_n = g_n$ $(n = 1, 2, \ldots)$.

The third step. Take a sequence $\{c_n\}$ of positive numbers such that

$$\operatorname*{s-lim}_{k\to\infty} \sum_{n=1}^{k} c_n f_n, \quad \operatorname*{s-lim}_{k\to\infty} \sum_{n=1}^{k} c_n T f_n$$

both exist. For instance, we may take $c_n = 2^{-n} (\|f_n\| + \|T f_n\|)^{-1}$. Since T is a closed operator, we have

$$x_0 = \sum_{n=1}^{\infty} c_n f_n \in D(T) \text{ and } T x_0 = y_0 = \sum_{n=1}^{\infty} c_n T f_n. \tag{13}$$

Hence, by the first step,

$$T x_0 = F(H)\, x_0 . \tag{14}$$

Let $B \in (H)'$ be a bounded self-adjoint operator. Then, by hypothesis, $B \in (T)'$. By Theorem 1, $F(H)$ is commutative with B. Hence

$$F(H)\, B x_0 = B F(H)\, x_0 = B T x_0 = T B x_0 . \tag{15}$$

Let $e_n(\lambda)$ be the defining function of the set $\{\lambda;\, |F(\lambda)| \leqq n\}$, and put

$$B = c_m^{-1} P_n H^k L_m, \quad \text{where} \quad P_n = e_n(H) .$$

Then we can show that

$$T P_n = F(H)\, P_n . \tag{16}$$

In fact, we have $L_m x_0 = \sum_{n=1}^{\infty} c_n L_m f_n = c_m f_m$, by (11) and $f_m \in M(f_m)$. Hence, by (15),

$$\begin{aligned} F(H)\, P_n H^k f_m &= F(H)\, c_m^{-1} P_n H^k L_m x_0 = F(H)\, B x_0 = T B x_0 \\ &= T c_m^{-1} P_n H^k L_m x_0 = T P_n H^k f_m , \end{aligned}$$

that is, for h spanned by $H^k f_m$ with fixed m, we have

$$F(H)\, P_n h = T P_n h . \tag{16'}$$

But such h's are dense in $M(f_m)$ and so, by (12), we see that if we let m take all positive integers then the h's are dense in X. Hence we have proved (16′) for those h's which are dense in X.

Now P_n is bounded by (4). By the operational calculus given below, the operator $F(H) P_n$ is equal to the function $F_n(H)$ where

$$\begin{aligned} F_n(\lambda) = F(\lambda)\, e_n(\lambda) &= F(\lambda), \text{ for } |F(\lambda)| \leqq n, \\ &= 0, \text{ for } |F(\lambda)| > n . \end{aligned}$$

Thus $F_n(H) = F(H) P_n$ is bounded.

Let $h^* \in X$ be arbitrary, and let $h^* = \text{s-}\lim_{j\to\infty} h_j$ where h_j's are linear combinations of elements of the form $H^k f_m$. Such a choice of h_j is possible as proved above. By the continuity of the operator $F(H) P_n$, we have

$$F(H) P_n h^* = \text{s-}\lim_{j\to\infty} F(H) P_n h_j .$$

Hence, by $\text{s-}\lim_{j\to\infty} P_n h_j = P_n h^*$ and (16′), we see that the closed operator T satisfies (16).

The fourth step. Let $y \in D(F(H))$ and set $y_n = P_n y$. Since $F(\lambda)$ is finite everywhere, we must have $\text{s-}\lim_{n\to\infty} P_n = I$. Thus $\text{s-}\lim_{n\to\infty} y_n = \text{s-}\lim_{n\to\infty} P_n y = y$. Hence, by (16), we obtain $T \supseteq F(H)$. For, $\text{s-}\lim_{n\to\infty} F(H)\, P_n y = \text{s-}\lim_{n\to\infty} F_n(H)$ $y = F(H)\, y$ whenever $y \in D(F(H))$.

Let $y \in D(T)$ and set $y_n = P_n y$. Then

$$T y_n = T P_n y = F(H) P_n y \text{ (by (16))},$$
$$T P_n y = P_n T y \text{ (by } P_n = e_n(H) \in (H)').$$

By the operational calculus given below, the function $F(H)$ of H is a closed operator. Hence, letting $n \to \infty$ in the above relations, we see that $F(H) \supseteq T$. We have thus proved that $T = F(H)$.

An Operational Calculus. We have

Theorem 3. (i) Let $\bar{f}(\lambda)$ be the complex-conjugate function of $f(\lambda)$. Then $D(\bar{f}(H)) = D(f(H))$ and, for $x, y \in D(f(H)) = D(\bar{f}(H))$, we have

$$(f(H)\, x, y) = (x, \bar{f}(H)\, y). \tag{17}$$

(ii) If $x \in D(f(H))$, $y \in D(g(H))$, then

$$(f(H)\, x, g(H)\, y) = \int_{-\infty}^{\infty} f(\lambda)\, \overline{g(\lambda)}\, d(E(\lambda)\, x, y). \tag{18}$$

(iii) $(\alpha f)(H)\, x = \alpha f(H)\, x$ if $x \in D(f(H))$. If $x \in D(f(H)) \cap D(g(H))$, then

$$(f + g)(H)\, x = f(H)\, x + g(H)\, x. \tag{19}$$

(iv) If $x \in D(f(H))$, then the condition $f(H)\, x \in D(g(H))$ is equivalent to the condition $x \in D(f \cdot g(H))$, where $f \cdot g(\lambda) = f(\lambda)\, g(\lambda)$, and we have

$$g(H)\, f(H)\, x = (g \cdot f)(H)\, x. \tag{20}$$

(v) If $f(\lambda)$ is finite everywhere, then $f(H)$ is a normal operator and

$$f(H)^* = \bar{f}(H). \tag{21}$$

In particular, $f(H)$ is self-adjoint if $f(\lambda)$ is real-valued and finite everywhere.

Proof. (i) $D(f(H)) = D(\bar{f}(H))$ is clear and

$$(f(H)\, x, y) = \int_{-\infty}^{\infty} f(\lambda)\, d(E(\lambda)\, x, y) = \int_{-\infty}^{\infty} f(\lambda)\, d(x, E(\lambda)\, y)$$
$$= \overline{(\bar{f}(H)\, y, x)} = (x, \bar{f}(H)\, y).$$

(ii) We know, by Theorem 1, that $E(\lambda)$ is commutative with $g(H)$. Thus

$$(f(H)\, x, g(H)\, y) = \int_{-\infty}^{\infty} f(\lambda)\, d(E(\lambda)\, x, g(H)\, y) = \int_{-\infty}^{\infty} f(\lambda)\, d(x, E(\lambda)\, g(H)\, y)$$
$$= \int_{-\infty}^{\infty} f(\lambda)\, d\, \overline{(g(H)\, E(\lambda)\, y, x)} = \int_{-\infty}^{\infty} f(\lambda)\, d\left(\int_{-\infty}^{\infty} \overline{g(\mu)}\, d\, \overline{(E(\mu)\, E(\lambda)\, y, x)}\right)$$
$$= \int_{-\infty}^{\infty} f(\lambda)\, d\left(\int_{-\infty}^{\lambda} \overline{g(\mu)}\, d\, \overline{(y, E(\mu)\, x)}\right) = \int_{-\infty}^{\infty} f(\lambda)\, \overline{g(\lambda)}\, d(E(\lambda)\, x, y).$$

(iii) is clear. (iv) Let x satisfy $\int_{-\infty}^{\infty} |f(\lambda)|^2 d\,||E(\lambda)\,x||^2 < \infty$. Then, by $E(\lambda)\,E(\mu) = E(\min(\lambda, \mu))$, the condition $\int_{-\infty}^{\infty} |g(\lambda)|^2 d\,||E(\lambda)\,f(H)\,x||^2 < \infty$ implies, by virtue of the commutativity of $E(\lambda)$ with $f(H)$,

$$\infty > \int_{-\infty}^{\infty} |g(\lambda)|^2 d\,||E(\lambda)\,f(H)\,x||^2 = \int_{-\infty}^{\infty} |g(\lambda)|^2 d\,||f(H)\,E(\lambda)\,x||^2$$

$$= \int_{-\infty}^{\infty} |g(\lambda)|^2 d\left(\int_{-\infty}^{\infty} |f(\mu)|^2 d\,||E(\mu)\,E(\lambda)\,x||^2\right)$$

$$= \int_{-\infty}^{\infty} |g(\lambda)|^2 d\left(\int_{-\infty}^{\lambda} |f(\mu)|^2 d\,||E(\mu)\,x||^2\right) = \int_{-\infty}^{\infty} |g(\lambda)\,f(\lambda)|^2 d\,||E(\lambda)\,x||^2.$$

Since the above calculation may be traced conversely, we see that, under the hypothesis $x \in D(f(H))$, the two conditions $f(H)\,x \in D(g(H))$ and $x \in D(f \cdot g(H))$ are equivalent and we have

$$(g(H)\,f(H)\,x, y) = \int_{-\infty}^{\infty} g(\lambda)\,d(E(\lambda)\,f(H)\,x, y)$$

$$= \int_{-\infty}^{\infty} g(\lambda)\,d\left(\int_{-\infty}^{\lambda} f(\mu)\,d(E(\mu)\,x, y)\right)$$

$$= \int_{-\infty}^{\infty} g(\lambda)\,f(\lambda)\,d(E(\lambda)\,x, y) = ((g \cdot f)\,(H)\,x, y).$$

(v) Let us put $h(\lambda) = |f(\lambda)| + \alpha$, $k(\lambda) = h(\lambda)^{-1}$, $g(\lambda) = f(\lambda)\,h(\lambda)^{-1}$, where α is any positive integer. Then $k(\lambda)$ and $g(\lambda)$ are both bounded functions. Hence $D(k(H)) = D(g(H)) = X$. Thus, by (iv),

$$f(H) = h(H)\,g(H) = g(H)\,h(H). \tag{22}$$

We have, by (i) and $D(k(H)) = X$, $k(H)^* = k(H)$, i.e., $k(H)$ is self-adjoint. By (iv) we have $x = h(H)\,k(H)\,x$ for all $x \in X$ and $x = k(H)\,h(H)\,x$ for all $x \in D(h(H))$. Hence $h(H) = k(H)^{-1}$. Thus, by Theorem 1 in Chapter VII, 3, $h(H)$ is self-adjoint. Therefore $D(f(H)) = D(h(H))$ is dense in X and so we may define $f(H)^*$. We shall show that $f(H)^* = \bar{f}(H)$. Let a pair $\{y, y^*\}$ of elements $\in X$ be such that $(f(H)\,x, y) = (x, y^*)$ for all $x \in D(f(H))$. Then, by $g(H)^* = \bar{g}(H)$ (implied by (i)) and (22),

$$(f(H)\,x, y) = (g(H)\,h(H)\,x, y) = (h(H)\,x, \bar{g}(H)\,y).$$

Hence, by $x \in D(f(H)) = D(h(H))$ and the self-adjointness of $h(H)$, we obtain

$$\bar{g}(H)\,y \in D(h(H)) \quad \text{and} \quad h(H)\,\bar{g}(H)\,y = y^*.$$

Thus, again by (22), we obtain $\bar{f}(H)\,y = y^*$ so that $f(H)^* = \bar{f}(H)$. Therefore, by (iv), we see that $f(H)$ is normal, that is, $f(H)^*\,f(H) = f(H)\,f(H)^*$.

Corollary. If $f(\lambda)$ is finite everywhere, then $f(H)$ is closed.

Proof. Clear from $f(H)^{**} = \bar{f}(H)^* = \bar{\bar{f}}(H) = f(H)$.

A historical note. Theorem 2 was first proved by J. VON NEUMANN [7] for the case of a bounded self-adjoint operator T. Cf. F. RIESZ [5]. The general case of a closed linear operator T was proved by Y. MIMURA [2]. The exposition given above is adapted from Y. MIMURA [2] and B. SZ. NAGY [1].

13. Stone's Theorem and Bochner's Theorem

As an example of functions of a self-adjoint operator, we give

Theorem 1 (M. H. STONE). Let $\{U_t\}$, $-\infty < t < \infty$, be a one-parameter group of class (C_0) of unitary operators in a Hilbert space X. Then

$$U_t = f_t(H), \text{ where } f_t(\lambda) = \exp(it\lambda) \text{ and } iH = A,\ H^* = H,$$

is the the infinitesimal generator of the group U_t. (1)

Conversely, for any self-adjoint operator H in X, $U_t = f_t(H)$ defines a one-parameter group of class (C_0) of unitary operators.

Proof. We have, by the representation theorem of the semi-group theory,

$$U_t x = \underset{n\to\infty}{s\text{-lim}} \exp(tiH(I - n^{-1}iH)^{-1})\, x.$$

Since the function $g(t) = \exp(ti\lambda(1 - n^{-1}i\lambda)^{-1})$ is smaller than $\exp((-nt\lambda^2)/(n^2 + \lambda^2))$ in absolute value, we have, for $H = \int \lambda\, dE(\lambda)$,

$$\exp(tiH(I - n^{-1}iH)^{-1}) = \int_{-\infty}^{\infty} \exp\left(\frac{ti\lambda}{1 - n^{-1}i\lambda}\right) dE(\lambda)$$

and, moreover,

$$\lim_{n\to\infty} \int_{-\infty}^{\infty} \left|\exp\left(\frac{ti\lambda}{1 - n^{-1}i\lambda}\right) - \exp(ti\lambda)\right|^2 d\,\|E(\lambda)\,x\|^2$$

$$= \lim_{n\to\infty} \int_{-\infty}^{\infty} \left|\exp\left(\frac{-t\lambda^2}{n - i\lambda}\right) - 1\right|^2 d\,\|E(\lambda)\,x\|^2 = 0.$$

This proves that $U_t = f_t(H) = \int_{-\infty}^{\infty} \exp(it\lambda)\, dE(\lambda)$.

For the converse part of Theorem 1, we observe that, by the operational calculus of the preceding section,

$$f_t(H)^* = f_{-t}(H) \quad \text{and} \quad f_t(H)\, f_s(H) = f_{t+s}(H),\ f_0(H) = I.$$

We also have the strong continuity of $f_t(H)$ at $t = 0$ by

$$\|f_t(H)\,x - x\|^2 = \int_{-\infty}^{\infty} |\exp(it\lambda) - 1|^2\, d\,\|E(\lambda)\,x\|^2 \to 0 \quad \text{as} \quad t \to 0.$$

Hence $U_t = f_t(H)$ is a one-parameter group of class (C_0) of unitary operators.

Remark. For the original proof, see M. H. STONE [2], Cf. also J. VON NEUMANN [8]. Another proof given by E. HOPF [1] is based on a theorem due to S. BOCHNER:

Theorem 2 (BOCHNER). A complex-valued continuous function $f(t)$, $-\infty < t < \infty$, is representable as

$$f(t) = \int_{-\infty}^{\infty} e^{it\lambda}\, dv(\lambda) \text{ with a non-decreasing,}$$

$$\text{right-continuous bounded function } v(\lambda), \tag{2}$$

iff $f(t)$ is *positive definite* in the following sense:

$$\int_{-\infty}^{\infty}\int_{-\infty}^{\infty} f(t-s)\, \varphi(t)\, \overline{\varphi(s)}\, dt\, ds \geqq 0$$

$$\text{for every continuous function } \varphi \text{ with compact support.} \tag{3}$$

The proof of Theorem 1 given by E. HOPF starts with the fact that $f(t) = (U_t x, x)$ satisfies condition (3) as may be seen from

$$\int_{-\infty}^{\infty}\int_{-\infty}^{\infty} (U_{t-s}x, x)\, \varphi(t)\, \overline{\varphi(s)}\, dt\, ds = \int_{-\infty}^{\infty}\int_{-\infty}^{\infty} (U_t x, U_s x)\, \varphi(t)\, \varphi(s)\, dt\, ds$$

$$= \left(\int_{-\infty}^{\infty} \varphi(t)\, U_t\, x\, dt, \int_{-\infty}^{\infty} \varphi(s)\, U_s x\, ds \right) \geqq 0.$$

We shall show that Bochner's theorem is a consequence of Stone's theorem.

Deduction of Theorem 2 from Theorem 1. Consider the totality $\mathfrak{F}$ of complex-valued functions $x(t)$, $-\infty < t < \infty$, such that $x(t) = 0$ except possibly for a finite set of values of t; the finite set may vary with x. $\mathfrak{F}$ is a pre-Hilbert space by

$$(x + y)(t) = x(t) + y(t), \ (\alpha x)(t) = \alpha x(t) \text{ and}$$

$$(x, y) = \sum_{-\infty<t,s<\infty} f(t-s)\, x(t)\, y(s) \text{ for } x, y \in \mathfrak{F}, \tag{4}$$

excepting the axiom that $(x, x) = 0$ implies $x = 0$. That $(x, x) \geqq 0$ for any $x \in \mathfrak{F}$ is a simple consequence of the positive definiteness of the function $f(t)$.

Let us set $\mathfrak{N} = \{x \in \mathfrak{F}; (x, x) = 0\}$. Then the factor space $\mathfrak{F}/\mathfrak{N}$ is a pre-Hilbert space with respect to the scalar product $(\bar{x}, \bar{y}) = (x, y)$ where $\bar{x}$ is the residue class mod $\mathfrak{N}$ containing $x \in \mathfrak{F}$. Let X be the completion of the pre-Hilbert space $X = \mathfrak{F}/\mathfrak{N}$. The operator U_τ defined by

$$(U_\tau x)(t) = x(t + \tau), \ x \in \mathfrak{F}, \tag{5}$$

surely satisfies the conditions

$$(U_\tau x, U_\tau y) = (x, y), \ U_\tau U_\sigma = U_{\tau+\sigma} \text{ and } U_0 = I. \tag{6}$$

Therefore, it is easy to see that $\{U_\tau\}$ naturally defines a unitary operator $\hat{U}_\tau$ in X in such a way that $\{\hat{U}_\tau\}$, $-\infty < \tau < \infty$, constitutes a one-parameter semi-group of class (C_0) of unitary operators in X; the strong continuity in t of $\hat{U}_t$ follows from the continuity of the function $f(t)$.

Hence, by Stone's theorem, $\hat{U}_t = \int_{-\infty}^{\infty} e^{it\lambda}\, dE(\lambda)$. Let $x_0(t) \in \mathfrak{F}$ be defined by $x_0(\tau) = 1$ and $x_0(t) = 0$ whenever $t \neq \tau$. Then, by (4) and (5), $f(\tau) = (U_\tau x_0, x_0)$. Therefore,

$$f(\tau) = \int_{-\infty}^{\infty} e^{i\tau\lambda}\, d\, \|E(\lambda)\, \bar{x}_0\|^2,$$

which proves Bochner's theorem.

Remark. The idea of using a positive definite function to define a pre-Hilbert space as in (4) was systematically applied by B. Sz. Nagy [3] to various interesting problems concerning the Hilbert space.

14. A Canonical Form of a Self-adjoint Operator with Simple Spectrum

Let $H = \int \lambda\, dE(\lambda)$ be a self-adjoint operator in a Hilbert space X with simple spectrum as defined in Chapter XI, 8. Thus there exists an element $y \in X$ such that the set $\{(E(\beta) - E(\alpha))\, y;\, \alpha < \beta\}$ spans a dense linear subspace of X. We put

$$\sigma(\lambda) = (E(\lambda)\, y, y). \tag{1}$$

Then $\sigma(\lambda)$ is monotone non-decreasing, right-continuous and bounded. We shall denote by $\sigma(B)$ the Baire measure determined on Baire sets on R^1 from $\sigma((a, b]) = \sigma(b) - \sigma(a)$. We denote by $L^2_\sigma(-\infty, \infty)$ the totality of complex-valued Baire-measurable functions $f(\lambda)$, $-\infty < \lambda < \infty$, such that $\|f\|_\sigma = \left(\int_{-\infty}^{\infty} |f(\lambda)|^2\, \sigma(d\lambda)\right)^{1/2} < \infty$. Then $L^2_\sigma(-\infty, \infty)$ is a Hilbert space by the scalar product $(f, g)_\sigma = \int_{-\infty}^{\infty} f(\lambda)\, \overline{g(\lambda)}\, \sigma(d\lambda)$ with the convention to consider $f = g$ in L^2_σ iff $f(\lambda) = g(\lambda)$ σ-a.e.

Theorem. With any $f(\lambda) \in L^2_\sigma(-\infty, \infty)$ we associate a vector $\hat{f}$ of X defined by

$$\hat{f} = \int_{-\infty}^{\infty} f(\lambda)\, dE(\lambda)\, y. \tag{2}$$

Then the correspondence $f(\lambda) \rightarrow \hat{f}$ is a one-one linear isometric mapping of $L^2_\sigma(-\infty, \infty)$ onto X. Let this mapping be denoted by V, i.e., $\hat{f} = Vf$. Then the operator $H_1 = V^{-1}HV$ in $L^2_\sigma(-\infty, \infty)$ is precisely the operator of multiplication by λ:

$$D(H_1) = D(V^{-1}HV) = \{f(\lambda);\, f(\lambda) \text{ and } \lambda f(\lambda) \text{ both } \in L^2_\sigma(-\infty,\infty)\}$$
$$\text{and } (H_1 f)(\lambda) = \lambda f(\lambda) \text{ whenever } f(\lambda) \in D(H_1). \tag{3}$$

Proof. We have, by $E(\lambda)\,E(\mu) = E(\min(\lambda,\mu))$,

$$(E(\lambda)\,y, \hat{f}) = \int_{-\infty}^{\infty} \overline{f(\mu)}\, d_\mu(E(\lambda)\,y, E(\mu)\,y) = \int_{-\infty}^{\infty} \overline{f(\mu)}\, d_\mu(E(\mu)\,E(\lambda)\,y, y)$$
$$= \int_{-\infty}^{\lambda} \overline{f(\mu)}\, d(E(\mu)\,y, y) = \int_{-\infty}^{\lambda} \overline{f(\mu)}\, \sigma(d\mu),$$

and so

$$(\hat{f}, \hat{g}) = \int_{-\infty}^{\infty} f(\lambda)\, d(E(\lambda)\,y, \hat{g}) = \int_{-\infty}^{\infty} f(\lambda)\, \overline{g(\lambda)}\, \sigma(d\lambda) = (f, g)_\sigma. \tag{4}$$

Therefore, V maps $D(V) = L^2_\sigma(-\infty,\infty)$ onto $\{\hat{f};\, \hat{f} = \int_{-\infty}^{\infty} f(\lambda)\, dE(\lambda)\,y,$ $f \in L^2_\sigma(-\infty,\infty)\}$ one-to-one, linearly and isometrically. Hence, in particular, $R(V)$ is a closed linear subspace of X. But $R(V)$ surely contains the elements of the form $\int_\alpha^\beta dE(\lambda)\,y = (E(\beta) - E(\alpha))\,y, -\infty < \alpha < \beta < \infty$, and so, by the hypothesis that the spectrum of H is simple, we see that $R(V) = R(V)^a = X$. Thus the first half of the Theorem is proved.

Next we have

$$E(\lambda)\,\hat{f} = E(\lambda) \int_{-\infty}^{\infty} f(\mu)\, dE(\mu)\,y = \int_{-\infty}^{\infty} f(\mu)\, d_\mu(E(\lambda)\,E(\mu)\,y)$$
$$= \int_{-\infty}^{\lambda} f(\mu)\, dE(\mu)\,y,$$

and so, by (4),

$$(E(\lambda)\,\hat{f}, \hat{g}) = \int_{-\infty}^{\lambda} f(\mu)\, \overline{g(\mu)}\, \sigma(d\mu). \tag{5}$$

Hence the condition $\int_{-\infty}^{\infty} \lambda^2 d(E(\lambda)\,\hat{f}, \hat{f}) < \infty$, equivalent to $\hat{f} \in D(H)$, is equivalent to the condition $\int_{-\infty}^{\infty} \lambda^2\, |f(\lambda)|^2\, \sigma(d\lambda) < \infty$. Moreover, in the last case, we have, by (20) in Chapter XI, 12,

$$HVf = H\hat{f} = \int_{-\infty}^{\infty} \lambda\, dE(\lambda)\,\hat{f},$$

and hence, by (4) and (5),

$$(H_1 f, g)_\sigma = (V^{-1}HVf, g)_\sigma = (HV \cdot f, V \cdot g) = (H \cdot \hat{f}, \hat{g})$$
$$= \int_{-\infty}^{\infty} \lambda\, d(E(\lambda)\,\hat{f}, \hat{g}) = \int_{-\infty}^{\infty} \lambda\, f(\lambda)\, \overline{g(\lambda)}\, \sigma(d\lambda).$$

On the other hand, we have

$$(H_1 f, g)_\sigma = \int_{-\infty}^{\infty} (H_1 f)(\lambda)\, g(\lambda)\, \sigma(d\lambda).$$

Therefore we must have

$$(H_1 f)(\lambda) = \lambda f(\lambda) \quad \sigma\text{-almost everywhere}.$$

Remark. There is a close connection between self-adjoint operators with simple spectrum and the *Jacobi matrices*. See M. H. STONE [1], p. 275. For a canonical form of a self-adjoint operator with not necessarily simple spectrum, see J. von Neumann's *reduction theory*: NEUMANN [9].

15. The Defect Indices of a Symmetric Operator. The Generalized Resolution of the Identity

Definition 1. Let $U = U_H = (H - iI)(H + iI)^{-1}$ be the Cayley transform of a closed symmetric operator H in a Hilbert space X. Let $X_H^+ = D(U_H)^\perp$, $X_H^- = R(U_H)^\perp$, and let $m = \dim(X_H^+)$, $n = \dim(X_H^-)$ be the dimension numbers of X_H^+, X_H^- respectively. Then H is said to be of the *defect indices* (m, n). H is self-adjoint iff it is of the defect indices $(0, 0)$ (see Chapter VII, 4).

Proposition 1. The defect indices (m, n) of a closed symmetric operator H may be defined as follows: m is the dimension number of the linear subspace $\{x \in X; H^* x = ix\}$; n is the dimension number of the linear subspace $\{x \in X; H^* x = -ix\}$.

Proof. Clear from Theorem 3 in Chapter VII, 4.

Example 1. Let $X = L^2(0, 1)$. Let D be the totality of absolutely continuous functions $x(t) \in L^2(0, 1)$ such that $x(0) = x(1) = 0$ and $x'(t) \in L^2(0, 1)$. Then the operator T_1 defined by $T_1 x(t) = i^{-1} x'(t)$ on $D = D(T_1)$ is of the defect indices $(1, 1)$.

Proof. As was shown in Example 4 of Chapter VII, 3, $T_1^* = T_2$ is defined by

$$T_2 x(t) = i^{-1} x'(t) \text{ on } D(T_2) = \{x(t) \in L^2(0, 1);\ x(t) \text{ is absolutely continuous such that } x'(t) \in L^2(0, 1)\}.$$

Thus the solution $y \in L^2(0, 1)$ of $T_1^* y = T_2 y = iy$ is a distribution solution of the differential equation

$$y'(t) = -y(t) \quad (y, y' \in L^2(0, 1)). \tag{1}$$

Then $z(t) = y(t) \exp(t)$ is a distribution solution of the differential equation

$$z'(t) = 0 \quad (z, z' \in L^2(0, 1)). \tag{2}$$

We shall show that there exists a constant C such that $z(t) = C$ for a.e. $t \in (0, 1)$. To this purpose, take any function $x_0(t) \in C_0^\infty(0, 1)$ such that

$\int_0^1 x_0(t)\,dt = 1$ and put

$$x(t) - x_0(t)\int_0^1 x(t)\,dt = u(t),\ w(t) = \int_0^t u(s)\,ds,$$

where $x(t)$ is an arbitrary function from $C_0^\infty(0,1)$. We have $w \in C_0^\infty(0,1)$ by $\int_0^1 u(s)\,ds = 0$. Therefore, by (2), we have

$$-\int_0^1 z(t)\,w'(t)\,dt = -\int_0^1 z(t)\,u(t)\,dt = 0,$$

that is,

$$\int_0^1 z(t)\,x(t)\,dt = C\int_0^1 x(t)\,dt, \quad \text{where} \quad C = \int_0^1 z(t)\,x_0(t)\,dt.$$

This proves, by the arbitrariness of $x(t) \in C_0^\infty(0,1)$, that $z(t) = C$ for a.e. $t \in (0,1)$.

Hence any solution of $T^*y = iy$ is of the form $y(t) = C\exp(-t)$. In the same manner we see that any solution of $T^*y = -iy$ is of the form $y(t) = C\exp(t)$. Thus T is of the defect indices (1, 1).

Definition 2. A symmetric operator H in a Hilbert space X is called *maximal symmetric* if there is no proper symmetric extension of H.

Proposition 2. A maximal symmetric operator H is closed and $H = H^{**}$. A self-adjoint operator H is maximal symmetric.

Proof. By Proposition 1 in Chapter VII, 3, H^{**} is a closed symmetric extension of H. Thus the first half of Proposition 2 is clear. Let H_0 be a symmetric extension of a self-adjoint operator H. Then, from $H \subseteqq H_0$, $H_0 \subseteqq H_0^*$, we obtain $H_0 \subseteqq H_0^* \subseteqq H^*$ and so, by $H = H^*$, $H \subseteqq H_0 \subseteqq H$. This proves that a self-adjoint operator H is maximal symmetric.

Corollary 1. Every maximal symmetric extension H_0 of a given symmetric operator H is also an extension of H^{**}.

Proof. The relation $H \subseteqq H_0$ implies the relation $H_0^* \subseteqq H^*$ and $H^{**} \subseteqq H_0^{**}$. Since, by the preceding proposition, $H_0 = H_0^{**}$, Corollary 1 is true.

Corollary 2. If H is a symmetric operator such that $H^* = H^{**}$, then the self-adjoint operator H^* is the only maximal symmetric extension of H.

Proof. Being self-adjoint, H^{**} is maximal symmetric. Thus any maximal symmetric extension H_0 of H, which is also a symmetric extension of H^{**} by Corollary 1, is identical to $H^{**} = H^*$.

We are thus in a position to state

Definition 3. A symmetric operator H such that $H^* = H^{**}$ is said to be *essentially self-adjoint*. A self-adjoint operator H is said to be *hypermaximal*. The latter terminology is due to J. von Neumann.

Example 2. Let $X = L^2(-\infty, \infty)$, and define an operator H by $Hx(t) = t \cdot x(t)$ for $x(t) \in C_0^0(-\infty, \infty)$. H is surely a symmetric operator in X. It is easy to see that H^* is the coordinate operator defined in Example 2 in Chapter VII, 3 so that H is essentially self-adjoint. The operator H defined by $Hx(t) = i^{-1}x'(t)$ for $x(t) \in C_0^1(-\infty, \infty)$ is also essentially self-adjoint in $X = L^2(-\infty, \infty)$. For, in this case, H^* is the momentum operator defined in Example 3 in Chapter VII, 3.

Theorem 1. Let the defect indices (m, n) of a closed symmetric operator H satisfy

$$m = m' + p, \; n = n' + p \quad (p > 0).$$

Then there exists a closed symmetric extension H' of H with the defect indices (m', n').

Proof. Let $\{\varphi_1, \varphi_2, \ldots, \varphi_p, \varphi_{p+1}, \ldots, \varphi_{p+m'}\}$, $\{\psi_1, \psi_2, \ldots, \psi_p, \psi_{p+1}, \ldots \ldots, \psi_{p+n'}\}$ be complete orthonormal systems of $X_H^+ = D(U_H)^\perp$, $X_H^- = R(U_H)^\perp$, respectively. Define an isometric extension V of U_H by

$$Vx = U_H x \quad \text{for} \quad x \in D(U_H),$$

$$V \cdot \sum_{i=1}^{p} \alpha_i \varphi_i = \sum_{i=1}^{p} \alpha_i \psi_i .$$

We have $R(I - U_H)^a = X$ by Theorem 1 in Chapter VII, 4. Thus, by $R(I - V) \supseteq R(I - U_H)$ and Theorem 2 in Chapter VII, 4, there exists a uniquely determined closed symmetric extension H' of H such that $V = (H' - iI)(H' + iI)^{-1}$. The defect indices of H' are (m', n') as may be seen from $\dim(D(V)^\perp) = m'$, $\dim(R(V)^\perp) = n'$.

Corollary. A closed symmetric operator H with the defect indices (m, n) is maximal symmetric iff either $m = 0$ or $n = 0$.

Proof. The "only if" part is clear from Theorem 1. If, for instance, $m = 0$, then, by $D(U_H) = X$, we must have $U_{H_0} = U_H$ for a closed symmetric extension H_0 of H. This proves $H_0 = i(I + U_{H_0})(I - U_{H_0})^{-1} = i(I + U_H)(I - U_H)^{-1} = H$. Also, in the case $n = 0$, we see that there is no proper closed symmetric extension of H.

Example 3. Suppose that $\{\varphi_1, \varphi_2, \ldots, \varphi_n, \ldots\}$ is a complete orthonormal system of a (separable) Hilbert space X. Then, by

$$U \cdot \sum_{i=1}^{\infty} \alpha_i \varphi_i = \sum_{i=1}^{\infty} \alpha_i \varphi_{i+1}, \quad \sum_{i=1}^{\infty} |\alpha_i|^2 < \infty$$

we define a closed isometric operator U such that $D(U) = X$ and $\dim(R(U)^\perp) = 1$. If $R(I - U)^a \neq X$, then there exists an $x \neq 0$ such that $x \in R(I - U)^\perp$. Consequently $((I - U)x, x) = 0$ and so $(Ux, x) = \|x\|^2 = \|Ux\|^2$. This implies that

$$\begin{aligned} \|(I - U)x\|^2 &= \|x\|^2 - (Ux, x) - (x, Ux) + \|Ux\|^2 \\ &= \|x\|^2 - \|x\|^2 - \|x\|^2 + \|x\|^2 = 0, \end{aligned}$$

that is, $Ux = x$ and so, by the above definition of U, x must be zero. This contradiction shows that $R(I-U)^a = X$. Hence, by Theorem 2 in Chapter VII, 4, U is the Cayley transform of a closed symmetric operator H. H is of the defect indices (0, 1) since $D(U) = X$ and $R(U)^{\perp}$ is spanned by φ_1. Thus H is maximal symmetric without being self-adjoint.

Theorem 2 (M. A. NAIMARK [3]). Let a closed symmetric operator H_1 in a Hilbert space X_1 be of the defect indices (m, n). Then we can construct a Hilbert space X, containing X_1 as a closed linear subspace, and a closed symmetric operator H in X with the defect indices $(m+n, m+n)$ such that

$$H_1 = P(X_1)\, H P(X_1), \text{ where } P(X_1) \text{ is the projection of } X \text{ onto } X_1.$$

Proof. Consider a Hilbert space X_2 of the same dimensionality as X_1. We construct a closed symmetric operator H_2 in X_2 with the defect indices (n, m). For instance, we may take $H_2 = -H_1$ supposing that X_2 coincides with X_1. Then we would have $\{x \in X_2;\, H_2^* x = ix\} = \{x \in X_1;\, H_1^* x = -ix\}$, $\{x \in X_2;\, H_2^* x = -ix\} = \{x \in X_1;\, H_1^* x = ix\}$ and so the defect indices of H_2 must be (n, m).

We then consider an operator H defined by

$$H\{x, y\} = \{H_1 x, H_2 y\} \text{ for } \{x, y\} \in X_1 \times X_2 \text{ with } x \in D(H_1),\, y \in D(H_2).$$

It is easy to see that H is a closed symmetric operator in the product Hilbert space $X = X_1 \times X_2$. The condition

$$H^*\{x, y\} = i\{x, y\} \quad (\text{or the condition } H^*\{x, y\} = -i\{x, y\})$$

means that $H_1^* x = ix$, $H_2^* y = iy$ (or $H_1^* x = -ix$, $H_2^* y = -iy$). Hence we see that the defect indices of H are $(m+n, m+n)$.

Corollary. Let, by Theorem 1, $\hat{H}$ be a self-adjoint extension of H, and let $\hat{H} = \int \lambda\, d\hat{E}(\lambda)$ be the spectral resolution of $\hat{H}$. Since H and a fortiori $\hat{H}$ are extensions of H_1 when H_1 is considered as an operator in X, we have the result:

$$\left.\begin{aligned} &\text{if } x \in D(H_1) \subseteqq X_1 = P(X_1)X, \text{ then } x = P(X_1)\,x \in D(\hat{H}) \text{ and} \\ &H_1 x = P(X_1)\,\hat{H} x = P(X_1)\,\hat{H} P(X_1)\,x = \int_{-\infty}^{\infty} \lambda\, dF(\lambda)\,x, \text{ where} \\ &F(\lambda) = P(X_1)\,\hat{E}(\lambda)\,P(X_1). \end{aligned}\right\} \quad (3)$$

The system $\{F(\lambda);\, -\infty < \lambda < \infty\}$ surely satisfies the conditions:

$$\left.\begin{aligned} &F(\lambda) \text{ is a self-adjoint operator in } X_1, \\ &\lambda_1 < \lambda_2 \text{ implies that } (F(\lambda_1)\,x, x) \leqq (F(\lambda_2)\,x, x) \text{ for every } x \in X_1, \\ &F(\lambda + 0) = F(\lambda), \\ &F(-\infty)\,x = \operatorname*{s-lim}_{\lambda \downarrow -\infty} F(\lambda)\,x = 0, \\ &F(\infty)\,x = \operatorname*{s-lim}_{\lambda \uparrow \infty} F(\lambda)\,x = x \quad \text{for all } x \in X. \end{aligned}\right\} \quad (4)$$

Remark 1. For the closed symmetric operator H_1, we have

$$H_1 x = \int_{-\infty}^{\infty} \lambda \, dF(\lambda) \, x \quad \text{for all} \quad x \in D(H_1),$$

where $\{F(\lambda); -\infty < \lambda < \infty\}$ satisfies (4). In this sense, H_1 admits a *generalized spectral resolution.*

Example 4. Let $X_1 = L^2(-\infty, 0)$, and let D_1 be the totality of absolutely continuous functions $x(t) \in L^2(-\infty, 0)$ such that $x(0) = 0$ and $x'(t) \in L^2(-\infty, 0)$. Then the operator H_1 defined by $H_1 x(t) = i^{-1} x'(t)$ on $D_1 = D(H_1)$ is a maximal symmetric operator with the defect indices (0, 1). This we see as in the above Example 1. Let $X_2 = L^2(0, \infty)$, and let D_2 be the totality of absolutely continuous functions $x(t) \in L^2(0, \infty)$ such that $x(0) = 0$ and $x'(t) \in L^2(0, \infty)$. Then the operator H_2 defined by $H_2 x(t) = i^{-1} x'(t)$ on $D_2 = D(H_2)$ is a maximal symmetric operator with the defect indices (1, 0). In this case, $X = X_1 \times X_2 = L^2(-\infty, \infty)$, and the operator H in Theorem 2 is given concretely by $Hx(t) = i^{-1} x'(t)$ for those $x(t) \in L^2(-\infty, \infty)$ for which $x(0) = 0$ and $x'(t) \in L^2(-\infty, \infty)$.

Remark 2. Since $\hat{H} = \int \lambda \, d\hat{E}(\lambda)$ is an extension of H_1, we have, for $x \in D(H_1)$,

$$\begin{aligned} \|H_1 x\|^2 = \|\hat{H} x\|^2 &= \int_{-\infty}^{\infty} \lambda^2 \, d\,\|\hat{E}(\lambda)\, x\|^2 = \int_{-\infty}^{\infty} \lambda^2 \, d(\hat{E}(\lambda)\, x, x) \\ &= \int_{-\infty}^{\infty} \lambda^2 \, d(\hat{E}(\lambda)\, P(X_1)\, x, P(X_1)\, x) = \int_{-\infty}^{\infty} \lambda^2 \, d(F(\lambda)\, x, x). \end{aligned}$$

However, the condition $\int_{-\infty}^{\infty} \lambda^2 \, d(F(\lambda)\, x, x) < \infty$ does not necessarily imply that $x \in D(H_1)$. Concerning this point we have

Theorem 3. In the case of a maximal symmetric operator H_1, we have, for the corresponding generalized spectral resolution $\int \lambda \, dF(\lambda)$ in X_1,

$$x \in D(H_1) \text{ is equivalent to the condition } \int_{-\infty}^{\infty} \lambda^2 \, d(F(\lambda)\, x, x) < \infty. \tag{5}$$

Proof. The reasoning in Remark 2 shows that $\int_{-\infty}^{\infty} \lambda^2 d(F(\lambda) x, x) < \infty$ implies that $\int_{-\infty}^{\infty} \lambda^2 \, d\,\|\hat{E}(\lambda)\, x\|^2 < \infty$, i.e., $x \in D(\hat{H})$. The operator

$$H' = P(X_1)\, \hat{H}\, P(X_1),$$

when considered as an operator in X_1, is a symmetric extension of H_1 and $D(H') = D(\hat{H}) \cap X_1$. Thus, by the maximality of H_1, we must have $H_1 = H'$. Thus $D(H_1) = D(H') = D(\hat{H}) \cap X_1$ and so, by $H' = P(X_1)\, \hat{H}\, P(X_1)$, the condition $\int_{-\infty}^{\infty} \lambda^2 \, d\,\|\hat{E}(\lambda)\, x\|^2 < \infty$ with $x \in X_1$ implies

$\int_{-\infty}^{\infty} \lambda^2 \, d(F(\lambda)\, x, x) < \infty$; and conversely, the condition $x \in D(H_1)$ implies
$\int_{-\infty}^{\infty} \lambda^2 \, d(F(\lambda)\, x, x) = \int_{-\infty}^{\infty} \lambda^2 \, d\, \|\hat{E}(\lambda)\, x\|^2 < \infty.$

Remark 3. Since, by Theorem 1, any closed symmetric operator can be extended to a maximal symmetric operator H_1, we can apply our Theorem 3 to the effect that (5) is valid. For a detailed exposition of the generalized spectral resolution, see N. I. ACHIESER-I. M. GLASMAN [1] or B. Sz. NAGY [3]. The spectral representation of a self-adjoint operator in a Hilbert space can be extended to a certain class of linear operators in a Banach space with an appropriate modification. This result is due to N. DUNFORD and may be considered as an "elementary divisor theory" in infinite dimensional spaces. N. DUNFORD-J. SCHWARTZ [6].

16. The Group-ring L^1 and Wiener's Tauberian Theorem

The Gelfand representation admits another important application in functional analysis, namely, an operator-theoretical treatment of the Tauberian theorem of N. WIENER.

The linear space $L^1(-\infty, \infty)$ is a ring with respect to the function sum and the product $\times$ defined by

$$(f \times g)(t) = (f * g)(t) = \int_{-\infty}^{\infty} f(t-s)\, g(s)\, ds. \tag{1}$$

For, by the Fubini-Tonelli theorem,

$$\int_{-\infty}^{\infty} \left| \int_{-\infty}^{\infty} f(t-s)\, g(s)\, ds \right| dt \leqq \int_{-\infty}^{\infty} |f(t-s)|\, dt \int_{-\infty}^{\infty} |g(s)|\, ds$$
$$= \int_{-\infty}^{\infty} |f(t)|\, dt \int_{-\infty}^{\infty} |g(t)|\, dt.$$

We have thus proved

$$\|f \times g\| \leqq \|f\| \cdot \|g\|, \text{ where } \|\ \| \text{ is the norm in } L^1(-\infty, \infty). \tag{2}$$

Therefore we have proved

Proposition 1. We can introduce formally a multiplicative unit e in the ring L^1 consisting of all elements $\tilde{z}$ given formally by

$$\tilde{z} = \lambda e + x, \ x \in L^1(-\infty, \infty). \tag{3}$$

In fact L^1 is a normed ring by the following rule:

$$\begin{gathered}(\lambda_1 e + x_1) + (\lambda_2 e + x_2) = (\lambda_1 + \lambda_2)\, e + (x_1 + x_2), \\ \alpha(\lambda e + x) = \alpha\lambda e + \alpha x, \\ (\lambda_1 e + x_1) \times (\lambda_2 e + x_2) = \lambda_1\lambda_2 e + \lambda_1 x_2 + \lambda_2 x_1 + x_1 \times x_2, \\ \|\lambda e + x\| = |\lambda| + \|x\|.\end{gathered} \tag{4}$$

This normed ring L^1 is called the *group-ring* of the group R^1 of real numbers written additively. We shall find all maximal ideals of this ring L^1. One is $I_0 = L^1(-\infty, \infty)$ from which L^1 is obtained by adjunction of the unit e. We shall find all maximal ideals $I \neq I_0$.

For any maximal ideal I of the normed ring L^1, we shall denote by $(\tilde{z}, I)$ the complex number which corresponds to the element $\tilde{z}$ by the ring homomorphic mapping $L^1 \to L^1/I$. Thus $(\tilde{z}, I_0) = \lambda$ for $\tilde{z} = \lambda e + x$, $x \in I_0$.

Let I be a maximal ideal of L^1 different from I_0. Then there exists a function $x \in L^1(-\infty, \infty) = I_0$ such that $(x, I) \neq 0$. We set

$$\chi(\alpha) = (x_\alpha, I)/(x, I), \text{ where } x_\alpha(t) = x(t+\alpha). \tag{5}$$

Then $\chi(0) = 1$, and, by $|(x_\alpha, I)| \leqq \|x_\alpha\| = \|x\|$, we see that $|\chi(\alpha)| \leqq \|x\|/|(x, I)|$. Thus the function $\chi(\alpha)$ is bounded in α. Moreover, by

$$|\chi(\alpha+\delta) - \chi(\alpha)| \leqq \|x_{\alpha+\delta} - x_\alpha\|/|(x, I)|,$$

the function $\chi(\alpha)$ is continuous in α. For, by (2) in Chapter X, 1, we have

$$\lim_{\delta \to 0} \int_{-\infty}^{\infty} |x(\alpha+\delta+t) - x(\alpha+t)|\, dt = 0.$$

On the other hand, we have, by $(x_{\alpha+\beta} \times x)(t) = (x_\alpha \times x_\beta)(t)$,

$$(x_{\alpha+\beta}, I)(x, I) = (x_\alpha, I)(x_\beta, I)$$

so that

$$\chi(\alpha+\beta) = \chi(\alpha)\,\chi(\beta). \tag{6}$$

From this we can prove that there exists a uniquely determined real number $\xi = \xi(I)$ such that

$$\chi(\alpha) = \exp(i \cdot \xi(I) \cdot \alpha). \tag{7}$$

In fact, by $\chi(n\alpha) = \chi(\alpha)^n$ and the boundedness of the function χ, we obtain $|\chi(\alpha)| \leqq 1$. Consequently, by $\chi(\alpha)\,\chi(-\alpha) = \chi(0) = 1$, we must have $|\chi(\alpha)| = 1$. Thus, as a continuous solution of (6) whose absolute value is one, the function $\chi(\alpha)$ must be given in form (7). That the value $\xi(I)$ is determined by I independently of the choice of $x \in L^1(-\infty,\infty)$ may be seen from $\chi(\alpha)(y, I) = (y_\alpha, I)$ which is implied by $x_\alpha \times y = x \times y_\alpha$.

Every continuous solution of (6) with the absolute value identically equal to one is called a *continuous unitary character* of the group R^1 of real numbers written additively.

We have thus constructed the (continuous unitary) character $\chi(\alpha)$ of the group R^1 with respect to the given maximal ideal $I \neq I_0$.

We next show how to reconstruct the ideal I with respect to this character or, what amounts to the same thing, to reconstruct the value (z, I) from χ.

For any $y(t) \in L^1(-\infty, \infty)$, we have

$$(x \times y)(t) = \int_{-\infty}^{\infty} x(t-s)\, y(s)\, ds = \int_{-\infty}^{\infty} x_{-s}(t)\, y(s)\, ds.$$

Hence, by (5) and the continuity of the ring homomorphism $L^1 \to L^1/I$, we obtain

$$(x \times y, I) = (x, I)\,(y, I) = (x, I) \int_{-\infty}^{\infty} \chi(-s)\, y(s)\, ds.$$

Thus, by $(x, I) \neq 0$, we obtain

$$(y, I) = \int_{-\infty}^{\infty} y(s) \exp(-i \cdot \xi(I) \cdot s)\, ds. \tag{8}$$

Therefore, for any $\tilde{z} = \lambda e + x$ with $x \in L^1(-\infty, \infty)$,

$$(\tilde{z}, I) = (\lambda e, I) + (x, I) = \lambda + \int_{-\infty}^{\infty} x(s) \exp(-i \cdot \xi(I) \cdot s)\, ds. \tag{9}$$

Conversely, any (continuous, unitary) character $\chi(\alpha) = \exp(i \cdot \xi \cdot \alpha)$ of the group R^1 defines a ring homomorphism

$$\tilde{z} \to \lambda + \int_{-\infty}^{\infty} x(s) \exp(-i \cdot \xi \cdot s)\, ds \quad (\tilde{z} = \lambda e + x,\ x \in L^1(-\infty, \infty)) \tag{10}$$

of L^1 onto the ring of complex numbers. For, we have, by the Fubini-Tonelli theorem.

$$\int_{-\infty}^{\infty} (x_1 \times x_2)(t) \exp(-i \cdot \xi \cdot t)\, dt$$
$$= \int_{-\infty}^{\infty} x_1(t) \exp(-i \cdot \xi \cdot t)\, dt \int_{-\infty}^{\infty} x_2(t) \exp(-i \cdot \xi \cdot t)\, dt.$$

We have thus proved

Theorem 1 (GELFAND [4] and RAIKOV [1]). There exists a one-to-one correspondence between the set of all maximal ideals $I \neq L^1(-\infty, \infty)$ of the group ring L^1 of the group R^1 and the set of all continuous, unitary characters $\chi(\alpha)$ of this group R^1. This correspondence is defined by formula (9).

We shall show that the normed ring L^1 is semi-simple or, what amounts to the same thing (see Chapter XI, 2), that the following theorem is true.

Theorem 2. The normed ring L^1 has no generalized nilpotent elements other than 0.

Proof. Let $x \in L^1(-\infty, \infty)$ and $y \in L^2(-\infty, \infty)$. Then, by Schwarz' inequality,

$$\left| \int_{-\infty}^{\infty} x(t-s)\, y(s)\, ds \right| \leq \left(\int_{-\infty}^{\infty} |x(t-s)|\, ds \int_{-\infty}^{\infty} |x(t-s)|\, |y(s)|^2\, ds \right)^{1/2},$$

and so, by the Fubini-Tonelli theorem, we see that the left hand side belongs to $L^2(-\infty, \infty)$ and

$$\|x \times y\|_2 \leqq \|x\| \, \|y\|_2, \text{ where } \| \ \|_2 \text{ is the norm in } L^2(-\infty, \infty). \tag{11}$$

Hence we can define a bounded linear operator T_x on $L^2(-\infty, \infty)$ into $L^2(-\infty, \infty)$ by

$$(T_x y)(t) = \int_{-\infty}^{\infty} x(t-s)\, y(s)\, ds \quad \text{whenever} \quad x \in L^1(-\infty, \infty); \tag{12}$$

and, moreover, we have

$$\|T_x\|_2 \leqq \|x\|, \tag{13}$$

$$T_x^* = T_{x^*} \quad \text{where} \quad x^*(t) = \overline{x(-t)}. \tag{14}$$

Thus, by applying the Fubini-Tonelli theorem again, we obtain

$$T_x T_x^* = T_{x \times x^*} = T_{x^* \times x} = T_x^* T_x, \text{ i.e., } T_x \text{ is a normal operator.} \tag{15}$$

Hence, by Chapter XI, 3, we have $\|T_x\|_2 = \lim_{n \to \infty} (\|T_x^n\|_2)^{1/n}$. Therefore, by

$$\|\underbrace{x \times x \times x \times \cdots \times x}_{n \text{ factor}}\| \geqq \|T_x^n\|_2,$$

we see that, if x is a generalized nilpotent element, then $\|T_x\|_2 = 0$. From this fact we easily prove that x must be a zero vector of $L^1(-\infty, \infty)$.

Let now $\tilde{z} = \lambda e + x$ with $x \in L^1(-\infty, \infty)$ be a generalized nilpotent element of the normed ring L^1. Then, for any maximal ideal I of L^1, we must have $(\tilde{z}, I) = \lambda + (x, I) = 0$. Hence the Fourier transform

$$(2\pi)^{-1/2} \int_{-\infty}^{\infty} x(t) \exp(-i \cdot \xi \cdot t)\, dt$$

must be identically equal to $-(2\pi)^{-1/2} \lambda$. From this we prove that λ must be 0. For, we have (the Riemann-Lebesgue theorem)

$$\left| \int_{-\infty}^{\infty} x(t) \exp(-i \cdot \xi \cdot t)\, dt \right| = \left| 2^{-1} \int_{-\infty}^{\infty} \left[x(t) - x\left(t + \frac{\pi}{\xi}\right) \right] \exp(-i \cdot \xi \cdot t)\, dt \right|$$

$$\leqq 2^{-1} \int_{-\infty}^{\infty} \left| x(t) - x\left(t + \frac{\pi}{\xi}\right) \right| dt \to 0 \quad \text{as} \quad \xi \to \infty.$$

Thus a generalized nilpotent element $\tilde{z} \in L^1$ must be of the form $\tilde{z} = x$ with $x \in L^1(-\infty, \infty)$ and so, by what we have proved above, $\tilde{z} = x = 0$.

We are now in a position to state and prove the *Tauberian theorem of N. Wiener* [2]:

Theorem 3. Let $x(t) \in L^1(-\infty, \infty)$ be such that its Fourier transform

$$(2\pi)^{-1/2} \int_{-\infty}^{\infty} x(t) \exp(-i \cdot \xi \cdot t)\, dt$$

does not vanish for any real ξ. Then, for any $y(t) \in L^1(-\infty, \infty)$ and $\varepsilon > 0$, we can find the real numbers β's, the complex numbers α's and a positive integer N in such a way that

$$\int_{-\infty}^{\infty} \left| y(t) - \sum_{j=1}^{N} \alpha_j x(t-\beta_j) \right| dt < \varepsilon. \tag{16}$$

Proof. It suffices to find $\tilde{z} \in L^1$ in such a way that

$$\|y - x \times \tilde{z}\| \leqq \varepsilon/2. \tag{17}$$

We first show that

$$\lim_{\alpha \to \infty} \int_{-\infty}^{\infty} |y(t) - y^{(\alpha)}(t)| \, dt = 0, \text{ where}$$
$$y^{(\alpha)}(t) = \frac{1}{\pi} \int_{-\infty}^{\infty} y(t-s) \frac{1 - \cos \alpha s}{\alpha s^2} ds. \tag{18}$$

For, by $\int_{-\infty}^{\infty} (1 - \cos \alpha s) (\alpha s^2)^{-1} ds = \pi \ (\alpha > 0)$, the left side is

$$\leqq (\pi)^{-1} \lim_{\alpha \to \infty} \int_{-\infty}^{\infty} \frac{(1 - \cos s)}{s^2} ds \left\{ \int_{-\infty}^{\infty} \left| y(t) - y\left(t - \frac{s}{\alpha}\right) \right| dt \right\} = 0.$$

Next we have

$$(2\pi)^{-1/2} \int_{-\infty}^{\infty} \left(\frac{2}{\pi}\right)^{1/2} \frac{1 - \cos \alpha u}{\alpha u^2} e^{-iu\xi} du = \begin{cases} 1 - |\xi|/\alpha & (|\xi| < \alpha), \\ 0 & (|\xi| \geqq \alpha). \end{cases} \tag{19}$$

For, we have

$$(2\pi)^{-1/2} \int_{-\alpha}^{\alpha} (1 - |\xi|/\alpha) e^{i\xi u} d\xi = \left(\frac{2}{\pi}\right)^{1/2} \int_0^{\alpha} \left(1 - \frac{\xi}{\alpha}\right) \cos u\xi \cdot d\xi$$
$$= \left(\frac{2}{\pi}\right)^{1/2} \frac{1}{u} \int_0^{\alpha} \left(1 - \frac{\xi}{\alpha}\right) d(\sin u\xi) = \left(\frac{2}{\pi}\right)^{1/2} \int_0^{\alpha} \frac{\sin u\xi}{u\alpha} d\xi$$
$$= \left(\frac{2}{\pi}\right)^{1/2} \frac{1 - \cos u\alpha}{u^2 \alpha},$$

and so we have only to apply Plancherel's theorem in Chapter VI, 2. Hence, by the Parseval relation of the Fourier transform,

$$y^{(\alpha)}(t) \text{ satisfies } \int_{-\infty}^{\infty} y^{(\alpha)}(t) \exp(-i \cdot \xi \cdot t) \, dt = 0 \text{ if } |\xi| \geqq \alpha. \tag{20}$$

Therefore, we may assume that $y \in L^1(-\infty, \infty)$ in (17) satisfies the condition

$$\int_{-\infty}^{\infty} y(t) \exp(-i \cdot \xi \cdot t) \, dt = 0 \text{ whenever } |\xi| \geqq \alpha. \tag{21}$$

Next we choose a positive number β and a sufficiently large positive number γ such that $[-\beta - \gamma, -\beta + \gamma]$ and $[\beta - \gamma, \beta + \gamma]$ both contain

$[-\alpha, \alpha]$. Let $C_1(\xi)$ and $C_2(\xi)$ be the defining functions of the intervals $[-\gamma, \gamma]$ and $[-\beta, \beta]$, respectively. Then

$$\begin{cases} u(\xi) = (2\beta)^{-1} \int\limits_{-\infty}^{\infty} C_1(\xi-\eta)\, C_2(\eta)\, d\eta = 1 \text{ for } \xi \in [-\alpha, \alpha], \\ \qquad\qquad\qquad\qquad\qquad\qquad\quad = 0 \text{ for sufficiently large } |\xi|, \\ 0 \leqq u(\xi) \leqq 1 \text{ for all real } \xi. \end{cases} \tag{22}$$

By the Parseval relation of the Fourier transform, we have

$$\hat{u}(t) = \frac{1}{2\beta} (2\pi)^{1/2}\, \hat{C}_1(t)\, \hat{C}_2(t).$$

Thus, by Plancherel's theorem, we see that $\hat{u}(t)$ belongs to $L^1(-\infty,\infty) \cap L^2(-\infty,\infty)$. Hence we may apply Theorem 1 so that

$$f(t) = \hat{u}(t) = (2\pi)^{-1/2}\,(u, I_t) \text{ with a certain maximal ideal } I_t \neq I_0. \tag{23}$$

Moreover, by Plancherel's theorem, the inverse Fourier transform of $f(t)$ is equal to $u(\xi)$, i.e., we have

$$u(\xi) = (2\pi)^{-1/2}\,(f, I_{-\xi}). \tag{24}$$

We next set $\tilde{g} = e - (2\pi)^{-1/2} f$. Then, by what was proved above,

$$\begin{aligned} &0 \leqq (\tilde{g}, I_\xi) \leqq 1;\ (\tilde{g}, I_\xi) = 0 \quad \text{whenever} \quad \xi \in [-\alpha, \alpha]; \\ &(\tilde{g}, I_\xi) = 1 \text{ for all sufficiently large } |\xi|. \end{aligned} \tag{25}$$

On the other hand, we have, for $x^*(t) = \overline{x(-t)}$, the relation $\overline{(x, I_\xi)} = (x^*, I_\xi)$. Thus, by hypothesis, we have $(x^* \times x, I_\xi) = |(x, I_\xi)|^2 > 0$ for all real ξ. Hence the element

$$\tilde{g} + x^* \times x \in L^1$$

satisfies the condition that $(\tilde{g} + x^* \times x, I) > 0$ for all maximal ideals I of L^1. Consequently, the inverse $(\tilde{g} + x^* \times x)^{-1}$ in L^1 does exist.

We define

$$\tilde{z} = (\tilde{g} + x \times x^*)^{-1} \times x^* \times y. \tag{26}$$

Then, by (4), the element $x \times \tilde{z}$ belongs to $L^1(-\infty, \infty)$. Moreover, we have for every real number ξ,

$$(x \times \tilde{z}, I_\xi) = (x, I_\xi)\,(\tilde{z}, I_\xi) = (x, I_\xi)\,\frac{(x^*, I_\xi)\,(y, I_\xi)}{(\tilde{g}, I_\xi) + (x, I_\xi)\,(x^*, I_\xi)}.$$

Hence, by (25) and the hypothesis that $(y, I_\xi) = 0$ whenever $|\xi| \geqq \alpha$, we obtain

$$(x \times \tilde{z}, I_\xi) = (y, I_\xi) \text{ for all real } \xi.$$

The normed ring L^1 is semi-simple by Theorem 2. Hence we must have $x \times \tilde{z} = y$. We have thus proved Theorem 3.

Corollary. Let $k_1(t)$ belong to $L^1(-\infty, \infty)$, and let its Fourier transform vanish for no real argument. Let $k_2(t)$ belong to $L^1(-\infty, \infty)$. Let

$f(t)$ be Baire measurable and bounded over $(-\infty, \infty)$. Let, for a constant C,

$$\lim_{t\to\infty} \int_{-\infty}^{\infty} k_1(t-s) f(s)\, ds = C \int_{-\infty}^{\infty} k_1(t)\, dt. \tag{27}$$

Then

$$\lim_{t\to\infty} \int_{-\infty}^{\infty} k_2(t-s) f(s)\, ds = C \int_{-\infty}^{\infty} k_2(t)\, dt. \tag{28}$$

Proof. We may plainly suppose that $C = 0$. And (28) holds for $k_2(t)$ of the form $k_2(t) = (x \times k_1)(t)$ with $x(t) \in L^1(-\infty, \infty)$. It is easy to prove that (28) holds for $k_2(t)$ if $k_2(t) = s\text{-}\lim_{n\to\infty} k^{(n)}(t)$ in $L^1(-\infty, \infty)$ with $k^{(n)}(t) \in L^1(-\infty, \infty)$ for which (28) holds $(n = 1, 2, \ldots)$. Hence, by Theorem 3, we see that (28) holds for every $k_2(t) \in L^1(-\infty, \infty)$.

Remark. N. WIENER ([1], [2] and [3]) has applied the above Corollary to a unified treatment of classical results concerning the limit relations in series and integrals which include a new proof of the prime number theorem. Cf. also H. R. PITT [1]. The above proof of Theorem 3 is adapted from M. FUKAMIYA [1] and I. E. SEGAL [1]. Cf. also M. A. NAIMARK [1] and C. E. RICKART [1] and the bibliographies cited in these books. For the sake of comprehension of the scope of the above Corollary, we shall reproduce Wiener's deduction of the special Tauberian theorem. The special Tauberian theorem, as formulated by J. E. LITTLEWOOD reads:

Theorem 4. Let $\sum_{n=0}^{\infty} a_n x^n$ converge to $s(x)$ for $|x| < 1$, and let

$$\lim_{x\to 1-0} s(x) = C. \tag{29}$$

Let, moreover,

$$\sup_{n\geq 1} n\,|a_n| = K < \infty. \tag{30}$$

Then

$$\sum_{n=0}^{\infty} a_n = C. \tag{31}$$

Proof. Put $f(x) = \sum_{n=0}^{[x]} a_n$. Then we see, by

$$|f(x) - s(e^{-1/x})| = \left| \sum_{n=1}^{[x]} a_n (1 - e^{-n/x}) - \sum_{[x]+1}^{\infty} a_n e^{-n/x} \right|$$

$$\leq \sum_{n=1}^{[x]} \frac{K}{n} \frac{n}{x} + \sum_{[x]+1}^{\infty} \frac{K}{n} e^{-n/x} \leq 2K + K \int_{[x]}^{\infty} e^{-u/x} u^{-1}\, du$$

$$\leq 2K + K \int_{1}^{\infty} e^{-u} u^{-1}\, du = \text{constant},$$

that $f(x)$ is bounded.

Hence, by partial integration,

$$s(e^{-x}) = \sum_{n=0}^{\infty} a_n e^{-nx} = \int_{-0}^{\infty} e^{-ux} df(u) = \int_0^{\infty} x e^{-ux} f(u)\, du.$$

Thus we have

$$C = \lim_{x\to 0} \int_0^{\infty} x e^{-ux} f(u)\, du = \lim_{\xi\to\infty} \int_{-\infty}^{\infty} e^{-\xi} e^{-e^{\eta-\xi}} f(e^{\eta})\, e^{\eta}\, d\eta. \tag{31'}$$

This may be written as

$$\lim_{t\to\infty} \int_{-\infty}^{\infty} k_1(t-s)\, f(e^s)\, ds = C \int_{-\infty}^{\infty} k_1(t)\, dt, \text{ where } k_1(t) = e^{-t} e^{-e^{-t}}, \tag{32}$$

since

$$\int_{-\infty}^{\infty} k_1(t)\, dt = \int_{-\infty}^{\infty} e^{-t} e^{-e^{-t}} dt = \int_0^{\infty} e^{-x} dx = 1.$$

Furthermore, we have

$$\int_{-\infty}^{\infty} k_1(t)\, e^{-iut} dt = \int_0^{\infty} x^{iu} e^{-x} dx = \Gamma(1+iu) \neq 0.$$

Thus we can apply the Corollary and obtain, for

$$k_2(t) = 0 \ (t<0); \ k_2(t) = e^{-t} \ (t>0),$$

the limit relation

$$C = C\int_0^{\infty} e^{-t} dt = C\int_{-\infty}^{\infty} k_2(t)\, dt = \lim_{t\to\infty} \int_{-\infty}^{\infty} k_2(t-s)\, f(e^s)\, ds$$

$$= \lim_{x\to\infty} x^{-1} \int_0^x f(y)\, dy.$$

It follows that, if $\lambda > 0$,

$$C = \frac{(1+\lambda)C - C}{\lambda} = \lim_{x\to\infty} \frac{1}{\lambda x} \left\{ \int_0^{(1+\lambda)x} f(y)\, dy - \int_0^x f(y)\, dy \right\}$$

$$= \lim_{x\to\infty} \frac{1}{\lambda x} \int_x^{(1+\lambda)x} f(y)\, dy = \lim_{x\to\infty} \left\{ f(x) + \frac{1}{\lambda x} \int_x^{(1+\lambda)x} [f(y) - f(x)]\, dy \right\}. \tag{33}$$

On the other hand, we have, by (30),

$$\left| \frac{1}{\lambda x} \int_x^{(1+\lambda)x} [f(y) - f(x)]\, dy \right| \leq \frac{1}{\lambda x} \int_x^{(1+\lambda)x} \sum_{[x]+1}^{[y]} \frac{K}{n}\, dy$$

$$\leq \sum_{[x]+1}^{[(1+\lambda)x]} \frac{K}{[x]} \leq \frac{[\lambda x]K}{[x]} \leq 2\lambda K$$

for sufficiently large value of x. Hence, by (33),

$$\overline{\lim_{x\to\infty}}\ |f(x) - C| \leq 2\lambda K,$$

and since λ is any positive number, we get

$$\lim_{x\to\infty} f(x) = C.$$

Thus we have proved (31).

XII. Other Representation Theorems in Linear Spaces

In this chapter, we shall prove three representation theorems in linear spaces. The first one, the *Krein-Milman theorem* says that a non-void convex compact subset K of a locally convex linear topological space is equal to the closure of the convex hull of the extremal points of K. The other two theorems concern the *representations of a vector lattice* as point functions and as set functions.

1. Extremal Points. The Krein-Milman Theorem

Definition. Let K be a subset of a real- or complex-linear space X. A non-void subset $M \subseteqq K$ is said to be an *extremal subset* of K, if a proper convex combination $\alpha k_1 + (1-\alpha) k_2$, $0 < \alpha < 1$, of two points k_1 and k_2 of K lies in M only if both k_1 and k_2 are in M. An extremal set of K consisting of just one point is called an *extremal point* of K.

Example. In a three dimensional Euclidean space, the surface of a solid sphere is an extremal subset of the sphere, and every point of the surface is an extremal point of the surface.

Theorem (Krein-Milman). A non-void compact convex subset K of a locally convex linear topological space X has at least one extremal point.

Proof. The set K is itself an extremal set of K. Let $\mathfrak{M}$ be the totality of compact extremal subsets M of K. Order $\mathfrak{M}$ by inclusion relation. It is easy to see that, if $\mathfrak{M}_1$ is a linearly ordered subfamily of $\mathfrak{M}$, the non-void set $\bigcap_{M\in\mathfrak{M}_1} M$ is a compact extremal subset of K which is a lower bound for the subfamily $\mathfrak{M}_1$.

Thus, by Zorn's lemma, $\mathfrak{M}$ contains a minimal element M_0. Suppose that M_0 contains two distinct points x_0 and y_0. Then there exists a continuous linear functional f on X such that $f(x_0) \neq f(y_0)$. We may assume that $Re\, f(x_0) \neq Re\, f(y_0)$. M_0 being compact, the set $M_1 = \{x \in M_0;\ Re\, f(x) = \inf_{y\in M_0} Re\, f(y)\}$ is a proper subset of M_0. On the other hand, if k_1 and k_2 are points of K such that $\alpha k_1 + (1-\alpha) k_2 \in M_1$ for some α with $0 < \alpha < 1$, then, by the extremal property of M_0, k_1 and k_2 both $\in M_0$. It follows from the definition of M_1 that k_1 and k_2 both $\in M_1$. Hence M_1 is a closed extremal subset properly contained in M_0. Since M_0 is a minimal element of $\mathfrak{M}$, we have arrived at a contradiction. Therefore M_0 must consist of only one point which is thus an extremal point of K.

Corollary. Let K be a non-void compact convex subset of a locally convex real linear topological space X. Let E be the totality of the extremal points of K. Then K coincides with the smallest closed set containing every convex combination $\sum_i \alpha_i e_i \left(\alpha_i \geqq 0,\ \sum_i \alpha_i = 1\right)$ of points $e_i \in E$, i.e., K is equal to the closure of the *convex hull* $\mathrm{Conv}(E)$ of E.

Proof. The inclusion $E \subseteqq K$ and the convexity of K imply that $\mathrm{Conv}(E)^a$ is contained in the compact set K. Suppose that there exists a point k_0 contained in $(K - \mathrm{Conv}(E)^a)$. We can then take a point $c \in \mathrm{Conv}(E)^a$ so that $(k_0 - c) \overline{\in} (\mathrm{Conv}(E)^a - c)$. The set $(\mathrm{Conv}(E)^a - c)$ being compact convex and $\ni 0$, there exists, by Theorem 3′ in Chapter IV, 6, a continuous real linear functional f on X such that

$$f(k_0 - c) > 1 \text{ and } f(k - c) \leqq 1 \text{ for } (k - c) \in (\mathrm{Conv}(E)^a - c).$$

Let $K_1 = \left\{x \in K;\ f(x) = \sup_{y \in K} f(y)\right\}$. Then, since $k_0 \in K$, the set $K_1 \cap E$ must be void. Moreover, since K is compact, K_1 is a closed extremal subset of K. On the other hand, any extremal subset of K_1 is also an extremal subset of K and hence any extremal point of K_1, which surely exists by the preceding Theorem, is also an extremal point of K. Since $K_1 \cap E$ is void, we have arrived at a contradiction.

Remark. The above Theorem and the Corollary were first proved by M. KREIN-D. MILMAN [1]. The proof given above is adapted from J. L. KELLEY [2]. It is to be noted that for the unit sphere $S = \{x \in X;\ \|x\| \leqq 1\}$ in a Hilbert space X the extremal points of S are precisely those on the surface of S, i.e., those of norm 1. This we easily see from (1) in Chapter I, 5. For applications of the notion of the extremal points to concrete function spaces, see, e.g., K. HOFFMAN [1].

A simple example. Let $C[0, 1]$ be the space of real-valued continuous functions $x(t)$ defined on $[0, 1]$ normed by $x = \max_t |x(t)|$. The dual space $X = C[0, 1]'$ is the space of real Baire measures on $[0, 1]$ of bounded total variations. The unit sphere K of X is compact in the weak* topology of X (see Theorem 1 in the Appendix to Chapter IV). It is easy to see that extremal points of K are in one-to-one correspondence with the linear functionals $f_{t_0} \in X$ of the form $\langle x, f_{t_0}\rangle = x(t_0)$, $t_0 \in [0, 1]$. The above Corollary says that any linear functional $f \in X$ is given as the weak* limit of the functionals of the form

$$\sum_{j=1}^{n} \alpha_j x(t_j), \quad \text{where } \alpha_j > 0,\ \sum_{j=1}^{n} \alpha_j = 1 \text{ and } t_j \in [0, 1].$$

Recently, G. CHOQUET [1] proved a more precise result: If X is a metric space, then E is a G_δ set and, for every $x \in K$, there exists a non-negative Baire measure $\mu_x(B)$ defined for Baire sets B of X such that $\mu_x(X-E) = 0$, $\mu_x(E) = 1$ and $x = \int_E y\mu_x(dy)$. As for the uniqueness of μ_x and further literature, see G. CHOQUET and P. A. MEYER [2].

2. Vector Lattices

The notion of "positivity" in concrete function spaces is very important, in theory as well as in application. A systematic abstract treatment of the "positivity" in linear spaces was introduced by F. RIESZ [6], and further developed by H. FREUDENTHAL [2], G. BIRKHOFF [1] and many other authors. These results are called the theory of *vector lattice*. We shall begin with the definition of the vector lattice.

Definition 1. A real linear space X is said to be a *vector lattice* if X is a lattice by a partial order relation $x \leqq y$ satisfying the conditions:

$$x \leqq y \text{ implies } x + z \leqq y + z, \tag{1}$$

$$x \leqq y \text{ implies } \alpha x \leqq \alpha y \text{ (or } \alpha x \geqq \alpha y) \text{ for every } \alpha \geqq 0 \text{ (or } \alpha \leqq 0). \tag{2}$$

Proposition 1. If, in a vector lattice X, we define

$$x^+ = x \vee 0 \text{ and } x^- = x \wedge 0, \tag{3}$$

then we have

$$x \vee y = (x - y)^+ + y, \quad x \wedge y = -((-x) \vee (-y)). \tag{4}$$

Proof. The one-one mappings $x \to x + z$ and $x \to \alpha x$ $(\alpha > 0)$ of X onto X both preserve the partial order in X.

Example. The totality $A(S, \mathfrak{B})$ of real-valued, σ-additive set functions $x(B)$ defined and finite on a σ-additive family $(S, \mathfrak{B})$ of sets $B \subseteq S$ is a vector lattice by

$$(x + y)(B) = x(B) + y(B), \quad (\alpha x)(B) = \alpha x(B)$$

and the partial order $x \leqq y$ defined by $x(B) \leqq y(B)$ on $\mathfrak{B}$. In fact, we have, in this case,

$$x^+(B) = \sup_{N \subseteq B} x(N) = \text{the positive variation } \overline{V}(x; B) \text{ of } x \text{ on } B. \tag{5}$$

Proof. We have to show that $\overline{V}(x; B) = (x \vee 0)(B)$. It is clear that $\overline{V}(x; B) \geqq 0$ and $x(B) \leqq \overline{V}(x; B)$ on $\mathfrak{B}$. If $0 \leqq y(B)$ and $x(B) \leqq y(B)$ on $\mathfrak{B}$, then, for any $N \subseteq B$, $y(B) = y(N) + y(B - N) \geqq x(N)$ and so $y(B) \geqq \overline{V}(x; B)$ on $\mathfrak{B}$.

Proposition 2. In a vector lattice X, we have

$$x \overset{\vee}{\wedge} y + z = (x+z) \overset{\vee}{\wedge} (y+z), \tag{6}$$

$$\alpha\,(x \overset{\vee}{\wedge} y) = (\alpha x) \overset{\vee}{\wedge} (\alpha y) \quad \text{for} \quad \alpha > 0, \tag{7}$$

$$\alpha\,(x \overset{\vee}{\wedge} y) = (\alpha x) \overset{\wedge}{\vee} (\alpha y) \quad \text{for} \quad \alpha < 0, \tag{8}$$

$$x \wedge y = -(-x) \vee (-y),\ x^- = -(-x)^+,\ x^+ = -(-x)^-. \tag{9}$$

Proof. Clear from (1) and (2).

Corollary.

$$x + y = x \vee y + x \wedge y, \text{ in particular, } x = x^+ + x^-. \tag{10}$$

Proof. $x \vee y - x - y = 0 \vee (y-x) - y = (-y) \vee (-x) = -(y \wedge x)$.

Proposition 3. We have

$$x \overset{\vee}{\wedge} y = y \overset{\vee}{\wedge} x \ \text{(commutativity)}, \tag{11}$$

$$\left.\begin{aligned} x \vee (y \vee z) &= (x \vee y) \vee z = \sup(x, y, z) \quad (12)\\ x \wedge (y \wedge z) &= (x \wedge y) \wedge z = \inf(x, y, z) \quad (13)\end{aligned}\right\}\ \text{(associativity)},$$

$$\left.\begin{aligned} (x \wedge y) \vee z &= (x \vee z) \wedge (y \vee z) \quad (14)\\ (x \vee y) \wedge z &= (x \wedge z) \vee (y \wedge z) \quad (15)\end{aligned}\right\}\ \text{(distributivity)}.$$

Proof. We have only to prove the distributivity. It is clear that $(x \wedge y) \vee z \leqq x \vee z,\ y \vee z$. Let $w \leqq x \vee z,\ y \vee z$. Then $w \leqq x \vee z = x + z - x \wedge z$ and so $x + z \geqq w + x \wedge z$. Similarly we have $y + z \geqq w + y \wedge z$. Hence

$$\begin{aligned} w + (x \wedge z) \wedge (y \wedge z) &= (w + x \wedge z) \wedge (w + y \wedge z)\\ &\leqq (x+z) \wedge (y+z) = x \wedge y + z, \end{aligned}$$

and so

$$w \leqq (x \wedge y) + z - (x \wedge y) \wedge z = (x \wedge y) \vee z.$$

We have thus proved (14). (15) is proved by substituting $-x, -y, -z$ for x, y, z, respectively in (14).

Remark 1. In a lattice, the *distributive identity*

$$x \wedge (y \vee z) = (x \wedge y) \vee (x \wedge z) \tag{16}$$

does not hold in general. The *modular identity*:

$$x \leqq z \text{ implies } x \vee (y \wedge z) = (x \vee y) \wedge z \tag{17}$$

is weaker than the distributive identity (15). Let G be a group. Then the totality of invariant subgroups N of G constitutes a *modular lattice*, that is, a lattice satisfying the modular identity (17), if we define $N_1 \vee N_2$ and $N_1 \wedge N_2$ as the invariant subgroup generated by N_1 and N_2 and the invariant subgroup $N_1 \cap N_2$, respectively.

Remark 2. A typical example of a *distributive lattice* is given by the *Boolean algebra*; a distributive lattice B is called a Boolean algebra if it satisfies the conditions: (i) there exist elements I and 0 such that $0 \leqq x \leqq 1$ for every $x \in B$, (ii) for any $x \in B$, there exists a uniquely determined *complement* $x' \in B$ which satisfies $x \vee x' = I$, $x \wedge x' = 0$. The totality B of subsets of a fixed set is a Boolean algebra by defining the partial order in B by the inclusion relation.

Proposition 4. We define, in a vector lattice X, the absolute value

$$|x| = x \vee (-x). \tag{18}$$

Then

$$|x| \geqq 0, \text{ and } |x| = 0 \text{ iff } x = 0, \tag{19}$$

$$|x+y| \leqq |x| + |y|, \ |\alpha x| = |\alpha|\,|x|. \tag{20}$$

Proof. We have

$$x^+ \wedge (-x^-) = x^+ \wedge (-x)^+ = 0. \tag{21}$$

For, by $0 = x - x = x \vee (-x) + x \wedge (-x) \geqq 2(x \wedge (-x))$ and the distributive identity (14), $0 = (x \wedge (-x)) \vee 0 = x^+ \wedge (-x)^+$. Thus

$$x^+ - x^- = x^+ + (-x)^+ = x^+ \vee (-x)^+.$$

On the other hand, from $x \vee (-x) \geqq x \wedge (-x) \geqq -((-x) \vee x)$, we have $x \vee (-x) \geqq 0$ and so by (21)

$$x \vee (-x) = (x \vee (-x)) \vee 0 = x^+ \vee (-x)^+ = x^+ + (-x)^+.$$

We have thus proved

$$x^+ - x^- = x^+ + (-x)^+ = x^+ \vee (-x)^+ = x \vee (-x). \tag{22}$$

We now prove (19) and (20). If $x = x^+ + x^- \neq 0$, then x^+ or $(-x^-)$ is > 0 so that $|x| = x^+ \vee (-x^-) > 0$. $|\alpha x| = (\alpha x) \vee (-\alpha x) = |\alpha|\,(x \vee (-x)) = |\alpha|\,|x|$. From $|x| + |y| \geqq x + y, -x - y$, we obtain $|x| + |y| \geqq (x+y) \vee (-x-y) = |x+y|$.

Remark. The decomposition $x = x^+ + x^-$, $x^+ \wedge (-x^-) = 0$, is called the *Jordan decomposition* of x; x^+, x^- and $|x|$ correspond to the *positive variation*, the *negative variation* and the *total variation*, of a function $x(t)$ of bounded variation, respectively.

Proposition 5. For any $y \in X$, we have

$$|x - x_1| = |x \vee y - x_1 \vee y| + |x \wedge y - x_1 \wedge y|. \tag{23}$$

Proof. We have

$$\begin{aligned} |a-b| &= (a-b)^+ - (a-b)^- = a \vee b - b - (a \wedge b - b) \\ &= a \vee b - a \wedge b. \end{aligned}$$

Hence the right side of (23) is, by (10), (14) and (15),

$$\begin{aligned} &= (x \vee y) \vee x_1 - (x \vee y) \wedge (x_1 \vee y) + (x \wedge y) \vee (x_1 \wedge y) - (x \wedge y) \wedge x_1 \\ &= (x \vee x_1) \vee y - (x \wedge x_1) \vee y + (x \vee x_1) \wedge y - (x \wedge x_1) \wedge y \\ &= x \vee x_1 + y - (x \wedge x_1 + y) = x \vee x_1 - x \wedge x_1 = |x - x_1|. \end{aligned}$$

Definition 2. A sequence $\{x_n\}$ of a vector lattice X is said to *O-converge* to an element $x \in X$, in symbol, $O\text{-}\lim\limits_{n\to\infty} x_n = x$ if there exists a sequence $\{w_n\}$ such that $|x - x_n| \leqq w_n$ and $w_n \downarrow 0$. Here $w_n \downarrow 0$ means that

$w_1 \geqq w_2 \geqq \cdots$ and $\bigwedge_{n\geqq 1} w_n = 0$. The $O\text{-}\lim_{n\to\infty} x_n$, if it exists, is uniquely determined. For, let $O\text{-}\lim_{n\to\infty} x_n = x$ and $O\text{-}\lim_{n\to\infty} x_n = y$. Then $|x - x_n| \leqq w_n$, $w_n \downarrow 0$ and $|y - x_n| \leqq u_n$, $u_n \downarrow 0$. Thus $|x - y| \leqq |x - x_n| + |x_n - y| \leqq w_n + u_n$ and $(w_n + u_n) \downarrow 0$ as may be seen from $w_n + \bigwedge_{n\geqq 1} u_n \geqq \bigwedge_{n\geqq 1} (w_n + u_n)$. This proves that $x = y$.

Proposition 6. With respect to the notion of O-lim, the operations $x + y$, $x \vee y$ and $x \wedge y$ are continuous in x and y.

Proof. Let $O\text{-}\lim_{n\to\infty} x_n = x$, $O\text{-}\lim_{n\to\infty} y_n = y$. Then $|x + y - x_n - y_n| \leqq |x - x_n| + |y - y_n|$ implies $O\text{-}\lim_{n\to\infty} (x_n + y_n) = x + y$. By (23), we have

$$\begin{aligned} |x \overset{\vee}{\wedge} y - x_n \overset{\vee}{\wedge} y_n| &\leqq |x \overset{\vee}{\wedge} y - x_n \overset{\vee}{\wedge} y| + |x_n \overset{\vee}{\wedge} y - x_n \overset{\vee}{\wedge} y_n| \\ &\leqq |x - x_n| + |y - y_n|, \end{aligned}$$

and so $O\text{-}\lim (x_n \overset{\vee}{\wedge} y_n) = \left(O\text{-}\lim_{n\to\infty} x_n\right) \overset{\vee}{\wedge} \left(O\text{-}\lim_{n\to\infty} y_n\right)$.

Remark. We have

$$O\text{-}\lim_{n\to\infty} \alpha \cdot x_n = \alpha \cdot O\text{-}\lim_{n\to\infty} x_n.$$

But, in general,

$$O\text{-}\lim_{n\to\infty} \alpha_n x \neq \left(\lim_{n\to\infty} \alpha_n\right) x.$$

The former relation is clear from $|\alpha x - \alpha x_n| = |\alpha|\,|x - x_n|$. The latter inequality is proved by the following counter example: We introduce in the two-dimensional vector space the *lexicographic partial order* in which $(\xi_1, \eta_1) \geqq (\xi_2, \eta_2)$ means that either $\xi_1 > \xi_2$ or $\xi_1 = \xi_2$, $\eta_1 \geqq \eta_2$. It is easy to see that we obtain a vector lattice. We have, in this lattice,

$$n^{-1}(1, 0) \geqq (0, 1) > 0 = (0, 0) \quad (n = 1, 2, \ldots).$$

Hence $O\text{-}\lim_{n\to\infty} n^{-1}(1, 0) \neq 0$. A necessary and sufficient condition for the validity of the equation

$$O\text{-}\lim_{\alpha_n\to\alpha} \alpha_n x = \alpha x \tag{24}$$

is the so-called *Archimedean axiom*

$$O\text{-}\lim_{n\to\infty} n^{-1} x = 0 \quad \text{for every} \quad x \geqq 0. \tag{25}$$

This we see by the Jordan decomposition $y = y^+ + y^-$.

Definition 3. A subset $\{x_\alpha\}$ of a vector lattice X is said to be *bounded* if there exist y and z such that $y \leqq x_\alpha \leqq z$ for all x_α. X is said to be *complete* if, for any bounded set $\{x_\alpha\}$ of X, $\sup_\alpha x_\alpha$ and $\inf_\alpha x_\alpha$ exist in X. Here $\sup_\alpha x_\alpha$ is the *least upper bound* in the sense of the partial order in X, and $\inf_\alpha x_\alpha$ is the *largest lower bound* in the sense of the partial order in X. A vector lattice X is said to be *σ-complete*, if, for any bounded sequence $\{x_n\}$ of X, $\sup_{n\geq 1} x_n$ and $\inf_{n\geq 1} x_n$ exist in X. We define, in a σ-complete vector lattice X,

$$O\text{-}\overline{\lim_{n\to\infty}}\, x_n = \inf_m \left(\sup_{n\geq m} x_n\right), \quad O\text{-}\underline{\lim_{n\to\infty}}\, x_n = \sup_m \left(\inf_{n\geq m} x_n\right). \tag{26}$$

Proposition 7. $O\text{-}\lim_{n\to\infty} x_n = x$ iff $O\text{-}\overline{\lim_{n\to\infty}}\, x_n = O\text{-}\underline{\lim_{n\to\infty}}\, x_n = x$.

Proof. Suppose $|x - x_n| \leqq w_n$, $w_n \downarrow 0$. Then $x - w_n \leqq x_n \leqq x + w_n$ and so we obtain $O\text{-}\overline{\lim_{n\to\infty}}\,(x - w_n) = x \leqq O\text{-}\overline{\lim_{n\to\infty}}\, x_n \leqq O\text{-}\overline{\lim_{n\to\infty}}\,(x + w_n) = x$, that is, $O\text{-}\overline{\lim_{n\to\infty}}\, x_n = x$. Similarly, we obtain $O\text{-}\underline{\lim_{n\to\infty}}\, x_n = x$.

We next prove the sufficiency. Put $u_n = \sup_{m\geq n} x_m$, $v_n = \inf_{m\geq n} x_m$, $u_n - v_n = w_n$. Then, by hypothesis, $w_n \downarrow 0$. Also, by $x_n \leqq u_n = x + (u_n - x) \leqq x + (u_n - v_n) = x + w_n$ and $x_n \geqq x - w_n$ obtained similarly, we prove $|x - x_n| \leqq w_n$. Thus $O\text{-}\lim_{n\to\infty} x_n = x$.

Proposition 8. In a σ-complete vector lattice X, αx is continuous in α, x with respect to O-lim.

Proof. We have $|\alpha x - \alpha_n x_n| \leqq |\alpha x - \alpha x_n| + |\alpha x_n - \alpha_n x_n| = |\alpha|\,|x - x_n| + |\alpha - \alpha_n|\,|x_n|$. Hence, if $O\text{-}\lim_{n\to\infty} x_n = x$, $\lim_{n\to\infty} \alpha_n = \alpha$, then the O-lim of the first term on the right is 0. Therefore, by putting $\sup_{n\geq 1} |x_n| = y$, $\sup_{m\geq n} |\alpha - \alpha_m| = \beta_n$, we have to prove $O\text{-}\lim_{n\to\infty} \beta_n y = 0$. But, by $y \geqq 0$ and $\beta_n \downarrow 0$, we see that $O\text{-}\lim_{n\to\infty} \beta_n y = z$ exists and $O\text{-}\lim_{n\to\infty} 2^{-1}\beta_n y = 2^{-1} z$. Since there exists, for any n, an n_0 such that $\beta_{n_0} \leqq 2^{-1}\beta_n$, we must have $z = 2^{-1} z$, that is, $z = 0$.

Proposition 9. In a σ-complete vector lattice X, $O\text{-}\lim_{n\to\infty} x_n$ exists iff

$$O\text{-}\lim_{n,m\to\infty} |x_n - x_m| = 0. \tag{27}$$

Proof. The necessity is clear from $|x_n - x_m| \leqq |x_n - x| + |x - x_m|$. If we set $|x_n - x_m| = y_{nm}$, then $O\text{-}\overline{\lim_{n\to\infty}}\, x_n \leqq x_m + O\text{-}\overline{\lim_{n\to\infty}}\, y_{nm}$, $O\text{-}\underline{\lim_{n\to\infty}}\, x_n \geqq x_m - O\text{-}\overline{\lim_{n\to\infty}}\, y_{nm}$. Hence

$$0 \leqq O\text{-}\overline{\lim_{n\to\infty}}\, x_n - O\text{-}\underline{\lim_{n\to\infty}}\, x_n \leqq O\text{-}\lim_{m\to\infty} \left(O\text{-}\overline{\lim_{n\to\infty}}\, y_{nm} - O\text{-}\underline{\lim_{n\to\infty}}\, y_{nm}\right) = 0.$$

Thus we have proved the sufficiency.

Proposition 10. A vector lattice X is σ-complete iff every monotone increasing, bounded sequence $\{x_n\} \subseteq X$ has $\sup_{n\geq 1} x_n$ in X.

Proof. We have only to prove the sufficiency. Let $\{z_n\}$ be any bounded sequence in X, and set $x_n = \sup_{m\leq n} z_m$. Then, by hypothesis, $\sup_{n\geq 1} x_n = z$ exists in X, and $z = \sup_{n\geq 1} z_n$. Similarly, we see that $\inf_{n\geq 1} z_n = \inf_{n\geq 1} \left(\inf_{m\leq n} z_m\right)$ also exists in X.

3. *B*-lattices and *F*-lattices

Definition. A real B-space (F-space) X is said to be a *B-lattice* (*F-lattice*) if it is a vector lattice such that

$$|x| \leqq |y| \quad \text{implies} \quad \|x\| \leqq \|y\|. \tag{1}$$

Examples. $C(S)$, $L^p(S)$ are B-lattices by the natural partial order $x \leqq y$ which means $x(s) \leqq y(s)$ on S ($x(s) \leqq y(s)$ a.e. on S for the case $L^p(S)$). $M(S, \mathfrak{B}, m)$ with $m(S) < \infty$ is an F-lattice by the natural partial order as in the case of $L^p(S)$. $A(S, \mathfrak{B})$ is a B-lattice by

$$\|x\| = |x|(S) = \text{the total variation of } x \text{ over } S.$$

We have, by (1) and $|x| = |(|x|)|$,

$$\|x\| = \|(|x|)\|. \tag{2}$$

In $A(S, \mathfrak{B})$, we have, moreover,

$$x \geqq 0, y \geqq 0 \quad \text{imply} \quad \|x + y\| = \|x\| + \|y\|. \tag{3}$$

S. Kakutani called a B-lattice satisfying (3) an *abstract L^1-space*. From (3) we have

$$|x| < |y| \quad \text{implies} \quad \|x\| < \|y\|. \tag{4}$$

The norm in $A(S, \mathfrak{B})$ is continuous in O-lim, that is

$$O\text{-}\lim_{n\to\infty} x_n = x \quad \text{implies} \quad \lim_{n\to\infty} \|x_n\| = \|x\|. \tag{5}$$

For, in $A(S, \mathfrak{B})$, $O\text{-}\lim_{n\to\infty} x_n = x$ is equivalent to the existence of $y_n \in A(S, \mathfrak{B})$ such that $|x - x_n|(S) \leqq y_n(S)$ with $y_n(S) \downarrow 0$. It is easy to see that $M(S, \mathfrak{B}, m)$ satisfies (4) and (5).

Proposition 1. A σ-complete F-lattice X satisfying (4) and (5) is a complete lattice. In particular, $A(S, \mathfrak{B})$ and $L^p(S)$ are complete lattices.

Proof. Let $\{x_\alpha\} \subseteq X$ be bounded. We may assume $0 \leqq x_\alpha \leqq x$ for all α, and we shall show that $\sup_\alpha x_\alpha$ exists. Consider the totality $\{z_\beta\}$ of z_β obtained as the sup of a finite number of x_α's: $z_\beta = \bigvee_{j=1}^{n} x_{\alpha_j}$. Set $\gamma = \sup_\beta \|z_\beta\|$. Then there exists a sequence $\{z_{\beta_j}\}$ such that $\lim_{j\to\infty} \|z_{\beta_j}\| = \gamma$. If we put $z_n = \sup_{j \leqq n} z_{\beta_j}$, then $O\text{-}\lim_{n\to\infty} z_n = w$ exists, and, by (5) and the definition of γ, we have $\|w\| = \gamma$. We shall prove that $w = \sup_\alpha x_\alpha$. Suppose that $x_\alpha \vee w > w$ for a certain x_α. Then, by (4), $\|x_\alpha \vee w\| > \|w\| = \gamma$. But, by $x_\alpha \vee w = O\text{-}\lim_{n\to\infty} (x_\alpha \vee z_n)$, $x_\alpha \vee z_n \in \{z_\beta\}$ and (5), we have $\|x_\alpha \vee w\| = \lim_{n\to\infty} \|x_\alpha \vee z_n\| \leqq \gamma$, which is a contradiction. Hence we must have $w \geqq x_\alpha$ for all x_α. Let $x_\alpha \leqq u$ for all x_α. Suppose $w \wedge u < w$.

Then, by (4), $||w \wedge u|| < \gamma$, contrary to the fact that $w \wedge u \geqq z_\beta$ for all z_β. Hence we must have $w = \sup_\alpha x_\alpha$.

Remark. In $C(S)$, $O\text{-}\lim_{n\to\infty} x_n = x$ does not necessarily imply $s\text{-}\lim_{n\to\infty} x_n = x$. In $M(S, \mathfrak{B}, m)$, $s\text{-}\lim_{n\to\infty} x_n = x$ does not necessarily imply $O\text{-}\lim_{n\to\infty} x_n = x$. To see this, let $x_1(s), x_2(s), \ldots$ be the defining functions of the intervals of $[0, 1]$:

$$\left[0, \frac{1}{2}\right], \left[\frac{1}{2}, \frac{2}{2}\right], \left[0, \frac{1}{4}\right], \left[\frac{1}{4}, \frac{2}{4}\right], \left[\frac{2}{4}, \frac{3}{4}\right], \left[\frac{3}{4}, \frac{4}{4}\right], \left[0, \frac{1}{8}\right], \left[\frac{1}{8}, \frac{2}{8}\right], \ldots$$

Then the sequence $\{x_n(s)\} \subseteq M([0, 1])$ converges to 0 asymptotically but does not converge to 0 a.e., that is, we have $s\text{-}\lim_{n\to\infty} x_n = 0$ but $O\text{-}\lim_{n\to\infty} x_n = 0$ does not hold.

Proposition 2. Let X be an F-lattice. Let a sequences $\{x_n\} \subseteq X$ satisfy $s\text{-}\lim_{n\to\infty} x_n = x$. Then $\{x_n\}$ **-converges to x relative uniformly*. This means that, from any subsequence $\{y_n\}$ of $\{x_n\}$, we can choose a subsequence $\{y_{n(k)}\}$ and a $z \in X$ such that

$$|y_{n(k)} - x| \leqq k^{-1} z \quad (k = 1, 2, \ldots). \tag{6}$$

Conversely, if $\{x_n\}$ *-converges relative uniformly to x, then $s\text{-}\lim_{n\to\infty} x_n = x$.

Proof. We may restrict ourselves to the case $x = 0$. From $\lim_{n\to\infty} ||y_n|| = 0$, we see that there exists a sequence $\{n(k)\}$ of positive integers such that $||k y_{n(k)}|| \leqq k^{-2}$. Then (6) holds by taking $z = \sum_{k=1}^{\infty} |k y_{n(k)}|$. Conversely, let condition (6) with $x = 0$ be satisfied. Then we have $||y_{n(k)}|| \leqq ||k^{-1} z||$ from $|y_{n(k)}| \leqq k^{-1} z$. Hence $s\text{-}\lim_{k\to\infty} y_{n(k)} = 0$. Therefore, there cannot exist a subsequence $\{y_n\}$ of $\{x_n\}$ such that $\lim_{n\to\infty} ||y_n|| > 0$.

Remark. The above proposition is an abstraction of the fact, that an asymptotically convergent sequence of $M(S, \mathfrak{B}, m)$ with $m(S) < \infty$ contains a subsequence which converges m-a.e.

4. A Convergence Theorem of Banach

This theorem is concerned with the almost everywhere convergence of a sequence of linear operators whose ranges are measurable functions. See S. Banach [2]. A lattice-theoretic formulation of the theorem reads as follows (K. Yosida [15]):

Theorem. Let X be a real B-space with the norm $|| \; ||$ and Y a σ-complete F-lattice with the quasi-norm $|| \; ||_1$ such that

$$O\text{-}\lim_{n\to\infty} y_n = y \quad \text{implies} \quad \lim_{n\to\infty} ||y_n||_1 = ||y||_1. \tag{1}$$

Let $\{T_n\}$ be a sequence of bounded linear operators $\in L(X, Y)$. Suppose that

$$\begin{aligned}&O\text{-}\lim_{n\to\infty} |T_n x| \text{ exists for those } x's \in X \text{ which form a set}\\ &G \text{ of the second category.}\end{aligned} \tag{2}$$

Then, for any $x \in X$, $O\text{-}\overline{\lim}_{n\to\infty} T_n x$ and $O\text{-}\underline{\lim}_{n\to\infty} T_n x$ both exist and the (not necessarily linear) operator $\tilde{T}$ defined by

$$\tilde{T}x = \left(O\text{-}\overline{\lim_{n\to\infty}}\, T_n x\right) - \left(O\text{-}\underline{\lim_{n\to\infty}}\, T_n x\right) \tag{3}$$

is continuous as an operator defined on X into Y.

Remark. The space $M(S, \mathfrak{B}, m)$ with $m(S) < \infty$ satisfies (1) if we write $y_1 \leqq y_2$ when and only when $y_1(s) \leqq y_2(s)$ m-a.e., the quasi-norm $\|y\|_1$ being defined by $\|y\|_1 = \int_S |y(s)|\,(1 + |y(s)|)^{-1}\, m(ds)$. Likewise $L^p(S, \mathfrak{B}, m)$ with $m(S) \leqq \infty$ also satisfies (1) by the same semi-order.

Proof of the Theorem. Set $T_n x = y_n$, $y_n' = \sup_{n \geqq m} |y_m|$, $y' = \sup_{n\geqq 1} |y_n|$, and consider the operators $V_n x = y_n'$ and $Vx = y'$ defined at least on G into Y. By (23) of the preceding section 2, each V_n is strongly continuous with T_k. Since $\lim_{n\to\infty} \|V_n x - Vx\|_1 = 0$ by (1), we have $\lim_{n\to\infty} \|k^{-1} V_n x\| = \|k^{-1} V x\|$ and further $\lim_{k\to\infty} \|k^{-1} V x\|_1 = 0$. These are implied by the continuity in α, y of αy in the F-space Y. Hence for any $\varepsilon > 0$,

$$G \subseteqq \bigcup_{k=1}^{\infty} G_k \text{ where } G_k = \left\{x \in X; \sup_{n\geqq 1} \|k^{-1} V_n x\|_1 \leqq \varepsilon\right\}. \tag{4}$$

By the strong continuity of V_n, each G_k is a strongly closed set of X. Thus some G_{k_0} contains a sphere of X, in virtue of the hypothesis that G is of the second category. That is, there exist an $x_0 \in X$ and a $\delta > 0$ such that $\|x_0 - x\| \leqq \delta$ implies $\sup_{n\geqq 1} \|k_0^{-1} V_n x\|_1 \leqq \varepsilon$. Hence, by putting $z = x_0 - x$, we see that

$$\sup_{n\geqq 1} \|k_0^{-1} V_n z\|_1 \leqq \sup_{n\geqq 1} \|k_0^{-1} V_n x_0\|_1 + \sup_{n\geqq 1} \|k_0^{-1} V_n x\|_1 \leqq 2\varepsilon,$$

that is, since $V_n(k_0^{-1} z) = k_0^{-1} V_n z$, we have

$$\sup_{n\geqq 1} \|V_n z\|_1 \leqq 2\varepsilon \text{ whenever } \|z\| \leqq \delta/k_0.$$

This proves that $s\text{-}\lim_{\|z\|\to 0} V_n \cdot z = 0$ uniformly in n.

G being dense in X, $V \cdot x$ is defined for all $x \in X$ and $V \cdot x$ is strongly continuous at $x = 0$ with $V \cdot 0 = 0$. Hence, by

$$|\tilde{T} \cdot x| \leqq 2V \cdot x \quad \text{and} \quad \|\tilde{T}x_1 - \tilde{T}x_2\|_1 \leqq \|\tilde{T}(x_1 - x_2)\|_1,$$

we see that $\tilde{T} \cdot x$ is strongly continuous at every $x \in X$.

Corollary. Under condition (1), the set $G = \{x \in X;\ O\text{-}\lim_{n\to\infty} T_n x \text{ exists}\}$ either coincides with X or is a set of the first category.

Proof. Suppose the set G be of the second category. Then, by our Theorem, the operator $\tilde{T}$ is a strongly continuous operator defined on X into Y. Thus $G = \{x \in X;\ \tilde{T}y = 0\}$ is strongly closed in X. Moreover, G is a linear subspace of X. Hence G must coincide with X. Otherwise, G would be non dense in X.

5. The Representation of a Vector Lattice as Point Functions

Let a vector lattice X contain a "*unit*" I with the properties:

$$I > 0, \text{ and, for any } f \in X, \text{ there exists an } \alpha > 0 \text{ such that } -\alpha I \leqq f \leqq \alpha I. \tag{1}$$

For such a vector lattice X, we can give an analogous representation similar to the representation of a normed ring as point functions.

An element $f \in X$ is called "*nilpotent*" if $n\,|f| \leqq I$ $(n = 1, 2, \ldots)$. The set R of all nilpotent elements $f \in X$ is called the "*radical*" of X. By (20) of Chapter XII, 2, R constitutes a linear subspace of X. Moreover, R is an "*ideal*" of X in the sense that

$$f \in R \text{ and } |g| \leqq |f| \text{ imply } g \in R. \tag{2}$$

Lemma. Let X_1 and X_2 be vector lattices. A linear operator T defined on X_1 onto X_2 is called a *lattice homomorphism* if

$$T(x \overset{\vee}{\wedge} y) = (Tx) \overset{\vee}{\wedge} (Ty). \tag{3}$$

Then T is a lattice homomorphism iff $N = \{x \in X;\ Tx = 0\}$ is an ideal of X_1.

Proof. Let T be a lattice homomorphism. Let $x \in N$ and $|y| \leqq |x|$. Then, by $T(|x|) = T(x \vee -x) = (Tx) \vee (T(-x)) = 0$, we obtain, $0 \leqq Ty^+ = T(y^+ \wedge |x|) = Ty^+ \wedge T\,|x| = 0$ and so $y^+ \in N$. Similarly we obtain $y^- \in N$ and thus $y = y^+ + y^- \in N$.

Next let a linear subspace $N = \{x \in X_1;\ Tx = 0\}$ be an ideal of X_1. Then the linear space X_2 is isomorphic to the factor space X_1/N. We have to show that $\overline{(x \overset{\vee}{\wedge} y)} = \bar{x} \overset{\vee}{\wedge} \bar{y}$, where $\bar{x}$ is the residue class mod N containing x. But, if $\bar{y} = \bar{z}$ i.e., if $y - z \in N$, then, by (23) in Chapter XII, 2, we obtain

$$|x \overset{\vee}{\wedge} y - x \overset{\vee}{\wedge} z| \leqq |y - z| \in N$$

so that the residue class $\overline{(x \overset{\vee}{\wedge} y)}$ is determined independently of the choice of the representative elements x, y from the residue classes $\bar{x}, \bar{y}$, respectively.

Remark. The above Lemma may be phrased as follows. Let N be a linear subspace of a vector lattice. Then the *linear-congruence* $a \equiv b \pmod N$ is also a *lattice-congruence*:

$$a \equiv b,\ a' \equiv b' \pmod N \quad \text{implies} \quad a \begin{smallmatrix}\vee\\ \wedge\end{smallmatrix} b \equiv a' \begin{smallmatrix}\vee\\ \wedge\end{smallmatrix} b' \pmod N,$$

iff N is an ideal of X.

Now an ideal N is called "*non-trivial*" if $N \neq \{0\}$, X. A non-trivial ideal N is called "*maximal*" if it is contained in no other ideal $\neq X$. Denote by $\mathfrak{M}$ the set of all maximal ideals N of X. The residual class X/N of X mod any ideal $N \in \mathfrak{M}$ is "*simple*", that is, X/N does not contain non-trivial ideals. It will be proved below that *a simple vector lattice with a unit is linear-lattice isomorphic to the vector lattice of real numbers, the non-negative elements and the unit I being represented by non-negative numbers and the number* 1. We denote by $f(N)$ the real number which corresponds to $f \in X$ by the linear-lattice-homomorphism $X \to X/N$, $N \in \mathfrak{M}$.

After these preliminaries we may state

Theorem 1. The radical R coincides with the intersection ideal $\bigcap_{N \in \mathfrak{M}} N$.

Proof. *The first step.* Let X be a simple vector lattice with a unit I. Then we must have $X = \{\alpha I; -\infty < \alpha < \infty\}$.

Proof. X does not contain a nilpotent element $f \neq 0$, for otherwise X would contain a non-trivial ideal $N = \{g; |g| \leqq \eta\, |f|$ with some $\eta < \infty\}$. Hence, by (1), X satisfies the *Archimedean axiom*:

$$\text{order-}\lim_{n\to\infty} n^{-1}\,|x| = 0 \quad \text{for all} \quad x \in X. \tag{4}$$

Suppose there exists an $f_0 \in X$ such that $f_0 \neq \gamma I$ for any real number γ. Let

$$\alpha = \inf_{f_0 \leqq \alpha' I} \alpha',\ \beta = \sup_{\beta' I \leqq f_0} \beta'.$$

Then, by (4), $\beta I \leqq f_0 \leqq \alpha I$ and $\beta < \alpha$. Hence $(f_0 - \delta I)^+ \neq 0$, $(f_0 - \delta I)^- \neq 0$ for $\beta < \delta < \alpha$. Thus, by $x^+ \wedge (-x^-) = 0$, the set $N_0 = \{g; |g| \leqq \eta\,(f_0 - \delta I)^+$ with some $\eta < \infty\}$ is a non-trivial ideal, contrary to the hypothesis.

The second step. For any non-trivial ideal N_0, there exists a maximal ideal N_1 containing N_0.

Proof. Let $\{N_0\}$ be the totality of non-trivial ideals containing N_0. We order the ideals of $\{N_0\}$ by inclusion relation, that is, we denote $N_{\alpha_1} \leqq N_{\alpha_2}$ if N_{α_1} is a subset of N_{α_2}. Suppose that $\{N_\alpha\}$ is a linearly ordered subset of $\{N_0\}$ and set $N_\beta = \bigcup_{N_\alpha \in \{N_\alpha\}} N_\alpha$. We shall then show that N_β is an upper bound of $\{N_\alpha\}$. For, if $x, y \in N_\beta$, there exist ideals N_{α_1} and N_{α_2} such that $x \in N_{\alpha_1}$ and $y \in N_{\alpha_2}$. Since $\{N_\alpha\}$ is linearly ordered, $N_{\alpha_1} \subseteqq N_{\alpha_2}$ (or $N_{\alpha_1} \supseteqq N_{\alpha_2}$) and so x and y both belong to N_{α_2}. This proves that

$(\gamma x + \delta y) \in N_{\alpha_2} \subseteqq N_\beta$ and that $|z| \leqq |x|$ implies $z \in N_{\alpha_1} \subseteqq N_\beta$. Since the unit I is contained in no N_α, I is not contained in N_β. Therefore N_β is a non-trivial ideal containing every N_α, i.e., N_β is an upper bound of $\{N_\alpha\}$. Thus, by Zorn's Lemma, there exists at least one maximal ideal containing N_0.

The third step. $R \subseteqq \bigcap_{N \in \mathfrak{M}} N$. Let $f > 0$ and $nf \leqq I$ $(n = 1, 2, \ldots)$. Then; for any $N \in \mathfrak{M}$, we have $nf(N) \leqq I(N) = 1$ $(n = 1, 2, \ldots)$, and hence $f(N) = 0$, that is, $f \in N$.

The fourth step. $R \supseteqq \bigcap_{N \in \mathfrak{M}} N$. Let $f > 0$ be not nilpotent. Then we have to show that there exists an ideal $N \in \mathfrak{M}$ such that $f \bar{\in} N$. This may be proved as follows.

Since $f > 0$ is not nilpotent, there exists an integer n such that $nf \not\leqq I$. We may assume that $nf \not\geqq I$, since otherwise $f \bar{\in} N$ for any $N \in \mathfrak{M}$ and so we have nothing to prove. Thus suppose $p = I - (n \cdot f) \cap I > 0$. Then, for any positive integer m, we do not have $m \cdot p \geqq I$. If otherwise, we would have $m^{-1} I \leqq I - (n \quad f) \cap I$ and hence

$$(n \cdot f) \wedge I = (n \cdot f) \wedge (1 - m^{-1})\, I.$$

Thus, by (6) in Chapter XII, 2,

$$(n \cdot f - (1 - m^{-1})\, I) \wedge m^{-1} I = (n \cdot f - (1 - m^{-1})\, I) \wedge 0 \leqq 0,$$

and so, by the distributivity of the vector lattice,

$$0 = \{(n \cdot f - (1 - m^{-1})\, I) \wedge m^{-1} I\} \vee 0 = (n \cdot f - (1 - m^{-1})\, I)^+ \wedge m^{-1} I,$$

that is, $(n \cdot f - (1 - m^{-1})\, I)^+ \wedge I = 0$. Put $b = (n \cdot f - (1 - m^{-1})\, I)^+$ and assume that $b > 0$. By hypothesis (1), we have $b < \alpha I$ with some $\alpha > 1$. Then $0 < b = b \wedge \alpha I$ and so $0 < (\alpha^{-1} b) \wedge I \leqq b \wedge I$, contrary to $b \wedge I = 0$. Therefore $b = 0$, i.e., $n \cdot f \leqq (1 - m^{-1})\, I$. This contradicts the fact that $n \cdot f \not\leqq I$. Hence the set $N_0 = \{g; |g| \leqq \eta\, |p| \text{ for some } \eta < \infty\}$ is a non-trivial ideal. N_0 is contained in at least one maximal ideal N, by the second step. Then $0 = p(N) = 1 - (n \cdot f(N)) \wedge 1$ which shows that $f(N) > 0$, that is, $f \bar{\in} N$.

We have thus proved our Theorem 1.

The vector lattice $\bar{X} = X/R$ is again a vector lattice with a unit $\bar{I}$. By Theorem 1, the intersection ideal $\bigcap_{\bar{N}} \bar{N}$ of all the maximal ideals $\bar{N}$ of $\bar{X}$ is the zero ideal and $\bar{X}$ contains no nilpotent element $\neq 0$. Hence $\bar{X}$ satisfies the Archimedean axiom

$$\underset{n \uparrow \infty}{\text{order-limit}}\; n^{-1}\, |\bar{f}| = 0 \quad \text{for all} \quad \bar{f} \in \bar{X}. \tag{5}$$

Let $\bar{N}$ be any maximal ideal of $\bar{X}$. Then the factor space $\bar{X}/\bar{N}$ is a simple vector lattice, and so, by the first step in the proof of Theorem 1, $\bar{X}/\bar{N}$ is linear-lattice-isomorphic to the vector lattice of real numbers; the non-negative elements and the unit being represented by non-negative numbers and 1. We denote by $\bar{f}(\bar{N})$ the real number which corresponds to $\bar{f}$ by this homomorphism $\bar{X} \to \bar{X}/\bar{N}$. We also denote by $\overline{\mathfrak{M}}$ the set of all maximal ideals of $\bar{X}$. We have thus

Theorem 2. By the correspondence $\bar{f} \to \bar{f}(\bar{N})$, $\bar{X}$ is linear-lattice-isomorphically mapped on the vector lattice $F(\overline{\mathfrak{M}})$ of real-valued bounded functions on $\overline{\mathfrak{M}}$ such that (i) $|\bar{f}| \to |\bar{f}(\bar{N})|$, (ii) $\bar{I}(\bar{N}) \equiv 1$ on $\overline{\mathfrak{M}}$ and (iii) $F(\overline{\mathfrak{M}})$ *separates the points* of $\overline{\mathfrak{M}}$ in the sense that

$$\begin{array}{l}\text{for two different points } \bar{N}_1, \bar{N}_2 \text{ of } \overline{\mathfrak{M}}, \text{ there exists at} \\ \text{least one } \bar{f} \in \bar{X} \text{ such that } \bar{f}(\bar{N}_1) \neq \bar{f}({}_2\bar{N}).\end{array} \tag{6}$$

Remark. We introduce a topology in $\overline{\mathfrak{M}}$ by calling the sets of the form

$$\{\bar{N} \in \overline{\mathfrak{M}};\ |\bar{f}_i(\bar{N}) - \bar{f}_i(\bar{N}_0)| < \varepsilon_i \quad (i = 1, 2, \ldots, n), \\ \text{where } -\bar{I} \leqq \bar{f} \leqq \bar{I} \quad \text{for all} \quad i\}$$

neighbourhoods of N_0. Then $\overline{\mathfrak{M}}$ is compact since it may be identified with a closed subset of a topological product (of the same potency as the cardinal number of the set of elements $\bar{f} \in X$ which satisfy $-\bar{I} \leqq \bar{f} \leqq \bar{I}$) of the closed intervals $[-1, 1]$. The proof is entirely similar to the case of the set of all maximal ideals of a normed ring in Chapter XI, 2. Moreover, each function $\bar{f}(\bar{N}) \in F(\overline{\mathfrak{M}})$ is continuous on the compact space $\overline{\mathfrak{M}}$ topologized in this way. Thus, by the Kakutani-Krein theorem in Chapter 0, 2, we see that $F(\overline{\mathfrak{M}})$ is dense in the B-space $C(\overline{\mathfrak{M}})$. The above two theorems are adapted from K. Yosida-M. Fukamiya [16]. Cf. also S. Kakutani [4] and M. Krein-S. Krein [2].

6. The Representation of a Vector Lattice as Set Functions

Let X be a σ-complete vector lattice. Choose any positive element x of X and call it a "unit" of X and write 1 for x; when not ambiguous we also write α for $\alpha \cdot 1$. A non-negative element $e \in X$ is called a "quasi-unit" if $e \wedge (1 - e) = 0$. A finite linear combination $\sum_i \alpha_i e_i$ of quasi-units e_i is called a "step-element", and we call the element $y \in X$ "absolutely continuous" (with respect to the unit 1) if y can be expressed as the O-lim of a sequence of step-elements. An element $z \in X$ is called "singular" (with respect to the unit 1) if $|z| \wedge 1 = 0$.

We shall give an abstract formulation of the Radon-Nikodym theorem in integration theory.

Theorem. Any element of X is uniquely expressed as the sum of an absolutely continuous element and a singular element.

Proof. *The first step.* If $f > 0$ and $f \wedge 1 \neq 0$, then there exist a positive number α and a quasi-unit $e_\alpha \neq 0$ such that $f \geqq \alpha e_\alpha$. We can, in fact, take

$$e_\alpha = \bigvee_{n \geqq 1} \{n(\alpha^{-1}f - \alpha^{-1}f \wedge 1) \wedge 1\}. \tag{1}$$

Proof. Put $y_\alpha = \alpha^{-1}f - \alpha^{-1}f \wedge 1$. Then we obtain

$$2e_a \wedge 1 = \left\{\bigvee_{n \geqq 1} (2ny_a \wedge 2)\right\} \wedge 1 = e_\alpha,$$

and so e_α is a quasi-unit. We obtain $f \geqq \alpha e_\alpha$ from

$$ny_\alpha \wedge 1 = n\alpha^{-1}f \wedge [1 + n(\alpha^{-1}f \wedge 1)] - n(\alpha^{-1}f \wedge 1)$$
$$\leqq (n+1)\alpha^{-1}f \wedge (n+1) - n(\alpha^{-1}f \wedge 1) \leqq \alpha^{-1}f \wedge 1 \leqq \alpha^{-1}f.$$

If we can show that $y_\alpha \wedge 1 > 0$ for some $\alpha > 0$, then $e_\alpha > 0$ for such α. Suppose that such positive α does not exist. Then, for every α with $0 < \alpha < 1$, we have

$$\alpha^{-1}(\alpha^{-1}f - \alpha^{-1}f \wedge 1) \wedge \alpha^{-1} = 0.$$

Hence $(f - f \wedge \alpha) \wedge 1 = 0$ and, letting $\alpha \downarrow 0$, we obtain $f \wedge 1 = 0$, contrary to the hypothesis $f \wedge 1 \neq 0$.

The second step. Let $f \geqq 0$, and $f \geqq \alpha e$ where $\alpha > 0$ and e is a quasi-unit. Then, for $0 < \alpha' < \alpha$, $e_{\alpha'} \geqq e$ and $f \geqq \alpha' e_{\alpha'}$ where $e_{\alpha'}$ is defined by (1).

Proof. For the sake of simplicity, we assume that $\alpha = 1$. For $0 < \delta < 1$

$$\frac{f}{1-\delta} + \frac{e}{1-\delta} \wedge 1 = \left(\frac{f}{1-\delta} + \frac{e}{1-\delta}\right) \wedge \left(1 + \frac{f}{1-\delta}\right)$$
$$\geqq \left(\frac{f}{1-\delta} + \frac{e}{1-\delta}\right) \wedge \left(1 + \frac{e}{1-\delta}\right)$$
$$= \frac{f}{1-\delta} \wedge 1 + \frac{e}{1-\delta}.$$

Since e is a quasi-unit, we have $2e \wedge 1 = e$, $e \wedge 1 = e$. Hence $me \wedge 1 = e$ when $m \geqq 1$. Thus, by $1 < (1-\delta)^{-1}$, we have $(1-\delta)^{-1} e \wedge 1 = e$. Therefore, we obtain, from the above inequality,

$$\frac{\delta}{1-\delta} e = \frac{e}{1-\delta} - \frac{e}{1-\delta} \wedge 1 \leqq \frac{f}{1-\delta} - \frac{f}{1-\delta} \wedge 1 = y_{1-\delta},$$

and so, by (1), $e \leqq e_{1-\delta}$.

The third step. The set of all quasi-units constitutes a *Boolean algebra*, that is, if e_1 and e_2 are quasi-units, then $e_1 \vee e_2$ and $e_1 \wedge e_2$ are also quasi-units and $0 \leqq e_i \leqq 1$. The quasi-unit $(1 - e)$ is the complement of e, and 0,1 are the least and the greatest elements, respectively in the totality of the quasi-units.

Proof. The condition $e \wedge (1-e) = 0$ is equivalent to $2e \wedge 1 = e$. Hence, if e_1, e_2 are quasi-units, then

$$2(e_1 \wedge e_2) \wedge 1 = (2e_1 \wedge 1) \wedge (2e_2 \wedge 1) = e_1 \wedge e_2,$$
$$2(e_1 \vee e_2) \wedge 1 = (2e_1 \wedge 1) \vee (2e_2 \wedge 1) = e_1 \vee e_2,$$

so that $e_1 \wedge e_2$ and $e_1 \vee e_2$ are also quasi-units.

The fourth step. Let $f > 0$, and set $\bar{f} = \sup \beta e_\beta$ where sup is taken for all positive rational numbers β. Because of the third step, the sup of a finite number of elements of the form $\beta_i e_{\beta_i}$ is a step-element. Hence $\bar{f}$ is absolutely continuous with respect to the unit element 1. We have to show that $g = f - \bar{f}$ is singular with respect to the unit element 1. Suppose that g is not singular. Then, by the first step, there exist a positive number α and a quasi-unit e such that $g \geqq \alpha e$. Hence $f \geqq \alpha e$, and so, by the second step, there exists, for $0 < \alpha_1 < \alpha$, a quasi-unit $e_{\alpha_1} \geqq e$ such that $f \geqq \alpha_1 e_{\alpha_1}$. We may assume that α_1 is a rational number. Thus $\bar{f} \geqq \alpha_1 e_{\alpha_1}$ and so $f = \bar{f} + g \geqq 2\alpha_1 e_{\alpha_1}$. Again, by the second step, there exists, for $0 < \alpha_1' < \alpha_1$, a quasi-unit $e_{2\alpha_1'} \geqq e_{\alpha_1}$ such that $f \geqq 2\alpha_1' e_{2\alpha_1'}$. We may assume that $2\alpha_1'$ is a rational number so that $\bar{f} \geqq 2\alpha_1' e_{2\alpha_1'}$. Hence $f = \bar{f} + g \geqq 3\alpha_1' e$. Repeating the process, we can prove that, for any rational number α_n with $0 < \alpha_n < \alpha$,

$$f \geqq (n+1)\,\alpha_n e \quad (n = 1, 2, \ldots).$$

If we take $\alpha_n \geqq \alpha/2$, we have $(n+1)\,\alpha_n e \geqq 2^{-1} n \alpha e$. Hence $f \geqq n\alpha e$ $(n = 1, 2, \ldots)$ with $\alpha > 0$, $e > 0$. This is a contradiction. For, by the σ-completeness of X, the Archimedean axiom holds in X.

The fifth step. Let $f = f^+ + f^-$ be the Jordan decomposition of a general element $f \in X$. Applying the fourth step to f^+ and f^- separately, we see that f is decomposed as the sum of an absolutely continuous element and a singular element. The uniqueness of the decomposition is proved if we can show that an element $h \in X$ is $= 0$ if h is absolutely continuous as well as singular. But, since h is absolutely continuous, we have $h = O\text{-}\lim_{n\to\infty} h_n$ where h_n are step-elements. h_n being a step-element, there exists a positive number α_n such that $|h_n| \leqq \alpha_n \cdot 1$. Since h is singular, we have $|h| \wedge |h_n| = 0$. Therefore $|h| = |h| \wedge |h| = O\text{-}\lim_{n\to\infty} (|h| \wedge |h_n|) = 0$.

Application to the Radon-Nikodym theorem. Consider the case $X = A(S, \mathfrak{B})$. We know already (Proposition 1 in Chapter XII, 3) that it is a complete lattice, and that in $A(S, \mathfrak{B})$,

$$x^+(B) = \sup_{B' \subseteq B} x(B') = \text{the positive variation } \overline{V}(x; B) \text{ of } x \text{ on } B. \tag{2}$$

We shall prepare a

Proposition. In $A(S, \mathfrak{B})$, let $x > 0$, $z \geqq 0$ in such a way that $x \wedge z = 0$. Then there exists a set $B_0 \in \mathfrak{B}$ such that $x(B_0) = 0$ and $z(S - B_0) = 0$.

Proof. Since $x \wedge z = (x - z)^- + z$, we know, from (2), that

$$(x \wedge z)(B) = \inf_{B' \subseteq B} (x - z)(B') + z(B) = \inf_{B' \subseteq B} [x(B') + z(B - B')]. \quad (3)$$

Hence, by the hypothesis $x \wedge z = 0$, we have

$$\inf_{B \in \mathfrak{B}} [x(B) + z(S - B)] = (x \wedge z)(S) = 0.$$

Hence, for any $\varepsilon > 0$, there exists a $B_\varepsilon \in \mathfrak{B}$ such that $x(B_\varepsilon) \leqq \varepsilon$, $z(S - B_\varepsilon) \leqq \varepsilon$. Put $B_0 = \bigwedge_{k \geqq 1} \left(\bigvee_{n \geqq k} B_{2^{-n}} \right)$. Then, by the σ-additivity of $x(B)$ and $z(B)$,

$$0 \leqq x(B_0) = \lim_{k \to \infty} x \left(\bigvee_{n \geqq k} B_{2^{-n}} \right) \leqq \lim_{k \to \infty} \sum_{n=k}^{\infty} 2^{-n} = 0,$$

$$0 \leqq z(S - B_0) = \lim_{k \to \infty} z \left(S - \bigvee_{n \geqq k} B_{2^{-n}} \right) \leqq \lim_{k \to \infty} z(S - B_{2^{-k}}) = 0.$$

Corollary. Let e be a quasi-unit with respect to $x > 0$ in $A(S, \mathfrak{B})$. Then there exists a set $B_1 \in \mathfrak{B}$ such that

$$e(B) = x(B \cap B_1) \quad \text{for all} \quad B \in \mathfrak{B}. \quad (4)$$

Proof. Since $(x - e) \wedge e = 0$, there exists a set $B_0 \in \mathfrak{B}$ such that $e(B_0) = 0$, $(x - e)(S - B_0) = 0$. Hence $e(S - B_0) = x(S - B_0) = e(S)$, and so $e(B) = x(B - B_0) = x(B \cap B_1)$ where $B_1 = S - B_0$.

We are now able to prove the Radon-Nikodym theorem in integration theory. In virtue of the above Corollary, a quasi-unit e with respect to $x > 0$ is, in $A(S, \mathfrak{B})$, a *contracted measure* $e(B) = x(B \cap B_e)$. Hence a step-element in $A(S, \mathfrak{B})$ is the integral of the form:

$$\sum_i \lambda_i \int_{B \cap B_i} x(ds),$$

that is, indefinite integral of a step function (= finitely-valued function). Hence an absolutely continuous element in $A(S, \mathfrak{B})$ is an indefinite integral with respect to the measure $x(B)$. The Proposition above says that a singular element g (with respect to the unit x) is associated with a set $B_0 \in \mathfrak{B}$ such that $x(B_0) = 0$ and $g(B) = g(B \cap B_0)$ for all $B \in \mathfrak{B}$. Such a measure $g(B)$ is the so-called *singular measure* (with respect to $x(B)$). Thus any element $f \in A(S, \mathfrak{B})$ is expressed as the sum of an indefinite integral (with respect to $x(B)$) and a measure $g(B)$ which is singular (with respect to $x(B)$). This decomposition is unique. The obtained result is exactly the Radon-Nikodym theorem.

Remark. The above Theorem is adapted from K. YOSIDA [2]. Cf. F. RIESZ [6], H. FREUDENTHAL [2] and S. KAKUTANI [5]. For further references, see G. BIRKHOFF [1].

XIII. Ergodic Theory and Diffusion Theory

These theories constitute fascinating fields of application of the analytical theory of semi-groups. Mathematically speaking, the ergodic theory is concerned with the "time average" $\lim_{t\uparrow\infty} t^{-1}\int_0^t T_s\,ds$ of a semi-group T_t, and the diffusion theory is concerned with the investigation of a stochastic process in terms of the infinitesimal generator of the semi-group intrinsically associated with the stochastic process.

1. The Markov Process with an Invariant Measure

In 1862, an English botanist R. BROWN observed under a microscope that small particles, pollen of some flower, suspended in a liquid move chaotically, changing position and direction incessantly. To describe such a phenomenon, we shall consider the transition probability $P(t,x;s,E)$ that a particle starting from the position x at time t belongs to the set E at a later time s. The introduction of the transition probability $P(t,x;s,E)$ is based upon the fundamental hypothesis that the chaotic motion of the particle after the time moment t is entirely independent of its past history before the time moment t. That is, the future history of the particle after the time moment t is entirely determined chaotically if we know the position x of the particle at time t. The hypothesis that the particle has no memory of the past implies that the transition probability P satisfies the equation

$$P(t, x; s, E) = \int_S P(t, x; u, dy)\, P(u, y; s, E) \quad \text{for} \quad t < u < s, \qquad (1)$$

where the integration is performed over the entire space S of the chaotic movement of the particle.

The process of evolution in time governed by a transition probability satisfying (1) is called a *Markov process*, and the equation (1) is called the *Chapman-Kolmogorov equation*. The Markov process is a natural generalization of the deterministic process for which $P(t, x; s, E) = 1$ or $= 0$ according as $y \in E$ or $y \bar{\in} E$; that is, the process in which the particle at the position x at the time moment t moves to a definite position $y = y(x, t, s)$ with probability 1 at every fixed later time moment s. The Markov process P is said to be *temporally homogeneous* if $P(t, x; s, E)$ is a function of $(s - t)$ independently of t. In such a case, we shall be concerned with the transition probability $P(t, x, E)$ that a particle at the position x is transferred into the set E after the lapse of t units of time. The equation (1) then becomes

$$P(t + s, x, E) = \int_S P(t, x, dy)\, P(s, y, E) \quad \text{for} \quad t, s > 0. \qquad (2)$$

In a suitable function space X, $P(t, x, E)$ gives rise to a linear transformation T_t:

$$(T_t f)(x) = \int_S P(t, x, dy) f(y), \quad f \in X, \tag{3}$$

such that, by (2), the *semi-group property* holds:

$$T_{t+s} = T_t T_s \quad (t, s > 0). \tag{4}$$

A fundamental mathematical question in statistical mechanics is concerned with the existence of the time average

$$\lim_{t\uparrow\infty} t^{-1} \int_0^t T_s f ds. \tag{5}$$

In fact, let S be the phase space of a mechanical system governed by the classical Hamiltonian equations whose Hamiltonian does not contain the time variable explicitly. Then a point x of S is moved to the point $y_t(x)$ of S after the lapse of t units of time in such a way that, by a classical theorem due to LIOUVILLE, the mapping $x \to y_t(x)$ of S onto S, for each fixed t, is an *equi-measure transformation*, that is, the mapping $x \to y_t(x)$ leaves the "phase volume" of S invariant. In such a deterministic case, we have

$$(T_t f)(x) = f(y_t(x)), \tag{6}$$

and hence *the ergodic hypothesis* of BOLTZMANN that

> the time average of any physical quantity = the space average of this physical quantity

is expressed, assuming $\int_S dx < \infty$, by

$$\lim_{t\uparrow\infty} t^{-1} \int_0^t f(y_s(x))\, ds = \int_S f(x)\, dx \Big/ \int_S dx \quad \text{for all} \quad f \in X,$$

dx denoting the phase volume element of S. (7)

A natural generalization of the equi-measure transformation $x \to y_t(x)$ to the case of a Markov process $P(t, x, E)$ is the condition of the existence of an *invariant measure* $m(dx)$:

$$\int_S m(dx)\, P(t, x, E) = m(E) \quad \text{for all} \quad t > 0 \quad \text{and all } E. \tag{8}$$

We are thus lead to the

Definition. Let $\mathfrak{B}$ be a σ-additive family of subsets B of a set S such that S itself $\in \mathfrak{B}$. For every $t > 0$, $x \in S$ and $E \in \mathfrak{B}$, let there be associated a function $P(t, x, E)$ such that

$$P(t, x, E) \geqq 0, \quad P(t, x, S) = 1, \tag{9}$$

for fixed t and x, $P(t, x, E)$ is σ-additive in $E \in \mathfrak{B}$, (10)

for fixed t and E, $P(t, x, E)$ is $\mathfrak{B}$-measurable in x, (11)

$$P(t + s, x, E) = \int_S P(t, x, dy)\, P(s, y, E)$$

(the Chapman-Kolmogorov equation). (12)

Such a system $P(t, x, E)$ is said to define a *Markov process* on the *phase space* $(S, \mathfrak{B})$. If we further assume that $(S, \mathfrak{B}, m)$ is a measure space in such a way that

$$\int_S m(dx)\, P(t, x, E) = m(E) \quad \text{for all} \quad E \in \mathfrak{B}, \tag{13}$$

then $P(t, x, E)$ is said to be a *Markov process with an invariant measure* $m(E)$.

Theorem 1. Let $P(t, x, E)$ be a Markov process with an invariant measure m such that $m(S) < \infty$. Let the norm in $X_p = L^p(S, \mathfrak{B}, m)$ be denoted by $\|f\|_p$, $p \geqq 1$. Then, by (3), a bounded linear operator $T_t \in L(X_p, X_p)$ is defined such that $T_{t+s} = T_t T_s$ $(t, s > 0)$ and

T_t is *positive*, i.e., $(T_t f)(x) \geqq 0$ on S m-a.e. if $f(x) \geqq 0$ on S m-a.e., (14)

$$T_t \cdot 1 = 1, \tag{15}$$

$\|T_t f\|_p \leqq \|f\|_p$ for $f \in X_p = L^p(S, \mathfrak{B}, m)$ with $p = 1, 2$ and ∞. (16)

Proof. (14) and (15) are clear. Let $f \in L^\infty(S, \mathfrak{B}, m)$. Then, by (9), (10) and (11), we see that $f_t(x) = (T_t f)(x) \in L^\infty(S, \mathfrak{B}, m)$ is defined and $\|f_t\|_\infty \leqq \|f\|_\infty$. Hence, by (9) and (13), we obtain, for $f \in L^\infty(S, \mathfrak{B}, m)$ with $p = 1$ or $p = 2$,

$$\begin{aligned} \|f_t\|_p &= \left\{\int_S m(dx) \left|\int_S P(t, x, dy)\, f(y)\right|^p\right\}^{1/p} \\ &\leqq \left\{\int_S m(dx) \left[\int_S P(t, x, dy)\, |f(y)|^p \cdot \int_S P(t, x, dy)\, 1^p\right]\right\}^{1/p} \\ &= \left\{\int_S m(dy)\, |f(y)|^p\right\}^{1/p} = \|f\|_p. \end{aligned}$$

We put, for a non-negative $f \in L^p(S, \mathfrak{B}, m)$ with $p = 1$ or $p = 2$,

$${}_nf(s) = \min(f(s), n), \text{ where } n \text{ is a positive integer}.$$

Then, by the above, we obtain $0 \leqq ({}_nf(s))_t \leqq ({}_{n+1}f(s))_t$ and $\|({}_nf)_t\|_p \leqq \|{}_nf\|_p \leqq \|f\|_p$. Thus, if we put $f_t(s) = \lim\limits_{n\to\infty} ({}_nf(s))_t$, then, by the Lebesgue-Fatou lemma, $\|f_t\|_p \leqq \|f\|_p$, that is, $f_t \in L^p(S, \mathfrak{B}, m)$. Again by the Lebesgue-Fatou lemma, we obtain

$$\begin{aligned} f_t(x) &= \lim_{n\to\infty} \int_S P(t, x, dy)\, ({}_nf(y)) \geqq \int_S P(t, x, dy) \left(\lim_{n\to\infty} ({}_nf(y))\right) \\ &= \int_S P(t, x, dy)\, f(y). \end{aligned}$$

Thus $f(y)$ is integrable with respect to the measure $P(t, x_0, dy)$ for those x_0's for which $f_t(x_0) \neq \infty$, that is, for m-a.e. x_0. Hence, by the Lebesgue-Fatou lemma, we finally obtain

$$\int_S P(t, x_0, dy) \left(\lim_{n\to\infty} ({}_nf(y))\right) = \lim_{n\to\infty} \int_S P(t, x_0, dy)\, ({}_nf(y)).$$

Therefore, $f_t(x_0) = \int_S P(t, x_0, dy) f(y)$ m-a.e. and $\|f_t\|_p \leqq \|f\|_p$. For a general $f \in L^p(S, \mathfrak{B}, m)$, we obtain the same result by applying the positive operator T_t to f^+ and f^-, separately.

Theorem 2 (K. YOSIDA). Let $P(t, x, E)$ be a Markoff process with an invariant measure m such that $m(S) < \infty$. Then, for any $f \in L^p(S, \mathfrak{B}, m)$ with $p = 1$ or $p = 2$, the mean ergodic theorem holds:

$$\underset{n\to\infty}{s\text{-}\lim}\, n^{-1} \sum_{k=1}^{n} T_k f = f^* \text{ exists in } L^p(S, \mathfrak{B}, m) \text{ and } T_1 f^* = f$$
$$\text{whenever } f \in L^p(S, \mathfrak{B}, m), \tag{17}$$

and we have

$$\int_S f(s)\, m(ds) = \int_S f^*(s)\, m(ds). \tag{18}$$

Proof. In the Hilbert space $L^2(S, \mathfrak{B}, m)$, the mean ergodic theorem (17) holds by (16) and the general mean ergodic theorem in Chapter VIII,3.

Since $m(S) < \infty$, we see, by Schwarz' inequality, that any $f \in L^2(S, \mathfrak{B}, m)$ belongs to $L^1(S, \mathfrak{B}, m)$ and $\|f\|_1 \leqq \|f\|_2 \cdot m(S)^{1/2}$. Hence the mean ergodic theorem $\lim_{n\to\infty} \int_S \left| f^*(s) - n^{-1} \sum_{k=1}^{n} (T_k f)(s) \right| m(ds) = 0$, together with $T_1 f^* = f^*$, hold for any $f \in L^2(S, \mathfrak{B}, m)$. Again, by $m(S) < \infty$ and $0 = \lim_{n\to\infty} \|f - {}_nf\|_1 = \lim_{n\to\infty} \int_S |f(s) - {}_nf(s)|\, m(ds)$ with ${}_nf(s) = \min(f(s), n)$, we see that $L^2(S, \mathfrak{B}, m)$ is L^1-dense in $L^1(S, \mathfrak{B}, m)$. That is, for any $f \in L^1(S, \mathfrak{B}, m)$ and $\varepsilon > 0$, there exists an $f_\varepsilon \in L^2(S, \mathfrak{B}, m) \cap L^1(S, \mathfrak{B}, m)$ such that $\|f - f_\varepsilon\|_1 < \varepsilon$. Hence, by (16), we obtain

$$\left\| n^{-1} \sum_{k=1}^{n} T_k f - n^{-1} \sum_{k=1}^{n} T_k f_\varepsilon \right\|_1 \leqq \|f - f_\varepsilon\|_1 \leqq \varepsilon.$$

The mean ergodic theorem (17) in $L^1(S, \mathfrak{B}, m)$ holds for f_ε and so, by the above inequality, we see that (17) in $L^1(S, \mathfrak{B}, m)$ must hold also for f.

Since the strong convergence implies the weak convergence, (18) is a consequence of (17).

Remark 1. The above Theorem 2 is due to K. YOSIDA [17]. Cf. S. KAKUTANI [6], where, moreover, the m-a.e. convergence $\lim_{n\to\infty} n^{-1} \sum_{k=1}^{n} (T_k f)(x)$ is proved for every $f \in L^\infty(S, \mathfrak{B}, m)$. We can prove, when the semi-group T_t is strongly continuous in t, that we may replace $s\text{-}\lim\, n^{-1} \sum_{k=1}^{n} T_k f$ in (17) by $\underset{t\uparrow\infty}{s\text{-}\lim}\, t^{-1} \int_0^t T_s f\, ds$. We shall not go into the details, since they are given in the books on ergodic theory due to E. HOPF [1] and K. JACOBS [1]. We must mention a fine report by S. KAKUTANI [8] on the development of ergodic theory between HOPF's report in 1937 and the 1950 International Congress of Mathematicians held at Cambridge.

In order to deal with the ergodic hypothesis (7), we must prove the *individual ergodic theorem* to the effect that

$$\lim_{n\to\infty} n^{-1} \sum_{k=1}^{n} (T_k f)(x) = f^*(x) \ m\text{-a.e.}$$

In the next section, we shall be concerned with the m-a.e. convergence of the sequence $n^{-1} \sum_{k=1}^{n} (T_k f)(x)$. Our aim is to derive the m-a.e. convergence from the mean convergence using Banach's convergence theorem in Chapter XII, 4.

2. An Individual Ergodic Theorem and Its Applications

We first prove

Theorem 1 (K. YOSIDA). Let X_1 be a real, σ-complete F-lattice provided with a quasi-norm $||x||_1$ such that

$$O\text{-}\lim_{n\to\infty} x_n = x \ \text{ implies } \ \lim_{n\to\infty} ||x_n||_1 = ||x||_1. \tag{1}$$

Let a linear subspace X of X_1 be a real B-space provided with a norm $||x||$ such that

$$s\text{-}\lim_{n\to\infty} x_n = x \ \text{ in } X \text{ implies } \ s\text{-}\lim_{n\to\infty} x_n = x \ \text{ in } X_1. \tag{2}$$

Let $\{T_n\}$ be a sequence of bounded linear operators defined on X into X such that

$$O\text{-}\overline{\lim_{n\to\infty}} |T_n x| \text{ exists for those } x\text{'s which form a set } S \text{ of the second category in } X. \tag{3}$$

Suppose that, to a certain $z \in X$, there corresponds a $\bar{z} \in X$ such that

$$\lim_{n\to\infty} ||T_n z - \bar{z}|| = 0, \tag{4}$$

$$T_n \bar{z} = \bar{z} \quad (n = 1, 2, \ldots), \tag{5}$$

$$O\text{-}\lim_{n\to\infty} (T_n z - T_n T_k z) = 0 \quad \text{for} \quad k = 1, 2, \ldots \tag{6}$$

Then

$$O\text{-}\lim_{n\to\infty} T_n z = \bar{z}. \tag{7}$$

Proof. Put $z = \bar{z} + (z - \bar{z})$. We define an operator $\tilde{T}$ on S into X_1 by

$$\tilde{T} x = O\text{-}\overline{\lim_{n\to\infty}}\, T_n x - O\text{-}\underline{\lim_{n\to\infty}}\, T_n x. \tag{8}$$

Then, by (5),

$$0 \leq \tilde{T} z \leq \tilde{T}(z - \bar{z}).$$

We have $\tilde{T}(z - T_k z) = 0$ $(k = 1, 2, \ldots)$ by (6), and $\lim_{k\to\infty} \|(z-\bar{z}) - (z - T_k z)\| = 0$ by (4). Hence, by Banach's theorem in Chapter XII, 4, we have $\tilde{T}(z - \bar{z}) = 0$. Hence $0 \leqq \tilde{T}z \leqq 0$, that is, $O\text{-}\lim_{n\to\infty} T_n z = w$ exists.

We have to show that $w = \bar{z}$. We have, by (4) and (2), $\lim_{n\to\infty} \|T_n z - \bar{z}\|_1 = 0$. Also we have, by (1) and $O\text{-}\lim_{n\to\infty} T_n z = w$, $\lim_{n\to\infty} \|T_n z - w\|_1 = 0$. Hence $w = \bar{z}$.

Specializing X_1 as the real space $M(S, \mathfrak{B}, m)$ and X as the real space $L^1(S, \mathfrak{B}, m)$, respectively, we can prove the following individual ergodic theorem.

Theorem 2 (K. Yosida). Let T be a bounded linear operator defined on $L^1(S, \mathfrak{B}, m)$ into $L^1(S, \mathfrak{B}, m)$ where $m(S) < \infty$. Suppose that

$$\|T^n\| \leqq C < \infty \quad (n = 1, 2, \ldots), \tag{9}$$

$$\overline{\lim_{n\to\infty}} \left| n^{-1} \sum_{m=1}^{n} (T^m x)(s) \right| < \infty \quad m\text{-a.e.} \tag{10}$$

Suppose that, for a certain $z \in L^1(S, \mathfrak{B}, m)$,

$$\lim_{n\to\infty} n^{-1} (T^n z)(s) = 0 \quad m\text{-a.e.}, \tag{11}$$

and the sequence $\left\{ n^{-1} \sum_{m=1}^{n} T^m z \right\}$ contains a subsequence which converges weakly to an element $\bar{z} \in L^1(S, \mathfrak{B}, m)$. (12)

Then

$$s\text{-}\lim_{n\to\infty} n^{-1} \sum_{m=1}^{n} T^m z = \bar{z}, \quad T\bar{z} = \bar{z}, \tag{13}$$

and

$$\lim_{n\to\infty} n^{-1} \sum_{m=1}^{n} (T^m z)(s) = \bar{z}(s) \quad m\text{-a.e.} \tag{14}$$

Proof. Consider the space $M(S, \mathfrak{B}, m)$, and take the F-lattice $X_1 = M(S, \mathfrak{B}, m)$ with $\|x\|_1 = \int_S |x(s)| (1 + |x(s)|)^{-1} m(ds)$, the B-lattice $X = L^1(S, \mathfrak{B}, m)$ with $\|x\| = \int_S |x(s)| m(ds)$ and $T_n = n^{-1} \sum_{m=1}^{n} T^m$. Then the conditions of Theorem 1 are satisfied. We shall, for instance, verify that (6) is satisfied. We have

$$T_n z - T_n T^k z = n^{-1}(T + T^2 + \cdots + T^n) z - n^{-1}(T^{k+1} + T^{k+2} + \cdots + T^{k+n}) z$$

and so, by (11), $\lim_{n\to\infty} (T_n z - T_n T^k z)(s) = 0$ m-a.e. for $k = 1, 2, \ldots$ Hence, by taking the arithmetic mean with respect to k, we obtain $\lim_{n\to\infty} (T_n z - T_n T_k z)(s) = 0$ m-a.e. for $k = 1, 2, \ldots$ Conditions (4) and (5), that is condition (13) is a consequence of the mean ergodic theorem in Chapter VIII, 3.

Remark. The above two theorems are adapted from K. YOSIDA [15] and [18]. In these papers there are given other ergodic theorems implied by Theorem 1.

As for condition (11) above, we have the following result due to E. HOPF [3]:

Theorem 3. Let T be a positive linear operator on the real $L^1(S, \mathfrak{B}, m)$ into itself with the L^1-norm $\|T\| \leqq 1$. If $f \in L^1(S, \mathfrak{B}, m)$ and $p \in L^1(S, \mathfrak{B}, m)$ be such that $p(s) \geqq 0$ m-a.e., then

$$\lim_{n\to\infty} (T^n f)(s) \Big/ \sum_{j=0}^{n-1} (T^j p)(s) = 0 \quad m\text{-a.e. on the set where } p(s) > 0.$$

If $m(S) < \infty$ and $T \cdot 1 = 1$, then, by taking $p(s) = 1$, we obtain (11).

Proof. It suffices to prove the Theorem for the case $f \geqq 0$. Choose $\varepsilon > 0$ arbitrarily and consider the functions

$$g_n = T^n f - \varepsilon \cdot \sum_{j=0}^{n-1} T^j p, \quad g_0 = f.$$

Let $x_n(s)$ be the defining function of the set $\{s \in S; g_n(s) \geqq 0\}$. From $x_n g_n = g_n^+ = \max(g, 0)$ and $g_{n+1} + \varepsilon p = T g_n$, we obtain, by the positivity of the operator T and $\|T\| \leqq 1$,

$$\int_S g_{n+1}^+ \cdot m(ds) + \varepsilon \int_S x_{n+1} p \cdot m(ds) = \int_S x_{n+1}(g_{n+1} + \varepsilon p)\, m(ds)$$

$$= \int_S x_{n+1} T g_n \cdot m(ds) \leqq \int_S x_{n+1} T g_n^+ \cdot m(ds) \leqq \int_S T g_n^+ \cdot m(ds) \leqq \int_S g_n^+ \cdot m(ds).$$

Summing up these inequalities from $n = 0$ on, we obtain

$$\int_S g_n^+ \cdot m(ds) + \varepsilon \int_S p \cdot \sum_{k=1}^{n} x_k \cdot m(ds) \leqq \int_S g_0^+ \cdot m(ds),$$

and hence

$$\int_S p \cdot \sum_{k=1}^{\infty} x_k \cdot m(ds) \leqq \varepsilon^{-1} \int_S g_0^+ \cdot m(ds).$$

Hence $\sum_{k=1}^{\infty} x_k(s)$ converges m-a.e. on the set where $p(s) > 0$. Thus, on the set where $p(s) > 0$, we must have $g_n(s) < 0$ m-a.e. for all large n.

Therefore, we have proved that $(T^n f)(s) < \varepsilon \sum_{k=0}^{n-1} (T^k p)(s)$ m-a.e. for all large n on the set where $p(s) > 0$. As $\varepsilon > 0$ was arbitrary, we have proved our Theorem.

As for condition (10), we have the following Theorem due to R. V. CHACON-D. S. ORNSTEIN [1]:

Theorem 4 (R. V. CHACON-D. S. ORNSTEIN). Let T be a positive linear operator on the real $L^1(S, \mathfrak{B}, m)$ into itself with the L^1-norm

$\|T\|_1 \leqq 1$. If f and p are functions in $L^1(S, \mathfrak{B}, m)$ and $p(s) \geqq 0$, then

$$\overline{\lim_{n\to\infty}} \left\{ \sum_{k=0}^{n} (T^k f)(s) \Big/ \sum_{k=0}^{n} (T^k p)(s) \right\} \text{ is finite } m\text{-a.e.}$$

on the set where $\sum_{n=0}^{\infty} (T^n p)(s) > 0$.

If $m(S) < \infty$ and $T \cdot 1 = 1$, then, by taking $p(s) = 1$, we obtain (10).

For the proof, we need a

Lemma (CHACON-ORNSTEIN [1]). If $f = f^+ + f^-$, and if $\overline{\lim_{n\to\infty}} \sum_{k=0}^{n} (T^k f)(s) > 0$ on a set B, then there exist sequences $\{d_k\}$ and $\{f_k\}$ of non-negative functions such that, for every N,

$$\int_S \sum_{k=0}^{N} d_k \cdot m(ds) + \int_S f_N \cdot m(ds) \leqq \int_S f^+ \cdot m(ds), \tag{15}$$

$$\sum_{k=0}^{\infty} d_k(s) = -f^-(s) \text{ on } B, \tag{16}$$

$$T^N f^+ = \sum_{k=0}^{N} T^{N-k} d_k + f_N. \tag{17}$$

Proof. Define inductively

$$d_0 = 0,\ f_0 = f^+,\ f_{-1} = 0,$$

$$f_{i+1} = (Tf_i + f^- + d_0 + \cdots + d_i)^+,\ d_{i+1} = Tf_i - f_{i+1}. \tag{18}$$

Note that

$$+ f^- + d_0 + \cdots + d_i \leqq 0, \tag{19}$$

and that the equality holds on the set where $f_i(s) > 0$, for

$$\begin{aligned} f_i &= (Tf_{i-1} + f^- + d_0 + \cdots + d_{i-1})^+ \\ &= (Tf_{i-1} - f_i + f^- + d_0 + \cdots + d_{i-1} + f_i)^+ \\ &= (d_i + f^- + d_0 + \cdots + d_{i-1} + f_i)^+. \end{aligned}$$

It follows from (18) that

$$T^j f^+ = \sum_{k=0}^{j} T^{j-k} d_k + f_j. \tag{20}$$

By the definition, f_i is non-negative, and so is d_i by the last two equations of (18) and (19). From (20), we have

$$\sum_{j=0}^{n} T^j f^+ = \sum_{j=0}^{n} \sum_{k=0}^{j} T^{j-k} d_k + \sum_{j=0}^{n} f_j. \tag{21}$$

We then prove

$$\sum_{j=0}^{n} T^j f^+ \leqq \sum_{j=0}^{n} d_j - \sum_{j=1}^{n} T^j f^- + \sum_{j=0}^{n} f_j. \tag{22}$$

To this purpose, we note that

$$\sum_{j=0}^{n} \sum_{k=0}^{j} T^{j-k} d_k = \sum_{j=0}^{n} T^j \left(\sum_{k=0}^{n-j} d_k \right)$$

and that, by (19) and the positivity of T,

$$-T^j f^- \geqq T^j\left(\sum_{k=0}^{n-j} d_k\right) \quad \text{for} \quad 1 \leqq j \leqq n. \tag{23}$$

Rewriting (22), we have

$$\sum_{j=0}^{n} T^j (f^+ + f^-) \leqq \sum_{j=0}^{n} (d_j + f_j) + f^-. \tag{24}$$

We will now prove that

$$\sum_{j=0}^{\infty} d_j(s) + f^-(s) \geqq 0 \ m\text{-a.e. on } \ B. \tag{25}$$

It is clear from the remark after (19) that (25) holds with equality m-a.e. on the set $C = \{s \in S;\, f_j(s) > 0 \text{ for some } j \geqq 0\}$. It remains to show that (25) holds on the set $B - C$. This is proved by noting that (24) implies that on B, and on $B - C$ in particular, we have the inequality

$$\sum_{j=0}^{\infty} (d_j + f_j)(s) + f^-(s) \geqq 0.$$

Now we note that (17) is exactly (20) and that (16) is implied from (19) and (25). To see that (15) holds, note that we have, by the assumptions on T and (18),

$$\int_S \left(\sum_{k=0}^{j} d_k + f_j\right) m(ds) \geqq \int_S \left(\sum_{k=0}^{j} d_k + T\cdot f_j\right) m(ds) = \int_S \left(\sum_{k=0}^{j+1} d_k + f_{j+1}\right) m(ds).$$

Hence we obtain (15) by induction on j, since $d_0 + f_0 = f^+$.

Proof of Theorem 4. It is sufficient to prove the Theorem under the hypothesis that $f(s) \geqq 0$ on S, and only to prove the finiteness of the indicated supremum when $p(s) > 0$. The first remark is obvious, and the second is proved as follows. By the hypothesis that the indicated supremum is finite at point s where $p(s) > 0$, we have

$$\varlimsup_{n\to\infty} \left\{\sum_{j=0}^{n} (T^{j+k} f)(s) \Big/ \sum_{j=0}^{n} (T^{j+k} p)(s)\right\} \text{ is finite } m\text{-a.e.}$$
$$\text{on the set where } (T^k p)(s) > 0, \tag{26}$$

and this implies that $\varlimsup_{n\to\infty} \left\{\sum_{j=0}^{n} (T^j f)(s) \Big/ \sum_{j=0}^{n} (T^j p)(s)\right\}$ is finite m-a.e. on the set where $(T^k p)(s) > 0$.

Now assume the contrary to the hypothesis. Then

$$\varlimsup_{n\to\infty} \left\{\sum_{j=0}^{n} (T^j f)(s) \Big/ \sum_{j=0}^{n} (T^j p)(s)\right\}$$

is infinite m-a.e. on a set E of positive m-measure, and $p(s) > \beta > 0$ on E for some positive constant β. We have therefore, that for any positive constant α,

$$\sup_n \left\{\sum_{j=0}^{n} (T^j((f - \alpha p)^+ + (f - \alpha p)^-))(s)\right\} > 0$$

m-a.e. on E. Applying the Lemma with f replaced by $(f - \alpha p)$, we obtain from (15) and (16) that

$$\int_S (f-\alpha p)^+ m(ds) \geqq \int_S \sum_{k=0}^{\infty} d_k m(ds) \geqq - \int_E (f - \alpha p)^- m(ds).$$

However, the extreme right term tends to ∞ as $\alpha \uparrow \infty$, and the extreme left term is bounded as $\alpha \uparrow \infty$. This is a contradiction, and so we have proved the Theorem.

We have thus proved

Theorem 5. Let T be a positive linear operator on $L^1(S, \mathfrak{B}, m)$ into itself with the L^1-norm $\|T\|_1 \leqq 1$. If $m(S) < \infty$ and $T \cdot 1 = 1$, then, for any $f \in L^1(S, \mathfrak{B}, m)$, the mean convergence of the sequence $\left\{n^{-1} \sum_{j=1}^{n} T^j f\right\}$ implies its m-a.e. convergence.

As a corollary we obtain

Theorem 6. Let $P(t, x, E)$ be a Markov process with an invariant measure m on a measure space $(S, \mathfrak{B}, m)$ such that $m(S) < \infty$. Then, for the linear operator T_t defined by $(T_t f)(x) = \int_S P(t, x, dy) f(y)$, we have i) the *mean ergodic theorem*:

$$\begin{aligned} &\text{for any } f \in L^p(S, \mathfrak{B}, m),\ \underset{n\to\infty}{\text{s-lim}}\, n^{-1} \sum_{k=1}^{n} T_k f = f^* \\ &\text{exists in } L^p(S, \mathfrak{B}, m) \text{ and } T_1 f^* = f^* \ (p = 1, 2), \end{aligned} \tag{27}$$

and ii) the *individual ergodic theorem*:

$$\begin{aligned} &\text{for any } f \in L^p(S, \mathfrak{B}, m) \text{ with } p = 1 \text{ or } p = 2, \\ &\text{finite} \lim_{n\to\infty} n^{-1} \sum_{k=1}^{n} (T_k f)(s) \text{ exists } m\text{-a.e. and is } = f^*(s)\ m\text{-a.e.}, \\ &\text{and moreover, } \int_S f(s)\, m(ds) = \int_S f^*(s)\, m(ds). \end{aligned} \tag{28}$$

Proof. By Theorem 2 in the preceding section, the mean ergodic theorem (27) holds and so we may apply Theorem 5.

Remark. If T_t is given by an equi-measure transformation $x \to y_t(x)$ of S onto S, the result (27) is precisely the mean ergodic theorem of J. VON NEUMANN [3] and the result (28) is precisely the individual ergodic theorem of G. D. BIRKHOFF [1] and A. KHINTCHINE [1].

A Historical Sketch. The first operator-theoretical generalization of the individual ergodic theorem of the Birkhoff-Khintchine type was given by J. L. DOOB [1]. He proved that $n^{-1} \sum_{k=1}^{n} (T_k f)(x)$ converges m-a.e. when T_t is defined by a Markov process $P(t, x, E)$ with an invariant measure m on a measure space $(S, \mathfrak{B}, m)$ such that $m(S) = 1$ and f is the defining function of a set $\in \mathfrak{B}$. It was remarked by S. KAKUTANI [6] that

Doob's method is applicable and gives the same result for f merely a bounded $\mathfrak{B}$-measurable function. E. HOPF [2] then proved the theorem assuming merely that f is m-integrable. N. DUNFORD-J. SCHWARTZ [4] extended Hopf's result by proving (28) for a linear operator T_t which increases neither the L^1- nor L^∞-norm, without assuming that T_t is positive but assuming that $T_k = T_1^k$ and $T_1 \cdot 1 = 1$. It is noted that Hopf's and the Dunford-Schwartz arguments use the idea of our Theorem 1. R. V. CHACON-D. S. ORNSTEIN [1] proved (28) for a positive linear operator T_t of L^1-norm $\leqq 1$, without assuming that T_t does not increase the L^∞-norm, and, moreover, without appealing to our Theorem 1. Here, of course, it is assumed that $T_k = T_1^k$ and $T_1 \cdot 1 = 1$. We will not go into details, since, for the exposition of this Chapter which deals with Markov processes, it sufficies to base the ergodic theory upon Theorem 6 which is a consequence of our Theorem 1.

3. The Ergodic Hypothesis and the H-theorem

Let $P(t, x, E)$ be a Markov process with an invariant measure m on a measure space $(S, \mathfrak{B}, m)$ such that $m(S) = 1$. We shall define the *ergodicity* of the process $P(t, x, E)$ by the condition:

$$\begin{aligned}\text{the time average } f^*(s) &= \lim_{n\to\infty} n^{-1} \sum_{k=1}^{n} (T_k f)(x) \\ &= \lim_{n\to\infty} n^{-1} \sum_{k=1}^{n} \int_S P(k, x, dy)\, f(y) \\ = \text{the space average } &\int_S f(x)\, m(dx) \quad m\text{-a.e.} \qquad (1)\end{aligned}$$

for every $f \in L^p(S, \mathfrak{B}, m)$ $(p = 1, 2)$.

Since $\int_S f^*(x)\, m(dx) = \int_S f(x)\, m(dx)$, (1) may be rewritten as

$$f^*(x) = \text{a constant } m\text{-a.e. for every } f \in L^p(S, \mathfrak{B}, m) \ (p = 1, 2). \qquad (1')$$

We shall give three different interpretations of the ergodic hypothesis (1) and (1').

1. Let $\chi_B(x)$ be the defining function of the set $B \in \mathfrak{B}$. Then, for any two sets $B_1, B_2 \in \mathfrak{B}$, *the time average of the probability that the points of B_1 will be transferred into the set B_2 after k units of time is equal to the product $m(B_1)\, m(B_2)$.* That is, we have

$$(\chi^*_{B_2}, \chi_{B_1}) = m(B_1)\, m(B_2). \qquad (2)$$

Proof. If $f, g \in L^2(S, \mathfrak{B}, m)$, then the strong convergence $n^{-1} \sum_{k=1}^{n} T_k f \to f^*$ in $L^2(S, \mathfrak{B}, m)$ implies, by (1'),

$$\begin{aligned}\lim_{n\to\infty} n^{-1} \sum_{k=1}^{n} (T_k f, g) &= (f^*, g) = f^*(x) \int_S g(x)\, m(dx) \\ &= \int_S f(x)\, m(dx) \cdot \int_S g(x)\, m(dx) \quad \text{for } m\text{-a.e. } x.\end{aligned}$$

By taking $f(x) = \chi_{B_2}(x)$, $g(x) = \chi_{B_1}(x)$, we obtain (2).

Remark. Since linear combinations of the defining functions $\chi_B(x)$ are dense in the space $L^p(S, \mathfrak{B}, m)$ for $p = 1$ and $p = 2$, we easily see that (2) is equivalent to the ergodic hypothesis (1'). (2) says that every part of S will be transferred uniformly into every part of S in the sense of the *time average*.

2. The mean ergodic theorem in Chapter VIII, 3 says that the mapping $f \to T^*f = f^*$ gives the eigenspace of T_1 belonging to the eigenvalue 1 of T_1; this eigenspace is given by the range $R(T^*)$. Hence the *ergodic hypothesis* (1)' *means precisely the hypothesis that* $R(T^*)$ *is of one-dimension*. Thus the ergodic hypothesis of $P(t, x, E)$ is interpreted in terms of the spectrum of the operator T_1.

3. The Markov process $P(t, x, E)$ is said to be *metrically transitive* or *indecomposable* if the following conditions are satisfied:

$$\begin{aligned}&S \text{ cannot be decomposed into the sum of disjoint sets}\\ &B_1, B_2 \in \mathfrak{B} \text{ such that } m(B_1) > 0,\ m(B_2) > 0\\ &\text{and } P(1, x, E) = 0 \text{ for every } x \in B_i,\ E \subseteqq B_j \text{ with } i \neq j.\end{aligned} \tag{3}$$

Proof. Suppose $P(t, x, E)$ is ergodic and that S is decomposed as in (3). The defining function $\chi_{B_1}(x)$ satisfies $T_1\chi_{B_1} = \chi_{B_1}$ by the condition

$$\begin{aligned}\int_S P(1, x, dy)\, \chi_{B_1}(y) = P(1, x, B_1) &= 1 = \chi_{B_1}(x), \quad \text{for } x \in B_1,\\ &= 0 = \chi_{B_1}(x), \quad \text{for } x \bar{\in} B_1.\end{aligned}$$

But, by $m(B_1)\, m(B_2) > 0$, the function $\chi_{B_1}(x) = \chi^*_{B_1}(x)$ cannot be equal to a constant m-a.e.

Next suppose that the process $P(t, x, E)$ is indecomposable. Let $T_1 f = f$ and we shall show that $f^*(x) = f(x)$ equals a constant m-a.e. Since T_1 maps real-valued functions into real-valued functions, we may assume that the function $f(x)$ is real-valued. If $f(x)$ is not equal to a constant m-a.e., then there exists a constant α such that both the sets

$$B_1 = \{s \in S; f(s) > \alpha\} \quad \text{and} \quad B_2 = \{s \in S;\ f(s) \leqq \alpha\}$$

are of m-measure > 0. Since $T_1(f - \alpha) = f - \alpha$, the argument under The Angle Variable of p. 391 implies that $T_1(f-\alpha)^+ = (f-\alpha)^+$, $T_1(f-\alpha)^- = (f-\alpha)^-$.

Hence we would obtain $P(1, x, E) = 0$ if $x \in B_i$, $E \subseteqq B_j$ $(i \neq j)$.

Remark. The notion of the metric transitivity was introduced by G. D. Birkhoff-P. A. Smith [2] for the case of an equi-measure transformation $x \to y_t(x)$ of S onto S.

An Example of the Ergodic Equi-measure Transformation. Let S be a torus. That is, S is the set of all pairs $s = \{x, y\}$ of real numbers x, y such that $s = \{x, y\}$ and $s' = \{x', y'\}$ are identified iff $x \equiv x' \pmod 1$ and $y \equiv y' \pmod 1$; S is topologized by the real number topology of its coordinates x, y. We consider a mapping

$$s = \{x, y\} \to T_t s = s_t = \{x + t\alpha, y + t\beta\}$$

of S onto S which leaves invariant the measure $dx\,dy$ on S. Here we assume that the real numbers α, β are integrally linearly independent mod 1 in the following sense: if integers n, k satisfy $n\alpha + k\beta \equiv 0$ (mod 1), then $n = k = 0$. Then the mapping $s \to T_t s$ is ergodic.

Proof. Let $f(s) \in L^2(S)$ be invariant by T_1, that is, let $f(s) = f(s_1)$ $dx\,dy$-a.e. on S. We have to show that $f(s) = f^*(s) =$ a constant $dx\,dy$-a.e. on S. Let us consider the Fourier coefficients of $f(s)$ and $f(s_1) = f(T_1 s)$, respectively.

$$\int_0^1\int_0^1 f(s) \exp(-2\pi i(k_1 x + k_2 y))\,dx\,dy,$$

$$\int_0^1\int_0^1 f(T_1 s) \exp(-2\pi i(k_1 x + k_2 y))\,dx\,dy.$$

By $T_1 s = \{x + \alpha, y + \beta\}$ and the invariance of the measure $dx\,dy$ by the mapping $s \to T_1 s$, the latter integral is equal to

$$\int_0^1\int_0^1 f(s) \exp(-2\pi i(k_1 x + k_2 y)) \exp(2\pi i(k_1 \alpha + k_2 \beta))\,dx\,dy.$$

Hence, by the uniqueness of the corresponding Fourier coefficients of $f(s)$ and $f(T_1 s)$, we must have

$$\exp(2\pi i(k_1\alpha + k_2\beta)) = 1 \quad \text{whenever}$$
$$\int_0^1\int_0^1 f(s) \exp(-2\pi i(k_1 x + k_2 y))\,dx\,dy \neq 0.$$

Hence, by the hypothesis concerning α, β, we must have

$$\int_0^1\int_0^1 f(s) \exp(-2\pi i(k_1 x + k_2 y))\,dx\,dy = 0 \quad \text{unless} \quad k_1 = k_2 = 0.$$

Therefore $f(s) = f^*(s)$ must reduce to a constant $dx\,dy$-a.e.

The Angle Variable. Let T_t be defined by the Markov process $P(t, x, E)$ with an invariant measure m on $(S, \mathfrak{B}, m)$ such that $m(S) = 1$. Let $f(s) \in L^2(S, \mathfrak{B}, m)$ be the eigenvector of T_1 pertaining to the eigenvalue λ of absolute value 1 of T_1:

$$T_1 f = \lambda f, \quad |\lambda| = 1.$$

Then we have $T_1 |f| = |f|$. For, we have, by the positivity of the operator T_1, $(T_1 |f|)(x) \geqq |(T_1 f)(x)| = |f(x)|$, that is, $\int_S P(1, x, dy)\,|f(y)| \geqq |f(x)|$. In this inequality, we must have the equality for m-a.e. x, as may be seen by integrating both sides with respect to $m(dx)$ and remembering the invariance of the measure m. Hence, if $P(t, x, E)$ is ergodic, then $|f(x)|$ must reduce to a constant m-a.e. Therefore, if we put

$$f(x) = |f(x)| \exp(i\Theta(x)), \quad 0 \leqq \Theta(x) < 2\pi,$$

we must have

$$T_1 \exp(i\Theta(x)) = \lambda \exp(i\Theta(x)).$$

In case $\lambda \neq 1$, such $\Theta(x)$ is called an *angle variable* of the ergodic Markov process $P(t, x, E)$.

The Mixing Hypothesis. Let T_t be defined by a Markov process $P(t, x, E)$ with an invariant measure m on $(S, \mathfrak{B}, m)$ such that $m(S) = 1$. A stronger condition than the ergodic hypothesis of $P(t, x, E)$ is given by

$$\lim_{k\to\infty} (T_k f, g) = (f^*, g) = \int_S f(x)\, m(dx) \cdot \int_S g(x)\, m(dx)$$
$$\text{for every pair } \{f, g\} \text{ of vectors} \in L^2(S, \mathfrak{B}, m). \tag{4}$$

This condition is called the *mixing hypothesis* of the Markov process $P(t, x, E)$. As in the case of the ergodic hypothesis, it may be interpreted as follows: *every part of S will be transferred into every part of S uniformly in the long run.* As for the examples of the mixing equi-measure transformation $x \to y_t(x)$, we refer the reader to the above cited book by E. Hopf.

H-theorem. Let $P(t, x, E)$ be a Markov process with an invariant measure m on $(S, \mathfrak{B}, m)$ such that $m(S) = 1$. Let us consider the function

$$H(z) = -z \log z, \quad z \geqq 0. \tag{5}$$

We can prove

Theorem (K. Yosida [17]). Let a non-negative function $f(x) \in L^1(S, \mathfrak{B}, m)$ belong to the *Zygmund class*, that is, let us assume that $\int_S f(x) \log^+ f(x)\, m(dx) < \infty$, where $\log^+ |z| = \log |z|$ or $= 0$ according as $|z| \geqq 1$ or $|z| < 1$. Then we have

$$\int_S H(f(x))\, m(dx) \leqq \int_S H((T_t f)(x))\, m(dx). \tag{6}$$

Proof. Since $H(z) = -z \log z$ satisfies $H''(z) = -1/z < 0$ for $z > 0$, $H(z)$ is a concave function. Hence we obtain

the weighted mean of $H(f(x)) \leqq H$ (the weighted mean of $f(x)$),

and so

$$\int_S P(t, x, dy)\, H(f(y)) \leqq H\Big(\int_S P(t, x, dy)\, f(y)\Big).$$

Integrating with respect to $m(dx)$ and remembering the invariance of the measure $m(E)$, we obtain (6).

Remark. By virtue of the semi-group property $T_{t+s} = T_t T_s$ and (6), we easily obtain

$$\int_S H((T_{t_1} f)(x))\, m\,(dx) \leqq \int_S H((T_{t_2} f)(x))\, m(dx) \text{ whenever } t_1 < t_2. \tag{6'}$$

This may be considered as an analogue of the classical *H-theorem* in statistical mechanics.

4. The Ergodic Decomposition of a Markov Process with a Locally Compact Phase Space

Let S be a separable metric space whose bounded closed sets are compact. Let $\mathfrak{B}$ be the set of all Baire subsets of S, and consider a Markov process $P(t, x, E)$ on $(S, \mathfrak{B})$. We assume that

$$f(x) \in C_0^0(S) \text{ implies } f_t(x) = \int_S P(t, x, dy)\, f(y) \in C_0^0(S). \tag{1}$$

The purpose of this section is to give the decomposition of S into *ergodic parts* and *dissipative part* as an extension of the Krylov-Bogolioubov case of the deterministic, reversible transition process in a compact metric space S (see N. Krylov-N. Bogolioubov [1]). The possibility of such an extension, to the case where S is a compact metric space and with condition (1), was observed by K. Yosida [17] and carried out by N. Beboutov [1], independently of Yosida. The extension to the case of a locally compact space S was given in K. Yosida [19]. We shall follow the last cited paper.

Lemma 1. Let a linear functional $L(f)$ on the normed linear space $C_0^0(S)$, normed by the maximal norm $\|f\| = \sup_{x \in S} |f(x)|$, be *non-negative*, viz., $L(f) \geqq 0$ when $f(x) \geqq 0$ on S. Then $L(f)$ is represented by a uniquely determined *regular measure* $\varphi(E)$, which is σ-additive and $\geqq 0$ for Baire subsets E of S, as follows:

$$L(f) = \int_S f(x)\, \varphi(dx) \text{ for all } f \in C_0^0(S). \tag{2}$$

Here the regularity of the measure φ means that

$$\varphi(E) = \inf \varphi(G) \text{ when } G \text{ ranges over all open sets } \supseteqq E. \tag{3}$$

Proof. We refer the reader to P. R. Halmos [1].

We shall call a non-negative, σ-additive regular measure $\varphi(E)$ defined on $\mathfrak{B}$ such that $\varphi(S) \leqq 1$ an *invariant measure* for the Markov process if

$$\varphi(E) = \int_S \varphi(dx)\, P(t, x, E) \text{ for all } t > 0 \text{ and } E \in \mathfrak{B}. \tag{4}$$

We have then

Lemma 2. For any $f \in C_0^0(S)$ and for any invariant measure $\varphi(E)$,

$$\lim_{n\to\infty} n^{-1} \sum_{k=1}^{n} f_k(x) = f^*(x) \left(f_k(x) = \int_S P(k, x, dy)\, f(y) \right) \text{ exists} \tag{5}$$

φ-a.e. and

$$\int_S f^*(x)\, \varphi(dx) = \int_S f(x)\, \varphi(dx). \tag{6}$$

This is exactly a corollary of Theorem 6 in the preceding section.

Now, let ${}_1f, {}_2f, {}_3f, \ldots$ be dense in the normed linear space $C_0^0(S)$. The existence of such a sequence is guaranteed by the hypothesis concerning the space S. Applying Lemma 2 to ${}_1f, {}_2f, {}_3f, \ldots$ and summing up the exceptional sets of φ-measure zero, we see that there exists a set N of φ-measure zero with the property:

$$\text{for any } x \overline{\in} N \text{ and for any } f \in C_0^0(S),\ \lim_{n\to\infty} n^{-1} \sum_{k=1}^{n} f_k(x) = f^*(x) \text{ exists}. \tag{7}$$

A Baire set $S' \subseteqq S$ is said to be of *maximal probability* if $\varphi(S - S') = 0$ for every invariant measure φ. Thus the set S' of all x for which $f^*(x) = \lim_{n\to\infty} n^{-1} \sum_{k=1}^{n} f_k(x)$ exists for every $f \in C_0^0(S)$ is of maximal probability. Hence if there exists an invariant measure with $\varphi(S) > 0$, then there exist a $g \in C_0^0(S)$ and a point x_0 such that

$$g^{**}(x_0) = \overline{\lim_{n\to\infty}}\, n^{-1} \sum_{k=1}^{n} g_k(x_0) > 0. \tag{8}$$

For, if otherwise, we would have $f^*(x) = 0$ on S for all $f \in C_0^0(S)$ and hence, by (6), $\int_S f(x)\, \varphi(dx) = 0$.

Let, conversely, (8) be satisfied for a certain $g \in C_0^0(S)$ and for a certain x_0. Let a subsequence $\{n'\}$ of natural numbers be chosen in such a way that $\lim_{n'\to\infty} (n')^{-1} \sum_{k=1}^{n'} g_k(x_0) = g^{**}(x_0)$. By a diagonal method, we may choose a subsequence $\{n''\}$ of $\{n'\}$ such that $\lim_{n''\to\infty} (n'')^{-1} \sum_{k=1}^{n''} ({}_jf)_k(x_0)$ exists for $j = 1, 2, \ldots$ By the denseness of $\{{}_jf\}$ in $C_0^0(S)$, we easily see that

$$\lim_{n''\to\infty} (n'')^{-1} \sum_{k=1}^{n''} f_k(x_0) = f^{***}(x_0) \text{ exists for every } f \in C_0^0(S).$$

If we put $f^{***}(x_0) = L_{x_0}(f)$, then

$$L_{x_0}(f) = L_{x_0}(f_1). \tag{9}$$

For, by condition (1), we have $f_1 \in C_0^0(S)$ and so

$$L_{x_0}(f) = \lim_{n''\to\infty} (n'')^{-1} \sum_{k=1}^{n''} f_{k+1}(x_0) = f_1^{***}(x_0) = L_{x_0}(f_1).$$

By Lemma 1, there exists a regular measure $\varphi_{x_0}(E)$ such that

$$L_{x_0}(f) = f^{***}(x_0) = \int_S f(x)\, \varphi_{x_0}(dx). \tag{10}$$

Surely we have

$$0 \leqq \varphi_{x_0}(E) \leqq 1, \tag{11}$$

and, by (9),

$$\int_S f(x)\, \varphi_{x_0}(dx) = \int_S \left(\int_S P(1, x, dy)\, f(y)\right) \varphi_{x_0}(dx).$$

By letting $f(x)$ tend to the defining function of the Baire set E, we see that $\varphi_{x_0}(E)$ is an invariant measure. We have $\varphi_{x_0}(S) > 0$, since, by (8) and (10), we have $L_{x_0}(g) = g^{***}(x_0) = \int_S g(x)\, \varphi_{x_0}(dx) > 0$. We have thus proved

Theorem 1. A necessary and sufficient condition for the non-existence of non-trivial invariant measures is

$$\lim_{n\to\infty} n^{-1} \sum_{k=1}^{n} f_k(x) = f^*(x) = 0 \text{ on } S \text{ for any } f \in C_0^0(S). \tag{12}$$

Definition. We will call the process $P(t, x, E)$ on $(S, \mathfrak{B})$ *dissipative* if condition (12) is satisfied.

Example. Let S be the half line $(0, \infty)$ and $\mathfrak{B}$ the set of all Baire sets of $(0, \infty)$. Then the Markov process $P(t, x, E)$ defined by

$$P(t, x, E) = 1 \text{ if } (x+t) \in E \text{ and } = 0 \text{ if } (x+t) \,\overline{\in}\, E$$

is dissipative.

We shall assume that $P(t, x, E)$ is not dissipative. Let D denote the set

$$\left\{x \in S;\ f^*(x) = \lim_{n\to\infty} n^{-1} \sum_{k=1}^{n} f_k(x) = 0 \quad \text{for all} \quad f \in C_0^0(S)\right\}.$$

Since it is equal to

$$\{x \in S;\ ({}_jf)^*(x) = 0 \quad \text{for} \quad j = 1, 2, \ldots\},$$

D is a Baire set. We shall call D the *dissipative part* of S. We can prove below that $\varphi(D) = 0$ for any invariant measure φ, and so, since the process $P(t, x, E)$ is assumed to be not dissipative, $S_0 = S - D$ is not void. We already know that there exists a Baire set $S_1 \subseteqq S_0$ with the property:

to any $x \in S_1$, there corresponds a non-trivial invariant measure

$\varphi_x(E)$ such that $f^*(x) = \lim_{n\to\infty} n^{-1} \sum_{k=1}^{n} f_k(x) = \int_S f(z)\, \varphi_x(dz)$

and $\varphi(S_0 - S_1) = 0$ for any invariant measure φ. (13)

For any invariant measure φ, we have (6) and hence, by (13),

$$\int_S f(y)\, \varphi(dy) = \int_{S_1} \left(\int_S f(z)\, \varphi_y(dz)\right) \varphi(dy).$$

Thus we obtain

$$\varphi(E) = \int_{S_1} \varphi_y(E)\, \varphi(dy), \tag{14}$$

and so we have proved

Theorem 2. Any invariant measure φ may be obtained as a convex combination of the invariant measure $\varphi_y(E)$ with y as a parameter.

We see, from (11) and (14), that the set

$$S_2 = \{x \in S;\, x \in S_1, \varphi_x(S_1) = 1\}$$

is of maximal probability. Hence S_0 is of maximal probability.

For any $f \in C_0^0(S)$ and for any invariant measure φ, we have

$$\int_{S_2} \varphi(dx) \Big(\int_{S_2} (f^*(y) - f^*(x))^2 \varphi_x(dy) \Big) = 0, \tag{15}$$

since the left hand side is equal to

$$\int_{S_2} \varphi(dx) \Big(\int_{S_2} f^*(y)^2 \varphi_x(dy) \Big) - 2 \int_{S_2} f^*(x)\, \varphi(dx) \Big(\int_{S_2} f^*(y)\, \varphi_x(dy) \Big)$$
$$+ \int_{S_2} f^*(x)^2 \varphi(dx) \cdot \int_{S_2} \varphi_x(dy)$$
$$= \int_{S_2} f^*(y)^2 \varphi(dy) - 2 \int_{S_2} f^*(x)^2 \varphi(dx) + \int_{S_2} f^*(x)^2 \varphi(dx),$$

by (13), (14), (6) and the definition of S_2. By applying (15) to ${}_1f, {}_2f, {}_3f, \ldots,$ we see that the set

$$S_3 = \Big\{ x \in S_2;\, \int_{S_2} (f^*(y) - f^*(x))^2 \varphi_x(dy) = 0 \quad \text{for all} \quad f \in C_0^0(S) \Big\}$$

is of maximal probability.

We shall give the *ergodic decomposition* of S. For any $x \in S_3$, put

$$E_x = \{y \in S_3;\, f^*(y) = f^*(x) \quad \text{for all} \quad f \in C_0^0(S)\}. \tag{16}$$

We can then show that each E_x contains a set $\hat{E}_x$ with the property:

$$\varphi_x(E_x) = \varphi_x(\hat{E}_x) \quad \text{and} \quad P(1, y, \hat{E}_x) = 1 \quad \text{for any} \quad y \in \hat{E}_x. \tag{17}$$

Proof. By the definition of S_3, we see that $f^*(y) = f^*(x)$ if the measure $\varphi_x(E)$ has variation at the point y. Hence $\varphi_x(E_x) = \varphi_x(S_3) = 1$. Thus, by the invariance of the measure φ_x, we obtain

$$1 = \varphi_x(E_x) = \int_{S_3} P(1, z, E_x)\, \varphi_x(dz) = \int_{E_x} P(1, z, E_x)\, \varphi_x(dz).$$

Since $0 \leqq P(1, z, E_x) \leqq 1$, there exists a Baire set $E^1 \subseteqq E_x$ such that

$$\varphi_x(E^1) = \varphi_x(E_x) \text{ and } z \in E^1 \text{ implies } P(1, z, E_x) = 1.$$

Put

$$E^2 = \{z \in E^1, P(1, z, E^1) = 1\}.$$

Since

$$\int_{E^1} P(1, z, E^1)\, \varphi_x(dz) = \varphi_x(E^1) = \varphi_x(E_x) = \int_{E^1} P(1, z, E_x)\, \varphi_x(dz),$$

we must have

$$\int_{E^1} (P(1, z, E_x) - P(1, z, E^1))\, \varphi_x(dz) = 0.$$

As $z \in E^2$ implies $P(1, z, E^1) = 1$ by the definition of E^2, we obtain $\varphi_x(E^1 - E^2) = 0$.

Next put

$$E^3 = \{z \in E^2;\ P(1, z, E^2) = 1\}.$$

Then, as above, $\varphi_x(E^3) = \varphi_x(E_x)$. In this way, we obtain a sequence $\{E^n\}$ such that

$$E_x \supseteqq E^1 \supseteqq E^2 \supseteqq \cdots,\ \varphi_x(E^n) = \varphi_x(E_x) \text{ and}$$
$$P(1, z, E^n) = 1 \text{ if } z \in E^{n+k}\ (n \geqq 0, k \geqq 1, E^0 = E_x).$$

We have thus obtained the set $\hat{E}_x = \bigcap_{n=1}^{\infty} E^n$ with the demanded property.

Therefore, we have

Theorem 3. $P(t, y, E)$ defines a Markov process in each $\hat{E}_x$ which is *ergodic*, in the sense that

$\hat{E}_x$ is not decomposable in two parts A and B such that $\varphi(A) \cdot \varphi(B) > 0$ for a certain invariant measure φ and $P(1, a, B) = 0$ when $a \in A$ and $P(1, b, A) = 0$ when $b \in B$. (18)

Proof. Let φ be any invariant measure in $\hat{E}_x$ with $\varphi(\hat{E}_x) = 1$. Then, by (14), $\varphi(E) = \int_{\hat{E}_x} \varphi(dz)\, \varphi_z(E)$ whenever $E \subseteqq \hat{E}_x$. Since $\varphi_z(E) = \varphi_x(E)$ for every $z \in \hat{E}_x$ by the definition of $\hat{E}_x$ we have

$$\varphi(E) = \varphi_x(E) \int_{\hat{E}_x} \varphi(dz) = \varphi_x(E).$$

Thus there is essentially only one invariant measure $\varphi_x(E)$ in $\hat{E}_x$. This proves the ergodicity. For, let $\hat{E}_x$ be decomposed as in (18). Then, by the invariance of φ,

$$\varphi(C) = \int_{\hat{E}_x} P(1, z, C)\, \varphi(dz) \quad \text{whenever} \quad C \subseteqq \hat{E}_x.$$

Thus the measure ψ defined by

$$\psi(C) = \varphi(C)/\varphi(A) \quad \text{if} \quad C \subseteqq A$$
$$= 0 \text{ if } C \subseteqq B$$

is invariant in $\hat{E}_x$ and differs from the unique invariant measure φ.

5. The Brownian Motion on a Homogeneous Riemannian Space

There is an interesting interplay between Markov processes and differential equations. Already in the early thirties, A. KOLMOGOROV proved, under a certain regularity hypothesis concerning the transition probability $P(t, x, E)$, that

$$u(t, x) = \int_S P(t, x, dy) f(y)$$

satisfies the equation of the *diffusion type*:

$$\frac{\partial u}{\partial t} = b^{ij}(x) \frac{\partial^2 u}{\partial x_i \partial x_j} + a^i(x) \frac{\partial u}{\partial x_i} = A u, \quad t > 0, \tag{1}$$

where the differential operator A is elliptic in the local coordinates $(x_1, x_2, \ldots, x_n)$ of the point x of the phase space S. Following Einstein's convention of the tensor notation, it is understood that, e.g., $a^i(x)\, \partial/\partial x_i$ means $\sum_{i=1}^{n} a^i(x)\, \partial/\partial x_i$. For the derivation of equation (1), see A. KOLMOGOROV [1]. The leit-motif in his research was the investigation of local characteristics of Markov processes.

Following work done by E. HILLE at Scandinavian Congress in 1949 (Cf. E. HILLE [9]) and K. YOSIDA [29] in 1948, W. FELLER [2] began in 1952 a systematic research of this new field in probability theory using the analytical theory of semi-groups. His investigations were further developed by E. B. DYNKIN [1], [2], K. ITÔ-H. MCKEAN [1], D. RAY, [1] G. A. HUNT [1], A. A. YUSHKEVITCH [1], G. MARUYAMA [1] and many younger scholars, especially in Japan, USSR and USA. These investigations are called the *theory of diffusion*. A comprehensive treatise on the diffusion theory by K. ITO-H. MCKEAN [1] is in the course of printing[1]. We shall sketch some of the salient analytical features of the theory.

Let S be a locally compact space such that $\mathfrak{B}$ is the totality of the Baire sets in S. To define the *spatial homogeneity* of a Markov process $P(t, x, E)$ on the space S, we suppose that S is an n-dimensional, orientable connected C^∞ Riemannian space such that the full group $\mathfrak{G}$ of the *isometries* of S, which is a Lie group, is transitive on S. That is, for each pair $\{x, y\}$ of points $\in S$, there exists an isometry $M \in \mathfrak{G}$ such that $M \cdot x = y$. Then the process $P(t, x, E)$ is *spatially homogeneous* if

$$P(t, x, E) = P(t, M \cdot x, M \cdot E) \text{ for each } x \in S, E \in \mathfrak{B} \text{ and } M \in \mathfrak{G}. \tag{2}$$

A temporally and spatially homogeneous Markov process on S is called a *Brownian motion* on S, if the following condition, known as the *continuity condition of Lindeberg's type*, is satisfied:

$$\lim_{t \downarrow 0} t^{-1} \int_{\mathrm{dis}(x,y) > \varepsilon} P(t, x, dy) = 0 \quad \text{for every} \quad \varepsilon > 0 \text{ and } x \in S. \tag{3}$$

[1] Berlin/Göttingen/Heidelberg: Springer.

Proposition. Let $C(S)$ be the B-space of all bounded uniformly continuous real-valued functions $f(x)$ on S normed by $\|f\| = \sup |f(x)|$.

Define

$$(T_t f)(x) = \int_S P(t, x, dy) f(y), \quad \text{if} \quad t > 0, \qquad (4)$$
$$= f(x), \text{ if } t = 0.$$

Then $\{T_t\}$ constitutes a contraction semi-group of class (C_0) in $C(S)$.

Proof. By $P(t, x, E) \geqq 0$ and $P(t, x, S) = 1$, we obtain

$$|(T_t f)(x)| \leqq \sup_y |f(y)|$$

and the positivity of the operator T_t. The semi-group property $T_{t+s} = T_t T_s$ is a consequence from the Chapman-Kolmogorov equation. If we define a linear operator M' by $(M'f)(x) = f(M \cdot x)$, $M \in \mathfrak{G}$, we obtain

$$T_t M' = M' T_t, \ t \geqq 0. \qquad (5)$$

For,

$$(M' T_t f)(x) = (T_t f)(M \cdot x) = \int_S P(t, M \cdot x, dy) f(y)$$
$$= \int_S P(t, M \cdot x, d(M \cdot y)) f(M \cdot y)$$
$$= \int_S P(t, x, dy) f(M \cdot y) = (T_t M' f)(x).$$

If $M \in \mathfrak{G}$ be such that $M \cdot x = x'$, we have

$$(T_t f)(x) - (T_t f)(x') = (T_t f)(x) - (M' T_t f)(x) = T_t(f - M'f)(x).$$

Hence by the uniform continuity of $f(x)$, we see that $(T_t f)(x)$ is uniformly continuous and bounded.

To prove the strong continuity in t of T_t, it suffices by the Theorem in Chapter IX, 1, to verify weak right continuity of T_t at $t = 0$. Therefore, it is surely enough to show that $\lim_{t \downarrow 0} (T_t f)(x) = f(x)$ uniformly in x. Now

$$|(T_t f)(x) - f(x)| = \left| \int_S P(t, x, dy) [f(y) - f(x)] \right|$$
$$\leqq \left| \int_{d(x,y) \leqq \varepsilon} P(t, x, dy) [f(y) - f(x)] \right| + \left| \int_{d(x,y) > \varepsilon} P(t, x, dy) [f(y) - f(x)] \right|$$
$$\leqq \left| \int_{d(x,y) \leqq \varepsilon} P(t, x, dy) [f(y) - f(x)] \right| + 2 \|f\| \int_{d(x,y) > \varepsilon} P(t, x, dy).$$

The first term on the right tends to zero uniformly in x as $\varepsilon \downarrow 0$ and, for fixed $\varepsilon > 0$, the second term on the right tends to zero uniformly in x as $t \downarrow 0$. The latter fact is implied by (3) and the spatial homogeneity. Thus $\lim_{t \downarrow 0} (T_t f)(x) = f(x)$ uniformly in x.

Theorem. Let x_0 be a fixed point of S. Let us assume that the *group of isotropy* $\mathfrak{G}_0 = \{M \in G; M \cdot x_0 = x_0\}$ is compact. Being a closed subgroup of a Lie group $\mathfrak{G}$, $\mathfrak{G}_0$ is a Lie group by a theorem due to E. CARTAN. Let A be the infinitesimal generator of the semi-group T_t. Then we have the following results.

[1]. If $f \in D(A) \cap C^2(S)$, then, for a coordinate system $(x_0^1, x_0^2, \ldots, x_0^n)$ at x_0,

$$(Af)(x_0) = a^i(x_0) \frac{\partial f}{\partial x_0^i} + b^{ij}(x_0) \frac{\partial^2 f}{\partial x_0^i \partial x_0^j}, \tag{6}$$

where

$$a^i(x_0) = \lim_{t \downarrow 0} t^{-1} \int_{d(x_0, x_i) \leqq \varepsilon} (x^i - x_0^i)\, P(t, x_0, dx), \tag{7}$$

$$b^{ij}(x_0) = \lim_{t \downarrow 0} t^{-1} \int_{d(x_0, x) \leqq \varepsilon} (x^i - x_0^i)(x^j - x_0^j)\, P(t, x_0, dx), \tag{8}$$

the limits existing independently of sufficiently small $\varepsilon > 0$.

[2]. The set $D(A) \cap C^2(S)$ is "big" in the sense that, for any $C^\infty(S)$ function $h(x)$ with compact support there exists an $f(x) \in D(A) \cap C^2(S)$ such that $f(x_0)$, $\partial f/\partial x_0^i$, $\partial^2 f/\partial x_0^i \partial x_0^j$ are arbitrarily near to $h(x_0)$, $\partial h/\partial x_0^i$, $\partial^2 h/\partial x_0^i \partial x_0^j$, respectively.

Proof. *The first step.* Let $h(x)$ be a C^∞ function with compact support. If $f \in D(A)$, then the "convolution"

$$(f \otimes h)(x) = \int_{\mathfrak{G}} f(M_y \cdot x)\, h(M_y \cdot x_0)\, dy \tag{9}$$

(M_y denotes a generic element of $\mathfrak{G}$ and dy a fixed right invariant Haar measure on $\mathfrak{G}$ such that $dy = d(y \cdot M)$ for every $M \in \mathfrak{G}$) is C^∞ and belongs to $D(A)$. The above integral exists since the isotropy group $\mathfrak{G}_0$ is compact and h has a compact support. By the uniform continuity of f and the compactness of the support of h we can approximate the integral by Riemann sums $\sum_{i=1}^{k} f(M_{y_i} \cdot x)\, C_i$ uniformly in x:

$$(f \otimes h)(x) = \text{s-}\lim_{k \to \infty} \sum_{i=1}^{k} f(M_{y_i} \cdot x)\, C_i.$$

Since $T_t M' = M' T_t$, M' commutes with A, i.e., if $f \in D(A)$, then $M' f \in D(A)$ and $A M' f = M' A f$. Putting $g(x) = (Af)(x)$, we obtain $g \in C(S)$ and

$$A\left(\sum_{i=1}^{k} f(M_{y_i} \cdot x)\, C_i\right) = \sum_{i=1}^{k} (A M'_{y_i} \cdot f)(x)\, C_i = \sum_{i=1}^{k} (M'_{y_i} A f)(x)\, C_i$$
$$= \sum_{i=1}^{k} g(M_{y_i} \cdot x)\, C_i,$$

and the right hand side tends to $(g \otimes h)(x) = (Af \otimes h)(x)$. Since the infinitesimal generator A is closed, it follows that $f \otimes h \in D(A)$ and $A(f \otimes h) = Af \otimes h$. Since S is a *homogeneous space* of the Lie group $\mathfrak{G}$, i.e., $S = \mathfrak{G}/\mathfrak{G}_0$, we can find a coordinate neighbourhood U of x_0 such that for each $x \in U$ there exists an element $M = M(x) \in \mathfrak{G}$ satisfying the conditions:

i) $M(x) \cdot x = x_0$,

ii) $M(x)\, x_0$ depends analytically on the coordinates $(x^1, x^2, \ldots, x^n)$ of the point $x \in S$.

This is so, since the set $\{M_y \in \mathfrak{G};\ M_y \cdot x = x_0\}$ forms an analytic submanifold of $\mathfrak{G}$; it is one of the cosets of $\mathfrak{G}$ with respect to the Lie subgroup $\mathfrak{G}_0$. Hence, by the right invariance of dy, we have

$$(f \otimes h)(x) = \int_G f(M_y M(x)\, x)\, h(M_y M(x)\, x_0)\, dy$$

$$= \int_G f(M_y \cdot x_0)\, h(M_y M(x) \cdot x_0)\, dy.$$

The right side is infinitely differentiable in the vicinity of x_0, and

$$D_x^s (f \otimes h)(x) = \int_G f(M_y \cdot x_0)\, D_x^s h(M_y M(x) \cdot x_0)\, dy. \tag{10}$$

The second step. Remembering that $D(A)$ is dense in $C(S)$ and choosing f and h appropriately, we obtain:

there exist C^∞ functions $F^1(x), F^2(x), \ldots, F^n(x) \in D(A)$ such that the Jacobian $\dfrac{\partial(F^1(x), \ldots, F^n(x))}{\partial(x^1, \ldots, x^n)} > 0$ at x_0, (11)

and, moreover

there exists a C^∞ function $F_0(x) \in D(A)$ such that

$$(x^i - x_0^i)(x^j - x_0^j) \frac{\partial^2 F_0}{\partial x_0^i \partial x_0^j} \geqq \sum_{i=1}^n (x^i - x_0^i)^2. \tag{12}$$

We can use $F^1(x), F^2(x), \ldots, F^n(x)$ as coordinate functions in a neighbourhood $\{x;\ \mathrm{dis}(x_0, x) < \varepsilon\}$ of the point x_0. We denote these new local coordinates by $(x^1, x^2, \ldots, x^n)$. Since $F^j(x) \in D(A)$, we have

$$\lim_{t \downarrow 0} t^{-1} \int_S P(t, x_0, dx)\, (F^j(x) - F^j(x_0)) = (AF^j)(x_0)$$

$$= \lim_{t \downarrow 0} t^{-1} \int_{\mathrm{dis}(x, x_0) \leqq \varepsilon} P(t, x_0, dx)\, (F^j(x) - F^j(x_0))$$

independently of sufficiently small $\varepsilon > 0$, by virtue of Lindeberg's condition (3). Hence there exists, for the coordinate function $x^1, x^2, \ldots, x^n$ $(x^j = F^j)$, a finite

$$\lim_{t \downarrow 0} t^{-1} \int_{\mathrm{dis}(x, x_0) \leqq \varepsilon} (x^j - x_0^j)\, P(t, x_0, dx) = a^j(x_0) \tag{13}$$

which is independent of sufficiently small $\varepsilon > 0$. Since $F_0 \in D(A)$, we have, again using Lindeberg's condition (3),

$$(A F_0)(x) = \lim_{t \downarrow 0} t^{-1} \int_S P(t, x_0, dx)(F_0(x) - F_0(x_0))$$

$$= \lim_{t \downarrow 0} t^{-1} \int_{\mathrm{dis}(x_0,x) \leqq \varepsilon} P(t, x_0, dx)(F_0(x) - F_0(x_0))$$

$$= \lim_{t \downarrow 0} \Bigg[t^{-1} \int_{\mathrm{dis}(x_0,x) \leqq \varepsilon} (x^i - x_0^i) \frac{\partial F_0}{\partial x_0^i} P(t, x_0, dx)$$

$$+ t^{-1} \int_{\mathrm{dis}(x_0,x) \leqq \varepsilon} (x^i - x_0^i)(x^j - x_0^j) \left(\frac{\partial^2 F_0}{\partial x^i \partial x^j} \right)_{x = x_0 + \theta(x - x_0)} P(t, x_0, dx) \Bigg],$$

where $0 < \theta < 1$. The first term on the right has a limit $a^i(x_0) \dfrac{\partial F_0}{\partial x_0^i}$. Hence, by the positivity of $P(t, x, E)$ and (12), we see that

$$\overline{\lim_{t \downarrow 0}}\, t^{-1} \int_{\mathrm{dis}(x_0,x) < \varepsilon} \sum_{i=1}^{n} (x^i - x_0^i)^2 P(t, x_0, dx) < \infty. \tag{14}$$

The third step. Let $f \in D(A) \cap C^2$. Then, by expanding $f(x) - f(x_0)$, we obtain

$$\frac{(T_t f)(x_0) - f(x_0)}{t} = t^{-1} \int_S (f(x) - f(x_0)) P(t, x_0, dx)$$

$$= t^{-1} \int_{\mathrm{dis}(x,x_0) > \varepsilon} (f(x) - f(x_0)) P(t, x_0, dx)$$

$$+ t^{-1} \int_{\mathrm{dis}(x,x_0) \leqq \varepsilon} (x^i - x_0^i) \frac{\partial f}{\partial x_0^i} P(t, x_0, dx)$$

$$+ t^{-1} \int_{\mathrm{dis}(x,x_0) \leqq \varepsilon} (x^i - x_0^i)(x^j - x_0^j) \frac{\partial^2 f}{\partial x_0^i \partial x_0^j} P(t, x_0, dx)$$

$$+ t^{-1} \int_{\mathrm{dis}(x,x_0) \leqq \varepsilon} (x^i - x_0^i)(x^j - x_0^j) C_{ij}(\varepsilon) P(t, x_0, dx)$$

$$= C_1(t, \varepsilon) + C_2(t, \varepsilon) + C_3(t, \varepsilon) + C_4(t, \varepsilon), \text{ say},$$

where $C_{ij}(\varepsilon) \to 0$ as $\varepsilon \downarrow 0$. We know, by (3), that $\lim_{t \downarrow 0} C_1(t, \varepsilon) = 0$ for fixed $\varepsilon > 0$, and that $\lim_{t \downarrow 0} C_2(t, \varepsilon) = a^i(x_0) \dfrac{\partial f}{\partial x_0^i}$, independently of sufficiently small $\varepsilon > 0$. By (14) and Schwarz' inequality, we see that $\lim_{\varepsilon \downarrow 0} C_4(t, \varepsilon) = 0$, boundedly in $t > 0$. The left side also has a finite limit $(Af)(x_0)$ as $t \downarrow 0$. So the difference

$$\overline{\lim_{t \downarrow 0}}\, C_3(t, \varepsilon) - \underline{\lim_{t \downarrow 0}}\, C_3(t, \varepsilon)$$

can be made arbitrarily small by taking $\varepsilon > 0$ sufficiently small. But, by (14), Schwarz' inequality and (3), the difference is independent of sufficiently small $\varepsilon > 0$. Thus a finite limit $\lim_{t \downarrow 0} C_3(t, \varepsilon)$ exists independently of sufficiently small $\varepsilon > 0$. Since we may choose $F \in D(A) \cap C^\infty(S)$ in such a way that $\partial^2 F / \partial x_0^i \, \partial x_0^j$ $(i, j = 1, 2, \ldots, n)$ is arbitrarily near α_{ij}, where α_{ij} are arbitrarily given constants, it follows, by an argument similar to the above, that

$$\text{a finite } \lim_{t \downarrow 0} t^{-1} \int_{\mathrm{dis}(x, x_0) \leqq \varepsilon} (x^i - x_0^i)(x^j - x_0^j) \, P(t, x_0, dx) = b^{ij}(x_0) \text{ exists} \tag{15}$$

and

$$\lim_{t \downarrow 0} C_3(t, \varepsilon) = b^{ij}(x_0) \frac{\partial^2 f}{\partial x_0^i \, \partial x_0^j} \, .$$

This completes the proof of our Theorem.

Remark. The above Theorem and proof are adapted from K. YOSIDA [20]. It is to be noted that $b^{ij}(x) = b^{ji}(x)$ and

$$b^{ij}(x_0) \, \xi_i \xi_j \geqq 0 \text{ for every real vector } (\xi_1, \xi_2, \ldots, \xi_n), \tag{16}$$

$$\text{because } (x^i - x_0^i)(x^j - x_0^j) \, \xi_i \xi_j = \left(\sum_{i=1}^{n} (x^i - x_0^i) \, \xi_i \right)^2 .$$

The Brownian Motion on the Surface of the 3-sphere. In the special case when S is the surface of the 3-sphere S^3 and $\mathfrak{G}$ is the group of rotations of S^3, the infinitesimal generator A of the semi-group induced by a Brownian motion on S is proved to be of the form

$$\begin{aligned} A = C \Lambda \ , & \text{ where } C \text{ is a positive constant and } \Lambda \text{ is the Laplacian} \\ = \frac{1}{\sin\theta} \frac{\partial}{\partial\theta} \sin\theta \frac{\partial}{\partial\theta} & + \frac{1}{\sin^2\theta} \frac{\partial^2}{\partial\varphi^2} \text{ on the surface of } S^3. \end{aligned} \tag{17}$$

Thus there exists essentially one Brownian motion on the surface of S^3. For the detail, see K. YOSIDA [27].

6. The Generalized Laplacian of W. Feller

Let S be an open interval (r_1, r_2), finite or infinite, on the real line, and $\mathfrak{B}$ be the set of all Baire sets in (r_1, r_2). Consider a Markov process $P(t, x, E)$ on $(S, \mathfrak{B})$ satisfying Lindeberg's condition

$$\lim_{t \downarrow 0} t^{-1} \int_{|x-y| > \varepsilon} P(t, x, dy) = 0 \quad \text{for every} \quad x \in (r_1, r_2) \quad \text{and} \quad \varepsilon > 0. \tag{1}$$

Let $C[r_1, r_2]$ be the B-space of real-valued, bounded uniformly continuous functions $f(x)$ defined in (r_1, r_2) and normed by $\|f\| = \sup_x |f(x)|$. Then

$$(T_t f)(x) = \int_{r_1}^{r_2} P(t, x, dy) \, f(y) \quad (t > 0); = f(x) \quad (t = 0) \tag{2}$$

defines a positive contraction semi-group on $C[r_1, r_2]$. We prove by (1) the weak right continuity of T_t at $t = 0$ so that T_t is of class (C_0) by the Theorem in Chapter IX, 1.

Concerning the infinitesimal generator A of the semi-group T_t, we have the following two theorems due to W. FELLER.

Theorem 1.

$$A \cdot 1 = 0; \tag{3}$$

A is of *local character*, in the sense that, if $f \in D(A)$ vanishes in a neighbourhood of a point x_0 then $(Af)(x_0) = 0$; (4)

If $f \in D(A)$ has a local maximum at x_0, then $(Af)(x_0) \leqq 0$. (5)

Proof. (3) is clear from $T_t \cdot 1 = 1$. We have, by (1),

$$(Af)(x_0) = \lim_{t \downarrow 0} t^{-1} \int_{|x_0 - y| \leqq \varepsilon} P(t, x_0, dy)\,(f(y) - f(x_0))$$

independently of sufficiently small $\varepsilon > 0$. Hence, by $P(t, x, E) \geqq 0$, we easily obtain (4) and (5).

Theorem 2. Let a linear operator A, defined on a linear subspace $D(A)$ of $C[r_1, r_2]$ into $C[r_1, r_2]$, satisfy conditions (3), (4) and (5). Let us assume that A does not *degenerate* in the sense that the following two conditions are satisfied:

there exists at least one $f_0 \in D(A)$ such that $(Af_0)(x) > 0$ for all $x \in (r_1, r_2)$, (6)

there exists a solution v of $A \cdot v = 0$ which is linearly independent of the function 1 in every subinterval (x_1, x_2) of the interval (r_1, r_2). (7)

Then there exists a strictly increasing continuous solution $s = s(x)$ of $A \cdot s = 0$ in (r_1, r_2) such that, if we define a strictly increasing function $m = m(x)$, not necessarily continuous nor bounded in (r_1, r_2), by

$$m(x) = \int^{x} \frac{1}{(Af_0)(t)}\, d(D_s^+ f_0(t)), \tag{8}$$

we obtain the representation:

$$(Af)(x) = D_m^+ D_s^+ f(x) \text{ in } (r_1, r_2) \quad \text{for any} \quad f \in D(A). \tag{9}$$

Here the right derivative D_p^+ with respect to a strictly increasing function $p = p(x)$ is defined by

$$\begin{aligned} D_p^+ g(x) &= \lim_{y \downarrow x} \frac{g(y+0) - g(x-0)}{p(y+0) - p(x-0)} \text{ at the point of continuity of } p, \\ &= \frac{g(x+0) - g(x-0)}{p(x+0) - p(x-0)} \text{ at the point of discontinuity of } p. \end{aligned} \tag{10}$$

Proof. *The first step.* Let $u \in D(A)$ satisfy $A \cdot u = 0$ and $u(x_1) = u(x_2)$, where $r_1 < x_1 < x_2 < r_2$. Then $u(x)$ must reduce to a constant in (x_1, x_2). If otherwise, $u(x)$ would, for example, assume a maximum at an interior point x_0 of (x_1, x_2) in such a way that $u(x_0) - \varepsilon \geqq u(x_1) = u(x_2)$ with some $\varepsilon > 0$. Let f_0 be the function $\in D(A)$ given by condition (6). Then, for a sufficiently large $\delta > 0$, the function $F(x) = f_0(x) + \delta u(x)$ satisfies $F(x_0) > F(x_i)$ $(i = 1, 2)$. Hence the maximum of $F(x)$ in $[x_1, x_2]$ is attained at an interior point x_0' although $(AF)(x_0') = (Af_0)(x_0') > 0$. This is a contradiction to condition (5).

The second step. By (7) and the first step, we see that there exists a strictly increasing continuous solution $s(x)$ of $A \cdot s = 0$. We reparametrize the interval by s so that we may assume that

$$\text{the functions 1 and } x \text{ are both solutions of } A \cdot f = 0. \tag{11}$$

We can then prove that

$$\text{if } (Ah)(x) > 0 \text{ for all } x \text{ in a subinterval } (x_1, x_2) \text{ of the interval } (r_1, r_2), \text{ then } h(x) \text{ is convex downwards in } (x_1, x_2). \tag{12}$$

For, the function $u(x) = h(x) - \alpha x - \beta$, which satisfies $(Au)(x) = (Ah)(x) > 0$ for all x in (x_1, x_2), can have no local maximum at an interior point of (x_1, x_2).

The third step. Let $f \in D(A)$. Then, by (12) and (6), we see that, for sufficiently large $\delta > 0$, the two functions $f_1(x) = f(x) + \delta f_0(x)$ and $f_2(x) = \delta f_0(x)$ are both convex downwards. Being the difference of two convex functions, $f(x) = f_1(x) - f_2(x)$ has a right derivative at every point x of (r_1, r_2). Let us put

$$A \cdot f = \varphi, \ A \cdot f_0 = \varphi_0, \ D_x^+ f_0(x) = \mu(x). \tag{13}$$

For $g_t = f - tf_0$, we have $A \cdot g_t = Af - tA \cdot f_0 = \varphi - t\varphi_0$. Hence, by (12),

$$\left.\begin{aligned} &t < \min_{x_1 < x < x_2} \varphi(x)/\varphi_0(x) \text{ implies that } g_t(t) \text{ is convex downwards} \\ &\text{in } (x_1, x_2) \text{ and so } D_x^+ g_t(x) \text{ is increasing in } (x_1, x_2). \end{aligned}\right\}$$

Applying the same argument for $t > \max_{x_1 < x < x_2} \varphi(x)/\varphi_0(x)$, we obtain:

$$\left.\begin{aligned} &\text{for any sub-interval } (x_1, x_2) \text{ of } (a, b), \text{ we have} \\ &\min_{x_1<x<x_2} \frac{\varphi(x)}{\varphi_0(x)} \times (D_x^+ f_0(x_2) - D_x^+ f_0(x_1)) \leqq D_x^+ f(x_2) - D_x^+ f(x_1) \\ &\qquad \leqq \max_{x_1<x<x_2} \frac{\varphi(x)}{\varphi_0(x)} \times (D_x^+ f_0(x_2) - D_x^+ f_0(x_1)). \end{aligned}\right\}$$

The continuity of the function $\varphi(x)/\varphi_0(x)$ implies, by the above inequality, that

$$D_x^+ f(x_2) - D_x^+ f(x_1) = \int_{x_1}^{x_2} \frac{\varphi(x)}{\varphi_0(x)} d(D_x^+ f_0(x)),$$

which is precisely the integrated version of $A \cdot f = D_m^+ D_x^+ f$.

Remark 1. We may consider the operator $D_m^+ D_s^+$ as a generalized Laplacian in the sense that D_s^+ and D_m^+ correspond to the *generalized gradiant* and the *generalized divergence* in one dimension, respectively. Feller called $s = s(x)$ the *canonical scale* and $m = m(x)$ the *canonical measure* of the Markov process in question. We easily see, from the first step above, that the function 1 and s constitute a *fundamental system of linearly independent solutions* of $A \cdot v = 0$ to the effect that any solution of $A \cdot v = 0$ can be expressed uniquely as a linear combination of 1 and $s(x)$. Thus the canonical scale of $P(t, x, E)$ is determined up to a linear transformation, i.e., another canonical scale s_1 must be of the form $s_1 = \alpha s + \beta$ with $\alpha > 0$; hence the canonical measure m_1 which corresponds to s_1 must be of the form $m_1 = \alpha^{-1} m$.

Remark 2. Theorem 2 gives the representation of the infinitesimal generator A at the *interior point* x of (r_1, r_2). To determine the operator A as the infinitesimal generator of a positive contraction semi-group T_t of class (C_0) on $C[r_1, r_2]$ into $C[r_1, r_2]$, we must consider the *lateral condition*, that is, the *boundary condition* of the operator A at both boundary points r_1 and r_2 in order to describe the domain $D(A)$ of A concretely and completely. According to Feller [2] and [6] (cf. E. Hille [6]), the boundary points r_1 (or r_2) are classified into the *regular-boundary*, the *exit-boundary*, the *entrance-boundary* and the *natural-boundary*. To this purpose, we introduce four quantities:

$$\sigma_1 = \iint_{r_1 < y < x < r_1'} dm(x)\, ds(y), \quad \mu_1 = \iint_{r_1 < y < x < r_1'} ds(x)\, dm(y),$$

$$\sigma_2 = \iint_{r_2 > y > x > r_2'} dm(x)\, ds(y), \quad \mu_2 = \iint_{r_2 > y > x > r_2'} ds(x)\, dm(y).$$

The boundary point r_i $(i = 1, 2)$ is called

$$\left.\begin{array}{ll} \textit{regular} \text{ in case} & \sigma_i < \infty,\ \mu_i < \infty \\ \textit{exit} \text{ in case} & \sigma_i < \infty,\ \mu_i = \infty \\ \textit{entrance} \text{ in case} & \sigma_i = \infty,\ \mu_i < \infty \\ \textit{natural} \text{ in case} & \sigma_i = \infty,\ \mu_i = \infty \end{array}\right\} \begin{array}{l} \text{(the conditions are} \\ \text{independent of the} \\ \text{choice of } r_1' \text{ and } r_2'\text{).} \end{array}$$

We shall illustrate by simple examples.

Example 1. $D_m^+ D_s^+ = d^2/dx^2$, $S = (-\infty, \infty)$. We can take $s = x$, $m = x$. Hence

$$\sigma_2 = \iint_{\infty > y > x > r_2'} dx\,dy = \int_{r_2'}^{\infty} (y - r_2')\,dy = \infty,$$

$$\mu_2 = \iint_{\infty > y > x > r_2'} dx\,dy = \infty,$$

and so ∞ is a natural boundary. Similarly $-\infty$ is also a natural boundary.

Example 2. $D_m^+ D_s^+ = d^2/dx^2$, $S = (-\infty, 0)$. We can take $s = x$, $m = x$. In this case, $-\infty$ is a natural boundary and 0 is a regular boundary.

Example 3. $D_m^+ D_s^+ = x^2 \dfrac{d^2}{dx^2} - \dfrac{d}{dx}$, $S = (0, \infty)$. A strictly increasing continuous solution $s = s(x)$ of $D_m^+ D_s^+ s = 0$ is given by $s(x) = \int^x e^{-1/t}\,dt$ so that

$$D_s = e^{1/x} \frac{d}{dx}, \quad D_m^+ D_s^+ = x^2 e^{-1/x} \frac{d}{dx} e^{1/x} \frac{d}{dx}.$$

Therefore $ds = e^{-1/x}\,dx$, $dm = x^{-2} e^{1/x}\,dx$. Hence

$$\sigma_1 = \iint_{0<y<x<1} x^{-2} e^{1/x}\,dx\,e^{-1/y} dy = \int_0^1 [-e^{1/x}]_y^1\, e^{-1/y}\,dy < \infty,$$

$$\mu_1 = \iint_{0<y<x<1} e^{-1/x}\,dx\,y^{-2} e^{1/y}\,dy = \int_0^1 e^{-1/x}\,[-e^{1/y}]_0^x\,dx = \infty.$$

Thus 0 is an exit boundary. Similarly we see

$$\sigma_2 = \iint_{\infty>y>x>1} e^{1/x} x^{-2}\,dx\,e^{-1/y}\,dy = \int_1^{\infty} [-e^{1/x}]_1^y \cdot e^{-1/y}\,dy = \infty,$$

$$\mu_2 = \iint_{\infty>y>x>1} e^{-1/x}\,dx\,e^{1/y} y^{-2}\,dy = \int_1^{\infty} e^{-1/x}\,[-e^{1/y}]_x^{\infty}\,dx = \infty.$$

That ∞ is a natural boundary.

Example 4. $D_m^+ D_s^+ = x^2 \dfrac{d^2}{dx^2} + \dfrac{d}{dx}$, $S = (0, 2)$. As above, we see that $ds = e^{1/x}\,dx$, $dm = x^{-2} e^{-1/x}\,dx$, and so we can verify that 0 is an entrance boundary and 2 is a regular boundary.

Feller's probabilistic interpretation of the above classification is as follows:

The probability that a particle, located at first in the interior of the open interval (r_1, r_2), will, after a finite lapse of time, reach a regular boundary or an exit boundary, is positive; while, the particle can, after a finite lapse of time, neither reach an entrance boundary nor a natural boundary.

References

FELLER's original paper was published in 1952: W. FELLER [2]. The proof of Theorem 2 given above is adapted from W. FELLER [3] and [4], who also delivered an inspiring address [5]. We also refer the readers to the forthcoming book by K. ITÔ-H. McKEAN [1] and E. B. DYNKIN [3], and the references given in these books.

7. An Extension of the Diffusion Operator

Let the possible state of a (not necessarily temporally homogeneous) Markov process be represented by the point x of an n-dimensional C^∞ Riemannian space R. We denote by $P(s, x, t, E)$, $s \leqq t$, the transition probability that a state x at a time moment s is transferred into a Baire set $E \subseteqq R$ at a later time moment t.

We shall be concerned with the possible form of the operator A_s defined by

$$(A_s f)(x) = \lim_{t \downarrow 0} t^{-1} \int_R P(s, x, s+t, dy)\,(f(y) - f(x)), \quad f \in C_0^0(R), \tag{1}$$

without assuming the Lindeberg type condition:

$$\lim_{t \downarrow 0} t^{-1} \int_{d(x,y) \geqq \varepsilon} P(s, x, s+t, dy) = 0 \quad \text{for all positive constants } \varepsilon$$

$$(d(x, y) = \text{the geodesic distance between two points } x \text{ and } y). \tag{2}$$

We can prove

Theorem. Let there exist an increasing sequence $\{k\}$ of positive integers such that, for a fixed pair $\{s, x\}$,

$$\lim_{a \uparrow \infty} k \cdot \int_{d(x,y) \geqq a} P(s, x, s + k^{-1}, dy) = 0 \text{ uniformly in } k, \tag{3}$$

$$k \int_R \frac{d(x, y)^2}{1 + d(x, y)^2} P(s, x, s + k^{-1}, dy) \text{ is uniformly bounded in } k. \tag{4}$$

Suppose that, for a function $f(x) \in C_0^2(R)$,

$$\text{a finite } \lim_{k \to \infty} k \left\{ \int_R P(s, x, s + k^{-1}, dy)\, f(y) - f(x) \right\} \text{ exists.} \tag{5}$$

We have then, in any fixed local coordinates $(x_1, x_2, \ldots, x_n)$ of the point $x \in R$,

$$\begin{aligned}(A_s f)(x) &= a_j(s, x) \frac{\partial f}{\partial x_j} + b_{ij}(s, x) \frac{\partial^2 f}{\partial x_i \partial x_j} \\ &\quad + \overline{\lim_{\varepsilon \downarrow 0}} \int_{d(x,y) \geqq \varepsilon} \left\{ f(y) - f(x) - \frac{\varrho(x, y)}{1 + d(x, y)^2} (y_j - x_j) \frac{\partial f}{\partial x_j} \right\} \\ &\quad \times \frac{1 + d(x, y)^2}{d(x, y)^2} G(s, x, dy),\end{aligned} \tag{6}$$

where

$$G(s, x, E) \text{ is non-negative and } \sigma\text{-additive for Baire sets } E \subseteq R \text{ and } G(s, x, R) < \infty, \tag{7}$$

$$\varrho(x, y) \text{ is continuous in } x, y \text{ such that } \varrho(x, y) = 1 \text{ or } 0 \text{ according as } d(x, y) \leqq \delta/2 \text{ or } > \delta \ (\delta \text{ is a fixed constant} > 0), \tag{8}$$

$$\text{the quadratic form } b_{ij}(s, x)\,\xi_i\xi_j \text{ is} \geqq 0. \tag{9}$$

Remark. Formula (6) is given in K. YOSIDA [26]. It is an extension of the *diffusion operator* of the form of the second order elliptic differential operator discussed in the preceding sections. The third term on the right of (6) is the sum of infinitely many difference operators. The appearence of such a term is due to the fact that we do not assume the condition (2) of Lindeberg's type. Formula (6) is an operator-theoretical counterpart of the P. Lévy-A. Khintchine-K. Itô *infinitely divisible law* in probability theory. About this point, see E. HILLE-PHILLIPS [1], p. 652.

Proof of Theorem. Consider a sequence of non-negative, σ-additive measures given by

$$G_k(s, x, E) = k \int_E \frac{d(x, y)^2}{1 + d(x, y)^2} P(s, x, s + k^{-1}, dy). \tag{10}$$

We have, by (3) and (4),

$$G_k(s, x, E) \text{ is uniformly bounded in } E \text{ and } k, \tag{11}$$

$$\lim_{a\uparrow\infty} \int_{d(x,y)\geqq a} G_k(s, x, dy) = 0 \text{ uniformly in } k. \tag{12}$$

Hence, for fixed $\{s, x\}$,the linear functional L_k given by

$$L_k(g) = \int_R G_k(s, x, dy)\, g(y), \quad g \in C_0^0(R),$$

is non-negative and continuous on the normed linear space $C_0^0(R)$ with the norm $\|g\| = \sup_x |g(x)|$; and the norm L_k is uniformly bounded in k.

Therefore, by the separability of the normed linear space $C_0^0(R)$, we can choose a subsequence $\{k'\}$ of $\{k\}$ such that $\lim_{k\to\infty} L_{k'}(g) = L(g)$ exists as a non-negative linear functional on $C_0^0(R)$. By virtue of a lemma in measure theory (P. R. HALMOS [1]) there exists a non-negative, σ-additive measure $G(s, x, E)$ such that $G(s, x, R) < \infty$ and

$$\lim_{k=k'\to\infty} \int_R G_k(s, x, dy)\, g(y) = \int_R G(s, x, dy)\, g(y) \quad \text{for all } g \in C_0^0(R). \tag{13}$$

We have

$$k\left[\int_R P(s, x, s+k^{-1}, dy)\, f(y) - f(x)\right]$$

$$= \int_R \left\{\left[f(y) - f(x) - \frac{\varrho(y, x)}{1 + d(y, x)^2}(y_j - x_j)\frac{\partial f}{\partial x_j}\right]\frac{1 + d(y, x)^2}{d(y, x)^2}\right\} G_k(s, x, dy)$$

$$+ \int_R \frac{\varrho(y, x)}{d(y, x)^2}(y_j - x_j)\frac{\partial f}{\partial x_j} G_k(s, x, dy). \tag{14}$$

The term $\{\}$ in the first integral on the right is, for sufficiently small $d(y, x)$,

$$= (y_j - x_j)\frac{\partial f}{\partial x_j} + (y_i - x_i)(y_j - x_j)\left(\frac{\partial^2 f}{\partial X_i \partial X_j}\right)\frac{1 + d(y, x)^2}{d(y, x)^2},$$

$$\text{where } X_j = x_j + \theta(y_j - x_j),\ 0 < \theta < 1.$$

Thus $\{\}$ is bounded and continuous in y. Hence, by (12) and (13), the first term on the right of (14) tends, as $k = k' \to \infty$, to $\int_R \{\}\, G(s, x, dy)$. Therefore, by (5),

$$\text{a finite } \lim_{k=k'\to\infty} \int_R \frac{\varrho(y, x)}{d(y, x)^2}(y_j - x_j)\frac{\partial f}{\partial x_j} G_k(s, x, dy) = a_j(s, x)\frac{\partial f}{\partial x_j}$$

exists. Hence, by (3), we obtain (6) by taking

$$b_{ij}(s, x) = \lim_{\varepsilon \downarrow 0}\lim_{k=k'\to\infty} k \int_{d(y,x)\leqq\varepsilon} (y_i - x_i)(y_j - x_j)\, P(s, x, s + k^{-1}, dy).$$

8. Markov Processes and Potentials

Let $\{T_t\}$ be an equi-continuous semi-group of class (C_0) on a function space X defined by $(T_t f)(x) = \int_S P(t, x, dy)\, f(y)$ where $P(t, x, E)$ is a Markov process on a measure space $(S, \mathfrak{B})$. Let A be the infinitesimal generator of T_t. Suggested by the special case where A is the Laplacian, an element $f \in X$ is called *harmonic* if $A \cdot f = 0$. Then f is harmonic iff $\lambda(\lambda I - A)^{-1} f = f$ for every $\lambda > 0$. An element $f \in X$ is called a *potential* if $f = \lim_{\lambda \downarrow 0}(\lambda I - A)^{-1} g$ with some g. Because, for such an element f, there holds by the closedness of A the *Poisson equation*

$$A \cdot f = \lim_{\lambda\downarrow 0} A(\lambda I - A)^{-1} g = \lim_{\lambda\downarrow 0}\{-g + \lambda(\lambda I - A)^{-1} g\} = -g.$$

Suppose that X is a vector lattice which is also a locally convex linear topological space such that every monotone increasing bounded sequence of elements $\in X$ converges weakly to an element of X which is greater than the elements of the sequence. We also assume that the resolvent $J_\lambda = (\lambda I - A)^{-1}$ is *positive* in the sense that $f \geqq 0$ implies $J_\lambda f \geqq 0$. The situation is suggested by the special case where A is the Laplacian considered in a suitable function space.

We may thus call *subharmonic* those elements $f \in X$ for which the inequality $A \cdot f \geqq 0$ holds. By virtue of the positivity of J_λ, a subharmonic element f satisfies $\lambda J_\lambda f \geqq f$ for all $\lambda > 0$. We shall prove an analogue of a well-known theorem of F. RIESZ concerning ordinary subharmonic functions (see T. RADO [1]):

Theorem. Any subharmonic element x is decomposed as the sum of a harmonic element x_h and a potential x_p, where the harmonic part x_h of x is given by $x_h = \lim_{\lambda \downarrow 0} \lambda J_\lambda x$ and x_h is the *least harmonic majorant* of x in the sense that any harmonic element $x_H \geqq x$ satisfies $x_H \geqq x_h$.

Proof (K. YOSIDA [4]). By the resolvent equation

$$J_\lambda - J_\mu = (\mu - \lambda) J_\lambda J_\mu, \tag{1}$$

we obtain

$$(I - \lambda J_\lambda) = (I + (\mu - \lambda) J_\lambda)(I - \mu J_\mu). \tag{2}$$

Since x is subharmonic, we see, by the positivity of J_λ, that

$$\lambda > \mu \quad \text{implies} \quad \lambda J_\lambda x \geqq \mu J_\mu x \geqq x.$$

Hence the weak-$\lim_{\lambda \downarrow 0} \lambda J_\lambda x = x_h$ exists by virtue of the hypothesis concerning bounded monotone sequences in X. Therefore, by the ergodic theorem in Chapter VIII, 4, we see that $x_h = \lim_{\lambda \downarrow 0} \lambda J_\lambda x$ exists and x_h is harmonic, i.e., $\lambda J_\lambda x_h = x_h$ for all $\lambda > 0$. We also have $x_p = (x - x_h) = \lim_{\lambda \downarrow 0} (I - \lambda J_\lambda) x = \lim_{\lambda \downarrow 0} (-A)(\lambda I - A)^{-1} x = \lim_{\lambda \downarrow 0} (\lambda I - A)^{-1}(-A x)$ which shows that x_p is a potential.

Let a harmonic element x_H satisfy $x_H \geqq x$. Then, by the positivity of λJ_λ and the harmonic property of x_H, we have

$$x_H = \lambda J_\lambda x_H \geqq \lambda J_\lambda x \quad \text{and hence} \quad x_H \geqq \lim_{\lambda \downarrow 0} \lambda J_\lambda x = x_h.$$

Researches by Hunt and by Doob. G. A. HUNT [1] gave a condition that a linear operator $\in L(C_0^0(R^n), C_0^0(R^n))$ becomes one of the resolvents of a semi-group induced by a Markov process $P(t, x, E)$ in R^n and discussed the potential theory associated with this Markov process. We refer the reader to an excellent exposition of Hunt's theory given in P. A. MEYER [1]. J. L. DOOB [2] discussed the boundary values of harmonic functions $f(x)$ when x tends to the boundary through the path of the corresponding Markov process. The essential tool in Doob's research is the limit theorem associated with the *martingale theory* developed systematically by him.

XIV. The Integration of the Equation of Evolution

The ordinary exponential function solves the initial value problem

$$dy/dx = \alpha y, \; y(0) = y.$$

We consider the diffusion equation

$$\partial u/\partial t = \Delta u, \quad \text{where} \quad \Delta = \sum_{j=1}^{m} \partial^2/\partial x_j^2 \text{ is the Laplacian in } R^m.$$

We wish to find a solution $u = u(x, t)$, $t > 0$, of this equation satisfying the initial condition $u(x, 0) = f(x)$, where $f(x) = f(x_1, \ldots, x_m)$ is a given function of x. We shall also study the wave equation

$$\partial^2 u/\partial t^2 = \Delta u, \quad -\infty < t < \infty,$$

with the initial data

$$u(x, 0) = f(x) \quad \text{and} \quad (\partial u/\partial t)_{t=0} = g(x),$$

f and g being given functions. This may be written in vector form as follows:

$$\frac{\partial}{\partial t}\begin{pmatrix} u \\ v \end{pmatrix} = \begin{pmatrix} 0 & I \\ \Delta & 0 \end{pmatrix}\begin{pmatrix} u \\ v \end{pmatrix}, \quad v = \frac{\partial u}{\partial t}$$

with the initial condition

$$\begin{pmatrix} u(x, 0) \\ v(x, 0) \end{pmatrix} = \begin{pmatrix} f(x) \\ g(x) \end{pmatrix}.$$

So in a suitable function space, the wave equation is of the same form as the diffusion (or heat) equation—differentiation with respect to the time parameter on the left and another operator on the right—or again similar to the equation $dy/dt = \alpha y$. Since the solution in the last case is the exponential function, it is suggested that the heat equation and the wave equation may be solved by properly defining the exponential functions of the operators

$$\Delta \text{ and } \begin{pmatrix} 0 & I \\ \Delta & 0 \end{pmatrix}$$

in suitable function spaces. This is the motivation for the application of the semi-group theory to Cauchy's problem. It is to be noted that the Schrödinger equation

$$i^{-1}\, \partial u/\partial t = H u = (\Delta + U(x))\, u, \text{ where } U(x) \text{ is a given function},$$

gives another example of the *equation of evolution* of the form

$$\partial u/\partial t = A u, \quad t > 0, \tag{1}$$

where A is a not necessarily continuous, linear operator in a function space.

The equation of the form (1) may be called a *temporally homogeneous equation of evolution*. We may integrate such an equation by the semi-group theory. In the following three sections, we shall give typical examples of such an integration. We shall then expound the integration theory of the *temporally inhomogeneous equation of evolution*

$$\partial u/\partial t = A(t)\, u, \quad a < t < b. \tag{2}$$

1. Integration of Diffusion Equations in $L^2(R^m)$

Consider a diffusion equation

$$\partial u/\partial t = A u, \ t > 0, \tag{1}$$

where the differential operator

$$A = a^{ij}(x) \frac{\partial^2}{\partial x_i \partial x_j} + b^i(x) \frac{\partial}{\partial x_i} + c(x) \quad (a^{ij}(x) = a^{ji}(x)) \tag{2}$$

is *strictly elliptic* in an m-dimensional euclidean space R^m. We assume that the real-valued coefficients a, b and c are $C^\infty(R^m)$ functions and that

$$\max\Big(\sup_x |a^{ij}(x)|, \sup_x |b^i(x)|, \sup_x |c(x)|, \sup_x |a^{ij}_{x_k}(x)|, \\ \sup_x |b^i_{x_k}(x)|, \sup_x |a^{ij}_{x_k x_s}(x)|\Big) = \eta < \infty. \tag{3}$$

The strictly elliptic hypothesis concerning A means that positive constants λ_0 and μ_0 exist such that

$$\mu_0 \sum_{j=1}^m \xi_j^2 \geqq a^{ij}(x)\, \xi_i \xi_j \geqq \lambda_0 \sum_{i=1}^m \xi_i^2 \text{ on } R^m \text{ for any real} \\ \text{vector } \xi = (\xi_1, \xi_2, \ldots, \xi_m). \tag{4}$$

Let $\hat{H}_0^1$ be the space of all real-valued $C_0^\infty(R^m)$ functions $f(x)$ normed by

$$\|f\|_1 = \Big(\int_{R^m} f^2\, dx + \sum_{j=1}^m \int_{R^m} f_{x_j}^2 dx\Big)^{1/2}, \tag{5}$$

and let H_0^1 be the completion of $\hat{H}_0^1$ with respect to the norm $\|f\|_1$. Let similarly H_0^0 be the completion of $\hat{H}_0^1$ with respect to the norm

$$\|f\|_0 = \Big(\int_{R^m} f^2\, dx\Big)^{1/2}. \tag{6}$$

We have thus introduced two real Hilbert spaces H_0^1 and H_0^0, and H_0^1 and $\hat{H}_0^1$ are $\|\ \|_0$-dense in H_0^0. We know, from the Proposition in Chapter I, 10, that H_0^1 coincides with the real Sobolev space $W^1(R^m)$; we know also that H_0^0 coincides with the real Hilbert space $L^2(R^m)$. We shall denote the scalar product in the Hilbert space H_0^1 (or in the Hilbert space H_0^0) by $(f, g)_1$ (or by $(f, g)_0$).

To integrate equation (1) in the complex Hilbert space $L^2(R^m)$ under condition (3) and (4), we shall prepare some lemmas. These lemmas will also play important roles in the following sections.

Lemma 1 (concerning partial integration). Let $f, g \in \hat{H}_0^1$. We have then

$$(Af, g)_0 = -\int_{R^m} a^{ij} f_{x_i} g_{x_j} dx - \int_{R^m} a^{ij}_{x_j} f_{x_i} g\, dx \\ + \int_{R^m} b^i f_{x_i} g\, dx + \int_{R^m} c f g\, dx, \tag{7}$$

that is, we may partially integrate, in $(Af, g)_0$, the terms containing the second order derivatives as if the integrated terms are zero.

Proof. By (3) and the fact that f and g both belong to $\hat{H}_0^1$, we see that $a^{ij} f_{x_i x_j} g$ is integrable over R^m. Hence, by the fact that f and g are both of compact support,

$$\int_{R^m} a^{ij} f_{x_i x_j} g\, dx = - \int_{R^m} a^{ij} f_{x_i} g_{x_j} dx - \int_{R^m} a^{ij}_{x_j} f_{x_i} g\, dx.$$

Remark. The formal adjoint A^* of A is defined by

$$(A^* f)(x) = \frac{\partial^2}{\partial x_i \partial x_j} (a^{ij}(x) f(x)) - \frac{\partial}{\partial x_i} (b^i(x) f(x)) + c(x) f(x). \tag{8}$$

Then, as above, we have the result: If $f, g \in \hat{H}_0^1$ then we may partially integrate, in $(A^* f, g)_0$, the terms containing the first and the second order derivatives as if the integrated terms are zero. That is, we have

$$\begin{aligned}(A^* f, g)_0 = &- \int_{R^m} a^{ij} f_{x_i} g_{x_j} dx - \int_{R^m} a^{ij}_{x_i} f g_{x_j} dx \\ &- \int_{R^m} b^i f g_{x_i} dx + \int_{R^m} c f g\, dx.\end{aligned} \tag{7'}$$

Corollary. There exist positive constants $\varkappa$, γ and δ such that, for all sufficiently small positive constants α,

$$\begin{aligned}&\alpha\delta \, \|f\|_1^2 \leqq (f - \alpha A f, f)_0 \leqq (1 + \alpha\gamma) \, \|f\|_1^2 \quad \text{when} \quad f \in \hat{H}_0^1, \\ &\alpha\delta \, \|f\|_1^2 \leqq (f - \alpha A^* f, f)_0 \leqq (1 + \alpha\gamma) \, \|f\|_1^2 \quad \text{when} \quad f \in \hat{H}_0^1,\end{aligned} \tag{9}$$

$$\begin{aligned}&|(f - \alpha A f, g)_0| \leqq (1 + \alpha\gamma) \, \|f\|_1 \, \|g\|_1 \quad \text{when} \quad f, g \in \hat{H}_0^1, \\ &|(f - \alpha A^* f, g)_0| \leqq (1 + \alpha\gamma) \, \|f\|_1 \, \|g\|_1 \quad \text{when} \quad f, g \in \hat{H}_0^1,\end{aligned} \tag{10}$$

$$|(Af, g)_0 - (f, Ag)_0| \leqq \varkappa \, \|f\|_1 \, \|g\|_0 \quad \text{when} \quad f, g \in \hat{H}_0^1. \tag{11}$$

Proof. (9) and (10) may be proved by (3), (4), (7) and (7′) remembering the inequality

$$2\alpha \, |ab| \leqq \alpha (\nu \, |a|^2 + \nu^{-1} \, |b|^2) \tag{12}$$

which is valid whenever α and ν are > 0. In fact, we can use an estimate such as

$$\left| \int_{R^m} a^{ij}_{x_j} f_{x_i} g\, dx \right| \leqq \sum_{i=1}^{m} m \eta (\nu \|f_{x_i}\|_0^2 + \nu^{-1} \|g\|_0^2).$$

We also obtain (11) from

$$(Af, g)_0 - (f, Ag)_0 = - \int_{R^m} (2 a^{ij}_{x_i} f_{x_j} g + a^{ij}_{x_i x_j} f g - 2 b^i f_{x_i} g - b^i_{x_i} f g)\, dx.$$

Lemma 2 (concerning the existence of solutions of $u - \alpha A u = f$). Let a positive number α_0 be so chosen that the above Corollary is valid for

$0 < \alpha \leqq \alpha_0$. Then, for any function $f(x) \in \hat{H}_0^1$, the equation

$$u - \alpha A u = f \quad (0 < \alpha \leqq \alpha_0) \tag{13}$$

admits a uniquely determined solution $u \in H_0^1 \cap C^\infty(R^m)$.

Proof. Let us define a bilinear functional $\hat{B}(u, v) = (u - \alpha A^* u, v)_0$ defined for functions $u, v \in \hat{H}_0^1$. From the above Corollary, we have

$$|\hat{B}(u, v)| \leqq (1 + \alpha\gamma)\, \|u\|_1 \, \|v\|_1, \ \alpha\delta \|u\|_1^2 \leqq \hat{B}(u, u). \tag{14}$$

Hence we may extend $\hat{B}(u, v)$, by continuity, to a bilinear functional $B(u, v)$ defined for $u, v \in H_0^1$ and satisfying

$$|B(u, v)| \leqq (1 + \alpha\gamma)\, \|u\|_1 \, \|v\|_1, \ \alpha\delta \|u\|_1^2 \leqq B(u, u). \tag{14'}$$

The linear functional $F(u) = (u, f)_0$ defined on H_0^1, is a bounded linear functional by $|(u, f)_0| \leqq \|u\|_0 \, \|f\|_0 \leqq \|u\|_1 \, \|f\|_0$. Hence, by F. Riesz' representation theorem in the Hilbert space H_0^1, there exists a uniquely determined $v = v(f) \in H_0^1$ such that $(u, f)_0 = (u, v(f))_1$. Thus, by the Milgram-Lax theorem applied to the Hilbert space H_0^1, we have

$$\begin{aligned} &(u, f)_0 = (u, v(f))_1 = B\,(u, S v(f)) \text{ for all } u \in H_0^1, \text{ where} \\ &S \text{ is a bounded linear operator on } H_0^1 \text{ onto } H_0^1. \end{aligned} \tag{15}$$

Let u run over $C_0^\infty(R^m)$, and let $v_n \in \hat{H}_0^1$ be such that $\lim_{n\to\infty} \|v_n - S v(f)\|_1 = 0$. Then

$$\begin{aligned} B(u, S v(f)) &= \lim_{n\to\infty} B(u, v_n) = \lim_{n\to\infty} \hat{B}(u, v_n) \\ &= \lim_{n\to\infty} (u - \alpha A^* u, v_n)_0 = (u - \alpha A^* u, S v(f))_0, \end{aligned}$$

because the norm $\| \ \|_1$ is larger than the norm $\| \ \|_0$. Hence

$$(u, f)_0 = (u - \alpha A^* u, S v(f))_0, \tag{15'}$$

that is, $S v(f) \in H_0^1$ is a distribution solution of the equation (13). Hence, by the strict ellipticity of $(I - \alpha A)$ and by the fact $f \in C_0^\infty(R^m)$, we see, from the Corollary of Friedrichs' theorem in Chapter VI, 9, that we may consider $u = S v(f) \in H_0^1$ to be a $C^\infty(R^m)$ solution of (13). Hence $u = S v(f) \in H_0^1 \cap C^\infty(R^m)$.

The uniqueness of such a solution u of (13) may be proved as follows. Let a function $u \in H_0^1 \cap C^\infty(R^m)$ satisfy $u - \alpha A u = 0$. Thus $A u \in H_0^1 \cap C^\infty(R^m) \subseteq H_0^0$ and so the expression $(u - \alpha A u, u)_0$ is defined and $= 0$. Let $u_n \in \hat{H}_0^1$ be such that $\lim_{n\to\infty} \|u - u_n\|_1 = 0$. We obtain, by partial integration as in (9),

$$0 = (u - \alpha A u, u)_0 = \lim_{n\to\infty} (u - \alpha A u, u_n)_0 \geqq \alpha\delta \, \|u\|_1^2, \quad \text{that is,} \quad u = 0.$$

Corollary 1. Positive constants $\hat{\alpha}_0$ and η_0 exist such that, for any $f \in \hat{H}_0^1$, the equation

$$\alpha u - A u = f \quad (0 < \hat{\alpha}_0 + \lambda_0 + \eta_0 \leqq \alpha) \tag{16}$$

admits a uniquely determined solution $u = u_f \in H_0^1 \cap C^\infty(R^m)$, and we have the estimate

$$||u_f||_0 \leqq (\alpha - \lambda_0 - \eta_0)^{-1} ||f||_0. \tag{17}$$

Proof. By Schwarz' inequality we have

$$||(\alpha I - A)\, u||_0 \cdot ||u||_0 \geqq |((\alpha I - A)\, u, u)_0| \quad \text{for} \quad u \in \hat{H}_0^1. \tag{18}$$

By partial integration, we have

$$((\alpha I - A)\, u, u)_0 = \alpha\, ||u||_0^2 + \int_{R^m} a^{ij} u_{x_i} u_{x_j} dx + \int_{R^m} a^{ij}_{x_j} u_{x_i} u\, dx$$
$$- \int_{R^m} b^i u_{x_i} u\, dx - \int_{R^m} c u u\, dx.$$

Hence, we have, by (3), (4) and (12),

$$((\alpha I - A)\, u, u)_0 \geqq \alpha\, ||u||_0^2 + \lambda_0 (||u||_1^2 - ||u||_0^2)$$
$$- m^2 \eta\, [\nu (||u||_1^2 - ||u||_0^2) + \nu^{-1} ||u||_0^2 + m^{-2} ||u||_0^2]$$
$$= (\alpha - \lambda_0 - m^2 \eta\, (\nu^{-1} - \nu + m^{-2}))\, ||u||_0^2 + (\lambda_0 - m^2 \eta\, \nu)\, ||u||_1^2.$$

Thus by (18) we have, for $\eta_0 = m^2 \eta\, (\nu^{-1} - \nu + m^{-2})$,

$$||(\alpha I - A)\, u||_0 \geqq (\alpha - \lambda_0 - \eta_0)\, ||u||_0 \quad \text{whenever} \quad u \in \hat{H}_0^1, \tag{17'}$$

by taking $\nu > 0$ so small that $(\lambda_0 - m^2 \eta\, \nu)$ and η_0 are > 0. We then take $\hat{\alpha}_0$ so large that for $\alpha \geqq \hat{\alpha}_0 + \lambda_0 + \eta_0$ we can apply Lemma 2 in solving (16).

Since the solution $u = u_f \in H_0^1 \cap C^\infty(R^m)$ of (16) is approximated by $||\ ||_1$-norm by a sequence of functions $\in \hat{H}_0^1$, we obtain the estimate (17) from (18) and (17').

Corollary 2. Consider A as an operator on $D(A) = (\alpha I - A)^{-1} \hat{H}_0^1 \subseteq H_0^0$ into H_0^0. Then the smallest closed extension $\hat{A}$ in H_0^0 of A admits, for $\alpha > \hat{\alpha}_0 + \lambda_0 + \eta_0$, the resolvent $(\alpha I - \hat{A})^{-1}$ defined on H_0^0 into H_0^0 such that

$$||(\alpha I - \hat{A})^{-1}||_0 \leqq (\alpha - \lambda_0 - \eta_0)^{-1}. \tag{19}$$

Proof. Clear from Corollary 1 remembering the fact that $D(A)$ and $\hat{H}_0^1$ both are $||\ ||_0$-dense in H_0^0.

Thus, by Corollary 2 in Chapter IX, 7, $\hat{A}$ is the infinitesimal generator of a semi-group T_t of class (C_0) in the B-space H_0^0 such that $||T_t||_0 \leqq e^{(\lambda_0 + \eta_0)t}$ for $t \geqq 0$.

Actually we can prove

Theorem 1. Let Complex-$\hat{H}_0^1$ be the space of all complex-valued functions $f \in C_0^\infty(R^m)$ with $||f||_1 = \left(\int_{R^m} |f|^2 dx + \sum_{j=1}^m \int_{R^m} ||f||^2 dx\right)^{1/2} < \infty$.

Let $\tilde{H}_0^1$ and $\tilde{H}_0^0$ be the completions of Complex-$\hat{H}_0^1$ with respect to the norm $\|f\|_1$ and $\|f\|_0$, respectively. We know that $\tilde{H}_0^1 =$ (complex) Sobolev space $W^1(R^m)$ (see Chapter I, 10). It is also clear that $\tilde{H}_0^0 =$ (complex) Hilbert space $L^2(R^m)$. We consider A as an operator defined on $D(A) = (\alpha I - A)^{-1}\tilde{H}_0^1 \subseteq L^2(R^m)$ into $L^2(R^m)$. Then the smallest closed extension $\tilde{A}$ in $L^2(R^m)$ of A is the infinitesimal generator of a holomorphic semi-group T_t of class (C_0) in $L^2(R^m)$ such that $\|T_t\|_0 \leqq e^{(\lambda_0+\eta_0)t}$ for $t \geqq 0$.

Proof. By the preceding Corollary 2 and the reality of the coefficients in the differential operator A, we see that the range $R(\alpha I - A) = (\alpha I - A)\cdot D(A)$ is $\|\ \|_0$-dense in $L^2(R^m)$ for $\alpha > \hat{\alpha}_0 + \lambda_0 + \eta_0$. Moreover, we have, for $(u + iv) \in L^2(R^m)$ such that $(u + iv) \in D(A)$,

$$\|(\alpha I - A)(u + iv)\|_0^2 = \|(\alpha I - A)u\|_0^2 + \|(\alpha I - A)v\|_0^2$$
$$\geqq (\alpha - \lambda_0 - \eta_0)^2 (\|u\|_0^2 + \|v\|_0^2).$$

Hence the inverse $(\alpha I - \tilde{A})^{-1}$ is a bounded linear operator on $L^2(R^m)$ into $L^2(R^m)$ such that $\|(\alpha I - \tilde{A})^{-1}\|_0 \leqq (\alpha - \lambda_0 - \eta_0)^{-1}$ for $\alpha > \hat{\alpha}_0 + \lambda_0 + \eta_0$.

Therefore, by the Theorem in Chapter IX, 10, we have only to show that

$$\overline{\lim_{|\tau|\uparrow\infty}} |\tau|\, \|((\alpha + i\tau) I - \tilde{A})^{-1}\|_0 < \infty. \tag{20}$$

We have, for $w \in D(A)$,

$$\|((\alpha + i\tau) I - A) w\|_0 \|w\|_0 \geqq |(((\alpha + i\tau) I - A) w, w)_0|. \tag{21}$$

By partial integration, we obtain, as in the proof of (17),

$$|\text{Real part } (((\alpha + i\tau) I - A) w, w)_0|$$
$$= \left| |\alpha|\, \|w\|_0^2 + \text{Real part} \left(\int_{R^m} a^{ij} w_{x_i} \overline{w}_{x_j} dx + a^{ij}_{x_i} w_{x_j} \overline{w}\, dx \right.\right.$$
$$\left.\left. - \int_{R^m} b^i w_{x_i} \overline{w}\, dx - \int_{R^m} c w \overline{w}\, dx \right)\right|$$
$$\geqq (\alpha - \lambda_0 - \eta_0) \|w\|_0^2 + (\lambda_0 - m^2 \eta \nu) \|w\|_1^2.$$

Similarly we have

$$|\text{Imaginary part } (((\alpha + i\tau) I - A) w, w)_0|$$
$$\geqq |\tau|\, \|w\|_0^2 - m^2 \eta (\|w\|_1^2 + m^{-2} \|w\|_0^2) | \geqq (|\tau| - \eta) \|w\|_0^2 - m^2 \eta \|w\|_1^2.$$

Suppose there exists a $w_0 \in D(A)$, $\|w_0\|_0 \neq 0$, such that

$$|\text{Imaginary part } (((\alpha I + i\tau) I - A) w_0, w_0)_0| < 2^{-1} (|\tau| - \eta) \|w_0\|_0^2$$

for sufficiently large τ (or for sufficiently large $-\tau$). Then, for such large τ (or $-\tau$), we must have

$$m^2 \eta \|w_0\|_1^2 > 2^{-1} (|\tau| - \eta) \|w_0\|_0^2$$

and so

$$|\text{Real part } (((\alpha + i\tau) I - A) w_0, w_0)_0| \geqq (\lambda_0 - m^2 \eta \nu) \frac{(|\tau| - \eta)}{2m^2 \eta} ||w_0||_0^2.$$

Hence we have proved Theorem 1 by (21).

Theorem 2. For any $f \in L^2(R^m)$, $u(t, x) = (T_t f)(x)$ is infinitely differentiable in $t > 0$ and in $x \in R^m$, and $u(t, x)$ satisfies the diffusion equation (1) as well as the initial condition $\lim_{t \downarrow 0} ||u(t, x) - f(x)||_0 = 0$.

Proof. We denote by $T_t^{(k)}$ the k-th strong derivative in $L^2(R^m)$ of T_t with respect to t. T_t being a holomorphic semi-group of class (C_0) in $L^2(R^m)$, we have $T_t^{(k)} f = \tilde{A}^k T_t f \in L^2(R^m)$ if $t > 0$ $(k = 0, 1, \ldots)$. Since $\tilde{A}$ is the smallest closed extension in $L^2(R^m)$ of A, we see that $A^k T_t f \in L^2(R^m)$ for fixed $t > 0$ $(k = 0, 1, 2, \ldots)$ if we apply the differential operator A^k in the distributional sense. Thus, by the Corollary of Friedrichs' theorem in Chapter VI, 9, $u(t, x)$ is, for fixed $t > 0$, equal to a $C^\infty(R^m)$ function after a correction on a set of measure zero.

Because of the estimate $||T_t||_0 \leqq e^{(\lambda_0 + \eta_0)t}$, we easily see that, if we apply, in the distributional sense, the elliptic differential operator

$$\left(\frac{\partial^2}{\partial t^2} + A\right)$$

any number of times to $u(t, x)$, then the result is locally square integrable in the product space $\{t; 0 < t < \infty\} \times R^m$. Thus again, by the Corollary of Friedrichs' theorem in Chapter VI, 9, we see that $u(t, x)$ is equal to a function which is C^∞ in (t, x), $t > 0$ and $x \in R^m$, after a correction on a set of measure zero of the product space. Thus we may consider that $u(t, x)$ is a genuine solution of (1) satisfying the initial condition $\lim_{t \downarrow 0} ||u(t, x) - f(x)||_0 = 0$.

Remark. The above obtained solution $u(t, x)$ satisfies the "forward and backward unique continuation property":

$$\begin{aligned} &\text{if, for a fixed } t_0 > 0,\ u(t_0, x) \equiv 0 \text{ on an open set } G \subseteqq R^m, \\ &\text{then } u(t, x) = 0 \quad \text{for every} \quad t > 0 \quad \text{and every} \quad x \in G. \end{aligned} \tag{22}$$

Proof. Since T_t is a holomorphic semi-group of class (C_0), we have

$$\lim_{n \to \infty} \left\| T_{t_0 + h} f - \sum_{k=0}^{n} (k!)^{-1} h^k A^k T_{t_0} f \right\|_0 = 0$$

for a sufficiently small h. Hence, as in the proof of the completeness of the space $L^2(R^m)$, there exists a sequence $\{n'\}$ of natural numbers such that

$$u(t_0 + h, x) = \lim_{n' \to \infty} \sum_{k=0}^{n'} (k!)^{-1} h^k A^k u(t_0, x) \text{ for a.e. } x \in R^m.$$

By hypothesis given in (22), we have $A^k u(t_0, x) = 0$ in G and so we must have $u(t_0 + h, x) = 0$ in G for sufficiently small h. Repeating the process we see that conclusion (22) is true.

References

The result of this section is adapted from K. Yosida [21]. As for the "forward and backward unique continuation property" (22), we have the more precise result that $u(t, x) = 0$ for every $t > 0$ and every $x \in R^m$. For, S. Mizohata [3] has proved a "space-like unique continuation property" of solutions $u(t, x)$ of a diffusion equation to the effect that $u(t, x) = 0$ for all $t > 0$ and all $x \in R^m$ if $u(t, x) = 0$ for all $t > 0$ and all $x \in G$. Concerning the holomorphic character of the semi-group T_t obtained above, see also R. S. Phillips [6]. It is to be noted here that the unique continuation property of the solutions of a heat equation $\partial u/\partial t = \Delta u$ was first proposed and solved by H. Yamabe-S. Itô [1].

There is a fairly complete discussion of parabolic equations from the view point of the theory of dissipative operators. See R. S. Phillips [7].

2. Integration of Diffusion Equations in a Compact Riemannian Space

Let R be a connected, orientable m-dimensional C^∞ Riemannian space with the metric

$$ds^2 = g_{ij}(x)\, dx^i dx^j. \tag{1}$$

Let A be a second order linear partial differential operator in R with real-valued C^∞ coefficients:

$$A = a^{ij}(x) \frac{\partial^2}{\partial x^i \partial x^j} + b^i(x) \frac{\partial}{\partial x^i}. \tag{2}$$

We assume that a^{ij} is a symmetric contravariant tensor and that $b^i(x)$ satisfies the transformation rule

$$\bar{b}^i = b^k \frac{\partial \bar{x}^i}{\partial x^k} + a^{kj} \frac{\partial^2 \bar{x}^i}{\partial x^k \partial x^j} \tag{3}$$

by the coordinate change $(x^1, x^2, \ldots, x^m) \to (\bar{x}^1, \bar{x}^2, \ldots, \bar{x}^m)$ so that the value $(Af)(x)$ is determined independently of the choice of the local coordinates. We further assume that A is *strictly elliptic* in the sense that there exist positive constants λ_0 and μ_0 such that

$$\mu_0 \sum_{j=1}^m \xi_j^2 \geqq a^{ij}(x)\, \xi_i \xi_j \geqq \lambda_0 \sum_{j=1}^m \xi_j^2 \text{ for every real vector } (\xi_1, \ldots, \xi_m) \text{ and every } x \in R^m. \tag{4}$$

We consider the Cauchy problem in the large on R for the diffusion equation: to find solution $u(t, x)$ such that

$$\partial u/\partial t = Au,\ t > 0,\ u(0, x) = f(x) \text{ where } f(x) \text{ is a given function on } R. \tag{5}$$

We shall prove

Theorem. If R is compact so that R is without boundary, then the equation (5) admits, for any initial function $f \in C^\infty(R)$, a uniquely determined solution $u(t, x)$ which is C^∞ in (t, x), $t > 0$ and $x \in R$. This solution can be represented in the form

$$u(t, x) = \int_R P(t, x, dy)\, f(y) \tag{6}$$

where $P(t, x, E)$ is the transition probability of a Markov process on R.

Proof. Let us denote by $C(R)$ the B-space of real-valued continuous functions $f(x)$ on R normed by $\|f\| = \sup_x |f(x)|$. We first prove

$$\begin{gathered}\text{for any } f \in C^\infty(R) \text{ and any } n > 0, \text{ we have}\\ \max_x h(x) \geqq f(x) \geqq \min_x h(x) \text{ where } h(x) = f(x) - n^{-1}(Af)(x).\end{gathered} \tag{7}$$

Suppose that $f(x)$ attains its maximum at $x = x_0$. We choose a local coordinate system at x_0 such that $a^{ij}(x_0) = \delta_{ij}$ ($= 1$ or 0 according as $i = j$ or $i \neq j$). Such a choice is possible owing to condition (4). Then

$$\begin{aligned} h(x_0) &= f(x_0) - n^{-1}(Af)(x_0)\\ &= f(x_0) - n^{-1} b^i(x_0) \frac{\partial f}{\partial x_0^i} - n^{-1} \sum_{i=1}^m \frac{\partial^2 f}{\partial (x_0^i)^2} \geqq f(x_0), \end{aligned}$$

since we have, at the maximum point x_0,

$$\frac{\partial f}{\partial x_0^i} = 0 \quad \text{and} \quad \frac{\partial^2 f}{\partial (x_0^i)^2} \leqq 0.$$

Thus $\max_x h(x) \geqq f(x)$. Similarly we have $f(x) \geqq \min_x h(x)$.

We shall consider A as an operator on $D(A) = C^\infty(R) \subseteq C(R)$ into $C(R)$. Then, by (7), we see that the inverse $(I - n^{-1}A)^{-1}$ exists for $n > 0$ and $\|(I - n^{-1}A)^{-1} g\| \leqq \|g\|$ for g in the range $R(I - n^{-1}A) = (I - n^{-1}A) \cdot D(A)$. This range is, for sufficiently large n, strongly dense in $C(R)$. For, as in the preceding section, we have the result: For any $g \in C^\infty(R)$ and for sufficiently large $n > 0$, the equation $u - n^{-1}Au = g$ admits a uniquely determined solution $u \in C^\infty(R)$. Because, by the compactness of the Riemann space R, Lemma 1 concerning the partial integration in the preceding section may be adapted to our Riemannian space R without boundary. Moreover, $C^\infty(R)$ is strongly dense in $C(R)$ as may be seen by the regularization of functions in $C(R)$ (see Proposition 8 in Chapter I, 1).

Hence the smallest closed extension $\bar{A}$ in $C(R)$ of the operator A satisfies the conditions:

$$\left.\begin{array}{l}\text{for sufficiently large } n > 0, \text{ the resolvent } (I - n^{-1}\bar{A})^{-1}\\ \text{exists as a bounded linear operator on } C(R) \text{ into } C(R)\\ \text{such that } \|(I - n^{-1}\bar{A})^{-1}\| \leqq 1,\end{array}\right\} \tag{8}$$

$$((I - n^{-1}\bar{A})^{-1}h)(x) \geqq 0 \text{ on } R \text{ whenever } h(x) \geqq 0 \text{ on } R, \tag{9}$$

$$(I - n^{-1}\bar{A})^{-1} \cdot 1 = 1. \tag{10}$$

The positivity of the operator $(I - n^{-1}\overline{A})^{-1}$ given in (9) is clear from (7). The equation (10) follows from $A \cdot 1 = 0$.

Therefore $\overline{A}$ is the infinitesimal generator of a contraction semi-group T_t in $C(R)$ of class (C_0). As in the preceding section, we see, by the strict ellipticity of A, that, for any $f \in C^\infty(R)$, the function $u(t, x) = (T_t f)(x)$ is a C^∞ function in (t, x) for $t > 0$ and $x \in R$ so that $u(t, x)$ is a genuine solution of (5).

Since the dual space of the space $C(R)$ is the space of Baire measures in R, we easily prove the latter part of the Theorem remembering (9) and (10).

3. Integration of Wave Equations in a Euclidean Space R^m

Consider a wave equation

$$\partial^2 u/\partial t^2 = A u, \quad -\infty < t < \infty, \tag{1}$$

where the differential operator

$$A = a^{ij}(x) \frac{\partial^2}{\partial x_i \partial x_j} + b^i(x) \frac{\partial}{\partial x^i} + c(x) \qquad (a^{ij}(x) = a^{ji}(x)) \tag{2}$$

is strictly elliptic in an m-dimensional Euclidean space R^m. We assume that the real-valued C^∞ coefficients a, b, and c satisfy conditions (3) and (4) in Chapter XIV, 1. As done there, we shall denote by $\hat{H}_0^1$ the space of all real-valued $C_0^\infty(R^m)$ functions $f(x)$ normed by

$$\|f\|_1 = \left(\int_{R^m} f^2\,dx + \sum_{j=1}^{m} \int_{R^m} f_{x_j}^2\,dx \right)^{1/2}$$

and let H_0^1 (and H_0^0) be the completion of $\hat{H}_0^1$ with respect to the norm $\|f\|_1$ $\left(\text{and with respect to the norm } \|f\|_0 = \left(\int_{R^m} f^2\,dx\right)^{1/2}\right)$.

Lemma. For any pair $\{f, g\}$ of elements $\in \hat{H}_0^1$ the equation

$$\left(\begin{pmatrix} I & 0 \\ 0 & I \end{pmatrix} - n^{-1} \begin{pmatrix} 0 & I \\ A & 0 \end{pmatrix} \right) \begin{pmatrix} u \\ v \end{pmatrix} = \begin{pmatrix} f \\ g \end{pmatrix} \tag{3}$$

admits, if the integer n be such that $|n^{-1}|$ is sufficiently small, a uniquely determined solution $\{u, v\}$, u and $v \in H_0^1 \cap C^\infty(R^m)$, satisfying the estimate

$$\begin{aligned} &((u - \alpha_0 A u, u)_0 + \alpha_0 (v, v)_0)^{1/2} \\ &\leq (1 - \beta\,|n|^{-1})^{-1}\,((f - \alpha_0 A f, f)_0 + \alpha_0 (g, g)_0)^{1/2}, \end{aligned} \tag{4}$$

with positive constants α_0 and β which are independent of n and $\{f, g\}$.

Proof. Let $u_1 \in H_0^1 \cap C^\infty(R^m)$ and $v_1 \in H_0^1 \cap C^\infty(R^m)$ be respectively the solutions of

$$u_1 - n^{-2} A u_1 = f \quad \text{and} \quad v_1 - n^{-2} A v_1 = g. \tag{5}$$

The existence of such solutions for sufficiently small $|n^{-1}|$ was proved in Lemma 2 of Chapter XIV, 1. Then

$$u = u_1 + n^{-1} v_1, \quad v = n^{-1} A u_1 + v_1 \tag{6}$$

satisfies (3), i.e., we have $u - n^{-1} v = f$, $v - n^{-1} A u = g$.

We next prove (4). We remark that

$$A u = n(v - g) \in H_0^1 \cap C^\infty(R^m) \subseteqq H_0^0 \text{ and hence, by } f, g \in C_0^\infty(R^m),$$
$$A v = n(A u - A f) \in H_0^1 \cap C^\infty(R^m) \subseteqq H_0^0.$$

We have thus, by (3),

$$\begin{aligned}(f - \alpha_0 A f, f)_0 &= (u - n^{-1} v - \alpha_0 A(u - n^{-1} v), u - n^{-1} v)_0 \\ &= (u - \alpha_0 A u, u)_0 - 2 n^{-1}(u, v)_0 + \alpha_0 n^{-1}(A u, v)_0 + \alpha_0 n^{-1}(A v, u)_0 \\ &\quad + n^{-2}(v - \alpha_0 A v, v)_0\end{aligned}$$

and

$$\begin{aligned}\alpha_0 (g, g)_0 &= \alpha_0 (v - n^{-1} A u, v - n^{-1} A u)_0 \\ &= \alpha_0 (v, v)_0 - \alpha_0 n^{-1}(v, A u)_0 - \alpha_0 n^{-1}(A u, v)_0 + \alpha_0 n^{-2}(A u, A u)_0.\end{aligned}$$

By a limiting process, we can prove that (9), (10) and (11) in Chapter XIV, 1 are valid for $f = u$ and $g = v$. Thus, by (12) in Chapter XIV, 1, there exists a positive constant β satisfying

$$\begin{aligned}((f - \alpha_0 A f, f)_0 + \alpha_0 (g, g)_0)^{1/2} &\geqq ((u - \alpha_0 A u, u)_0 + \alpha_0 (v, v)_0 \\ &\quad - \alpha_0 |n^{-1}| \, |(A u, v)_0 - (A v, u)_0| - 2 |n^{-1}| \, |(u, v)_0|)^{1/2} \\ &\geqq (1 - \beta |n^{-1}|)((u - \alpha_0 A u, u)_0 + \alpha_0 (v, v)_0)^{1/2}\end{aligned}$$

for sufficiently large $|n|$.

The above estimate for the solutions $\{u, v\}$ belonging to $H_0^1 \cap C^\infty(R^m)$ shows that such solutions are uniquely determined by $\{f, g\}$.

Corollary. The product space $H_0^1 \cap H_0^0$ of vectors

$$\begin{pmatrix} u \\ v \end{pmatrix} = \{u, v\}', \quad \text{where } u \in H_0^1 \text{ and } v \in H_0^0, \tag{7}$$

is a B-space by the norm

$$\left\| \begin{pmatrix} u \\ v \end{pmatrix} \right\| = \|\{u, v\}'\| = (B(u, u) + \alpha_0 (v, v)_0)^{1/2}, \tag{8}$$

where $B(f, g)$ is the extension by continuity with respect to the norm $\| \ \|_1$ of the bilinear functional

$$\hat{B}(f, g) = (f - \alpha_0 A f, g)_0 \text{ defined for } f, g \in \hat{H}_0^1.$$

We know that $B(u, u)^{1/2}$ is equivalent to the norm $\|u\|_1$ (see Chapter XIV, 1):

$$\alpha_0 \delta \|u\|_1^2 \leqq B(u, u) \leqq (1 + \alpha_0 \gamma) \|u\|_1^2. \tag{9}$$

Let the domain $D(\mathfrak{A})$ of the operator

$$\mathfrak{A} = \begin{pmatrix} 0 & I \\ A & 0 \end{pmatrix} \tag{10}$$

be the vectors $\{u, v\}' \in H_0^1 \times H_0^0$ such that $u, v \in H_0^0$ are given by (6). Then the Lemma shows that the range of the operator

$$\mathfrak{J} - n^{-1}\mathfrak{A}, \quad \text{where } \mathfrak{J} = \begin{pmatrix} I & 0 \\ 0 & I \end{pmatrix},$$

contains all the vectors $\{f, g\}'$ such that $f, g \in \hat{H}_0^1$. Thus the smallest closed extension $\overline{\mathfrak{A}}$ in $H_0^1 \times H_0^0$ of the operator $\mathfrak{A}$ is such that the operator $(\mathfrak{J} - n^{-1}\overline{\mathfrak{A}})$ with integral parameter n admits, for sufficiently large $|n|$, an everywhere in $H_0^1 \times H_0^0$ defined inverse $(\mathfrak{J} - n^{-1}\overline{\mathfrak{A}})^{-1}$ satisfying

$$\|(\mathfrak{J} - n^{-1}\overline{\mathfrak{A}})^{-1}\| \leqq (1 - \beta\,|n^{-1}|)^{-1}. \tag{11}$$

We are now prepared to prove

Theorem. For any pair $\{f(x), g(x)\}$ of $C_0^\infty(R^m)$ functions, the equation (1) admits a C^∞ solution $u(t, x)$ satisfying the initial condition

$$u(0, x) = f(x), \quad u_t(0, x) = g(x) \tag{12}$$

and the estimate

$$(B(u, u) + \alpha_0 (u_t, u_t)_0)^{1/2} \leqq \exp(\beta\,|t|)\,(B(f, f) + \alpha_0 (g, g)_0)^{1/2}. \tag{13}$$

Remark. Formula (9) shows that $B(u, u)$ is comparable to the potential energy of the *wave* (= the solution of (1)) $u(t, x)$, and $(u_t, u_t)_0$ is comparable to the kinetic energy of the wave $u(t, x)$. Thus (13) means that the total energy of the wave $u(t, x)$ does not increase more rapidly than $\exp(\beta\,|t|)$ when the time t tends to $\pm\infty$. This is a kind of *energy inequality* which governs wave equations in general.

Proof of Theorem. Estimate (11) shows that $\overline{\mathfrak{A}}$ is the infinitesimal generator of a *group* T_t of class (C_0) in $H_0^1 \times H_0^0$ such that

$$\|T_t\| \leqq \exp(\beta\,|t|), \quad -\infty < t < \infty. \tag{14}$$

By hypothesis we have, for $k = 0, 1, 2, \ldots,$

$$\overline{\mathfrak{A}}^k \begin{pmatrix} f \\ g \end{pmatrix} = \mathfrak{A}^k \begin{pmatrix} f \\ g \end{pmatrix} \in C_0^\infty(R^m) \times C_0^\infty(R^m) \subseteqq H_0^1 \times H_0^0.$$

Hence, if we put

$$\begin{pmatrix} u(t, x) \\ v(t, x) \end{pmatrix} = T_t \begin{pmatrix} f(x) \\ g(x) \end{pmatrix},$$

then, by the commutativity of $\overline{\mathfrak{A}}$ with T_t, we have

$$\frac{\partial^k}{\partial t^k} \begin{pmatrix} u(t, x) \\ v(t, x) \end{pmatrix} = \frac{\partial^k T_t}{\partial t^k} \begin{pmatrix} f(x) \\ g(x) \end{pmatrix} = \overline{\mathfrak{A}}^k \begin{pmatrix} u(t, x) \\ v(t, x) \end{pmatrix} \in H_0^1 \times H_0^0$$

for $k = 0, 1, 2, \ldots$ Here we denote by $\partial^k T_t/\partial t^k$ the kth strong derivative in $H_0^1 \times H_0^0$. Therefore, by $H_0^1 \subsetneqq H_0^0 = L^2(R^m)$ and the strict ellipticity of the operator A, we see, as in the proof of Theorem 2 in Chapter XIV, 1, that $u(t, x)$ is C^∞ in (t, x) for $-\infty < t < \infty$, $x \in R^m$, and satisfies equation (1) with (12) and estimate (13).

Remark. The result of the present section is adapted from K. Yosida [22]. Cf. J. L. Lions [1]. P. D. Lax has kindly communicated to the present author that the method of integration given in this section is very similar to that announced by him in Abstract 180, Bull. Amer. Math. Soc. **58**, 192 (1952). It is to be noted here that our method can be modified to the integration of wave equations in an open domain of a Riemannian space. There is another approach to the integration of wave equations based upon the theory of dissipative semi-groups. See R. S. Phillips [8] and [9]. The method is closely connected with the theory of symmetric positive systems by K. Friedrichs [2]. Cf. also P. Lax-R. S. Phillips [3].

4. Integration of Temporally Inhomogeneous Equations of Evolution in a Reflexive *B*-space

We shall be concerned with the integration of the equation

$$dx(t)/dt = A(t)\, x(t), \quad a \leqq t \leqq b. \tag{1}$$

Here the unknown $x(t)$ is an element of a B-space X, depending on a real parameter t, while $A(t)$ is a given, in general unbounded, linear operator with domain $D(A(t))$ and range $R(A(t))$, both in X, depending also on t.

T. Kato [3], [4] was the first to make a successful attack on the problem of integration of (1). He assumed the following four conditions:

(i) The domain $D(A(t))$ is independent of t and is strongly dense in X such that, for $\alpha > 0$, the resolvent $(I - \alpha A(t))^{-1}$ exists as a bounded linear operator $\in L(X, X)$ with the norm $\leqq 1$.

(ii) The operator $B(t, s) = (I - A(t))(I - A(s))^{-1}$ is uniformly bounded in norm for $t \geqq s$.

(iii) $B(t, s)$ is, at least for some s, of bounded variation in t in norm, that is, for every partition $s = t_0 < t_1 < \cdots < t_n = t$ of $[a, b]$,

$$\sum_{j=0}^{n-1} \|B(t_{j+1}, s) - B(t_j, s)\| \leqq N(a, b) < \infty.$$

(iv) $B(t, s)$ is, at least for some s, weakly differentiable in t and $\partial B(t, s)/\partial t$ is strongly continuous in t.

Under these conditions, Kato proved that the limit

$$U(t, s)\, x_0 = \underset{\max|t_{j+1}-t_j|\to 0}{\text{s-lim}} \prod_{j=n-1}^{0} \exp\left((t_{j+1} - t_j)\, A(t_j)\right) x_0$$

exists for every $x_0 \in X$ and gives the unique solution of (1) with the initial condition $x(s) = x_0$ at least when $x_0 \in D(A(s))$.

Kato's method is thus an abstraction of the classical polygon method of Cauchy for ordinary differential equations $dx(t)/dt = a(t)\,x(t)$. Although the method is very simple and natural in its idea, the proof is somewhat lengthy because it is concerned with a general B-space X. KATO [3] has shown that the proof can be simplified if the space X is reflexive.

Another method of integrating equation (1) has been devised by J. L. LIONS [2]. He assumes the operator $A(t)$ to be an elliptic differential operator with smooth coefficients depending on t, and seeks distributional solutions $x(t)$ by transforming equation (1) into an integrated form in concrete function spaces such as the Sobolev spaces $W^{k,2}(\Omega)$ and their variants.

We shall not give details of the methods of KATO and LIONS, but refer the reader to T. KATO [3], [4] and J. L. LIONS [2]. We also refer the reader to a paper by O. A. LADYZHENSKAYA-I. M. VISIK [1] which is motivated by a similar idea as Lions'. It is to be noted here that the cited book by Lions contains a fairly complete list of references concerning the integration of the equations of evolution up to 1961.

In the present section, we shall give a method of integration of (1) which is based upon a uniqueness lemma and the local sequential weak compactness of a reflexive B-space. Our motivation may be described as follows.

If $A(t) = A$ is independent of t and A is the infinitesimal generator of a semi-group T_t of class (C_0) in X, then the solution $x(t)$ of

$$dx(t)/dt = A\,x(t),\ a \leqq t \leqq b,\ x(a) = x_0 \in D(A) = \text{the domain of } A\,, \tag{2}$$

is given by

$$\begin{aligned} &x(t) = T_t \cdot x_0 = \exp((t-a)\,A)\,x_0 = \underset{n\to\infty}{s\text{-lim}} \exp((t-a)\,A_n)\,x_0, \\ &\text{where}\quad A_n = A(I - n^{-1}A)^{-1}. \end{aligned} \tag{3}$$

Suggested by this time independent case, let $A(t)$ satisfy two conditions:

$$\left.\begin{array}{l} A(t) \text{ is, for } a \leqq t \leqq b, \text{ a closed linear operator with} \\ \text{dense domain } D(A(t)) \subseteqq X \text{ and range } R(A(t)) \subseteqq X \\ \text{such that, for } \lambda \geqq 0, \text{ the resolvent } (\lambda I - A(t))^{-1} \text{ exists} \\ \text{with the estimate } \|(I - \lambda^{-1}A(t))^{-1}\| \leqq (1 - \lambda^{-1}M)^{-1} \\ \text{for } \lambda \geqq M, \text{ where } M \text{ is independent of } \lambda \text{ and } t; \end{array}\right\} \tag{4}$$

$$\begin{array}{l} \text{the strong derivative } dA(t)^{-1}/dt = B(t) \text{ exists and is} \\ \text{strongly continuous in } t,\ a \leqq t \leqq b. \end{array} \tag{5}$$

We then consider the initial value problem:

$$\begin{array}{l} dx_n(t)/dt = A_n(t)\,x_n(t),\ \ a \leqq t \leqq b,\ \ x_n(a) = A(a)^{-1}y, \\ \text{where } y \text{ is an arbitrary element of } X \text{ and } A_n(t) = A(t) \\ \times (I - n^{-1}A(t))^{-1} = n(J_n(t) - I),\ J_n(t) = (I - n^{-1}A(t))^{-1}. \end{array} \tag{6}$$

Because of (5), $J_n(t)$ is strongly continuously differentiable with respect to t, and so $J_n(t)$ is uniformly continuous in t in the sense of the norm of operator. Thus problem (6) has a uniquely determined solution $x_n(t)$ which may be obtained, for instance, by successive approximation starting with the first approximation $\exp((t-a)\,A_n(a))\,A(a)^{-1}y$. Hence we are lead to the problem of finding conditions for $A(t)$ under which the sequence $\{x_n(t)\}$ converges, strongly or weakly, to the solution $x(t)$ of the initial value problem

$$dx(t)/dt = A(t)\,x(t),\ a \leqq t \leqq b,\ x(s) = A(a)^{-1}y. \tag{1'}$$

The purpose of the present section is to give a partial answer to this problem.

Lemma. The solution of (1′), if it exists, must satisfy

$$\|x(t)\| \leqq \|x(a)\| \cdot \exp((t-a)\,M). \tag{7}$$

Proof. We follow an argument essentially given in T. Kato [3]. We have, for $\delta > 0$,

$$\begin{aligned} x(t+\delta) &= x(t) + \delta A(t)\,x(t) + o(\delta) \\ &= (I + \delta A(t))\,(I - \delta A(t))\,(I - \delta A(t))^{-1}\,x(t) + o(\delta) \\ &= (I - \delta A(t))^{-1}\,x(t) - \delta^2 A(t)\,(I - \delta A(t))^{-1}\,A(t)\,x(t) + o(\delta) \\ &= (I - \delta A(t))^{-1}\,x(t) - \delta((I - \delta A(t))^{-1} - I)\,A(t)\,x(t) + o(\delta). \end{aligned}$$

Because of (4), we have $s\text{-}\lim_{\delta\downarrow 0}(I - \delta A(t))^{-1}z = z$ for any $z \in X$ (see (2) in Chapter IX, 7). Hence, by the estimate in (4), we have

$$\|x(t+\delta)\| \leqq (1-\delta M)^{-1}\,\|x(t)\| + o(\delta)$$

and so $d^+\|x(t)\|/dt \leqq M\,\|x(t)\|$. This proves the Lemma.

Theorem 1. We assume that X is a reflexive B-space so that any bounded sequence in X contains a subsequence which converges weakly to an element of X. Besides (4) and (5), we also assume that

$$A(t)\,B(t) = A(t)\,[dA(t)^{-1}/dt] \text{ is strongly continuous in } t. \tag{8}$$

Then there exists a uniquely determined solution of (1′) which is given by $x(t) = \text{weak-}\lim_{n\to\infty} x_n(t)$.

Proof. Since $A_n(t) = n(J_n(t) - I)$ and $\|J_n(t)\| \leqq (1 - n^{-1}M)^{-1}$, we have, for $\delta > 0$ and $n > M$,

$$\begin{aligned} \|(I - \delta A_n(t))\,x\| &= \|(I - \delta n(J_n(t) - I))\,x\| \geqq (1 + n\,\delta)\,\|x\| \\ &\quad - \delta n(1 - n^{-1}M)^{-1}\,\|x\| = (1 - \delta M(1 - n^{-1}M)^{-1})\,\|x\|. \end{aligned}$$

Hence for sufficiently small $\delta > 0$, the bounded inverse $(I - \delta A_n(t))^{-1}$ exists with the estimate $\|(I - \delta A_n(t))^{-1}\| \leqq (1 - \delta M(1 - n^{-1}M)^{-1})$. Therefore, by the Lemma, we see that the solution $x_n(t)$ of (6) satisfies

$$\|x_n(t)\| \leqq \|x_n(a)\| \cdot \exp((t-a)\, M_1),\, M_1 \geqq M\,(1-n^{-1}\,M)^{-1}. \tag{9}$$

Thus $x_n(t)$ is uniquely determined by the initial condition $x_n(a) = A\,(a)^{-1}\,y$ and so we may put

$$\begin{aligned} x_n(t) &= U_n(t, a)\, x_n(a) = U_n(t, a)\, A\,(a)^{-1}\, y, \\ \text{where} \quad &\|U_n(t, a)\| \leqq \exp((t-a)M_1). \end{aligned} \tag{10}$$

We next put

$$y_n(t) = A_n(t)\, U_n(t, a)\, A\,(a)^{-1}\, y = A_n(t)\, x_n(t) = dx_n(t)/dt. \tag{11}$$

We see, by (5), that $A_n(t)$ is strongly continuously differentiable in t and

$$dA_n(t)/dt = -A_n(t)\, B\,(t)\, A_n(t), \tag{12}$$

because we have

$$\delta^{-1}(A_n(t+\delta) - A_n(t)) = -A_n(t+\delta)\,\{[A_n(t+\delta)^{-1} - A_n(t)^{-1}]/\delta\}\, A_n(t),$$
$$A_n(t)^{-1} = A\,(t)^{-1} - n^{-1} I.$$

Thus $y_n(t)$ satisfies the initial value problem

$$dy_n(t)/dt = A_n(t)\, y_n(t) - A_n(t)\, B\,(t)\, y_n(t),\ a \leqq t \leqq b,\ y_n(a) = J_n(a)\, y. \tag{13}$$

We see, by (8), that $A_n(t)\, B\,(t) = J_n(t)\, A\,(t)\, B\,(t)$ is strongly continuous in t, and thus uniformly bounded in t and n by (4). Hence

$$\|A_n(t)\, B\,(t)\| \leqq C \quad \text{for} \quad a \leqq t \leqq b \quad \text{and} \quad n > M. \tag{14}$$

We have, for $\delta > 0, y_n(t+\delta) = y_n(t) + \delta A_n(t)\, y\,(t) - \delta A_n(t)\, B\,(t)\, y_n(t) + o\,(\delta)$.

Hence, as in the proof of the Lemma, we obtain

$$d^+ \|y_n(t)\|/dt \leqq (M_1 + C)\, \|y_n(t)\|,$$

and so

$$\begin{aligned} \|y_n(t)\| &\leqq \|y_n(a)\| \exp((t-a)\,(M+C)) = \|J_n(a)\, y\| \cdot \exp((t-a)\,(M_1+C)) \\ &\leqq (1-n^{-1}M)^{-1} \exp((t-a)\,(M_1+C))\, \|y\|. \end{aligned} \tag{15}$$

Hence, by $dx_n(t)/dt = y_n(t)$, we see that $x_n(t)$ is bounded and strongly continuous in t, uniformly in t and n. Thus, by the reflexivity of X, there exists a subsequence $\{n'\}$ of $\{n\}$ such that $\text{weak-}\lim_{n' \to \infty} x_{n'}(t) = x\,(t)$ exists simultaneously for all t, $a \leqq t \leqq b$.

We next prove that

$$\begin{aligned} &x\,(t) \in D\,(A\,(t)) \text{ and } A\,(t)\, x\,(t) = \text{weak-}\lim_{n' \to \infty} A_{n'}(t)\, x_{n'}(t) \\ &\text{is bounded and strongly measurable in } t. \end{aligned} \tag{16}$$

Indeed, we can prove that $A\,(t)\, x\,(t)$ is strongly continuous in t. In the proof given below, we shall make use of the fact that, for the dual operators, we have (see Theorem 1 in Chapter VIII, 6)

$$J_n(t)' = ((I - n^{-1}A(t))^{-1})' = (I' - n^{-1}A(t)')^{-1},\ (A(t)^{-1})' = (A(t)')^{-1}. \tag{17}$$

Since X is reflexive, $D(A(t)')$ is strongly dense in the dual space X' as will be proved below. Hence, by $\|J_n(t)'\| \leqq (1 - n^{-1}M)^{-1}$ which follows from (4) and (17), we see that $s\text{-}\lim_{n\to\infty} J_n(t)' f' = f'$ for every $f' \in X'$. Thus, by $\text{weak-}\lim_{n'\to\infty} x_{n'}(t) = x(t)$, we get

$$\text{weak-}\lim_{n'\to\infty} J_{n'}(t)\, x_{n'}(t) = x(t). \tag{18}$$

We have

$$\langle A(t)\, J_n(t)\, x_n(t),\ (A(t)^{-1})' f'\rangle = \langle J_n(t)\, x_n(t), f'\rangle$$

by 17) and so, by (18),

$$\lim_{n\to\infty} \langle A(t)\, J_{n'}(t)\, x_{n'}(t),\ (A(t)^{-1})' f'\rangle = \langle x(t), f'\rangle. \tag{19}$$

Now, by (17), the range $R((A(t)^{-1})') = D(A(t)')$ and it is strongly dense in X'. If otherwise, the reflexivity of X assures the existence of a $w_0 \in X$, $w_0 \neq 0$, such that $0 = \langle w_0, (A(t)^{-1})' f'\rangle = \langle A(t)^{-1} w_0, f'\rangle$. This proves that $A(t)^{-1} w_0 = 0$, i.e., $w_0 = 0$ which is a contradiction. Therefore, by the boundedness in t and n of $y_n(t) = A_n(t)\, x_n(t)$, we see that $\text{weak-}\lim_{n'\to\infty} A_{n'}(t)\, x_{n'}(t) = u(t)$ must exist. We have, by (19), $\langle u(t), (A(t)^{-1})' f'\rangle = \langle x(t), f'\rangle$, that is, $A(t)^{-1} u(t) = x(t)$.

Since the strong continuity in t of $A_n(t)\, x_n(t)$ implies that the closure of the set $\{A_n(t)\, x_n(t);\ a \leqq t \leqq b,\ n = 1, 2, \ldots\}$ is separable, the weak measurability of $u(t) = \text{weak-}\lim_{n'\to\infty} A_{n'}(t)\, x_{n'}(t)$ in t implies, by Pettis' theorem in Chapter V, 4, its strong measurability. To prove that $u(t)$ is strongly continuous in t, we integrate (13), obtaining

$$y_n(t) = U_n(t, a)\, y_n(a) - \int_a^t U_n(t, s)\, A_n(s)\, B(s)\, y_n(s)\, ds. \tag{20}$$

The second term on the right is strongly continuous in t, uniformly in t and n, because of estimate (10), (14) and the boundedness of $y_n(t)$ in t and n. The first term on the right, $U_n(t, a)\, y_n(a)$, is strongly continuous in t, uniformly in t and n, if $A_n(a)\, y_n(a) = A(a)\, J_n(a)\, J_n(a)\, y$ is bounded in n. This we prove as in the above case for the strong continuity in t of $x_n(t) = U_n(t, a)\, x_n(a)$. But $A_n(a)\, y_n(a) = A(a)\, J_n(a)\, J_n(a)\, y$ is bounded in n if y is in the domain $D(A(a))$ of $A(a)$, which is strongly dense in X. Thus, by virtue of the Lemma, we see that, for any $y \in X$, the first term on the right of (20) is strongly continuous in t, uniformly in t and n. Thus $y_n(t)$ is strongly continuous in t, uniformly in t and n. This proves that $u(t) = A(t)\, x(t) = \text{weak-}\lim_{n\to\infty} y_{n'}(t)$ is strongly continuous in t.

Therefore, by letting $n' \to \infty$ in the relation

$$x_{n'}(t) = x_{n'}(a) + \int_a^t A_{n'}(s)\, x_{n'}(s)\, ds = A(a)^{-1} y + \int_a^t A_{n'}(s)\, x_{n'}(s)\, ds,$$

we obtain

$$x(t) = A(a)^{-1} y + \int_a^t A(s)\, x(s)\, ds. \tag{21}$$

Hence, by the strong continuity in t of $u(t) = A(t)\, x(t)$, we see that (1) is satisfied. Since the solution $x(t)$ is uniquely determined by the initial value $x(a)$ in virtue of the Lemma, we see that the original sequence $\{x_n(t)\}$ itself must converge weakly to $x(t)$.

Remark. The above result is adapted from K. YOSIDA [28]. Our condition (8) implies that the domain $D(A(t))$ of $A(t)$ is independent of t as T. KATO did remark to the present author. For, set $A(t)\,(dA(t)^{-1}/dt) = C(t)$ and let $W(t)$ be the solution of the equation $dW(t)/dt = -C(t)\, W(t)$, $W(a) = I$. Then $d(A(t)^{-1} W(t))/dt = A(t)^{-1} C(t)\, W(t) - A(t)^{-1} C(t)$ $W(t) = 0$, and so

$$A(a)^{-1} W(a) = A(a)^{-1} = A(t)^{-1} W(t).$$

This proves that $D(A(t)) =$ the range $R(A(t)^{-1})$ is independent of t.

Theorem 2. Let X be a Hilbert space, and let $A(t)$ satisfy, beside (4) and (5), the further conditions:

$$A(t) \text{ is self-adjoint and } (A(t)\, x, x) \leqq -\|x\|^2 \text{ for } x \in D(A(t)), \tag{22}$$

$$\text{there exists a constant } \alpha \text{ with } 2^{-1} \leqq \alpha \leqq 1 \text{ such that } (-A(t))^{\alpha} \cdot B(t) \text{ is bounded in } t \text{ in the operator norm.} \tag{23}$$

Here the fractional power $(-A(t))^{\alpha}$ is defined, by virtue of the spectral resolution $A(t) = \int_{-\infty}^{-1} \lambda\, dE_t(\lambda)$, through

$$(-A(t))^{\alpha} = \int_{-\infty}^{-1} (-\lambda)^{\alpha}\, dE_t(\lambda). \tag{24}$$

Then the sequence $\{x_n(t)\}$ converges weakly to the uniquely determined strongly continuous solution of the integral equation (21).

Proof. The proof of Theorem 1 shows that we need only prove that $\{y_n(t)\}$ is uniformly bounded in t and n.

We have

$$\begin{aligned} \frac{d}{dt}\|y_n(t)\|^2 &= 2\, Re\left(\frac{dy_n(t)}{dt}, y_n(t)\right) \\ &= 2\, Re\,(A_n(t)\, y_n(t), y_n(t)) - 2\, Re\,(A_n(t)\, B(t)\, y_n(t), y_n(t)). \end{aligned} \tag{25}$$

Since $-A_n(t) = -A(t)(I - n^{-1}A(t))^{-1}$ is self-adjoint and non-negative with $-A(t)$, we have, by Schwarz' inequality,

$$\begin{aligned} |Re(A_n(t)B(t)x, x)| &\leqq |((-A_n(t))^{\alpha} B(t)x, (-A_n(t))^{1-\alpha}x)| \\ &\leqq \|(-A_n(t))^{\alpha} B(t)x\| \cdot \|(-A_n(t))^{1-\alpha}x\| \\ &\leqq 2^{-1}\varepsilon \|(-A_n(t))^{\alpha} B(t)x\|^2 + 2^{-1}\varepsilon^{-1}\|(-A_n(t))^{1-\alpha}x\|^2 \end{aligned}$$

for any $\varepsilon > 0$. Since $A_n(t) = A(t)J_n(t) = \int_{-\infty}^{-1} \lambda(1 - n^{-1}\lambda)^{-1} dE_t(\lambda)$, the spectrum of $-A_n(t)$ lies in the interval $[(1 + n^{-1})^{-1}, \infty)$. Hence, by $2^{-1} \leqq \alpha \leqq 1$, we see that

$$2^{-1}\varepsilon^{-1}\|(-A_n(t))^{1-\alpha}x\|^2 \leqq \|(-A_n(t))^{1/2}x\|^2 = (-A_n(t)x, x)$$

for all t and n, by taking $\varepsilon > 0$ appropriately.

Therefore we see that the right hand side of (25) is, by (22) and (23), smaller than

$$\begin{aligned} 2^{-1}\varepsilon \|(-A_n(t))^{\alpha} B(t) y_n(t)\|^2 &= 2^{-1}\varepsilon \|J_n(t)^{\alpha}(-A(t))^{\alpha} B(t) y_n(t)\|^2 \\ &\leqq K\|y_n(t)\|^2, \text{ where } K \text{ is independent of } t \text{ and } n. \end{aligned}$$

Hence $\|y_n(t)\| \leqq \|y_n(a)\| \cdot \exp((t-a)K^{1/2})$ and so $y_n(t)$ is bounded in t and n.

Remark. Let $X = L^2(0, 1)$ and $A(t)$ be the multiplication operator in $L^2(0, 1)$:

$$A(t) \cdot x(s) = -(1 + |t - s|^{-\beta})\, x(s).$$

If $\beta \geqq 2$, then, by taking $\alpha \geqq 1/2$ such that $\beta(1-\alpha) - 1 \geqq 0$, we see that the conditions of Theorem 2 are satisfied. A similar case was recently treated by H. TANABE [1]. It seems that his method cannot be applied to the case $\beta = 1$. Here we have $B(t) = (1 + |t-s|)^{-2}$ or $= -(1 + |t-s|)^{-2}$ according as $t > s$ or $t < s$. Thus the right side of (25) in this case is $\leqq 0$ and so $\|y_n(t)\| \leqq \|y_n(a)\|$. Hence Theorem 2 applies to this case also.

References. The idea of approximating equation (1) by equations (6) was first devised by the present author. See K. YOSIDA [23]. In this paper, the B-space X need not be reflexive. Actually, equation (1) is treated in the space L^1 under the assumption that $A(t)$ is a second order elliptic differential operator with very smooth coefficients. It is proved that the approximate solution $x_n(t) \in L^1$ will tend, as $n \uparrow \infty$, to a distributional solution of the differential equation (1). That $x(t)$ itself is a genuine solution of the differential equation (1) is proved by using a suitable *parametrix* of the *adjoint equation* to (1):
$-dy(s)/ds = A^*(s)\,y(s)$, where $A^*(s)$ is the formal dual operator of $A(s)$. See e.g., K. YOSIDA [24] and [25]. Cf. also S. ITÔ [1]. According to J. KISYŃSKI [1], published during the proof reading, the assumption of the reflexivity of the space X in Theorem 1 can be omitted.

5. The Method of Tanabe and Sobolevski

Let X be a complex B-space, and consider the equation of evolution in X with a given *inhomogeneous term* $f(t)$:

$$dx(t)/dt = Ax(t) + f(t), \quad a \leqq t \leqq b. \tag{1}$$

Then the solution $x(t) \in X$ with the initial condition $x(a) = x_0 \in X$ is given formally by the so-called *Duhamel Principle* from the solution $\exp((t-a)A)x$ of the *homogeneous equation* $dx/dt = Ax$:

$$x(t) = \exp((t-a)A)x_0 + \int_a^t \exp((t-s)A) \cdot f(s)\,ds. \tag{2}$$

This suggests that a temporally inhomogeneous equation in X:

$$dx(t)/dt = A(t)x(t), \quad a \leqq t \leqq b, \tag{3}$$

may be solved formally as follows. We rewrite equation (3) in the form

$$dx(t)/dt = A(a)x(t) + (A(t) - A(a))x(t). \tag{4}$$

By virtue of the formalism (2), the solution $x(t)$ of (4) with the initial condition $x(a) = x_0$ will be given as the solution of an abstract integral equation

$$\begin{aligned} x(t) &= \exp((t-a)A(a))x_0 \\ &\quad + \int_a^t \exp((t-s)A(s))(A(s) - A(a))x(s)\,ds. \end{aligned} \tag{5}$$

Solving (5) formally by successive approximation, we obtain approximate solutions:

$$\begin{aligned} x_1(t) &= \exp((t-a)A(a))x_0, \\ &\dots \\ x_{n+1}(t) &= \exp((t-a)A(a))x_0 \\ &\quad + \int_a^t \exp((t-s)A(s))(A(s) - A(a))x_n(s)\,ds. \end{aligned}$$

Hence the solution $x(t)$ of (5) would be given formally by

$$x(t) = \exp((t-a)A(a))x_0 + \int_a^t \exp((t-s)A(s))R(s,a)x_0\,ds, \tag{6}$$

where

$$\begin{aligned} R(t,s) &= \sum_{m=1}^{\infty} R_m(t,s), \\ R_1(t,s) &= (A(t) - A(s))\exp((t-s)A(s)), \quad s < t, \\ &= 0, \quad s \geqq t, \\ R_m(t,s) &= \int_s^t R_1(t,\sigma)R_{m-1}(\sigma,s)\,d\sigma \quad (m = 2, 3, \dots). \end{aligned} \tag{7}$$

Justification for the above formal procedure of integration has been given by H. TANABE [2] using the theory of holomorphic semi-groups as given in Chapter IX, 10. We shall follow Tanabe's approach, and assume the following conditions:

For each $t \in [a, b]$, $A(t)$ is a closed linear operator with domain dense in X and range X such that the resolvent set $\varrho(A(t))$ of $A(t)$ contains a fixed angular domain Θ of the complex λ-plane consisting of the origin 0 plus the set $\{\lambda; -\theta < \arg \lambda < \theta$ with $\theta > \pi/2\}$. The resolvent $(\lambda I - A(t))^{-1}$ is strongly continuous in t uniformly in λ on any compact set $\subseteqq \Theta$. (8)

There exists positive constants M and N such that, for $\lambda \in \Theta$ and $t \in [a, b]$, we have $\|(\lambda I - A(t))^{-1}\| \leqq N (|\lambda| - M)^{-1}$ whenever $|\lambda| > M$ with $N = 1$ for real λ. (9)

The domain $D(A(t))$ of $A(t)$ is independent of t so that, by the closed graph theorem in Chapter II, 6, the operator $A(t) A(s)^{-1}$ is in $L(X, X)$. It is assumed that there exists a positive constant K such that $\|A(t) A(s)^{-1} - A(r) A(s)^{-1}\| \leqq K |t - r|$ for s, t and $r \in [a, b]$. (10)

Under these conditions, we can prove

Theorem. For any $x_0 \in X$ and s with $a \leqq s < b$, the equation

$$dx(t)/dt = A(t)\, x(t), \quad x(s) = x_0, \quad s < t \leqq b \tag{3'}$$

admits a uniquely determined solution $x(t) \in X$. This solution is given by

$$x(t) = U(t, s)\, x(s) = U(t, s)\, x_0, \quad \text{where} \tag{11}$$

$$\left.\begin{aligned} &U(t, s) = \exp((t - s) A(s)) + W(t, s), \\ &W(t, s) = \int_s^t \exp((t - \sigma) A(\sigma))\, R(\sigma, s)\, d\sigma \quad \text{with} \quad R(t, s) \quad \text{given} \\ &\text{by } (7). \end{aligned}\right\} \tag{12}$$

For the proof, we prepare three lemmas.

Lemma 1. $R(t, s)$ satisfies, with a constant C,

$$\|R(t, s)\| \leqq KC \cdot \exp(KC(t - s)), \tag{13}$$

and $R(t, s)$ is strongly continuous in $a \leqq s < t \leqq b$.

Proof. By (8) and (9), we see that each $A(s)$ generates a holomorphic semi-group which is given by (see Chapter IX, 10)

$$\left.\begin{aligned} &\exp(tA(s)) = (2\pi i)^{-1} \int_{C'} e^{\lambda t} (\lambda I - A(s))^{-1}\, d\lambda, \quad \text{where } C' \text{ is} \\ &\text{a smooth contour running from } \infty\, e^{-i\theta} \text{ to } \infty\, e^{i\theta} \text{ in } \Theta. \end{aligned}\right\} \tag{14}$$

Hence, by $A(s)(\lambda I - A(s)) = \lambda(\lambda I - A(s))^{-1} - I$, we have, for $(b-a) > t > 0$,

$$\|\exp(tA(s))\| \leqq C \quad\text{and}\quad \|A(s)\exp(tA(s))\| \leqq Ct^{-1},$$
$$\text{where the positive constant } C \text{ is independent of } t > 0 \text{ and } s \in [a, b]. \tag{15}$$

We have, by (7),

$$R_1(t, s) = (A(t) - A(s))\,A(s)^{-1}A(s)\exp((t-s)A(s)),\ t > s,$$

and so, by (10) and (15),

$$\|R_1(t, s)\| \leqq KC. \tag{16}$$

It is also clear, from (8) and (14), that $R_1(t, s)$ is strongly continuous in $a \leqq s < t \leqq b$. Next, by induction, we obtain

$$\begin{aligned}\|R_m(t, s)\| &\leqq \int_s^t \|R_1(t, \sigma)\| \cdot \|R_{m-1}(\sigma, s)\|\,d\sigma \\ &\leqq \int_s^t (KC)^m (\sigma - s)^{m-2}(\underline{|m-2})^{-1} d\sigma \\ &= (KC)^m (t-s)^{m-1} (\underline{|m-1})^{-1},\end{aligned}$$

and hence (13) is obtained. In the same way, we see that $R(t, s)$ is strongly continuous in $a \leqq s < t \leqq b$.

Lemma 2. For $s < \tau < t$, we have

$$\|R(t, s) - R(\tau, s)\| \leqq C_1\left(\frac{t-\tau}{t-s} + (t-\tau)\log\frac{t-s}{t-\tau}\right), \tag{17}$$

where C_1 is a positive constant which is independent of s, τ and t.

Proof. We have, by (7),

$$\begin{aligned}R_1(t, s) - R_1(\tau, s) &= (A(t) - A(\tau))\exp((t-s)A(s)) \\ &\quad + (A(\tau) - A(s))\,[\exp((t-s)A(s)) - \exp((\tau - s)A(s))].\end{aligned}$$

By (10) and (15), the norm of the first term on the right is dominated by $KC(t-\tau)(t-s)^{-1}$. The second term on the right is

$$= (A(\tau) - A(s))\int_{\tau-s}^{t-s} \frac{d}{d\sigma}\exp(\sigma A(s)) = (A(\tau) - A(s))\,A(s)^{-1}$$
$$\times \int_{\tau-s}^{t-s} A(s)^2 \exp(\sigma A(s))\,d\sigma,$$

and we have, by (15),

$$\left\| \int_{\tau-s}^{t-s} A(s)^2 \exp(\sigma A(s))\, d\sigma \right\| \leqq \int_{\tau-s}^{t-s} \|(A(s) \exp(2^{-1}\sigma A(s)))^2\|\, d\sigma$$

$$\leqq \int_{\tau-s}^{t-s} (2C/\sigma)^2\, d\sigma = 4C^2 \left[\frac{-1}{\sigma}\right]_{\tau-s}^{t-s} = 4C^2 (t-\tau)(t-s)^{-1}(\tau-s)^{-1}.$$

Therefore

$$\|R_1(t, s) - R_1(\tau, s)\| \leqq KC(1 + 4C)\frac{t-\tau}{t-s}. \tag{18}$$

On the other hand, by (7),

$$\sum_{m=2}^{\infty} R_m(t, s) - \sum_{m=2}^{\infty} R_m(\tau, s) = \int_s^t R_1(t, \sigma)\, R(\sigma, s)\, d\sigma$$

$$- \int_s^{\tau} R_1(\tau, \sigma)\, R(\sigma, s)\, d\sigma = \int_{\tau}^{t} R_1(t, \sigma)\, R(\sigma, s)\, d\sigma$$

$$+ \int_s^{\tau} (R_1(t, \sigma) - R_1(\tau, \sigma))\, R(\sigma, s)\, d\sigma.$$

The norm of the first term on the right is dominated by

$$\int_{\tau}^{t} \|R_1(t, \sigma)\|\, \|R(\sigma, s)\|\, d\sigma \leqq K^2 C^2 \exp(KC(b-a))(t-\tau).$$

The norm of the second term on the right is, by (13) and (18), dominated by

$$\int_s^{\tau} \|R_1(t, \sigma) - R_1(\tau, \sigma)\|\, \|R(\sigma, s)\|\, d\sigma$$

$$\leqq K^2 C^2 (1 + 4C) \exp(KC(b-a)) \int_s^{\tau} (t-\tau)(t-\sigma)^{-1}\, d\sigma$$

$$= K_1 (t-\tau) \log\frac{t-s}{t-\tau}.$$

Therefore we obtain (17).

Lemma 3. For $s < t$, we have

$$\|A(t)\{\exp((t-s)A(t)) - \exp((t-s)A(s))\}\| \leqq C_2, \text{ where } C_2 \text{ is a positive constant independent of } s \text{ and } t. \tag{19}$$

Proof. We obtain, from (14),

$$A(t)\{\exp((t-s)A(t)) - \exp((t-s)A(s))\}$$

$$= (2\pi i)^{-1} \int_C e^{\lambda(t-s)} A(t)(\lambda I - A(t))^{-1}(A(t) - A(s))(\lambda I - A(s))^{-1} d\lambda.$$

On the other hand, we have $A(t)(\lambda I - A(t))^{-1} = \lambda(\lambda I - A(t))^{-1} - I$, and so, by (9),

$$\|A(t)(\lambda I - A(t))^{-1}\| \leqq \frac{|\lambda|}{|\lambda| - M} + 1 \text{ for } \lambda \in \Theta \text{ and } t \in [a, b]. \tag{20}$$

Hence we obtain (19) by (10) and

$$\begin{aligned} &\|(A(t) - A(s))(\lambda I - A(s))^{-1}\| \\ &\quad \leqq \|(A(t) - A(s)) A(s)^{-1}\| \, \|A(s)(\lambda I - A(s))^{-1}\|. \end{aligned}$$

Proof of the Theorem. We rewrite $W(t, s)$ given in (12) as follows:

$$\begin{aligned} W(t,s) &= \int_s^t \exp((t-\tau) A(t)) R(t, s)\, d\tau \\ &+ \int_s^t \{\exp((t-\tau) A(\tau)) - \exp((t-\tau) A(t))\} R(\tau, s)\, d\tau \\ &+ \int_s^t \exp((t-\tau) A(t)) (R(\tau, s) - R(t, s))\, d\tau. \end{aligned}$$

By approximating the integrals by Riemann sums and making use of the closure property of the operator $A(t)$, we see that we can apply $A(t)$ to each term of the right side of the above equality. For, by (19), we can apply $A(t)$ to the second term of the right side; and also to the third term of the right side by (15) and (17); we also have, by $A(t)\exp((t-\tau)A(t)) = -d\exp((t-\tau)A(t))/d\tau$,

$$A(t) \int_s^t \exp((t-\tau) A(t)) R(t, s)\, d\tau = \{\exp((t-s) A(t)) - I\} R(t, s).$$

Hence we obtain

$$\begin{aligned} A(t)\, U(t, s) &= A(t) \exp((t-s) A(s)) + \{\exp((t-s) A(t)) - I\} R(t, s) \\ &+ \int_s^t A(t) \{\exp((t-\tau) A(\tau)) - \exp((t-\tau) A(t))\} R(\tau, s)\, d\tau \\ &+ \int_s^t A(t) \exp((t-\tau) A(t)) (R(\tau, s) - R(t, s))\, d\tau. \end{aligned} \tag{21}$$

The above proof shows that $A(t)\, U(t, s)$ is strongly continuous in $a \leqq s < t \leqq b$ and

$$\|A(t)\, W(t, s)\| \leqq C_3 \text{ and } \|A(t)\, U(t, s)\| \leqq C_3 (t-s)^{-1}, \tag{22}$$

where C_3 is a positive constant independent of s and t.

We next define, for $s < (t-h) < t$,

$$U_h(t, s) = \exp((t-s) A(s)) + \int_s^{t-h} \exp((t-\tau) A(\tau)) R(\tau, s)\, d\tau. \tag{23}$$

Since a holomorphic semi-group $\exp(tA(u))$ is differentiable in $t > 0$, we have

$$\frac{\partial}{\partial t} U_h(t, s) = A(s) \exp((t-s) A(s)) + \exp(hA(t-h)) R(t-h, s) + \int_s^{t-h} A(\tau) \exp((t-\tau) A(\tau)) R(\tau, s) \, d\tau.$$

Hence we have, by (7),

$$\frac{\partial}{\partial t} U_h(t, s) - A(t) U_h(t,s) = \exp(hA(t-h)) R(t-h, s) - R_1(t, s) - \int_s^{t-h} R_1(t, \tau) R(\tau, s) \, d\tau. \tag{24}$$

By (8), (13) and (14), $\exp(hA(t-h)) R(t-h, s)$ tends strongly to $R(t, s)$ as $h \downarrow 0$. Thus we have

$$\underset{h\downarrow 0}{s\text{-}\lim} \left(\frac{\partial}{\partial t} U_h(t, s) - A(t) U_h(t, s)\right) x_0 = \left(R(t, s) - R_1(t, s) - \int_s^t R_1(t, \sigma) R(\sigma, s) \, d\sigma\right) x_0, \quad x_0 \in X. \tag{25}$$

The right side must be 0 as may easily be proved by (7). Since we have

$$\underset{h\downarrow 0}{s\text{-}\lim}\, A(t) U_h(t, s) x_0 = A(t) U(t, s) x_0$$

by the reasoning used in proving (21), we obtain from (25)

$$\underset{h\downarrow 0}{s\text{-}\lim} \frac{\partial}{\partial t} U_h(t, s) x_0 = A(t) U(t, s) x_0 \text{ for } t > s \text{ and } x_0 \in X. \tag{26}$$

The right side of (26) being strongly continuous in $t > s$, we see, by integrating (26) and remembering $\underset{h\downarrow 0}{s\text{-}\lim}\, U_h(t, s) x_0 = U(t, s) x_0$, that

$$\frac{\partial}{\partial t} U(t,s) x_0 = A(t) U(t, s) x_0 \text{ for } t > s \text{ and } x_0 \in X. \tag{27}$$

Therefore, $x(t) = U(t, s) x_0$ is the desired solution of (3'). The uniqueness of the solution may be proved as in the preceding section.

Comments and References

The above Theorem and proof are adapted from H. Tanabe [2]. To illustrate his idea we have made Tanabe's conditions somewhat stronger. Thus, e.g., the condition (9) may be replaced by a weaker one:

$$\|A(t) A(s)^{-1} - A(r) A(s)^{-1}\| \leq K_1 |t - r|^\varrho \text{ with } 0 < \varrho < 1.$$

For the details, see the above cited paper by H. TANABE which is a refinement of H. TANABE [3] and [4]. It is to be noted that the Russian school independently has developed a similar method. See, e.g., P. E. SOBOLEVSKI [1] and the reference cited in this paper.

Komatsu's work. H. KOMATSU [1] made an important remark regarding Tanabe's result given above. Let Δ be a convex complex neighbourhood of the real segment $[a, b]$ considered as embedded in the complex plane. Suppose that $A(t)$ is defined for $t \in \Delta$ and satisfies (8) and (9) in which the phrase "$t \in [a, b]$" is replaced by "$t \in \Delta$". Assume, moreover, that there exists a bounded linear operator A_0 which maps X onto D, the domain of $A(t)$ which is assumed to be independent of $t \in \Delta$, in a one-to-one manner and such that $B(t) = A(t) A_0$ is strongly holomorphic in $t \in \Delta$. Under these assumptions, KOMATSU proved that the operator $U(t, s)$ constructed as above is strongly holomorphic in $t \in \Delta$ if

$$|\arg(t-s)| < \theta_0 \text{ with a certain } \theta_0 \text{ satisfying } 0 < \theta_0 < \pi/2.$$

The result may be applied to the "forward and backward unique continuation" of solutions of temporally inhomogeneous diffusion equations as in Chapter XIV, 1. In this connection, we refer to H. KOMATSU [2], [3] and T. KOTAKÉ-M. NARASIMHAN [1].

Kato's work. To get rid of the assumption that the domain $D(A(t))$ is independent of t, T. KATO [6] proved that, in the above Theorem, we may replace condition (10) by the following:

$$\left.\begin{array}{l}\text{For a certain positive integer } k, \text{ the domain } D((-A(t))^{1/k}) \\ \text{is independent of } t. \text{ (Here } (-A(t))^{1/k} \text{ is the fractional} \\ \text{power as defined in Chapter IX, 11.) Further, there} \\ \text{exist constants } K_2 > 0 \text{ and } \gamma \text{ with } 1 - k^{-1} < \gamma \leqq 1 \\ \text{such that, for } s, t \in [a, b], \ \|(-A(t))^{1/k} (-A(s))^{1/k} - I\| \\ \leqq K_2 |t-s|^{\gamma}.\end{array}\right\} \quad (10)'$$

Tanabe's and the Kato-Tanabe recent work. With the same motivation as KATO, H. TANABE [1] devised a method to replace condition (10) by

$$\left.\begin{array}{l}A(t)^{-1} \text{ is once strongly differentiable in } a \leqq t \leqq b \\ \text{and such that, for positive constants } K_3 \text{ and } \alpha, \\ \left\|\dfrac{dA(t)^{-1}}{dt} - \dfrac{dA(s)^{-1}}{ds}\right\| \leqq K_3 |t-s|^{\alpha}. \\ \text{Further, there exist positive constants } N \text{ and } \varrho \text{ with} \\ 0 < \varrho \leqq 1 \text{ such that} \\ \left\|\dfrac{\partial}{\partial t}(\lambda I - A(t))^{-1}\right\| \leqq N |\lambda|^{\varrho-1}, \ \varrho < 1.\end{array}\right\} \quad (10)''$$

For details, see KATO-TANABE [8]. Their point is to start with the first approximation $\exp((t-a) A(t)) x_0$ instead of $\exp((t-a) A(a)) x_0$.

Nelson's work on Feynman integrals. The semi-group method of integration of Schrödinger equations gives an interpretation of Feynman integrals. See E. NELSON [2].

The Agmon-Nirenberg work. Agmon-Nirenberg [1] discussed the behaviour as $t \uparrow \infty$ of solutions of the equation $\frac{1}{i}\frac{du}{dt} - Au = 0$ in some B-spaces.

Bibliography

ACHIESER, N. I.
[1] (with I. M. GLAZMAN) Theorie der linearen Operatoren im Hilbert-Raum, Akademie-Verlag 1954.

AGMON, S.
[1] (with L. NIRENBERG) Properties of solutions of ordinary differential equations in Banach space. Comm. P. and Appl. Math. **16**, 121—239 (1963).

AKILOV, G.
[1] See KANTOROVITCH-AKILOV [1].

ALEXANDROV, P.
[1] (with H. HOPF) Topologie, Vol. I, Springer 1935.

ARONSZAJN, N.
[1] Theory of reproducing kernels. Trans. Amer. Math. Soc. **68**, 337—404 (1950).

BALAKRISHNAN, V.
[1] Fractional powers of closed operators and the semi-groups generated by them. Pacific J. Math. **10**, 419—437 (1960).

BANACH, S.
[1] Théorie des Opérations Linéaires, Warszawa 1932.
[2] Sur la convergence presque partout de fonctionnelles linéaires. Bull. Sci. Math. France **50**, 27—32 and 36—43 (1926).

BEBOUTOV, M.
[1] Markoff chains with a compact state space. Rec. Math. **10**, 213—238 (1942).

BERGMAN, S.
[1] The Kernel Function and the Conformal Mapping. Mathematical Surveys, No. 5 (1950).

BERS, L.
[1] Lectures on Elliptic Equations. Summer Seminar in Appl. Math. Univ. of Colorado, 1957.

BIRKHOFF, G.
[1] Lattice Theory. Colloq. Publ. Amer. Math. Soc., 1940.

BIRKHOFF, G. D.
[1] Proof of the ergodic theorem. Proc. Nat. Acad. Sci. USA **17**, 656—660 (1931).
[2] (with P. A. SMITH) Structure analysis of surface transformations. J. Math. Pures et Appliq. **7**, 345—379 (1928).

BOCHNER, S.
[1] Integration von Funktionen, deren Wert die Elemente eines Vektorraumes sind. Fund. Math. **20**, 262—276 (1933).
[2] Diffusion equations and stochastic processes. Proc. Nat. Acad. Sci. USA **35**, 369—370 (1949).
[3] Vorlesungen über Fouriersche Integrale, Akademie-Verlag 1932.

[4] Beiträge zur Theorie der fastperiodischen Funktionen. Math. Ann. **96**, 119—147 (1927).

BOGOLIOUBOV, N.
[1] See KRYLOV-BOGOLIOUBOV [1].

BOHR, H.
[1] Fastperiodische Funktionen, Springer 1932.

BOURBAKI, N.
[1] Topologie Générale. Act. Sci. et Ind., nos. 856, 916, 1029, 1045, 1084, Hermann 1940—42.
[2] Espaces Vectoriels Topologiques. Act. Sci. et Ind., nos. 1189, 1229, Hermann 1953—55.

BROWDER, F. E.
[1] Functional analysis and partial differential equations, I—II. Math. Ann. **138**, 55—79 (1959) and **145**, 81—226 (1962).

CHACON, R. V.
[1] (with D. S. ORNSTEIN) A general ergodic theorem. Ill. J. Math. **4**, 153—160 (1960).

CHOQUET, G.
[1] La théorie des représentations intégrales dans les ensembles convexes compacts. Ann. Inst. Fourier **10**, 334—344 (1960).
[2] (with P.-A. MEYER) Existence et unicité des représentations intégrales dans les convexes compacts quelconques. Ann. Inst. Fourier **13**, 139—154 (1963).

CLARKSON, J. A.
[1] Uniformly convex spaces. Trans. Amer. Math. Soc. **40**, 396—414 (1936).

DIEUDONNÉ, J.
[1] Recent advances in the theory of locally convex vector spaces. Bull. Amer. math. Soc. **59**, 495—512 (1953).

DOOB, J. L.
[1] Stochastic processes with an integral-valued parameter. Trans. Amer. Math. Soc. **44**, 87—150 (1938).
[2] Probability theory and the first boundary value problem. Ill. J. Math. **2**, 19—36 (1958).
[3] Probability methods applied to the first boundary value problem. Proc. Third Berkeley Symp. on Math. Statist. and Prob. **II**, 49—80 (1956).

DUNFORD, N.
[1] (with J. SCHWARTZ) Linear Operators, Vol. I, Interscience 1958.
[2] Uniformity in linear spaces. Trans. Amer. Math. Soc. **44**, 305—356 (1938).
[3] On one-parameter groups of linear transformations. Ann. of Math. **39**, 569—573 (1938).
[4] (with J. SCHWARTZ) Convergence almost everywhere of operator averages. J. Rat. Mech. Anal. **5**, 129—178 (1956).
[5] (with J. SCHWARTZ) Linear Operators, Vol. II, Interscience 1963.
[6] (with J. SCHWARTZ) Linear Operators, Vol. III, Interscience, to appear.

DYNKIN, E. B.
[1] Markoff processes and semi-groups of operators. Teorya Veroyatn. **1** (1956).
[2] Infinitesimal operators of Markoff processes. Teorya Veroyatn. **1** (1956).

EBERLEIN, W. F.
[1] Weak compactness in Banach spaces. Proc. Nat. Acad. Sci. USA **33**, 51—53 (1947).

EHRENPREIS, L.
[1] Solutions of some problems of division. Amer. J. Math. **76**, 883—903 (1954).
ERDÉLYI, A.
[1] Operational Calculus and Generalized Functions, Reinhart 1961.
FELLER, W.
[1] On the generation of unbounded semi-groups of bounded linear operators. Ann. of Math. **58**, 166—174 (1953).
[2] The parabolic differential equation and the associated semi-group of transformations. Ann. of Math. **55**, 468—519 (1952).
[3] On the intrinsic form for second order differential operators. Ill. J. Math. **2**, No. 1, 1—18 (1958).
[4] Generalized second order differential operators and their lateral conditions. Ill. J. Math. **1**, No. 4, 459—504 (1957).
[5] Some new connections between probability and classical analysis. Proc. Internat. Congress of Math. 1958, held at Edinburgh, pp. 69—86.
[6] On differential operators and boundary conditions. Comm. Pure and Appl. Math. **8**, 203—216 (1955).
[7] Boundaries induced by non-negative matrices. Trans. Amer. Math. Soc. **83**, 19—54 (1956).
[8] On boundaries and lateral conditions for the Kolmogoroff differential equations. Ann. of Math. **65**, 527—570 (1957).
FOIAS, C.
[1] Remarques sur les semi-groupes distributions d'opérateurs normaux. Portugaliae Math. **19**, 227—242 (1960).
FREUDENTHAL, H.
[1] Über die Friedrichssche Fortsetzung halbbeschränkter Hermitescher Operatoren. Proc. Acad. Amsterdam **39**, 832—833 (1936).
[2] Teilweise geordnete Modulen. Proc. Acad. Amsterdam **39**, 641—651 (1936).
FRIEDMAN, A.
[1] Generalized Functions and Partial Differential Equations, Prentice-Hall 1963.
FRIEDRICHS, K.
[1] Differentiability of solutions of elliptic partial differential equations. Comm. Pure and Appl. Math. **5**, 299—326 (1953).
[2] Symmetric positive systems of differential equations. Comm. Pure and Appl. Math. **11**, 333—418 (1958).
[3] Spektraltheorie halbbeschränkter Operatoren, I—III. Math. Ann. **109**, 465—487, 685—713 (1934) and **110**, 777—779 (1935).
FUKAMIYA, M.
[1] Topological methods for Tauberian theorem. Tohoku Math. J. 77—87 (1949).
[2] See YOSIDA-FUKAMIYA [16].
GÅRDING, L.
[1] Dirichlet's problem for linear elliptic partial differential equations. Math. Scand. **1**, 55—72 (1953).
[2] Some trends and problems in linear partial differential equations. Internat. Congress of Math. 1958, held at Edinburgh, pp. 87—102.
GELFAND, I. M.
[1] (with I. M. ŠILOV) Generalized Functions, Vol. I—III, Moscow 1958.
[2] Normierte Ringe. Rec. Math. **9**, 3—24 (1941).

[3] (with N. Y. VILENKIN) Some Applications of Harmonic Analysis (Vol. IV of Generalized Functions), Moscow 1961.
[4] (with D. RAIKOV) On the theory of characters of commutative topological groups. Doklady Akad. Nauk SSSR **28**, 195—198 (1940).
[5] (with D. A. RAIKOV and G. E. ŠILOV) Commutative Normed Rings, Moscow 1960.

GLAZMAN, I. M.
[1] See ACHIESER-GLAZMAN [1].

GROTHENDIECK, A.
[1] Espaces Vectoriels Topologiques, seconde éd., Sociedade de Mat. de São Paulo 1958.
[2] Produits Tensoriels Topologiques et Espaces Nucléaires. Memoirs of Amer. Math. Soc. No. 16 (1955).

HAHN, H.
[1] Über Folgen linearer Operatoren. Monatsh. für Math. und Phys. **32**, 3—88 (1922).
[2] Über lineare Gleichungssysteme in linearen Räumen. J. reine und angew. Math. **157**, 214—229 (1927).
[3] Über die Integrale des Herrn Hellinger und die Orthogonalinvarianten der quadratischen Formen von unendlich vielen Veränderlichen. Monatsh. für Math. und Phys. **23**, 161—224 (1912).

HALMOS, P. R.
[1] Measure Theory, van Nostrand 1950.
[2] Introduction to Hilbert Space and the Theory of Spectral Multiplicity, Chelsea 1951.

HAUSDORFF, F.
[1] Mengenlehre, W. de Gruyter 1935.

HELLINGER, E.
[1] Neue Begründung der Theorie quadratischer Formen von unendlichvielen Veränderlichen. J. reine und angew. Math. **136**, 210—271 (1909).

HELLY, E.
[1] Über Systeme linearer Gleichungen mit unendlich vielen Unbekannten. Monatsh. für Math. und Phys. **31**, 60—91 (1921).

HILBERT, D.
[1] Wesen und Ziele einer Analysis der unendlich vielen unabhängigen Variablen. Rend. Circ. Mat. Palermo **27**, 59—74 (1909).

HILLE, E.
[1] (with R. S. PHILLIPS) Functional Analysis and Semi-groups. Colloq. Publ. Amer. Math. Soc., 1957. It is the second edition of the book below.
[2] Functional Analysis and Semi-groups. Colloq. Publ. Amer. Math. Soc., 1948.
[3] On the differentiability of semi-groups of operators. Acta Sci. Math. Szeged **12B**, 19—24 (1950).
[4] On the generation of semi-groups and the theory of conjugate functions. Proc. R. Physiogr. Soc. Lund **21**, 1—13 (1951).
[5] Une généralization du problème de Cauchy. Ann. Inst. Fourier **4**, 31—48 (1952).
[6] The abstract Cauchy problem and Cauchy's problem for parabolic differential equations. J. d'Analyse Math. **3**, 81—196 (1954).

[7] Perturbation methods in the study of Kolmogoroff's equations. Proc. Internat. Congress of Math. 1954, held at Amsterdam, Vol. III, pp. 365—376.
[8] Linear differential equations in Banach algebras. Proc. Internat. Symposium on Linear Analysis, held at Jerusalem, 1960, pp. 263—273.
[9] Les probabilités continues en chaine. C. R. Acad. Sci. **230**, 34—35 (1950).

HOFFMAN, K.
[1] Banach Spaces of Analytic Functions, Prentice-Hall 1962.

HOPF, E.
[1] Ergodentheorie, Springer 1937.
[2] The general temporally discrete Markoff processes. J. Rat. Mech. and Anal. **3**, 13—45 (1954).
[3] On the ergodic theorem for positive linear operators. J. reine und angew. Math. **205**, 101—106 (1961).

HOPF, H.
[1] See ALEXANDROV-HOPF [1].

HÖRMANDER, L.
[1] On the theory of general partial differential operators. Acta Math. **94**, 161—248 (1955).
[2] Lectures on Linear Partial Differential Equations, Stanford Univ. 1960.
[3] Linear partial differential equations without solutions. Math. Ann. **140**, 169—173 (1960).
[4] Local and global properties of fundamental solutions. Math. Scand. **5**, 27—39 (1957).
[5] On the interior regularity of the solutions of partial differential equations. Comm. Pure and Appl. Math. **9**, 197—218 (1958).
[6] Linear Partial Differential Operators, Springer 1963.

HUNT, G. A.
[1] Markoff processes and potentials, I—II. Ill. J. of Math. **1** and **2** (1957 and 1958).

ITÔ, K.
[1] (with H. P. MCKEAN) Diffusion Processes and Their Sample Paths, Springer, to appear.

ITÔ, S.
[1] The fundamental solutions of the parabolic differential equations in differentiable manifold. Osaka Math. J. **5**, 75—92 (1953).
[2] (with H. YAMABE) A unique continuation theorem for solutions of a parabolic differential equation. J. Math. Soc. Japan **10**, 314—321 (1958).

JACOBS, K.
[1] Neuere Methoden und Ergebnisse der Ergodentheorie, Springer 1960.

JOHN, F.
[1] Plane Waves and Spherical Means Applied to Partial Differential Equations, Interscience 1955.

KAKUTANI, S.
[1] Iteration of linear operations in complex Banach spaces. Proc. Imp. Acad. Tokyo **14**, 295—300 (1938).
[2] See YOSIDA-MIMURA-KAKUTANI [10].
[3] Weak topology and regularity of Banach spaces. Proc. Imp. Acad. Tokyo **15**, 169—173 (1939).
[4] Concrete representation of abstract (M)-spaces. Ann. of Math. **42**, 994—1024 (1941).

[5] Concrete representation of abstract (*L*)-spaces and the mean ergodic theorem. Ann. of Math. **42**, 523—537 (1941).
[6] Ergodic theorems and the Markoff processes with a stable distribution. Proc. Imp. Acad. Tokyo **16**, 49—54 (1940).
[7] See YOSIDA-KAKUTANI [7].
[8] Ergodic Theory. Proc. Internat. Congress of Math. 1950, held at Cambridge, Vol. 2, pp. 128—142.

KANTOROVITCH, L.
[1] (with G. AKILOV) Functional Analysis in Normed Spaces, Moscow 1955.

KATO, T.
[1] Remarks on pseudo-resolvents and infinitesimal generators of semi-groups. Proc. Japan Acad. **35**, 467—468 (1959).
[2] Note on fractional powers of linear operators. Proc. Japan Acad. **36**, 94—96 (1960).
[3] Integration of the equation of evolution in a Banach space. J. Math. Soc. of Japan **5**, 208—234 (1953).
[4] On linear differential equations in Banach spaces. Comm. Pure and Appl. Math. **9**, 479—486 (1956).
[5] Fractional powers of dissipative operators. J. Math. Soc. of Japan **13**, 246—274 (1961); II, ibid. **14**, 242—248 (1962).
[6] Abstract evolution equations of parabolic type in Banach and Hilbert spaces. Nagoya Math. J. **19**, 93—125 (1961).
[7] Fundamental properties of Hamiltonian operators of Schrödinger type. Trans. Amer. Math. Soc. **70**, 195—211 (1950).
[8] (with H. TANABE) On the abstract evolution equation. Osaka Math. J. **14**, 107—133 (1962).

KELLEY, J. L.
[1] General Topology, van Nostrand 1955.
[2] Note on a theorem of Krein and Milman. J. Osaka Inst. Sci. Tech., Part I, **3**, 1—2 (1951).

KELLOGG, O. D.
[1] Foundations of Potential Theory, Springer 1929.

KHINTCHINE, A.
[1] Zu Birkhoffs Lösung des Ergodenproblems. Math. Ann. **107**, 485—488 (1933).

KISYŃSKI, J.
[1] Sur les opérateurs de Green des problèmes de Cauchy abstraits. Stud. Math. **23**, 285—328 (1964).

KODAIRA, K.
[1] The eigenvalue problem for ordinary differential equations of the second order and Heisenberg's theory of *S*-matrices. Amer. J. of Math. **71**, 921—945 (1949).

KOLMOGOROV, A.
[1] Über analytische Methoden in der Wahrscheinlichkeitsrechnung. Math. Ann. **104**, 415—458 (1931).

KOMATSU, H.
[1] Abstract analyticity in time and unique continuation property of solutions of a parabolic equation. J. Fac. Sci. Univ. Tokyo, Sect. 1, **9**, Part 1, 1—11 (1961).
[2] A characterization of real analytic functions. Proc. Japan Acad. **36**, 90—93 (1960).

[3] A proof of Kotaké-Narasimhan's theorem. Proc. Japan Acad. **38**, 615—618 (1962).
[4] Semi-groups of operators in locally convex spaces, to appear.

KOTAKÉ, T.
[1] Sur l'analyticité de la solution du problème de Cauchy pour certaines classes d'opérateurs paraboliques. C. R. Acad. Sci. Paris **252**, 3716—3718 (1961).
[2] (with M. S. NARASIMHAN) Sur la régularité de certains noyaux associés à un opérateur elliptique. C. R. Acad. Sci. Paris **252**, 1549—1550 (1961).

KÖTHE, G.
[1] Topologische lineare Räume, Vol. I, Springer 1960.

KREIN, M.
[1] (with D. MILMAN) On extreme points of regularly convex sets. Stud. Math. **9**, 133—138 (1940).
[2] (with S. KREIN) On an inner characteristic of the set of all continuous functions defined on a bicompact Hausdorff space. Doklady Akad. Nauk SSSR **27**, 429—430 (1940).

KREIN, S.
[1] See KREIN-KREIN [2].

KRYLOV, N.
[1] (with N. BOGOLIOUBOV) La théorie générale de la mesure dans son application à l'étude des systèmes de la mécanique non linéaires. Ann. of Math. **38**, 65—113 (1937).

LADYZHENSKAYA, O. A.
[1] (with I. M. VISIK) Problèmes aux limites pour les équations aux dérivées partielles et certaines classes d'équations opérationnelles. Ousp. Mat. Nauk **11**, 41—97 (1956).

LAX, P. D.
[1] (with A. N. MILGRAM) Parabolic equations, in Contributions to the Theory of Partial Differential Equations, Princeton 1954.
[2] On Cauchy's problem for hyperbolic equations and the differentiability of solutions of elliptic equations. Comm. Pure and Appl. Math. **8**, 615—633 (1955).
[3] (with R. S. PHILLIPS) Local boundary conditions for dissipative system of linear partial differential operators. Comm. Pure and Appl. Math. **13**, 427—455 (1960).

LERAY, J.
[1] Hyperbolic Differential Equations, Princeton 1952.

LEWY, H.
[1] An example of a smooth linear partial differential equation without solutions. Ann. of Math. **66**, 155—158 (1957).

LIONS, J. L.
[1] Une remarque sur les applications du théorème de Hille-Yosida. J. Math. Soc. Japan **9**, 62—70 (1957).
[2] Equations Différentielles Opérationnelles, Springer 1961.
[3] Espaces d'interpolation et domaines de puissance fractionaires d'opérateurs. J. Math. Soc. Japan **14**, 233—241 (1962).
[4] Les semi-groups distributions. Portugaliae Math. **19**, 141—164 (1960).

LUMER, G.
[1] Semi-inner product spaces. Trans. Amer. Math. Soc. **100**, 29—43 (1961).
[2] (with R. S. PHILLIPS) Dissipative operators in a Banach space. Pacific J. Math. **11**, 679—698 (1961).

MAAK, W.
[1] Fastperiodische Funktionen. Springer 1950.
MALGRANGE, B.
[1] Existence et approximation des solutions des équations aux dérivées partielles et des équations de convolution. Ann. Inst. Fourier **6**, 271—355 (1955—56).
[2] Sur une classe d'opérateurs différentielles hypoelliptiques. Bull. Soc. Math. France **58**, 283—306 (1957).
MARUYAMA, G.
[1] On strong Markoff property. Mem. Kyushu Univ. **13** (1959).
MAZUR, S.
[1] Sur les anneaux linéaires. C. R. Acad. Sci. Paris **207**, 1025—1027 (1936).
[2] Über konvexe Mengen in linearen normierten Räumen. Stud. Math. **5**, 70—84 (1933).
MCKEAN, H.
[1] See ITÔ-MCKEAN [1].
MEYER, P. A.
[1] Séminaire de Théorie du Potentiel, sous la direction de M. BRELOT-G. CHOQUET-J. DENY, Faculté des Sciences de Paris, 1960—61.
[2] See CHOQUET-MEYER [2].
MIKUSIŃSKI, J.
[1] Operational Calculus, Pergamon 1959.
MILGRAM, A. N.
[1] See LAX-MILGRAM [1].
MILMAN, D.
[1] On some criteria for the regularity of spaces of the type (B). Doklady Akad. Nauk SSSR **20**, 20 (1938).
[2] See KREIN-MILMAN [1].
MIMURA, Y.
[1] See YOSIDA-MIMURA-KAKUTANI [10].
[2] Über Funktionen von Funktionaloperatoren in einem Hilbertschen Raum. Jap. J. Math. **13**, 119—128 (1936).
MINLOS, R. A.
[1] Generalized stochastic processes and the extension of measures. Trudy Moscow Math. **8**, 497—518 (1959).
MIYADERA, I.
[1] Generation of a strongly continuous semi-group of operators. Tohoku Math. J. **4**, 109—121 (1952).
MIZOHATA, S.
[1] Hypoellipticité des équations paraboliques. Bull. Soc. Math. France **85**, 15—50 (1957).
[2] Analyticité des solutions élémentaires des systèmes hyperboliques et paraboliques. Mem. Coll. Sci. Univ. Kyoto **32**, 181—212 (1959).
[3] Unicité du prolongement des solutions pour quelques opérateurs différentiels paraboliques. Mém. Coll. Sci. Univ. Kyoto, Sér. A **31**, 219—239 (1958).
[4] Le problème de Cauchy pour les équations paraboliques. J. Math. Soc. Japan **8**, 269—299 (1956).
[5] Systèmes hyperboliques. J. Math. Soc. Japan **11**, 205—233 (1959).
MORREY, C. B.
[1] (with L. NIRENBERG) On the analyticity of the solutions of linear

elliptic systems of partial differential equations. Comm. Pure and Appl. Math. **10**, 271—290 (1957).

NAGUMO, M.

[1] Einige analytische Untersuchungen in linearen metrischen Ringen. Jap. J. Math. **13**, 61—80 (1936).

[2] Re-topologization of functional spaces in order that a set of operators will be continuous. Proc. Japan Acad. **37**, 550—552 (1961).

NAGY, B. VON SZ.

[1] Spektraldarstellung linearer Transformationen des Hilbertschen Raumes, Springer 1942.

[2] See RIESZ-NAGY [3].

[3] Prolongements des transformations de l'espace de Hilbert qui sortent de cet espace. Akad. Kiado, Budapest 1955.

NAIMARK, M. A.

[1] Normed Rings, P. Noordhoff 1959.

[2] Lineare differentiale Operatoren, Akad. Verlag 1960.

[3] Über Spektralfunktionen eines symmetrischen Operators. Izvestia Akad. Nauk SSSR **17**, 285—296 (1943).

NAKANO, H.

[1] Unitärinvariante hypermaximale normale Operatoren, Ann. of Math. **42**, 657—664 (1941).

NARASIMHAN, M. S.

[1] See KOTAKÉ-NARASIMHAN [2].

NELSON, E.

[1] Analytic vectors. Ann. of Math. **670**, 572—615 (1959).

[2] Feynman integrals and the Schrödinger equations. J. of Math. Physics **5**, 332—343 (1964).

NEUMANN, J. VON

[1] Allgemeine Eigenwerttheorie Hermitescher Funktionaloperatoren. Math. Ann. **102**, 49—131 (1929).

[2] On rings of operators, III. Ann. of Math. **41**, 94—161 (1940).

[3] Zur Operatorenmethode in der klassischen Mechanik. Ann. of Math. **33**, 587—643 (1932).

[4] Almost periodic functions in a group. I. Trans. Amer. Math. Soc. **36**, 445—492 (1934).

[5] Über adjungierte Funktionaloperatoren. Ann. of Math. **33**, 249—310 (1932).

[6] Über die analytischen Eigenschaften von Gruppen linearer Transformationen und ihrer Darstellungen. Math. Z. **30**, 3—42 (1929).

[7] Zur Algebra der Funktionaloperatoren und Theorie der normalen Operatoren. Math. Ann. **102**, 370—427 (1929—30).

[8] Über einen Satz von Herrn M. H. Stone. Ann. of Math. **33**, 567—573 (1932).

NIRENBERG, L.

[1] Remarks on strongly elliptic partial differential equations. Comm. Pure and Appl. Math. **8**, 643—674 (1955).

[2] On elliptic partial differential equations. Ann. Scuola Norm. Sup. Pisa **13**, 115—162 (1959).

[3] See MORREY-NIRENBERG [1].

[4] See AGMON-NIRENBERG [1].

ORNSTEIN, D. S.

[1] See CHACON-ORNSTEIN [1].

PALEY, R. E. A. C.
[1] (with N. WIENER) Fourier Transforms in the Complex Domain, Colloq. Publ. Amer. Math. Soc., 1934.
PEETRE, J.
[1] A proof of the hypoellipticity of formally hypoelliptic differential operators. Comm. Pure and Appl. Math. **16**, 737—747 (1961).
PETER, F.
[1] (with H. WEYL) Die Vollständigkeit der primitiven Darstellungen einer geschlossenen kontinuierlichen Gruppe. Math. Ann. **97**, 737—755 (1927).
PETROWSKY, I. G.
[1] Sur l'analyticité des solutions d'équations différentielles. Rec. Math. **47** (1939).
PETTIS, B. J.
[1] On integration in vector spaces. Trans. Amer. Math. Soc. **44**, 277—304 (1938).
PHILLIPS, R. S.
[1] See HILLE-PHILLIPS [1].
[2] The adjoint semi-group. Pacific J. Math. **5**, 269—283 (1955).
[3] An inversion formula for Laplace transform and semi-groups of linear operators. Ann. of Math. **59**, 325—356 (1954).
[4] See LUMER-PHILLIPS [2].
[5] On the generation of semi-groups of linear operators. Pacific J. Math. **2**, 343—369 (1952).
[6] On the integration of the diffusion equation with boundaries. Trans. Amer. Math. Soc. **98**, 62—84 (1961).
[7] Dissipative operators and parabolic partial differential operators. Comm. Pure and Appl. Math. **12**, 249—276 (1959).
[8] Dissipative hyperbolic systems. Trans. Amer. Math. Soc. **86**, 109—173 (1957).
[9] Dissipative operators and hyperbolic systems of partial differential equations. Trans. Amer. Math. Soc. **90**, 193—254 (1959).
[10] See LAX-PHILLIPS [3].
PITT, H. R.
[1] Tauberian Theorems. Tata Inst. of Fund. Research, 1958.
PONTRJAGIN, L.
[1] Topological Groups, Princeton 1939.
RADO, T.
[1] Subharmonic Functions, Springer 1937.
RAIKOV, D. A.
[1] See GELFAND-RAIKOV [4].
[2] See GELFAND-RAIKOV-ŠILOV [5].
RAY, D.
[1] Resolvents, transition functions and strongly Markovian processes. Ann. of Math. **70**, 43—72 (1959).
RICKART, C. E.
[1] General Theory of Banach Algebras, van Nostrand 1960.
RIESZ, F.
[1] Zur Theorie des Hilbertschen Raumes. Acta Sci. Math. Szeged **7**, 34—38 (1934).
[2] Über lineare Funktionalgleichungen. Acta Math. **41**, 71—98 (1918).

[3] (with B. VON SZ. NAGY) Leçons d'Analyse Fonctionelle, Akad. Kiado, Budapest 1952.
[4] Some mean ergodic theorems. J. London Math. Soc. **13**, 274—278 (1938).
[5] Sur les fonctions des transformations hermitiennes dans l'espace de Hilbert. Acta Sci. Math. Szeged **7**, 147—159 (1935).
[6] Sur la Décomposition des Opérations Linéaires. Proc. Internat. Congress of Math. 1928, held at Bologna, Vol. III, 143—148.

RYLL-NARDZEWSKI, C.
[1] See J. MIKUSIŃSKI [1].

SAKS, S.
[1] Theory of the Integral, Warszawa 1937.
[2] Addition to the note on some functionals. Trans. Amer. Math. Soc. **35**, 967—974 (1933).

SCHATTEN, R.
[1] A Theory of Cross-spaces, Princeton 1950.

SCHAUDER, J.
[1] Über lineare, vollstetige Funktionaloperationen. Stud. Math. **2**, 1—6 (1930).

SCHWARTZ, J.
[1] See DUNFORD-SCHWARTZ [1].
[2] See DUNFORD-SCHWARTZ [4].
[3] See DUNFORD-SCHWARTZ [5].
[4] See DUNFORD-SCHWARTZ [6].

SCHWARTZ, L.
[1] Théorie des Distributions, vol. I et II, Hermann 1950, 1951.
[2] Transformation de Laplace des distributions. Comm. Sém. Math. de l'Univ. de Lund, tome suppl. dédié à M. Riesz, 196—206 (1952).
[3] Lectures on Mixed Problems in Partial Differential Equations and the Representation of Semi-groups. Tata Inst. Fund. Research, 1958.
[4] Les équations d'évolution liées au produit de compositions. Ann. Inst. Fourier **2**, 165—169 (1950—1951).
[5] Exposé sur les travaux de Gårding, Séminaire Bourbaki, May 1952.

SEGAL, I. E.
[1] The span of the translations of a function in a Lebesgue space. Proc. Nat. Acad. Sci. USA **30**, 165—169 (1944).

SHMULYAN, V. L.
[1] Über lineare topologische Räume. Math. Sbornik, N. S. **7** (49), 425—448 (1940).

ŠILOV, G.
[1] See GELFAND-ŠILOV [1].
[2] See GENFAD-RALKOV-ŠILOV [5].

SMITH, P. A.
[1] See BIRKHOFF-SMITH [2].

SOBOLEV, S. L.
[1] Sur un théorème d'analyse fonctionnelle. Math. Sbornik **45**, 471—496 (1938).
[2] Certaines Applications de l'Analyse Fonctionnelle à la Physique Mathématique, Leningrad 1945.

SOBOLEVSKI, P. E.
[1] Parabolic type equations in Banach spaces. Trudy Moscow Math. **10**, 297—350 (1961).

STONE, M. H.
[1] Linear Transformations in Hilbert Space and Their Applications to Analysis. Colloq. Publ. Amer. Math. Soc., 1932.
[2] On one-parameter unitary groups in Hilbert space. Ann. of Math. **33**, 643—648 (1932).
SZEGÖ, G.
[1] Orthogonal Polynomials. Colloq. Pub. Amer. Math. Soc., 1948.
TANABE, H.
[1] Evolution equations of parabolic type. Proc. Japan Acad. **37**, 610—613 (1961).
[2] On the equations of evolution in a Banach space. Osaka Math. J. **12**, 365—613 (1960).
[3] A class of the equations of evolution in a Banach space. Osaka Math. J. **11**, 121—145 (1959).
[4] Remarks on the equations of evolution in a Banach space. Osaka Math. J. **12**, 145—166 (1960).
[5] See KATO-TANABE [8].
TANNAKA, T.
[1] Dualität der nicht-kommutativen bikompakten Gruppen. Tohoku Math. J. **53**, 1—12 (1938).
TAYLOR, A.
[1] Introduction to Functional Analysis, Wiley 1958.
TITCHMARSH, E. C.
[1] Introduction to the Theory of Fourier Integrals, Oxford 1937.
[2] Eigenfunction Expansion Associated with Second-order Differential Equations, Vol. I—II, Oxford 1946—1958.
TRÈVES, F.
[1] Lectures on Partial Differential Equations with Constant Coefficients. Notas de Matematica, No. 7, Rio de Janeiro 1961.
TROTTER, H. F.
[1] Approximation of semi-groups of operators. Pacific J. Math. **8**, 887—919 (1958).
VILENKIN, N. Y.
[1] See GELFAND-VILENKIN [3].
VISIK, I. M.
[1] See LADYZHENSKAYA-VISIK [1].
VITALI, G.
[1] Sull'integrazioni per serie. Rend. Circ. Mat. di Palermo **23**, 137—155 (1907).
WATANABE, J.
[1] On some properties of fractional powers of linear operators. Proc. Japan Acad. **37**, 273—275 (1961).
WECKEN, F. J.
[1] Unitärinvariante selbstadjungierte Operatoren. Math. Ann. **116**, 422—455 (1939).
WEIL, A.
[1] Sur les fonctions presque périodiques de von Neumann. C. R. Acad. Sci. Paris **200**, 38—40 (1935).
WEYL, H.
[1] The method of orthogonal projection in potential theory. Duke Math. J. **7**, 414—444 (1940).

[2] Über gewöhnliche Differentialgleichungen mit Singularitäten und die zugehörigen Entwicklungen willkürlicher Funktionen. Math. Ann. **68**, 220—269 (1910).
[3] See PETER-WEYL [1].

WIENER, N.
[1] See PALEY-WIENER [1].
[2] Tauberian theorems. Ann. of Math. **33**, 1—100 (1932).
[3] The Fourier Integral and Certain of Its Applications, Cambridge 1933.

YAMABE, H.
[1] See ITô-YAMABE [2].

YOSIDA, K.
[1] Lectures on Differential and Integral Equations, Interscience 1960.
[2] Vector lattices and additive set functions. Proc. Imp. Acad. Tokyo **17**, 228—232 (1940).
[3] Mean ergodic theorem in Banach spaces. Proc. Imp. Acad. Tokyo **14**, 292—294 (1938).
[4] Ergodic theorems for pseudo-resolvents. Proc. Japan Acad. **37**, 422—425 (1961).
[5] On the differentiability and the representation of one-parameter semi-groups of linear operators. J. Math. Soc. Japan **1**, 15—21 (1948).
[6] Holomorphic semi-groups in a locally convex linear topological space. Osaka Math. J., **15**, 51—57 (1963).
[7] (with S. KAKUTANI) Operator-theoretical treatment of Markoff process and mean ergodic theorems. Ann. of Math. **42**, 188—228 (1941).
[8] Fractional powers of infinitesimal generators and the analyticity of the semi-groups generated by them. Proc. Japan Acad. **36**, 86—89 (1960).
[9] Quasi-completely continuous linear functional operators. Jap. J. Math. **15**, 297—301 (1939).
[10] (with Y. MIMURA and S. KAKUTANI) Integral operators with bounded measurable kernel. Proc. Imp. Acad. Tokyo **14**, 359—362 (1938).
[11] On the group embedded in the metrical complete ring. Jap. J. Math. **13**, 7—26 (1936).
[12] Normed rings and spectral theorems. Proc. Imp. Acad. Tokyo **19**, 356—359 (1943).
[13] On the unitary equivalence in general euclid spaces. Proc. Jap. Acad. **22**, 242—245 (1946).
[14] On the duality theorem of non-commutative compact groups. Proc. Imp. Acad. Tokyo **19**, 181—183 (1943).
[15] An abstract treatment of the individual ergodic theorems. Proc. Imp. Acad. Tokyo **16**, 280—284 (1940).
[16] (with M. FUKAMIYA) On vector lattice with a unit, II. Proc. Imp. Acad. Tokyo **18**, 479—482 (1941).
[17] Markoff process with a stable distribution. Proc. Imp. Acad. Tokyo **16**, 43—48 (1940).
[18] Ergodic theorems of Birkhoff-Khintchine's type. Jap. J. Math. **17**, 31—36 (1940).
[19] Simple Markoff process with a locally compact phase space. Math. Japonicae **1**, 99—103 (1948).
[20] Brownian motion in a homogeneous Riemannian space. Pacific J. Math. **2**, 263—270 (1952).
[21] An abstract analyticity in time for solutions of a diffusion equation. Proc. Japan Acad. **35**, 109—113 (1959).

[22] An operator-theoretical integration of the wave equation. J. Math. Soc. Japan **8**, 79—92 (1956).
[23] Semi-group theory and the integration problem of diffusion equations. Internat. Congress of Math. 1954, held at Amsterdam, Vol. 3, pp. 864—873.
[24] On the integration of diffusion equations in Riemannian spaces. Proc. Amer. Math. Soc. **3**, 864—873 (1952).
[25] On the fundamental solution of the parabolic equations in a Riemannian space. Proc. Amer. Math. Soc. **3**, 864—873 (1952).
[26] An extension of Fokker-Planck's equation. Proc. Japan Acad. **25**, 1—3 (1949).
[27] Brownian motion on the surface of 3-sphere. Ann. of Math. Statist. **20**, 292—296 (1949).
[28] On the integration of the equation of evolution. J. Fac. Sci. Univ. Tokyo, Sect. 1, **9**, Part 5, 397—402 (1963).
[29] An operator-theoretical treatment of temporally homogeneous Markoff processes. J. Math. Soc. Japan **1**, 244—253 (1949).

YUSHKEVITCH, A. A.
[1] On strong Markoff processes. Teorya Veroyatn. **2** (1957).

Index

Druck: Julius Beltz, Weinheim/Bergstraße